RENEWALS 458-4574

DATE DUE

GAYLORD			PRINTED IN U.S.A.

The Logic of Thermostatistical Physics

Springer
Berlin
Heidelberg
New York
Barcelona
Hong Kong
London
Milan
Paris
Tokyo

Physics and Astronomy

ONLINE LIBRARY

http://www.springer.de/phys/

Gérard G. Emch   Chuang Liu

The Logic
of Thermostatistical
Physics

With 19 Figures

 Springer

Professor Gérard G. Emch
Department of Mathematics
University of Florida
358 Little Hall, P.O. Box 11810
Gainesville 32611-8105, FL, USA

Professor Chuang Liu
Department of Philosophy
University of Florida
330 Griffin-Floyd Hall, P.O. Box 118545
Gainesville 32611-8545, FL, USA

Library of Congress Cataloging-in-Publication Data applied for.

Die Deutsche Bibliothek - CIP-Einheitsaufnahme
Emch, Gérard G.: The logic of thermostatistical physics : with 4 tables / Gérard G. Emch ; Chuang Liu. -
Berlin ; Heidelberg ; New York ; Barcelona ; Hong Kong ; London ; Milan ; Paris ; Tokyo : Springer, 2002
(Physics and astronomy online library) ISBN 3-540-41379-0

ISBN 3-540-41379-0 Springer-Verlag Berlin Heidelberg New York

Springer-Verlag Berlin Heidelberg New York
a member of BertelsmannSpringer Science+Business Media GmbH

http://www.springer.de

Typesetting: Camera copy by the authors
Cover design: *design & production* GmbH, Heidelberg

Printed on acid-free paper   SPIN 10553005   55/3141/mf   5 4 3 2 1 0

Preface

This book addresses several of the foundational problems in thermophysics, i.e. thermodynamics and statistical mechanics. It is an interdisciplinary work in that it examines the philosophical underpinning of scientific models and theories; it also refines the analysis of the problems at hand and delineates the place occupied by various scientific models in a generalized philosophical landscape.

Hence, our philosophical – or *theoretical* – inquiry focuses sharply on the concept of models; and our empirical – or *laboratory* – evidence is sought in the model–building activities of scientists who have tried to confront the epistemological problems arising in the thermophysical sciences.

Primarily for researchers and students in physics, philosophy of science, and mathematics, our book aims at informing the readers – with all the indispensable technical details made readily available – about the nature of the foundational problems, how these problems are approached with the help of various mathematical models, and what the philosophical implications of such models and approaches involve. Some familiarity with elementary thermophysics and/or with introductory–level philosophy of science may help, but neither is a prerequisite. The logical and mathematical background required for the book are introduced in the Appendices. Upon using the Subject Index, the readers may easily locate the concepts and theorems needed for understanding various parts of the book. The Citation Index lists the authors of the contributions we discuss in detail.

Since ours is not a textbook on statistical mechanics, readers should not expect the usual comprehensive coverage of standard textbook topics; for instance, they will not find any discussion of topics such as the Onsager relations or computer-assisted proofs; neither of which would inform our discusssion of the roles of models.

We especially confront a pervasive trend in today's philosophical discussions of contemporary scientific theories: rather than *telling* what the theories amount to, we *demonstrate* how they are constructed and why they should be believed. We let our readers see in detail the rigor of the results in thermophysics, how such rigor is achieved, and what it signifies. By so doing, we may enable philosophers of science to form independent judgment on foundational problems in thermophysics.

We also alert and inform scientists concerning the philosophical issues and the nature of the reasoning pertinent to the praxis of thermophysics; some of the stones uncovered by our analysis are offered as building materials for wider syntheses in the natural sciences.

We eschew a rather widespread temptation in the science literature on modeling or mathematical modeling to identify these concepts with a listing of specific procedures or techniques in problem solving, however extensive the latter may be.

Since the material in this book and its intended audience are diverse, a few words on how the book may be used are in order.

For physicists and mathematicians who are familiar with the standard presentation of thermophysics but are curious about the history and the fundamentals of probability and probabilistic reasoning, Chap. 4 may provide an interesting and useful point of entry. There we first summarize dual traditions in the interpretation of the concept of probability and then give a detailed study of some of the milestones, from Bernoulli's binomial distribution and Pascal's triangle to Laplace's systematic rationalizations. We include also some historical details that illustrate the meaning of the theorems; meanwhile, these illustrations deepen our understanding of probability and provide an appreciation of their implications. We then show in Chap. 5 how a clear separation between the syntax and the semantics – which one can already witness in Hilbert's work in geometry – made it possible for Kolmogorov to systematize the formal aspects of probability. The importance of such a separation can be appreciated further when one gets to Chap. 6, where our discussion of the competing semantics shows that waiting for a settlement of the semantics for probability would have greatly hampered the understanding of its formal properties. Our examination of the notion of randomness in Chaps. 5 and 6 (and the references cited and briefly introduced therein) should well inform those who are curious about the foundations of probability theories.

If our readers from physics and mathematics are, moreover, curious about the philosophical implications of the separation of the syntax and semantics of a theory, and about how the conception of scientific theories is analyzed in the context of such a separation, they will find in Chap. 1 an in-depth discussion of the philosophical landscape. Our readers should look there for the identification of the components involved in bridging the gap between the logician's (or philosopher's) conception of models and that of scientists; see also Appendix E. In particular, we explain how the recognition of the syntax/semantics distinction helps to introduce general features of theory construction in thermophysics.

For physicists and mathematicians who have a working knowledge of thermophysics but are concerned with the validity of statistical mechanics, Chap. 7 is the place to start reading. We articulate first in this chapter Boltzmann's original ergodic hypothesis, and we evaluate what he could have meant by this in the light of the limitations imposed by his finitist philo-sophical position, and by the contemporary mathematics (i.e. the absence of properly formulated theories of measure and integration) and physics (i.e. the scarcity of worked-out models beyond the ideal gas). We then discuss two improved versions of ergodicity, Birkhoff's and von Neumann's ergodic theorems, which rendered the notion of ergodicity mathematically rigorous but made its application to real physical systems a problem that is still with us today. Chapters 8 and 9 examine finer properties and models of the er-godic hierarchy. An in-depth critical discussion of the relevance of several models to the foundations of statistical mechanics is given there as well. One of our purposes is to provide specific anchors for the discussion of the de-liberately controversial assertion according to which ergodic conditions are neither necessary nor sufficient for a proper understanding of the physical theory.

Students of formal logic will find that their field enters in at least three ways in our enterprise: (1) in the sustained distinction between the properties of language (syntax) and its interpretations (semantics); (2) in the discussion of the very notion of models; (3) in the use of recursive function theory and algorithmic complexity to discuss randomness and, consequently, entropy.

For philosophers of science in general, and of physics in particular, this is a book about models and model-building in physics. For those who are familiar with the literature of the semantic (or structural or architectonic) view of scientific theories, Chap. 1 offers a critique of the existing views and a proposal for a new, hybrid view. The problem was flagged in the title quo-tation from van Fraassen and the chapter endeavors to address that problem. For those who are not familiar with this literature, the chapter will serve as an introduction to a general philosophical account of models and theories in science; these readers will find here a list of different types of theories and models, and then how these fit into a general scheme of theory construction and testing in the praxis of science. Chapters 2 and 3 contain introductory material for thermophysics. Chapter 2, on Thermostatics, not only informs our readers about the essentials of a phenomenological theory of macroscopic systems but also demonstrates how the separation of syntax and semantics works for the construction and analysis of such a theory. Chapter 3 provides a simple but thoroughly worked-out tour of some of the most significant microscopic models of gases; see also Chap. 11.

Those philosophers of physics who are anxious to examine some real mod-els in physics – and also to see how these models are conceived and used – may want to go directly to Chaps. 7–15. The models in Chaps. 7–10 are all directed towards the foundations of thermophysics; and the models in Chaps. 11–15 are mostly chosen for their ability to deliver microscopic explanations for well tested thermal phenomena. For those readers who are familiar with the philosophical literature on the foundations of statistical mechanics and problems of ergodicity, a wealth of carefully worked–out models is presented which exhibits the different properties listed in the ergodic hierarchy. These readers will find out not only what these models mean to the foundations of statistical mechanics, but also why. Enough technical steps are detailed to allow an informed appreciation of the mathematical workings of the models. Specific applications of the formalism of classical and quantum statistical mechanics are the focus of Chaps. 11–14; we do not shy away from showing step-by-step to our readers how the complex phenomena of phase transition and criticality are explained in (classical) lattice models via some of the intricate and powerful tools that mathematics offers to physics. The renormalization semi-group method is one of them. In Chap. 15 we illustrate with the help of specific (quantum) models some of the questions non-equilibrium statistical mechanics has begun to answer.

A very prominent feature of model–building is the relative places of idealization and approximation. Throughout the book we have laid emphasis on the various manifestations of these two closely related concepts. We give an integrated summary of such discussions in the last chapter of the book, Chap. 16. Since the entire range of applications of these two concepts in science is relatively understudied in the philosophy of science literature, part of our intention for that chapter is to stimulate further philosophical investigations of these concepts. Simulation is another understudied concept in philosophy, although it is becoming more and more important in scientific research. We give a brief account of various aspects of the concept without, however, pretending to do it full justice. Nevertheless our intention, here also, is to stimulate further studies of the concepts in connection with the practice of model–building and model–testing in the physical sciences.

In most of our book, the microscopic world is viewed through the lenses of classical rather than quantum mechanics. This is done for two reasons. The first is that we believe that the main problem posed by the attempt to reconcile macroscopic and microscopic theories is a reduction process involving the interplay of various scales, and that the understanding of this interplay is not primarily dependent on whether the microscopic picture is classical or quantum. In this context, our readers steeped in the traditional philosophic discourse may want to pay attention to the issues involved in the pragmatic meaning of the mathematical notion of limit; see for example Sects. 1.6 and 16.4. These issues are brought forward specifically in connection with the use of: (1) the thermodynamical limit (described Sects.10.2 and 12.1. and implemented, for instance in Sect. 12.2 and the Models 2–4 of Sect. 15.3; see also its role in Chap. 14); (2) the high–temperature limit (Sect. 10.3); (3) the critical scaling in phase transitions (Chap. 13); and (4) the van Hove limit (described in Sect. 15.2 and illustrated in detail with Models 3 and 4 of Sect. 15.3). These limits are presented here in a manner that ought to illuminate such reflections.

The second reason behind our taking the classical rather than the quantum option is that we want to avoid the danger that the conceptual problems involved in the reduction be masked by the conceptual difficulties inherent in a fully consistent presentation of quantum premises. Nevertheless, we also consider explicitly quantum situations: at the end of Sect. 5.3; in the course of Sect. 6.3; in several parts of Chap. 8; in Sect. 10.2; and in Chaps. 14 and 15.

Finally, we should mention that we systematically verified – and modified when necessary – the translations of all the original texts we quote.

Written and oral comments on various parts of the book, as well as discussions on related materials, are gratefully acknowledged. We were indeed privileged to engage in stimulating conversations with several colleagues, among whom we wish to mention the following: Douglas Cenzer, Irene Hueter, Henry Kyburg, Jean Larson, Mark Pinsky, Erhard Scheibe, Abner Shimony, Rick Smith, Steve Summers and Alexander Turull; Giovanni Boniolo, Jim Brown, Craig Callender, Nancy Cartwright, Robert D'Amico, Pierre Extermann, Steven French, Carl Hoefer, Risto Hilpinen, Wladyslav Krajewsky, Peter Lewis, Elliott Lieb, Philippe Martin, James McAllister, Heide Narnhofer, Michael Readhead, Miklós Rédei, Geoffrey Sewell, Walter Thirring, and Jakob Yngvason.

We are indebted to Prof. Wolf Beiglböck for suggesting that we "write a book on the foundations", for endorsing readily our idea to focus on the roles of models in science, and for his willingness to take time during the Summer 2000 London IAMP Congress and examine closely the first version of this work. All along, the stimulating suggestions of our Scientific Editor helped us significantly to complete our task.

Antoinette Emch-Dériaz read through the typescript, for some parts several times over, shooting down obscure sentences and typographical errors; however, the responsibility lies with us for those that escaped the hunt.

Lastly, Antoinette Emch–Dériaz and Mingmin Zhu cannot be thanked enough for their unfailing patience and steadfast support.

Gainesville, FL, USA   *Gérard Emch*
June 2001   *Chuang Liu*

Contents

List of Figures and Tables

Figures

Tables

1. Theories and Models:
a Philosophical Overview

In connection with the development of modal logic ..., we have seen the development of a very rich formal semantics. There, a language is characterized ... by specifying the structure of models for theories formulated in that language. ... in philosophy of science much attention has been given to the characterization of the structure of models as they appear in the scientific literature. The first central problem is to bring these two efforts together, because at first sight, the model structures found in semantics, and the models of scientific theories ... are totally different.

[Van Fraassen, 1983]

1.1 Introduction

In this chapter we lay out some philosophical fundamentals for our later logical analysis of thermostatistics. As an overview, we proceed rather dogmatically: making precise definitions and sharp distinctions; building up somewhat simplistic scaffoldings which delineates our later philosophical analyses. Modifications will have to be made in later chapters when we see the intricate details of thermostatistics, but we trust that it is useful to begin with a clear direction, charted by the fundamentals representing the best understanding reached to-date by philosophers on the nature of scientific theories.

As van Fraassen correctly pointed out almost twenty years ago (cf. the opening quote of this chapter), the central problem of providing a precise and accurate understanding of the scientific theories is to bring together two apparently separate strains: the logicians' use of models in formal semantics and the scientists' use of them in the practice of theory construction. Since (and even before) that remark of van Fraassen's, works have been steadily pouring out, most of which defended and/or developed a 'semantic view' (or 'semantic approach'). In the following, we review and build upon what we consider to be the important developments in this literature. However, it will soon become clear that even today, the problem posed by van Fraassen has not been resolved with complete satisfaction. A gap still exists in the two conceptions of models from the two communities – the philosophers' and the scientists'; and while the former has been consolidating and entrenching its results in explicating the formal theory of the semantic (or structural or architechtonic) view (or approach), the latter has never relented in its effort to evoke models for explaining the phenomena at hand, which often defy classification.

We begin (in Sect. 1.2) with an account of the two views of the nature of scientific theories: the syntactic vs. the semantic. Emphasis is given to explaining how the concept of models in mathematical logic is used in both of the two views, and addressing the question of whether they can be construed as complementary, despite their advocates' insistence on its being otherwise. The truth may be that they represent the two sides of the same coin. In Sect. 1.3, we turn to an in-depth discussion of the different conceptions of models. We discover there that the conception used by scientists to denote models furnished to either divine a theory or explain a phenomenon is usually quite different from the logical conception. We do not claim exhaustiveness in this inquiry, but rather a rough classification of different kinds of models used in various fields and noticed previously by philosophers. In Sect. 1.4, we bring the two conceptions – the logician's and the scientist's – of models together and examine the possibility of accommodating them in the semantic approach. As it turns out, it is not possible to do that without changing the approach in some fundamental ways. This leads us into the proposal for an enriched semantic view for scientific theories. Then come three sections (Sects. 1.5–7) which discuss respectively how the theories of confirmation, of reduction, and of explanation may differ under the two approaches.

1.2 The Syntactic vs. the Semantic View

The division between syntax and semantics in mathematical logic makes it possible to define two different but related notions of a 'theory'. A theory may be defined *syntactically* as a set of assertions $\mathcal{T}$ that is a derivational closure of a set of assertions $\mathcal{S}$, i.e.

$$\mathcal{T} = \{\varphi \in L \mid \mathcal{S} \vdash \varphi\} \,, \tag{1.2.1}$$

where $\mathcal{S}$ consists of $\mathcal{T}$'s axioms. And a theory may be defined *semantically* as a set of assertions that are true in a set of models $\mathcal{A}$, i.e.

$$\mathcal{T} = \{\varphi \in L \mid \mathcal{A} \models \varphi\} \,. \tag{1.2.2}$$

A typical syntactic theory is the Euclidean geometry which consists of five axioms and all the theorems derivable from the axioms via rules of geometrical construction. And an example for a semantic theory would be group theory which consists of all the first-order assertions that a set of groups satisfies; the groups are definable by set-theoretic entities. (For the syntactic nature of Euclidean geometry, see the opening remarks in [Hilbert, 1899]; and for

group theory, see [Pontrjagin, 1939, Kurosh, 1955-56, Hamermesh, 1962]).
Furthermore, when a theory is axiomatizable, one only needs to find out
whether a structure satisfies the axioms. If it does, it is a model of the theory;
cf. "completeness" in Appendix A. Hence,

$$\mathcal{A} \models \mathcal{T} \quad \text{if and only if} \quad \mathcal{A} \models \mathcal{S} \tag{1.2.3}$$

where $\mathcal{S}$ is the set of $\mathcal{T}$'s axioms. It is helpful to keep the notions of these two
kinds of theories in mind when we consider the concept of scientific theories
from either the syntactic or the semantic view.

Ideally, $\mathcal{T}$'s models are determined uniquely by the given interpretation.
Once the L-structures are given, we can discuss items such as the consistency of a theory, one theory being equivalent or reducible to another, and
one structure being isomorphic to or an extension of another structure, etc.
(cf. Appendix A). Even though the consistency, equivalence and reduction of
– or among – theories are primarily syntactic notions and can in principle
be handled syntactically, it is much easier and more elegant (because of the
completeness of first-order theories) to use the structures. For instance, to
prove *syntactic* consistency of a theory one must make sure that no contradiction is derivable from it, which could be a daunting task when the theory
is complex. But *semantically*, one only needs to make sure that at least one
structure exists which satisfies the theory. And again, because of the completeness, the equivalence or reduction of theories can be formulated in terms
of the isomorphism or extensions of their models.

One caveat should be observed at this point. The schemes reviewed above
are only known to apply to first-order languages. Therefore, we know neither
if they can be applied to theories which are not first-order but close to it,
nor if any of our actual scientific languages can be made into a first-order
language, even though originally, it may be more involved.

Let us first turn to the **syntactic view** (or 'the received view') of scientific theory; it is usually [Suppe, 1977] attributed to logical positivism –
a philosophical tradition whose manifesto was written in 1928 by the members of the Vienna Circle, in which the intellectual heritage is traced back
to [Mach, 1886], and even to [Hume, 1748]. For the positivists, a scientific
theory is, or can be reconstructed as, a set of interpreted assertions some of
which are axioms, while the other assertions in the theory are derivable from
the axioms through syntactic (including mathematical) rules. The language
is divided into two mutually exclusive kinds:

$$L_\theta = \{V_\theta, C_\theta\} \quad \text{and} \quad L_0 = \{V_0, C_0\} , \tag{1.2.4}$$

where V_θ consists of a vocabulary of non-observational terms, and C_θ is a calculus in terms of V_θ; while V_0 consists of a vocabulary of observational terms,
and C_0 is a calculus in terms of V_0. (C_0 not only contains singular assertions
in terms of V_0, but also quantified and modal assertions as well.)

Accordingly, there are two types of interpretations. At the first (or lower) level, one has the simple interpretation, $\mathcal{I}$, which relates terms in V_0 to observational results. It supplies the semantics for C_0 just as what one does in mathematical logic. The second (or higher) level of interpretation is a partial interpretation via rules of correspondence, $\mathcal{C}$, which relates terms in V_θ to terms in V_0.

One must note that interpretations via $\mathcal{C}$ are radically different from $\mathcal{I}$: they are partial definitions or reductional assertions so that the meaning of terms in V_θ and of assertions/inferences in C_θ can be given ultimately in terms of V_0 and C_0. The positivists consider

$$\{\, C_\theta \,(\text{and } V_\theta)\, ,\, \mathcal{C}\,\} \qquad (1.2.5)$$

to be a scientific theory proper. A theory acquires its meaning and confirmation (positivists consider them one and the same) exclusively through matching the observational assertions with data under some interpretations in $\mathcal{I}$. The rest (especially terms in V_θ and assertions in C_θ) gain their meaning and/or confirmation when they are partially defined by, and/or connected with (via $\mathcal{C}$) C_0 and V_0.

In the positivist view, models – especially models for higher theoretical postulates which only use V_θ – have little significance in formulating the relationships among various types of assertions in a theory. While the models for assertions in V_0 are deemed obvious, the models for assertions in V_θ are generally discouraged – see Duhem [1954] for a strong denunciation of the use of theoretical models. Instead of interpreting theoretical terms directly and discuss theoretical models, one first reduces them to V_0. In this sense, theories are like domes of interconnected knots – i.e. assertions – which are semantically 'sealed', so to speak, around their lower edges – areas consisting exclusively of observational assertions. The meanings of assertions high up in such domes are transmitted to them from the edges via the correspondence rules.

Take the theory of thermodynamics known to Gibbs or Planck, for example. The theory consists of two different kinds of variables: such as, on the one hand, pressure, volume or temperature; and, on the other hand, entropy, enthalpy, or Gibbs potential. The former kind are observational, the latter non-observational; and the latter are explicitly defined in terms of the former. The axioms of the theory are the two laws, and every other assertions about any thermodynamic properties of a system can be derived from those and the system's state-equations. (The state-equations are necessary because the two laws are completely general and so applicable to all thermo-systems.)

One can now see that 'the syntactic view', if not solely in contrast to 'the semantic view', is somewhat of a misnomer. It gives the impression that syntacticists only consider the uninterpreted and highly abstract calculus (such as C_θ) as a theory, which is in fact not the case. Interpretations and correspondence rules are also essential components of a theory. The right way of conceiving this as a syntactic view ought to be that it emphasizes the importance of a theory's linguistic structure at the expense of other elements. From now on, we use the term only in this more circumscribed sense.

We now turn to the **semantic view.** [Suppes, 1972] is among the pioneers of this view and perhaps the least dogmatic. His early attempt at an axiomatization of mechanics (classical and quantum) champions an approach to scientific theories that includes both the syntactic and the semantic apparatus of mathematical logic. Moreover, in defending the semantic view, he maintained,

> *The view we want to support* [a prototype of the semantic view] *in this essay is not that this standard sketch* [the syntactic view] *is wrong, but rather that it is far too simple.* [Suppes, 1972]

A theory, according to Suppes, is given by a predicate (such as *"being a mechanical system"*) which admits both an intrinsic and an extrinsic characterization: the former by a set of uninterpreted assertions and the latter by the set of models that satisfies those assertions. The syntactic view, which emphasizes the former, is not *"wrong"* but *"far too simple"* (ibid.). By incorporating a theory's models, such issues as the theory's axiomatization, its consistency and its various symmetry properties can be discussed with much simpler means. This shows that Suppes' theory is imitating very closely the relation between the syntax and the semantics in the first-order logic (see Appendix A). The theory-model pair may also be applied to different levels of theorization. For instance, a theory of measurement of a particular kind may have axioms and theorems explicating the experimental procedures and sets of numbers (i.e. data) as models that they satisfy.

The flow-chart in Fig. 1.1, which is a slightly modified version of Giere's, seems to capture the core of the semantic view. The first box of Fig. 1.1 (con-

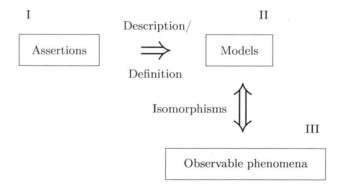

Fig. 1.1. The semantic view

taining assertions in a language) would be where a syntacticist says a theory lies. Since the syntactic features of assertions and their interdependent relations are no longer of primary importance to a semanticist, the fine linguistic structure touted by the syntacticists is no longer appreciated by the semanticists. Instead, which set of assertions should suffice depends entirely on what kind of models is under consideration. The predicate is described or defined by the assertions, but the property that it picks up is specified by the structure of the models. So, the assertions should describe necessary and sufficient conditions for picking out the models. Hence the two descriptions are equivalent and exchangeable so long as they are co-extensive (i.e. they pick out the same set of models).

The following might be one of the simplest example of equivalent descriptions. Imagine that, in a two-dimensional space with cartesian coordinates (x, y), there is a trajectory which is a straight line going through $(0,0)$ and has an angle q to $0x$ such that $tan\, q = a/b$, where $a, b > 0$. The assertion

$$A \text{ straight line goes through } (0,0) \text{ and has an angle} \atop q \text{ to } 0x \text{ such that } tan\, q = a/b, \text{ where } a, b > 0 \qquad (1.2.6)$$

is a theory that the trajectory satisfies, but the trajectory also makes true the assertion

$$(x)(y) \quad (ax - by = 0) \, , \qquad (1.2.7)$$

which is another theory that is equivalent to the previous one.

Here, we assume that the notion of 'models' is used in the logical sense as defined earlier in this section and in Appendix A, namely, models are interpretations, i.e. $m = (\mathbf{S}, z)$, that make true the assertions in the first box. Depending on which kind of models one evokes, the semantic view can be formulated in two different versions: the state-space vs. the system version.

V.1 : The state-space version: *the domain of any set of models only contains trajectories (understood in the broadest and most abstract sense of the term) in a state-space defined by the values of the variables.*

Such models are certainly semantic models, for the trajectories are exactly those that satisfy the quantified assertions (e.g. algebraic equations).

V.2 : The system version: *the domain of any set of models contains physi cally possible systems. And such a system provides the extension for the predicate, "being a system of ...".*

Such models can also be taken as semantic models if we consider the structures to be isomorphic to some first-order relational structures. Giere certainly believes that they are semantic models when he says that *"this terminology even overlaps nicely with the usage of logicians for whom a model of a set of axioms is an object, or a set of objects, that satisfies the axioms"*

[Giere, 1988]. Surely, the object must be an interpretation in the logical sense
if it is to satisfy any axioms.

The relation between the second and the third boxes is understood as an
empirical (or theoretical) hypothesis, which is taken in the literature to say
that the phenomena (or phenomenal systems) are isomorphic, or at least,
similar, to the empirical substructures in the set of models. Note that the
term "isomorphism" already has a precise definition in mathematics: an isomorphism is a map from one set to another (not necessarily a different from
the first) which is "iso" – i.e. 1–1 and onto – and a "morphism" – i.e. preserves
the *chosen* relations and structures among the elements. We need here a more
sophisticated notion, akin to one that is familiar to differential geometers: an
atlas of consistent local isomorphisms. To explain this in non-technical terms,
consider one of the best and most familiar examples of an isomorphism, the
one between a city and its map. The two sets contain radically different elements; namely, the city: roads, buildings, parks, etc.; and the map: lines of
different colors and widths, dots of different colors and sizes, etc. But as far as
the two-dimensional spatial relationships among the elements are concerned,
the map is isomorphic to the city. An atlas of the Earth is a collection of maps
or charts *each of which* is isomorphic (with respect to the 2-dimensional spatial relations) to the local region it represents. The charts must be consistent:
on all subregions that are described by more than one chart, the corresponding charts must be locally isomorphic. It is the atlas that gives the whole
picture. Note in particular that an atlas consisting of at least two charts is
necessary to describe smoothly a 2-sphere, e.g. to avoid singularities at the
poles; for the mathematical side of this analogy, see Sect. B.3.

Now, models in the second box of Fig. 1.1 are theoretical constructs which
are mapped isomorphically onto a manifold, which we call the *model manifold*;
and the empirical submanifold of this manifold is then isomorphic to the
phenomena (i.e. in the third box). To see these relations in more detail, we
should note that the model manifold consists of open areas each of which is
isomorphic to a model and, on all subregions described by more than one
model, the corresponding regions of these models are isomorphic.

The relation between the second and the third boxes can now be understood as a hypothesis assertion that there is an isomorphism between the
empirical submanifold – i.e. those areas in the model manifold that are isomorphic to the observable parts of the models – and the phenomena. If we
use the state-space for model constructions – assuming that all states are
reachable – and if we think of the phenomena as collections of states of some
systems, then the empirical hypothesis relating the second and the third
boxes would simply be the following claim: the models which are trajectories
in the state-space correspond to the states of the systems in question as an
atlas of local isomorphisms.

Suppose, for instance, that the state-space is the pV-space of thermodynamics, where $p(t)$, $V(t)$ are variables representing, respectively, the pressure

and the volume of a thermo-system. A model could simply be a set of trajectories: $\{p(t), V(t) \mid t \in [t_o, t_1]\}$, which represents the time-evolution of the system. To see if an actual system is one to which this predicate is attributable, one only needs to compare the measured data of that system with the values in the models (provided all conditions of calibration and approximation are taken into account). Not all hypotheses are this simple, and so the comparison might be much more involved and indirect. But the semanticists believe that in the end, all such comparisons are essentially of the same kind. However, this claim is too naïve, as we argue in Sect. 1.4.

Furthermore, a mathematical theory usually contains two types of basic equations which correspond to laws of co-existence and of succession, the former of which determine the states, and the latter, the time-evolution of a system in the state-space. As linguistic items, the equations or the lawlike assertions are among those in the first box that, together with other assertions, define models or systems. It is not clear what exactly are laws of nature in the semantic view. It seems that they should be sets of models, but how can such sets be determined independently of the assertions we call lawlike? We come back to this point in Sect. 1.5 when we entertain an improvement of the current version of the view.

Now, two extra points call for our immediate attention.

First, Suppe uses the counterfactuals to capture the idealized nature of all models. If that is allowed, the possible-world talk lends itself conveniently to the semantic view, whether or not one is a realist about such worlds. A non-viscous fluid is simply a fluid in a possible world where frictional forces do not exist. Similarly, it seems that all conditions of idealization and/or approximation can be thus understood or analyzed, and the models of an assertion are simply the possible worlds in which it is true. Possible worlds may differ from the actual one by having different initial/boundary conditions, or they may differ by having or lacking some contingent but lawlike features. The latter worlds are more distant from the actual one than the former, but they are the ones that most likely account for conditions of idealization and approximation. Second, the semanticists are not entirely clear on what the two arrows in the flow-chart in Fig. 1.1 really are. These arrows seem to be semantic relations since they are relations between words and the world; but are they? Not really.

If the above possible-world talk holds, the first arrow is certainly semantic. The models in the second box are the possible worlds in which the assertions in the first box are true, but the second arrow between the second and the third box does not represent a semantic relation, at least, not simply so. If Suppe's possible worlds include the actual one, the relation is simply a membership relation. The set of observable phenomena which is the actual world is in turn a member of the set in the second box. But if Suppe meant to include only abstract or mathematical worlds or structures in the second box, then we have no precise idea what that relation is. Therefore, the relationship between the second and third box depends on whether one believes in the reality of possible worlds: it is one of similarity for the believers and one of isomorphism for the non-believers. (We take it that similarity is a relation holding between two structures of the same kind, while isomorphism is one that can hold between two structures of radically different compositions.)

At this juncture, we must turn our attention to an approach to scientific theories that is closely related to the semantic view – the *structuralist approach*, which is popular in Europe but seldom discussed in the United States. It began with Sneed's seminal book [Sneed, 1971] which, according to its author's own admission, is a project of extending Suppes' then roughly sketched theory. It launched a program with a paradigmatic vision and a strong following – see e.g. [Stegmüller, 1973] – which culminated in [Balzer, Moulines, and Sneed, 1987].

Logical reconstruction is the professed aim of structuralism, which is not necessarily shared by other semanticists. To put it in a nutshell, the task is to discover the logical structure of scientific theories by constructing set-theoretic predicates which are descriptions of set-theoretic entities – models, so that empirically testable claims can be made by applying the predicates to appropriate objects or systems. For instance, to have a theory of mechanics is to give a set-theoretic structure to the predicate, *"being a mechanical system"*, which in turn defines a set of abstract models of mechanics. Then one can use it to make claims such as 'the planetary system is a mechanical system', and experiments or observations can be devised to test the claim. The main virtue of structuralism is its clarity and transparency. While other versions of semanticism are rather shy in answering questions such as whether everything must be rendered in first-order terms or whether the models must have a purely set-theoretic structure, the structuralists are forthcoming. They believe that the structure of any scientific theory is given by its first-order set-theoretic structures, and is defined using a metalanguage in which the structures are accurately and plainly described.

In this program, however, it is not always clear what the relation is between a scientific theory and a corresponding set-theoretic description of its models. It may be easy to see in simple mathematical theories, such as group theory, but this is often rather difficult for the empirical sciences. And some set-theoretic reconstructions of theories in physics are criticized as being neither necessary nor sufficient for the scope of the original theory. For an acerbic review, see [Truesdell, 1984].

When the structuralists claim, for instance, that a set-theoretic description picks out the models for classical particle mechanics, they give no proof that the models are the ones satisfied by and only by classical particle mechanics, nor any proof that the set-theoretic description is equivalent to classical particle mechanics. Moreover, the insistence on first-order structures and a set-theoretic description of them makes the view too abstract for, and distant from, any real concern of the structure of a scientific theory. Nevertheless, it does provide a clear mirror in which the details of the other versions are reflected. While skeptical about the adequacy of the structuralist approach, we by no means share the sentiment of absolute disgust that Truesdell showed over what he called the "Suppesian Stew", nor do we follow the argument that if structuralism does not work, Suppes' pioneering work in the semantic approach should be thrown out of the window as well.

Structuralist or not, the semantic view is said to have several advantages over the syntactic one, and we list below four of these, following [Suppes, 1972, Giere, 1988, Van Fraassen, 1989, Suppe, 1989].

(a) Scientific theories should not be made dependent on their linguistic formulations. Otherwise, it does not only force an unnecessary proliferation of theories but also obscures their structures. The aim of theorizing is to represent the properties and behaviors of the systems in question, not the ways in which they are linguistically rendered. The syntactic view endorses such a dependence but the semantic view denies it. One should note that in criticizing the syntactic view, the semanticists are not clear as to whether the syntacticists have the option of using propositions, rather than assertions, as essential components of a scientific theory. Even though there might be difficulties in defending the use of propositions in general (as a problem in philosophy of language), the use of propositions helps one to get rid of the trivial problem of having unnecessary proliferation of assertions. If by propositions we mean beliefs (or mental states) that are expressed by certain assertions, or that they are the content (or the meaning) of the assertions, then we can identify a theory with a set of propositions about a certain phenomenon with the appropriate logical and theoretical constraints. We go no further into the details of this possibility beyond raising the flag that the dichotomy between the syntactic and the semantic view might be a misplaced one.

(b) The model-building activity is common in both the natural and the social sciences. This is most conspicuous in those cases where scientists construct in their imagination systems of objects whose properties or behaviors resemble familiar systems. This is where reasoning by analogy figures prominently and the models come before their precise description in a language. The semantic view does make sense of this practice, but the syntactic view does not. However, it is not clear whether such models are the ones that are available to the semantic view. If the semanticist models have to be state-space trajectories, as one of the versions suggests, then they are certainly not the type that scientists build in their imagination, such as Bohr's model for hydrogen atoms and Kekule's model for benzene rings. Even worse are the set-theoretic descriptions: scientists almost never construct that kind of models in their praxis. We say more about this in the next section.

(c) The syntactic view is riddled with difficulties such as the unattainable distinction between the theoretical and the observable (e.g. the distinction between V_θ and V_0). One of the main reasons for such a distinction to fail is that in empirical science, there does not seem always to be a privileged set of terms (V_0) and assertions, the understanding and evaluation of which depend on no theoretical reasoning. Such difficulties do not affect the semantic view, for it does not recognize those distinctions.

(d) The syntactic view greatly oversimplifies the structure of theory-testing by trying to directly connect experimental data with the assertions in a theory. One may see this point in, for instance, [Popper, 1935] scheme for hypothesis (or theory) testing. Let 'all $\mathcal{F}$s are $\mathcal{G}$s' be a theoretical claim. One can then derive an assertion such as

$$\text{if } a \text{ is } \mathcal{F}, \text{ then } a \text{ is also } \mathcal{G}.$$

Suppose that, though highly theoretical,

$$\mathcal{F} \text{ and } \mathcal{G} \text{ are reducible to } \mathcal{F}' \text{ and } \mathcal{G}'$$

which are sets of observable terms. One can then test whether

$$\text{some } a, \text{ which is } \mathcal{F}', \text{ is also } \mathcal{G}'$$

by directly comparing it with experimental findings. For Suppes, this does not even remotely resemble what actually goes on in the actual theory-testing. With the semantic view, however, theories of different levels of idealization (and/or approximation) may be characterized by the theory-model pairs and their relations by those among the models. A more detailed discussion on idealization/approximation and other related concepts in theory testing is conducted in Sect. 1.6 below.

1.3 Conceptions of Models

The term, 'model', is widely used in the English language and bear several distinct meanings. According to *OED*, as a noun, it has fifteen distinct meanings grouped under three categories: as representation of structure, as type of design, and as object of imitation, all of which have something in common with our scientific usage of the term, whose philosophical nature has been extensively discussed in the literature [Apostel, 1961, Freudenthal, 1961, Braithwaite, 1962, Achinstein, 1964, 1965, 1968, McMullin, 1968, Hesse, 1970, Leatherdale, 1974, Wartofsky, 1979, Redhead, 1980, Kroes, 1989, Hughes, 1990, Horgan, 1994]. As we did in the previous section, we proceed by first organizing the different concepts from the literature into a coherent system; this requires minimal but necessary re-baptizing of some of the terms.

I. Mathematical Logic

1. *Theory*: see Appendix A.

2. *Interpretation and Model*: see Appendix A.

3. *Theory Testing*

i   *Test of Consistency*: to construct a structure and show that the theory is true under the intended interpretation.

ii *Direct testing of a set of assertions*: to see if a model satisfies a set of assertions.

iii *Indirect testing of a set of assertions*: to conduct a direct testing of a set A of assertions, and see if the set of assertions being tested are derivable from A.

For instance, from the hypothesis that pure water boils at different temperatures under different atmospheric pressures we can derive that samples of water placed at different altitudes boil at different temperatures. Testing the samples and measuring their boiling temperatures constitutes a testing of our hypothesis. One should note that 'truth' here is a two-place predicate: ... is true in ..., where the first blank is to be filled by an assertion and the second by a model.

II. Natural Sciences

1. *Theory*

i   *Structural theory*: a theory that describes a system by characterizing its structural properties, such as a theory of atomic nuclei or of mammals.

ii *Behavioral theory*: a theory that describes the behavior of a system by characterizing the possible states it may occupy, such as the paths of an subatomic particle in a cloud chamber or the graphs of a mammal's blood pressure as a function of time.

2. *Interpretation*

i   *The semantic interpretation*: to give meanings to quantities in a theory so that, directly or indirectly, they can be given values.

ii *The philosophical interpretation*: to understand those terms in a theory that refer to fundamental entities and processes so that we know to what they refer.

For instance, an interpretation of 'electron' as a corpuscle, as a wave, or as a wave-particle is a philosophical interpretation, while specifying what range of values each of an electon's properties admits is a semantic interpretation.

3. *Models*:

i   *Causal/Structural models*: true representations of systems whose causal and/or structural properties provide the explanation for observable phenomena.

ii *Causal/Structural analog models*: analogical representations of the systems mentioned above.

iii *Substantial models*: true representations of postulated substances.
iv *Substantial analog models*: analogical representations of
   a substantial kind.
v   *Functional models*: true representations of the functions of systems
   while ignoring their real structures and substance.
vi *Mathematical models*: true representations of the mathematical
   (geometrical or algebraic) structures of the phenomena
   (disregarding the true physical substance and structures).
vii *Logical models*: true representations of the logical relations (e.g.
   inductive support) among assertions in a theory. It is questionable
   whether such models exist in actual scientific praxis, but we list
   them for the sake of completeness.
viii *State-space models*: subsets of states that a system occupies. These
   are behavioral models; trajectories of mechanical particles are best
   examples, but if understood in its most general sense, we may even
   simply call models of this type trajectories, namely, a linearly
   ordered set of points in a state-space.

For instance, taking molecules in a gas as pointlike objects would be giving the
gas a substantial model, and thinking of them during collisions as billiard balls
would be using a substantial analog model. A mechanical model for an electric
circuit or for the metabolism of an organism would be a functional model,
and if the functions can be expressed in precise mathematical formulas, it
may be said to be a mathematical model.

One must be careful with the state-space models. For classical systems,
they are configuration spaces, such as the position-momentum space of an N-
particle system. For quantum systems, they are neither configuration spaces
nor Hilbert spaces but spaces of quantum states whose trajectories are traced
out by the systems in accordance with the Schrödinger equation. More precisely, a pure state is described by a wave function, $\Psi \in \mathcal{H}$, (satisfying
$\|\Psi\|^2 = 1$), which is interpreted through the expectation value of the observables A :

$$\psi : A \in \mathcal{B}(\mathcal{H}) \mapsto < \psi; A >= (\Psi, A\Psi) \in C \quad .$$

And a general state is described by a positive linear map

$$\psi : A \in \mathcal{B}(\mathcal{H}) \mapsto < \psi; A >= Tr(\varrho A) \in C \quad ,$$

where the *density matrix* ϱ is a positive linear operator of trace 1.

Below is a category of models that are entirely different from the above
and do not even resemble semantic models in logic. Here the logical distinction
between theory and model is ignored.

4. *Epistemic models*: theories or models that are distinguished merely by
their epistemic status. Here, a theory is a set of assertions that truly and

completely describes its phenomena, while a model is a *potential* theory that approximately describes them.

i   *Idealized models*: theories that are true only under idealization conditions.

ii *Exploratory models*: theories that are true only of known aspects of the phenomena while it is clear that other essential aspects may exist.

iii *Analog models*: theories that are true of well-known phenomena that closely resemble the unknown intended ones.

iv *Instantiation models*: theories that are true of specific systems which instantiate theories that consist of universal/general laws/principles.

For instance, the theory of ideal gas uses an idealized model for gas and the model is also exploratory because relaxing some of the idealizations gives us a better theory (e.g. the van der Waals theory). The theory is also an instantiation model for thermodynamics.

5. *Theory testing*

i   *Test of consistency*: to find a model of any kind of theories listed above.

ii *Direct testing of a set of assertions*: to see if a model satisfies a set of assertions under the intended interpretation and if the model resembles the real system.

iii *Indirect testing of a set of assertions*: to see if a set of assertions is derivable from a set of directly tested assertions.

iv *Suitability of models*: to see if using the model which is tested, one can derive those results that are obtained from a theory without the model (e.g. the molecular model for a thermodynamic theory of gases).

v   *Model reduction*: to see if a phenomenological model can be explained without remainder by a (or more than one) causal/structural model(s).

vi *Model testing*: to see if a model represents within reasonable limits the intended phenomena.

(5.vi) is needed for the sake of (5.ii) above, for without it (5.ii) would be defining an empty set: what use could there be of a theory which is satisfied by a model that does not represent the intended phenomena? For the purpose of testing aeronautic phenomena, model devices have to be built and to have their suitability tested. It is interesting to note, as Emch [1993] pointed out, that while theoretical physicists construct models mostly for approximation, such as we have in the category of epistemic models, mathematical physicists use them mostly for testing the consistency of their theories, such as we have in 5.i above.

In the literature, other notions of models and theories are also used. Let us now review what they are and how they are related to the ones given above.

First, for Achinstein and Hesse (AH) [Achinstein, 1964, 1965, Hesse, 1970] – see, also [McMullin, 1968, 1978, Hughes, 1990] – a theoretical model is an abstract and nonlinguistic entity that bears analogy to some other familiar theoretical systems; it is equipped with an exactly describable *inner* structure or *mechanism* which *explains* the system's observable behavior, and is treated as an *approximation* useful for certain purposes: either theoretical (e.g. giving explanations) or practical (e.g. making predictions). This concept is unfortunately indifferent to how the models are specified: whether by a set of assertions in some language or by a computer generated picture or by some other means. This seems to imply that models can have multiple equivalent descriptions.

Second, one can formulate what we call Earman* concept [Earman, 1989, Beth, 1949, 1961, Van Fraassen, 1967, 1970], according to which a model is simply that which satisfies a solution or set of solutions of the (differential) equations of a scientific theory (under specific initial/boundary conditions). Furthermore, a theory should be identified with its model closure (i.e. all of the things that its solutions represent). One should of course distinguish between the solutions which are either algebraic formulae or geometric figures (essentially linguistic items) and the models (non-linguistic) that they are used to represent. Given the state-space version (van Fraassen), these models are trajectories (understood in the broadest and most abstract sense of the term) in a state (or phase) space (limited by the parameters and given by the values of the variables).

Third, Redhead [1980] takes models, especially those in physics, to be impoverished or enriched versions of theories, the latter being sets of interpreted assertions (which include, *inter alia*, laws and definitions). A simplified version of a theory is one of its models (i.e. an impoverished version), and so is a further articulated (e.g. with more details) version of a more general theory (i.e. an enriched version). Theories and models are entities of the same logical category; the difference being that they serve different purposes in the practice of theory-building – see also [Emch, 1993]. These offer sharper delineations of the relations between a theory and its models than is offered in [Truesdell and Muncaster, 1980] they may be said to be along the same line of approach. Nevertheless, we reproach [Truesdell and Muncaster, 1980] for often equating models with theories in science.

All of these concepts are undoubtedly instantiated in science and daily life; the question is whether they are indeed distinct concepts. For scientists, the notion of models of whatever kind bears at least some connotation of their being instruments by which an access to the phenomena in question may be more expediently gained than otherwise. There may be in practice many ways to gain such access, the easiest and most common of which would be to use an actual and *well-understood* system whose similarity with the object under study is either known or easily predicted. Then, one formulates a theory of such a system and takes it to be roughly describing the mechanism of the target object. It is easier to understand and control the degrees of approximation and the difference between the object and its model when one can evaluate them quantitavively, so that the differences and their possible reduction (which may improve the approximation) can be precisely represented and predicted. Many variations of such a strategy exist: one may, for instance, drop the 'actualness' requirement and use a purely mathematical structure, if one only aims at capturing some structural properties of a system or proving the consistency of a theory. 'Well-understood' does not have to mean 'already familiar'; to be a well-known system might mean being a 'well-defined' system, the discovery of whose structures being only a matter of time and diligence.

To see how the latter notions are related to the ones in the main list, let us first take $\mathcal{M}^A$ to be the class of interpretations whose systems or structures are similar to the target system or structure $\mathcal{A}$, and $\mathcal{T}^A$ a theory of $\mathcal{M}^A$, i.e. $\mathcal{T}^A = \{\varphi \in \mathcal{L} \mid \mathcal{M}^A \models \varphi\}$, such that $\mathcal{T}^A$ is an approximation of a true theory of $\mathcal{A}$. The same is true if $\mathcal{M}^A$ is a set of trajectories or behaviors that approximate the true trajectories or behaviors of the target object $\mathcal{A}$. Now, $\mathcal{T}^A$ could be any one of the epistemic models except the instantiating models, and they are models in Redhead's conception. If $\mathcal{M}^A$ contains causal/structural models or substantial models, it may be identified with the AH models; and if it contains state-space models, Earman* models are most appropriate for it.

Suppes [1972] believes that the different conceptions of models used by scientists and mathematicians alike share the same meaning as the logical conception: they only differ in uses [Apostel, 1961, Wartofsky, 1979]. Suppes is, and in a sense must be, right. Otherwise, *either* the semantic view fails to capture the use of models in sciences (if its models are different in kind from the ones scientists use), *or* it cannot use the results in model theory, namely, the notions of isomorphism, embedding, etc., among L-structures may not even apply (if its models are different in kind from the ones logicians use). This is a dilemma that semanticists cannot afford to be straddled on. When asked with what a scientific theory should be identified, the semanticists must make a choice: the dilemma is forced on them.

1.4 Models and Semantics: a Hybrid View

In principle, the only way to avoid the dilemma – either one's semantic view fails to capture model-building in science or it cannot use the results of model theory in logic – is to identify all models used in science as semantic models. But if semantic models are restricted to first-order structures, such identification is clearly impossible, because many models in science are not merely first-order L-structures. In other words, the structuralists are hung on one of the horns of the dilemma: their models find no correspondence in the model-building practice in science. But there is no reason why we should adhere to the restriction of first-order structures only. Since our scientific languages are usually not first-order, they should not be expected to be satisfied only by first-order structures, and yet no one could deny that theories in science obey in general the semantic rules given in mathematical logic.

As we explained in Sect. 1.2, both of the two alternative versions of the semantic view, [V.1] and [V.2], are compatible with the notion of semantic models. However, each version picks out a different type of L-structures: [V.1] picks out the state-space models and [V.2] the causal/structural and substantial models. Had the semantic view been construed to include both types, then it would have been nearly correct. A scientific theory would be then a set of assertions that defines a class of models of the [V.1] or the [V.2] type. But the received view takes [V.1] and [V.2] to be alternative formulations of the semantic view [Lloyd, 1988, pp. 14–20] [Hausman, 1992, pp. 72–78]. For those who choose [V.1], all theories are (ultimately) sets of state-space structures which satisfy a set of assertions thus 'defining' them. And these authors seem to think that law assertions are a part, and indeed the core, of such definitions since the state-space structures are what make the law assertions true: the causal/structural models are then unnecessary, since everything is eventually reduced to a state-space structure. For those who choose [V2], the situation is less clear. Giere seems to think that once one obtains the right set of systems, the behaviors of these systems as represented by the state-space structures are automatically given; so one should not worry; and Suppes does not even mention the state-space.

To illustrate this point, let us use an example which we discuss again in Sect. 2.4: the ideal gas. From the account given there, it is considered the result of applying the more general theory of thermodynamics of equilibrium and homogenous fluids (and so, as an aside, it should *ipso facto* be a model according to Redhead). Besides the assertions that define the object, we have the two laws of thermodynamics, (2.4.5) and (2.4.6), and the Gay-Lussac law, (2.4.14). In other words, the laws that govern the dynamic process of the ideal gas are

$$\eta = \mathrm{d}U + \tau \qquad \text{and} \qquad \eta = T \mathrm{d}S \,, \tag{1.4.1}$$

while

$$pV = RT \quad ; \quad C_v = c \tag{1.4.2}$$

are sufficient conditions under which the two general laws (1.4.1) are satisfied. From the derivation given in Chap. 2, one sees that the solution of these equations is given in (2.4.18) and (2.4.19):

$$\begin{aligned} U(T) &= C_v \left(T - T_o \right) + U(T_o) \\ S(V,T) &= C_v \ln(T/T_o) + R \ln(V/V_o) + S(V_o, T_o). \end{aligned} \tag{1.4.3}$$

Here, we see an application of the inverse function theorem (cf. Appendix B) : the usual extrinsic variables, V and T, can be replaced by the intrinsic ones, U and S.

Obviously, both pairs of equations define trajectories in the V-T state-space – here, level curves, e.g. $U(T) = u$ or $S(V, T) = s$, where u and s are constants. Strictly speaking, the models that satisfy the laws (1.4.1) (and the definition of the object) belong to the causal/structural or substantial category and thus fit [V.2], while those satisfying equations (1.4.3) belong to the state-space category and thus fit [V.1]. But those who choose [V.1] take the state-space level curves to be models of equations (1.4.1) (i.e. the general laws), and for those who choose [V.2], the models are systems or ideal gas (abstract objects) which satisfy (1.4.1). And these systems are not identical to the level curves, even though the latter do represent the behavior of the former. To further clarify this point in general, let us consider the following points.

First, there is a dangerous equivocation in the above use of the term 'satisfaction'. An L-structure **satisfies** an assertion in L, and a set of values *satisfies* a law assertion (e.g. an equation) as the latter's solution. Both are legitimate usages, but with very different meanings. The former which we call sat_1 is a relation between an assertion and a L-structure that makes it true, while the latter – call it sat_2 – is a relation between two sets of assertions such that the one set is derivable from the other. That a trajectory satisfies a law assertion (most likely a differential equation) would have to mean that it sat_1 a set of assertions (expressing obtained values) which, in turn, sat_2 the law assertion.

Second, the law assertions when interpreted 'explain' the presence of the trajectories, in the sense that the structures which satisfy the former are causally responsible for the latter. Hence, the equations (for laws) and the solutions (for trajectories) cannot be descriptions of the same L-structures (or models). The two types of models are certainly related, but one cannot be the model for the theory of the other.

Third, many law assertions or equations only yield – as solutions – observable values under special circumstances. Hence, the structures that satisfy the equations cannot be identified with the trajectories which are models of such sets of observable values. But perhaps they can be identified with the possible solutions in the following sense. Whenever a law assertion is given, all of its possible solutions are determined, whether we know them or not. Why cannot we identify those with the law? In principle we indeed can, but practically such identification is meaningless. Suppose that we have as our system a fluid whose lawlike structure is given by a set of differential equations on its molecules which is so complex that no solutions are known. If we insist on obtaining the set of trajectories that satisfies those solutions as the equations' models, we are not able to say anything about the fluid, even though we can in fact say a lot about it at least as long as its Reynold's number is not reached. Furthermore, we cannot tell whether some of the fluid's properties which are represented by the set of equations are reducible to some other properties which may be represented by another set of equations, etc. That includes all the benefits that the semantic view is supposed to offer. In other words, if the set of solutions that 1–1 corresponds to the law-assertions is known in detail, it is harmless (even though a mistake) to take their models to be the models of those laws. But if not, one must go back to the laws' own models, namely, those structures that sat_1 the laws.

Fourth, microscopic models defined by postulates of theoretical entities and their behavior are widely used in science to explain macroscopic (or observable) phenomena represented in the form of sets of recorded experimental data (or fitted graphs resulting from those data). In this case, it is even more egregious to identify the structures that satisfy such graphs to be the same models that satisfy the laws governing the microscopic entities in whichever sense of 'satisfaction'. Moreover, although the causal laws of the microscopic entities are essential in our attempt to explain the macroscopic phenomena, the state-space structures of individual entities are rarely considered or even considerable. The most obvious example would be the causal explanation of thermodynamic behavior of ideal gas by evoking its molecular structure. Although the equations of motion for those molecules are essential in the explanation, no one dreams or should dream of solving them for each molecule in the gas. In this case, the causal/structural models of the molecules of ideal gas are important but not the state-space models. The state-space models only come into play at the macroscopic level.

We are of course not claiming that ontologically a system (e.g. an ideal gas) is not extensionally identical to its entire state-space instantiation. But ontology is not the issue here. At issue is what should be identified as the *theory* of a phenomenon.

Now, the foregoing discussion indicates in which direction modifications should be made on the original version(s) of the semantic view. We now see that [V.1] and [V.2] should not be taken as two alternative versions of the view but rather as illustrating two aspects or components of a single view. Keeping in mind the critique of the concepts we gave in our scheme in Sect. 1.2, we now present the flow-chart of Fig. 1.2 that offers a hybrid view. Some observations are in order.

1. While the *explanatory* relation between boxes II and IV can be seen as exceptionless on the level of their members, the *derivational* relation between boxes I and III should only be seen as between the two classes because it is clearly not always true between their members, for the two classes may share many assertions; for example, they may share those that assert the presence of the constituents of the system in question.

2. Models in the second box may also be called theoretical or explanatory models: their domains contain idealized systems whose structures or internal mechanisms – mostly lawlike and representing the essential properties

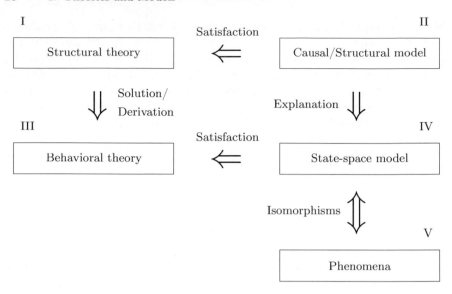

Fig. 1.2. The enhanced (or hybrid) semantic view

of the real systems of which they are idealizations – are described by law, or lawlike, assertions (or differential equations). Suppose that $\mathcal{M}$ is a set of causal/structural models for a structural theory, $\mathcal{T}$. $\mathcal{M}$ may contain structures of radically different compositions that are nonetheless homomorphic (even isomorphic) to one another *with respect to* $\mathcal{T}$. Some of them may be well-known, others may not. This is where analogy among models comes in and causal/structural analog models play a role. Moreover, a realist may take some models to be real systems, while an empiricist may consider them only as heuristic devices.

3. Models in box IV may also be called observational models. These are the state-space trajectories that satisfy either the solutions of those equations in a structural theory or an extension of an actual data collection (e.g. the best fitted graph through the raw data points). That is why we always consider observational models to be idealized data. Analogical reasoning does not apply to these models.

4. As far as comparing a model with an actual system is concerned, one can only use models in box IV. According to van Fraassen, all models which may be isomorphic to a phenomenon must be the ones in a state-space. Empirically, that is the only way in which we have access to actual systems; for it does not make much sense to compare an idealized system in box II with an actual system, except in the indirect sense of how much idealization has been introduced. What corresponds to a model in box II is empirically shielded from us so that we can know such a model – if it exists – only through

indirect means. Hence, Giere's theoretical hypothesis, which says that an actual pendulum, for instance, is similar to the one in our theory, can only mean that a curve plotted by observing the movement of a real pendulum is similar to one of its state-space models. This point, which concerns the issue for establishing a confirmation theory for the semantic view, is discussed in detail in Sect. 1.6.

5. Boxes I and II and boxes III and IV may form pairs to serve on any theoretico-experimental landscape. At the theoretical end, for instance, we have mathematical physics whose theories are quite distant from possible measurements. Still, one may find an axiomatized theory, $\mathcal{T}$, which is satisfied by some highly idealized systems (i.e. Boxes I and II) (e.g. the Ising model of ferromagnetism), and also a theory, $\mathcal{T}^*$, of some Gedanken experiments whose models (in the set of state-space models) supply idealized data (i.e. Boxes III and IV). At the experimental end, we may have a theory of the experimental set-up (which can be structural) whose models are similar to the actual experimental apparatus (i.e. Boxes I and II), and we also may have an extension of the collected data (i.e. Boxes III and IV); the former is what Suppes calls *"models of experiment"* and the latter *"models of data"* [Suppes, 1962, p. 259]. The problem with the syntactic view is that its scheme can only be applied across the whole vertical landscape.

6. In combating the syntactic view, semanticists went overboard in denying the significant role language (especially a mathematical language) plays in scientific theories. After decades of 'language bashing', the original assessment of the situation in [Suppes, 1972] seems to be the right one (see the quote from Suppes in Sect. 1.2). Therefore, we not only distinguish *theories* from *models* (either structural or behavioral) but also put them on a par; and we would not mind calling items in boxes I and III theories as long as one knows that items in boxes II and IV must be given to complete the story. The real problem with the syntactic view is that models of highly theoretical claims are never considered; theoretical reasoning is imagined to be carried out exclusively through syntactic rules until one obtains solutions that can be experimentally tested. So construed, the view is naturally plagued by the problem of underdetermination. What mitigates the problem is not that we should abandon language, but that we should take theoretical models (e.g. the AH models) seriously, so that the relation between the highly theoretical and the experimental can be studied through the relation between their models.

7. The syntactic view makes realism of unobservable and theoretical laws a very implausible, though not impossible, position. If the calculus is not even interpreted until one reaches the level of experimentally obtainable results, it does not seem tenable to ask whether or what the theoretical terms denote. The semantic view turns the table towards realism, although not completely. Van Fraassen's constructive empiricism extols the utility, but denies the reality, of theoretical models, Suppe's quasi-realism takes them to be real but

non-actual worlds, which may be viewed as realism along [Cartwright, 1983] line, because Cartwright's realism is an entity realism in that she regards entities postulated in theoretical models as possibly real even though claims about those entities may be simply untrue. [Giere, 1988] is probably the strongest realist position stemming from the semantic view.

1.5 Semantic View and Theory Testing (Confirmation)

Neither the index in Sneed's classical work [Sneed, 1971] nor that of the most recent and ambitious milestone of structuralism [Balzer, Moulines, and Sneed, 1987] contains any of the following entries: confirmation, justification, testing, and experiment; nor is there any name associated with works in any area of confirmation theory or scientific methodology. Works on this side of the Atlantic are less neglectful. Giere [1988] has chapters on justification and experiment, so does Lloyd [1988]; and in Suppe [1989] the relation between theories and phenomena is given a detailed account. But if one looks up the relevant parts of those books on confirmation or theory testing, one either finds a discussion disconnected from the semantic view (e.g. Giere) or one that is limited to only one species of theory testing (e.g. Lloyd and Suppe).

To us, this neglect seems surprising and unjustified. Confirmation of scientific theories ought to be a major part of a philosophy of science and most approaches in the literature, such as [Hempel, 1965], [Salmon, 1989] and the Bayesian [Howson and Urbach, 1993], are in a syntactic orientation.

If the semantic view is to replace the syntactic view, should not it address this issue and provide revisions to the old accepted approaches?

According to [Van Fraassen, 1989] succinct formulation of the semantic view, to test a theory is to test *"a theoretical hypothesis, which asserts that certain (sorts of) real systems are among (or related in some way to) members of that class"* (222–3). So stated, the theoretical hypothesis (TH) gives no general guidance as to how one may go about testing it. For instance, suppose the domain of a class of models contains non-viscous fluids, and so a typical TH would be that *"this tank of pure liquid helium is among or related to one of the fluids in the models"*. It is of no use if it does not inform us how we can find out whether one of the model fluids is or resembles the liquid helium sample.

Now, what are the possible ways of substantiating a TH ? The simplest would be when the semantic models of a theory – e.g. aeronautics – include simple mechanical systems – e.g. *model* airplanes. According to Apostel [1961], all one needs to do to test whether airplanes work precisely according to our theory is to build model airplanes and fly them under the controlled environment. If a model plane performs well within pre-assigned limits and we know that merely altering its absolute scale without changing the proportions of its parts does not affect its performance, then the theory is confirmed.

This is perhaps also true for more complex but artificially constructable systems, e.g. chemical compounds (this may be technically complicated but still simple with respect to the logic of testing). Other kinds of models, such as those of natural biological organisms and abstract microscopic cooperative systems, are not so easily tested.

Suppe [1989, p. 84] speaks of a theory being empirically true if *"the class of theory-induced physical systems are identical with the class of causally possible physical systems"*. So, to test a theory is to find out empirically whether such an identity relation holds. However, in the class of causally possible physical systems, Suppe clearly meant to include actual as well as non-actual (yet physically possible) systems. Hence, the testing could not be a matter of ostensibly identifying through empirical means the members of that class in the actual world. Suppe suggests that we develop an *"experimental methodology [that] enables us to determine the states of a causally possible physical system by observing and measuring, and so on, the corresponding phenomenal system."*[Suppe, 1989, p. 103]. Despite its special terminology, Suppe's claim in fact captures the general idea of theory testing for semanticists, namely: to test a theory is to find out by empirical means if the behavior (i.e. sets of states) of an element in the model set *matches* (with some degree of approximation) that of the phenomenon in question (i.e. the actual states of some actual system).

Suppe's method seems most applicable to behavioral theories and state-space models since it is essentially a search for a match between the latter and the experimentally obtained data. The laws that are satisfied (in the sense of sat_1, see Sect. 1.4) by the state-space models are laws of coexistence; or in practice, they are the equations by which a theoretical graph of consecutive states can be plotted. For instance, the Gay-Lussac law of ideal gas undergoing a quasi-static process

$$t \mapsto (p(t), V(t), T(t)) \tag{1.5.1}$$

is such that at time t, $\{p(t), V(t)\}$ gives a complete description of the state of the gas, i.e. $T(t)$ is itself a function of $p(t)$ and $V(t)$ alone:

$$T = T(p, V) \quad \text{specifically} \quad T(p, V) = \frac{1}{R} pV \tag{1.5.2}$$

which is the state-equation of the gas.

To put it roughly, the testing of a set of state-space models defined by a behavioral theory is to construct experiments on the intended phenomenal system in which the system's states – through time – are obtained and plotted as measurement results. We then compare the experimentally obtained trajectory with the one predicted by the law and see if they match. If they do, the theory is confirmed; otherwise, it is invalidated.

This kind or level of testing or confirmation corresponds to two items in our scheme under *"Theory Testing"*, namely, *"direct test of a set of assertions"* and *"model testing"* [(5.ii) and (5.vi) in Sect. 1.3]. In our present context, the assertion could either be an algebraic equation, e.g. again $pV = RT$ or a long conjunction of simple assertions in the form 'T at time t has value $T(t)$'. And the models should all be state-space models (i.e. the testing of such models are direct testing).

In order for this methodology to work, the TH in our case should probably be read as saying that the set of actual states of some systems when mapped to the real numbers matches some of our state-space models. If we identify the actual system with the collection of all of its actual states, which seems to be a sound idea in ontology, the sense in which the two systems of different compositions resemble each other seems clear. However, the sense of a comparison between two radically different kinds of systems is only metaphorical; what is really tested here are individual assertions of the form: for a system S, and time t, the number attributable to the measured property of S at time t matches the number attributable to the corresponding property of M at t, where S is an actual system and M a model. The possibility of such a direct comparison is given by a theory of measurement whose foundations run deep in logic and mathematics.

One should note of course that there is no perfect match and that approximations of some degree are always involved. In most cases two kinds of factors play a role in such approximation processes. First, most theories, even if they are at the level of phenomenological laws, are only true under idealized conditions. Such conditions must be taken into consideration when one handles the problems of how predictions and data may ideally fit together. This is known by scientists as sorting out *intrinsic deviations*. Second, interference and random noise always create some error in our experiments, which must also be estimated (e.g. through statistical means). This is known as considering *extrinsic fluctuations* . But there are some cases in which the degrees of approximation are not precisely knowable because either the idealization is too radical or the system too complex. Hence, we distinguish *controllable* from *uncontrollable* approximations. The former consists of approximations in which the deviations in the initial/boundary conditions of a process are connected by some laws to those of the predicted outcomes. The laws do not have to be precisely known, but some assurance of their existence is necessary. When no such laws appear to exist between those deviations, the approximations are uncontrollable. Of course, specific experimental testings require specific models of approximation, but we do not see any general philosophical problems arising in this respect.

Is this in any way significantly different from a syntacticist version of theory testing or confirmation? One of the earliest and most popular theories of confirmation is the Hypothetical-Deductive (HD) method, which though discredited in its simple form still constitutes the core of many more sophisticated theories. In its simple form it can be given by a micro-scheme: for any hypothesis H and auxiliary assumption A, if H & A yield an assertion P

which is capable of being tested in an experiment, and P is tested positive, then H & A survive the test. If P is tested negative (i.e. if P disagrees with the experimental results), then either H or A is false. Since A is usually already better confirmed or tested, H must be rejected. $\{H, A\} \models P$ is purely a logical deduction (or a mathematical derivation). So stated, the HD method seems a perfect theory for syntacticists: it is a theory about logical relations among assertions. Models need not be invoked at all.

As Suppes rightly pointed out (see Sect. 1.2), the syntactic view is not wrong but too simplistic, omitting the rich details of measurements and experiments where the semantic view provides much illumination. And [Suppe, 1989, pp. 96–99, 118–151] exposition of his theory of experimental testing could certainly fill the lacuna. When one asks how P is tested in the above HD method, the semanticist but not the syntacticist has the right answer. But confirmation does not happen only at the level of measurements and for phenomenological laws. Suppose that H is a highly abstract or theoretical hypothesis and P one of $H's$ measurable consequences. It looks as if the syntactic view has a better hand in characterizing how H is confirmed from P up, while the semantic view gives a better picture for how P is tested. There are many examples for this point throughout our book, for the first of which, see the discussion of Brownian motion in Sect. 3.1.

Hence, not all models in science are state-space models, nor are all theories (as descriptions of sets of models) behavioral theories (cf. Fig. 1.2, the hybrid view). Scientists not only propose theories that describe highly abstract structural properties of microscopic systems but also construct models for such systems. The question is whether it makes sense to talk about confirming or testing a structural theory by comparing its models to real systems in the world. Does it make sense to say that a microscopic system S of electrons is P, where P is a set-theoretical predicate defining a kind of physical systems, such as a 'Fermion gas'? Since S here is only remotely accessible to our observation, it is not clear whether a comparison would make sense between S and some other models that satisfy P since none of these models can be observational. Moreover, one would be ill-advised to construct state-space models for such systems and think that they can be used for testing the TH, because those models are most likely to 'exist' in some abstract mathematical spaces which have no direct connection to the kind of state-spaces that are directly testable. For example, the set of light waves which Arago, Fresnel and Young postulated as a linear space is not the space in which they observed the interference patterns of light beams. Whereas the latter is the space we all know, the reality of the former was then far from a settled question.

Besides the types of testing or confirmation discussed above, which are widely examined in the literature, there are other kinds of testing in science – see, (5.i– vi) in Sect. 1.3. The testing of consistency (5.i) is an important enterprise in fields ranging from mathematical physics to econometrics. To see whether a theory is consistent is to find out whether there is (at least) one structure (specified in a language other than the one for the theory) that satisfies the theory. This activity is especially made intelligible by the semantic view because if a theory is a description of a set of models: what could be more important initially than finding out that the set is not empty?

What is involved in the suitability of models is not as straightforward as briefly stated earlier (cf. 5.iv in Sect. 1.3). Let us consider the case of using a molecular model for a thermodynamic theory of gases. The molecular model can be understood as an epistemic model or a semantic one. If it is epistemic, it may be an instantiation model (4.iv in Sect. 1.3) in the sense that the kinetic theory of gases can be considered as an instance of the general theory of thermodynamics: while the kinetic theory is sensitive to the structural constraints of various thermo-systems, thermodynamics is not. But this does not seem right, for the kinetic theory is better regarded as an instantiation of statistical mechanics. For one law to be an instance of another (under specific constraints), the two laws must be of the same type (i.e. at least using the same terms). This is true between kinetic theory and statistical mechanics but not between kinetic theory and thermodynamics. Between the molecular models and thermodynamics, there is a relation of *reduction* involved which lies beyond the scope of testing or confirmation. Similar considerations apply if the model is a semantic one. A molecular model cannot be said to satisfy (in the sense of sat_1 in Sect. 1.4) a thermodynamic theory.

Thus, we conclude that a successful theory of confirmation must be a combination of syntactic and semantic methods (such a combination also results in the hybrid semantic view in Sect. 1.4). To test a highly theoretical theory, one must take its law-assertions and relate them via some sort of derivation to those lower level (phenomenological) generalizations. Whatever general features and conditions of adequacy such a derivation should meet, they could not possibly be characterized solely in terms of relations among different models. The semantic perspective makes its most significant contribution at the level of phenomenological laws. This matches very well with our hybrid view, namely, one should distinguish two different types of theories, both of which have their own models. And since the models of structural theories (i.e. causal/structural models) are almost never accessible to empirical testing, such theories have to be tested indirectly, through some syntactic means.

1.6 Semantic View and Theory Reduction

Closely related to the notion of confirmation is that of *theory reduction* (or equivalence, a degenerate case of reduction), which defines a logical relation between two theories about the same phenomena.

We first examine a largely syntactic approach to reduction, which has been in dominance for the better part of the 20th century. Then we articulate an approach which is consistent with, but not strictly following from, the hybrid view (cf. Sect. 1.4 above). The conceptual structure and the main elements of this approach are borrowed from Scheibe's theory of reduction [Scheibe, 1988, 1993, 1997], because we find the theory most congenial to our hybrid view; but we give the core concepts of Scheibe's theory in a somewhat different formulation so as to show how a semanticist can benefit from this approach. For a criticism from Scheibe of how the traditional semanticists went to the other extreme, see [Scheibe, 1997, pp. 45–47].

Let us begin with some obvious examples for theory reduction (and equivalence) and a few naïve observations upon them. The simplest and perhaps trivial reductional relation holds between two logically equivalent theories for exactly the same set of phenomena. Take the example of the two laws of thermodynamics formulated in integral or differential forms (cf. 2.4.3–4 vs. 2.4.5–6). The integral equations are equivalent (cf. 2.4.7) to the differential equations on $D = \{(V, T) \in R^2 \mid V > 0,\ T > 0\}$ as a consequence of Green's theorem (cf. Thm. 3, Appendix B). It also follows from this equivalence relation that every system which satisfies one set of equations also satisfies the other set. Next consider, in the theory of gases, the relation between the Gay-Lussac law [2.2.9–10] and the van der Waals law [11.3.1]. They concern the same set of phenomena, but are clearly not logically equivalent. There are more adjustable (or independent) parameters in the latter than in the former. It seems reasonable to say the Gay-Lussac law for the "ideal gas" is *reducible* to the van der Waals law in that the latter makes more accurate predictions on the behavior of "real gases". In the same vein, Galileo's relativity is said to be reducible to Einstein's (special) theory of relativity. They can be considered to be about the same set of phenomena – the free particle dynamics – and special relativity does not contain any more adjustable parameters. When an object's speed v becomes comparable to the speed of light, the two theories give different predictions under the same circumstances. Nevertheless, there is a reduction relation because Galileo's relativity is recoverable from special relativity under an easily specified condition: when the speed v of a system becomes negligible against the speed c of light:

$$v << c\,. \tag{1.6.1}$$

The relation between classical mechanics and quantum mechanics is more strenuous. The functions that one uses to determine a quantum state are no longer the same kind of functions of spacetime that one uses to describe a classical state. But the classical theory stands with respect to the quantum theory in an analogous reducing relation as Galileo's relativity stands to Einstein's theory; we have here that the classical theory is sufficient when the energy E of a system is much greater than $\hbar$, the Planck constant; one writes:

$$\hbar << E\quad. \tag{1.6.2}$$

A simple consequence of the above reduction relation is a theorem, which can be attributed to Ehrenfest. It states that at high temperature quantum statistical mechanics gives results that are in agreement with classical statistical mechanics. One of the first triumphs of quantum statistical mechanics was to show, in particular, that while at ordinary temperatures, the heat capacity (or specific heat) differs widely from one solid to another, and for each substance depends on the temperature, in manners that the classical theory cannot predict, the quantum theory can make explicit how these differences subside as the temperature is raised, and that the heat capacity of each solid reaches (albeit only asymptotically) a value that is equal to the universal value beyond which classical theory could not discriminate – see (10.3.44–45).

The reduction between different theories, such as thermodynamics and statistical mechanics, biology and physics, and sociology and biology are even more complex and/or tentative. The only obvious sense in which the former theory in each of those pairs could be reducible to the latter is that the latter is expected to provide a better or a deeper understanding of the phenomena than the former; however such a conception is not well defined. These prototypes and observations are crude and naïve, and the hope is that our analysis in this section brings useful refinement and clarity to these and other similar examples.

With few exceptions (e.g. the structuralist approach) and until recent years, the philosophical account of reduction has been discussed exclusively in the traditional, largely syntactic, approach to scientific theories. In this tradition (see our discussion of the syntactic view in Sect. 1.2), the reducing and the reduced theories are characterized as axiomatized or semi-axiomatized sets of assertions or assertions. To see if theory $\mathcal{T}$ so described is reducible to theory $\mathcal{T}'$, one looks at how the terms and assertions appearing in the two theories are related. Among the early accounts [Nagel, 1961], such relations are taken to be those of definability (with respect to terms) and derivability (with respect to lawlike assertions). Such an account, if successfully construed, is believed to be universal, namely, that it applies to all cases of theory reduction in natural and social sciences. The two D's (i.e. definability and derivability) may be construed in different ways: different accounts of reduction may result. But however one manipulates them, the syntactic orientation is kept intact.

For such an account, various formal (i.e. logical) and informal (i.e. pragmatic) constraints must be introduced and the result is a set of criteria for the reducibility of one theory to another. For example, in [Nagel, 1961, p. 336ff] account of the general features of the thermodynamics-to-statistical-mechanics reduction, the reduction is characterized primarily as derivability of the reduced theory (*thermodynamics*) from the reducing one (*statistical mechanics*). The formal conditions Nagel requires for such derivations are roughly the following. (a) All assertions in both theories, such as law assertions and special hypotheses, must be formulated with syntactic transparency and have unambiguous meanings; in other words, the constituent terms of the assertions must be clearly recognizable from their syntactic formulations. (b) *"Every assertion of a science S must be analyzed as a linguistic structure, compounded out of more elementary expressions in accordance with tacit or explicit rules of composition"*(p. 349). These rules are syntactic rules of expression formation. (c) The two theories must share a large number of terms and assertions with the same meanings and truth assignments. This is a semantic requirement. Informal conditions, according to Nagel (p. 358ff), include (d) no or few ad hoc assumptions should be allowed in the reducing theory for the purpose of reduction lest any theory can be reduced to any other theory if the later is gerrymandered to serve the purpose; (e) the reducing theory must be significantly more general than the reduced one; (f) the reduction should produce an effect of unification, e.g. the reducing theory should entail a great variety of previously disconnected experimental generalizations. When these conditions are met, a theory, $\mathcal{T}$, is reduced to another theory, $\mathcal{T}'$, just in case the terms in $\mathcal{T}$ are in principle definable in terms of $\mathcal{T}'$'s terms; and assertions in $\mathcal{T}$, again, are in principle derivable from those in $\mathcal{T}'$ under specified limits and boundaries.

As with the syntactic view of theories, this approach is much too simplistic and, in some sense, even wrong. It is naïve to think that one can come up with a single account that encompasses all the reduction relations obtainable between any two theories in science (the diversity and complexity of these relations are partially revealed in the examples given above). It is worse than naïve to think that if there is a unified account of reduction, it is to be found in how scientific theories are formulated syntactically. Since reductive relations among theories most essentially concern the understanding of the phenomena which theories standing in such relations compete to describe, it is unlikely that the improvement of this understanding can necessarily be had by formally reducing one theory to another. Also, we should remember that it is not a good idea to use only assertions in analyzing theories and their relations: interpretations and models play an equally if not more important role. Hence, independently of any interest in seeing whether a syntactic approach along the Nagelian line can work, we should try to determine in what way models may help to construct a better account of reduction (and equivalence). These considerations should suffice in explaining why a search for a semantic account of reduction is desirable. (For a more thorough criticism of the Nagelian approach, see [Suppe, 1977]).

Before we delineate a *semantic* account of reduction, let us see what possible apparatuses might be available for this purpose in model theory (cf. Appendix A). Two concepts from that appendix are particularly relevant in this connection: one is the notion of 'substructure' (or 'extension') and the other is that of 'reduct' (or 'expansion'). Accompanying them are the notions of homomorphism, embedding and isomorphism, the second is an 'injection' (i.e. a 1–1 into mapping) and the third a 'bijection' (i.e. a 1–1 onto mapping). For one structure, $\mathcal{A}$, to be the substructure of another, $\mathcal{B}$, (or $\mathcal{B}$ an extension of $\mathcal{A}$), it must be of the same type as $\mathcal{B}$; and the universe of $\mathcal{A}$ is a subset of the universe of $\mathcal{B}$. And an injection from $\mathcal{A}$'s elements into the $\mathcal{B}$'s which is defined on all items in the type (e.g. functions and relations) must exist. Reduct (or expansion) requires that the reduced structure, $\mathcal{A}$, have the same universe as the reducing one, $\mathcal{B}$, and its type a subset of the type of $\mathcal{B}$; and an injection must exist from its elements into those of $\mathcal{B}$'s which is defined on all items in its type. Hence, a substructure of a structure, $\mathcal{B}$, may have fewer objects in its universe, but has the same number of constants, functions and relations as in $\mathcal{B}$; whereas, a reduct of $\mathcal{B}$ must have the same number of objects in its universe, but may have more constants, functions, and relations than $\mathcal{B}$ has. Now, if $s\,\mathcal{B}$ is a substructure of $\mathcal{B}$, and $r\,\mathcal{B}$ one of $\mathcal{B}$'s reducts, then $[s \circ r]\,\mathcal{B}$ yields a structure which may have fewer objects, fewer functions, etc. When a structure $\mathcal{A}$ satisfies a theory $\mathcal{T}$ (i.e. becomes $\mathcal{T}$'s model), every assertion in $\mathcal{T}$ is true in $\mathcal{A}$. Then, one can intuitively see that a reduct of $\mathcal{A}$ satisfies $\mathcal{T}$ if $\mathcal{A}$ is a model of $\mathcal{T}$; but this is not necessarily true for the substructure(s) of $\mathcal{A}$.

Unfortunately, none of these notions (or other notions such as isomorphism and elementary equivalence) can be directly applied in formulating or examining the semantic view of reduction. Scientific theories and models are much too complex for such simple first-order notions to apply. Nevertheless, the logical notions usually direct us towards the right direction.

We would like to follow Scheibe's example in trying not to get our hopes too high and projects too ambitious. The reductional relations in science, as pointed out by Scheibe [1997, pp. 40–41] and briefly indicated in the examples given at the beginning of this section, are multifarious and thus such *diversity* should be reflected in our account. However, systematicity is also a virtue since a motley of descriptions each for a different reductional relation could hardly be called a view or account of reduction. With this in mind, let us survey the main ideals of Scheibe's theory [Scheibe, 1993, 1997]. What Scheibe is trying to do in this new account is

> [T]o develop ... a synthetic theory of reduction. In this theory we start from a couple of special and widely differing concepts of reduction and gain the most general cases by their recursive [repeated] combination. [Scheibe, 1993]

The key idea is *recombination* and *decomposition*. According to Scheibe, one should distinguish and separate which reductional relations among scientific theories are *complete* and which are *partial*, and further for a complete reduction, whether it is *exact* or *approximate*. Both exact and approximate reductions may be *elementary* or *non-elementary*; the latter being constructions from combinations of the former. Although the complexity of the actual practice of science makes it almost impossible to list, once for all, all the elementary types, we have good reasons to believe that most complex reductional relations are *decomposible* to a combination of simpler, elementary relations; and some elementary reductions have almost ubiquitous presence in science [Scheibe, 1997, pp. 35–44, 109, 169–171]

Now, suppose that $\mathcal{T}$ is a theory that is reduced to another theory $\mathcal{T}'$. An elementary type of reduction between $\mathcal{T}$ and $\mathcal{T}'$ is represented by the notation: $R(\mathcal{T}, \mathcal{T}'; v)$, which reads: *theory $\mathcal{T}$ is i-reduced by v to theory $\mathcal{T}'$* [Scheibe, 1993, p. 251]. Here v is what Scheibe calls the vehicle of the reduction. If the reduction is elementary and of an identified kind, then it is expressed as $R_i(\mathcal{T}, \mathcal{T}'; v)$, where i is an index that traces the kind of reduction invoked. One may notice that this notation for reductional relations is no longer present in [Scheibe, 1997]; but for our purpose of giving a brief account of reduction under semanticism, the notation renders good services. Hence, following Scheibe, we take every reduction to be a ternary, rather than a binary, relation that is transitive. Two elementary relations are *composible* if the following is satisfied.

[R]: For any three theories, $\mathcal{T}$, $\mathcal{T}'$, and $\mathcal{T}''$:
    if $R_i(\mathcal{T}, \mathcal{T}'; v)$ and $R_i(\mathcal{T}', \mathcal{T}''; w)$ exist among them,
    then $R_{i,j}(\mathcal{T}, \mathcal{T}''; t)$ exists in which $t = v \circ w$.

Hence, two reductions are composible depending on whether it is possible to find a $t = v \circ w$ for the two relations. Such compositions can be applied repeatedly to produce any number of reduction relations as long as each item within the chain (except the two at both ends) is at the same time a reducing and a reduced theory.

Now, the selection of elementary reduction relations is crucial to the success of this approach. One may fail to see how a certain reduction is a combination of the elementary relations not because no such combination exists but because the elementary relations are not correctly chosen. And further, in using the scheme satisfying [R], one may require that the theories at the two ends of a reduction sequence be theories that actually appeared in history, but it would be unreasonable to require this for every theory within the sequence. So, one may have to use possible theories to complete the sequence; this should be harmless as long as such theories could have appeared in the past. Scheibe [1997] examined in great detail two of the most common elementary reductional relations: *(direct) generalization* and *refinement*. A theory $\mathcal{T}'$ is a direct *generalization* of $\mathcal{T}$ if and only if they are formulated in the same language and the model set of $\mathcal{T}'$ is strictly larger than that of $\mathcal{T}$. To specify the vehicle of reduction in generalization, one isolates the axioms of the two theories; the vehicle is an assertion which when added to the axioms of $\mathcal{T}'$ reduces them to those of $\mathcal{T}$. More precisely, suppose that Σ' is the set of axioms for $\mathcal{T}'$, Σ that for $\mathcal{T}$, and v is the vehicle, *all in the same language*. Let $\mathcal{M}(X)$ denote the set of models satisfying X. Then, we have $\mathcal{M}(\Sigma' v) = \mathcal{M}(\Sigma)$. For example, in the description of the motion of a single planet around the sun, Newton's theory of gravitation is a direct generalization of Kepler's three laws; the *vehicle* that makes this reduction possible is the condition that one considers in Newton's theory only those solutions that are bounded. When this condition is relaxed, Newton's theory allows in addition the description of parabolic trajectories, and the hyperbolic trajectories of the comets.

A theory $\mathcal{T}'$ is a *refinement* of $\mathcal{T}$ by means of D if and only if they share the same set of physical systems in the universe of their models and if σ' is a predicate in $\mathcal{T}'$ of a type of a system's property, then $D(\sigma') = \sigma$ is a less complete or refined predicate of the same type [Scheibe, 1993, pp. 253–254]. D is the vehicle for this type of reduction, and although the details for D are not yet well-defined, the idea seems right: $\mathcal{T}'$ is a refinement of $\mathcal{T}$ if they deal with the same set of physical systems or phenomena while $\mathcal{T}'$ accounts for it in more sophisticated predicates. For example, Planck's law of blackbody radiation is a refinement of the Stephan–Boltzmann law. The two laws obviously deal with exactly the same set of physical systems, and while the Stephan–Boltzmann law gives the total energy of the radiation, Planck's law offers the energy density – or distribution. More precisely, we have Planck's law (10.3.29):

$$\varrho(\nu, T) = \frac{8\pi}{c^3} h\nu^3 [exp(h\nu/kT) - 1]^{-1} \quad ,$$

where T is the temperature, h the Planck constant, ν the frequency of the radiation, and k the Boltzmann constant; it, when integrated, i.e. $\int_0^\infty d\nu \, \varrho(\nu, T)$, gives us the total energy of the form: $E = \sigma T^4$, where σ is a constant – the Stephan–Boltzmann law (10.3.34). According to the above definition of refinement, we see that $D(\varrho) = E$ is a less refined property than, but of the same kind – i.e. energy – as, $\varrho(\nu, T)$. We see in the next paragraph a related case of this example.

Now, the notion of equivalence between two theories can be seen as a sub-kind of refinement [Scheibe, 1993, pp. 255–256]. $\mathcal{T}$ and $\mathcal{T}'$ are equivalent if and only if they are refinements of each other. In other words, the vehicle D that transforms $\mathcal{T}'$ to $\mathcal{T}$ has a inverse, D^{-1}, which transforms $\mathcal{T}$ back to $\mathcal{T}'$.

It would be nice if all *exact* reductional relations in physics could be accountable as combinations via [R] of these two elementary relations (including equivalence). But it is very unlikely that this is true [Scheibe, 1997, Chap. 1]. However, unless one can list all the elementary reductions or give a general characterization of what constitutes an elementary reduction, the claim that all exact reductions are decomposible becomes a somewhat vacuous claim. Nevertheless, the complexity of a reductional relation does not necessarily come from the compositional complexities of exact reductions. There are in physics also reductional relations that are distinctly not exact, namely, by adding the vehicle, $\mathcal{T}'$ only *approximately* returns to $\mathcal{T}$. For example, adding the condition, $v \ll c$ (v being the speed of the system and c that of light), Einstein's special relativity only returns to Galilean relativity approximately (in other words, the latter is false if the former is true no matter how slow the system is moving). In fact, in the former, space and time are separately absolute; while in the latter, they are observer-dependent, and the remaining absolute being that of space-time. Another example of this can be seen in the relationship between Planck's law of blackbody radiation and either Wien's law or Rayleigh–Jeans' law (10.3.38). Here the vehicles are naturally the conditions under which Wien's law and the Rayleigh–Jeans' law are *recoverable* from Planck's law. In Wien's case, we have $T \ll h\nu/k$, and in the Rayleigh–Jeans' case, it is $T \gg h\nu/k$.

What one needs is to add to the two elementary relations the notion of *approximation* which gives us the notion of *asymptotic reduction*. The intuitive idea of such a reduction is that for $\mathcal{T}$ to be asymptotically reducible to $\mathcal{T}'$, there exists a family $\{\mathcal{T}^\varepsilon \mid \varepsilon > 0\}$ such that $\mathcal{T}^\varepsilon$ is exactly reducible to $\mathcal{T}'$ and

$$d\left(P(\mathcal{T}^\varepsilon), P(\mathcal{T})\right) < \varepsilon \quad, \tag{1.6.3}$$

where $P(X)$ is any observable prediction derivable from theory X, but for a different formulation, and d is visualized as some distance function on a space describing the range of all predictions. An asymptotic reduction can thus in general be decomposed into combinations of elementary asymptotic reductions, just as it would if it were an exact reduction.

Now, returning to the examples given at the beginning of this section, it seems clear that the first one – the integral (I) and differential (F) description of thermodynamic systems – are equivalent theories since there exists a D and D^{-1}, such that

$$R_r\,(I,\ F;\ D) \quad \text{and} \quad R_r\,(F,\ I;\ D^{-1}) \quad,$$

where R_r is a refinement reduction. And the relation between the Gay-Lussac theory (GL) and the Van der Waals (W) theory of gases is an asymptotic reduction which comprise a refinement and an approximation in that there exist a refinement vehicle D and a theory W^a such that $R_r\,(W, W^a; D)$ and W^a is an approximation of GL. Further, the relation between Galileo's (space/time) relativity (TR) and Einstein's (space-time) special relativity (SR) seems to be an asymptotic reduction consisting of a generalization and an approximation. In other words, there is

a generalization vehicle v and a theory SR^a such that $R_g\,(SR,\ SR^a;\ v)$ and SR^a is an approximation of TR.

The mechanism of Scheibe's theory has its roots in the model theory of mathematical logic (cf. Appendix A). Recalling our two notions introduced earlier, *substructure* and *reduct*, generalization is formally similar to substructure and refinement to reduct. The important difference lies in the introduction – in Scheibe's theory – of the notion of reductional vehicles which has no use in model theory. It is only natural to introduce new elements when one applies model theory to account for certain substantive relations – such as reduction – among scientific theories. In other words, we find Scheibe's theory appealing and ringing true precisely because: (a) it has its foundations in the mathematical theory of models, and (b) the added constraints accommodate the right features of reduction.

Once again, this section is not intended to be a summary and/or a critique of Scheibe's theory which appears in its latest form in [Scheibe, 1997, 1999]. Such an undertaking would be urgently needed since the works are currently only available in German, but unfortunately, our present book is not the place to do it. Here we can only adopt Scheibe's theory in spirit, so to speak.

1.7 Semantic View and Structural Explanation

When Hempel [1965] laid out his deductive-nomological (D-N) scheme for deterministic explanation and the inductive-statistical (I-S) scheme for statistical explanation, logical positivism had waned, but the syntactical approach which rode to power with logical positivism was still going strong. So, it is natural that Hempel takes an explanation to be an argument relating *explanans* – assertions that explain – as its premises with the *explanadum* – the assertion to be explained – as its conclusion. One should note that Hempel is clear in the very beginning that a distinction should be made between an explanation of the occurrence of an event and that of an empirical generalization. This distinction has been observed, by and large, by the subsequent contributors to the subject. To explain an event's occurrence, one provides an account of the circumstances under which it must, or is very likely to, occur. And this, according to Hempel, is accomplished when all the relevant initial/boundary conditions are given in conjunction to the necessary covering laws. Whereas, to explain an empirical generalization, which is often called 'theoretical explanation', one invokes laws of higher order or of theoretical character from which the empirical generalizations are derivable. To understand is to be able to anticipate, and the latter is guaranteed by the D-N and I-S type arguments.

Under Hempel's scheme, confirmation seems to be the exact opposite process of explanation and a kind of symmetry between the two prevails. If $\{L, \alpha\}$ implies β, then β is explained by L and α, and L is (partially) confirmed by β. It is not difficult to imagine that when a new piece of datum is recorded and recognized, it provides new support for the theory that anticipates it at the same time it is explained by that theory as to why it should be as it is. This symmetry holds roughly on both levels of explanation, i.e., whether the explanadum is a single event or an empirical law. When it is the latter, we say that theoretical laws that explain it are further confirmed by a newly discovered empirical generalization.

Hempel's account of explanation has been criticized, modified and improved in the past four decades – see [Salmon, 1989] for a retrospective – and all but a few of these contributions are given along the same syntactic line. The approach of *structural explanation* is one of those few that takes a different path. So far two different versions of this approach are present in the literature. It was first given a voice by [McMullin, 1978] and then modified by [Hughes, 1989a,b, 1990]. In his argument, McMullin pitched his proposal for structural explanation against what he calls 'nomothetic explanation' which purportedly refers to those models of explanation developed by Hempel, Salmon, and others. We discuss first McMullin's account, keeping an eye at the same time on its real differences with the Hempelian paradigm.

McMullin [1978] defines the structural explanation at the very beginning of his article: *"[w]hen the properties or behavior of a complex entity are explained by alluding to the structure of that entity, the resultant explanation may be called a structural one"*(p. 139). The account or model is better called 'hypothetico-structural' (HS), because the entities and their structures are in most cases postulated as something remotely connected to observations. It is clear from the article that McMullin considers the postulation of 'hidden' or 'unobservable' or 'theoretical' structures as crucial to HS explanation. We find this requirement unnecessarily stringent. Perhaps most of the historical cases of HS explanation happen to conform to this stipulation, but it is not necessarily so. One may presumably explain the moving of a clock's hands by the structure of the parts inside the clock, all of which are directly visible. Why should such an explanation be excluded from the group of HS explanations?

Otherwise, the concept seems clearly defined. The ways in which a structure of a system provides explanation for its behavior presumably have to be specified for each specific case, but examples abound in science and daily life that show how this can be done. For the simplest case, we have the swing of a pendulum being explained by the very structure of that contraption. More complex would be the case of explaining the behavior of a glass of water by invoking the molecular structure of the liquid that is relevant to such behavior. For these cases, the notion of models comes immediately to mind. *Prima facie*, the semantic view and the use of models should most befit structural explanation, but before we see if this is indeed so, we need to clarify a related issue.

What is needed, and McMullin does not supply, is an account of how exactly a structural explanation differ from a Hempelian one. McMullin takes *"a theory to be a set of propositions, explanatory in its intent.... part of the theory will be a specification of the structure"* (1978, p. 139). But how is this different from Hempel's largely syntactic schemes? If the explanadum is a type of behavior of a system, then according to the Hempelian schemes (either the I-N or the I-S), the generalization that describes the behavior is explained by theoretical laws that govern (deterministically or indeterministically) the microscopic components of the system together with the initial/boundary conditions, if it is derivable from them. Nothing is said to exclude a specification of the system's structure in the set of initial/boundary conditions. If this is right, McMullin's structural explanation is threatened to collapse into a species of Hempel's account of scientific explanation in general:

it is a set of Hempelian explanations where assertions about a system's structure at a certain time, t, appear in the initial conditions at t. However, we do think that there is a significant difference between McMullin and Hempel if McMullin's account is interpreted in the direction of the semantic view. But we need to discuss Hughes' version of structural explanation beforehand, as it offers the needed element to argue for that difference.

In Hughes' version (1989) of structural explanation, McMullin's account is modified and extended. The modification is introduced by way of broadening the notion of theoretical structures. Besides physical structures of material objects, Hughes takes facts and their relations also as legitimate structures for explanations.

> *A structural explanation displays the elements of the models the theory uses and shows how they fit together. More picturesquely, it dissembles the black box, shows the working parts, and puts it together again. 'Brute facts' about the theory are explained by showing their connections with other facts, possibly less brutish.*   [Hughes, 1989a]

To see what Hughes really means by these claims, let us look at his *"structural explanation for the EPR correlations"* [Hughes, 1989a, pp. 202–207]. The explanadum in that case is the correlation among pairs of particles that are prepared in a singlet state prior to their being measured at some distance apart from each other, and the failure of attributing classical probability distributions to the measured results. And the explanans, according to Hughes, consists of four essential principles of quantum theory: representability, entanglement, non-orthodoxy, and projectability. These principles may also be seen as facts about quantum phenomena that support a non-classical probability structure. And it is this structure that Hughes argues to be capable of explaining the EPR correlations.

One may object, especially if one believes in McMullin's conception, that this is no structural explanation since it does not tell us how the physical structures of the systems (i.e. the particles and the measurement apparatuses) account for the non-classical correlations. But we take the spirit of Hughes' concept to be that logical, mathematical, or some other abstract models (cf. Sect. 1.3) should equally be counted as capable of providing explanations. It is in this sense that we take Hughes' account to be an extension of McMullin's and to have opened the door to the semantic view.

Moreover, whether or not a structural explanation should involve physical or some more abstract structures seems to depend on the nature of the explanadum. Hughes' structural explanation of the EPR correlations, for instance, uses some pure mathematical structures of quantum theory because what is explained there is not the existence of those correlations but why they deviate from the classical ones. Some philosophers – such as [Fine, 1989] – even argue against the belief that the non-classical nature of the EPR correlations is something that needs an explanation; they seem to view it as an unjustified prejudice towards the classical world that lures people into thinking that quantum correlations are odd properties. However, it is clear, at least, that Hughes' account is not for explaining the occurrence of some singular events or repeated phenomena.

For another example, we have two harmonic oscillators: a pendulum swinging in small angles in nearly vacuous space and an electric circuit whose current oscillates. If the explanation is why they behave similarly, we may cite the formal similarity between the laws that govern their movements. However, this is not an explanation of why a pendulum or the current in an electric circuit oscillates. For this, the specific physical structure of neither is needed.

Now, let us see whether in light of the semantic view, structural explanation provides substantial improvement over the Hempelian and the later largely syntactic schemes. Let us suppose that it is harmless to extend the notion of structures in structural explanation to that of models in the semantic view (Hughes' modification, as we mentioned above, makes this plausible). If a scientific theory is essentially a description of a set of models (i.e. structures), it should be obvious that structures are the one responsible for the success or failure of scientific explanations, not the law-assertions that describe the relations among their properties. Moreover, from our previous discussion of the kinds of models included in the semantic view, it is easy to see that the structures McMullin has in mind are mostly those advocated by Achinstein and Hesse (the AH models, cf. Sect. 1.3). And the arrow in Fig. 1.2 between boxes II and IV, which we labeled as 'Explanation', seems to be adequately captured by this extended version of structural explanation. Let us examine below the benefits this account may have over the Hempelian.

First, the notion of structural explanation (when properly extended) provides a way of avoiding the defects of Hempelian schemes which take explanations to be arguments. Explanations have to be formulated as linguistic entities because they have to be deductively valid or inductively strong arguments. But to explain an empirical generalization about a certain system structurally, one does not necessarily need to derive that generalization from theoretical laws. It is sufficient if one knows the physical or biological structure of the system. Take, for instance, the case of explaining the thermobehavior of a system by invoking the system's molecular structure (see our discussion in Sect. 3.2). Knowing that a tank of gas consists of molecules that move randomly in the tank is sufficient for explaining why pumping more heat into it makes it warmer, given that we know that heat is a form of transferred energy and a system's temperature is proportional to the kinetic energy of each of its molecules. For this, one does not need to know any theoretical law that connect the behavior of the molecules and that of the gas, nor does one need to construct any deductive argument. To generalize a bit further, an understanding of a system's structure, especially when it is complex such as in the case of a biological organism, usually gives us an idea about the essential dispositions of the system, which in turn tells us how the system behaves under certain idealized conditions. More often than not, it fails to provide accurate predictions of what actually happens in most real circumstances. According to Hempel, no explanation is to be had in these circumstances, but that seems mistaken. Intuition tells us that we do have explanations in such cases, and the structural account provides the justification for such an intuition.

Second, one may provide an explanation for or understanding of all the behavior patterns of a system if one knows its physical or biological structure. So, an explanation may exist on a level where explicit laws and their deductive relations to lower-level generalizations are not known or do not exist. For instance, we think with good reason that we can explain the running of a car or a clock if we know enough of its physical composition and structure. We may not be able to derive every possible behavior of the car or the clock based on our current knowledge of its composition and structure, but in each case, we know at least the direction towards which such a derivation can be had. So is also the case with providing a structural explanation of thermodynamic phenomena by using the molecular structures of the thermo-system in question. When one considers the details of Boltzmann's H–theorem, one could not help but realize that a straightforward derivation of every thermodynamical behavior of a tank of gas from the mass and position/momentum of every one of its molecules is not really possible. But that does not prevent us from recognizing that Boltzmann's understanding of the structure of gases provides excellent structural explanations for those behavior patterns. In other words, to explain, one should not always require derivability.

Third, even for those theoretical explanations in which the empirical laws are derivable from the theoretical ones, producing the derivations may not be what explanations are about. Rather than saying that Kepler's laws are explained by the Newtonian laws of mechanics because the former can be derived from the latter, as the syntacticists would have it, it is probably better to say that different mechanical phenomena such as the movement of the planets, of a block on a smooth incline, and of pendulum in a near vacuum are explained by Newtonian mechanics because they are all Newtonian systems as accurately captured by these, among the many models that satisfy the Newtonian laws of mechanics.

2. Thermostatics

Les causes primordiales ne nous sont point connues mais elles sont assujetties à des lois simples et constantes que l'on peut découvrir par l'observation et dont l'étude est l'objet de la Philosophie naturelle.

[Fourier, 1822]

2.1 Introduction

Thermostatistical physics was developed to describe properties of matter in bulk. In this study we distinguish three levels. Note that we write *distinguish* not *separate*: we will have many occasions to explore the passages between these levels, and we do not want to let ourselves be confined within any particular level. Yet, we must start somewhere, even though we anticipate some traveling back and forth, treading over old roads with new shoes. Succinctly, these three levels are: the *gathering of data and/or information* and their presentation in ways that are recognizable enough to participate in the second level, namely *theory building*, while the results obtained at both of these levels are to be formulated in a manner that is verifiable/falsifiable enough to allow, at the third level, their *being tested and evaluated*. The dynamics and organization of this book is provided by our efforts to show how models thus arise as scaffoldings erected by scientists to organize their judgments at all three of these levels. In doing so we are often confronted with questions of consistency, either within individual fields of research or across diverse contexts. A superficial commentator might be tempted to advance that some – or even much of – pure mathematics is established in a "context-free" manner; we argue, in contrast to this caricatural view, that much of the appeal and integrity of the sciences, of which thermo-physics is a particularly instructive representative, depends on the fact that their practitioners have to take into account the necessity of traveling back and forth between different "contexts".

At the first level we collect and choose the classes of phenomena that constitute the purview of the field. This goes from the most rudimentary feelings we have on what is cold(er) or warm(er), to the much more sophisticated experimental evidence of the melting and recrystalization of solids, and of the existence of transport phenomena (such as the fact that metals

conduct electricity; see 15.3 Model 4); or (see Sect. 14.1) the appearance
of superconductive materials at very low temperatures [Kamerlingh Onnes,
1911]; for reminiscences on the conditions under which this discovery actually occurred, see [De Nobel and Lindenfeld, 1996]. In 1986 superconductivity
was discovered to be possible in a range of temperatures that are much easier to reach, and for materials (ceramics) that apparently share very little
else with the traditional carriers (metals or metal alloys) of superconductivity; for an account of their discovery, see [Bednorz and Müller, 1987]; and
for an update [Nowotny, 1997]. Note already here that some of the above
"more sophisticated experimental findings" are "evidence" only by reference to
thought-patterns or "models" provided by the prior existence of some structures and/or class of phenomena of the kind we associate with the second
level.

At the second level we consolidate and rationalize the corpus of experience
collected at level one; compare, e.g. with Aphorism XIX in Book I of [Bacon,
1560]. Even more importantly however new experiments are suggested, at
this level, in order to refine the primitive notions acquired at the first level;
to mention here only one example of the latter type of activity was the establishment of a temperature scale which was first [Newton, 1701, Amontons,
1702, Fahrenheit, 1724, Fourier, 1822] "reasonable and practical" and then of
a better scale of such significant importance that it came to be referred to as
"absolute" [Thomson, 1848, 1852]. Thermodynamics is the collective name of
the phenomenological theory resulting at this level. The theoretical determination [Carnot, 1824] of the limit of the efficiency of a heat engine was one of
the first signs the theory had come of age in France; in the Anglo-Saxon coun-
tries pragmatic, but momentous accomplishments occurred around that time
with the adaptation of the steam engines of Savery and Watt to transporta-
tion (on water by Fulton, 1807; and on land by Stephenson, 1829). The fact
that the latter happened apparently quite independently of Carnot's work has
led to some interesting sociological speculations; we must however resist engaging these, lest we venture too much out of focus; for similar reasons most
of the technological achievements deriving from thermodynamics are only
given scant consideration. Note that the examples we gave for the activities
involved at this level were chosen to emphasize that modeling here does *not*
necessarily appeal to microscopic mechanical models – although, even at this
level, the prevalent statistical aspects of the theorizing pertaining to thermal
phenomena has become intimately linked with atomistic models, a point we
have the occasion to develop later.

It is at the third level that occurs much of the physical modeling activity with which we are concerned in this book. To a large extend, although
not exclusively, this activity starts out of an openly reductionist attempt that
aims at "interpreting" thermodynamical quantities, and their relations, in mechanical terms (see, Sect. 1.6). Two types of models should be distinguished
here again: *macroscopic* and *microscopic*. To the class of macroscopic models

belong the attempts to correlate among themselves quantities such as the "pressure", "volume", "elasticity" of a gas ... all these quantities adhering to the usual discourse of classical mechanics. Still in the class of macroscopic models belong some early attempts – see Sect. 11.3 – to formulate certain phase transitions. In contrast, the primary categories for microscopic models are the fundamental interactions between elementary constituents, albeit what scientists have held to be "elementary constitutents" has changed in the course of time: from molecules and atoms, to nuclei and electrons, photons or phonons, to quarks and strings. Difficulties however soon mount when trying to deal consistently with the second type of models, as the discourse becomes pervaded with probabilistic notions that do not primarily pertain to the domain of pure mechanics; we have to move here into the province of statistical mechanics. The recourse to this new element – randomness – totally absent from thermodynamics itself, as it was from classical mechanics as well, is motivated to a large extend by the limited information we have on the details of what is actually going on, such as (but not only) the conditions that enter in the determination of the initial state from which a microscopic dynamical evolution is postulated to proceed according to the deterministic laws of mechanics.

It is primarily in this context that logical questions discussed in abstraction in Chap. 1 – e.g. questions of self-consistency and completeness – occur at every turn. While models in statistical mechanics play various roles, not limited to testing hypotheses, this aspect of models is one we want to consider more than it is usually done in the context of physical theories. In the remaining sections of the present chapter, we review briefly some of the materials in level one and two, upon which the theory of thermostatistics rests.

2.2 Thermometry

The first notion proper to thermostatistical physics is that of *temperature*. We distinguish between the primary notion itself and its quantification, i.e. the establishment of a temperature scale. Unless one wishes to call the heat equation itself a model rather than a theory, the modeling activity in this section is limited to the emphasis placed here on the theoretical delineation of the object of the study through its laboratory apprehension.

The so-called *zeroth law of thermodynamics* collects our crudest notions of warm and cold in the empirical statement that when a warmer body, say A, is brought into contact with a cooler body, say B, heat flows from the warmer to the cooler one until a point where no observable variation occurs anymore; A and B are then said to be in thermodynamical equilibrium with one another. Moreover this reflexive property is recognized to be transitive, i.e. if A is in thermodynamical equilibrium with B, and if B is in thermodynamical equilibrium with C, then A is in thermodynamical equilibrium with C. This equivalence relation is expressed also by saying that the two bodies A and B are at the same temperature.

The semantic issues here should not be underestimated. A consensus on the meaning of statements to the effect that the heat flows in a certain direction depends on the acceptance that such a flow is detected through specific events such as: the lowering of the temperature of the warmer body and/or the rising of the temperature of the cooler body, the condensation of a vapor and/or the evaporation of a liquid, the freezing of a liquid and/or the melting of a solid. To reduce the discomfort inherent to the initial compromises one must make when dealing with the ill-defined concepts of a nascent field of research, one should remark here that the heart of the zeroth law is that temperature appears as a marker of an equilibrium situation, one in which no heat flows anymore. It is important to realize that this does not require anything beyond the ability to determine that a stage has been reached where no heat is transferred; in particular, it does not require to compare and/or measure anything like non-zero quantities of heat.

The zeroth law, as just circumscribed, will get us started on the project of giving a tentative description of the primary notion of temperature. Our point in this section is that this description is sufficient to allow us to proceed towards quantifying it, i.e. to move towards the establishment of a temperature scale. We should immediately note that if Θ were a numerical evaluation of *temperature*, any function $T = f(\Theta)$ would do as well, provided only that it be strictly monotonically increasing. The removal of that ambiguity is the aim of thermometry, the measure of temperature (even though, due to historical confusion the etymology seems to indicate that it is a measure of *heat*; in fact, well into the 19th century, papers on the subject of thermometry would write about the "degree of heat", a terminology that has now been abandoned in favor of speaking of a degree of temperature).

The construction of thermometers has a long, somewhat tortuous history: see e.g. [Newton, 1701] (in connection with his cooling law), [Amontons, 1702] (construction of an air pressure thermometer), [Fahrenheit, 1724] (construction of mercury thermometers consistent with one another over a larger range of temperatures). The reading of these papers (and even later ones) points to the fact that the proper definition of *what* was measured – let alone *how* to measure "it" consistently – was not clear even by the beginning of the 19th century.

To stake the territory, let us consider what Fourier [1822] had to say on the subject. We choose to comment on this book for several reasons. Firstly, the author consolidates here some 15 years of researches which took place right at the time historians often consider as the true beginning of the 19th century. Secondly, Fourier's discourse is articulated very much on the hinge between the first two levels we mentioned in the introduction to this chapter. Thirdly (and especially in the re-edition produced by Darboux, which we cite here), the ideas are discussed with remarkable clarity. Fourier states there his famous "equation of motion of heat" in a solid:

$$\partial_t T = K\{\partial_x{}^2 + \partial_y{}^2 + \partial_z{}^2\}T \qquad (2.2.1)$$

where T is the temperature, treated as a function $T(x, y, z; t)$ of the position (x, y, z) where the temperature is evaluated, and of the time t when it is evaluated. K is a structural constant, independent of $(x, y, z; t)$, but the value of which depends on the solid considered (this structural dependence is of no concern to us in this chapter, although its understanding is ultimately in the purview of quantum statistical mechanics). As this equation occurs only in Art.127 (p. 105) of Fourier's treatise, our Reader could legitimately wonder what the author had done before that.

First of all, we should reassure the Reader that, in spite of the term *"heat"* equation for an equation governing the spacio-temporal behavior of *"temperature"*, Fourier is quite clear that an essential distinction should be made between the two terms, although he is rather vague on the nature of "heat" which he explains only tentatively in terms of analogies. Nevertheless, he is quite on the mark when he realizes that:

(1) what he is about to do is to reach quite outside the realm of the other "analytic" theories of his time, preponderantly "rational mechanics"; and

(2) to have a chance of success, his theory must deal with quantities that are phenomenologically defined well enough to be measured; therefore it is "temperature" which he chooses to single out as the marker of the propagation of "heat". Consequently his "analytic theory" produces and studies a "heat equation" which has to govern explicitly "temperature", not "heat". The transport equation known today as the Fourier law – or *heat equation* – consists of the following generalization of (2.2.1):

$$C_V(T)\,\partial_t T = -\nabla[\kappa \nabla T]$$

where $C_V(T)$ is the specific heat at constant volume (Sect. 2.3), ∇ denotes the differential operator $(\partial_x, \partial_y, \partial_z)$, κ is the thermal conductivity; the quantity $J = \kappa \nabla T$ is called the *heat flux*. The derivation of this Fourier law from microscopic premises has been a perennial challenge throughout the history of thermophysics. We discuss in this section only the ingredients and presuppositions that entered in the original Fourier equation (2.2.1). Nevertheless, we may already mention here that in the kinetic theory of Clausius, Maxwell and Boltzmann – Chap. 3, and in particular formulas (3.3.21) – the notion of mean-free path λ, together with the fact that the speed of the particles of a gas is proportional to $\sqrt{T}$, imply that κ is proportional to $\sqrt{T}$ and to d^{-2} (where d stands for the diameter of the molecules), while it is independent of the density; the first and third predictions are born out quite well for gases near equilibrium; and a theoretical estimate of the constant of proportionality allows to derive from the measurement of κ an estimate for d, thus giving quantitative evidence for the existence and size of molecules. While the kinetic theory was – and still largely is – based on ill-controlled stochastic assumptions, a purely stochastic, but exactly controlled "random walk" model is presented in Sect. 3.1 for a related equation, the diffusion equation of Brownian motion.

Let us now return to Fourier's original equation. For this equation to be mathematically unambiguous, it must refer to a "temperature" that is measured on a scale defined at worst up to an affine transformation; indeed this equation governs as well a quantity that is measured on two scales that differ at most by their origin and their unit, i.e. $\tilde{T}$ also satisfies equation (2.2.1) if and only if

$$\tilde{T} = a + bT \quad . \tag{2.2.2}$$

In every day parlance, it should not matter whether the temperature is measured in °C or in °F.

As did most constructors of thermometers before him, Fourier gets rid of the easy ambiguities by choosing two fixed points for his temperature scale: 0° is chosen to be the melting temperature of ice, and $1°$ [Celsius later said $100°$] to be the boiling temperature of water under an atmospheric pressure which he chose to correspond to a column of 760 mm of mercury, the mercury of the barometer being held at $0°$. The remaining problem is the hard one, namely how to divide the basic interval of the scale in equal segments. Newton [1701] satisfied himself that he had achieved this in connection with his experimental confirmation of what is known today as *Newton's cooling law*, namely

$$\frac{\mathrm{d}}{\mathrm{d}t} T = -k(T - T_o) \quad . \tag{2.2.3}$$

Note that it is no accident that this law – too – is invariant under an affine transformation (2.2.2). For this task of interpolating and extrapolating between the two fixed temperatures chosen on the scale, Fourier prefers to use the observation that most metals – not just mercury as in [Fahrenheit, 1724] thermometers – contract proportionally for the same decrease of temperature. Fourier thus divides the scale proportionally to the volume of the bar of any metal he wants to use as his thermometers. Remarkably enough, this was good enough for his purposes, although Fourier is well aware that this dilation law is only good in a limited range of temperatures, far away from the melting point of the expanding material of which the thermometer is made – water, for instance, does not contract as the freezing temperature is approached from above, but in fact it dilates from about $4^o\,C$ down to $0^o\,C$, – a circumstance from which Fourier can choose to stay away, although this freedom is not granted to the frogs that have to survive in a freezing pond.

Before leaving Fourier [1822], it is relevant to our subsequent purposes to note an important observation he made later in his book [Art. 376, p. 435], namely that his heat equation can be used to describe the "diffusion of heat"; in particular the function

$$T(x, y, z; t) = (4\pi Kt)^{-\frac{3}{2}} \exp{-\frac{x^2 + y^2 + z^2}{4Kt}} \qquad (\text{with} \quad t > 0) \tag{2.2.4}$$

satisfies equation (2.2.1). The normalization factor $(2\pi)^{-\frac{3}{2}}$ ensures that

$$\iiint dx\,dy\,dz\, T(x, y, z; t) = 1 \quad . \tag{2.2.5}$$

Note the time-dependence of the spatial width

$$\langle T(t) \rangle \equiv \left[\iiint dx\,dy\,dz\, x^2\, T(x, y, z; t) \right]^{\frac{1}{2}} \tag{2.2.6}$$

of the distribution T, namely

$$\langle T(t) \rangle = \sqrt{(t/t_o)}\, \langle T(t_o) \rangle \tag{2.2.7}$$

i.e. the width of T grows proportionally to the square-root of the elapsed time, a relation that Fourier recognizes to be indeed typical of a *diffusion process*.

What we have at this point is an *empirical temperature*, namely a linear temperature scale, the unit of which can be agreed upon, so that the readings of various thermometers can give consistent results. This means that we can handle the b in (2.2.2); yet we still do not know what meaning, if any, one should attach to a. The existence of a privileged value of a would lead to consider the corresponding temperature scale as allowing the definition and/or measurement of an *absolute temperature*. One might wonder why we need an absolute temperature: why would not temperature be like location, or potential energy, where only differences do matter (after all, we no longer think the "ground zero", i.e. the center of the universe a viable concept). A historically accurate explanation aside, we are aware now that temperature is a measure of internal energy, a notion that requires a zero point; the latter should indeed mark the state of a body which has no internal (or thermal) energy. Hence it is a, not b, that should concern us.

Let us remark, before we move on, that the above shows the importance and complexity of model-building behind the simple acts of collecting temperatures. As we discussed in Sects. 1.4–5, the semantic view of theory construction and testing gives us a better philosophical reconstruction of what goes on in thermometry (e.g. Suppes' *"models of experiments"* and *"models of data"*). Here we see, how much theoretical reasoning, such as the exchange of heat – leading to equilibrium – and the separation between heat and one of its manifestations, temperature, has to take place before a quantitative model can be built in which temperature data can be collected and interpreted. We also see how the thorny problem of interpretation is already present in thermometry; for instance, whether the quantity we measure is the quantity of heat or something else. Further, we see a nice illustration of the difference between a structural and a behavioral theory/description (cf. Sect. 1.4) in the distance which had to be run between Fourier's heat equation (2.2.1) and the final determination of T, with the absolute temperature scale.

The root for the next progress is to be found in earlier works, by Boyle [1660, 1662] and Mariotte [1676 and 1679] to whom we owe careful observations establishing that under the same equilibrium conditions, i.e. *at constant temperature (!),* the pressure p of the air enclosed in a vessel of variable volume V, follows the Boyle–Mariotte law

$$pV = C \qquad (2.2.8)$$

where C is a constant, the relation of which with the temperature is not investigated further than what is just said, namely that the value of C does not vary as long as the temperature does not change. Notice that, so far, one does not need a temperature scale, but only the zeroth law of thermodynamics.

Over a century later Gay-Lussac [1802] – see also [Gay-Lussac, 1807] – improves on these observations in three ways, with the stated purpose to determine whether one could trust the *division* of the scales on thermometers. First, he repeats the experiments with a variety of gases as diverse as: air, oxygen, hydrogen, nitrogen, nitrous oxide, ammonia, muriatic acid, sulphurous acid and carbonic acid. Second, he compares for different temperatures the constants obtained in the right-hand side of (2.2.8). Third, he normalizes the results of his observations by taking account of the quantity of gas present in his trials, measured in "molecular" units, i.e. as M/m where M is the mass of the gas, measured in grams, and m is its "molecular mass" (as determined from the ratios appearing in chemical reactions: 32 grams for oxygen, 2 grams for hydrogen, etc.; note that this does not presuppose any allegiance to the existence of "molecules" or "atoms" in the sense understood today). He concludes that all these substances "have the same expansion between the same degrees of heat." Specifically, his findings on gases expanding while maintained at constant pressure, amount to saying that there exists a universal constant R, such that

$$p\,(V - V_o) = R\,\frac{M}{m}(T - T_o) \quad . \qquad (2.2.9)$$

Hence, one could choose the zero of the temperature scale – i.e. a in (2.2.2) – in such a manner that (2.2.9) reads

$$pV = R\,\frac{M}{m}T \qquad (2.2.10)$$

which is known as the *Gay-Lussac law;* in the sequel, we consider mostly the normalized case where $M = m$.

Both of the main concerns on the establishment of an acceptable temperature scale have thus been met. Yet, a few remarks are still necessary.

First, a limitation on the universality of the law: Gay-Lussac already makes note of the fact that the relation (2.2.9) is not strictly satisfied by a "vapor" that is too near to the temperature at which it becomes a liquid; the following discussion is tacitly kept within the limits imposed by this restriction.

Second, Gay-Lussac also conducted experiments on the "free" expansion of gases, where he confirmed that gases do have similar properties independently of their chemical identity; there, his laboratory procedures left room for improvement, especially in terms of controlling the exchange of heat with the outside; this was done quite elaborately by Joule and Thomson (Lord Kelvin) [Joule, 1845, Thomson, 1848, Joule and Thomson, 1854 and 1862].

Third, the meticulous experimental investigations pursued by Joule and Thomson were borne out of the correct perception that the results were of fundamental importance in the derivation of the mechanical equivalent of heat – [Mayer, 1842, Joule, 1843, 1847, 1849, 1850, Von Helmoltz, 1847, Colding, 1851, Regnault, 1853]; for a survey, see [Kuhn, 1959] – and we thus defer to the next section our report on these experiments. Fourth, the qualification of "absolute" given to the pragmatic temperature scale suggested by (2.2.10), while justified in part by Gay-Lussac's observation that (2.2.9) holds for such a wide variety of gases, finds later on in the history of our subject an even more solid theoretical corroboration as it coincides with the "absolute temperature" proposed by Thomson [1848] who thus turned on its head Carnot's formula on the efficiency of ideal heat engines.

Our vehicle must change gears at this point, since it has to negotiate a constructive escalation from temperature to the concept of heat, which we must now engage.

2.3 The Motive Power of Heat vs. Conservation Laws

What is heat? The short answer is: "A form of energy that is convertible from and to other forms of energy, such as mechanical work." The courts of physics successively accepted various empirical evidences, first for conservation of the quantity of heat, and then for its convertibility. Some of these evidences are presented in this section.

A longer answer is encapsulated in the "first" and the "second laws of thermodynamics", which we enunciate in Sect. 2.4.

Nowadays, we understand the concept of heat as a contribution that has to be introduced, in certain circumstances that obtain when one studies thermodynamical processes, in order to preserve the law of conservation of energy. Much of the teaching of physics proceeds today in terms of conservation laws, and these have proven such a reliable guide in research that one can view them as the "laws which the laws of nature have to obey"[Wigner, ca. 1980]. A few examples should suffice: the conservation of linear momentum (Newton's "quantity of motion"), and of *vis viva*, in classical mechanics; the conservation of mass in chemical reactions; the Maxwell equations of electromagnetism; the equivalence of mass and energy in relativity; the use of symmetry groups in quantum spectroscopy; the conservation of charges, and more generally the conservation laws attached to internal symmetries in elementary particle physics.

Initially however, much of the difficulty in conceiving of heat as an ingredient participating in the conservation of energy laid in that the law itself had not been formulated beyond the confines of mechanics – and even there the conservation of *vis viva* had not been a simple affair. The full realization that an extension of this principle of conservation of energy was called for, in the context of the study of heat, did not take shape without controversy: more than a dozen authors (beginning with the 1840s) can be read to have some claim in a burst of independent and almost simultaneous discoveries, as Kuhn [1959] shows. But Kuhn is also quick to point out that it is rather doubtful whether the core of the theory, as we describe it today, would have been recognized by these authors through the various avatars in which it is manifested in their diverse writings. Yet, these pioneers are in large part responsible for the fact that modern physics relies so "naturally" on conservation laws.

Hence, in order to sort out the fundamental concepts behind the theory of heat, we should first try and reconstruct some of the causes for the delays in the formulation and acceptance of a principle of conservation of energy in connection with thermal phenomena, and the concomitant "mechanical equivalent of heat". We propose to argue that this counter-influence was a powerful, largely successful, but ultimately misguided *model*, called the *caloric*; and we examine below some of the reasons why it had such a strong hold on the scientific community that it effectively retarded the construction of thermodynamics.

Even a casual glance shows Carnot [1824] writing of the *"mechanical effects"* [read "work"] of the *"fall of heat"* in a style that cannot but evoke the water mills or steam pumps of his engineering contemporaries. In the first quarter of the 19th century, heat and light are widely viewed as substances which are therefore conserved, as is the total mass in the chemical reactions. Lavoisier [1789], who summarized this conservation law in the famous aphorism "Rien ne se perd, rien ne se crée", had listed explicitly the *caloric* among the *elementary substances* – i.e. those that cannot be decomposed by any known process of analysis – together with oxygen, azote, hydrogen ... and light. Consequently, we have to consider here some of the evidences pointing to, and then away from, this *misguided* conclusion, namely: if heat behaves so much as a substance then it *is* a substance – the *caloric* – the quantity of which can indeed be measured in terms of a well-defined unit – the *calorie*; see also [Fox, 1971].

From a philosophical perspective, the question is whether caloric is a substantial model for heat or an analogue model (cf. Sect. 1.3, 3.iii–iv). We should learn from this historic battle over caloric – the story of which unfolds in the rest of this section – that, while it is easy to distinguish conceptually the basic notions behind the two models, it is not always clear which model should be used with respect to a particular phenomenon.

After reassessing carefully quantitative experiments repeated since at least the 17th century (e.g. by Carlo Renaldi and by Hermann Boerhaave), Black

[1803] implements the following situation. Two bodies A and B possibly made of different substances, and of masses m_A and m_B, are initially at temperatures T_A and T_B with $T_A > T_B$. The two bodies are then put into contact until they reach mutual equilibrium at the (same !) temperature T_{AB}. The observations made by Black indicate that there exist two constants C_A and C_B, such that

$$C_A \, m_A \, (T_A - T_{AB}) \; = \; C_B \, m_B \, (T_{AB} - T_B) \qquad (2.3.1)$$

a formula still known as the fundamental law of calorimetry. One can rewrite this formula as follows, putting into the LHS all terms describing the initial situation, and in the RHS all terms corresponding to the final situation:

$$C_A \, m_A \, T_A + C_B \, m_B \, T_B \; = \; C_A \, m_A \, T_{AB} + C_B \, m_B \, T_{AB} \quad . \qquad (2.3.2)$$

The analogy between this equality and Newton's law of conservation of the momentum (which for reasons that are pertinent here, he had called *quantity of motion*), namely

$$m_A \, v_A + m_B \, v_B \; = \; m_A \, v_{AB} + m_B \, v'_{AB} \qquad (2.3.3)$$

helps understand why Black felt justified in calling $C_A \, m_A \, T_A$ the *quantity of the matter of heat* initially contained in the body A, and C_A the *capacity for the matter of heat* (we would speak of "specific heat") of the substance A; similar interpretations are evidently made for the other three terms in (2.3.2). Note that this is indeed an *interpretation* that goes further than the experimental evidence recorded in (2.3.1). The law of calorimetry is more restrictive in that it compares the quantities of heat *exchanged:* the quantity of heat that flows from A to B [i.e. the LHS of (2.3.1)] is equal to the quantity of heat that is received by B from A [i.e. the RHS of (2.3.1)].

To the objection that for each material A, the proportionality C_A is constant only over a range of temperatures sufficiently far away from the transition temperature of the material considered, Black had to respond with the introduction of a new concept: *latent heat*. He defines the latent heat of fusion – or vaporization – as the quantity of heat one must transfer to a solid, once it has reached its melting temperature, in order to melt it thoroughly – or the quantity of heat one must transfer to a liquid, once it has reached its boiling temperature, in order to turn it completely into its vapor. One could argue that the word "latent" is an adroit use of language that might conceal some uneasiness with regard to the nature of heat or perhaps with the conservation of the caloric; however, we do not wish to pursue this line now.

Rather, we want to notice that, for gases, it is necessary to distinguish between the specific heat at constant volume, C_V, and the specific heat at constant pressure, C_p. An unexpected application of these notions is the explanation given by Laplace [1816] – see the presentations by Finn [1964] and

Truesdell [1980] – for the discrepancy (about 20%) between the experimental value of the speed of sound, say in the air, and the value computed by Newton [1687, Book II, Prop. XLVIII]; see also [Chandrasekhar, 1995, pp. 588–591]. Newton had seen correctly that the speed v of sound – a compression wave – should be given by the formula

$$v = (\varrho\beta)^{-\frac{1}{2}} \qquad \text{with} \qquad \beta = \frac{1}{\varrho}\frac{d\varrho}{dp} \qquad (2.3.4)$$

where ϱ is the density of the air, and β its compressibility. But Newton had derived his value of β from the Boyle–Mariotte law (2.2.8), i.e. in its differential form along an isothermal curve

$$p dV + V dp = 0 \quad , \qquad (2.3.5)$$

to get that the isothermal compressibility β_T is given by

$$\varrho\,\beta_T = (\partial_p\varrho)_T = \frac{\varrho}{p} \quad . \qquad (2.3.6)$$

This is where Laplace locates the flaw that had vexed his predecessors, among them no less than Lagrange [1759]: in a phenomenon as rapid as a sound wave surely is, the compression and expansion of the air should *not be assumed* to be isothermal, but rather to be adiabatic – i.e. to proceed without exchange of heat – the underlying idea being that such rapid compression and expansions should not have to involve exchanging any quantity of caloric. He then argues that an adiabatic process is described by a curve, now called an adiabat, given by

$$pV^{\gamma} = C' \qquad \text{with} \qquad \gamma = \frac{C_p}{C_V} = \text{const} \qquad (2.3.7)$$

where C' is a constant (the value of which is irrelevant here). This is known today as the *Laplace–Poisson law for adiabatic change*. We use it here in its differential form

$$\gamma\,p\,dV + V dp = 0 \qquad (2.3.8)$$

to get that, along an adiabat, the compressibility β_S satisfies

$$\varrho\,\beta_S = (\partial_p\varrho)_S = \frac{\varrho}{p}\gamma^{-1} \quad . \qquad (2.3.9)$$

Hence (see (2.3.4)), Laplace obtains

$$v = \sqrt{\frac{p}{\varrho}\,\gamma} = v_{Newt}\sqrt{\gamma} \qquad (2.3.10)$$

which is his famous claim

> the real speed of sound equals the product of the speed according to the New-tonian formula by the square-root of the ratio of the specific heat of the air subject to the constant pressure of the atmosphere at various temperatures to its specific heat when its volume remains constant.

The experimental value $\gamma = 1.40$ thus gives nearly perfectly the needed 20% correction. This pragmatic value was later understood from the kinetic theory of diatomic ideal gases for which $C_p = 7/2$ and $C_V = 5/2$. In an interesting twist of history, a century and a half later, one determines the value of the adiabatic compressibility β_S in imperfect fluids – especially fluids in the vicinity of their critical points – by measuring the speed of sound in these fluids, and using (2.3.4) in the form $\beta_S = \varrho v^{-2}$; [Levelt-Sengers, 1966].

While Laplace's arguments were explicitly rooted in a caloric theory of heat, hindsight makes it not hard to see that the argument is in fact not dependent on this model, as Laplace [1823, Tome 5, Livre XII] and Poisson [1823] came close to make clear by 1823. This is what we want to examine now.

One who believes that the gas contains a quantity of caloric would assume naturally that a "heat function" Q of the state variables (p, V) exists; as one assumes an underlying constitutive equation $p = \varpi(V, T)$ (such as the Gay-Lussac law for the ideal gas), one would write Q as well as a smooth function of (V, T) or of (p, T). Along any smooth process that consists of a continuous succession of equilibrium states: $(p(0), V(0)) \mapsto (p(t), V(t))$, one would have then $Q(t) = Q(p(t), V(t))$ and thus

$$\frac{dQ}{dt} = (\partial_p Q)_V \frac{dp}{dt} + (\partial_V Q)_p \frac{dV}{dt} \quad . \tag{2.3.11}$$

Then the specific heats C_p (at constant pressure) and C_V (at constant volume) would be defined by

$$C_p = (\partial_T Q)_p \quad \text{and} \quad C_V = (\partial_T Q)_V \quad . \tag{2.3.12}$$

Provided that the constitutive equation (e.g. the Gay-Lussac law (2.2.9), linking p, V and T satisfy the usual conditions of inevitability the following changes of variables – from (p, V) to (p, T) [resp. (V, T)] – are mathematically legitimate and would give:

$$C_p = (\partial_V Q)_p (\partial_T V)_p \quad \text{and} \quad C_V = (\partial_p Q)_V (\partial_T p)_V \quad . \tag{2.3.13}$$

Upon restricting one's attention to the perfect gas, one would obtain from the Gay-Lussac constitutive equation (2.2.10):

$$C_p = (\partial_V Q)_p \frac{R}{p} \quad \text{and} \quad C_V = (\partial_p Q)_V \frac{R}{V} \quad . \tag{2.3.14}$$

Together with (2.3.11), this would give

$$dQ = C_V \frac{V}{R} dp + C_p \frac{p}{R} dV \quad . \tag{2.3.15}$$

Along an adiabat (i.e. with $dQ = 0$) this expression reduces to the differential form (2.3.8) of the Laplace Poisson law. To be able to integrate (2.3.8) to give

(2.3.7) one needs to assume that γ is constant (at least over each adiabatic process separately). Laplace and Poisson knew that the ratio $\gamma = C_p/C_V$ is constant for a perfect gas from the experiments of Gay-Lussac (where p and T were allowed to range over a full order of magnitude). Notice incidentally that the Laplace formula (2.3.10) for the speed of sound is valid without this constancy assumption since it follows already from (2.3.8), the validity of which does not require this assumption.

The above derivation assumes a caloric model in as much as it is explicitly based on the existence of a heat function $Q(p, V)$. With the hindsight provided in part – but only in part – by a reflexion on the work of Carnot, one knows that such a function *does not exist*, and we proceed to explain why. Before we do so, nevertheless, it is useful to investigate how much of the derivation of (2.3.8) can be salvaged when the caloric assumption is omitted.

Let us first pick up the above derivation to push it one step further. We can rewrite (2.3.15) as

$$dQ = C_V \frac{V\,dp + p\,dV}{R} + \frac{C_p - C_V}{R} p\,dV \tag{2.3.16}$$

or equivalently

$$dQ = C_p \frac{V\,dp + p\,dV}{R} - \frac{C_p - C_V}{R} V\,dp \quad . \tag{2.3.17}$$

Upon using again the constitutive equation for a perfect gas, one obtains

$$dQ = C_V dT + J^{-1} p\,dV \tag{2.3.18}$$

or equivalently

$$dQ = C_p dT - J^{-1} V\,dp \tag{2.3.19}$$

where J is defined by

$$J(C_p - C_V) = R \quad . \tag{2.3.20}$$

From equations (2.3.18–19) alone, the expression (2.3.8) of the Laplace–Poisson law follows, and thus so does Laplace's formula for the speed of sound.

The problem is then reduced to that of reinterpreting equations (2.3.18–19) without appeal to the caloric theory. This is done by reinterpreting the *differential form* dQ in the LHS of (2.3.18) (which we denote by the symbol η) as the *primary* object, rather than Q itself; indeed, we do not need a *heat function* Q to describe a *heating differential* η ; mathematically, we say that η does not need to be an exact differential, i.e. that its integral over a process may depend – and it does – on the process itself and not just on the end points of the process. Physically, we first notice the empirical fact that, for an ideal gas, J is a constant; we can then assume without loss of generality that $J = 1$; indeed this is accomplished by choosing properly the units in which we measure the heating η. Under these circumstances, (2.3.20) is known as the Mayer relation. We can now rewrite (2.3.18) as

$$\eta = dU + p\,dV \qquad \text{with} \qquad dU = C_V\,dT \qquad (2.3.21)$$

which we reinterpret by stating that the heating that goes into the gas is equal to the sum of the rate of change – dU – of the internal energy U of the gas (a true function of its state (p, V)), plus the rate of work – $p\,dV$ – that is produced by the gas. Hence the Laplace–Poisson equation can indeed be derived without reference to an underlying caloric theory.

We should remark here that the passage from (2.3.15) to (2.3.21) surmizes that J is constant. In the light of the already accepted fact that γ is constant, this implies that C_p and C_V be *separately* constant. The latter implication, in itself, is however *not compatible* with the caloric theory on which was based the original derivation of (2.3.10) given by Laplace. Indeed, if C_p and C_V were constant, $\partial_p(\partial_V Q) = \partial_V(\partial_p Q)$ in (2.3.15) would imply $C_p = C_V$, i.e. $\gamma = 1$ instead of $\gamma > 1$. In particular, this is not an option in Laplace's theory, as it contradicts both its basic motivation – the isotherms and the adiabats are distinct – and its conclusion – the correction (2.3.10) to Newton's formula for the speed of sound. Nevertheless, since Laplace's sleek prose emphasizes the interpretation of his elaborate (and sometimes impenetrable) mathematical arguments in terms of the caloric theory, and since (2.3.21) is tantamount to the recognition of the law of conservation of energy (for which his readership was not yet prepared), Laplace's spectacular correction (2.3.10) to Newton's theoretical computation of the speed of sound was considered, for some time, as yet another triumph of the caloric. Moreover, with the progresses of calorimetry, – see already [Regnault, 1842; 1847, 1853] – specific heats and latent heats can nowadays be measured with great precision, and their properties still provide powerful information on important aspects of the structure of materials.

There are therefore pertinent circumstances where heat, by itself, is conserved. Hence the temptation to identify it as some kind of "substance", maybe with some reservations to the effect that it is a particularly subtle substance, imponderable or intangible, but nevertheless one to which some reality must [*sic*] be imparted. The unit in which this quantity can be expressed was given a pragmatic definition in terms of calorimetry: a *calorie* is the quantity of heat needed to raise by 1^oC the temperature of 1 gram of water (under ordinary atmospheric pressure; moreover, since this quantity of heat varies by about 1% between 0^oC and 100^oC, one must also specify the temperature, say 20^oC around which the measurement has to be made; such precautions are not directly relevant to our purpose). Incidentally, this official definition persisted until 1928; nowadays, it is more convenient to define the unit of heat, still called the calorie, as being equal to 4.1840 *joules*, a statement that subsumes the recognition of a mechanical equivalent of heat. As we shall presently see, this recognition however is associated with the demise of the caloric itself as a proper model for heat. The problem with that model of "heat as substance" is that it impedes a consistent extension of thermodynamics, beyond calorimetry, as the model compulsively fails to warn against connotations that are too intimately associated with the limited circumstances for which it was devised.

The above case study gives another example of the benefit of adopting the semantic view of scientific theories. The caloric model of heat is certainly a semantic model for the quantity Q and $dQ(=\eta)$ and its relation with other quantities in thermodynamics. Excluding such models from one's consideration of theories, current or historical, would have resulted in a failure to understand the rationale of the "Newton–Laplace argument" on specific heat.

One last remark before we review the reasons that led to the abandonment of the view that heat can be understood *as if* it were merely a substance. It is indeed interesting to note that these reasons did not go unchallenged, at least on methodological grounds. For instance, in [Mach, 1872, 1894] we are exquisitely shown that summarily walking away from viewing heat as a substance, simply on the cognizance that it is not universally conserved, would be an accident of language, or history. He argues by transposing the standard argument for heat to a similarly fraught discussion of the "quantity of electricity".

As might be expected, Mach has an even mightier horn to toot: he wants to discredit the atomistic tenet that "heat" should be viewed as a kind of "motion" since it is could not be held anymore as a kind of "substance". The idea that heat might be indeed a kind of motion was once advanced by Ampère [1832] – see also [Ampère, 1835] – who proposed to replace the "quantity of caloric" by the *vis viva* of vibrating molecules". A similar conclusion was to be advanced later from radically different premises, by Clausius [1857].

In Ampère's work, the proposal is presented as a new avatar of the old analogy between heat and light. Such an updating seemed then to be necessary as a response to laboratory evidences – the diffraction/interference patterns of *light* – accumulated over the past twenty years and leading to the replacement of the particle theories (Descartes, Newton) of light by a wave theory (Arago, Fresnel, Young). Brush [1983, Sect. 1.8] makes an impressive case that the caloric's demise might be due to this – now discredited – *wave theory of heat* rather than to the canonical influences that are widely recognized today. Nevertheless, our purpose being to investigate how models partake in the ideas that form the corpus of thermostatics, we choose to turn now to Carnot's reflections and the arguments that sprung from there.

After discussing some preliminary models – that show. incidentally, his command of the functioning of actual steam engines – Carnot devises the *gedanken experiment* which is now known as the Carnot cycle; and then he proceeds to prove that this model is universal. *This is the model which*

Clausius [1850], Thomson [1848], *and others of the pioneers, cite among the sources of their thinking.*

Three assumptions enter explicitly Carnot's argument. The first is that heat is a substance – the caloric – and hence is conserved. We present the model itself without any reference to this assumption. As we shall see, the caloric however enters explicitly in the second part of Carnot's argument – the universality of the model – in conjunction with a second assumption, namely the absence of *perpetuum mobile*. The third assumption is a weakened version of the assertion that the adiabatic expansion and contraction of a perfect gas are governed by (2.3.7) (with $\gamma > 1$!); Carnot only wants to assume that "the temperature of a gaseous fluid is raised by [adiabatic] compression and lowered by [adiabatic] expansion". As a point of history – rather than one of principle – it is conceivable that Carnot did not even know (2.3.7), then quite a recent discovery. It is another matter to determine why Carnot's expositor, Clapeyron [1834], asserts that the analytic form of the adiabats for an ideal gas is not known. One can speculate that both disliked the idea that γ be a constant, for reasons that could be suspected from our discussion of Laplace's correction to the formula for speed of sound: in all these papers the underlying caloric theory mars the computation of the specific heats.

There are also two idealizations implied in the construction of this gedanken experiment. One is the tacit acceptance that a cycle (such as the one to be described in the first part of the argument) can be implemented by perfectly reversible transformations. The other is the tacit use (in the first part of the argument) of the Gay-Lussac law, or at least the Boyle–Mariotte law, for the isotherms characteristic of a perfect gas; this again is dispensable – as the second part (universality) of the argument will show.

For the nascent thermodynamics, such idealizations were eminently acceptable; see however the end of this section, and note in anticipation that the existence of *irreversible* transformations later turned out to be inherent to the core of thermodynamics.

On may also easily see in the following description that Carnot's system is one of the typical examples for the kind of models which Suppe and Giere believe to be the core of any scientific theories (i.e. what Suppe calls "physical systems") (cf. Sect. 1.2, V.2). They are idealized systems whose dispositional properties essential to the phenomena in question are given an accurate description (such as in the Carnot cycle).

The gedanken experiment goes as follows. A gas is encapsulated in a vessel closed by a piston. One tags the gas through a cycle of *reversible* transformations that proceed in a continuous succession of equilibrium states. In the beginning the pressure, volume and temperature of the gas are $\{p_A, V_A, T_A\}$. The gas is first let to expand adiabatically – i.e. without any exchange of heat with its surroundings, of which it is thus thermally isolated – until it reaches the state $\{p_B, V_B, T_B\}$. The third of the three Carnot's assumptions mentioned above amounts to say:

$$V_B > V_A \quad , \quad T_B < T_A \quad \text{and thus} \quad p_B < p_A \frac{V_A}{V_B} < p_A \quad . \tag{2.3.22}$$

In a second step, the gas is compressed isothermally – its temperature being maintained constant by keeping the gas (in its vessel!) in thermal contact with an infinite external reservoir at temperature T_B until the gas reaches the state $\{p_C, V_C, T_C\}$. We have now:

$$V_C < V_B \quad , \quad T_C = T_B \quad \text{and thus} \quad p_C = p_B \frac{V_B}{V_C} > p_B \quad . \tag{2.3.23}$$

During this, heat is flowing from the gas to the reservoir at temperature T_B. The third step is again adiabatic, but is now a compression pursued until the gas reaches the state $\{p_D, V_D, T_D\}$ with

$$V_D < V_C \quad , \quad T_D = T_A \quad \text{and thus} \quad p_D > p_C \frac{V_C}{V_D} > p_C \quad . \tag{2.3.24}$$

Finally, the gas is let to expand isothermally – its temperature being maintained constant by keeping the gas in thermal contact with an infinite external reservoir at temperature T_A until it reaches back its initial state $\{p_A, V_A, T_A\}$; so that we have

$$V_A > V_D \quad , \quad T_D = T_A \quad \text{and thus} \quad p_A = p_D \frac{V_D}{V_A} < p_D \quad . \tag{2.3.25}$$

During this, heat is flowing to the gas from the reservoir at temperature T_A.
A graphical summary of the cycle just described is given in Fig. 2.1.

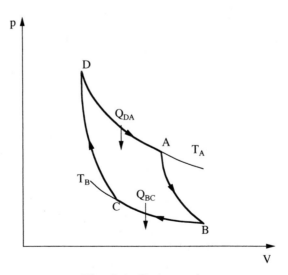

Fig. 2.1. Carnot cycle

From his assumption on adiabatic compression and expansion, Carnot concludes first that along the above cycle of transformations: "at equal volume – that is for similar positions of the piston – the temperature is higher during the expansions than during the compressions", these being in all cases either isothermal or adiabatic. From this simple fact, he concludes that the net result of the cycle is that "there remains an excess of motive power", since the net motive power is equal to the area of the curvilinear parallelogram of Fig. 2.1 (see also the quantitative discussion presented below). In other words, Carnot's gedanken experiment is a *model* for an engine that produces mechanical work by transferring heat from a hotter heat reservoir at constant temperature (here T_A) to a colder heat reservoir at constant temperature (here T_B, with $T_B < T_A$).

In his "Réflexions" [see e.g. footnote (1) to page 37 of the original edition (the pagination is preserved in Fox' critical edition)], Carnot subscribes to the caloric model of heat in as much as he thinks that the *whole* quantity of heat taken from the body A (the hotter body) is transferred *completely* to the body B (the cooler one):

> We assume implicitly in our demonstration that, when a body has gone through any changes whatever and that after any number of transformations it is brought back identically to its initial state, ... [then] this body is found to contain the same quantity of heat as it contained at first, or in other words, the quantities of heat absorbed or released in these various transformations compensate exactly.

This reading is explicitly confirmed in [Clapeyron, 1834] mathematical exegesis of Carnot's work:

> all the heat given up by body A to the gas during its expansion, in contact with it has flowed into the body B during the compression of the gas which took place in contact with that body.

This quote is authoritative for the purpose of the subsequent history since it is only through the work of Clapeyron that Clausius and Thomson knew of Carnot's. While the assumption is crucial to make sense of the following proof as Carnot gave it, it is this very assumption that is disallowed by Clausius and Thompson.

Having constructed a model, call it M_1, Carnot proceeds then to show that this model is *universal* in the sense that any model, say M_2, that would function between the same temperatures T_A and T_B and absorb the same quantity of heat from the reservoir at higher temperature would not produce more work than M_1 would. To show this, Carnot constructs a *reductio ad absurdum* argument. Suppose M_2 produces more work that M_1; then use part of the work produced by M_2 to run the cycle M_1 in reverse; i.e. M_1 functions then as a heat pump, transferring heat from the colder reservoir to the hotter one, the transfer being for equal but opposite quantities of heat since, by definition, all the transformations, that occur as M_1 runs its course, are reversible. The net effect of this coupling of the two machines would be that after completing one cycle together, they would both be back in their initial state, *without* leaving behind any net balance of heat absorbed or expanded, *and yet* having *produced* a net amount of work. This would contradict the postulate that a *perpetuum mobile* is impossible. By the same token, if the model M_2 were also made up only of reversible transformations, the two models would produce exactly the same quantity of work. Carnot can therefore conclude

> *The motive power of heat is independent of the agents employed to develop it; its quantity is determined solely by the temperatures of the bodies between which, in the final result, the transfer of the caloric occurs.*

He has produced "the" universal model for a heat engine that is functioning between the temperature T_A and T_B. Except for the recourse to the "caloric", the above quote from Carnot is still part of today's understanding of the situation. To emphasize the universality of the model, we underline in particular that it is not limited to ideal gases, nor steam, nor any vapor; for instance in [Wannier, 1966, 1987, pp. 122–124], an isomorphic "Carnot cycle" is constructed in which the agent is, instead, a paramagnetic substance placed in a variable external magnetic field: the value m of "magnetization" plays the role of the "pressure" in the initial model M_1 and the value H of the exterior magnetic field the role of the volume V.

To improve on Carnot's remarkable qualitative result, let us review the argument quantitatively, following "more or less" a line that could be traced back to [Clapeyron, 1834], but which was clarified by Clausius [1850]. Moreover, to make this analytic presentation as unambiguous as possible, we assume that we work with a perfect gas: the isotherms satisfy the Gay-Lussac law (2.2.10), and we further assume that the adiabats satisfy the Laplace–Poisson law (2.3.7). This specificity is only an apparent loss of generality, as the second part of Carnot's argument can be used to show that the resulting model is in fact equivalent to any other reversible model operating between the same temperatures. We take advantage of this remark to simplify the presentation below.

The work produced along any of the processes occurring in the cycle M_1 is given by the mechanical definition of work, namely

$$W = \int_{V_1}^{V_2} p(V) \, \mathrm{d}V \quad . \tag{2.3.26}$$

Upon using (2.2.10) for the isotherms, and (2.3.7) for the adiabats, one finds by elementary calculus:

$$\left. \begin{aligned} W_{AB} &= R\,(T_A - T_B)/(\gamma - 1) \; ; \; W_{BC} = RT_B \,(\ln V_C - \ln V_B) \\ W_{CD} &= R\,(T_B - T_A)/(\gamma - 1) \; ; \; W_{DA} = RT_A \,(\ln V_A - \ln V_D) \end{aligned} \right\} . \tag{2.3.27}$$

A little bit of algebra gives furthermore:

$$\frac{V_B}{V_C} = \frac{V_A}{V_D} \quad . \tag{2.3.28}$$

These computations imply immediately two consequences of utmost importance for what follows.

The first of these consequences is that the quantities of work done along each of the adiabats (i.e. when there are, by definition, no accompanying exchange of heat) compensate one another exactly:

$$W_{AB} + W_{CD} = 0 \tag{2.3.29}$$

i.e. it takes exactly as much work to bring the system back from T_B to T_A as the work one gained by letting the system go from T_A to T_B, when both operations are done adiabatically.

One defines the efficiency of the engine as

$$\varepsilon = 1 - |\frac{W_{BC}}{W_{DA}}| \quad . \tag{2.3.30}$$

This is a function of Q_A, T_A and T_B, independent of the substance made to undergo the cycle. The second consequence of the computations (2.3.27–28) is that the efficiency of the engine does satisfy:

$$\varepsilon = \frac{T_A - T_B}{T_A} \quad . \tag{2.3.31}$$

It is important to realize that (2.3.30) is the efficiency of any ideal, reversible heat engine operating between two temperatures T_A and T_B with $T_A > T_B$. As Thomson [1852] emphasizes, this shows that the *absolute temperature* entering the RHS of (2.2.10), the Gay-Lussac's law, is much more absolute than it appeared at first: its significance is now not limited anymore to ideal gases in equilibrium far from their critical temperatures; indeed, as (2.3.17) shows, this temperature makes sense in a much wider context, namely as a measure of the efficiency of the *universal* Carnot engine (recall that the above conclusions, drawn from this class of models, are independent of any particular agent used in the construction of a particular representative in the class). The zero of the temperature scale is essential; its unit is irrelevant; as we emphasized in Sect. 2.2.

Although one could argue directly for the definition (2.3.30) of the efficiency of "the" ideal heat engine functioning between the temperatures T_A and T_B, it is more germane to our purposes to take advantage here of the reinterpretation Clausius [1850] gives of the Carnot cycle. He first notes that Carnot's interpretation of heat as a substance that is conserved, i.e. exchanged or transferred only among itself, is not fully satisfactory and/or consistent throughout the description of the functioning of the cycle. Where Mayer [Mayer, 1842] cites explicitly the classical injunctions "*causa aequat ef-*

fectum" and that "causes" are both indestructible and convertible, some form
of these injunctions seems to support also, if only implicitly, Clausius' understanding of Carnot's argument. In essence, Clausius' summary and criticism
are

(a) in *agreement* with Carnot: whenever work is done by heat and no permanent change occurs in the condition of the working body, a certain
   quantity of heat passes from a hotter to a colder body;
(b) in *contradistinction* to Carnot: no experimental evidence has ever been
   presented for an equality between the quantity of heat restituted by the
   working body to the cooler reservoir and the quantity of heat received
   by this body from the hotter reservoir.

While some of the heat is transferred, not all the heat necessarily is. Clausius
then expounds the modern view that, while heat can be exchanged against
heat, and work exchanged against work; heat can also be converted to or from
work. In the Carnot cycle, Clausius proposes to consider that the work done
by, or on, the system should be divided into two parts: the work that causes
changes "exterior" to the system – such as the work done along isotherms of
a perfect gas – and the work that causes changes "interior" to the system –
such as the work done along the adiabats of this gas.

In so re-interpreting the Carnot engine, Clausius is fully subscribing to
the view that heat and work are two mutually convertible forms of energy, i.e.
that there is a mechanical equivalent of heat. Upon assuming $dU = 0$ along
the isotherms, i.e. $U = U(T)$ or $\partial_V U = 0$ – which is not generally valid, but
is correct for a perfect gas – Clausius saw that (2.3.30) can be rewritten as

$$\varepsilon = 1 - |\frac{Q_{BC}}{Q_{DA}}| \qquad (2.3.32)$$

where Q_{DA} is the heat given to the system by the reservoir at temperature
T_A, and Q_{BC} is the heat rejected by the system into the reservoir at temperature T_B; recall that the system does not absorb nor reject heat along
the adiabats AB and CD! Thus, in the whole process, heat is not merely
transferred, as Carnot would have it, but part of it is transformed into work.

To salvage Carnot's result (specifically on the universality of his model of
an ideal heat engine) from sinking with the wreck of the supporting caloric
model requires the replacement of Carnot's appeal to the absence of *perpetuum mobile* by a stronger premise; this is provided by the second law of
thermodynamics (see Sect. 2.4).

At the logical point at which we have already arrived in this section, note
that it is not premature to infer from Clausius' interpretation, in accord with
(2.3.29), that the internal energy of an ideal gas is a function of its temperature only, a conclusion that was indeed drawn by Thomson as a consequence
of his work with Joule.

One might perhaps wonder with Clausius why Carnot did not put things
that way. It is in fact quite remarkable that Carnot himself – in Notes [see

[Picard, 1927]] written between 1824 and his death in 1832, and published only in 1872 – pursued the query in a radically different direction which could now be read as a harbinger of the law of conservation of energy. Why then, were these insights not poured by Carnot in the crucible of the rich contemporary scientific exchanges? Is it possible that he could never envisage that the time could have become ripe for a speculative theory that he, himself, considered to be in open contradiction with the thrust of his *opus magnum* [Carnot, 1824] ?

Part of the answer may be that the caloric was not then the discredited model it was rapidly becoming in the 1850s [Clausius, 1857]. We must nevertheless leave to the historian the formulation of a more convincing explanation of why Carnot did not publish the thoughts he delineated in his Notes; one such explanation is offered in the Introduction of [Fox, 1978] critical edition of Carnot's "Réflections". We merely want to point out at this juncture that although some weird speculations on the nature of heat can be found in [Fourier, 1822], Fourier had stated also therein (Art.52) some very definite and strongly-phrased convictions:

> *The effects of heat are in no way comparable to those of a fluid ... It would be useless to attempt and derive from such a hypothesis the laws on the propagation [of heat] which we expound in this Ouvrage ... the behavior of this element [heat] is thus entirely different of the behavior of any air-like substance.*

Fourier's treatise appeared in 1822 (i.e. ten years before Carnot died) and it was read widely, certainly in France.

That Clausius' view took shape concurrently in several other minds is the object of the paper [Kuhn, 1959] cited earlier. Mach [1872] – see also [Mach, 1894] – views much of this development as only a convenient displacement of the problem: "energy is the term that has come into use for that indestructible something of which the measure is mechanical work". We prefer to say that we have a new interpretation of heat, displacing an *old model*, according to which heat is a substance that is hosted in a neutral manner by the system; the latter is viewed as a mere vehicle – or as Carnot had written "the steam is here only as a mean of transporting the caloric."

This view is replaced by a *new model* representing heat as a contributing factor that participates to the changes in the state of the system. The shift in paradigms here is only in part required for consistency; but its success is also largely due to its allowing for a more economical description (or nascent *theory*) of the class of phenomena at hand: from the conservation of heat in calorimetry, to the conservation of the sum "heat plus work" in engines, and the Laplace–Poisson law governing the adiabatic expansion of an ideal gas (see the discussion presented above).

Moreover, a view of heat and work as convertible currencies encompasses more than the study of engines, in which heat is exchanged for work. The convertibility must include also the exchange of work into heat. The *heat pump* obtained by running a heat engine in reverse was indeed part of Carnot's proof of the universality of his model. As a more immediate experience, we all know that work is spent in the production of heat in all phenomena involving friction. A study of the latter had been explored by Rumford, Hirn and others [see [Kuhn, 1959]] who started distinguishing there something like a principle of conservation of energy. In any case, this path had been reconnoitered well enough to be mentioned as one of his motivations by Clausius for his reinterpretation of Carnot's cycle. It is friction indeed that led to the ultimate quantification necessary to the acceptance of the conservation of energy in thermodynamical processes.

The exchange rate between heat and mechanical work – the so-called mechanical equivalent of heat – was thus measured quantitatively by Joule in a series of varied experiments over a period of several years, staring with [Joule, 1843]; the results of these experiments are collected in his *great memoir* [Joule, 1850] . The archetype quoted today in this connection consists in the measurement of the elevation of temperature produced in a viscous fluid by a paddle wheel driven by a motor expanding for this purpose a quantity of mechanical work that can be reliably measured. Joule's own conclusions from this type of experiments are:

> *1st. That the quantity of heat produced by the friction of bodies, whether solid or liquid, is always proportional to the quantity of force expanded;*
> *2nd. That the quantity of heat capable of increasing the temperature of a pound of water (weighed in vacuo, and taken between 55° and 60°) by 1° Fahr. requires for its evolution the expenditure of a mechanical force* [i.e. work] *represented by the fall of 772 lb. through the space of one foot.*

As a postscript to this section, we should remark that we went too casually over two qualifiers as we wrote about *reversible* processes and *perfect* gases; while these are useful idealizations, it is nevertheless quite important to know that there *exist* irreversible phenomena and imperfect gases.

An example of an irreversible phenomenon is the free expansion of gases. In the experimental study of this process, it is instructive to distinguish three successive stages of sophistication, as it was pioneered by Gay-Lussac and refined by Joule [1845] and by Joule and Thomson [1854 and 1862]. As a fringe benefit from the recounting these experiments on an irreversible process, we get our first glimpse at what it can mean for a gas not to be perfect.

In the first stage, one simply has two containers of equal volume, linked by a pipe equipped with a stopcock. In the beginning, the stopcock is closed, the first container is filled with a gas, the second container is empty; both containers are placed in a calorimeter at temperature T_1 . Then, the stopcock is open, and the gas flows from the first to the second container, doing no work since the expansion is done into a vacuum. When thermal equilibrium has been reached between all parts of the gas in both containers, the temperature T_2 of the calorimeter is read again; it is found that $T_2 = T_1$. Hence no heat has been exchanged between the gas and its surroundings; since no work has been exchanged either, one concludes that the internal energy of the gas depends only on its temperature. All this is thus in agreement with what one had learned from reversible processes. This free expansion however is not reversible: the gas is not going to flow back spontaneously (i.e without exchanging work nor heat) into the first container.

Now comes the second stage of sophistication. Joule repeats his experiment, this time by putting the two containers in *different* calorimeters. He notes that whereas the temperature of the first calorimeter increases – slightly, but surely – the temperature of the second decreases in such a manner that the quantity of heat absorbed by the first calorimeter is very closely equal to the quantity of heat released by the second calorimeter; hence, Joule concludes with satisfaction that the total amount of heat exchanged between the gas and its surroundings is again zero. Yet ... why did the two temperatures not remain separately constant?

The third stage examines this question with the help of a second refinement. A steady flow of gas is passed through a long pipe (that is not conducting heat) partially constricted by a porous plug. The result confirms what one could have expected from Joule's second experiment. The temperature on the gas upstream from the plug increases, while the temperature down-stream decreases. This fact establishes what is called the Joule–Thomson effect. It is of great technological interest since it allows to refrigerate gases to their liquefaction point. Joule and Thomson were lucky however to have worked with air; had they worked instead with hydrogen, or with Helium, they would have been unable to observe this cooling effect at room temperature. The effect itself, and its dependence on the nature of the gas considered, are understood today as a reflection of the deviations of the equation of state from the ideal Gay-Lussac law (2.2.10). These deviations, encoded to a good approximation by the van der Waals equation – (11.3.1-2) – have been explained later from an Ansatz on the space dependence of the molecular forces; we are not yet ready for that in this section, beyond taking cognizance of the Joule Thomson effect as an indication of a limit inherent – even at the macroscopic level – to the idealization involved in considering only perfect gases.

2.4 Energy vs. Entropy

Throughout the previous section, we saw a corpus of understandings emerge from diverse reflections and experiments on the nature of heat. In the present section, we examine how these variously colored strands could be braided into a theory that we know today as the thermodynamics of homogeneous fluids. Accounting for this development becomes easier when one realizes that the founders of the theory were speaking a language formalized nowadays – see Appendix B – as the calculus of differentials, albeit they surely did so with their own diction and idiosyncratic spellings. As the previous section was articulated around Carnot's work, the present one builds up from the foundations laid by [Clausius, 1850, 1857, 1865, 1879].

Clausius' frame of mind was more in tune with the emerging scientific method than Carnot's; yet, the impact of both of these pioneers suffered from some drawbacks in the presentation of the physical motivation as well as of the mathematical formulation of their theories.

On the mathematical side, we noted that Carnot's apparent lack of sophistication was bridged, although still precariously, by the reformulation Clapeyron gave of Carnot's main ideas. On the physical side, the flaw was certainly much more serious: we saw that Carnot's model, and in particular his universality argument, was based on a wrong physical premise, namely the caloric; nevertheless, he managed not only to unearth the correct conclusion on the efficiency of the heat engine, but also his methodology was truly inspired: it laid an essential emphasis on *cyclic* processes.

Clausius took advantage of the latter, substituted the correct ingredient for the caloric, and intuited the general picture, namely the first and second laws – or principles – of thermodynamics. One does not detract from Clausius' presentation by noting that the revamped discussion of the Carnot cycle can be read in Joule's work, and that Thomson proposed a more vernacular, and thus more widely accessible, formulation of the principles; we review also both of these contributions later in the present section.

On the shadowy side, Clausius explicitly predicated his own presentation – especially in its original version [Clausius, 1850] – on overly restrictive assumptions, among which the somewhat ambiguous role played by the constitutive equation for the perfect gas; that he might have chosen to follow that track for rhetoric's sake is historically arguable, yet the didactic reasons for such an approach surely have faded away today. Moreover, in spite of Clausius' awareness, and in fact his quite explicit recognition – especially in his systematic re-exposition of the theory [Clausius, 1879] – of the mathematical concept of non-integrability of differentials, he still couched his general principles in a now antiquated mathematical notation; this unfortunately was not without contributing to the confusion later found in secondary sources, where the issues have been clouded and unnecessary left hanging out of the simple focus which Clausius had achieved. In order to present Clausius' ideas in this focus, we resort to a logical device examined in Chap. 1 – also see, Appendix A – namely, *separating a theory's syntax from its semantics*.

We view the following discussion – starting with (2.4.1) - (2.4.4) – to be a case which illustrates the importance of the syntax in the formation of a theory. Traditional semanticists maintain that the form of the linguistic description of a set of models has no significance because the task of that description is maximally fulfilled once the right set of models is picked out. This is certainly not true in the formulation of the following theory. Hence, we see in this case a tacit support for the enhanced semantic view we proposed in Sect. 1.4.

The syntactic part of Clausius' formulation of the first and second laws, governing the reversible thermodynamics of homogenous fluid, can be stated nowadays very concisely.

2.4.1 The Syntax of Thermodynamics

It subsumes the rules of manipulation encoded in the calculus of differentials (a mathematical primer on this language is given in Appendix B, and we use it below without further ado). The variables span a simply connected two-dimensional region $D \subset \mathsf{R}^2$; usually $D = \{(V,T) \in \mathsf{R}^2 \mid V > 0, T > 0\}$. Two symbols are designated, namely the two smooth differentials η and τ defined on D, having the form

$$\eta = \Lambda_V \, dV + C_V \, dT \tag{2.4.1}$$

$$\tau = p \, dV \tag{2.4.2}$$

(where Λ_V, C_V, and p are smooth real-valued functions on D) and satisfying the following two axioms:

$$\int_\Gamma (\eta - \tau) = 0 \tag{2.4.3}$$

$$\int_\Gamma \frac{1}{T} \eta = 0 \tag{2.4.4}$$

for every simple closed path $\Gamma \subset D$.

The syntax is now completely specified.

Before engaging the semantics, we want to point out that the two axioms, written above in their integral form, can be alternatively presented in differential form, namely

$$\exists \text{ a smooth function } U : D \to \mathsf{R} \quad \text{such that}: \qquad \eta = dU + \tau \tag{2.4.5}$$

$$\exists \text{ a smooth function } S : D \to \mathsf{R} \quad \text{such that}: \qquad \eta = T \, dS \quad . \tag{2.4.6}$$

From a logical point of view, it is important to realize that this is only an equivalent reformulation of the syntax: the sentences are formally distinct, but they refer to the very same propositions, expressed in either integral or differential terms. Indeed , as a consequence of Green's Theorem (see Thm. B.1.3), we have

$$\begin{array}{ll} (4.3) \vdash (4.5) \quad , \quad (4.5) \vdash (4.3) \\ (4.4) \vdash (4.6) \quad , \quad (4.6) \vdash (4.4) \end{array} \tag{2.4.7}$$

where (see Appendix A) for any pair of assertions $\{\Sigma, \sigma\}$ we write $\Sigma \vdash \sigma$ to indicate that σ is deductible (i.e. can be inferred) from Σ in accordance with the inference rules of the syntax of the calculus of differentials (see Appendix B). Moreover, granted (2.4.1) and (2.4.2):

$$(4.5) \vdash (4.9) \quad , \quad (4.9) \vdash (4.5) \atop (4.6) \vdash (4.10) \quad , \quad (4.10) \vdash (4.6) \Bigg\} \qquad (2.4.8)$$

where the assertions (2.4.9) and (2.4.10) are

$$\partial_T (\Lambda_V - p) = \partial_V C_V \qquad (2.4.9)$$

$$\Lambda_V = T(\partial_T \Lambda_V - \partial_V C_V) \quad . \qquad (2.4.10)$$

Finally,

$$\{ (2.4.9), (2.4.10) \} \vdash (2.4.12) \qquad (2.4.11)$$

with

$$\Lambda_V = T\partial_T p \quad . \qquad (2.4.12)$$

A prerequisite for the above syntax to become the formal structure of a physical theory is that it be equipped with an *interpretation* of the formal objects introduced so far, and of their relations: we thus now proceed with these identifications.

2.4.2 The Semantics of Thermodynamics

D is the *state-space* of a homogeneous fluid in equilibrium; each state is uniquely specified by its coordinates V and T. The variables V and T stand respectively for the *volume* and the uniform *temperature* of the fluid. A path $\Gamma \subset D$ traces a *reversible process*. The differentials η and τ are interpreted as the *heating* on, and the *working* by the fluid. Specifically, along any path $\Gamma \subset D$,

$$Q_\Gamma = \int_\Gamma \eta \qquad W_\Gamma = \int_\Gamma \tau \qquad (2.4.13)$$

give a numerical value, respectively, to the total *heat received* and the total *work performed* by the fluid in the course of the process Γ. Finally the functions that appear in the differentials [see (2.4.1) and (2.4.2)] are interpreted as the uniform *pressure* p of the fluid, its *specific heat* C_V at constant volume, and the *latent heat* Λ_V at constant volume.

The conditions (2.4.3) and (2.4.4) are recognized as the first and second laws governing the (reversible) thermodynamics of uniform fluids in equilibrium. Their formally equivalent expressions (2.4.5) and (2.4.6) in differential form physically introduce two new functions defined on the state-space D : the *internal energy* U and the *entropy* S of the fluid.

The interpretative part of the semantics is now completed. Yet, a *model* must still be exhibited to show the compatibility of the structures so far introduced (cf. Sect. 1.3). The ideal gas provide this model; it is uniquely specified by two *constitutive equations,* the first of which incorporates the Gay-Lussac phenomenological law:

$$p(V, T) = R \frac{T}{V} \quad \text{with} \quad R \text{ a positive constant} \quad . \qquad (2.4.14)$$

As an immediate consequence of this constitutive equation, we receive:

$$\Lambda_V = p \qquad \text{and thus} \qquad \partial_V U = 0 \quad ; \quad \partial_V C_V = 0 \tag{2.4.15}$$

which is to say that the internal energy U of an ideal gas and its specific heat C_V are functions of the temperature only. This is a particular feature of the ideal gas, one that accounts for much of its theoretical appeal.

So far, we have that the putative functions U and S must satisfy the differential equations:

$$dU = C_V \, dT \qquad ; \qquad dS = C_V \frac{dT}{T} + R \frac{dV}{V} \quad . \tag{2.4.16}$$

The second constitutive equation defining the ideal gas is now:

$$\partial_T C_V = 0 \tag{2.4.17}$$

so that C_V is now a constant [see (2.4.15)].

For the purpose of building models which demonstrate the self-consistency of the syntax, (2.4.15) and (2.4.17) present the advantage that we can now straightforwardly integrate the differential equations (2.4.16) for U and S to obtain:

$$U(T) = C_V \cdot (T - T_o) + U(T_o) \tag{2.4.18}$$

and

$$S(V, T) = C_V \ln(\frac{T}{T_o}) + R \ln(\frac{V}{V_o}) + S(V_o, T_o) \quad . \tag{2.4.19}$$

The argument leading to (2.4.18–19) shows that the functions U and S can be well-defined, thus proving the internal consistency of the syntax.

One can push further the logical analysis of the syntax, and inquire whether the axioms are independent of one another and whether any of them is trivial. For this purpose, we modify the ideal gas model and introduce the two-parameter family of variations defined by the constitutive equations

$$p = R \frac{T}{V} \quad , \quad C_V = c + \alpha R \ln V \quad , \quad \Lambda_V = \frac{1}{1 + \beta} p \tag{2.4.20}$$

where R, c, α and β are constants. Note now that the first axiom [in its differential form (2.4.9)] is satisfied *iff*

$$\alpha = \beta \tag{2.4.21}$$

and that the second axiom [in its differential form (2.4.10)] is satisfied *iff*

$$\alpha = 0 \quad . \tag{2.4.22}$$

Hence the two axioms are compatible, since there exists a model with

$$\{ \alpha - 0 \quad \text{and} \quad \beta = 0 \} \tag{2.4.23}$$

(namely the ideal gas) in which both of the assertions are true. The two axioms are independent, since there are models with

$$\{\,\alpha \neq 0 \quad \text{and} \quad \beta = \alpha\,\} \quad \text{or} \quad \{\,\alpha = 0 \quad \text{and} \quad \beta \neq \alpha\,\} \tag{2.4.24}$$

in which one of the assertions is true and the other is not. Finally both axioms are non-trivial since there are models with

$$\{\,\alpha \neq 0 \quad \text{and} \quad \beta \neq \alpha\,\} \tag{2.4.25}$$

where neither of the assertions are true.

Beyond the confirmation of the syntax, we should now turn to the semantics. First, we recall from Sect. 2.3 that (2.4.17) is incompatible with the tenets of the caloric theory. Even worse, at the time Clausius proposed that the specific heats could be constant, this seemed to contradicts the "then prevailing views" based on laboratory evidence that was later found wanting; this subsequent vindication of his original proposal is documented in [Clausius, 1879]. Let it suffice here to mention with Clausius that the constancy of the specific heats was observed [Regnault, 1842; 1847] to hold (within about 1 part in 1000) for air and hydrogen over one order of magnitude in pressure, from about 1 atm. to about 10 atm., and over a range of temperature extending from $0°C$ to $200°C$.

Next, upon using (2.4.14) for the change of variables $(V, T) \to (p, V)$, and the notation $\gamma = (C_V + R)/C_V$, we can rewrite (2.4.19) in the form

$$S(p, V) = C_V \ln\left[\frac{p\,V^\gamma}{p_o\,V_o{}^\gamma}\right] + S(p_o, V_o) . \tag{2.4.26}$$

The equations (2.4.14), (2.4.18) and (2.4.26) have a simple interpretation. Indeed, they describe (a mole of) an ideal gas: (2.4.14) is its constitutive equation; (2.4.18) gives its energy, and (2.4.26) – or (2.4.19) – its entropy. Note that the energy is indeed a function only of the temperature; so that the isotherms $pV = $ const are the paths along which the internal energy is constant. Similarly the adiabats, defined as the paths along which the entropy is constant, are of the form $pV^\gamma = $ const ; kinetic theory (see Chap. 3) gives the value $\gamma = 5/3$ for a mono-atomic gas; compare with the value $\gamma = 7/5$ encountered in Sect. 2.3 in the course of our discussion of the Laplace–Poisson derivation of the speed of sound in air, a di-atomic gas.

As it was Clausius [1850] who proposed to name *"entropy"* – after $\eta\,\tau\varrho o\pi\eta$ (= a transformation) – the function S on which he called attention. Accordingly, he argues in [Clausius, 1879] that the curves describing in state-space processes along which the entropy is constant should be called "isentropic curves" (a name for which he credits Gibbs [1873]); Rankine had called them "adiabats" – again according to Clausius, from the primitive form of the Greek $\delta\iota\alpha\beta\alpha\iota\nu\epsilon\iota\nu$ (= to pass through).

Whatever this etymological flexing of classical scholarly muscles may evoke today, it reflects, on the part of Clausius at least, the recurring suggestion that, upon inverting the pair of functions $U(V,T)$ and $S(V,T)$, the energy and entropy could be taken as a parametrization (U,S) of the state-space of the fluid, rather than the perhaps less intrinsic parametrization (V,T) by the volume and temperature:

> *I have intentionally formed the word* entropy *so as to be as similar as possible to the word* energy, *since both of these quantities, which are to be known by these names, are so nearly related to each other in their physical significance that a certain similarity in their names seemed to me advantageous.*

[Clausius, 1865]

To explore further the physical scope of the theory proposed so far, we go back to the generic case, and we review the Carnot cycle, now in the frame provided solely by the two laws of thermodynamics. With Clausius, we draw the Reader's attention to the fact that the discussion given below does *not* depend on any further restriction that would reduce its validity to the simplest model, the ideal gas.

Recall from Sect. 2.3 that a general Carnot cycle is a closed path, operating between the temperatures $T_A > T_B$, and constituted of four consecutive smooth segments, namely

$$\left. \begin{array}{lll} \text{an adiabatic expansion :} & \text{from} & (V_A, T_A) \text{ to } (V_B, T_B) \\ \text{an isothermic contraction :} & \text{from} & (V_B, T_B) \text{ to } (V_C, T_B) \\ \text{an adiabatic contraction :} & \text{from} & (V_C, T_B) \text{ to } (V_D, T_A) \\ \text{an isothermic expansion :} & \text{from} & (V_D, T_A) \text{ to } (V_A, T_A) \end{array} \right\} \quad . \quad (2.4.27)$$

The first law (2.4.3), applied to this cycle, gives

$$Q_{AB} + Q_{BC} + Q_{CD} + Q_{DA} = W \tag{2.4.28}$$

where $Q_{AB} = 0 = Q_{CD}$ as, by definition, the fluid is thermally isolated along any adiabatic path; Q_{DA} is the heating done on the fluid by the reservoir at temperature T_A (i.e. the heat absorbed by the fluid from the reservoir at temperature T_A); $(-Q_{BC})$ is the cooling done on the fluid by the reservoir at temperature T_B, (i.e. the heat returned by the fluid to the reservoir at temperature T_B); and W is the total work performed by the fluid during the whole cycle. We thus have:

$$W = Q_{DA} - (-Q_{BC}) \quad . \tag{2.4.29}$$

The second law (2.4.4) gives in turn, when applied to the same cycle:

$$\frac{1}{T_B} Q_{BC} + \frac{1}{T_A} Q_{DA} = 0 \quad . \tag{2.4.30}$$

From these two expressions follows immediately the value of the efficiency of the Carnot engine, defined as the ratio of the work W performed by the engine and the quantity of heat Q_{DA} one has to provide to the engine at the temperature T_A ; note that the heat $-Q_{BC}$ returned by the fluid to the reservoir at the lower temperature T_B is discounted as useless for the purpose of the engine. We obtain:

$$\varepsilon \equiv \frac{W}{Q_{DA}} = 1 - \frac{T_B}{T_A} \quad . \tag{2.4.31}$$

We have thus recovered the basic result (2.3.31) of Carnot's theory; over the latter, this derivation presents several improvements. Firstly, it replaces the untenable premise of the caloric by two simply and clearly formulated laws (or principles). Secondly, the derivation claims to apply to the general theory of homogenous fluids, not just to the ideal gas. Thirdly, the derivation itself is more direct and transparent.

Yet, the general discussion – as presented so far – still begs an important question. Indeed, whereas the first law (2.4.3) expresses an immediate physical concern – the conservation of energy – the second law is left to appear more removed from a motivation that could be given an immediately apprehensible, pragmatic meaning. Here again, the key to the answer resides in an update of the proof of the universality of the Carnot cycle. In the discussion which follows, the second law (2.4.4) is purposefully *not* be assumed, since we now propose to show how Clausius [1879] was led to this condition from a more intuitive premise, which he phrases thus (see p. 78):

> Heat cannot, of itself, pass from a colder to a hotter body.

There is evidently some vagueness in the undefined terms "of itself"; but what this statement is meant to say will hopefully become sharper as the context is brought into focus, e.g. by the use made of it immediately after (4.33) below. We refer to this statement as *Clausius' heuristic condition*.

Consider again two reversible Carnot engines M and M' operating between the temperatures $T_A = T'_A$ and $T_B = T'_B$ with $T_A > T_B$. During each cycle, they respectively absorb, from a reservoir at temperature T_A , the quantities of heat Q_A and Q'_A , while they return, to the reservoir at temperature T_B , the quantities of heat Q_B and Q'_B . The two engines are matched in such a way that they perform the same quantity of work

$$Q_A - Q_B = W = W' = Q'_A - Q'_B \quad . \tag{2.4.32}$$

We want to prove that one has necessarily $Q_B = Q'_B$ (and hence $Q_A = Q'_A$). Suppose this is not the case, and that $Q_B > Q'_B$. Since M is assumed to be reversible, the quantity of work W performed by M' could be used to operate M in reverse, i.e as a heat pump which we label $(-M)$. The total system $(-M) \cup M'$ would thus transfer from the reservoir at temperature T_B to the reservoir at temperature T_A a quantity of heat [see (2.4.32)]

$$Q_B - Q'_B = Q_A - Q'_A \tag{2.4.33}$$

that would be positive by assumption. This is what Clausius' intuitive statement quoted above does intend to prevent. Hence $Q_B > Q'_B$ is thus ruled out.
A similar argument, with now $(-M') \cup M$ rules out $Q'_B > Q_B$. Consequently
$Q_B = Q'_B$ and $Q_A = Q'_A$ so that

$$\frac{Q_B}{Q_A} = \frac{Q'_B}{Q'_A} \qquad . \tag{2.4.34}$$

As this ratio depends only on T_A and T_B and not on the specific fluid of
which M and M' are made, we can compute its value for M from the value
it takes for M' when the fluid of the latter is a perfect gas. We have then
that U is constant along isotherms; hence, upon using (2.4.14), we have

$$Q'_B = W'_B = \int_{V'_B}^{V'_C} p \, dV = RT_B \ln(\frac{V'_C}{V'_B}) \tag{2.4.35}$$

$$Q'_A = W'_A = \int_{V'_D}^{V'_A} p \, dV = RT_A \ln(\frac{V'_A}{V'_D}) \qquad . \tag{2.4.36}$$

Upon using now (2.4.19), the quotient of these two equations reduces to:

$$\frac{Q'_B}{Q'_A} = \frac{T_B}{T_A} \qquad . \tag{2.4.37}$$

Hence, because of (2.4.34), we have for the *general* Carnot cycle:

$$\frac{Q_B}{T_B} = \frac{Q_A}{T_A} \qquad . \tag{2.4.38}$$

To account for all heats as "heat received" by the fluid, we write $Q_{DA} \equiv Q_A$
and $Q_{BC} \equiv -Q_B$. With this notation, (2.4.38) reads:

$$\frac{Q_{BC}}{T_B} + \frac{Q_{DA}}{T_A} = 0 \qquad . \tag{2.4.39}$$

This is equation (2.4.30), now derived from the above "Clausius heuristic
condition", rather than from the formal, mathematically unambiguous condition (2.4.4). One of the great intuitions of Clausius was to have realized
that the two equations are heuristically equivalent. Indeed, upon considering
an arbitrary closed, simple path Γ, Clausius approximates it as the exterior
boundary of a collection of abutting Carnot cycles, from which he infers:

$$\sum_n \frac{Q_n}{T_n} = 0 \qquad \text{and then} \qquad \int_\Gamma \frac{dQ}{T} = 0 \quad . \tag{2.4.40}$$

One might think that Clausius is making here the damning mistake of
considering Q as a function of which one can take the derivative dQ. A more
attentive reading of this expression, and of its context, shows however that
this is not the case, **precisely because the dQ is written here along
a path**. Clausius [1879, p. 112] leaves no doubt on what he means:

The magnitudes W and Q thus belong to that class which was described in the mathematical introduction [of the book], of which the peculiarity is that, although their differential coefficients are determinate functions of the two independent variables, yet they themselves cannot be expressed as such functions, and can only be determined when a further relation between the variables is given, and thereby the way in which the variations took place is known.

This may not be the most elegant English prose – as it is a translation from late nineteenth-century German – but it is lucid enough: the integrals of what Clausius keeps writing dW and dQ depend on the whole path, and not only on its initial and final points. Because of the lasting confusion Clausius' antiquated notation has contributed to harbor, rather than keeping at bay, modern authors make the distinction explicit, and write (as we did) τ and η for plain differentials, reserving the notations dU and dS for the cases when there are differentiable functions on D (here U and S) for which this notation makes sense: the emphasis is thus kept on the fact that the very *negation of the existence* of "functions" such as W and Q that would correspond to the differentials τ and η is at the core of the understanding of thermodynamics.

To a large extend, this emphasis on the core of the doctrine is the reason why we decided to present it according to its logical structure: the corrected syntax first, and then the semantics (interpretation and model). While this approach is not true to the bibliographical chronology, we feel that it is faithful to Clausius' central achievement.

Concerning the syntax. In his own summary, at the beginning of Chap. V, Clausius states:

In the general treatment of the subject hitherto adopted we have succeeded in expressing the two main principles of the Mechanical Theory of Heat by two very simple equations

which he writes

$$\mathrm{d}Q = \mathrm{d}U + \mathrm{d}W \qquad\qquad \mathrm{d}Q = T\mathrm{d}S \quad . \qquad\qquad (2.4.41)$$

Except for the notational distinction between differentials as noted above, these are the equations we render by (2.4.5) and (2.4.6). We stated that these are equivalent – through Green's theorem – to (2.4.3) and (2.4.4). Clausius knew that also. Indeed, he writes at the end of his Chap. III, right after the demonstration reproduced above and leading to (2.4.40):

If the integral $\int \frac{\mathrm{d}Q}{T}$, corresponding do any given succession of variations of a body, be always equal to zero provided the body returns finally to its original condition, whatever the intervening conditions may be, then it follows that the expression under the integral sign, viz. $\frac{\mathrm{d}Q}{T}$, must be a perfect differential of a quantity, which depends only on the present condition of the body, and is altogether independent of the way in which it has been brought into that condition. If we denote this quantity by S , we may put

$$\frac{\mathrm{d}Q}{T} = \mathrm{d}S \qquad or \qquad \mathrm{d}Q = T\mathrm{d}S$$

an equation which forms another expression, very convenient in the case
of certain investigations, for the second main principle of the Mechanical
Theory of Heat.

The Reader will have recognized here the equation we wrote as (2.4.4), again
with the necessary adjustment of notation; similarly for (2.4.3).

As for the semantics, nothing needs to be changed from the presentation
we gave, except for the order in which things are done, especially in view of
our providing Clausius' heuristic derivation leading to (2.4.40). Moreover we
mentioned from the beginning that the syntax given so far was that of the
reversible thermodynamics of homogeneous fluids. Clausius makes this *caveat*
shortly after he had summarized the two principles in the beginning of Chap.
V in the manner quoted above. He then delayed taking up the matter of
"non-reversible processes" to Chap. X , where he proposes his "Completion of
the Mathematical Expression for the second Main Principle". We thus retain
here the motivation he had for his formulation of this "second principle"
of a thermodynamics that is not necessarily supposed to be reversible, as
he stated in [Clausius, 1879]; and we give only a brief review of what the
tradition represents to be his main result. Nevertheless, we found it proper
to present a critical discussion of some of the tenets of this extension.

Such processes occur in very different forms, although in their substance
they are nearly related to each other. One case of this kind has already
been mentioned ... , that in which the force under which a body changes its
condition, e.g. the force of expansion of a gas, does not meet with a resis-
tance equal to itself, and therefore does not perform the whole amount of
work which it might perform during the change in condition. Other cases
of the kind are the generation of heat by friction and by the resistance of
air, and also the generation of heat by a galvanic current in overcoming
the resistance of the wire. Lastly the direct passage of heat from a hot to
a cold body, by conduction or radiation, falls into this class.

Let us reconsider indeed the free expansion of an ideal gas discussed briefly
at the end of Sect. 2.3. The internal energy U of the final state (V_2, T_2) is the
same as it was for the initial state (V_1, T_1), since $T_2 = T_1$ and we assumed
we have an ideal gas. For the entropy however, we have from (2.4.19) (since
$T_2 = T_1$ and $V_2 \geq V_1$):

$$S(V_2, T_2) - S(V_1, T_1) = R \ln \frac{V_2}{V_1} \geq 0 \qquad (2.4.42)$$

and thus

$$S(V_2, T_2) \geq S(V_1, T_1) \quad . \qquad (2.4.43)$$

Stated in a manner similar to his heuristic formulation (as reproduced above)
of the second principle for reversible processes, Clausius' contribution was to
intuit, from the fact that a free expansion occurs whereas no free contraction
does, that *in any transformation in which a body is left to itself, the entropy*
of the final state is larger or equal to its value in the initial state.

The language must be used with some caution, as can already be illustrated with the model of the free expansion of an ideal gas. Indeed while the initial and final states are *equilibrium* states, so that U and S make sense for these states, this is not the case for the intermediary states of the gas: whenever the two vessels – assumed to be of equal volume, and in communication with one another through however small an aperture – contain unequal amounts of the gas, the state of the whole gas is not an equilibrium state.

This is true even if the aperture is assumed to be so small that one can approximate the situation *in each container separately* by viewing it in a state of equilibrium where, in particular, its density is uniform throughout each vessel, and the pressure and temperature are well-defined for each vessel. Since the whole gas is not in equilibrium, the transformation under study – the free expansion of the gas – is not representable as a path in the space D, the points of which are equilibrium states of the whole gas. Hence our distinction between a "process" (that is a path entirely contained in D) and "a transformation" (which is only a map associating a final state to an initial state, both in D, and for which the intermediary situations, if any, are not required to be describable by points in D).

This being the case, a new interpretation has to be found for the differentials η ("heating") and τ ("working") and even more importantly for the quantities "heat" and "work" that have been defined only in terms of paths in D, if we want to be able to say in a precise manner what is meant by the heuristic condition that "the body is left to itself".

One way out could be to look for a generalization of the theory that would encompass *"local"* equilibrium states. This would allow to deal properly with the thermodynamics of situations such as those considered by Fourier. This however is not within the purview of this section; moreover, this was mastered only much later in the 20th century.

Another way, which we can accommodate here, at least for orientation purposes, is to take advantage of the fact that the experiment is done within a calorimeter, precisely to verify that no heat has passed from or to the gas. During the free expansion of the gas, the water of the calorimeter can be assumed to experience a reversible process: in fact, in this particular case of the ideal gas model, since the temperature does not change, the state of the water does not change. As for the work done by the gas, the expansion being free means that it meets no external resistance, so that no work is performed. Hence, "left to itself" means that in the transformation considered the balance of the internal energy between initial and final state is zero, and that no work has been performed; one can equivalently invoke the first principle and state that no heat and no work has been exchanged between the gas and its surroundings. The remarkable feature is that under such circumstances, the balance of the entropy between the final and initial state must be non-negative: no free contraction is observed. The principle raises this to an axiom, or a dogma; any situation where it could be violated would not conform to the laws of thermodynamics, which means that it would not be described by thermodynamics as we know it. More conservative minds would take solace in the hope that the second principle will not be challenged by laboratory evidence.

Clausius then allowed himself [Clausius, 1865] to speculate rather freely on the entropy of the Universe, which would be bound to increase always, whereas its energy would remain constant. We cannot follow him in this matter here, at least as long as the entropy is defined only in terms of the thermodynamics of finite and homogeneous fluids. Yet, we could not completely ignore that this extrapolation had so fired some imaginations that the speculative "heat death" of the Universe is the only trace left by Clausius' work in the popular folklore.

For the purpose of delineating the logics presiding over thermo-statistical physics, we gave in this section a rather thorough description of Clausius' contributions to the foundations of one aspect of the field, thermodynamics. Among the Founders, he focused indeed most systematically on formulating thermodynamics independently of any underlying microscopic assumptions. Although we all expect such assumptions to play ultimately an important role, this assessment is simply postponed, as molecular or atomic models is discussed at length later in this book. We also have to postpone dealing with inhomogeneous mixtures.

Before leaving the classical thermodynamics of homogeneous fluids however, we should avoid leaving the impression that Clausius was working in isolation. We encountered Joule and Thomson, Rankine and others during our discussion of the conservation of energy in Sect. 2.3, and they will reappear later. Reading Rankine is made harder nowadays by the appeals he made to some of the now strange conceptions he had about the microscopic world. Thomson was often quite close to the concerns of the present section. He contributed mightily indeed to the formulation of the second principle of thermodynamics in a way that seems to have appealed best to those researchers with empirical minds, synthetic longings, and a taste for the vernacular. In one of his influential contributions, Thomson [1851b] sets out the two propositions on which *"the whole theory of the motive power of heat is founded"*. These are:

> PROP. 1 (Joule). – *When equal quantities of mechanical effect are produced by any means whatever from purely thermal sources, or lost in purely thermal effects, equal quantities of heat are put out of existence or are generated.*
>
> PROP. 2 (Carnot and Clausius). – *If an engine be such that, when it is worked backwards, the physical and mechanical agencies in every part of its motions are all reversed, it produces as much mechanical effect as can be produced by any thermodynamic engine, with the same temperatures of source and refrigerator, from a given quantity of heat.*

Thompson's own version of the Second Law of Thermodynamics has the expected Victorian prohibitive twinge:

> *It is impossible, by means of inanimate material agency, to derive mechanical effect from any portion of matter by cooling it below the temperature of the coldest of the surrounding objects.*

Whether these alternate formulations add much to Clausius' theory is not a matter of taste only: one should take into account that at the time he wrote these, Thomson had only the original version of Clausius work, while we used Clausius' presentation of the ideas as it emerged after a long digestive period, in which he also had plenty of time to absorb Thomson contribution, and to formulate responses to his criticisms. The latter are sometimes hard to comprehend. For instance, while Thomson certainly did recognize that Clausius had enunciated an axiom (subsequently the Second Law of Thermodynamics) equivalent to his own, he still wrote, as late as 1864 in a letter to Tait [see [Smith and Wise, 1989, pp. 343–4]], concerning Clausius 1850 paper:

> *it is this* [paper] *that gives Clausius his greatest claim. In it he takes Carnot's theory & adapts it to true thermodynamics. But he does so mixedly with a hypothesis* [Mayer's] *justly assumed as probably APPROXIMATE for gases. It is only by separating out what depends on this hypoth. from what does not that I was able to see that Clausius in his gaseous part does really use the Carnot pure & simple.*

In our own presentation we were careful to separate the general theory from the particular form it takes in the ideal gas model, and we showed which general conclusions can be deducted (not just intuited) from this particular model. As we drew from the revised presentation [Clausius, 1879], we felt that this transient matter of priorities was better let to rest in favor of focusing on the permanent achievements of the theory: where improvements were made was not on the matter in dispute between Thomson and Clausius, but on the generalization of the scope of the theory necessitated to encompass more than homogeneous fluids in equilibrium. For added perspective on the trends that started developing at the beginning of the last quarter of the 19th century, consult [Clausius, 1879, Chap. XIII]: it offers various elements that could be instructive although (or perhaps because) some are not exempt of polemic overtones; in particular, Clausius chose to review there various aspects of the contributions, objections and/or rejoinders by several of those who harbored "different views": Thomson, Rankine, Boltzmann, Decher, Zeuner, Hirn, Wand, Tait.

We mentioned earlier that the equivalence between the two forms of Clausius' syntax for thermostatic (2.4.3–2.4.4) and (2.4.5–2.4.6) is but an expression of a general mathematical fact, Green's theorem, that was very much in the air at the time. Many of the recountings of its history are not exempt from some national bending: the French (with Ampère), the Germans (with Gauss), and the Russians (with Ostrogradski) do indeed have legitimate claims to an independent discovery, if not always to scientific priority or mathematical rigor and generality. Even the Swiss can claim part of the action: while we presented this as a corollary to Green's theorem, Euler "knew" of it (and it is not in our purview here to discuss Euler's notion of what constitutes a proof). In addition, many thermodynamics textbooks refer to one of the equivalent conditions of the theorem [condition (a) in the corollary of Green's theorem we listed as Thm. 3 in Appendix B] as "Maxwell's integrability condition". In the Anglo-Saxon world, the history of the theorem and its corollary is usually traced back to George Green whose life and work are surrounded by darkness; [Grattan-Guiness, 1995, Cannell, 1999]. His main contribution, [Green, 1828], was ignored by the scientific community until 1845 when Thomson rediscovered it and recognized its value in connection with his own research on electricity and magnetism, where he felt Green had indeed anticipated him [see [Smith and Wise, 1989, p. 204]].

Thomson was thus instrumental in getting Green's work republished in the early 1850s. Moreover, he had seen that the 3-dimensional extension of this theorem was what he needed; this form of the theorem has become known as ... Stokes' theorem, by one of the twists by which the superposition of various mode of scientific communication is prone to confuse later historians. The first appearance of Thomson's version of this theorem seems to be in a letter that he wrote to Stokes in 1847; Stokes apparently liked it so much that he posed it as an examination question in the Smith's prize competition held in February 1854; one of the two winning papers was by no less than James Clerk Maxwell (!), the other by E.J. Routh; see [Everitt, 1974] and also [Cross, 1985].

To complicate even further the intellectual history of the theorem, Thomson himself did not use this result for his own work on thermodynamics, while he used it most extensively in his work on electricity and magnetism, e.g. [Thomson, 1851a], thus seemingly ignoring the interest Green's theorem could have in the formulation of the principles of thermodynamics, and this in spite of the fact that he published his version of the first and second laws the very same year [Thomson, 1851b]. Recall also that – just at this time – Thomson was sponsoring the re-publication of [Green, 1828]!

The explanation is certainly not to be found in Thompson not having read Clausius' work: he refers to it explicitly in his published work, he mentions him in his notebook (on 15 August, 1850; incidentally, the same day he enters the name of 'young Clerk Maxwell for the first time) and recommends the reading of Clausius' work privately to some of his correspondents. For these epistolary information, see the very well-documented intellectual biography [Smith and Wise, 1989] of Thomson. In particular, we read there:

> In August 1850: Rankine wrote to Thomson to thank him for "calling my attention to the paper of Clausius, in Poggendorff's Annalen, on the Mechanical Theory of Heat. I approve your suggestion to send a copy of my paper either to Clausius or Poggendorff". Thomson was therefore aware of Rudolph Clausius' ... first paper, published by April 1850, 'On the motive power of heat' in which Clausius first enunciated and established his version of the second law of thermodynamics. This fact, however, does not imply that Thomson had assimilated its contents, and so the critical period

*between August, 1850 and March, 1851, when Thomson published his own
theory, requires careful interpretation.*

[Smith and Wise, 1989, p. 320]

It should be repeated here that Thomson wanted to see things his own way,
as already evidenced by the protracted resistance he had offered to Joule's
equivalence principle: while he had accepted that mechanical work could be
transformed into heat (e.g. in the instance of friction), he was stuck for several
years with ideas close to Carnot's analogy between the 'fall of heat' and
the fall of water in watermills, and according to which mechanical work is
produced by the mere *transfer* of heat:

*The heat contained in a body was for him [Thomson] purely a function of
the body's physical condition ... heat ... was therefore what would now be
called a 'state function'; any gain or loss of heat depended only on initial
and final states and not on the path between them ... Indeed, he seems
to have regarded conversion of mechanical effect to heat as an irreversible
process of dissipation ... in which available work is lost to man. Hence,
conversion of work to heat did not imply reconversion to work, or* mutual
convertibility

[Smith and Wise, 1989, pp. 314 and 311]

Compare the above argument to the view expressed explicitly even in the
title of [Joule, 1847] paper, where the "equivalence" is referred to as "the
identity between the caloric and the mechanical force"). Eventually, Thomson
rallied to Joule's complete equivalence. The difficulty he might have had in
espousing other peoples' ideas is surely a manifestation of the obverse side
to his originality, although we would not allow ourselves to go quite as far
as claiming that *"it was Thompson's peculiar genius to generate powerful
disconnected insights rather than complete theories."* [Everitt, 1974]

A somewhat more generous version of this "peculiar" trait of Thomson might nevertheless still help understand that the critical assessment on
Thompson's assimilation of the contents of Clausius' theory – reproduced
above from [Smith and Wise, 1989] – seems directed mainly to the differences Thomson had with Clausius on the proper physical formulation of the
principles. This very assessment might also throw some light on Thompson's
reticence to formulate the basic laws of thermodynamics in *integral rather
than differential form*. It is a strange thing indeed to see Thomson failing
here to take advantage of the perspective offered by a straight application
of a theorem he had pulled out of the obscurity in which Green's mode of
publication had condemned it, and which Thomson himself had grasped intimately enough to provide a generalization suitable to his own pursuits in
the theory of electricity and magnetism. Who would believe that this failure
could be deliberate?

Once again, we emphasize that the existence of an *entropy* function *is*
(the essential content of) the *second law of thermodynamics*.

We presented in this section a discussion of the syntax and semantics of
thermodynamics, following one main strand of the historical development,

namely the one associated most consistently with the name of Clausius. This presentation, however, is more contingent than meets the eye. For one thing, the notion of temperature enters the theory twice: first as a primary notion, and second as an integrating factor in $\eta = T \, dS$. It might therefore be more natural to parametrize the equilibrium state-space D with more intrisic variables, such as U, the internal energy, and V, the working coordinate(s). Also, the classical understanding of Carnot cycles involves "quasi-static" processes (so slow and mild as to be traceable as trajectories entirely contained in D) rather than more violent transformations of which one requires only that the initial and the final states belong to D, which is all one should need to contruct a function S the value of which $S(X)$ depends only on the state $X = (U, V) \in D$ and not on the history of that state.

One way out of these embarrassments – so that T and S appear as derived, rather than primary, concepts – was pioneered by [Carathéodory, 1909, Planck, 1926, Giles, 1964, Buchdahl, 1966] and firmly consolidated by [Lieb and Yngvason, 1998]. This approach focuses on an axiomatization of a (partial) ordering $\prec$, defined for "sufficiently many" pairs $(X, Y) \in D \times D$, with $X \prec Y$ to be interpreted as "Y is adiabatically accessible from X", a reading which is specified by the axioms – 17 of them at the last count. Ultimately, the axiomatization is couched in such a way that it leads to a function $S : X \in D \to S(X) \in \mathsf{R}$ with the property $X \prec Y \models S(X) \leq S(Y)$; hence, *if* one restricts oneself to any class $C \subseteq D$ such that $\{ \ X, Y \in C \models$ either $X \prec Y$, or $Y \prec X$, or both $\}$, *then* $\{ \ X \prec Y$ iff $S(X) \leq S(Y) \ \}$. Moreover, the axioms ensure that S is unique up to a scale factor, concave, and extensive.

3. Kinetic Theory of Gases

Der Mondschein geht wie ein langer Blitz vorbei, und die reglose Fahne hat unruhige Schatten. Sie träumt.

. . .

Aber die Fahne ist nicht dabei.

[Rilke, 1906]

The purpose of this chapter is to convey that certain aspects of the macroscopic theory called thermodynamics do admit simple statistical models that point towards a microscopic, mechanical theory; one of the first successful examples of the latter was [Boltzmann, 1896–1898] kinetic theory of gases, which we review and assess in Sect. 3.3. Before we do that, however, we present – in Sect. 3.1 – a simple stochastic model for diffusion, and – in Sect. 3.2 – the derivation of Maxwell's equilibrium velocity distribution in a gas. Finally, we try and capture in Sect. 3.4 the light thrown by the Ehrenfest urn model – affectionately known as the "dog-flea" model – on the controversies raised by Boltzmann's work. The common language of routine probability praxis is used throughout this chapter, as is the underlying concept of randomness. An approach, more careful with conceptual matters, is presented in the following chapters; the present chapter is therefore to be considered as a motivational primer. It also shows a procession of models from simple and frankly unrealistic to somewhat more complex and realistic, and back – which gives illustrations to the process of model building in the search for a microscopic theory of macroscopic phenomena.

3.1 A Random Walk Model for Diffusion

Recall that the very first equation written in Chap. 2 was Fourier's heat equation (2.2.1). It has the same form as the diffusion equation for colloïdal particles in a fluid. A visual example of the latter is provided by the spreading of the particles of tobacco smoke in the air. When we take into account the possibility of a drift v, this phenomenological equation governs the (one-particle) density $f(x_1, \ldots, x_d; t)$ of the colloïdal particles at time t and reads:

$$\partial_t f = -\sum_{i=1}^{d} v_i \partial_{x_i} f + D \sum_{i=1}^{d} \partial_{x_i}^2 f \qquad (3.1.1)$$

where $d = 1, 2,$ or 3 is the dimension of the space in which the diffusion takes place. Note that even though equation (3.1.1) was formulated for colloïdal particles, it is still phenomenological – meaning macroscopic – and that this equation could also admit a continuous micro-model. For nothing *in the equation* necessarily demands that the parts of the system in question be particles. And yet, if a model of colloïdal particles is to be constructed – as the Brown observations described below seem to demand – one must bridge the gap between the smooth and deterministic appearance of the macroscopic diffusion and the non-smoothness and randomness observed at the microscopic level. When one observes the motion of the colloïdal particles under a microscope, the particles appear to follow random, jagged trajectories; this phenomena was known since the 18th century, but tentatively cast at first in a mistaken vitalist perspective; it was studied systematically in [Brown, 1828, 1829]. The same jagged patterns were found to persist over several orders of magnitude, provided the time-scale was adjusted according to the square of the space-scale; carried to its limit, such a universal scaling law would imply that the trajectories, while surely continuous, should look as if they were nowhere differentiable; direct observations have not shown evidence to falsify this view; the name of Perrin [1910] is associated with the delicate measurements associated with this observation.

From equation (3.1.1), we see that – for whatever microscopic corpuscular model of the diffusion process one is to give – the average position of the particles, and the corresponding mean-square deviations must satisfy:

$$\langle x_i \rangle_t = v_i\, t \qquad \text{and} \qquad \langle\, (x_i - \langle x_i \rangle_t)^2\, \rangle_t = 2D\, t \quad . \qquad (3.1.2)$$

A model also should provide microscopic interpretations for v_i and D. Theoretically, (3.1.2) will be seen to reflect the fact that the solution of the diffusion equation (3.1.1), when one takes for initial condition the situation where all the particles are concentrated at the origin, is given by:

$$f(x_1, \ldots, x_d; t) = (4\pi Dt)^{-\frac{d}{2}} \exp[-\frac{1}{4Dt} \sum_{i=1}^{d} (x_i - v_i t)^2] \quad . \qquad (3.1.3)$$

Hence, we model the link between the diffusion equation (3.1.1) and the Brownian motion (3.1.2), such that $f(x_1, \ldots, x_n; t)$ admits a probabilistic interpretation under the constraints of equation (3.1.2). Upon noticing that, in (3.1.3):

$$f(x_1, \ldots, x_d; t) = \prod_{i=1}^{d} f_i(x_i; t) \quad \text{with}$$

$$f_i(x_i; t) = (4\pi Dt)^{-\frac{1}{2}} \exp[-\frac{1}{4Dt}(x_i - v_i t)^2] \qquad (3.1.4)$$

we conclude that the motions in the d directions are *independent* of one another. For our model building purposes, this implies that we could search for understanding already in an analysis of the one-dimensional case. We therefore assume henceforth that $d = 1$; accordingly, we drop the index i.

To model the one-dimensional diffusion equation, we start by considering a particle restricted to moving on a one-dimensional lattice

$$\{\, x = sa \mid s \in \mathsf{Z} \,\}$$

where a, the distance between two successive sites, is constant. The time is also chosen to be discrete, so that the clock indicates the instants

$$\{\, t = n\tau \mid n \in \mathsf{Z}^+ \,\}$$

where again τ is constant. The motion is then described by the condition that at every instant t the particle moves with a probability p (resp. q) by one step to the right (resp. left), with

$$p + q = 1$$

to ensure that the particle has no other choices. We denote by X_t the function – over the space of trajectories – which gives the position of the particle at time t – probabilists refer to X_t as a *random variable*. The above condition translates mathematically as a statement that the following conditional probabilities entirely govern the motion:

$$\left.\begin{array}{l} Prob\,\{\, X_{t+\tau} = x + a \mid X_t = x \,\} = p \\ Prob\,\{\, X_{t+\tau} = x - a \mid X_t = x \,\} = q \end{array}\right\} \quad . \tag{3.1.5}$$

From this law of motion, we want to determine the position of the particle at any time $t > 0$, given that the particle is starting at the site $x = 0$ at time $t = 0$; specifically, we want to evaluate the function f defined by:

$$f(x,t) \equiv Prob\,\{\, X_t = x \mid X_0 = 0 \,\} \quad . \tag{3.1.6}$$

Since the law of motion is entirely described by (3.1.5), the position of the particle at time $t+\tau$ depends only on its position at the immediately previous instant, namely t. This *Markovian* property translates here as the difference equation:

$$f(x, t + \tau) = pf(x - a, t) + qf(x + a, t) \tag{3.1.7}$$

which we have to solve for the initial condition

$$f(x, 0) = \begin{cases} 1 & \text{if} \quad x = 0 \\ 0 & \text{otherwise} \end{cases} \tag{3.1.8}$$

In order to solve the difference equation (3.1.7) with initial condition (3.1.8), note that to reach the position $x = sa$ at time $t = n\tau$ the particle has to make exactly j (resp.$n - j$) jumps to the right (resp. left) subject to the condition

$$s = j - (n - j) \quad \text{i.e.} \quad j = \frac{1}{2}(n + s) \quad . \tag{3.1.9}$$

Since there are exactly

$$\binom{n}{j} = \frac{n!}{j!\,(n - j)!} \tag{3.1.10}$$

ways to do this, we conclude that

$$f(sa, n\tau) = \begin{cases} \binom{n}{\frac{1}{2}(n+s)} p^{\frac{1}{2}(n+s)}\, q^{\frac{1}{2}(n-s)} & \text{when} \quad |s| \le n \\[2mm] 0 & \text{otherwise} \end{cases} \tag{3.1.11}$$

and we verify that this is the solution of (3.1.7–8) by a straightforward substitution.

We can now discuss whether this is a good enough model for our diffusion process. For this purpose we can compute the average position of the particle at time t, $< x >_t \equiv \sum (sa)\, f(sa, n\tau)$ and its mean-square deviation $< (x - < x >_t)^2 >$, which is defined similarly. We find

$$< x >_t = (p - q)\frac{a}{\tau}\, t \quad \text{and} \quad < (x - < x >_t)^2 >_t = 4pq\,\frac{a^2}{\tau}\, t \quad . \tag{3.1.12}$$

Hence our model *can* be made to satisfy (3.1.2) by adjusting its parameters p, a and τ in such a manner that

$$(p - q)\frac{a}{\tau} = v \quad \text{and} \quad 4pq\,\frac{a^2}{\tau} = 2D \quad . \tag{3.1.13}$$

Once this is done, *the model satisfies the constraints and is therefore on the right track. However,* we still have to turn the difference equation (3.1.7) into a differential equation of the form (3.1.1) – with $d = 1$ – describing a diffusion. To achieve this, we consider a limit in which the spacing between the sites of our lattice goes to zero. To preserve the relations (3.1.13), and thus to keep (3.1.2) satisfied, we must impose a link between several limiting procedures, specifically we must require that

$$\lim_{a \to 0} \frac{a^2}{\tau} = 2D \quad \text{and} \quad \lim_{a \to 0} \frac{1}{a}(p - q) = \frac{v}{2D} \tag{3.1.14}$$

which implies in particular that $\tau \to 0$ and $p, q \to \frac{1}{2}$; specifically that $p \approx \frac{1}{2}(1 + \frac{v}{2D}a)$. We can now appeal to the Taylor expansion theorem and write – upon taking into account the first of the above two conditions, namely $a^2 \approx \tau$:

$$\begin{aligned} f(x; t + \tau) &= f(x; t) + \tau(\partial_t f)(x; t) + \cdots \\ f(x \pm a\,; t) &= f(x; t) \pm a(\partial_x f)(x; t) + \tfrac{1}{2}a^2(\partial_x^2 f)(x; t) + \cdots \end{aligned} \tag{3.1.15}$$

Upon rewriting (3.1.7) we now obtain (via the above linked limits):

$$lim_{\tau \to 0} \tfrac{1}{\tau}[f(x; t + \tau) - f(x; t)] =$$
$$lim_{a \to 0}[(q - p)\tfrac{a}{\tau}(\partial_x f)(x; t) + \tfrac{1}{2}\tfrac{a^2}{\tau}(\partial_x^2 f)(x; t)] \tag{3.1.16}$$

i.e.

$$\partial_t f = -v \partial_x f + D \partial_x^2 f \quad . \tag{3.1.17}$$

This is precisely the one-dimensional version of our diffusion equation (3.1.1), thus showing that we obtained a stochastic model for the diffusion process. It can be shown also that, in the limit (3.1.14), the probability distribution (3.1.11) of the discrete model approaches (3.1.3) with $d = 1$ which is a solution of the differential equation (3.1.17). After the material to be presented in Chap. 4, this will be seen as just another manifestation of the Poisson–Laplace law of large numbers. The simpler case where $v = 0$ (no drift) is covered in the second section of the classic paper [Kac, 1947b]; the next three sections of that paper present more sophisticated models, in the presence of external forces, with or without barriers; and the fifth section contains Kac's original contribution to the dog-flea model to be discussed in our Sect. 3.4 below.

These examples provide a good illustration of one of the most salient features in the praxis of model-building – which is often neglected in the literature, perhaps with the exception of Emch [1993]. Our one-dimensional model is no doubt a highly idealized mathematical structure. The model does not even pretend to describe any actual process or system.

Nevertheless, it satisfies the phenomenological theory – at least, its core equation – which is well tested by experiments. It is often the case that when a phenomenological theory is established, its core components, at least, are sufficiently general so that its microscopic models are radically underdetermined. To exploit such a theoretical opportunity, scientists often construct, at first without much regard to their *empirical* plausibility, models as simple as possible so that the limits of building an explanatory micro-model for the theory can be clearly seen. In the present case, some guidelines from the phenomenon of Brownian motion are observed in the one-dimensional model, but otherwise, it is made as simple as one can imagine, and yet all macroscopic constraints are shown to be satisfied by it, and hence, *the theory is proven to admit a model.* Similar comments can be made about the dog-flea model to be discussed in Sect. 3.4 below: although admitedly unrealistic, the dog-flea model was able to pierce through some of the conceptual smoke blown around Boltzmann's ideas regarding the origin of irreversibility. In other words, even if the world out there is not nearly as simple microscopically as in the model of the present section, such a simple world would have exhibited macro-phenomena of diffusion very much the same as in ours. Two other attractive features of these examples are worth noting. One is that before the model of one-dimensional diffusion is produced, the phenomenological theory is *dismantled* as much as possible (e.g. to independent dimensions) so that the micro-model is necessary only for the smallest macro-component; and the other is that it shows nicely how the structural theory, i.e. (3.1.1) is related to the behavioral theory (3.1.4) by the latter being the *solution* of the former (cf. Sect. 1.4, the enhanced view).

The detailed analysis of the underlying Brownian motion has had a seminal influence on the elaboration of the mathematical theory of stochastic processes. The scientific literature (in mathematics, physics, chemistry, biology) on the subject is enormous and continues to grow. An informative and concise presentation – including a brief historical introduction – can be found in [Nelson, 1967]. Let us simply point out here three specific areas: (1) the historical development, and ultimate acceptance, of the molecular view of matter; (2) the study of transport equations; and (3) the recent awakening of stochastic financial analysis.

To the first of these areas are associated the names of Einstein and Smoluchovski. In particular, we want to mention the first [Einstein, 1905a] of a series of papers – collected in [Fürth, 1956] – which presents a more realistic model – certainly more realistic than our one-dimensional model of Brownian motion – for the diffusion process, and in particular provides a physical interpretation of the diffusion constant D; as well as another series of papers starting with [Smoluchowski, 1906a,b] on whose insights we will have occasion to comment later on; several of Smolukowski's papers have been translated and collected in [Ingarden, 1999], with informative introductions by M. Kac and S. Chandrasekhar.

Consider a system, submitted to a constant field of force, and consisting of identical, non-interacting, smooth spheres of radius r (modeling the colloïdal particles), suspended in a homogeneous fluid characterized by its viscosity η and its temperature T. Upon taking into account: (a) the external forces to which the particles are submitted; (b) the friction force experienced by these particles in their motion through the fluid, and (c) the fact that their mean motion is governed by a diffusion equation, one can compute that the diffusion constant must satisfy the Einstein diffusion formula:

$$D = k\frac{T}{6\pi r\eta} \qquad (3.1.18)$$

where k is a constant such that kT has the dimension of energy. In his derivation of (3.1.18), Einstein showed that k can be identified with the Boltzmann constant $k = (N_{mole}/N)R$ where for a macroscopic gas, N_{mole} (resp. N) is the number of moles (resp. molecules) it contains, and R is the gas constant in $pV = N_{mole}RT = NkT$; (see below the work of Boltzmann and Maxwell for the introduction of k). Note that (3.1.18) is not merely a prediction of some exclusively macroscopic quantity, but is connected to the microscopic corpuscularity of diffusion systems.

To verify this prediction, one must prepare a uniform collection of spherical particles of colloïdal dimension; this was achieved by [Perrin, 1910] and his collaborators, and provided a direct measure of k, and thus, indirectly of the Avogadro number, the number of molecules in a mole, within better than .5% of the present day value $6.02 \cdot 10^{23}$. Their results was in such a good agreement with the expected value of the Boltzmann constant, that it could be considered as an argument for the existence of molecules, an argument that, in its time, was hoped to be strong enough to convince the last sceptics, such as the positivists Ostwald and Mach; yet, it is a sad comment on the development of science that old theories seem never to die: only their proponents do.

Our second point concerns an important lesson to be learned from the model discussed in this section, namely the role of rescalling of the space and time parameters; this idea was to guide much of the subsequent attempts to derive hydrodynamic transport equations from deterministic Hamiltonian law governing the microscopic evolution. As to this writing, the latter aim has not been achieved yet, even for the Fourier law of Sect. 2.2, although some remarkable progress has been made recently with hybrid models involving Hamiltonian dynamical systems coupled to stochastic reservoirs, see e.g. [Jaksic and Pillet, 1998, Eckmann, Pillet, and Rey-Bellet, 1999b, Eckmann and Hairer, 2000]; for an up-to-date review, see [Bonetto, Lebowitz, and Rey-Bellet, 2000].

Finally, in our days, when the scientific establishment is submitted again to the pressure of fiscal relevance, it is amusing to note that, essentially contemporary to the pioneering explorations of Einstein and Smoluckowski, a student of Poincaré defended a thesis [Bachelier, 1900] in which he compared the fluctuations of the stock market – in his case the Paris Exchange – to the jagged appearance of Brownian motion. Since the stock index can hardly be negative, a crude analogy would not work: Brownian motion is unbounded, below as well as above. To avoid this, one can try and identify the composite stock index with the random variable obtained as the exponential of our process: $\Xi_t \equiv \alpha \exp\{X_t\}$. Under this identification, the drift term v can be interpreted as the exponential increase attributed to the necessity for the stock market to keep up with such things as the "fixed" interest rates in the bond market. It seems that, just in the course of the recent years, this rather naïve stroke of insight has stimulated some interesting developments; see e.g. [Baxter and Rennie, 1996, Case, 1998].

3.2 The Maxwell Distribution

As we go back to our exploration of how statistical arguments found a place in thermophysics, we turn now to the derivation of the Maxwell velocity distribution. While the idea that matter is made up of small parts – atoms or molecules – goes back to Democritus (5th Century B.C.), the first quantitative microscopic model for a gas was proposed by D. Bernoulli [1738]; the model was taken over by Krönig [1856], and shortly thereafter, improved in major ways – both technical and conceptual – in [Clausius, 1857, 1858]. The

original aim, as stated by Bernoulli [1738], was to give a theoretical account
for the air-pressure thermometer proposed in [Amontons, 1702]. The following is a review where we emphasize the modeling assumptions under which
Bernoulli's reasoning can be boiled down to its essentials.

Consider a cubic box Λ with rigid walls. Let $V = L^3$, be the volume of the
box. This box contains N independent point-like particles of mass m. Assume
that the particles are uniformly distributed in space among six beams, each
one of which impinges perpendicularly on one of the walls of the box. The
particles are supposed to move with uniform speed v and to collide elastically
with the walls: hence, in such a collision, the momentum of a particle changes
by an amount equal to $2mv$. Let ν be the number of particles that hit a wall
during a small time interval τ; clearly, since the particles all move at constant
speed v between the collisions, only those particles that are within a distance
$v\tau$ from the wall reach it during the time interval τ; since the particles are
distributed uniformly in space, and since there are six beams:

$$\nu = \frac{1}{6} N \frac{L^2 v \tau}{V} \quad . \tag{3.2.1}$$

Newton's definition of force, as the agent of a momentum change, gives the
value of the force F, and thus of the pressure p, exercised by these particles
on the walls:

$$p = F \frac{1}{L^2} = \{\frac{1}{\tau} 2mv \cdot \nu\} \frac{1}{L^2} \quad . \tag{3.2.2}$$

Upon substituting (3.2.1) into (3.2.2), we obtain:

$$pV = \frac{2}{3} N \frac{1}{2} mv^2 \quad . \tag{3.2.3}$$

This is the first quantitative result in the kinetic theory of gases, namely
the interpretation of the *product pressure × volume* in terms of the *kinetic
energy* of the individual particles in the gas. Let us moreover compare this
with what we know as the Gay-Lussac law $pV = NkT = N_{mole}RT$ (see
1.2.10) – recall that Bernoulli was interested in the crude version of the law
due to Amontons, namely that a change of temperature ΔT is accompanied
by a change of pressure Δp proportional to ΔT. We obtain in this manner
the second quantitative result in the kinetic theory of gases, namely the
interpretation of the *absolute temperature* in terms of the *kinetic energy* of
the individual particles in the gas:

$$\frac{3}{2} kT = \frac{1}{2} mv^2 \quad . \tag{3.2.4}$$

As was to be understood later, the factor 3 in this relation refers to the fact
that we treated only point-like particles in a 3−dimensional space, so that
the molecules of this gas have 3 degrees of freedom.

Evidently this model of a gas is very crude, with its six beams of molecules
moving uniformly perpendicularly to the walls of the container. Nevertheless,

this illustrates an important feature of model-making in general and the making of microscpic models in particular. We see here a tight connection between the degrees of qualitative description of a phenomenon and the degrees of idealization. In order to cross the macro-micro divide and obtain precise quantitative relations between macro-quantities and micro-quantities, one first approaches the problem with highly idealized microscopic models from which a qualitative estimate of the relations can be obtained, which helps us to divine which quantities to relate. The validity of such model-buildings with a high degree of idealization usually rests on some preconceived notions of the kind of substance which comprises the models and the kind of micro-quantities to which the macro-quantities in question are supposed to be reduced. In the Bernoulli case, a gas consists of a large number of small particles the movements of which obey Newton's laws of mechanics, and that a quantity such as temperature should be proportional to such motions. The derivation of (3.2.3–4), therefore, acts as a qualitative justification of the molecular models for gases.

It was [Maxwell, 1860] self-imposed task to derive from first principles a more realistic equilibrium velocity distribution $\varphi^{(o)}$ for the molecules of a gas. He required this distribution to satisfy the following conditions:

(a) the velocity distribution should be properly normalized;
(b) the velocity distributions in the three directions should be mutually independent;
(c) the velocity distribution should be isotropic.

Specifically, with N denoting the total number of molecules in the gas; $\boldsymbol{v} = (v_1, v_2, v_3)$ their velocity; and V the volume occupied by the gas, these conditions read:

$$\left.\begin{array}{l} (a) \ \int_{R^3} dv_1 \, dv_2 \, dv_3 \, \varphi^{(o)}(\boldsymbol{v}) \ = \ n = \frac{N}{V} \\[2ex] (b) \ \varphi^{(o)}(\boldsymbol{v}) = \varphi_1^{(o)}(v_1) \, \varphi_2^{(o)}(v_2) \, \varphi_3^{(o)}(v_3) \\[2ex] (c) \ \varphi^{(o)}(\boldsymbol{v}) = \Phi(\|\boldsymbol{v}\|^2) \end{array}\right\} \qquad (3.2.5)$$

From (3.2.5.b and c) we see that for each $i = 1, 2, 3$:

$$\frac{1}{2v_i}\partial_{v_i} \ln \varphi_i^{(o)}(v_i) \ = \ \partial_{\|\boldsymbol{v}\|} \ln \Phi(\|\boldsymbol{v}\|^2) \quad . \qquad (3.2.6)$$

In fact, instead of (c), Maxwell required that the functions $\varphi_i^{(o)}$ (with $i = 1, 2, 3$) be the same. He then realized that this can be derived (see 3.2.7 below) as a consequence of the invariance of $\varphi^{(o)}$ under rotation. But, for each i separately, the LHS of the equations (3.2.6) is independent of v_j with $j \neq i$; and they are all equal (to the RHS). Hence, neither can depend on any v_i, i.e. they must be constant. The most general function Φ, such that its logarithmic derivative is constant, is of the form:

$$\Phi(\|\boldsymbol{v}\|^2) = Z \, \exp\{A\|\boldsymbol{v}\|^2\} \qquad (3.2.7)$$

where A and Z are two constants. Finally, condition (3.2.5.a) imposes that $A < 0$. Hence, we can write without loss of generality

$$\varphi^{(o)}(\boldsymbol{v}) = n \{\frac{1}{2\pi\alpha^2}\}^{\frac{3}{2}} \exp\{-\frac{1}{2\alpha^2}\|\boldsymbol{v}\|^2\} \qquad (3.2.8)$$

where α is a constant that remains to be adjusted by a simple adaptation of the reasoning that led to (3.2.3) which should now read

$$p = \frac{1}{V} \int_{v_1>0} dv_1 \int_{R^2} dv_2 \, dv_3 \; \varphi^{(o)}(v_1, v_2, v_3) \, (2mv_1) \cdot v_1 = nm\alpha^2 \qquad (3.2.9)$$

i.e.

$$pV = Nm\alpha^2 \, .$$

Again, upon comparing this with the Gay-Lussac law $pV = NkT$ we obtain $\alpha^2 = kT/m$. Hence:

$$\varphi^{(o)}(\boldsymbol{v}) = n \{\frac{m}{2\pi kT}\}^{\frac{3}{2}} \exp\{-\frac{m}{2kT}\|\boldsymbol{v}\|^2\} \qquad (3.2.10)$$

which is the *Maxwell distribution law*. In particular, it implies that the *energy is equally distributed among the degrees of freedom of the particles of the gas*, in the sense that averages computed with respect to the distribution (3.2.10) give:

$$\langle \frac{1}{2}mv_1^2 \rangle = \langle \frac{1}{2}mv_2^2 \rangle = \langle \frac{1}{2}mv_3^2 \rangle = \frac{1}{2}kT \quad . \qquad (3.2.11)$$

In the same vein, one can rewrite the total energy of the particles in the gas in the form

$$U = N < \frac{1}{2}m\|\boldsymbol{v}\|^2 >= \frac{3}{2}NkT \qquad (3.2.12)$$

which gives for the specific heat per mole of the gas, the classical value

$$C_V = \frac{3}{2}R \qquad (3.2.13)$$

where we used again the defining relation $Nk = N_{mole}R$. As the energy is to be distributed equally between the degrees of freedom of the system, one would infer from this model that for an ideal gas formed of rigid diatomic molecules – where each molecule has five degrees of freedom – the specific heat would be $C_V = \frac{5}{2}R$, which is indeed the case: see our discussion of the Laplace computation of the speed of sound in air in Sect. 2.3.

These, however, are only indirect tests of the Maxwell distribution: by themselves, they do not give it any advantage over Bernoulli's model, as they concern only averaged quantities; these tests do not discriminate between a gas made of particles which move with different velocities according to a certain smooth, isotropic, distribution; and a gas made of particles which all move along six main directions with the same speed, numerically equal to the average speed of the same distribution. In fact, they treat all velocity distributions as empirically equivalent as long as the distributions give the same observable averages.

Therefore, Maxwell's motivation for constructing his model as an improvement of Bernoulli's must have come from his metaphysical conviction of what seems more realistic – specifically, the three equations (3.2.5) – and his faith in the eventual availability of *direct tests* for the distribution. Such tests were devised and implemented much after Maxwell's time, first by O. Stern in the 1930s, but quite spectacularly in the late 1950s with the advent of plasma physics; see e.g. [Zemansky, 1964, pp. 93–96]. There, the gas is formed of ionized particles that emit a light-ray as they return to their ground state; the wave-length λ_o of the light-ray emitted by a stationary particle is sharply specified by the law of atomic physics. As now the particles move away from the observer with velocity v, the wave-length λ of the light-ray, as seen by the a stationary observer, is shifted according to the *Doppler law*:

$$\frac{\lambda - \lambda_o}{\lambda_o} = \frac{v}{c} \tag{3.2.14}$$

where c is the velocity of light. The velocity distribution is therefore observable as the distribution of the Doppler shifted wave-length. The available optical resolution is good enough to produce a spectacular normal (i.e. Gaussian) spectral distribution, as predicted by the Maxwell law (3.2.10).

Here, we see the sketch of an example of what Suppes [1962] calls "*models of experiment*" (cf. Sect. 1.4): before one can test in a laboratory whether Maxwell's model is more realistic than, say, Bernoulli's, one has to contruct another model which involves the interplay between the elements in Maxwell's model and the light rays so that the variation – or the lack of it – of velocities of Maxwell's particles can be recorded and observed as the distribution of Doppler shifts. As a fringe benefit, notice from (3.2.14) and (3.2.10) that the width $\Delta\lambda$, at half-height, of the spectral distribution is given by

$$\frac{\Delta\lambda}{\lambda_o} = \alpha\sqrt{T} \quad \text{with} \quad \alpha = 2\sqrt{2\ln 2}\sqrt{\frac{R}{Mc^2}} \tag{3.2.15}$$

where R ($= 8.314 \cdot 10^7$ ergs per Kelvin per mole) is the gas constant, M is the molar mass of the particles (e.g. 12 moles for Carbon), and c ($= 3 \cdot 10^{10}$ cm) is the velocity of light. Hence (3.2.15) can be used to infer the temperature of the plasma; as this runs anywhere from several hundred thousand to several millions degrees Kelvin, it would be out of question to measure such a temperature with ordinary material thermometers. Conversely, the values of R and c imply that one needs this kind of temperature to observe an appreciable Doppler broadening in the procedure just described.

Maxwell himself was quite aware that his derivation of (3.2.10) belonged squarely in statistics: not only did he correctly impose the independence condition (3.2.5.b), but he insisted on the analogy between (3.2.8) and what we discuss in Sect. 3.2 below as de Moivre–Laplace's law of large numbers, another avatar of which is Gauss' law of errors; here is Maxwell [1860]:

> *It appears from this proposition* [i.e. (3.2.8)] *that the velocities are distributed among the particles, according to the same law as the errors are distributed among the observations in the theory of the method of least squares.*

Concepts and methods pertaining genuinely to the statistical discourse thus made their entry in the world of thermophysics. A modern probabilist would only want to rephrase the above quote from Maxwell by asserting that the components of the velocities are independent, identically distributed, Gaussian random variables. The mechanism of the interactions (e.g. collisions) between the particles of the gas is completely absent from the Maxwell model.

3.3 The Boltzmann Equation

Although Clausius [1858] had spoken of a "mechanical theory of heat", and some of Boltzmann's opponents were to attack him for similar alleged pretenses, Boltzmann's work showed clearly how and why the lofty aim of a truly mechanical theory of heat had not been achieved as yet. It also provides an example for our argument towards the hybrid semantic view of scientific theories (cf. 3.1.4).

[Boltzmann, 1872] equation is a differential law designed to model the future evolution of the one-particle density function

$$f : (\boldsymbol{x}, \boldsymbol{v}; t) \in \Lambda \times \mathsf{R}^3 \times \mathsf{R}^+ \mapsto f(\boldsymbol{x}, \boldsymbol{v}; t) \in \mathsf{R}^+ \qquad (3.3.1)$$

for a gas of N particles of mass m moving in a container Λ of finite volume V. In particular for every $\Lambda' \subseteq \Lambda$

$$N_{\Lambda'}(t) = \int_{\Lambda'} \mathrm{d}\boldsymbol{x} \int_{\mathsf{R}^3} \mathrm{d}\boldsymbol{v}\, f(\boldsymbol{x}, \boldsymbol{v}; t) \qquad (3.3.2)$$

is the average number of particles in Λ' at time t. Note that $N_\Lambda(t) = N$ for all t.

Two mechanisms are brought into play: the free motion of the particles under an external field of forces $\boldsymbol{F} : \boldsymbol{x} \in \Lambda \mapsto \mathsf{R}^3$; and the *binary collisions* between the particles; a third mechanism, namely the interaction of the particles with the walls of Λ, is neglected except for the fact that these keep the gas inside the container. To simplify the exposition, some hypotheses are made on the inter-particle interactions: the particles are point centers of force, with the force acting along the line of centers; the forces are short-range; and the particles are without internal structure. These conditions imply in particular: that the collisions are symmetric with respect to their initial and final conditions; and that, except in the immediate vicinity of a collision, the particles move independently of one another, solely under the influence of the external forces.

The *Boltzmann equation* reads

$$(\partial_t f)(\boldsymbol{x}, \boldsymbol{v}; t) = (\partial_t f)_{\text{streaming}}(\boldsymbol{x}, \boldsymbol{v}; t) + (\partial_t f)_{\text{collision}}(\boldsymbol{x}, \boldsymbol{v}; t) \qquad (3.3.3)$$

where

$$(\partial_t f)_{\text{streaming}}(\boldsymbol{x}, \boldsymbol{v}; t) = -[(\boldsymbol{v} \cdot \nabla_{\boldsymbol{x}} + \boldsymbol{F} \cdot \nabla_{\boldsymbol{v}})f](\boldsymbol{x}, \boldsymbol{v}; t) \qquad (3.3.4)$$

and

$$
\begin{aligned}
(\partial_t f)_{\text{collision}}(\boldsymbol{x}, \boldsymbol{v}; t) &= \int d\boldsymbol{u} \int d\Omega \, \|\boldsymbol{v} - \boldsymbol{u}\| \, \tilde{\sigma} \, \Gamma(\boldsymbol{x}, \boldsymbol{u}', \boldsymbol{v}', \boldsymbol{u}, \boldsymbol{v}; t) \\
\text{where} & \\
\Gamma(\boldsymbol{x}, \boldsymbol{u}', \boldsymbol{v}', \boldsymbol{u}, \boldsymbol{v}; t) &= f(\boldsymbol{x}, \boldsymbol{u}'; t) \, f(\boldsymbol{x}, \boldsymbol{v}'; t) - f(\boldsymbol{x}, \boldsymbol{u}; t) \, f(\boldsymbol{x}, \boldsymbol{v}; t) \, ;
\end{aligned}
\qquad (3.3.5)
$$

in these equations $\boldsymbol{u}'$ and $\boldsymbol{v}'$ denote the final velocities of two particles that enter a collision with initial velocities $\boldsymbol{u}$ and $\boldsymbol{v}$ respectively, so that the laws of conservation of momentum and of energy imply

$$\boldsymbol{u}' + \boldsymbol{v}' = \boldsymbol{u} + \boldsymbol{v} \quad \text{and} \quad \|\boldsymbol{u}'\|^2 + \|\boldsymbol{v}'\|^2 = \|\boldsymbol{u}\|^2 + \|\boldsymbol{v}\|^2 \qquad (3.3.6)$$

and thus

$$\|\boldsymbol{u}' - \boldsymbol{v}'\| = \|\boldsymbol{u} \quad \boldsymbol{v}\| \quad . \qquad (3.3.7)$$

The term $(\partial_t f)_{\text{streaming}}$ in (3.3.4) represents the contribution to the evolution in the absence of collisions: when $(\partial_t f)_{\text{collision}} = 0$, the Boltzmann equation reduces to the Liouville equation of classical mechanics.

The term $(\partial_t f)_{\text{collision}}$ in the full Boltzmann equation (3.3.3) therefore gives the contribution of the collisions, and its derivation involves a combination of the kinematics of collisions with some delicate statistical assumptions. There are two parts to this term: the *depletion* of $f(\boldsymbol{x}, \boldsymbol{v}; t)$ resulting from particles having velocity $\boldsymbol{v}$ before a collision, and thus a final velocity $\boldsymbol{v}' \neq \boldsymbol{v}$ after this collision; and the *accretion* of $f(\boldsymbol{x}, \boldsymbol{v}; t)$ resulting from particles having a velocity $\boldsymbol{v}' \neq \boldsymbol{v}$ before the collision, and ending up with the velocity $\boldsymbol{v}$ after the collision.

We compute first the depletion. Choose a system of coordinates moving with the initial velocity $\boldsymbol{u}$ of one of two particles about to enter into a collision; in this system, this particle is at rest; call it the "target". The other of the two particles, which has an initial laboratory velocity v moves, in this system of coordinates with the initial velocity $v - u$ relative to the target. Due to the inter-particle interaction-potential, the target offers a total cross-section σ_{tot} to the moving particle. Let it suffice here to say that this quantity can be explicitly computed from the (microscopic) interaction-potential. Hence the particle moving with velocity $v - u$ reaches the target in a time interval τ if and only if it is contained initially in a right-cylinder of base σ_{tot} and volume $\sigma_{\text{tot}} \|v - u\| \tau$. Since the target is distributed according to $f(x, u; t)$, and the second particle according to $f(x, v; t)$, it seems reasonable ($\spadesuit$) therefore to write the depletion as:

$$(\partial_t f)_{\text{out}}(x, v; t) = - f(x, v; t) \int du\, f(x, u; t) \|v - u\| \sigma_{\text{tot}} \quad . \qquad (3.3.8)$$

[The marker ($\spadesuit$) denotes a sequence of related statements on which we defer comment until the derivation has run its course.]

With Ω denoting the solid angle between the relative velocity $w' = v' - u'$ after the collision, and the relative velocity $w = v - u$ before the collision, we have $d\Omega = \sin\theta\, d\theta\, d\varphi$; and we can write the total cross-section

$$\sigma_{\text{tot}} = \int d\Omega\, \tilde{\sigma} \qquad (3.3.9)$$

with $\tilde{\sigma}$ a function of the azimutal angle θ, and in general of $\|v - u\|$, but not of the polar angle φ. Upon taking this into account, and rearranging the order of the terms, we can rewrite (3.3.8) as

$$(\partial_t f)_{\text{out}}(x, v; t) = - \int du \int d\Omega\, \tilde{\sigma} \|v - u\| f(x, u; t) f(x, v; t) \quad . \quad (3.3.10)$$

The symmetry of the collision kinematics implies that this expression of the depletion – due to the collisions $(u, v) \rightarrow (u'v')$ – translates immediately to an expression of the accretion due to the inverse collisions $(u', v') \rightarrow (uv)$:

$$(\partial_t f)_{\text{in}}(x, v; t) = \int du' \int d\Omega\, \tilde{\sigma} \|v' - u'\| f(x, u'; t) f(x, v'; t) \quad . \quad (3.3.11)$$

Upon adding (3.3.10) and (3.3.11), and taking into account (3.3.7), we obtain (3.3.5), i.e. the collision term $(\partial_t f)_{\text{collision}}$ in the Boltzmann equation (3.3.3), all the terms of which have now been described.

This is a clear – perhaps the purest – example of a structural theory which is equipped with a structural model (cf. II.3 in Sect. 1.3 and Sect. 1.4), namely, it gives a hypothetical but quite realistic account of the (micro-) structural properties of a fluid in a general state and a dynamical equation which portents to describe the law that governs, under the idealization of molecular chaos ($\spadesuit$) , the fluid's movement as the causal consequence (in time) of those structural properties. The equation is about the time-derivative $\partial_t f$, so that it only gives the disposition but not yet the behavior, especially not yet the observable behavior, of the fluid. And we shall see in a moment that the precise behavior is not forthcoming from the equation; so Boltzmann's model and equation are not easily replaceable by any behavioral theory or a state-space model.

Before we examine the pragmatic as well as the foundational problems of such a putatively fundamental equation, it seems salutary to examine whether the implications of this model warrant the effort of a critique. The most famous consequence of the Boltzmann equation is the so-called H-theorem:

Theorem 3.3.1. *Suppose that the Boltzmann equation, in the absence of external forces i.e. with $F = 0$, admits a space-homogeneous solution, i.e. $f(x, v; t) = \varphi(v; t)$. Then the function*

$$H_\varphi : t \in \mathsf{R}^+ \mapsto \int dv\, \varphi(v; t)\, \ln \varphi(v; t) \in \mathsf{R}$$

is always non-increasing; specifically $\frac{dH_\varphi}{dt} \leq 0$ with the equality sign holding if and only if $\varphi(u'; t)\, \varphi(v'; t) - \varphi(u; t)\, \varphi(v; t) = 0$.

Proof: Under the assumptions of the theorem, the Boltzmann equation reduces to

$$(\partial_t \varphi)(v; t) = \int du \int d\Omega \, \|v - u\| \, \tilde{\sigma} \{\varphi(u'; t)\, \varphi(v'; t) - \varphi(u; t)\, \varphi(v; t)\} \quad . \tag{3.3.12}$$

Because of (3.3.2)

$$\int dv\, (\partial_t \varphi)(v; t) = 0 \quad ; \tag{3.3.13}$$

hence

$$\frac{dH_\varphi}{dt} = \int dv\, (\partial_t \varphi)(v; t)\, \ln \varphi(v; t) \quad . \tag{3.3.14}$$

Upon inserting (3.3.12) into (3.3.14), we receive

$$\frac{dH_\varphi}{dt} = \int dv \int du \int d\Omega\, \tilde{\sigma}\, \|v - u\| \{\varphi(u'; t)\varphi(v'; t) - \varphi(u; t)\varphi(v; t)\} \ln \varphi(v; t) \tag{3.3.15}$$

which, owing to its symmetry properties under the exchanges $u \leftrightarrow v$ and $(u, v) \leftrightarrow (u', v')$, can be rewritten as

$$\frac{dH_\varphi}{dt} = -\frac{1}{4} \int dv \int du \int d\Omega\, \tilde{\sigma}\, \|v - u\|)\, (x - y)(\ln x - \ln y) \tag{3.3.16}$$

with

$$x = \varphi(u'; t)\, \varphi(v'; t) \quad \text{and} \quad y = \varphi(u; t)\, \varphi(v; t) \quad . \tag{3.3.17}$$

The H-theorem now follows immediately from (3.3.16–17) and the elementary remark that, since the logarithm is a strictly increasing function of its argument, $(x - y)(\ln x - \ln y) \geq 0 \quad \forall \; x, y \in \mathsf{R}^+$, with equality holding if and only if $x = y$.   **q.e.d.**

Let us hasten to add an observation to our philosophical view of models and theories, which we could not have realized in our earlier dogmatic slumber (i.e. in Chap. 1). Structural models of highly theoretical entities, such as Boltzmann's model here, not only may – if solvable – provide trajectories of the systems they model as solutions of their differential (or integro-differential) equations; but more importantly, *they may yield, without solving the equations,* general theorems *concerning the* tendencies *or* limits *of their behavior, such as the* H-*theorem.* This case already lends support to our argument in Sect. 1.4) that behavioral theories and state-space models are not sufficient for describing, or even reconstructing, the model-building activities in science; *hence,* a reliable philosophical account needs the hybrid view. Models can be built from one's theoretical commitments and intuitions and far-reaching results derived from them before, or without, any equations are solved.

Corollary 3.3.1. *The only space-homogeneous and stationary solution of the Boltzmann equation with* $\boldsymbol{F} = 0$ *is the Maxwell distribution (3.2.10).*

Proof: Let $\varphi^{(o)}(\boldsymbol{v}) = \varphi(\boldsymbol{v}, t)$ be such a solution. From the theorem, we have that $\varphi^{(o)}(\boldsymbol{u})\varphi^{(o)}(\boldsymbol{v})$ is invariant under the collisions $(\boldsymbol{u}, \boldsymbol{v}) \to (\boldsymbol{u}', \boldsymbol{v}')$; thus $\ln \varphi^{(o)}(\boldsymbol{v})$ is an additive invariant under collisions, i.e.

$$\ln \varphi^{(o)}(\boldsymbol{u}') + \ln \varphi^{(o)}(\boldsymbol{v}') = \ln \varphi^{(o)}(\boldsymbol{u}) + \ln \varphi^{(o)}(\boldsymbol{v}) \quad . \tag{3.3.18}$$

Since there are exactly five linearly independent additive invariants: the constant function, the three components of the momentum and the kinetic energy – see (3.3.6) – $\ln \varphi^{(o)}(\boldsymbol{v})$ must be a linear combination of these, namely:

$$\ln \varphi^{(o)}(\boldsymbol{v}) = a + \boldsymbol{b} \cdot \boldsymbol{v} + c\|\boldsymbol{v}\|^2 \tag{3.3.19}$$

or equivalently:

$$\varphi^{(o)}(\boldsymbol{v}) = Z \exp\{A \|\boldsymbol{v} - \boldsymbol{v_o}\|^2\} \quad . \tag{3.3.20}$$

Since $\boldsymbol{v_o}$ would imply a uniform drift of the gas, the condition that the container be at rest imposes $\boldsymbol{v_o} = 0$. With this, (3.3.20) reduces exactly to (3.2.7), which is the distribution Maxwell derived from first principles. The constants A and Z are finally adjusted, as in the passage from (3.2.1) to (3.2.10), by imposing a normalization – here (3.1.35) – and invoking the Gay-Lussac law.   **q.e.d.**

While Maxwell's derivation of (3.2.10) deals only with the equilibrium distribution itself, the Boltzmann model proposes a mechanism describing a temporal approach to this distribution; the latter being obtained now as the space-homogeneous, stationary solution of the differential equation that constitutes the model.

This special solution of the Boltzmann equation illustrates the relation between a structural theory and a behavioral one. However, the initial conditions are so special that it is almost a "*trivial*" solution, and yet, considering that most thermo-systems spend most of their time in such conditions, the solution may also be said to be the "*most probable*" one; this statement will have to be justified.

Two quantities can now be given meaning. We call *free path* the trajectory of a particle between two successive collisions; the average time between collisions, i.e. the average duration τ of a free path; and mean *free-path* its average length λ; the averages are understood with respect to the Maxwell–Boltzmann distribution (3.2.10). These are defined by:

$$\tau \equiv (\overline{n_c})^{-1} \quad \text{and} \quad \lambda \equiv \overline{v}\tau$$

where $\overline{n_c}$ is the average number of collisions of one molecule with other molecules per unit time:

$$\overline{n_c} = \left[\int d\boldsymbol{u}\, \varphi(\boldsymbol{u}) \right]^{-1} \int d\boldsymbol{u}\, \varphi(\boldsymbol{u}) \int d\boldsymbol{v}\, \varphi(\boldsymbol{v}) \, \|\boldsymbol{u} - \boldsymbol{v}\| \, \sigma \quad ;$$

and $\overline{v}$ is the average of the absolute value of the velocity along a fixed, but arbitrary, direction:

$$\left[\int dv_1\, \varphi_1(v_1) \right]^{-1} \int dv_1\, \varphi_1(v_1) \, |v_1| \quad .$$

For a gas the molecules of which are hard spheres of diameter d, leading to a constant cross-section $\sigma = \pi d^2$, a straightforward computation gives then:

$$\tau = \left[4d^2 \left(\frac{N}{V}\right) \sqrt{\frac{\pi kT}{m}} \right]^{-1} ; \quad \overline{v} = \sqrt{\frac{8kT}{\pi m}} \text{ and } \lambda = \left[\sqrt{2}\pi d^2 \left(\frac{N}{V}\right) \right]^{-1} . \quad (3.3.21)$$

As usual, (N/V) denotes the number of molecules per unit volume, k is the Boltzmann constant, T is the absolute temperature, and m is the mass of a molecule.

Note that our averaging with respect to the Maxwell–Boltzmann distribution is not essential to a qualitative estimate of the mean-free path λ. Clausius [1858], who first introduced the concept, evaluated λ with respect to a uniform velocity distribution, i.e. $\|\boldsymbol{v}\|$ is constant, and the direction of $\boldsymbol{v}$ is uniformly distributed; this only required an innocuous adjustment, namely to replace in (3.3.21) Clausius' coefficient 4/3 by the above $\sqrt{2}$.

Note further that, with $D^3 \equiv (V/N)$, D gives an estimate of the average distance between "neighboring" molecules; we therefore have:

$$\frac{\lambda}{D} \sim \left(\frac{D}{d}\right)^2 . \quad (3.3.22)$$

From this we can quantify the condition that the theory deals with a *dilute gas*. Typically:

$$\tau \approx 4 \cdot 10^{-10}\,\mathrm{s} \quad ; \quad \bar{v} \approx 5 \cdot 10^4\,\mathrm{cm/s} \quad ;$$

$$\lambda \approx 2 \cdot 10^{-5}\,\mathrm{cm} \quad ; \quad d \approx 4 \cdot 10^{-8}\,\mathrm{cm} \quad ; \quad D \approx 5 \cdot 10^{-7}\,\mathrm{cm} \quad .$$

Hence, (3.3.22) emphasizes the fact that for a dilute gas – defined by the condition that the average distance D between neighboring molecules be reasonably large compared to their diameter d – the mean free path is *very large* in comparison with the distance D. The latter condition plays an essential role in the discussion of the derivation of the H-theorem, as was repeatedly emphasized; see e.g. [Boltzmann, 1895b], where Boltzmann asserts that he could reasonably expect ($\spadesuit$) [in the form of (3.3.8) or (3.3.41)] *to hold at all times* only when the condition $\lambda/D \gg 1$ is satisfied – together with the condition that the system be large, so that he could ignore the collisions of the molecules with the walls.

Theorem 3.3.1 suggests to try and relate the Boltzmann H-function and the thermodynamical entropy. To this effect let us make the change of variable $\boldsymbol{v} \to \boldsymbol{p} = \frac{1}{m}\boldsymbol{v}$ and normalize the distributions f defined on the phase space $V \times R^3 = \{(\boldsymbol{x}, \boldsymbol{p})\}$, by

$$\int_V \mathrm{d}\boldsymbol{x} \int_{R^3} \mathrm{d}\boldsymbol{p}\, f(\boldsymbol{x}, \boldsymbol{p}) = N \tag{3.3.23}$$

where N denotes the number of molecules contained in a container of volume V. Let us then define

$$S = -k \int_V \mathrm{d}\boldsymbol{x} \int_{R^3} \mathrm{d}\boldsymbol{p}\, f(\boldsymbol{x}, \boldsymbol{p}) \ln f(\boldsymbol{x}, \boldsymbol{p}) \quad . \tag{3.3.24}$$

In particular, for the space-homogeneous Boltzmann–Maxwell distribution

$$f(\boldsymbol{x}, \boldsymbol{p}) = \frac{N}{V}(2\pi mkT)^{-3/2} \exp\{-\frac{1}{kT}\frac{1}{2m}\|p\|^2\} \quad , \tag{3.3.25}$$

we find

$$S = kN\left\{\ln\left[(\frac{V}{N})\,T^{3/2}\right] + c_1\right\} \tag{3.3.26}$$

where it is important to note that the constant $c_1 = \frac{3}{2}[1 + \ln(2\pi mk)]$ is just a plain number, i.e. is separately independent of N, V and T. Except for a correction in the numerical value of c_1 – discussed after formula (3.3.31) below – (3.3.26) is known as the *Sackur–Tetrode formula*. Irrespective of the numerical value of c_1, the above formula (3.3.26) calls for two comments.

First, when $N = N_{Avogadro}$, i.e. when the container contains one mole of the gas, *the equilibrium entropy (3.3.26) coïncides with the equilibrium thermodynamical entropy of the ideal gas* (compare with 2.4.19).

Second, *the equilibrium entropy* (3.3.26) *is an extensive quantity.* In particular, suppose that a wall separates two containers of volume V_1 and V_2 containing gases of the same composition, each in equilibrium, and both at the same spatial density $N_1/V_1 = N_2/V_2$ the same temperature $T_1 = T_2$; let S_1 and S_2 be the corresponding entropies, calculated from (3.3.26). Let us now remove the wall separating the two containers: the composite system has uniform density $N/V = (N_1 + N_2)/(V_1 + V_2) = N_1/V_1 = N_2/V_2$ and temperature $T = T_1 = T_2$: its entropy, S_{1+2} computed again from (3.3.26), satisfies

$$S_{1+2} = S_1 + S_2 \quad . \tag{3.3.27}$$

As the kinetic theory of the ideal gas is a prototype for the more general statistical mechanics, it is interesting to note that in this generalized context, the extensivity of the entropy follows *only if* one introduces "the correct Boltzmann counting"; this counting moreover ensures that this *classical* formula is compatible – at high temperature – with the tenets of *quantum* statistical mechanics. Hence the analogy that leads from the Boltzmann H-function to the expression (3.3.24) has proven to be more than a superficial guide to the development of the general theory of thermodynamical equilibrium.

Let us thus briefly summarize here the route prescribed by statistical mechanics; for more systematic presentation, see Chap. 10. One first defines the *N-particle partition function* as the integral of the N-particle phase-space $\Gamma = V^N \times R^{3N}$:

$$Z^{(N)} = C_N \int_\Gamma \mathrm{d}^N x \, \mathrm{d}^N p \, \exp\{-\frac{1}{kT} H^{(N)}\} \tag{3.3.28}$$

where $H^{(N)}$ is the total Hamiltonian (mechanical energy) of the system. If one neglects the interactions between the molecules of the gas, this Hamiltonian is a sum of 1-particle Hamiltonians, and the presence of the exponential implies that the partition function becomes then a product of 1-particle partition functions; and if the particles are identical, then all these are equal:

$$Z^{(N)} = C_N (Z^{(1)})^N \quad . \tag{3.3.29}$$

Rules are then given to compute the thermodynamical variables from the partition function $Z^{(N)}$; for instance the pressure, energy and entropy are obtained from:

$$\left. \begin{aligned} p &= (\partial_V \{kT \ln Z^{(N)}\})_{N,T} \\[1em] U &= kT^2 (\partial_T \{\ln Z^{(N)}\})_{N,V} \\[1em] S &= k \ln Z^{(N)} + \tfrac{1}{T} U \end{aligned} \right\} \tag{3.3.30}$$

With the one-particle Hamiltonian $H(1) = \frac{1}{2m} \|p\|^2$, corresponding to the free mono-atomic gas discussed earlier, we recover immediately

$$p = \frac{1}{V} NkT \quad and \quad U = \frac{3}{2} NkT \quad . \tag{3.3.31}$$

Due to the presence of logarithmic derivatives, these expressions are independent of the choice of the coefficient C_N in (3.3.29); this is however *not* the case for S. For instance, the simplest guess, namely $C_N = 1$ leads to an entropy that is not extensive; this is called the *Gibbs paradox*.

However, the entropy becomes extensive – for large values of N, when $\ln(N!) \approx N \ln N$ – if one chooses $C_N = (N! \, h^3)^{-1}$; here h is an adjustable constant; as for $N!$, it is to be understood as the number of permutations of N indistinguishable objects. This guess for C_N is called the *correct Boltzmann counting* and it leads again to our formula (3.3.26), except for the replacement of the constant c_1 by $c_h = \frac{3}{2} \left[1 + \ln(2\pi mk/h^2)\right]$. The high temperature limit of the quantum entropy shows that the constant h has to be identified with the Planck constant, a result which is supported by precise calorimetric experiments; this is spectacular enough an accomplishment to warrant mention even in Britannica [online]: the entropy of Argon, a noble gas, measured at $T = 298.2$ K is reported as $S = 154.8$ J/degree K mol, against the computed *Sackur–Tetrode* value $S = 154.7$ J/degree K mol.

This exhausts what we wanted to report on the pioneering contribution of the Boltzmann H-theorem to the elaboration of *equilibrium* thermo-physics. We now turn to its *non-equilibrium* aspects, where some of the short-comings of the Boltzmann model glare us in the face.

Inasmuch as the Boltzmann equation could be seen as a consequence of mechanics, it would provide a rudimentary "mechanical theory of heat". However, we now want to review some of the reasons – pragmatic as well as epistemological – why it is not a satisfactory mechanical model.

The first of the pragmatic limitations of the model is that the Boltzmann equation is a non-linear, integro-differential equation: no general theorem insures that it admits a solution over an extended time interval, except for the stationary Maxwell–Boltzmann distribution. With much effort [Lanford, 1976], it has been possible to show that such solutions do exist, however for times that are a fraction of the collision time – note that this is not quite as bad as it sounds, since τ is an *average* time, around which the dispersion may be large enough to allow very many collisions to take place.

Except for a handful of remarkable papers collected in [Lebowitz and Montroll, 1983], physicists seem to have largely ignored this mathematical problem, which is nevertheless serious from a fundamental point of view; there are moreover other serious and immediate difficulties, as we shall see.

This problematic situation vividly illustrates one of our key arguments for our hybrid semantic view of theories (cf. Sect. 1.4). Here is a deep – and arguably not too idealized – model of a classical fluid, which specifies the fluid's structures and dynamics; it also yields profound theorems about its general behavior, e.g. the H-theorem; and yet, no non-stationary solution of the equation, valid at least on some mesoscopic time scale has been exhibited so far. Unlike some other cases in which the existence of solutions can be proven, although some of these may be too complex to obtain in explicit forms, it is not clear that the Boltzmann equation admits any convincing non-stationary solutions at all. While the equation has been the subject of heavy criticisms, no one would dare to condemn it outright, nor should one dismiss it entirely. It appears to be an exemplar of a theoretical model which bridges the microscopic and the macroscopic level of a system in terms of the system's microscopic dispositions rather than explicit behavior. It captures some – but not all – of the characteristic features that a good and reasonable microscopic model must possess.

Even at a more immediately pragmatic level, the difficulties one encounters in extracting quantitative predictions from the Boltzmann model already start in the simplest of all cases, the derivation of the evolution equations for the velocity-average of quantities that are additively conserved in the course of collisions:

$$\langle \chi \rangle (\boldsymbol{x}, t) = \int d\boldsymbol{v} \, f(\boldsymbol{x}, \boldsymbol{v}, t) \, \chi(\boldsymbol{x}, \boldsymbol{v}) \tag{3.3.32}$$

where $\chi(\boldsymbol{x}, \boldsymbol{u}) + \chi(\boldsymbol{x}, \boldsymbol{v}) = \chi(\boldsymbol{x}, \boldsymbol{u}') + \chi(\boldsymbol{x}, \boldsymbol{v}')$ for any collision $(\boldsymbol{u}, \boldsymbol{v}) \rightarrow (\boldsymbol{u}', \boldsymbol{v}')$. As to be expected:

$$\int d\boldsymbol{v} \, (\partial_t f)_{\text{collision}} (\boldsymbol{x}, \boldsymbol{v}, t) \, \chi(\boldsymbol{x}, \boldsymbol{v}) = 0 \quad . \tag{3.3.33}$$

Hence only the streaming term of the Boltzmann equation is relevant to the evolution of the velocity-averages of these quantities. Starting with

$$\varrho(\boldsymbol{x}, t) = \int d\boldsymbol{v} \, f(\boldsymbol{x}, \boldsymbol{v}, t) \tag{3.3.34}$$

we receive the conservation law:

$$\partial_t \varrho + \nabla \cdot (\varrho \boldsymbol{u}) = 0 \quad \text{where} \quad u^i(\boldsymbol{x}, t) = \langle v^i \rangle (\boldsymbol{x}, t) = \int d\boldsymbol{v} \, f(\boldsymbol{x}, \boldsymbol{v}, t) \, v^i \tag{3.3.35}$$

i.e. the u^i are the three first moments of the distribution f. Hence, solving the differential equation for ϱ requires that we know about the evolution of the u^i. Since the v^i are also additive conserved quantities, we proceed as above to find the differential equation that governs their evolution:

$$\varrho(\partial_t + u^j \partial_j) u^i = \frac{\varrho}{m} F^i - \partial_j P^{ij} \quad \text{where} \quad P^{ij} = \varrho \langle (v^i - u^i)(v^j - u^j) \rangle \, . \tag{3.3.36}$$

Here again, unfortunately, the components of the pressure tensor $\boldsymbol{p}$ are yet higher moments of the distribution f. This unending escalation in the hierarchy of higher moments of f makes the solution of any of these equations depend on the solution of the next one: not a matter that can be handled without further ado. Note that the same escalation occurs when we write down the equation for the evolution of the local temperature, defined by analogy with (3.2.12) as:

$$\Theta(\boldsymbol{x},t) \;=\; \langle\,\|\boldsymbol{v}-\boldsymbol{u}\|^2\,\rangle(\boldsymbol{x},t) \;=\; \int d\boldsymbol{v} \,-\, f(\boldsymbol{x},\boldsymbol{v},t)\,\|\boldsymbol{v}-\boldsymbol{u}\|^2 \quad . \qquad (3.3.37)$$

The differential equation for Θ reads then:

$$\varrho(\partial_t + \boldsymbol{u}\cdot\nabla)\Theta \;=\; -\frac{2}{3}(\nabla\cdot\boldsymbol{q} - \boldsymbol{p}:\Lambda) \qquad (3.3.38)$$

where

$$q^i = \frac{1}{2}m\varrho\langle\,\|\boldsymbol{v}-\boldsymbol{u}\|^2(v^i-u^i)\,\rangle \quad \text{and} \quad \Lambda_{ij} = \frac{1}{2}m(\partial_i u_j + \partial_j u_i) \quad . \qquad (3.3.39)$$

To break these hierarchies, one usually resorts to some idealization scheme which produces approximate results (or for short, approximation scheme). The scheme is based typically on the assumption that if the initial state of the gas is "close" to "local equilibrium", then it remains so in the course of time. It is not our purpose here to show how this approximation leads, for instance, to the Navier–Stokes equation of viscous hydrodynamics:

$$\varrho(\partial_t + \boldsymbol{u}\cdot\nabla)\boldsymbol{u} \;=\; \frac{\varrho}{m}\mathbf{F} \,-\, \nabla(P - \frac{1}{3}\mu\nabla\cdot\boldsymbol{u}) + \mu\nabla^2\boldsymbol{u} \quad , \qquad (3.3.40)$$

but we needed to note the fact. Similarly, one obtains a heat conduction equation which, in the static case $\boldsymbol{u}=0$, reduces to the Fourier heat equation (2.2.1).

These developments and other elaborate approximation schemes are a staple of the physicist's and physico-chemist's traditional education, see [Chapman and Cowling, 1932], [Grad, 1963], [Wannier, 1966, 1987], [Huang, 1963, 1987], [Mazo, 1967]; they are usually found under the heading "Chapman–Enskog method". These approximation schemes cannot be ignored: we mention this line of research here to emphasize two points: (1) even if one could prove the *existence* of solution, the Boltzmann equation does not provide answers – beyond the H-theorem – that are rigorous consequences of the equation; and (2) this equation can be manipulated to model some "real physics" transport phenomena. That the challenge has not yet been met is due to the fact that, however successful their praxis may be, these approximation schemes do not rest on a satisfactory foundation nor do we have any satisfactory explanation for their success. In other words, drastic simplifications are made without the support of *a priori* error-estimates that would delineate the domain of their computational consequences. On the brighter side, the situation can be clarified when the proper distinctions are made between the *time-scales* pertinent to the phenomena under discussion; some references are indicated later in this section; and a suggestive illustration obtains from the dog-flea model discussed in the next section; this line of ideas is pursued in Chap. 15.

In spite – or in part perhaps because – of its successes, even if admittedly partial, the Boltzmann model has been exposed immediately to ferocious criticisms to demonstrate that it *cannot* be derived from mechanics alone. Asides from the acrimony, it is true that if it were so derived, then it would not show an irreversible approach to equilibrium; this is usually rendered as the *Umkehreinwand* (or reversibility objection) and the *Wiederkehreinwand* (or recurrence objection).

The first objection [[Loschmidt, 1876,1877] vs. [Boltzmann, 1877, 1878]] is that mechanical systems are reversible, whereas thermodynamics is not; to this objection, Boltzmann purportedly replied:

Go ahead reverse them

(*them* being the velocity of each of the molecules of a gas).

The second objection [[Zermelo, 1896] vs. [Boltzmann, 1896]] is based on a theorem by Liouville, asserting that the measure of every subset of phase-space is invariant under the time evolution generated by the symplectic equations of Hamiltonian mechanics: from this, Poincaré showed that any system having a finite energy and confined in a finite region of space must return arbitrarily close to its initial state, again and again, although the average recurrence time can be computed to be exorbitantly large; thus, the irreversible approaching of a certain terminal state would be impossible for such a system. And again, Boltzmann's equally apocryphal answer to this was

Do you want to wait that long?

These two objections have been repeatedly discussed from a great variety of angles, for which we refer the interested Reader to an abundant literature that extends from the physics textbooks mentioned earlier in this commentary to the recent philosophical studies by [Sklar, 1993] or [Von Plato, 1994], and the historical analyses that are presented, e.g. in [Brush, 1976, 1983].

Boltzmann himself, for the rest of his life, insisted on how clear it was to him that his theory subsumed more than a straight application of the principles of mechanics:

> *... my Minimum Theorem, as well as the so-called Second Law of Ther-modynamics, are only theorems of probability. The Second Law can never be proven mathematically from the equations of dynamics* [read reversible mechanics] *alone ... What I proved in my papers is as follow: It is ex-tremely probable that* H *is very near to its minimum value; if it is greater, it may increase or decrease, but the probability that it increases is always greater. Thus, if I obtain a certain value for* dH/dt, *this result does not hold for every time element* dt, *but is only an average value. But the greater the number of molecules, the smaller is the time-interval* dt *for which the results hold good.*   [Boltzmann, 1895a]

> *Clausius, Maxwell, among others, have repeatedly insisted that the theorem of gas theory has the character of a statistical truth ... I have also empha-sized that the Second Law of Thermodynaics is, from a molecular point of view, merely a statistical law. Zermelo's paper shows that my writings have been misunderstood.*   [Boltzmann, 1896]

A purely stochastic model was proposed by [Ehrenfest and Ehrenfest, 1907] which fully vindicates Boltzmann in his attempts to convince the Physics establishment of his time that the discussion should focus on the nature of the statistical assumptions to be made rather than on whether any such assumptions were logically necessary: the above quote shows that, for Boltzmann, assumptions beyond straight mechanics were needed. This model is presented in Sect. 3.4 below.

The additional, somewhat hidden, assumption of statistical nature in Boltzmann's work is often referred to as Boltzmann's *Stosszahlansatz*. When Boltzmann introduced it, it might have been somewhat burried in the argument, but there can be no doubt that Boltzmann later recognized the probabilistic – i.e. non-mechanistic – component in the theory, as we hope to have provided witnesses in the preceding quotes; see also further confirmations of (3.3.41) below in [Boltzmann, 1894].

As we presented earlier the standard derivation of the Boltzmann equation, we marked with the sign (♠) the place where this assumption is made; it consists in assuming that two particles entering a collision are independent and identically distributed, i.e. we assumed that we could "reasonably write" the two-particle distribution function in the form

$$f^{(2)}(\boldsymbol{x}, \boldsymbol{v}, \boldsymbol{x}', \boldsymbol{v}'; t) = f(\boldsymbol{x}, \boldsymbol{v}; t)\ f(\boldsymbol{x}', \boldsymbol{v}'; t) \quad \text{for all} \quad t \in \mathsf{R}^+ \quad . \qquad (3.3.41)$$

This neglect of inter-particle correlations cannot be justified on the basis of mechanics alone [Burbury, 1894]; in fact it is easy to construct an initial condition with three particles, in which the first and the second collide, and then the second collide with the third, in such a manner that the first and second collide again. In this second collision of particles 1 and 2, the particles are certainly not independent. To take this into account carries beyond the Boltzmann equation itself, and involves the appearance of divergences that have then to be tamed; for systematic presentations of this program, see [Ernst, Haines, and Dorfman, 1969] and [Cohen and Thirring, 1973].

Another way out of the difficulties has been along the route proposed by [Van Hove, 1955, 1957, 1959] and [Grad, 1958], where several limits are taken simultaneously and played against each other to define an asymptotic time scale; for e.g.: [Lanford, 1976, Martin, 1979, Spohn, 1991, Cercignani, Illner, and Pulvirenti, 1994, Cercignani and Sattinger, 1998] and Chap. 15.

These limits are somewhat in consonance with the reasons Boltzmann offered for his use of probability in the kinetic theory of gases: the probabilities represent some averages over a time-scale intermediate between the microscopic time characteristic of collisions, and the macroscopic time proper to the thermodynamical description of transport phenomena; this separation of time-scales is all the more neat when the gas is dilute. Beyond the Boltzmann equation itself, the definition of the time-scale proper to the extraction of the evolution of macroscopic quantities is one of the perennial problems of statistical mechanics; see e.g. [Lebowitz, 1983, 1999a,b].

Boltzmann's views on the use of probabilistic notions – distributions obtained from time-averages – contributed mightily, beyond the bounds of the *kinetic theory of gases* or even thermophysics, to the opening of a new field of inquiry in mathematical physics, namely *ergodic theory*; see Chaps. 7–9. We also present, in Chap. 14, some quantum models for the approach to equilibrium.

At the end of a section on Boltzmann's work, the Reader might question why we did not even mention the notion of *ensemble*, for the introduction of which Gibbs [1902], in his preface, credited [Boltzmann, 1871a] (the explicit reference being to the first section of that paper). One can also find already in [Boltzmann, 1868] an interpretation of probabilities as time-averages. Maxwell, writing purportedly *à propos* Boltzmann's ideas, described an ensemble as

> *a large number of systems similar to each other in all respects except in the initial circumstances of the motion, which are supposed to vary from system to system, the total energy being the same in all*
>
> Maxwell [1879b]

(thus focusing on what we call the microcanonical ensemble).

To this, Boltzmann responded

> *There is a difference in the conceptions of Maxwell and Boltzmann in that the latter characterizes the probability of a state by the average time during which the system is in that state, whereas the former assumes an infinity of equal systems with all possible initial states.*
>
> [Boltzmann, 1881]

If that sounds confusing, there may reside one of the reasons that made Hilbert wonder; see Sect. 5.2. In the present chapter, we chose to follow Gibbs in either thinking of *"representative ensembles"* – which is somewhat equivocal – or simply accepting to *"avoid any reference to an ensemble of systems"* altogether, recognizing with him that in this context ensembles are *"creations of the imagination"*. When we say "representative ensembles" we are not very far from the very name by which Boltzmann refers to them: *"Inbegriff von Systeme"*. He is also reticent about the reality of such embodiment of systems; [Von Plato, 1994] quotes him as thinking the notion of ensemble to be a *"Kunstgriff"* – that is an artifice or an artificial device – in need of interpretation.

Such concerns – albeit of certain epistemological value – may have diverted the focus of Boltzmann's contemporaries away from his profound insight: the role of time-averages in bringing probabilistic reasoning into what was to become statistical mechanics. We come back in Chaps. 7–10 to the problematic involved in the use of ensembles.

3.4 The Dog-Flea Model

The model was invented by [Ehrenfest and Ehrenfest, 1907] with the explicit purpose to clear some of the clouds still hanging from the controversy about Boltzmann's H–theorem. It has provided stimulating playgrounds to several generations of physicists and mathematicians; among the physicists [Kohlrausch and Schroedinger, 1926, Ter Haar and Green, 1953, Klein, 1956], and among the mathematicians [Steinhaus, 1938, Kac, 1946, 1947a,b]. The model indeed is praised as:

> ... *probably one of the most instructive models in the whole of Physics and although merely an example of a finite Markov chain, it is of considerable independent interest.*

> [Kac, 1959b]

Authoritative accounts of the model can be found in the texts by Ter Haar [1954], Kac [1959b] and Klein [1970], each of which has its own accent evoking the involvement of the authors with different aspects of the model.

Two dogs, A and B, share a population of N fleas; although this is not essential, we assume that N is even in order to allow for an even sharing where each dog as an equal and integer number of fleas. The fleas are labeled by an index running from 1 to N. The "observable" of the model is x, the number of fleas on dog A. The time evolution is prescribed as follows: at every click of the clock, a number from 1 to N is drawn "at random" from a bag, and announced. Upon hearing this the flea, the label of which has just been drawn, jumps immediately from the dog it pestered to the other dog, and it stays there until called again. *Problem:* Discuss the time-dependence x_t of the population x, with the time t measured in units of the interval between two successive clicks of the clock.

Intuitively, it seems clear that this process should lead to an approach of an "equilibrium" where each dog has $N/2$ fleas. Indeed, if at time t, there are many more fleas on dog A than on dog B the probability that the flea that is called at that time is on A (rather than on B) should be much bigger than $1/2$, so that the flea jumps to B, thus diminishing x by one unit; if on the contrary the flea population is about evenly divided between the two dogs, the population on a given dog is about as likely to increase as it is to decrease. We now verify that this intuition is correct on the average, and we discuss the evolution with sufficient details to show the sense in which it is irreversible, nevertheless exhibits recurrences, and leads to trajectories, the points of which are most likely to be local maxima as long as $x_t > N/2$ – and local minima when $x_t < N/2$ – in line with the property Boltzmann assigned to his H–curve [Boltzmann, 1898].

The model is Markovian with transition probability $P_{m,n}$ defined as the probability that there are n fleas on dog A at time $t + 1$, when it is known that there were m fleas on A at time t; from the very description of the time evolution just given, this probability is:

$$P_{m,n} \equiv P(x_{t+1} = n \mid x_t = m) = \frac{1}{N} m\, \delta_{n,m-1} + \frac{1}{N}(N-m)\, \delta_{n,m+1} \quad (3.4.1)$$

i.e.

$$P_{m,n} = \begin{cases} \frac{1}{N} m & \text{for} \quad n = m-1 \\[2mm] \frac{1}{N}(N-m) & \text{for} \quad n = m+1 \\[2mm] 0 & \text{otherwise} \end{cases}$$

Let $p(x_t = n)$ denote the probability that there are n fleas on A at time t, so that $\langle x_t \rangle = \sum_{n=1}^{N} n\, p(x_t = n)$ is the average population of fleas on A at time t. Similarly, upon using $p(x_{t+1} = n) = \sum_{m=1}^{N} p(x_t = m) P_{m,n}$ and (3.4.1), one finds that the evolution of $\langle x_t \rangle$ is given by the following difference equation:

$$\langle x_{t+1} \rangle = 1 + (1 - \frac{2}{N}) \langle x_t \rangle \quad . \qquad (3.4.2)$$

This equation admits a unique solution, provided an initial condition $\langle x_{t=0} \rangle = x_o$ be assigned; we thus find that the deviation from the equilibrium value always decays exponentially:

$$[\langle x_t \rangle - \frac{N}{2}] = [x_o - \frac{N}{2}] \cdot \exp(-\gamma t) \quad \text{with} \quad \gamma = -\ln(1 - \frac{2}{N}) > 0 \qquad (3.4.3)$$

independently of the initial distribution $\{p(x_{t=0} = n) \mid n = 1, 2, ..., N\}$, thus justifying the statement that, *in the average, the model shows an approach to equilibrium* similar to Newton's cooling law; see Fig. 3.1.

Our next remark is that the model admits *exactly one time-invariant probability distribution*; i.e. there exists exactly one distribution $\{p(n) \mid n = 1, 2, ..., N\}$ with $p(n) \geq 0 \; \forall \, n = 1, 2, ..., N$ and $\sum_{n=1}^{N} p(n) = 1$ such that $p(n) = \sum_{m=1}^{N} p(m)\, P_{m,n}$, where $P_{m,n}$ is the transition probability (3.4.1). This invariant distribution is

$$p(n) = \frac{1}{2^N} \binom{N}{n} \qquad n = 0, 1, ..., N \qquad . \qquad (3.4.4)$$

(Compare it with (3.1.11) when $p = q = \frac{1}{2}$; an instant of reflexion shows that this coïncidence is not accidental). This distribution is familiar – from the beginning of classical probability theory – in an apparently different context: it gives the probability that in a sequence of N trials of a fair coin, exactly n heads come up; see e.g. (3.2.17). Its most important properties for us here are: (i) it is symmetric around the value $n = N/2$; and (ii) it is very peaked at that value, all the more so that N is large. Being invariant under the time-evolution, it plays for this model the role played by the Liouville measure in the Hamiltonian situation considered in the Poincaré recurrence theorem.

With respect to the canonical invariant distribution (3.4.4), we write $P(x_{t-1} = n, x_t = m)$ in two equivalent ways:

$$P(x_{t-1} = n \mid x_t = m) \, P(x_t = m) = P(x_t = m \mid x_{t-1} = n) \, P(x_{t-1} = n) \quad ; \tag{3.4.5}$$

upon substituting for $P(x_t = m)$ and $P(x_{t-1} = n)$ the stationary values $p(m)$ and $p(n)$, given by (3.4.4), and for $P(x_t = m \mid x_{t-1} = n)$ the time-invariant, forward transition probability given by (3.4.1), we find from (3.4.4):

$$P(x_{t-1} = n \mid x_t = m) = P(x_{t+1} = n \mid x_t = m) \tag{3.4.6}$$

which gives the sense in which this model is *time-reversible*: the backward transition probability is equal to the forward transition probability.

We show now that on the average – with respect to the invariant measure (3.4.4) – the deviations from the equilibrium value $x = N/2$ exhibited by individual trajectories tend to die out. For this purpose, let us now compare the four possible configurations around $x_t = n$:

$$\left.\begin{array}{ccccccc}
n-1 & \rightarrow & n & \rightarrow & n-1 & \text{denoted} & (\wedge) \\
n-1 & \rightarrow & n & \rightarrow & n+1 & \text{denoted} & (/) \\
n+1 & \rightarrow & n & \rightarrow & n-1 & \text{denoted} & (\backslash) \\
n+1 & \rightarrow & n & \rightarrow & n+1 & \text{denoted} & (\vee)
\end{array}\right\} \quad . \tag{3.4.7}$$

Let $P^{(\wedge)}{}_n$ be the probability of the configuration $(\wedge)$ given $x_t = n$; i.e.

$$P^{(\wedge)}{}_n \equiv P(x_{t-1} = n-1, \, x_{t+1} = n+1 \mid x_t = n) \quad .$$

Upon using (3.4.1), (3.4.4) and (3.4.6), we compute:

$$P^{(\wedge)}{}_n \equiv P(x_{t-1} = n-1, \, x_{t+1} = n-1 \mid x_t = n) =$$

$$P(x_{t-1} = n-1, \, x_t = n, \, x_{t+1} = n-1) \, P(x_t = n)^{-1} =$$

$$\frac{P(x_{t-1} = n-1)}{P(x_t = n)} \, P_{n-1,n} \, P_{n,n-1} = \frac{p(n-1)}{p(n)} \frac{N-(n-1)}{N} \frac{n}{N} = \frac{1}{N^2} m^2$$

and similarly for the other three conditional probabilities $P^{(/)}{}_n$, $P^{(\backslash)}{}_n$ and $P^{(\vee)}{}_n$. The results so obtained can be summarized in the quotients:

$$P^{(\wedge)}{}_n : P^{(/)}{}_n : P^{(\backslash)}{}_n : P^{(\vee)}{}_n = \left[\frac{n}{N-n}\right] : 1 : 1 : \left[\frac{N-n}{n}\right] . \tag{3.4.8}$$

The interpretation of this comparison is twofold: (a) in particular, it gives

$$P^{(/)}{}_n = P^{(\backslash)}{}_n \tag{3.4.9}$$

which confirms the reversibility already established in (3.4.6); (b) it implies moreover that

$$\begin{array}{llll}
P^{(\wedge)}{}_n > P^{(/)}{}_n = P^{(\backslash)}{}_n > P^{(\vee)}{}_n & \text{when} & n > N/2 \\
P^{(\vee)}{}_n > P^{(/)}{}_n = P^{(\backslash)}{}_n > P^{(\wedge)}{}_n & \text{when} & n < N/2
\end{array} \tag{3.4.10}$$

which is to say that, in probability, every point of the trajectory $\{x_t \mid t \in \mathbb{Z}^+\}$ is a local maximum (resp. a local minimum) according to whether $x_t > N/2$ (resp. $x_t < N/2$). Indeed, for any n, $P^{(\wedge)}{}_n$ is the probability that if dog A is pestered by n fleas at time t then it had $n-1$ fleas at time $t-1$ and it also has $n-1$ fleas at time $t+1$, in other words, that $x_t = n$ is a local maximum for the population of fleas on dog A; similar interpretations go for the other three probabilities compared in (3.4.10). In fact (3.4.8) shows how much stronger these peaks are far away from equilibrium than they are close to the value $n = N/2$.

Hence the dog-flea model vindicates Boltzmann's notion that under certain statistical assumptions it is possible to obtain the properties he expected his H−curve to exhibit. Fig. 3.1 shows the computer-generated graph of individual histories on four timescales. The two short-time pictures – 200 and 500 iterations – are qualitatively similar; they also show the exponential decay of the average $< x_t >$, as given by (3.4.3).

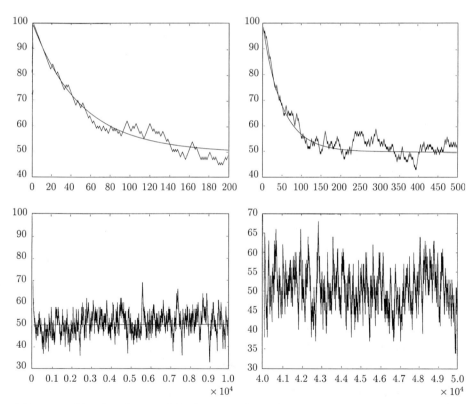

Fig. 3.1. The dog-flea model: short- vs. long-time behavior. Individual histories for time-scales of 10^2 and 10^4 iterations. (the changes in vertical time-scale is to obtain better resolution)

The long term behavior is quite different as illustrated in the last two windows of Fig.3.1: the model does exhibit recurrences. Although the theoretical derivation of the results is too involved to be properly presented here, we want to indicate the sense in which recurrences are to be expected to appear.

Let $f_k(n)$ be the probability that a trajectory, starting at n, comes back to n *for the first time* at time $t = k$. The mean recurrence time Θ_n for n, and its variance Λ_n, are then defined by

$$\Theta_n = \sum_{k=1}^{\infty} k\, f_k(n) \quad , \quad \Lambda_n^2 = \sum_{k=1}^{\infty} (k - \Theta_n)^2\, f_k(n) \quad . \tag{3.4.11}$$

Kac proved the following three results. First:

$$\sum_{k=1}^{\infty} f_k(n) = 1 \tag{3.4.12}$$

which means that every n certainly recurs. Second:

$$\Theta_n = \frac{1}{p(n)} = 2^N \cdot \frac{(N-n)!\, n!}{N!} \tag{3.4.13}$$

which becomes huge as soon as n differs significantly from $N/2$, while it is reasonably small when $n \approx N/2$. Kac gives the following illustration: if the clock clicks every second, and $N = 20,000$, then Θ varies from $2^{20,000}$ sec. i.e. $\approx 10^{6,000}$ years when $n = N$, down to ≈ 175 sec. when $n = N/2$. We note in passing that (3.4.12) and the first equality in (3.4.13) were established in [Kac, 1947a] for a class of Markov processes that is much more general than the model studied in this section.

Kac's third result emphasizes the fact that the mean recurrence time is an average, and hence its significance depends on its variance Λ being small when compared to Θ itself; this unfortunately is not the case for deviations far away from equilibrium; indeed for N large and $n \approx N$: Λ_n is commensurable with Θ_n, so that the immensity of mean recurrence times, for large N and far away from equilibrium, can be misleading.

The last two windows of Fig.3.1 show the long-time behavior traced in our computer experiment. Note that the compression of the time-scale in this Fig. makes the peaks appear to be quite narrow, so that the resolution of the graphics makes them look somewhat smaller than they really are; we also digitally recorded – see Table 3.1 – their actual height and we checked that the statistics of the resulting time-sequences conforms with the theoretical predictions sketched below. We gratefully acknowledge the help of our colleague, Prof. Rick Smith of the Mathematics Department of the University of Florida, for the efficient conduct of this computer experiment; this also gave us an opportunity to appreciate the tenacity that must have been involved in the data collection and analysis in [Kohlrausch and Schroedinger, 1926].

Table 3.1. The dog-flea model: Recurrences, predicted and measured: p_n^N is the theoretical probability that Dog A is visited by n fleas at any given time; $N = 100$ is the total number of fleas in each run; $x_n = \frac{1}{5}X_n$ where X_n is the number of times Dog A was visited simultaneously by n fleas during an actual run X consisting of $5 \cdot 10^4$ iterations; y_n and Y_n are defined similarly for another run Y of $5 \cdot 10^4$ iterations

n	58	59	60	61	62	63	64	65	66	67
$p_n^N \cdot 10^4$	222	159	109	71	45	27	16	8.6	4.6	2.3
x_n	210	167	131	93	61	38	21	12	7.6	5
x_{N-n}	238	160	113	78	48	27	17	10	5	2.4
y_n	224	170	136	97	61	35	16	11	5.6	5.4
y_{N-n}	215	153	131	92	56	33	22	14	5	3.8

Table 3.1 shows a generally good agreement between the computed recurrence times $\Theta_n^N = \{p_n^N\}^{-1}$ – see (3.4.4),(3.4.13) – and the corresponding empirically obtained frequencies, within the fluctuations observed from one to another run of $5 \cdot 10^4$ iterations. On the basis of the evidence presented in the table, the theoretical equality $\Theta_n^N = \Theta_{N-n}^N$ may appear to be empirically violated; yet, the average occupancy over the two $5 \cdot 10^4$ runs reported in Table 3.1 are 49.97 and 49.93; these would favor a bias in the opposite direction. However, a partition of the runs into segments of length 10^4 shows that even on this long-scale there are fluctuations; these can be so large as to reverse the above bias in some segments even at the borderline where $n = 66$ or 67. Recurrence times, after all, are stochastic objects.

The core of the lesson offered by the dog-flea model is to be looked in the contrast between the apparently irreversible approach to equilibrium observed in the time-scales [a few hundreds of iterations] reported in Fig. 3.1 (a and b) vs. the chaotic equilibrium fluctuations that empirically prevail over the time-scales [tens of thousands of iterations] of Fig. 3.1 (c and d). Hence the Ehrenfest dog-flea model throws useful light on the role of statistics which Boltzmann tried to emphasize during the long controversy he had with some of his detractors. The darker side of the coin is that the model being intrinsically stochastic, the question of the connection between statistics and the mechanical aspects of the physical problem at hand is strictly speaking out of its purview.

Yet, the validity of the conclusions derived from the Ehrenfest model extends significantly beyond the model itself, as was shown by the analysis of Brownian motion observed in colloïdal suspensions. In the latter context [Smoluchowski, 1916] could already conclude from his experience with the analysis of actual laboratory data:

A process appears irreversible – [resp.] reversible – when the initial state has a recurrence time which is long - [resp.] short – compared to the time of observation.

With remarkable perspicacity, Smoluchowski immediately commented on the remaining problem facing the applications to laboratory situations:

> *Since, in general, one does not know how to calculate the recurrence time, this is not particularly useful for practical calulations, however, a purely conceptual clarification of these concepts* [could be] *valuable.*

4. Classical Probability

presque toutes nos connaissances ne sont que probables
[Laplace, 1795]

Statistical concepts entered thermophysics with the *kinetic theory of gases* in a manner we began to describe in Chap. 3. To go any further into the discussion of the foundations of statistical mechanics depends on a delineation of the different roles played by the several concepts that reside at the core of probability theory. As we try and refine the philosophical analysis – see Chap. 1 – of various modeling activities, we find that the successive phases in the development of probability theory provide illustrations of the process by which *ad hoc* solutions of somewhat disparate problems compete, and – even if they are ultimately discarded – do contribute to the elaboration of a calculus we refer to as *Classical probability*. These notions are reviewed in the present chapter. Long traditions in betting and assessing opinions contributed to the birth of Probability in the 17th century. During the next two centuries the theory reached a level of maturity sufficient to cover effectively diverse applications to actuarial, medical, judicial and demographic concerns, as well as to the natural sciences; the theoretical developments that occurred in these formative years are also highlighted in Sect. 4.2 (up to the end of the 18th century) and 4 (for the 19th century). Yet, the early probabilists – from the middle of the 17th century to almost the end of the 19th century – seem to have been mainly motivated and/or concerned with solving concrete problems: it is only well into the 20th century that sufficient power of abstraction became available to address and formulate the foundations of probability theory, bringing thus the enterprise back into contact with the main stream of contemporary mathematics; see e.g. [Kac, 1985, Chung, 1998].

4.1 Different Models for Probability

Probability concerns have been traced back to the origins of civilization. Our purpose here is to retrace some general lines or tendencies in a long history; thus we thought it may be useful in this chapter to give the biographical dates of the authors cited. The didactic device we chose in order to place in some reasoned perspective the various contributions to the field is to begin with a long–standing duality revived by Hacking [1975] in the motivations of probability theory.

On the one hand, betting and games of chance were common among Ancient Egyptians, Sumerians and Assyrians, or closer to us among Jews, Greeks or Romans alike, not to mention the ample literary evidence from India or China. All exploited the empirical facts that long sequences of trials (whether they involved talis, astragalus or coin tossing) are individually irregular (in modern terms: the unpredictability that makes betting entertaining), and yet present enough global regularities (in modern terms again: the limiting relative frequencies, the purported knowledge of which makes betting tantalizing). These two basic characteristics of randomness, rooted as they are in easily accessible and oft-repeated observations, proved astonishingly hard to formalize mathematically. This tenacious unyielding persists indeed today, in lotteries, annuities or life insurance, and such medical practices as inoculation and vaccination which draw on the analytic tools developed in the 18th century.

On the other hand, the rendering of probable opinions was traditionally inseparable from an explicitly recognized authority. The Latin root *probabilis* covers more than what meets the modern eye, namely *probare*, "to test"; it also carries such connotation as "to approve", "to prove", or "to be trustworthy", a meaning that has survived in the modern institution of the "probate court", although it is was once of wider currency, e.g. in the ancient expression "a probable judge".

Hacking [1975] illustrates the medieval prominence of the latter, lasting well into the 18th century, as evidenced by a quote from Gibbon [1737–1794]: *"Such a fact is probable but undoubtedly false."* [Gibbon, 1776/1788, Chap. xxiv, fn.No.120]. That type of probability can hardly be identified as a strict frequency interpretation; it is much more in line with [Pascal, 1670, No. 233] famous wager on the existence of a unique and personal God: *"the God of Abraham, Isaac and Jacob – not of the philosophers nor of the scholars."* This is all the more remarkable that Pascal [1623–1662] and Fermat [1601–1665] are often placed [Todhunter, 1865] at the beginning of the period – from the middle 17th century to the early 19th century – during which probability became a tool of science, culminating in [Laplace, 1820]; note incidentally that Laplace [1749–1827], in sharp contrast with Pascal's outlook, is reputed to have replied to Napoleon's query about the absence of God in [Laplace, 1796] *"Sire, I did not need that assumption."*

Thus, almost from the beginnings, one could discern a dual motivation for the introduction of probability: a *frequentist* interpretation and an often controversial interpretation as *degree of belief (or confidence)*. This simple opposition, however useful it may be for a first orientation into the semantic issues, needs to be refined as our analysis proceeds further. Fig. 4.1 offers

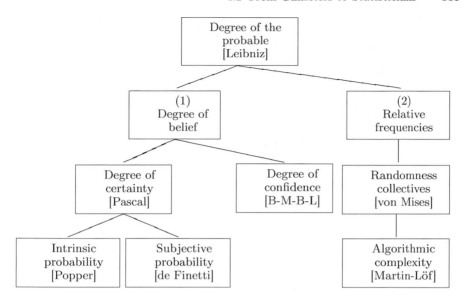

Fig. 4.1. Interpretations of the probable

a preview of the developments we consider, first in this chapter, and then in Chap. 6 once the syntax has been clarified (Chap. 5).

4.2 From Gamblers to Statisticians

The first steps in the elaboration of a theory of probability were taken in the Renaissance when the Humanists, the Reformers and the early Scientists alike realized that there were just too many conflicting authorities to invoke for *probable opinions*. Rejecting the multiple filiations of the Casuists' commentaries, they proposed to substitute the direct study of the Original Texts and of the Book of Nature. Thus, the art to make deductions from (or at least corroborated by) *incomplete data* had to be developed. To this line belong Pascal, Fermat and the early precursors of decision theory. Observations *a posteriori* were to be supported by reliable *a priori* reasons and/or computations. Examples are to be found especially in the Italy of the 16th and 17th centuries: Pacioli [1445–1514], Cardano [1501–1576], Tartaglia [1499–1557], Peverone [1500–1559], or even Galileo [1564–1642], are some of these early figures. Typically, they were concerned with games of chance: how to divide the stakes in an interrupted game; *or* how to enumerate and weigh possible outcomes. To be specific, we start with a brief discussion of one case for each of these two problems [Hacking, 1975]; we should mention, nevertheless that Galileo elsewhere computed *averages* over measurements: another of the perennial motivations for probability theory.

In the first example, a fixed amount of money is to be won by the first of two players A and B who obtains n successes when A is betting "head" against B betting "tail" in a sequence of fair coin-tossing trials. Suppose the game has to be interrupted, and the money must be allocated at that point in a manner that takes fairly into account the current state of the game, knowing that the successive trials have effectively produced $n-1$ heads and $n-k$ tails (with $k \geq 1$). A similar problem was posed by Pacioli in 1494; Cardano [1539] claimed to have discovered an error in the then prevalent solutions (although the publication of his own solution was delayed until 1564), and Tartaglia [1556] gave a tentative answer of 2:1 in favor of A for the case $k = 3$. For this same case, the solution 7:1 follows from an argument given by Peverone [1558]. It goes essentially as follows. If the game were to continue as first arranged, and if "head" were to obtain at the next trial, A would win; in contrast, B would need "tail" to occur in each of the next k trials; from this follows that A should stake 1 unit in the game, whereas B should stake

$$(1 + 2 + \ldots + 2^{k-1}) = 2^k - 1 \tag{4.2.1}$$

units. In modern terms, the probability that B wins is $\left(\frac{1}{2}\right)^k$, i.e. 1 out of 2^k cases, e.g. for $k = 3$, the first of the eight cases:

$$TTT, TTH, THT, THH, HTT, HTH, HHT, HHH. \tag{4.2.2}$$

From this follows, in general, the allocation $(2^k - 1) : 1$, i.e. indeed $7 : 1$ when $k = 3$. Note that, even in the presence of the uneven distribution effectively obtained at the time when the game was interrupted, all parties concerned accept that the coin is fair. We see later in this section that the detection of the unfairness of a coin, using only the result of a sequence of trials, does require some sophistication and considerable perseverance; moreover, devising a strategy to guard one's chances – by contract before the sequence of trials is carried on – against a possibly unfair coin is indeed a very hard problem; these two problems can begin to be approached satisfactorily only when modern mathematical tools are brought in; for some vistas into that territory, see Chap. 5.

The second example is an argument by Galileo starting from the elementary fact that $9, 10, 11$ or 12 can be broken up in six distinct sums of three integers n_i with $1 \leq n_i \leq 6$; specifically, for 9:

$$\left. \begin{array}{l} 1+2+6 \quad ; \quad 1+3+5 \quad ; \quad 2+3+4; \\ \quad 1+4+4 \quad ; \quad 2+2+5; \\ \quad \quad 3+3+3 \end{array} \right\} \tag{4.2.3}$$

and for 10:

$$1+3+6 \quad ; \quad 1+4+5 \quad ; \quad 2+3+5 \,; \atop 2+2+6 \quad ; \quad 2+4+4 \quad ; \quad 3+3+4 \quad \Big\} \qquad (4.2.4)$$

and similarly for 11 and 12. Galileo notices *however* that if these numbers are obtained by throwing three dice *"it is known from long observation that dice players consider* 10 *and* 11 *to be more advantageous than* 9 *and* 12.*"* Having thus set up his reader, Galileo offers the following *"very simple explanation, namely that some numbers are more easily and more frequently made from others."* Explicitly, the entries in the above lists can be made in a number of permutations; for instance $1+2+6$ can be obtained in six permutations, while $1 + 4 + 4$ can be obtained in three permutations, and $3 + 3 + 3$ in only one. Upon counting each entries with its multiplicity, we find that in a three-dice throw

$$9 \quad \text{obtains in} \quad 6 + 6 + 6 + 3 + 3 + 1 = 25 \quad \text{ways} \qquad (4.2.5)$$

while

$$10 \quad \text{obtains in} \quad 6 + 6 + 6 + 3 + 3 + 3 = 27 \quad \text{ways}, \qquad (4.2.6)$$

thus justifying the odds 27:25 for obtaining 10 over 9; and similarly for the odds of obtaining 11 over 12.

Five remarks can be made concerning the significance of Galileo's argument.

(i) The argument tacitly assumes that all permutations are *equally probable.*

(ii) The "long observation" to which Galileo refers is a clear-cut case of an early attempt towards a *limiting relative frequency interpretation*: 25/216 for the chance of obtaining 9 in a throw of three dice, *vs.* 27/216 for obtaining 10 (similarly for 12 *vs.* 11).

(iii) As pointed out by Hacking [1975], it proposes an *empirical refutation of a statistical hypothesis,* namely that each of the 6 partitions which we enumerated for 9 (or for 10) are equally probable. Incidentally, Hacking points out, tongue-in-cheek, that the latter hypothesis – which one could offer under Leibniz' [1646–1716] *Principle of Identity of the Indiscernible* – amounts to assuming Bose–Einstein (rather than Boltzmann) statistics for unmarked dice.

(iv) The above example illustrates a nascent trend to reduce probability to *combinatorics,* counting the equally probable possible outcomes; for a witness to the emergence of the latter, see [Leibniz, 1666]. In the sequel, as we observe the emergence of a more systematic theory, we try to recognize the distinction between what belongs to combinatorics and what to probability proper – see e.g. our discussion of the Bernoulli theorem (4.3.12).

(v) More generally, this episode illustrates Galileo's uncanny abilities to perceive important features beyond what everybody could see, and to tell his story well. Another example is provided by [Holton, 1995] recounting – in text as well as in pictures – the different reactions Harriot (1560–1621) and Galileo had in the same year 1609 when they both observed the Moon through a telescope. A mathematician and astronomer of some note, but perhaps all too aware of the classical myth attached to the supposed perfection of the Selenean disk, Harriot clearly sees, draws, but makes nothing of the jaggedness of the line separating light from darkness. [Galileo, 1610] seeks for its causes, thinks of the shadows projected by the mountains of the Earth, and offers a check for the analogy he is proposing: he derives from his observations an estimate of the heights of the mounts on the Moon. In Galileo's sepias, the Moon – one of the celestial bodies – appears, not as a myth, but as the real physical object demanded by a forthcoming new system of the World. Much later, and in the context of medical sciences Claude Bernard [1813–1878] expresses a fundamentally similar conviction:

> *Empiricism can only serve to accumulate facts, not to build up Science. The experimentalist who does not know what he is looking for will not understand what he finds.*

[Bernard, 1865]

If we are to think of the above two simple examples as attempts to articulate the meaning of probability, we may say that the moves are in some sense reductional: they are schemes of *reducing* the probability of a *compound* event to the probabilities of the constituent simple or singular events. Such a move is very similar to the truth-functional semantics in 20th-century logic, i.e. the truth-value of a compound sentence is reduced to the truth-functional combination of the truth-values of the constituent atomic sentences. In our case, the reduction scheme involved the nascent mathematical theory of *combinatorics* (which is again very similar to the scheme of truth-functional semantics). Once this is clearly understood, we see that the entire interpretive burden is laid on the notion of "*being equally probable*", as mentioned in remark (i) above. As we shall see in this and the next two chapters, this notion has defied a clear and complete logical definition until this day, and the failure of a logical definition was largely responsible for other interpretations of probability, outside the realm of formal logic.

Back to the main topic of this chapter, it seems reasonable to view *classical probability* as the doctrine embodied in [Laplace, 1820]. This treatise offers a synthesis remarkable by its broad coverage, its systematic presentation, and the incisive mathematical analysis it displays. These qualities came to fruition in the course of several avatars. The work first appeared in 1812; in its second edition, an *Introduction* was added, consisting of a revised version of a position paper outlining a research program, [Laplace, 1795]; the fourth edition of the latter was incorporated in the 1820 edition of the *Théorie analytique*. In fact, Laplace's interest in probability had started much earlier, with [Laplace, 1774] and [Laplace, 1781], in part under Condorcet's aegis – for some direct evidence of the parallelism of their concerns, see e.g. [De Condorcet, 1784]; for the full story, consult [Baker, 1975]. Condorcet proposed deep insights into the principles on which probability theory was to be built, but he did not have Laplace's technical acuity; some of the latter is evidenced for us by quite sophisticated papers on infinite series and the asymptotic behavior of certain integrals: these papers were strongly motivated by Laplace's sustained efforts to invest in the calculus of probability his command of the most advanced mathematical analysis; see *Oeuvres Complètes* [Laplace, 1878-1912].

As for the beginnings of the classical period in probability theory, there are good reasons to limit our search for the precursors to the middle of the 17th century, specifically to [Huygens, 1657] [1629–1695] and to the epistolary exchange between Pascal [1623–1662] and Fermat [1601–1665], centered around 1654, and on which Poisson [1781–1840] commented :

> *A problem concerning games of chance, proposed to an austere Jansenist by a man of the world* – Antoine Gombault, Chevalier de Méré [1610–1685] *– was at the origin of the calculus of probability.*
>
> [Poisson, 1837]

This oft-quoted and admittedly well-crafted characterization unfortunately does not do justice to Pascal in limiting his contribution to the *'calculus'* of probability. A current of scholarship has recently surfaced to show that Pascal's scope extended to a proposal for a track towards a theory of knowledge that would be an alternative to that followed by Descartes' disciples; see for instance [Van Fraassen, 1989] or [Carraud, 1992]. The thesis seems to be that Pascal saw fluctuations and uncertainties, not as accidents due to an imperfect knowledge, but as intrinsic components of what there is to know about natural phenomena. In [Chevalley, 1995], a compelling case is made for the necessity to elaborate on the undergirdings of Pascal's project as it differs from the Cartesian approach that is spanning the classical doctrine, especially when it comes to apply probabilistic ideas in the physical sciences.

As we examine the dialogue between Pascal and Fermat, it is clear that the new elements with respect to the earlier discussions (e.g. by Cardano and Galileo; see above) are not to be found in the problems themselves – still mostly games of chance – but rather in the systematic methodological scope in which the solutions were approached and/or presented. Many of the questions were reduced to particular cases of general rules that were derived for *counting* the number of favorable outcomes against the number of all possible outcomes. The *calculus* of probability came thus into existence, and one of the main mathematical areas that were developed for the purpose was the field of *combinatorics*. This however characterizes only the first of the three panels of a triptych that ultimately depicts classical probability. As it appears later in this section, the central panel of this triptych is a purported resolution of the duality between the two approaches to probability theory degree of certainty and long-run frequencies; the third panel is the bearing of the differential and integral calculus on the control and evaluation of limits.

A typical scene of the **first panel** depicting the use of combinatorial techniques in our context is the *arithmetic triangle*. It consists of an unending array of numbers, the first 12 rows of which are reported in Table 4.1. This array was known to many of Pascal's predecessors, among them Tartaglia [1556] – although not in immediate connection with games of chance – and many others dating back to the 13th, or even the 11th century. It became known as the Pascal triangle by reason of the systematic study of its properties, apparently drafted by Pascal at the time of his correspondence with Fermat, and published posthumously [Pascal, 1665]. Three properties stand out: a rule of formation (4.2.7–8), a summation formula (4.2.9), and a unimodal distribution (Table 4.1).

Table 4.1. The Pascal triangle

$n\backslash k$	0	1	2	3	4	5	6	7	8	9	10	11	12	2^n
0	1													1
1	1	1												2
2	1	2	1											4
3	1	3	3	1										8
4	1	4	6	4	1									16
5	1	5	10	10	5	1								32
6	1	6	15	20	15	6	1							64
7	1	7	21	35	35	21	7	1						128
8	1	8	28	56	70	56	28	8	1					256
9	1	9	36	84	126	126	84	36	9	1				512
10	1	10	45	120	210	252	210	120	45	10	1			1024
11	1	11	55	165	330	462	462	330	165	55	11	1		2048
12	1	12	66	220	495	792	924	792	495	220	66	12	1	4096

First, the rule of formation, already known to [Stifel, 1544]. Let us denote by $b(n, k)$ the entry in the nth row and the kth column; then

$$b(n + 1, k) = b(n, k - 1) + b(n, k) \qquad (4.2.7)$$

which together with the boundary conditions

$$b(n, 0) = 1 = b(n, n) \quad \forall \quad n = 1, 2, 3, \ldots \qquad (4.2.8)$$

determine the $b(n, k)$ uniquely, i.e. give a recursive definition of these numbers.

The second of the properties of the arithmetic triangle, already entered in the right column of Table 4.1, is

$$\sum_{k=0}^{n} b(n, k) = 2^n \quad . \qquad (4.2.9)$$

The third property is introduced later to serve as a heuristic motivation for the theorem due to Jakob Bernoulli [1654–1705]; it is the (weak) law of large numbers; see (4.3.12) below.

Note that the second property, when written in the form

$$\sum_{k=0}^{n} b(n,k) = (1+1)^n \qquad (4.2.10)$$

suggests to examine the expansion of the polynomial $(x+y)^n$:

$$\sum_{k=0}^{n} \beta(n,k)\, x^k\, y^{n-k} = (x+y)^n \qquad (4.2.11)$$

where the unknown coefficients $\beta(n,k)$ are determined by comparing, term-by-term the expansions of $(x+y)^n(x+y)$ and $(x+y)^{n+1}$; one indeed verifies that the $\beta(n,k)$ satisfy also the recursive definition (4.2.7–8) of the entries of the Pascal triangle.

These relations are satisfied again by the numbers defined by:

$$\binom{n}{k} = \frac{n!}{k!(n-k)!} \qquad (4.2.12)$$

Consequently, the three kinds of numbers just introduced coincide exactly:

$$b(n,k) = \beta(n,k) = \binom{n}{k} \qquad (4.2.13)$$

The above results, in part or in total, are credited to a variety of authors – besides Pascal [1623–1662] – e.g. [Leibniz, 1666] [1646–1716], [Gregory, 1670] [1638–1675], [Wallis, 1685] [1616–1703], [Newton, 1711] [1642–1727], [Bernoulli, 1713, Jakob] [1654–1705]. All of them, from Pascal to Bernoulli, saw the relevance of these combinatorial results for the nascent calculus of probability. The key towards their interpretations – see (4.2.16) and (4.2.17) below – is to go back to (4.2.10), and to *define* for any $0 \le p \le 1$, and $q = 1 - p$:

$$\boxed{p_n(k) = \binom{n}{k} p^k\, q^{n-k}} \qquad (4.2.14)$$

so that (4.2.10–13) imply:

$$\sum_{k=0}^{n} p_n(k) = 1 \quad \text{and} \quad 0 \le p_n(k) \le 1 \qquad (4.2.15)$$

In modern parlance, (4.2.15) says that, at fixed n, the collection $\{\, p_n(k) \mid k = 0, 1, \ldots, n \,\}$ is a probability distribution. The remaining question for us is thus to determine the events, the probability of which is so distributed.

To position ourselves back in the times when these formulas were discovered, we note first the bare combinatoric fact that $b(n,k)$ is the number of ways to choose k objects among n identical objects. For instance, let

$x = (x_1, x_2, \ldots, x_n)$, with $x_k \in \{0, 1\}$, $k = 1, 2, \cdots, n$, be the string of length $l(x) = n$, *representing* the outcome of n successive throws of a coin, with the representational convention $x_m = 1$ if head comes out at the mth throw, and $x_m = 0$ if tail comes out instead. Then $b(n, k)$ is the number of strings of length n along which we count exactly k heads:

$$\binom{n}{k} = \text{card}\{ x \mid l(x) = n ; \sum_{m=1}^{n} x_m = k\} \quad . \tag{4.2.16}$$

As we used here for the first time the notation card, let us specify that card A denotes the cardinality of the set A; when this set is finite, card A is simply the finite integer $N = \text{card}\, A$ equal to the number of the elements of A; when no such integer exist, we say that the set is infinite, and we write card $A = \infty$. A fair coin is one in which the probable certainty for "head" is the same as that for "tail": $p = q = 1/2$. Upon inserting this equal *a priori* probability condition into (4.2.14), we find in this particular case:

$$p_n(k) = \binom{n}{k} (\frac{1}{2})^k (\frac{1}{2})^{n-k} = \frac{1}{2^n} b(n, k) \quad . \tag{4.2.17}$$

Hence, in this case, $p_n(k)$ is the ratio of two numbers: the number $b(n, k)$ of favorable (i.e. "head") outcomes in a string of n trials, over the number 2^n of all possible outcomes – i.e. the number of n–strings in which 1 appears k times divided by the number of all possible strings of length n.

All of our authors, from Pascal and Fermat to Bernoulli, agree in view of (4.2.17) that it is reasonable to call $p_n(k)$ the probability for the event "k heads in n throws", and that this interpretation extend to (4.2.14) when the coin is biased (or "loaded"), so that the probable certainty p for "head" may differ from $1/2$. There is one assumption that was made tacitly all along, namely that the successive throws are "independent" of the previous outcomes, a notion that was not formalized until the concept of conditional probability was formulated (see below in our discussion of Bayes' contributions). Leaving this aside for the moment, our authors also seem to agree that the composite probability (4.2.14) is a reasonable *prediction,* one that would be borne out in the long run, i.e. if the coin were thrown sufficiently many times: the early discussions leading to probability theory could be subtitled "conversations with a gambler". It is thus quite remarkable that this tentative statement was soon to take the form of a theorem – see (4.3.12) below – to which Bernoulli's name is attached, and which we call today the (weak) law of large numbers. We discuss Bernoulli's theorem in some detail shortly. For the time-being, let it suffice to say that it is based on a simple observation, namely the third property we announced for the arithmetic triangle: along any given row of Table 4.1, the $b(n, k)$ become very peaked around values of k such that $k/n \approx 1/2$, and this tendency is increasingly marked as n increases. Bernoulli's theorem states that the same is true for the general $p_n(k)$ in (4.2.14).

To see a modern illustration of this fact, let us briefly present a model of a coin that is *not fair*, i.e. one for which the *a priori* probability p is different from $1/2$. Consider for this purpose, the "Weldon data" reproduced in [Fisher, 1968]:

Table 4.2. The Weldon data

k	$N(k)$	$\nu(k)$	$p_{12}(k)$	$\tilde{p}_{12}(k)$
0	185	.007 033	.007 707	.007 123
1	1149	.043 678	.046 244	.043 584
2	3265	.124 116	.127 171	.122 224
3	5475	.208 127	.211 952	.207 736
4	6114	.232 418	.238 446	.238 324
5	5194	.197 445	.190 757	.194 429
6	3067	.116 589	.111 275	.115 660
7	1331	.050 597	.047 689	.050 549
8	403	.015 320	.014 903	.016 109
9	105	.003 991	.003 312	.003 650
10	14	.000 532	.000 497	.000 558
11	4	.000 152	.000 045	.000 051
12	0	.000 000	.000 002	.000 002
Σ	26,306	1.000 003	1.000 000	.999 999

These data report on $N = 26,306$ repetitions of a sequence of independent trials, each of which consists of the throwing of the same die 12 times, each time counting as success the appearance 5 *or* 6 on the face of the die. The corresponding $b(12, k)$ are listed in the last row of Table 4.1. The first two columns of Table 4.2 give the raw experimental data: the number $0 \leq k \leq 12$ of successes in each trial, and the number $N(k)$ of trials in which k successes were counted. In the third column we list the relative frequencies of success: $\nu(k) = N(k)/N$. The corresponding $p_{12}(k)$ from (4.2.14), with $p = 2 \times 1/6 = 1/3$, are listed in column 4. The agreement between the experimental $\nu(k)$ and the computed $p_{12}(k)$ would be pretty good, were it not for the fact that one could have hoped for an even better fit, in view of the large data base, which consists of $12 \times 26,306 = 315,672$ throws of a single die. This suggests to compute, from the experimental data, a *revised* or *updated* attribution to the *a priori* probability of success for a single throw of the die, namely:

$$\tilde{p} = 106,602/315,672 = .337\,698\,6 \neq .333\,333\,3 = p,$$

where the total number of successes $106,602 = \sum_{k=1}^{12} k\,N(k)$ is computed from the experimental data. We recomputed for the fifth column of Table 4.2, the correspondingly revised values of the $\tilde{p}_{12}(k)$; the better agreement with the individual measured values of the relative frequencies $\nu(k)$ thus may suggest a slight bias in the die serving to collect the data. Naturally, although the sample is quite large, it is still finite; hence the question arises as to whether a measure – a "degree of confidence"– could be defined, allowing one to claim that the difference between the probable assessment p and the experimental value $\tilde{p}$ is significant or not. The search for an answer to that question motivates the development of modern statistical methods, starting with Bayes [1702–1761].

Prior to our discussion of Bayes' contribution, we still have to analyze how Bernoulli's result leads to the better credence usually granted – all else being equal – to trials that are repeated a large number of times. Yet, before we turn to a general discussion of Bernoulli's theorem, we should sound a brief warning on methodology, and propose a few historical vignettes.

The word "theorem" presupposes that a definition has been agreed upon. In the present case, the definition is for the mathematical object $p_n(k)$, namely (4.2.14). Neither the statement of Bernoulli's theorem, nor its proof, require the interpretation of $p_n(k)$ as a probability; as a sad reflection upon the times – theirs and ours – no explicit definition of probability was found to be a compelling demand for close to a century, until Laplace advanced one; and even so, it took yet another century until Kolmogorov boiled down the ingredients and rules of probability calculus to a syntax that could satisfy a modern mathematician.

In the time-frame that is ours in this chapter, probabilistic reasoning was still very much in the "fact finding" empirical phase of its development. Indeed, when studying the nature and modes of probabilistic reasoning in the 18th century one can easily be baffled by the profusion of particular problems, most of them of a combinatorial nature. These can be understood as models of a discourse, conducted in an as yet unstructured language, or as

corollaries to the great meta-theorems which remained unformulated
[Bahar and Spencer, 1998]

to quote (!) from an article written at the occasion of the departure of Paul Erdös, a towering figure in 20th-century combinatorics. Thus the account of the main stream in the development of probabilities from Pascal to Laplace [Todhunter, 1865] is a mostly faithful description of each separate solution to each separate problem. Note nevertheless that this useful – albeit piece-wise – *retrospective,* is contemporary to the opening of a systematic statistical *prospective* in the natural sciences: e.g. Maxwell's derivation of the normal velocity distribution, and Boltzmann's kinetic theory; see Chap. 3.

Asides from the perennial problems relative to game of chance, and the multiplicity of models of the "urns-and-balls" variety, probabilistic reasoning was applied also to the evaluation of a whole panoply of ventures. Since these are of marginal interest for our purposes here, we mention only a few of these risks. Lotteries and annuities – by which the lender would receive rents on his or some other person's head for as long as this person lived – became a favorite mean of the Enlightened despots to raise funds without raising a more obvious taxation of the rich and privileged; the famous astronomer

Halley [1656–1742], the mathematicians de Moivre [1667–1754], Euler [1707–1783], Waring [1736–1798] – Lucasian chair at Cambridge from 1760 to 1798 – and Lagrange [1736–1813] were consulted or otherwise chose to write on the subject: [Halley, 1693], [De Moivre, 1725], [Euler, 1767], [Waring, 1792], [Lagrange, 1798].

Also, as the very title of several of these papers may suggest, the Enlightenment did reach into the 18th-century practice of medicine; see e.g. [Bradley, 1971, Emch-Dériaz, 1982, Fagot, 1982]. Broad issues of public health were addressed, birth and mortality tables were more systematically collected for the purpose of evaluating cures [Black, 1789] and controlling the spreading of diseases and/or the working conditions in urban concentrations. The difficulty of introducing statistical arguments in matters where human intervention could drastically modify life expectancies is vividly illustrated by the smallpox controversy that pitched against one another Daniel Bernoulli [1700–1782] and d'Alembert [1717–1783]; see: [Bernoulli, 1760–1765] and [D'Alembert, 1760]; these battles continued well into the 19th century (see the end of Sect. 4.4).

Political science also became more quantitative, with various proposals for alternate forms of decision making – whether they be membership in learned societies or the composition and administration of the courts of law; one of the most prominent proponents of this line was De Condorcet [1784] [1743–1794]. His disciple, Lacroix [1765–1841] was also quite influential, perhaps mostly for his textbooks; he had taught mathematics at the Lycée under Condorcet's supervision, and he produced a famous *"Traité du calcul différentiel et intégral"* (3 volumes, 1797) from which generations learned calculus, first in France, and after 1816, in England – see [Struik, 1948]. He also wrote an introduction to probability theory [Lacroix, 1816 and 1833], deliberately tilted towards social phenomena; its enduring fame came in no small part from the fact that it was chosen by Quetelet [1796–1874], one of the 19th-century pioneers in social mathematics, as the textbook for the first probability course he taught at the Museum in Brussels.

For the assessment of the probable in the courts of law, the position of an influential precursor, Beccaria [1738–1794], is succinctly presented in [Beccaria, 1764, art. XIII and XIV]. It is also telling that the author of the collection of reprints [Maseres, 1795] was Francis Maseres [1731–1824], Attorney General of Québec until 1769, who has left his own mark with his writings on the powers of juries as*"judges of law, as well as judges of fact"*. Two threads were braided in the contemporary judicial literature; the first was the discussion of the effects of the manner in which juries are constituted on the decisions to which they arrive; and the second was the perennial problem of weighing the evidence presented by witnesses, against the jurisprudence of precedents. Along both threads qualitative probable reasoning was turned slowly into quantitative probabilistic computations; in the realm of the classical probabilists, these efforts culminated with the publication of [Poisson, 1837]; note nevertheless that [Laplace, 1820] in the *Avertissement* to the second edition makes a point of the fact that

> the theory of probability of [statements by] *witnesses, omitted in the first edition, is presented here with the development required by its importance.*

This rapid overview of some of the applications should suffice as motivating remarks, and we return to applied probability at the end of Sect. 4.4.

The quantification and early solution of a great variety of problems relative to games of chance – dice, cards, lotteries, annuities, and their extensions to medicine and law – is so impressive that, even in retrospect, it can be overwhelming (see e.g. [Todhunter, 1865]): the temptation may arise to view the history of probability as a succession of vignettes documenting its *practice*. One of [Hacking, 1975] merits is to have emphatically drawn attention to the fact that, across the web of applications, the classical probabilists wove into their *theory* another thread, spun out of contemporary philosophical concerns. The duality he recognized between degree of belief and relative frequencies pertains to these motivations; as we already pointed out in Fig. 4.1, this is a simplification that must be sharpened to account for what the 18th-century practitioners actually did, even though often only tacitly. Thus two different philosophical approaches should be discerned: one is rooted in the objective view of probability, and the other in its subjective aspects. That the potential for tensions did not flare up at the time may be attributed in a large part to the success the Encyclopedists met in their advocacy of the ultimate power of human reason: for a picture of the 18th-century intellectual background before which the probability play unfolded, see the informative monographs by Baker [1975] and Daston [1988].

Anchored on the rationalist tradition that belongs to the philosophers of continental Europe – such as Descartes [1596–1650] and Leibniz [1646–1716] – the former view held that knowing the "objective" probability intrinsic to a system – dice, cards, etc – mathematical reasoning, mostly combinatorics, would prescribe in terms of relative frequencies the outcome of a sequence of trials.

The second view was inherited from the British empiricists (the "associassionists" in [Daston, 1988]): Locke [1632–1704], Berkeley [1685–1753], Hartley [1705–1757], and Hume [1711–1778]; it deemed that human minds in their "natural state" [Locke, 1690] should be able to infer correctly the probability relative to a system from the data acquired while registering the outcome of a sequence of trials. Even though they were by no means directly involved in the mathematicians' work, the empiricists' epistemological views vividly contributed to the solution of the problems that culminated in the Bayes–Laplace theorem; the latter appeared then as the complement to the de Moivre–Laplace form of the Bernoulli theorem known as the law of large numbers. See below (4.3.12) for the Bernoulli theorem, (4.3.24) for the de Moivre–Laplace theorem, and (4.3.43) for the Bayes–Laplace theorem.

4.3 From Combinatorics to Analysis

In this section, we concentrate on a review of the contributions of Jakob Bernoulli [1664–1705], de Moivre [1667–1754] and Bayes [1702–1761] to the problems of the prediction of effects and the determination of the probability of causes, as they had coalesced at the time when [Laplace, 1820] and [Poisson, 1837] were published. In line with the tone and aspirations of our authors, we present the issues from the vantage point afforded by the mathematical tools they contributed to create.

We begin with the observation that, at fixed n and p, the function $p_n(.)$ satisfies a condition that generalizes the third property of the entries $b(n,.)$ of the arithmetic triangle, namely that they define a *unimodal distribution* peaked around a central value $k = m$ where m is the unique integer such that

$$\left.\begin{array}{l} (n+1)p - 1 < m \leq (n+1)p \quad i.e. \\ m = np + \delta \quad \text{with} \quad p-1 < \delta \leq p \end{array}\right\} \qquad . \tag{4.3.1}$$

Like the $b(n,.)$ (see Table 4.1), the $p_n(.)$ are monotonically increasing (resp. decreasing) for $k < m$ (resp. $k \geq m$): to show this, it is sufficient to verify, from the definition (4.2.14), that:

$$\frac{p_n(k)}{p_n(k-1)} = 1 + \frac{(n+1)p - k}{kq} \; ; \; \frac{p_n(k+1)}{p_n(k)} = 1 + \frac{(n+1)p - (k+1)}{(k+1)q} . \tag{4.3.2}$$

Hence, on one side of m, i.e. when $k < m$, the second of the conditions (4.3.1), namely $m \leq (n+1)p$ implies $(n+1)p - k > 0$ and thus, from the first of the identities (4.3.2): $p_n(k-1) < p_n(k)$; on the other side of m, i.e. when $k \geq m$, the first of the conditions (4.3.1), namely $(n+1)p - 1 < m$ implies $(n+1)p - (k+1) < 0$ and thus, from the second of the identities (4.3.2): $p_n(k) > p_n(k+1)$. Note also, from (4.3.1), that

$$\left.\begin{array}{ll} \text{either} & 0 \leq \frac{m}{n} - p \leq \frac{1}{n}p \\ \text{or} & 0 < p - \frac{m}{n} < \frac{1}{n}q \end{array}\right\} \quad \text{i.e.} \quad \left|\frac{m}{n} - p\right| \leq \frac{1}{n} \times \text{Max}\{p,q\} \tag{4.3.3}$$

so that, at p fixed, m/n approaches p as n becomes large.

Having shown that the function $p_n(.)$ is unimodal, and peaked around $k = m$, with m/n approximating p, we are ready for the derivation of the Bernoulli theorem which states that, for very large values of n, the $p_n(.)$ becomes very peaked around m; so much so, in fact, that given $\varepsilon > 0$, the ratio

$$\frac{\text{card}\{x \mid l(x) = n \, ; \, |\frac{1}{n}\sum_{j=1}^{n} x_j - p| \geq \varepsilon\}}{\text{card}\{x \mid l(x) = n\}} \tag{4.3.4}$$

can be made arbitrarily small by letting n be large enough. Hence, as n increases indefinitely, the *degree of confidence* P_n, that the observed relative frequency of heads $\frac{1}{n}\sum_{j=1}^{n} x_j$ deviates from the probable assessment p by less than an arbitrary quantity ε, i.e. is asymptotically approaching 1 (hence the name, degree of confidence, which we choose for this aspect of the original notion, the "degree of belief").

Bernoulli's proof, although correct, is quite convoluted, and yet it gives only a pessimistic estimate on how large n needs to be, in order to make P_n closer to 1 than a prescribed amount η. As we are interested in the content of the theorem rather than in the historical accidents of its proof, we choose to follow the much more transparent path proposed in [Feller, 1968, 1971]. From (4.2.14), the probability of having *at least* r heads in n throws of the coin (i.e. the relative number of strings of length n with the desired property) is

$$P_n\{\sum_{j=1}^{n} x_j \geq r\} = \sum_{\nu=0}^{n-r} p_n(\nu + r) \quad . \tag{4.3.5}$$

We consider separately the two cases $r > np$ and $r < np$. For the former case, we have:

$$\frac{p_n(\nu + r)}{p_n(\nu + r - 1)} < a \quad \text{with} \quad a = 1 - \frac{r - np}{rq} \quad . \tag{4.3.6}$$

Upon comparing the RHS of (4.3.5) with the sum of the geometric series $\sum_{\nu} a^{\nu}$, we obtain:

$$\left.\begin{array}{c} \sum_{\nu=0}^{n-r} p_n(\nu + r) \leq p_n(r) \sum_{\nu=0}^{n-r} a^{\nu} \leq p_n(r) \sum_{\nu=0}^{\infty} a^{\nu} = \\[2mm] = p_n(r) \frac{1}{1-a} = p_n(r) \frac{rq}{r-np} \end{array}\right\} . \tag{4.3.7}$$

Moreover,

$$1 = \sum_{k=0}^{n} p_n(k) \geq \sum_{k=m}^{r} p_n(k) \geq (r - np)\, p_n(r) \tag{4.3.8}$$

where the last inequality comes from two trivial facts: (1) there are less that $(r - np)$ terms in the central sum; and (2) $p_n(k)$ is a decreasing function of k for $k \geq m$. Upon inserting (4.3.7–8) into (4.3.5) we find:

$$P_n\{\sum_{j=1}^{n} x_j \geq r\} \leq \frac{rq}{(r - np)^2} \qquad \text{for} \quad r > np \quad . \tag{4.3.9}$$

Upon choosing $r = n(p + \varepsilon)$, (4.3.9) reduce to

$$P_n\left\{\left(\sum_{j=1}^{n} x_j - np\right) \geq n\varepsilon\right\} \leq \frac{1}{\varepsilon^2}\frac{1}{n} \qquad \text{for} \quad r > np \quad . \tag{4.3.10}$$

Upon noticing now that the substitution of x_j by $1 - x_j$ corresponds to the exchange of the roles of heads and tails, and thus to substituting p by q and r by $(n - r)$ we obtain from (4.3.10):

$$P_n \left\{ \left(np - \sum_{j=1}^{n} x_j \right) \geq n\varepsilon \right\} \leq \frac{1}{\varepsilon^2} \frac{1}{n} \qquad \text{for} \quad r < np \quad . \qquad (4.3.11)$$

Hence (4.3.10) and (4.3.11) can be compounded to give:

$$P_n \left\{ \left| \frac{1}{n} \sum_{j=1}^{n} x_j - p \right| \leq \varepsilon \right\} \geq 1 - \frac{1}{\varepsilon^2} \frac{1}{n} \qquad (4.3.12)$$

Note that $1/(\varepsilon^2 n)$ is very small exactly when n is much larger than ε^{-2}. The precise mathematical statement (4.3.12) is purely a result in *combinatorics*. It becomes a *probabilistic statement,* called the **Bernoulli theorem,** or the *(weak) law of large numbers,* only when one interprets it as stating that P_n measures the degree of confidence that, in n independent trials, the frequency $\nu_n = (1/n) \sum_j x_j$ of the successes falls no further than the prescribed bound ε from the value p of the probable assessment.

The strength of a general result had now been added to the mass of evidence provided by many particular applications: the temptation must have been great then to identify probability and combinatorics. For instance, towards the end of his article on *Combinaison* in the *Encyclopédie*, d'Alembert added:

> This theory is indeed very useful in the calculus of games of chance, and
> it is from it that the whole science of probability flows.
>   [Diderot and d'Alembert, 1751–1780, vol. III (1753), p. 663]

In this passage, d'Alembert distinguishes the part (combinatorics) from the whole (probability theory); some of his readers, however, were not so subtle; ironically, even the author of the "reasoned summary", in the supplement to the *Encyclopédie* [Table, vol. 1 (1780), p. 350], omits the strict inclusion allowed by d'Alembert, thus betraying the persistent misconception that still, towards the end of the 18th century, tended to identify probability and combinatorics. One of the reasons for this superficial account is that Bernoulli's theorem (4.3.12), *stricto sensu* a result in combinatorics, was given what could have seemed to be a compelling interpretation as an estimate of how large n must be to make one "acceptably confident" that, after n trials, the observed relative frequency is close to the probable assessment made before the trials. One must realize that as long as a definition of probability is not delineated precisely, it is far from unambiguous to say why, beyond the mathematical correctness of (4.3.12), this degree of confidence is deemed acceptable as a statement about the conclusion(s) one can draw from the observation of a (finite) number n of actual trials.

Yet, Bernoulli's theorem was regarded by some – e.g. Hartley [1749] – to be a genuine solution to Hume's skeptical problem of inductive reasoning (which includes probable reasoning). Along the line already suggested by Hume in his skeptical solution to the problem, Hartley argued that Bernoulli's theorem assured that in a long run of repeated trials, the association of impressions or ideas allows to prescribe the right proportion (read *a priori* probability) that governs the outcome of subsequent trials. One of the significant contributions of the empiricists to the classical theory was the idea that in the long run all chancy elements in a sequence eventually cancel one another and leave the stable *a priori* probability manifested in a reasonable mind. This was clearly part of de Moivre's thinking – see quote in our discussion of (4.3.17) below – although it was left for Bayes to supply most of the much needed mathematical precision necessary to clarify these ideas.

To begin dealing with this problem, we consider the other two panels of the probability triptych.

The **second panel** is concerned with the nature of probable reasoning: Leibniz' project of a treatise on the mathematical logic of the probable had not materialized in spite of his expertise in combinatorics [Leibniz, 1666]. Curiously enough, Leibniz seems to have anticipated Hilbert

> *I have more than once insisted on the necessity for a new kind of logic that would treat thedegrees of probability... I wish a skillful mathematician would undertake a comprehensive and systematic reasoned work on all sorts of games, something which would be of great utility for perfecting the art of discovery, since the human mind displays itself better in games than in the most serious matter.*

[Leibniz, 1703–1705]

Note in passing that the tongue-in-cheek qualifier on the human mind's attraction for games of chance was to disappear in Hilbert's version; see Chap. 5.

This systematic synthesis, while started by Jacob Bernoulli, as Baker suggests, was to become the central panel of the classical probability triptych, even though only after the **third panel** was deployed. Indeed, as *limits* were beginning to be considered seriously (law of large numbers, or long-run relative frequencies), as well as for more technical reasons (some rather delicate estimates had to be controlled) a then quite novel mathematical alley was opening to the development of probabilistic analysis: *differential and integral calculus*. One of the prominent practitioners of the new discipline was Abraham de Moivre [1667–1754]. A Huguenot from Champagne (France), compelled to emigrate in England by the Revocation of the Edit de Nantes (1685), he studied avidly – s.v. *Vitry* in [Diderot and d'Alembert, 1751–1780, vol. XVII, p. 363] – the just published Newton's *Principia* (1687); and he became so proficient in the new calculus that he gained the friendship of Newton himself, who is reported [Todhunter, 1865] saying *"Go to Mr. de Moivre, he knows these things better than I do."* Elected a member of the Royal Society in 1697, de Moivre made diverse contributions to several areas of mathematics, among which his *Doctrine of chance* [De Moivre, 1738], first published in 1718 – five years after [Bernoulli, 1713] *Ars conjectandi*. There is also a third edition (1756) which, although posthumous, was made following de Moivre's own revisions. We chose to refer throughout to the second edition because it publishes *for the first time,* and with only a few additions, the material printed in [De Moivre, 1733], a 7-page note, only privately distributed; this contains the first known derivation of the normal distribution which de Moivre obtains as a limit of the binomial distribution; and for which he offers his own version of the Stirling formula (see below). The original printing had been lost and was rediscovered by Pearson [1924]; see also for a bibliographical precis [Daw and Pearson, 1973], and for an article that could have been entitled 'Bernoulli and de Moivre compared' [Pearson, 1925].

Let us now see what the new calculus could do to improve on Bernoulli's theorem (4.3.12). Recall that the function $p_n(.)$ (see 4.2.14) is very peaked around a "central" term occuring for $k = m$ with $m \approx np$. The key to the improvement is to show that, when n is very large, $p_n(.)$ can be approximated by:

$$p_n(m + k) \approx \frac{1}{\sqrt{2\pi npq}} \exp\{-\frac{1}{2npq} k^2\} \quad . \tag{4.3.13}$$

This is done in two steps. First, one evaluates the value of the function at its maximum:

$$p_n(m) \approx \frac{1}{\sqrt{2\pi npq}} \quad ; \tag{4.3.14}$$

and second, one argues that

$$p_n(m + k) \approx p_n(m) \exp\{-\frac{1}{2npq} k^2\} \quad . \tag{4.3.15}$$

Each of these results is remarkable in itself, and for each the derivation has its own features which we want to indicate.

The first result is an immediate consequence of the so-called Stirling formula, which de Moivre discovered independently, and perhaps even first; this formula gives an asymptotic expansion for $n!$ as n becomes very large:

$$n! \approx n^n e^{-n} \sqrt{2\pi n} \tag{4.3.16}$$

by which one means

$$n! = n^n e^{-n} \sqrt{2\pi n} + o(n!) \quad \text{i.e.} \quad \lim_{n \to \infty} \frac{1}{n!} n^n e^{-n} \sqrt{2\pi n} = 1 \quad .$$

This formula does *not belong to combinatorics* anymore. In its original derivation, it is a remarkably astute application of differential and integral calculus – involving playing back and forth with the exponential function and its inverse, the logarithmic function – to show, firstly, that as n becomes very large, $[\ln(n!) - (n + \frac{1}{2}) \ln n + n]$ approaches a constant, and secondly, that this constant happens to be $\ln \sqrt{2\pi}$, i.e.

$$\lim_{n \to \infty} [\ln(n!) - n \ln n + n - \frac{1}{2} \ln(2\pi n)] = 0$$

of which the Stirling formula (4.3.14) is the exponentiation. The factor $\sqrt{2\pi}$ makes the normal distribution – in the RHS of (4.3.24) – a true probability distribution, i.e. one that is properly normalized to have total unit weight; see (4.3.25–26). Yet, rather than a lucky accident, the Stirling formula is a particular case of a more general – and also more transparent – asymptotic formula, as we want to indicate briefly here, since we also make repeated use of this material elsewhere.

Scholium 4.3.1. *(Laplace): Let $a > 0$, $f : t \in [0, a) \mapsto f(t) \in$ R be twice continuously differentiable, with $f(0) = f'(0) = 0$, $f''(0) > 0$; and let*

$$I(x) = \int_o^a dt \, e^{-x f(t)} .$$

Then

$$I(x) \approx \left(\frac{\pi}{2x f''(0)} \right)^{\frac{1}{2}} \quad \text{as} \quad x \to \infty \quad .$$

To obtain the central idea of both the statement of this scholium and its proof, consider the function $f(t) = bt^2/2$ with $b > 0$. Let Γ be the function defined on $(0, \infty)$ by

$$\Gamma(x) = \int_o^\infty dt \, e^{-t} t^{x-1}$$

which satisfies $\Gamma(x + 1) = x\Gamma(x)$ and $\Gamma(1) = 1$: for all positive integers n

$$n! = \Gamma(n + 1) \quad .$$

Now, simple changes of variables give:

$$\left[x^{x+\frac{1}{2}} e^{-x} \right]^{-1} \Gamma(x + 1) = x^{\frac{1}{2}} \left[I_-(x) + I_+(x) \right]$$

where

$$I_-(x) = \int_o^1 dt \, e^{-x f_-(t)} \quad \text{and} \quad I_+(x) = \int_o^\infty dt \, e^{-x f_+(t)}$$

with

$$f_\pm(t) = \pm t - \ln(1 \pm t) \ ,$$

both of which satisfy the conditions of the lemma. Hence

$$\left[x^{x+\frac{1}{2}} e^{-x} \right]^{-1} \Gamma(x + 1) \approx x^{\frac{1}{2}} 2 \left(\frac{\pi}{2x} \right)^{\frac{1}{2}} = \sqrt{2\pi}$$

or equivalently

$$\Gamma(x + 1) \approx x^x e^{-x} \sqrt{2\pi x} \tag{4.3.17}$$

thus proving the Stirling formula (4.3.16) (in fact, for all positive real x). Upon writing (4.2.14) for $m = np$, we have:

$$p_n(m) = \frac{\Gamma(n+1)}{\Gamma(np+1)\Gamma(nq+1)}p^{np}q^{nq}$$

which (4.3.17) reduces indeed to (4.3.14).

To give a simple idea of where (4.3.15) comes from, we consider the particular case where $p = \frac{1}{2} = q$ with n even, i.e. m an integer. We have:

$$p_{2m}(m+k) = \frac{(2m)!}{(m+k)!(m-k)!}2^{-2m} = p_{2m}(m)\frac{m!}{(m+k)!}\frac{m!}{(m-k)!}$$

$$= p_{2m}(m)\prod_{l=1}^{k}\left[\frac{m-(l-1)}{m+l}\right] = p_{2m}(m)\prod_{l=1}^{k}\left[\frac{1-(l-1)/m}{1+l/m}\right] \quad .$$

Notice that, at fixed k, and when n becomes very large, $(l-1)/m$ becomes very small and we can approximate $1 - (l-1)/m$ by $\exp\{(l-1)/m\}$ and similarly for the term $1 - l/m$; hence the above expression can be rewritten as

$$p_{2m}(m+k) = p_{2m}(m)\exp\{-\sum_{l=1}^{k}\frac{2l-1}{m}\} = p_{2m}(m)\exp\{-\frac{1}{m}k^2\} \quad .$$

In the symmetric case $p = \frac{1}{2} = q$ and n even, this proves (4.3.15) – thus (4.3.13) since we already have proven (4.3.14). The same idea can be adapted to the general case, at the cost of a somewhat heavier notation which does not throw any further light on the basic mechanism by which, for large n the binomial distribution (4.2.14) reduces to the distribution (4.3.13).

This result is all the more remarkable when one considers that calculus was still a fledgling discipline in the early 1730s – as we have seen, Newton knew de Moivre – so that our early probabilists were indeed well-informed, up-to-date and competent mathematicians. When he discovered (4.3.13) de Moivre fully realized that it represents a very significant improvement over the Bernoulli theorem; today, we know the RHS of (4.3.13) as the universal *normal distribution*; see also (1.3.24). de Moivre went further, and speculated that he had established the necessary link between probable assessments (or *a priori* probabilities) and relative frequencies:

> *although chance produces irregularities, still the odds will be infinitely great that in the process of time* [i.e. in the limit of large values of n] *those irregularities will bear no proportion to the recurrence of that order which naturally results from original design.*
>
> [De Moivre, 1738]

Even if this still sounds as the dream of a visionary, one must recognize that de Moivre had begun raising the mathematical scaffolding that would be necessary to support the central panel of the classical probability triptych: the interpretation of the theory in observational terms, specifically the purported resolution of the duality between the two approaches to probability theory (probable certainty, and observed frequencies). We said that this was only the *beginning* of the scaffolding, and we have two reasons for saying so at this point. Firstly, while de Moivre was proceeding from the probable assessments to the limiting relative frequencies, the return path still had to be cleared, namely to go from observed frequencies to the (updating of the) probability of causes. For this task, Bayes was to receive some deserved credit and much censure (some of the latter undeserved), as we shall see below. Our second reason for seeing de Moivre as a pioneer, rather than a settler, is that the latter role was to be fulfilled by the immensely resourceful Laplace, who brought de Moivre's and Bayes' results to more transparent and symmetric forms; see (4.3.24) and (4.3.43) below.

Lest the Reader be worried that the approximations involved in (4.3.13) be not sufficiently controlled, we can rewrite this as

$$p_n(m+k) = \frac{1}{\sqrt{2\pi npq}} \exp\{-\frac{1}{2npq} k^2\} \, \eta_n(k) \tag{4.3.18}$$

as an upper bound can be found – see e.g. [Feller, 1968, 1971] – for the error term:

$$\varrho_n(k) = |\ln \eta_n(k)| < \frac{k^2}{npq} + \frac{2k}{npq} \quad ; \tag{4.3.19}$$

consequently, as $n \to \infty$ and k is constrained to an interval $k < K_n$ where $K_n{}^3/n^2 \to 0$, we have that there exists, for every $\varepsilon > 0$, a N_ε such that $n > N_\varepsilon$ implies

$$1 - \varepsilon \le p_n(m+k) \left[\frac{1}{\sqrt{2\pi npq}} \cdot \exp\left\{-\frac{1}{2npq} k^2\right\} \right]^{-1} \le 1 + \varepsilon \tag{4.3.20}$$

so that the approximation is *uniform in k*. Hence, one can use (4.3.20) to estimate the degree of confidence for the number of occurrences of "head" to fall between α and β; namely:

$$P_n\left\{\alpha \le \sum_{k=1}^{n} x_k \le \beta\right\} = \sum_{\nu=\alpha}^{\beta} p_n(\nu) \tag{4.3.21}$$

can be approximated by

$$\frac{1}{\sqrt{2\pi npq}} \sum_{k=\alpha-m}^{\beta-m} \exp\left\{-\frac{1}{2npq} k^2\right\} \quad . \tag{4.3.22}$$

Upon using the definition of the integral as a limit of sums, and the notations

$$\sigma_n = \sqrt{npq} \quad \text{and} \quad \nu_n = \frac{1}{n}\sum_{k=1}^{n} x_k \quad , \tag{4.3.23}$$

we finally obtain the following asymptotic relation between the probable assessment p and the distribution of relative frequencies ν_n:

$$\lim_{n\to\infty} P_n\{p + a\frac{\sigma_n}{n} \leq \nu_n \leq p + b\frac{\sigma_n}{n}\} = \frac{1}{\sqrt{2\pi}} \int_a^b du \exp\{-\frac{1}{2}u^2\} \qquad (4.3.24)$$

which is known as **de Moivre–Laplace's theorem** or *de Moivre–Laplace's version of Bernoulli's (weak) law of large numbers.*

Before we discuss the meaning of (4.3.24), notice that we can now interpret the factor $\sqrt{2\pi}$ appearing in the limit (4.3.15): it ensures the normalization $N = 1$ in

$$N = \frac{1}{\sqrt{2\pi}} \int_{-\infty}^{+\infty} du \, \exp\{-\frac{1}{2}u^2\} \qquad . \qquad (4.3.25)$$

Indeed, upon changing to polar coordinates, we see that

$$N^2 = \frac{1}{2\pi} \iint_{-\infty}^{+\infty} du \, dv \, \exp\{-\frac{1}{2}(u^2 + v^2)\}$$
$$= \frac{1}{2\pi} \int_0^{2\pi} d\theta \int_0^{+\infty} dr \, r \exp\{-\frac{1}{2}r^2\} = 1 \qquad . \qquad (4.3.26)$$

The de Moivre–Laplace's theorem (4.3.24) involves no less than three quantities which could – *each one of them !* – lay some claim to be called a "probability". There is, firstly, the *a priori* "probability" (or probable assessment) p, determined before the trials have begun; this is an assessment *from original design*, to borrow de Moivre's own words. Secondly, there is the *a posteriori* "probability" (or relative frequency) ν_n, i.e. the average number of favorable outcomes computed as the ratio of the number of favorable outcomes in a string of trials, and the number n of trials. And, thirdly, there is the "probability" P_n (or degree of confidence) that ν_n be within certain "acceptable" bounds:

$$p + a\sqrt{\frac{pq}{n}} \leq \nu_n \leq p + b\sqrt{\frac{pq}{n}} \qquad . \qquad (4.3.27)$$

From the reasoning that lead to (4.3.24), P_n is precisely (i.e. *this and no more*) the ratio of the number N_n^+ of strings that satisfy condition (4.3.27), and the total number N_n of strings of length n (here $N_n = 2^n$ since we decided, for didactic reasons, to limit our attention to throws of a coin; but the reasoning can trivially be extended to the throw of one or several dices, or any other sequence of independent trials, each with the same, finite, number of possible outcomes.) This again shows that the concept of probability is still somewhat ambiguous, covering, as it seems to do, a variety of statements referring to different circumstances, namely: making an assessment of probability (p), observing a relative frequency (ν_n), and computing the degree of confidence (P_n), of a prediction that ν_n approaches p within a prescribed margin of error.

In light of the above quote from de Moivre – see the quote before (4.3.18) – the classical probabilists' understanding of probability seems to go as follows. Given the *a priori* probability, p, of a type of events, if chance had no hand in nature, perfect regularity would prevail and the frequencies of any sequence of independent occurrences of the same events should match p, or at least be very close to it. However, chance is present and for short sequences, the frequencies fluctuate and appear irregular. Fortunately, orders and regularities return in sequences whose lengths becomes very large, and the approaching of the relative frequencies to p can be precisely estimated by (4.3.24). But first, what exactly is the nature of the *a priori* probability? And second, what is the relation between it and *chance*, whose existence seems to jeopardize regularity? While Laplace's ignorance interpretation of p gives a consistent answer to the first question, it does not even begin to address the second.

While the above is a conceptual difficulty, there is yet another difficulty, now in the operational meaning to be given to (4.3.24). This has to do with the determination of its domain of usefulness: what should we make of a string of trials that has resulted in a relative frequency ν_n way-out of the bounds we had deemed "acceptable" ? Two attitudes seem to be possible. *Either* we can claim that we know for sure the value of our probable assignment p: we had examined the physical properties of the coin, and determined that they show the proper relation (or symmetry, in case $p = 1/2$) between head and tail. We would then be tempted simply to ignore this string of trials as somehow irrelevant, uncharacteristic or abnormal, a fluke of some sort. *Or* we may begin entertaining doubts about the thoroughness of our preliminary examination of the coin, or maybe of the persistence of their conditions through, or relative to, the string of trials just performed. Instead of dismissing this string as merely accidental, should we not take it seriously, and try to infer, from the observed frequencies ν_n , some information to re-evaluate p? If we were to choose the latter strategy, how reliable then would this re-evaluation be?

This is the germ of what is sometimes referred to as the *inverse problem*, a germ that is already discernible in Bernoulli and de Moivre, although the emphasis on the importance of the question, and the discussion of its solution are usually and justifiably attributed first to Bayes, and then to Laplace; see, e.g. [Dale, 1988]. In fact, this problem pertain to the philosophical line of inquiry we had mentioned earlier, namely the justification of inductive reasoning; in line with [Locke, 1690] constructive skepticism or [Hume, 1740] defense of probable reasoning, we have no choice: we *must* make some kind of probabilistic inference. In 1763 Richard Price [1723–1791] communicated to the Royal Society, with his own commentary, some notes [Bayes, 1763] left by "the Rev. Thomas Bayes, FRS" [1702–1761]; Bayes was a Presbyterian minister, and his election to the Royal Society in 1742 may have had much to do with his being the generally recognized author of a famous, if anonymous, tract defending Newton's fluxions against the attacks of Bishop Berkeley. The title of the work that interests us here *"An essay towards solving a problem in the doctrine of chances"* is a rather transparent reference to [De Moivre,

1738]. The subject of Bayes' paper belongs to what we would call today the doctrine of *statistical inference*, which de Moivre had mentioned as the *hardest problem that can be proposed on the subject of chance*. Bayes' paper leads to two somewhat different results, now known respectively as the "Bayes' rule" (4.3.31), and the "Bayes' theorem" (4.3.38). *Both have drawn thunder.* It seems therefore fair to first inquire of the avowed aim of the work. In his philosophically informed preface, Price writes of Bayes:

> *In an introduction he has written to this Essay, he says, that his design ... was to find out a method by which we may judge concerning the probability that an event has to happen, in given circumstances, upon supposition that we know nothing concerning it but that, under the same circumstances, it has happened a certain number of times, and failed a certain number of times ... to estimate the chance that the probability for the happening of an event perfectly unknown, should lie between any two named degrees of probability, antecedently to any experiments made about it; and that it appeared to him that the rule must be to suppose the chance the same that it should lie between any two equidifferent degrees; which, if it were allowed, all the rest might be easily calculated in the common method of proceeding in the doctrine of chance ... [T]he problem now mentioned is by no means merely a curious speculation in the doctrine of chances, but necessary to be solved in order to [provide] a sure foundation for all our reasoning concerning past facts, and what is likely to be hereafter ... [W]e cannot determine, at least not to any nicety, in what degree repeated experiments confirm a conclusion, without the particular discussion of the aforementioned problem; which, therefore, is necessary to be considered by any one who would give a clear account of the strength of analogical or inductive reasoning.*

Bayes is more concise and mathematically precise:

> *PROBLEM:* Given *the number of times in which an unknown event has happened and failed.* Required *the chance that the probability of its happening in a single trial lies somewhere between any two degrees of probability that can be named.*

Bayes' paper provoked many controversies; one of them concerns the use (and misuses) that originated in the concept of conditional probability, which Bayes himself put into a precise formulation, and uses correctly. The *conditional probability* $P(A|B)$ denotes the probability that an event A occurs, given that an event B has occurred; i.e. the joint event $A \cap B$ occurs with probability $P(A \cap B)$ satisfying

$$P(A \cap B) = P(A|B)\, P(B) \quad . \tag{4.3.28}$$

In particular, since $A \cap B = B \cap A$,

$$P(B|A)\, P(A) = P(A|B)\, P(B) \quad . \tag{4.3.29}$$

Let now $\{A_i | i = 1, 2, \ldots, n\}$ be an exhaustive collection of mutually exclusive events (i.e. exactly one of the events A_i, must occur exactly once in every trial – e.g. "head" or "tail" in the throwing of a coin) and B be any event; we have then:

$$P(B) = \sum_{j=1}^{n} P(B|A_j) P(A_j) \tag{4.3.30}$$

and thus (3.29) and (3.30) imply:

$$P(A_i|B) = \frac{P(B|A_i) P(A_i)}{\sum_{j=1}^{n} P(B|A_j) P(A_j)} \tag{4.3.31}$$

When one interprets the A_i's as the probable causes for the event B, the equality (4.3.31) is known as **Bayes' rule for the probability of causes**. As pointed out in [Dale, 1982], this rule appears nowhere in Bayes' essay, but was recognized by Laplace; moreover, Bayes favors continuous rather than discrete distributions, which seem to have been introduced in this context for didactic purposes by Laplace.

Still another twist is given to the interpretation of (4.3.31) when one wishes to consider it as a probabilistic statement between claims (or conclusions) drawn from available evidences.

We now briefly turn to the concept of *statistical independence*. It is intimately related to the notion of conditional probability, and appears with this connection explicitly in the work of Bayes, although it had been used more or less tacitly before. This concept was destined to play a strong motivating role in the 20th-century resurgence of probability theory *as a mathematical discipline;* for an application that could not be suspected in the 18th century, see in Sect. 5.3: the Erdös–Kac theorem on the number of prime divisors of a large integer.

Two events are said to be *statistically independent* whenever any (and thus all) of the following equivalent conditions is satisfied:

$$P(A|B) = P(A) \; ; \; P(A \cap B) = P(A) P(B) \; ; \; P(B|A) = P(B). \tag{4.3.32}$$

This formalizes a notion we already used several times, for instance when we said that we were considering a string $x = \{x_1, \ldots, x_n\}$ of n "independent" throws of a coin. It is explicitly on that basis – rather than simply carrying on a tacit or *ad hoc* custom – that Bayes accepted and used Bernoulli's probability distribution (4.2.14). Bayes' own discussion emphasizes indeed that the successive trials are to be *statistically independent* – see (4.3.32) – for the product law to hold; in Bayes words: *"The probability that several independent events shall all happen is a ratio compounded of the probabilities of each."* In particular, we have

$$\varrho_n(k) = p^k q^{n-k} \tag{4.3.33}$$

for the probability for a string which shows, in a specified order, exactly k successes and $(n - k)$ failures; this depends evidently only on n and k but not on the order in which these outcomes appear. Hence, when the order is deemed irrelevant, $\varrho_n(k)$ should be multiplied by the number of strings of length n that show exactly k successes and $(n-k)$ failures. This is the probabilistic part of the argument; it is then completed by the usual combinatorial result namely that there are $b_n(k)$ such strings. This gives indeed (4.2.14).

Another famous – or sometimes infamous – use of the concept of conditional probability is in *Laplace's law of succession;* according to Keynes [1921]:

> No other formula in the alchemy of logic has exerted more astonishing powers. It has established the existence of God from the basis of total ignorance and it has measured precisely the probability that the sun will rise tomorrow.

Indeed, stated in the vernacular, this law suggests that you can bet with odds 5000×365.25 to 1 that the sun will rise tomorrow knowing that it has done so for the last 5000 years ... although, as Feller points out tongue-in-cheek, the certainty that the sun rose on Feb.5 3123 BC is not really better than the certainty that it will do so tomorrow ! Moreover, as the following derivation shows, the application of the law to the case at hand would require an averaging over a large number of universes, all identically prepared.

These criticisms not withstanding, let us consider, in a manner typical of 18th-century probability discussions, an urn-and-ball model of the situation. Given: a collection of $(N+1)$ urns, numbered $1, 2, \ldots, N+1$; each urn containing exactly N balls, with urn number k containing k red balls and $(N-k)$ white balls. An urn is chosen at random. From this specific urn, n drawings are made, the color the ball drawn at each drawing is noted, and the ball is replaced in its urn immediately after each of the drawings. Let A_n be the event in which the n successive drawings result together in a total of exactly n red balls; A_{n+1} is similarly defined, i.e it is the event in which the next drawing results also in a red ball. We ask for $P(A_{n+1}|A_n)$.

Since the probability for drawing exactly n red balls from the kth urn is $(k/N)^n$, we obtain, upon averaging over the $N+1$ urns :

$$P(A_n) = \frac{1}{N+1} \sum_{k=0}^{N} (\frac{k}{N})^n \quad . \tag{4.3.34}$$

Note that, for each of the independent drawings *separately,* the odds are even that a red ball be drawn:

$$P(A_1) = \frac{1}{2} \quad . \tag{4.3.35}$$

For an arbitrary number n of independent drawings, and in the presence of a large number $(N+1)$ of urns, the sum can be approximated by

$$\frac{1}{N^n} \int_0^N dk\, k^n = N\frac{1}{n+1}$$

so that (4.3.34) can be approximated by

$$P(A_n) \approx \frac{N}{N+1} \cdot \frac{1}{n+1} \qquad (4.3.36)$$

Hence

$$P(A_{n+1}|A_n) = \frac{P(A_{n+1})}{P(A_n)} \approx 1 - \varepsilon_n \quad \text{with} \quad \varepsilon_n = \frac{1}{n+2} \qquad (4.3.37)$$

which indeed becomes very close to 1 as the number n of drawings become very large. This should be compared to (4.3.35).

The odds were similarly evaluated in the case of the raising sun metaphor discussed above, and we saw how they provide a healthy warning against the merits of any "plausible reasoning" that would take too seriously the explanatory power of Laplace's law of succession, a close cousin of Bayes' rule, itself the subject of much abuse.

The collection of significant samples is a delicate art that can hardly accommodate the brutal use of a sledge hammer, however reliable the latter could be by criteria foreign to the situation at hand: a theorem has assumptions and these must be respected.

Let us now turn to Bayes's theorem. The question is to evaluate the degree of certainty p from the observed frequency ν_n. Specifically, if in a sequence of n throws of a coin, we observed k heads, what can we infer on p? **Bayes' theorem** is said to state that – at least for large n:

$$P_n\{\alpha \le p \le \beta \mid k\} = \frac{(n+1)!}{k!\,(n-k)!} \int_\alpha^\beta \mathrm{d}p\, p^k\, q^{n-k} \qquad (4.3.38)$$

is the degree of confidence that p is between α and β, n and k being given.

To understand what enters in that claim, we follow Laplace's approach – which is more general than what Bayes actually wrote – so that we are better able to finger a restrictive assumption that seems to be tacit in Bayes' paper; in defense of Bayes, we hasten to say that he does take care to emphasize its justification *for the specific problem he is considering*. In general, the retrodiction (or postdiction, as opposed to prediction) should now take the form:

$$P_n\{\alpha \le p \le \beta \mid k\} = \left[\int_0^1 P_n(k|p)\,\mathrm{d}\mu(p) \right]^{-1} \int_\alpha^\beta P_n(k|p)\,\mathrm{d}\mu(p) \qquad (4.3.39)$$

where $\mathrm{d}\mu(p) = M(p)\mathrm{d}p$ corresponds to the probability density $M(p)$ of the degree of certainty p; and $P_n(k|p)$ is the conditional probability, given a putative probable assessment p, that k favorable outcomes occur in n trials, i.e. $P_n(k|p) = p_n(k)$ with $p_n(k)$ as in Bernoulli's formula (4.2.14) which – as we saw above – Bayes rederived with an emphasis on the statistical independence of the successive trials. So far, (4.3.39) appears as a continuous analog to (4.3.31).

Now comes the restrictive assumption tacitly made in the printed version of Bayes' work. In the particular case where we profess *no* bias on how p is distributed, i.e. when we restrict ourselves to the uniform distribution:

$$M(p) \equiv \frac{d\mu(p)}{dp} = 1 \quad \forall \quad p \in [0,1] \quad , \tag{4.3.40}$$

(4.3.39) becomes

$$P_n\{\alpha \le p \le \beta \mid k\} =$$
$$\left[\int_0^1 dp \; b(n,k) \, p^k \, q^{n-k}\right]^{-1} \int_\alpha^\beta dp \; b(n,k) \, p^k \, q^{n-k} \tag{4.3.41}$$

since the coefficients $b(n,k)$ are constant in p, they can be taken out of the integrals and thus cancel each other; finally, the denominator, which we know as a classical Euler integral, is evaluated by recursive use of integration-by-parts:

$$\int_0^1 dp \, p^k \, q^{n-k} = \frac{k! \, (n-k)!}{(n+1)!} \quad . \tag{4.3.42}$$

Hence (4.3.41) reduces indeed to Bayes's theorem, in the form given in (4.3.38).

From a purely mathematical standpoint, therefore, Bayes' theorem is a rather innocent compound of Bernoulli's combinatorics (more elementary in fact than what enters in the Bernoulli theorem itself) and Eulerian integral calculus. Moreover, from the point of view of probabilistic analysis, recall Bayes' explicit emphasis on the role of statistical independence, here (4.3.32–33), in the derivation of (4.2.14).

Yet, Bayes' result has raised much controversy – some of which rather ill-tempered, which we therefore do not cite – even into the 20th century; see e.g., for the statistician's account: [Fisher, 1956], and most recently the commemorative papers [Berger, Boukai, and Wang, 1997]; and for the philosopher's account: [Earman, 1996].

The primary reason seems to be the interpretation and justification of the restrictive assumption (4.3.40). Why should one take a uniform distribution? One can argue [Fisher, 1956] that Bayes himself may have been so uncomfortable with it that he could not resolve to send the paper for publication (it is indeed posthumous). What appears to have made Bayes uncomfortable about his assumption of the *a priori* uniform distribution (4.3.40) might not be the worry that it is not true (since it seems justifiable in many particular cases), but that it brings in his argument a compromise towards non-statistical reasoning. Bayes knew of de Moivre's announcement that he had an "inverse" to Bernoulli's theorem and it may well be that the motivation for his own theorem was his dissatisfaction with the argument in [De Moivre, 1738, *Remark II*, pp. 251–254] which is entirely *a priori* and largely unmathematical as it invokes the permanence of natural laws that govern the proportions (or ratios) of true causes. In contrast, Bayes' argument appears thoroughly mathematical and statistical, except for the part in *the* scholium (in Prop. 9) which assumes uniform *a priori* distribution of prior probabilities.

What is certainly clear, from the text we have (albeit it went through a reworking by Price), is that the author(s) felt obligated not to present the argument in its general form, but to couch it in terms of a specific model where the restrictive assumption seems unassailable. We only want to remark here that in this respect, Bayes was proposing a continuous version of something which had been in use since Pascal and Fermat, or Leibniz, and became the tacit *modus vivandi* in 18th-century probability, namely the *principle of indifference* or *principle of insufficient reason*. For instance, this principle is clearly undergirding the entry *Probabilité* in the *Encyclopédie*:

> this principle is used when one considers that the various cases are equally possible. And this is indeed only as an assumption relative to our limited knowledge that we say, for instance, that all the points [= faces] of a die are equally likely to come out; it is not that when they roll out of the cup, the points that must come out have not already the disposition which – combined with that of the cup, the tablecloth, the strength and manner with which the die is thrown – will make them surely arrive, but that these being entirely unknown to us, we have no reason to prefer one point to another; we will therefore assume them all equally easy to produce.
> [Diderot and d'Alembert, 1751–1780, Vol. XIII pp. 393–400],

We may prefer today to see this as a symmetry argument. The Encyclope-dist proceeds then to discuss an example akin to the one presented in the beginning of Sect. 4.2, and offered by Galileo, namely that the computation of probabilities must take into account that some events are conjunctions of more elementary events which are the ones that are equally probable. This was given a concise and explicit form by Laplace

> The theory of probability consists in reducing all events that could happen in a given situation, to a number of equally possible [read "probable"] cases, i.e. such that we would be equally undecided on their existence, [read "occurrence"] and to determine among these cases, the number of those that are favorable to the event the probability of which we want.
> [Laplace, 1820]

It is therefore not surprising that Bayes' theorem, with its uniform distribution, obtained Laplace's approval. Nevertheless, as a consequence of the fact that $p_n(.)$ becomes very peaked as n becomes very large, one may suspect that *in that limit* the a priori distribution (4.3.40) should become irrelevant – provided that μ is restricted to be a sufficiently smooth function. Laplace indeed derived for $P\{\alpha \leq p \leq \beta\}$ an expression similar to the de Moivre–Laplace theorem (4.3.24), namely, with the value $r = k/n$ given by the *observed* frequency:

$$\lim_{n \to \infty} P_n\{\nu_n + a\frac{\tau_n}{n} \leq p \leq \nu_n + b\frac{\tau_n}{n}\} = \frac{1}{\sqrt{2\pi}} \int_a^b dv \exp\{-\frac{1}{2}v^2\} \quad (4.3.43)$$

where we used the following notation, to complete the analogy with (4.3.23) in (4.3.24):

$$\tau_n = \sqrt{nr(1-r)} \qquad \text{and} \qquad \nu_n = \frac{1}{n}\sum_{j=1}^{n} x_j = r \qquad (4.3.44)$$

(4.3.43) is referred to as the **Bayes–Laplace theorem.** This theorem could be discussed almost *verbatim* along the lines followed for the de Moivre–Laplace form (4.3.24) of the law of large numbers. These two theorems allow to compare the assessment of probability (p) made on the fairness of the coin *before* a string of trials has begun, and the observed frequency (ν_n) registered *after* the string of trials has been performed; both comparisons allow only a certain *degree of confidence* (P_n); neither pretends to decide whether a significant discrepancy is due to the coin being intrinsically biased, or to the string of trials being unfairly rigged. It is eventually the primacy we choose to give to one of the determinations $(p$ or $\nu_n)$ over the other that calls for the use of one of the theorems rather than the other. In the practice of quality control, both theorems ought to be used: firstly, one calibrates the testing device with reliable products; then, and then only, does one use the testing device to pass or reject questionable products.

One may also note that [De Finetti, 1972] criticized and reconstructed the Bayes–Laplace theorem based on his subjectivist interpretation of probability; for an update, see [Regazzini, 1997]. We discuss de Finetti's approach in Chap. 6.

Evidently, the usefulness of both the de Moivre–Laplace theorem and the Bayes–Laplace theorem can be severely limited by the finiteness of the size of the sample. Hence the necessity to develop a craft of statistical inference to deal with practical situations that cannot afford unlimited luxuries. The Weldon data – summarized in Table 4.2 – provide an illustration where the large size of the sample drawn in an actual experiment makes it likely that some useful information obtains from the Bayes–Laplace theorem: 315,672 individual throws of an individual die, resulting in a relative frequency $\nu_n = .337\,699$ for the event consisting of *either* 5 *or* 6 showing up. At sake is to evaluate whether ν_n is *significantly* different from the outcome one would have predicted from a fair (or ideal) die, namely $p_o = .333\,333$. For this purpose, let us use, as a demarking line the average $\varrho \equiv (\nu_n + p_o)/2$. The Bayes–Laplace theorem gives

$$P\{\varrho \leq p \leq 1\} = .995 \qquad \text{and} \qquad P\{0 \leq p \leq \varrho\} = .005 \qquad (4.3.45)$$

so that the odds for the actual die to have $p \geq \varrho$ rather than $p \leq \varrho$ – which is the case for p_o – are 200 : 1. The Bayes–Laplace statistician accepts this as a strong evidence that the real die was indeed loaded. Incidentally, the same computation shows that the odds are about 200 : 1 for $.3353 \leq p \leq .3401$. If instead of throwing the die $n = 315,672$ times, the same relative frequency

$\nu_n = .3377$ had been already obtained for a smaller number n of trials, conventional wisdom has it that the corresponding probability P_n – i.e. the degree of confidence that p be still in the same limits $.3353 \leq p \leq .3401$ – would decrease with n; we indicate in Table 4.3 how well the Bayes–Laplace theorem supports this presumption.

Table 4.3. A Bayes–Laplace confidence evaluation

n	$3 \cdot 10^5$	10^5	$3 \cdot 10^4$	10^4	$3 \cdot 10^3$	10^3	$3 \cdot 10^2$	10^2
P_n	.995	.89	.62	.39	.22	.12	.06	.03

Remarkably enough, it so happens that odds of $200 : 1$ is a close approximation of the odds $[212 : 1$ corresponding to $-\frac{a}{\sqrt{2}} = \frac{b}{\sqrt{2}} = 2.00$ and thus $P_n = .99532]$ which Poisson did consider to be acceptable – so that the probability P_n *"differs almost imperceptibly from certitude"* – in matters of concern for his treatise [Poisson, 1837], namely judicial and/or medical practices; hence the total number of trials $n \approx 3 \cdot 10^5$ used for the Weldon data is not quite as arbitrary as it might have first appeared: it is essentially what is necessary to get the degree of confidence required by Poisson. For an analysis of the contributions to probability made by the remarkable polymath Poisson seems to have been, see [Sheynin, 1977/78], although a sceptic Reader may want to take in counterpoint [Daston, 1987].

4.4 From Here to Where?

To close this chapter on the emergence of classical probability, we weigh Laplace's consolidation of the work of de Moivre and Bayes. His asymptotic results (4.3.24) and (4.3.43) – the de Moivre–Laplace and Bayes–Laplace theorems – intimate some intrinsic linkage between the aspect of the degree of belief we isolated as probable assessment, namely p, and the relative frequency ν_n : provided the number of trials is large enough, one can safely offer a *prediction* of ν_n from the knowledge of p, or a *postdiction* of p from the observed value of ν_n ; note again the symmetry between the two formulas. This linkage seemed very attractive from a pragmatic point of view. Nonetheless, two kinds of objections appeared that originated from two directions: the demands for a deeper pragmatic analysis, emanating from the statisticians; and the renewed epistemological interest of the mathematicians; both of which matured, however, only in the 20th century.

In its time, classical probability theory seemed to be a discourse free from conflicting preconceptions. The world (in particular the physical world) was seen as deterministic, and the introduction of probability was made necessary only by the frailties of the human mind: (i) our incomplete knowledge of the initial conditions that enter in the determination of the *probable assessment* p; and (ii) our limitations on the conduct of the repeated trials needed to establish reliably the *observed* relative frequency ν_n.

The license to identify notions of different origins was used – with fruitful consequences – in the introduction of *expectation values, standard deviations and higher-order correlations.* These were conceived as averages over *ensembles;* for instance, in our archetypal string of coin-tossings, the expected number E_n of heads showing up in n trials is a (weighed) average over the ensemble of the 2^n binary strings of length n :

$$E_n \equiv \left\langle \sum_{k=1}^{n} x_k \right\rangle \equiv \sum_{k=1}^{n} k \, p_n(k) = np \qquad (4.4.1)$$

and the standard deviation is similarly written as: σ_n

$$\sigma_n{}^2 \equiv \left\langle \left(\sum_{k=1}^{n} x_k - \left\langle \sum_{j=1}^{n} x_j \right\rangle \right)^2 \right\rangle = \left\langle \sum_{k=1}^{n} x_k{}^2 \right\rangle - \left\langle \sum_{k=1}^{n} x_k \right\rangle^2 = npq \quad . \quad (4.4.2)$$

However, as we saw in our discussion of the Weldon data (Tables 4.2 and 4.3), a very large sample is necessary to warrant a sharp conclusion; pragmatic considerations, such as the size of the sample and the choice of sampling techniques, are issues that were poorly addressed in the main body of classical probability.

Assessments of probability and relative frequencies are objects of different natures. Fusing them together – i.e. confusing them – led to serious philosophical, and then mathematical, difficulties. We saw a benign case with Laplace's law of succession; Condorcet's voting paradox is another witness to the awkwardness inherent to the extension of probable reasoning to applications of classical probability calculus beyond the purview of elementary statistics. These and a host of other epistemological problems erupted in the early 20th century when distinctions started to be made cogently between the *syntax* and the *semantics* of probability.

The 19th century largely carried along – much within the conceptual mold cast by Laplace – the traditions of classical probability; the domain of its applications was certainly enlarged, in particular to encompass a quantitative approach to the social and moral sciences [Lacroix, 1816 and 1833, Poisson, 1837, Quetelet, 1835 and 1869]. Much more than the actuarial predictions of 18th-century probabilists, the systematic collection of statistical data did gain the patronage of governments attempting to measure and control socio-economic activities. The statisticians were recognized – and did cooperatively recognize themselves [Hacking, 1983] – as an important arm of the state bureaucracy. Yet, their official pursuits contributed little to the elaboration of a calculus which could be formulated *independently* of the specific interpretation appearing in applications.

While this flood of mathematical modeling did not hinder the survival of a fundamentally deterministic view of the world, it largely submerged it. The epistemological question became moot and/or irrelevant, as to whether the models modeled the real world or only our ignorance of its inner workings.

Well into the 19th century, the separate interpretation to be given to probable assessments (i.e. to individual probabilities, or "degree of certainty") and the results of extensive data collections (or "relative frequencies") remained largely unexamined, not merely unresolved. The field of medical statistics offers good illustrations of the semantic tensions that were felt under the pragmatic veneer. As we have just seen, these tensions could be largely ignored as long as the purpose of data collection was to inform public health policies such as the control of epidemics or the evaluation of the salubrity of urban collectives. For private health, however, the questions were more delicate, as they involved individuals, and thus required that attention be paid to the mentalities as well as to the mathematics. Already in the 17th and 18th centuries attempts had been made to establish – "arithmetically", if not yet statistically – a taxonomy of diseases and of the efficiency of specific medications and medical practices. This trend spread forth into the 19th century; see e.g. P. Louis [1787–1872], who is often called the founder of medical statistics in France: he introduced the practice of the 'control group' in the Paris hospitals. Yet these efforts were soon to meet with renewed criticisms from those who maintained that medicine, the *art* of healing individuals, was intrinsically *not* amenable to the general formulas of *science* (geometry and astronomy were often taken as references). As the debate was raging among physicians, some of the luminaries of the Polytechnique School of applied mathematics tried to step in and to sort out the uses or abuses involved in appealing to probability. Navier [1785–1836] recognized explicitly – see e.g. [Navier, 1835] – that the observations (i.e. the collection of data), as well as their analysis, had to be informed by medical knowledge, and he proposed that the role of mathematicians be to codify the methods of reasoning: mathematics should be used only to update *a priori* judgments on probable causes. Poisson [1837] warned that the very nature of the law of large numbers is to require *large* numbers! He gave specific illustrations to show explicitly the limits of the degree of confidence one could place in statistical data. Upon reading some of the classics, for instance the work of the clinician Louis [1829], the modern reader cannot but be surprised of how necessary was Poisson's admonition that one should go beyond a rhetorical appeal to mere averages.

Such readings are all the more amazing in view of the fact that the physicians and the mathematicians often were using the same words, but in such different meanings that they were in effect talking through each other. This might explain why even a graduate from Polytechnique, soon after to turn physician (and later government official), argued again Poisson's case on the *"degree of variability of averages"* with so little echoes that even his name – J. Gavarret [1809–1890] – is now largely forgotten although he published not only a treatise on medical statistics but also several books on the impact of the physical sciences (in particular electricity and thermodynamics) on medicine [Gavarret, 1840]; a notable exception is the recent monograph by Matthews [1995] who examines this debate in France, and then follows its avatars through Germany, England and the US, thus tracing the development of scientific medical statistics well into the 20th century.

In closing this section, we should mention a connnection between probability theory and the theory of errors. In the latter, the basic problem is to determine what should be the "best" value to assign to a quantity, repeated measurements of which have resulted in a string of numbers. This problem attracted the attention of several 18th century mathematicians. Among them, we want to mention [Simpson, 1757] whose probabilistic approach seems the closest to our present concerns; indeed, he may have been the first author to use explicitly an invariance argument, namely that the probability of a positive error is equal to the probability of a negative one; moreover, his presentation repeatedly appeals to analogies with coin tossing and dice throwing, evoking the binomial distribution.

Upon deciding that the "best" value should be the most probable, Gauss [1809] identified a set of properties – or axioms – that determines uniquely

$$\varphi(x) = \sqrt{\frac{1}{2\pi\sigma^2}} \; e^{-\frac{1}{2\sigma^2}(x-x_o)^2} \qquad\qquad (4.4.3)$$

as the probability distribution of the "errors", i.e. of the differences $(x_k - x_o)$ between each of the measured numbers x_k and the value x_o to be computed from them. For the case of n independent measurements carried out with the same precision – as measured by σ – he obtained then:

$$\Omega(x_1,\ldots,x_n) = (\frac{1}{2\pi\sigma^2})^{\frac{n}{2}} \; e^{-\sum_{k=1}^{n}\frac{1}{2\sigma^2}(x_k-x_o)^2} \quad . \qquad (4.4.4)$$

He finally noticed that

$$\{\, \Omega \quad \text{maximum} \,\} \qquad \text{iff} \qquad \{\, \sum_{k=1}^{n} \frac{1}{2\sigma^2}(x_k - x_o)^2 \quad \text{minimum} \,\} \quad , \qquad (4.4.5)$$

a condition which he interpreted as the claim that the most probable value x_o of the unknown is the one for which the sum of the squares of the differences from the observed values is the smallest, hence as a justification of the (simplest form of) the *least squares method*. Note moreover that if one has reasons to believe that the successive measurements are not done all with the same precision, this can be taken care of in (4.4.4) by replacing σ by σ_k, and thus obtain from this the weighted average:

$$x_o = (\sum_{k=1}^{n} \frac{1}{\sigma_k^2})^{-1} \sum_{k=1}^{n} \frac{1}{\sigma_k^2} \, x_k^2 \quad .$$

The so-called χ^2–method is obtained from similar premises. For didactic purposes we illustrate it here for a linear law $y = a + bx$. Suppose this law is submitted to an experimental test, the repeated performance of which gives a string of data $\{(x_n, y_n) \mid n = 1, \cdots, N\}$; each of the pairs (x_n, y_n) deviates from the linear law by an "error" $\varepsilon_n = y_n - a - bx_n$. The most probable values of the coefficients a and b compatible with the data collected obtain when $\mathcal{E} \equiv \sum_{n=1}^{N} \varepsilon_n^2$ is minimum, i.e. when $\partial_a \mathcal{E} = 0$ and $\partial_b \mathcal{E} = 0$. Upon substituting the definition of ε_n in these two conditions, we find in the system of "normal" equations

$$\left. \begin{array}{l} N a \quad + \left(\sum_n x_n\right) b \; - \; \left(\sum_n y_n\right) \quad = 0 \\ \left(\sum_n x_n\right) a + \left(\sum_n x_n^2\right) b \; - \; \left(\sum_n x_n y_n\right) = 0 \end{array} \right\} \qquad (4.4.6)$$

which can be solved uniquely for a and b, since $N\left(\sum_n x_n^2\right) - \left(\sum_n x_n\right)^2 > 0$ except for the trivial case when all the x_n are identical to one another. The χ^2–method generalizes naturally for laws that are more sophisticated than the linear law considered here; it ensures that the number of normal equations matches the number of parameters in the law that is tested.

The main idea of the original derivation of the *Gaussian distribution* (4.4.3) anticipated by a half-century that offered by Maxwell for the velocity distribution in gases; in particular, note the assumptions Gauss listed: the distribution being symmetrical, normalized to 1, differentiable, unimodal, and vanishing at infinity; complemented by the requirement of statistical independence, namely

$$\Omega(x_1, \ldots, x_n) = \prod_{k=1}^{n} \varphi(x_k - x_o) \quad .$$

He concluded then – compare with the argument presented in our Sect. 3.2:

$$\frac{1}{x} \frac{\mathrm{d} \ln \varphi}{\mathrm{d}x}(x) = k$$

where k must be a negative constant; from this (4.4.3) follows by integration and normalization.

Analytically, the Gaussian distribution is the same function as that obtained in the RHS of the de Moivre–Laplace and Bayes–Lapace theorems (4.3.24 and 4.3.43); however, it should be noted that, while the latter are asymptotic formulas valid under the assumption of a large number of measurements, Gauss' distribution – and later Maxwell's – is derived under different and more intrinsic assumptions. Nevertheless, once it is agreed, for whatever reasons, that the errors are so distributed, the mathematical form of the distribution can serve – as Laplace noticed also – as a justification for the least squares method in the theory of errors.

The name for the *least square method* was proposed in [Legendre, 1805] on pragmatic grounds that did not invoke probability considerations. Note that Simpson, Legendre and Gauss – even in the titles of the treatises in which they offered their version of the theory of errors – are motivated by the computational problems of position astronomy for the solar system. In fact a bitter priority battle was waged between Legendre and Gauss; Legendre clearly published first, namely in 1805; but Gauss, writing in 1809, claimed he had used the method as far back as 1795; it even appears [Hall, 1970, Reich, 1985] that entries in Gauss' diary show that in 1794 he had believed the method was due to Tobias Mayer who had written on errors in the 1750s. Gauss certainly could claim for himself the probabilistic derivation; and he subsequently kept refining his theory of errors [Gauss, 1823]. As for priority, the story of Ceres is worth considering.

On Jan. 1, 1801 the astronomer Piazzi in Palermo first sighted a very faint celestial object of the 8th order of magnitude. His observations extended over a span of 41 nights, of which some were unavailable due to unclement weather; they were fatally interrupted after February 11 as the faint object came too close to the halo of the Sun. These data covered too short a piece of the orbit – about 3° – to allow Piazzi to determine which of the conic sections it was, i.e. whether he had discovered a new planet or just seen another comet. Piazzi was then unable to locate the object again, and had to satisfy himself with the publication of his inconclusive raw data, first as a speculative announcement in June/July, and then as a complete account in September/October. Close to one year had elapsed, and in spite of the best efforts of several astronomers, Piazzi's observations could not be confirmed: the new "planet" was lost! It was under these circumstances that Gauss announced he had been able to compute, from Piazzi's data, the orbit of a planet and he predicted its current position where, indeed, it was found by Zach from Gotha on Dec. 7, 1801 and then by Olbers from Bremen on Jan. 1, 1802. How did Gauss succeed where so many had failed? It is only in (Gauss 1809) that he published his detailed account, but he wrote in his preface that

> scarcely any trace of resemblance remains between the method in which the orbit of Ceres was first computed and the form of this work.

This quote is taken over in [Teets and Whitehead, 1999] where the Reader will find an account of the original computations, based on a 1802 letter from Gauss to Olbers. Our modern authors emphasize that Gauss had reduced the problem to the following general formulation:

> from three geocentric observations (longitude and latitude) of a planet, determine two heliocentric vectors approximating the planet's position at two different times. From these two heliocentric vectors, the six orbital elements for the planet can be determined.

There are moreover two further aspects to the problem, the first of which explains why Gauss chose to propose his derivation of the least square method in a book on astronomy, besides the fact that so did Legendre and Simpson. By then, the astronomy community had recognized that while the solution of the problem indeed required only three observations, these data would have to be free of experimental errors; Gauss states explicitly in this connection that his aim is to remedy this insufficiency by using the data accumulated in the course of as many observations as were available. As the discovery of Ceres played a central motivating role in the writing of (Gauss 1809), it was imperative for him to master whatever unavoidable observational errors remained in Piazzi's data – recall the span in time and space: 41 nights and $3°$ – in order to reach the precision necessary to locate the celestial body almost a year later after it had been lost from sight. Gauss' biographers seem to agree that this is the first significant use by Gauss of the method of least squares, which he later exploited thoroughly in his geodetic work as Land Surveyor of the King of Hannover.

The third aspect of the problem of computing the orbit of a planet is that Kepler's laws cover only a two-body problem: the Sun and the planet, while the solar system consists of several mutually interacting bodies. Gauss developed enough of a perturbation theory to convince himself that, to the approximation necessary for his computations, the effect of the nearest heavy planet – Jupiter – on the orbit of the much smaller Ceres would modify the latter's orbit in ways that could be observed only after several revolutions. Nevertheless, he discusses in the last sections of [Gauss, 1809] how observations extended over a long time would be able to detect modifications of the orbit from which one could compute the mass of Jupiter. For more about the difficulties inherent already in the three-body problem, and the landscape surrounding Poincaré's treatment of secular perturbations, see Sect. 9.4.

The last question of interest to us is why did the Piazzi–Gauss discovery create such a stir. The reason is that it happened to be the first of the "missing planets" that "ought to be" between Mars and Jupiter according to a curious empirical formula proposed by Titius and Bode in 1776–1782. According to this formula, the difference of the mean distance between the Sun and the successive planets should double at each step:

$$\frac{A_{n+1} - A_n}{A_n - A_{n-1}} = 2 \qquad \text{i.e.} \qquad A_n = a + b\, 2^n \tag{4.4.7}$$

where a and b are *two* numerical constants. A choice of units, e.g, $A_{Earth} = 1$ leaves only *one* free parameter in the formula for A_n. With $a = .4$ and $b = .3$ a fit within 5% or better obtains for *all* of the following *eight* planets: Mercury $(n = -\infty)$, Venus $(n = 0)$, Earth $(n = 1)$, Mars $(n = 2)$, Jupiter $(n = 4)$, Saturn $(n = 5)$; the discovery of Uranus, which still satisfies (4.4.6), within the same 5% fit for $(n = 6)$, by Herschel in 1781 made the formula even more appealing; the formula again is satisfied, with the same precision, and $(n = 3)$, by the mean distance between Ceres and the Sun, as computed by Gauss: his success could not have gone unnoticed. Now that the minute size and faintness of Ceres had made it clear*er* why the missing planet had not been found before, other minor planets were discovered in the same range $(n = 3)$ shortly thereafter: Pallas (1802); Juno (1804) and Vesta (1807); Gauss computed their orbits and mentioned those computations repeatedly in [Gauss, 1809]. Tens of thousands of other small planets or asteroids – to use the modern generic name – were discovered since the photographic plate made it much easier to identify objects that move against a steady celestial background. This abundance led to speculations that *the* missing planet might have exploded, leaving behind the small fragments that are now sighted. Another theory holds that the asteroïds are primordial material prevented by Jupiter's strong gravity from collecting into a planet.

We should hasten to add, however, that while Pluto – to be discovered in 1930 – corresponds to $(n = 7)$, there is a planet between Uranus and Pluto, namely Neptune – discovered in 1846 – thus irrevocably falsifying the purported Titius–Bode law as it would demand $6 < n < 7$ [the actual astronomical data would only fit $(n \approx 6.7)$]. From this point of view, the Titius–Bode formula played a role similar to that of the Balmer formula in the elaboration of Bohr's quantum theory and the development of spectroscopy: a *wrong model* – here the Titius–Bode formula – had contributed to the acceptance of, and enthusiastic interest in, the discovery of *true facts* – here the planet Ceres – and unexpected advances in deductive sciences – here the theory of errors – as well as in experimental sciences – here the field of asteroïd research.

In the first third of the 19th century, statistical regularities were to gain explanatory power much beyond the theory of errors. The *homme moyen* [Quetelet, 1835 and 1869], whom nobody had seen roaming the streets of Bruxelles or Paris, emerged from Quetelet's demographic tables, begotten by the spirit announced in [Lacroix, 1816 and 1833] and [Poisson, 1837]. The Gaussian distribution – which we saw appear in various guises, cf. (4.3.24), (4.3.43) and (4.4.3) – was so effectively promoted by Quetelet under the name of *normal* distribution that it became normative. Deviations from the center of this normal distribution were used by social analysts to delimit *deviant* behavior. At about the same time, the physical scientists who followed Maxwell – see Chap. 3 – interpreted the standard deviation of the molecular velocity in a gas at equilibrium, in terms of a thermodynamical variable of heretofore independent standing, the *absolute temperature.*

Needless to say, these interpretations had very different motivations. While the social scientists were reporting the "normal distribution" of observed *relative frequencies,* the physicists who accepted the "Maxwell distribution" saw it as an inspired speculation on probable assessments that embodied minimal assumptions of symmetry and statistical independence. Yet, both groups of scientists plotted a distribution curve, that coïncides formally with those we discussed in connection with the asymptotic laws of large numbers, and which they knew from Laplace – (4.3.24) and (4.3.43) – or from Gauss (4.4.3) as the law of errors in physical measurement practice.

The least one can say about these usages is that there was a confluence; and the hardest commentary may be that they left room for much confusion.

Nevertheless, this is not to say that there were no 19th–century dissenters from the commonly held Laplacian faith, especially on the question of whether empirical inference from limiting frequencies is to be given precedence over deductions from the valuations of degrees of certainty, i.e. *a priori* probabilities. Daston [1994] recognized here a " *1840 cluster of revisionists"* – S.D. Poisson [1781–1840], B. Bolzano [1781–1848], R.L. Ellis [1817–1859], J.K. Fries [1773–1843], J.S. Mill [1806–1873] and A.A. Cournot [1801-1877] – whose pertinent criticisms were however so divergent that, for another half–century, they could not begin to break the coexistence tacitly accepted by the classical probabilists.

On quite another front, the methodology sketched already by G. Boole [1815–1864], and even more markedly by J. Venn [1834–1923] and C.S. Peirce [1839–1914] makes these authors now appear as precursors in the opposite epistemological approaches that led R. von Mises [1883–1953] and B. de Finetti [1906-1985] to their investigations, which we examine in Chap. 6.

5. Modern Probability: Syntax and Models

> *What epistemology intends is to construct thinking pro-*
> *cesses in a way in which they ought to occur. Episte-*
> *mology thus considers a logical substitute rather than the*
> *real process.*[Hence] *the distinction between the context of*
> *discovery and the context of justification. Epistemology*
> *cannot be concerned with the first but only with the latter*
> *... the rational reconstruction of knowledge.*
>
> [Reichenbach, 1938]

5.1 Quiet and Quaint No More

The articulation between classical and modern probability hinges on unresolved ambiguities in the very definition of the concept of probability – *de re*
[properties of things, e.g. stability of frequencies] or *de dicto* [properties of
propositions, e.g. insufficient reason] :

> *The complete definition of probability is ... a petitio principii: how are we*
> *to recognize that all cases are equally probable? A mathematical definition*
> *here is not possible; in each application we will have to make some con-*
> *ventions, to say that we consider this and that cases as equally probable.*
> *These conventions are not quite arbitrary, but they escape the purview of*
> *the mathematician who will not have to examine them once they are spec-*
> *ified.*
>
> [Poincaré, 1912]

The resolution came in the form of a recognition of the different roles of syntax
and semantics. In this respect [Poincaré, 1912] lecture notes are an instructive
witness. Most of the materials consist of mathematically precise, elegant and
quite didactic accounts of various chapters in classical probability theory,
with some distinct sharpenings of Gauss' theory of errors. The first edition
appeared in 1896, the second in 1912; the latter begins with an *Introduction,*
which Poincaré lifted from a chapter in his popular book [Poincaré, 1908a].
While the main body is concerned with the "hows" of probability theory,
the introduction addresses the "whys": the emphasis is almost entirely on the
semantics.

Although he places himself squarely in the Laplace tradition, via the work
of Bertrand – see e.g. [Bertrand, 1888] – he refines the Laplacian *de dicto*

indifference interpretation of probability as ignorance of the causes, and he adds a profound *de re* justification that plays a central role in modern ergodic theory, namely the importance of the relative size of causes and effects. If the effect of a small cause is small, we may ignore both the cause and the effect; if, however, a cause – or a difference in the causes – is too small for us to be able to assess it, while its consequence is so large to make for a directly observed difference, we call the latter a random effect – Poincaré says *"fortuit"*. So far, this is certainly worth saying, but it does not go very much beyond common sense. From this anodyne qualitative remark, Poincaré passes to a quantitative one; and it is of special interest to us that he draws support for his thesis from the kinetic theory of gases on which he had written earlier [Poincaré, 1894]. Specifically, if we denote by Λ the factor by which a small difference in the impact parameters is amplified by a collision between hard sphere molecules, then the effect of n successive collisions is Λ^n. We illustrate this *exponential sensitivity* to the initial conditions and discuss the centrality of this remark for ergodic theory in Models 5 and 6 of Sect. 8.4 , and in Sects. 9.4 and 9.5.

Upon coming back to the limitations due to the theorist – rather than to the things themselves – Poincaré also notices that one of the justifications for the use of probability is the complexity of the systems encountered in the kinetic theory of gases, which he contrasts with the tamer situation encountered with integrable systems, where

> the differential equations ... are too simple if they conserve anything, if they admit a uniform integral; if something from the initial conditions is kept unaltered, it is clear that the final situation will not be independent of the initial situation.
>
> [Poincaré, 1912]

The Poincaré of celestial mechanics [Poincaré, 1892–1907] is certainly in the background of this insight; see our Sects. 8.1 and 8.4. Furthermore, to this effect of the interactions, Poincaré adds the effect of the geometry of the boundary conditions, even in the absence of interactions:

> If the shape of the container is sufficiently complicated, the distribution of the molecules and that of their velocities will soon get uniform; this would not happen if the container where spherical ... the distance of an orbit from the center would remain constant.
>
> [Poincaré, 1912]

We will see that, in many of the models of statistical mechanics, the complexity of the equations can itself be traced to a significant degree to boundary conditions governing collisions between hard-core molecules – see e.g. Fig. 8.5 – so that we have a somewhat ambiguous region in the justification for the use of probability: again between the realm of things and the realm of their effective descriptions.

As Poincaré's name appears here for the first time in our book, we may refer to the following evaluations of a multifaceted mathematical and philosophical heritage that extends far beyond the confines of this chapter: [Danzig, 1954, Giedymin, 1982, Browder, 1983, Mette, 1986]. No introduction, however, could be nicer than the reading of his popular trilogy [Poincaré, 1903, 1904, 1908a]; for explorative vistas into his practice as a mathematician, see [Poincaré, 1892–1907].

Let us also introduce an eminent and immediate heir to Poincaré whom we shall meet several times in this chapter. From 1928 to his death in 1956, Emile Borel was the Director of the Institut Henri Poincaré which he helped establish. He is indeed one of the few mathematicians who managed to have a political career: elected to the French Parliament, he was for a time Ministre de la Marine, and was instrumental in the foundation of the CNRS. We know him nevertheless for his influence on mathematical thinking and his teaching, first at the University of Lille, then at the Ecole Normale Supérieure and, from 1909, at the Sorbonne. He belongs to our picture as a transitional figure in French probability theory. His lineage in the classical school still shows strongly in his book [Borel, 1909a], but in the same year, his paper [Borel, 1909b] broke new ground assuring Borel of a place as a precursor in the axiomatic approach to probability [Barone and Novikov, 1977/78]. We state later in this chapter [Thm. 5.2.3] the strong form of the law of large numbers, the original version of which is due to Borel [1909b]. His concept of denumerable probability is indeed a step in the direction of a purely measure theoretical formulation such as [Kolmogorov, 1933]. In his Sect. 2.2 on Borel, Von Plato [1994] locates the precise point where Borel stops and Kolmogorov forges ahead: it is Kolmogorov's extension theorem – his *Hauptsatz*, see Thm. 5.2.1 below – a mathematical statement involving a limiting procedure that goes beyond what Borel was willing to consider as being empirically acceptable. According to this reading, the question – see however [Barone and Novikov, 1977/78] – is not, therefore, whether Borel realized that denumerable probability could be summed up in measure theory: he had philosophical reasons not to want to go that way. Compare in this respect models 2 and 4 in Sect. 5.2 below; and notice the second of the comments offered after the description of these models. Mathematically, some of these models are isomorphic; the empirical ways by which one comes about them are however very different. As we show in Sect. 5.2, Kolmogorov [1933] offers the general syntactic frame, while Borel is still adhering to a specific semantics.

In a book such as ours, we must also note that Borel's writings on probability theory were informed not only from his mathematical contributions to the theory of sets in analysis but also from his interests in the physical sciences; even a cursory look in Borel's collected works [Borel, 1972] quickly locates papers, too numerous to be cited here, on kinetic theory of gases, statistical mechanics and irreversibility, including an *exposé* in French of the epoch-making *Encyklopädie* article of P. & T. Ehrenfest; see our Chap. 3 and Sect. 7.1. Borel enjoyed the gift of a fine pen which enabled him to address the non-specialized audiences; e.g. in [Borel, 1913, 1914b,c].

Some of the ambiguous mathematical status Probability theory still had at the dawn of the 20th century is captured for the record in Hilbert's address to the 1900 International Congress of Mathematicians, specifically in his *6th problem* which is discussed in our Sect. 5.2 below: among the responses this problem received, the [Kolmogorov, 1933] *syntax* is the best recognized; his "axioms" can be seen in the watermark through the pages of all modern texts, subsumed as they are by most practicing probabilists. We also present in Sect. 5.2 an array of models for Kolmogorov's axioms and we include there our first mathematical example of an ergodic theorem: it appears in the study of Weyl's sequences modulo 1.

In Sect. 5.3 of the present chapter, we consider the concept of *informational entropy*, introduced by Shannon and Khinchin; and we underline its similarities with the Boltzmann *H-function* discussed in the Sect. 3.3.

Finally, we explore in Sect. 5.4 the presumption that most sequences are random, and yet elusive in the sense that it is hard to determine whether a specific sequence is random. We confront Borel's notion of statistical randomness with what is empirically known about the decimal expansion of π. This section is tentative and is meant to prepare the grounds for the discussion of the competing semantics (von Mises and de Finetti) to be presented in Chap. 6. Although these semantics were developed almost contemporaneously with Kolmogorov's syntax, we felt that the postponement of their study to a separate chapter would help delineating the essence and limits of both the syntax and the semantics of Probability theory.

5.2 From Hilbert's 6th Problem to Kolmogorov's Syntax

At the 1900 International Congress of Mathematicians in Paris, Hilbert presented a series of problems that marked the frontiers of contemporary mathematical science [Hilbert, 1902]. His 6th problem concerns us in this section; it reads:

> Mathematical Teatment of the Axioms of Physics. *The investigations on the foundations of geometry suggest the problem: To treat in the same manner, by means of axioms, those physical sciences in which mathematics plays an important role; in the first rank are the theory of probabilities and mechanics. As to the axioms of the theory of probabilities, it seems to me desirable that their logical investigation should be accompanied by a rigorous and satisfactory development of the method of mean values in mathematical physics, and in particular in the kinetic theory of gases.*

Hilbert's plea for an axiomatization of physics is reinforced by his paper [Hilbert, 1918]. In the written version of his 1900 lecture, Hilbert gives the following five references to specify the range of his inquiry. The first four concern mechanics: [Mach, 1883]; [Hertz, 1894]; [Boltzmann, 1897]; [Volkmann, 1900]. These are perennial world classics, and they all show explicit concerns about the foundations.

Hilbert's fifth reference, the only one specifically pertaining to everyday statistics, is to a text by Bohlmann [1901]; here, Hilbert's choice could perhaps appear more accidental, yet it may be worth noting with [Von Plato, 1994], who cites [Schneider, 1988], that Bohlmann had given a course on the "Mathematics of Insurance" at Hilbert's own Göttingen in the Spring 1900. The collection by Schneider [1988] is indeed quite informative; it consists in a choice of texts, with introductions, that present a panorama extending from the origins up to some of the responses Hilbert's sixth problem generated. Significantly it announces, even in its title, that the coverage ends in 1933: the publication of [Kolmogorov, 1933] marked the official birth of modern probability theory. For generations to come, it established that the language of probability was to be measure theory, with expectation values to be defined as

$$\langle f \rangle \ = \ E(f) \ = \ \int \mathrm{d}\mu \ f \quad . \tag{5.2.1}$$

While the idea of an intrinsic connection between measure theory and probability was not original with Kolmogorov, the immediate and universal recognition of the importance of his work is to be understood as the winning combination of two ingredients: a crisp starting point – which we now review – and the richness of its consequences. In particular, Kolmogorov's sharp definition of random variables, as measurable functions on probability spaces, placed this notion at the cuting edge leading to general limit theorems - some of which are mentioned towards the end of this section. Furthermore, the notion of random variables has become the corner-stone on which much of the subsequent theory of stochastic processes is built.

Kolmogorov's first move is to announce the color of his cards, reflecting the syntactic concerns of Hilbert; he states:

> The theory of probability, as a mathematical discipline, can and should be developed from axioms in exactly the same way as Geometry and Algebra. This means that after we have defined the elements to be studied and their basic relations, and have stated the axioms by which these relations are to be governed, all further exposition must be based exclusively on these axioms, independent of the usual concrete meaning of these elements and their relations.

As if to make sure that we understand his deliberate take, Kolmogorov refers to the precedent provided by [Hilbert, 1899]. While Kolmogorov's was the most immediately successful, there were several other attempts at an axiomatization of probability theory; in particular, Kolmogorov mentions R. von Mises and S. Bernstein, although he warns that his aim is different from theirs. We discuss von Mises' semantic approach in Sect. 6.1 below; for the purpose of the present chapter, it seems fair to note that in the main body of his 1933 paper, Kolmogorov only pays lip service to the "limiting frequencies" interpretation which he professes not to need. As for [Bernstein, 1917], although it ultimately received the accolade of being the first axiomatization of probability [Aleksandrov, Akhiezer, Gnedenko, and Kolmogorov, 1969], its early circulation was severely limited by its remote venue in spite of the contacts Bernstein maintained with mathematicians not only in the Soviet Union but also in Western Europe, in particular in France, where he lectured at the Sorbonne.

Among other precursors whom Kolmogorov [1933] cites, note also [Lomnicki, 1923] and [Lévy, 1925]. For a detailed study of the early precedents to an axiomatic approach to probability theory, cf. [Barone and Novikov, 1977/78].

As [Kolmogorov, 1933] first section of his first chapter is to set the first five of his axioms, we follow him in this; but we immediately complete the list with two axioms: the first is Kolmogorov's sixth axiom, the appearance of which is delayed in Kolmogorov's text until the beginning of the second chapter; the next one is not labelled as such by Kolmogorov, but is introduced by him to define what he calls *Borel field of probability*, with the comment that *"we obtain full freedom of action"* only when this is required.

The Kolmogorov axioms for $(\Omega, \mathcal{F}, P)$:

I. Ω is a set, and $\mathcal{F}$ is a ring of subsets of Ω, i.e. $\mathcal{F}$ is closed under finite unions and differences;

II. $\Omega \in \mathcal{F}$;

III. $P : A \in \mathcal{F} \mapsto P(A) \in \mathsf{R}^+$;

IV. $P(\Omega) = 1$;

V.   Whenever $A, B \in \mathcal{F}$ with $A \cap B = \emptyset$:

$$P(A \cup B) = P(A) + P(B) ;$$

VI. For every sequence $\{A_n \in \mathcal{F} \mid A_n \supseteq A_{n+1}\}$ for which $\bigcap_n A_n = \emptyset$:

$$\lim_{n \to \infty} P(A_n) = 0 ;$$

VII. $\mathcal{F}$ is a σ−ring, i.e., in addition to I, it is required that for all countable collections $\{A_n \in \mathcal{F}\}$ of mutually disjoint elements of $\mathcal{F}$:

$$\bigcup_n A_n \in \mathcal{F} .$$

For the systematic introduction of these concepts, Kolmogorov cites [Hausdorff, 1927] and [Lebesgue, 1928]; both refer to [Borel, 1898] for the origin of the theory (concepts, examples and theorems). The modern student learns these from classic textbooks such as [Halmos, 1950], [Rudin, 1964], [Ash, 1972], or [Cohen, 1980]. We briefly review the basic definitions, and give a few examples.

Firstly, the union $A \cup B$ of two subsets of a set Ω is the subset of all elements $\omega \in \Omega$ that are in A *or* B. The difference $A - B$ is the set of elements in A which are not in B : $A - B = \{\omega \in A \mid \omega \notin B\}$. To say that $\mathcal{F}$ is closed under unions and differences, means that if A and B belong to the collection $\mathcal{F}$, then so do $A \cup B$ and $A - B$; in particular since the empty set is of the form $\emptyset = A - A$, $\emptyset \in \mathcal{F}$. Also, the intersection of two subsets A and B of Ω, namely $A \cap B = \{\omega \in A \mid \omega \in B\}$, can be rewritten as $A \cap B = A - [A - B]$; hence $\mathcal{F}$ is closed under intersections as well.

Secondly, since $\Omega \in \mathcal{F}$, $\mathcal{F}$ is closed under the complementation: $A \to \neg A = \Omega - A$, thus making the ring $\mathcal{F}$ an algebra, and when VII is added, a σ-algebra.

Thirdly, because of IV and V: $P(A) + P(\neg A) = 1$, and thus III can be read, without loss of generality, as $P : A \in \mathcal{F} \mapsto P(A) \in [0,1]$; and we have $P(\emptyset) = 0$.

Fourthly, as a consequence of the first five axioms only, for every finite string of mutually disjoint elements in $\mathcal{F}$ – i.e. $\{A_n \in \mathcal{F} \mid n = 1, 2, \cdots, N\}$ with $A_n \cap A_m = \emptyset$ whenever $n \neq m$ – we have:

$$P\left(\bigcup_{n=1}^{N} A_n\right) = \sum_{n=1}^{N} P(A_n).$$

Lastly, the purpose of axions VI and VII is to extend the above additivity result from finite strings to infinite sequences of subsets of $\mathcal{F}$. To see this, note first that, by axiom VII, the infinite union of countably many, mutually disjoint subsets of $\mathcal{F}$ still belongs to $\mathcal{F}$, so that

$$P\left(\bigcup_{n=1}^{\infty} A_n\right)$$

is defined. Moreover, for any finite positive integer N :

$$\bigcup_{n=1}^{\infty} A_n = \left(\bigcup_{n=1}^{N} A_n\right) \bigcup R_N \quad \text{where} \quad R_N = \bigcup_{n>N} A_n$$

so that, from the finite additivity obtained from the first five axioms, we have:

$$P\left(\bigcup_{n=1}^{\infty} A_n\right) = \sum_{n=1}^{N} P(A_n) + P(R_N) \quad .$$

Hence, it is sufficient to prove that

$$\lim_{N \to \infty} P(R_N) = 0 \quad ,$$

but this follows from the continuity axiom VI, since

$$R_N \supseteq R_{N+1} \quad \text{and} \quad \bigcap_{N=1}^{\infty} R_N = \emptyset \quad .$$

Hence, for any countable subcollection $\{A_n\}$ of elements of $\mathcal{F}$ where the A_n are mutually disjoint, we have:

$$P(\bigcup_{n=1}^{\infty} A_n) = \sum_{n=1}^{\infty} P(A_n) \quad . \tag{5.2.2}$$

This property is referred to by saying that the set function P defined on $\mathcal{F}$ is *countably additive*; many authors, including [Kolmogorov, 1933] (at least in its Engl. transl.) and [Lebesgue, 1928], say *completely* additive. In the sequel, we call *probability space* any triple $(\Omega, \mathcal{F}, P)$ that satisfies the seven Kolmogorov axioms.

Here we should note that the very language by which Kolmogorov's theory is summarized is the language of the semantic view of theories. The theory *defines* a *space* of mathematical discourse in which properties prescribed by the axioms obtain. Recalling the controversy over the syntactic and the semantic view we discussed in Chap. 1, one may then wonder whether we have blundered in calling Kolmogorov's theory the syntax – rather than the semantics – of probability. It is the syntax precisely because of the following reason. Even though the axioms define a set-theoretic object [the probability space, $(\Omega, \mathcal{F}, P)$], the object has no meaning in that neither the elements in Ω nor those in $\mathcal{F}$ and P are realized. Until such realization is given, as we see in the following four models, we do not have any systems against which assertions of probability can be evaluated as being satisfied (i.e. true) or not satisfied (i.e. false) by these systems.

This is in fact a case which enriches our understanding of the semantic view. It should by now be clear that the *syntax* or the syntactic part of a theory does not necessarily refer only to the symbols used in the theory and the sentences which comprise the theory. Rather, it may better be taken as a mathematical space so abstract and general that none of the objects in the space is *yet* given an interpretation or meaning. And the semantics is not supplied until an interpretation or meaning is given, which results in specific models as the ones we present below. Such a notion of syntax and the distinction between it and the semantics of a theory applies both to mathematical and to scientific theories: the syntax of thermodynamics, for instance, is a *space* defined by the two axioms given in Chap. 2 (5.4.1–2) or (5.4.3–4), whose magnitudes are not yet identified with anything physical, such as work, heat and temperature. As far as logic is concerned, the space so defined could be about the "dynamics" of a person's emotion rather than about his body heat.

Therefore, we now see even more clearly how misleading the dichotomy can be between the terms, the "semantic" view and the "syntactic" view. It is not that the syntax of a theory is not significant and so it should be ignored in a semantic construal of the theory, but rather semanticism, if correctly understood, requires that we locate the syntax of a theory on a different level, a level where the trivial symbolic differences of sentential formations play no role while the theory is not yet constrained by the meaning of each abstract object.

Although it would be tempting, on epistemological grounds, to omit the continuity axiom, the resulting theory would lack in predictive power; in particular, this axiom is essential in proving the strong law of large numbers – Thm. 5.2.3 below.

Theorem 5.2.1 (Kolmogorov's extension theorem). *Let $\mathcal{F}_o$ be an algebra of subsets of a set Ω. Let $P_o : A \in \mathcal{F}_o \mapsto P_o(A) \in \mathbb{R}$, satisfy:*

$P_o(\Omega) = 1$;

$\forall A \in \mathcal{F}_o : P_o(A) \geq 0$;

for every countable partition of any set $A \in \mathcal{F}_o$ into subsets $\{A_k \in \mathcal{F}_o\}$:
$$P_o(A) = \sum_k P_o(A_k).$$

Then there exists a unique extension of P_o to a probability measure P on the smallest $\sigma-$algebra $\mathcal{F}$ containing $\mathcal{F}_o$, i.e.. a countably additive, non-negative set function on $\mathcal{F}$ such that for all $A \in \mathcal{F}_o : P(A) = P_o(A)$.

This extension theorem, with its uniqueness, is stated and proved in [Kolmogorov, 1933]. It admits a generalization to consistent systems of random variables; Kolmogorov also gives this in the paper, and calls it the *"Hauptsatz"*, i.e. the fundamental theorem; it is sometimes this generalization which is referred to as Kolmogorov's extension theorem. The form given above is sufficient for our discussion. At the end of the section in which the extension theorem is stated, Kolmogorov adds the following semantic remark:

> *Even if the sets (events) A of $\mathcal{F}_o$ can be interpreted as actual and (perhaps only approximately) observable events, it does not, of course, follow from this that the sets of the extended field $\mathcal{F}$ admit such an interpretation ... the sets of $\mathcal{F}$ are generally merely ideal events to which nothing corresponds in the outside world. However, if reasoning which utilizes the probabilitiy of such ideal events leads us to a determination of the probability of an actual event of $\mathcal{F}_o$, then, from an empirical point of view also, this determination will automatically fail to be contradictory.*
>
> [Kolmogorov, 1933]

The extension theorem seems to mark the place where Borel did not want to go, as he insisted that all terms in the syntax be empirically meaningful. Kolmogorov's trump card – as he clearly let it be known in the above quote – was that, not only does the extension exist, but it is unique. Evidently, the theorem does not create countable additivity *ex nihilo*: the extension refers to that from $\mathcal{F}_o$ to $\mathcal{F}$, and countable additivity of P_o is assumed, albeit only where it makes sense (see the third condition in the hypothesis of the theorem). We now need to give a few models to show that the axioms are consistent, and allow a variety of distinct realizations.

Model 1: Probabilities for finitely many possible events. Let $\Omega = \{\omega_1, \omega_2, \cdots \omega_n\}$ be any finite set, with n (distinct) elements. Let $\mathcal{F}$ be the collection of all (i.e. 2^n) subsets of Ω. Let $\{p_1, p_2, \cdots, p_n\}$ be any $n-$tuple of positive numbers adding up to 1. Finally, for any non-empty subset $A =$

$\{\omega_{k_1}, \omega_{k_2}, \cdots, \omega_{k_m} \mid 1 \leq m \leq n\} \subseteq \Omega$,   let $P(A) = \sum_{i=1}^{m} p_{k_i}$. This is the generic example of a finite field of probability. It covers all the special cases discussed in Chap. 4. In the special case where all p_k are equal, P reduces to the counting measure $P(A) = \mathrm{card}(A)/n$ [where, as usual, $\mathrm{card}(A)$ is the number of elements of Ω appearing in A] and in particular, for every $\omega \in \Omega : P(\{\omega\}) = 1/n$. This is the uniform probability measure of Laplace: equal weight to all elementary events.

Model 2: Probabilities on a compact interval. Let $\Omega = [0, 1]$; $\mathcal{F}$ be the Borel field generated by the semi-open intervals $[a, b) \subseteq [0, 1]$, and P_L the countably additive set function P_L determined by $P_L([a, b)) = b - a$. The fact that this function exists on $\mathcal{F}$, and is unique, needs to be proven: Kolmogorov [1933] knew the relevant 'extension theorem' – see Thm. 5.2.1 above – the origin of which he traces back to [Hausdorff, 1927]. We refer to this P_L as the *Lebesgue measure* on $[0, 1]$, by a common abuse of nomenclature which we discuss in the comments following the descriptions of our four main models; we also defer to those comments for the sense in which the probability space $([0, 1], \mathcal{F}, P_L)$ is canonical. In the meantime note that, since any point $x \in [0, 1]$ is the intersection of countably many semi-open intervals of length decreasing to zero, every singleton $\{x\}$ belongs to $\mathcal{F}$, with $P_L(\{x\}) = 0$. Consequently every countable $X \subset [0, 1]$ is in $\mathcal{F}$, with $P(X) = 0$; in particular, since the set Q of the rational is countable, $P_L(\mathsf{Q} \cap [0, 1]) = 0$.

Model 3: Probabilities associated to Bernoulli trials. Let $\Omega = \Xi$ be the set of all infinite binary *sequences,* i.e. the elements of Ξ are of the form $\xi : k \in \mathsf{Z}^+ \mapsto \xi_k \in \{0, 1\}$ or $\xi = (\xi_1, \xi_2, \cdots, \xi_n, \cdots)$ with $\xi_i = 0$ or 1. The set Ξ is clearly infinite; in fact it is uncountable. By contrast, the set $\{0, 1\}^{(n)}$ of all binary *strings* $x = (x_1, x_2, \cdots x_n)$ of finite length $l(x) = n$ has 2^n elements, and the set $\Xi_o = \cup_n \{0, 1\}^{(n)}$ of all finite strings is infinite, but countable. [Note that we follow with this model the convention, already adopted earlier in this work, to distinguish *sequences* (which are infinite) from *strings* (which are finite).] For every $x \in \Xi_o$, let

$$A(x) = \{\xi \in \Xi \mid \xi_k = x_k \ \forall \ 1 \leq k \leq l(x)\} \qquad (5.2.3)$$

consist of all binary sequences that start with the string x ; $A(x)$ is called the *cylinder set* (or fiber) with base x . Let now $\mathcal{F}$ be the Borel field generated from this (countable) collection of subsets of Ξ . Finally P is prescribed by the value it takes on the cylinder sets. For this we choose a number p with $0 \leq p \leq \frac{1}{2}$, and let

$$P_p[A(x)] = \prod_{k=1}^{l(x)} p(x_k) \quad \text{with} \quad p(x_k) = \begin{cases} p & \text{if} \quad x_k = 1 \\ 1 - p & \text{if} \quad x_k = 0 \end{cases} .$$

Models with $\frac{1}{2} \leq p \leq 1$ are obtained by replacing in the above p by $(1 - p)$.

Model 4: Probabilities on Cantor sets. The construction of these models needs a few preliminary materials. We first let Ξ be, as in model 3 above, the set of all binary sequences; we then choose some $\lambda \in (0, \frac{1}{2}]$; and we define:

$$C_\lambda = \left\{ \pi_\lambda(\xi) = \frac{1-\lambda}{\lambda} \sum_{k=1}^\infty \xi_k \lambda^k \mid \xi \in \Xi \right\} . \qquad (5.2.4)$$

We claim: (i) $C_\lambda \subseteq [0,1]$, with equality holding if and only if $\lambda = \frac{1}{2}$, so that each $\lambda \in (0, \frac{1}{2}]$, C_λ is uncountable; and yet, (ii) for all $\lambda \in (0, \frac{1}{2})$, $P_L(C_\lambda) = 0$ with respect to the Lebesgue measure P_L.

To see (i), note first that for $\xi = (0,0,\cdots)$, $\pi_\lambda(\xi) = 0$; similarly, for $\xi = (1,1,\cdots)$, $\pi_\lambda(\xi) = 1$; and thus $C_\lambda \subseteq [0,1]$. Moreover, every $x \in [0,1]$ has a dyadic expansion, which is to say that there exists $\xi \in \Xi$ such that $\pi_{\frac{1}{2}}(\xi) = x$, i.e. $C_{\frac{1}{2}}$ covers $[0,1]$. Hence $C_{\frac{1}{2}}$ is uncountable, and thus so is every C_λ. Note however that $\pi_{\frac{1}{2}}$, which is surjective, is not injective; indeed for every finite binary string of length $n-1$, the infinite sequences

$$\xi_k = \begin{cases} x_k & k < n \\ 0 & k = 0 \\ 1 & k > n \end{cases} \quad \text{and} \quad \eta_k = \begin{cases} x_k & k < n \\ 1 & k = 0 \\ 0 & k > n \end{cases}$$

satisfy $\pi_{\frac{1}{2}}(\xi) = \pi_{\frac{1}{2}}(\eta)$. For any other value of $\lambda \in (0, \frac{1}{2})$ this degeneracy is lifted, but the map π_λ, while now injective, badly fails to be surjective. Indeed, note in a first stage, that the sequence $\xi_{11} = (0,1,1,\cdots)$ gives $\pi_\lambda(\xi_{11}) = \lambda$ while the sequence $\xi_{12} = (1,0,0,\cdots)$ gives $\pi_\lambda(\xi_{12}) = 1-\lambda$, so that

$$C_\lambda \subseteq [0,\lambda] \cup [1-\lambda, 1]$$

where the length of the missing interval $(\lambda, 1-\lambda)$ is $(1-2\lambda)$; similarly, in a second stage, the sequence $\xi_{21} = (0,0,1,1,\cdots)$ gives $\pi_\lambda(\xi_{21}) = \lambda^2$ while the sequence $\xi_{22} = (0,1,0,0,\cdots)$ gives $\pi_\lambda(\xi_{22}) = \lambda(1-\lambda)$, so that

$$C_\lambda \cap [0,\lambda] \subseteq [0,\lambda^2) \cup [1-\lambda, \lambda]$$

where the length of the missing interval $(\lambda^2, \lambda(1-\lambda))$ is $\lambda(1-2\lambda)$; similarly for $C_\lambda \cap [1-\lambda, 1]$. By successive iterations, we obtain in this manner that C_λ is the complement of a set H_λ which is the countable union of open sets, and has total measure

$$P_L(H_\lambda) = \sum_{k=0}^\infty 2^k [\lambda^k(1-2\lambda)] = 1 \quad i.e. \quad P_L(C_\lambda) = 0$$

which thus holds indeed for all $\lambda \in (0, \frac{1}{2})$.

With $\lambda \in (0, \frac{1}{2}]$ and $0 \le p \le \frac{1}{2}$ fixed, we can now define the model $\{\Omega_\lambda, \mathcal{F}_\lambda, P_{\lambda,p}\}$ as follows. First $\Omega_\lambda = C_\lambda$. Second, for each $A \subset \Xi$, let $\pi_\lambda(A) \equiv A_\lambda \equiv \{\pi_\lambda(\xi) \mid \xi \in A\}$; $\mathcal{F}_\lambda$ is then the Borel field generated by the cylinder sets $A_\lambda(x) \equiv A(x)_\lambda$. Finally the probability measure $P_{\lambda,p}$ is determined by $P_{\lambda,p}[A_\lambda(x)] = P_p[A(x)]$.

Comments on the above models. For each $\lambda \in (0, \frac{1}{2})$, the set C_λ is referred to as the *Cantor set* with deleted middle $l = (1-2\lambda)$. It is self-similar on a scale λ; in particular, for every positive integer $n : \{C_\lambda \cap [0, \lambda^n]\} = \lambda^n C_\lambda$. While the Lebesgue measure does not carry information on the respective sizes of the Cantor sets corresponding to different values of $\lambda \in (0, \frac{1}{2})$, there is a more discriminating concept – the *Hausdorff dimension*:

$$\dim_H(C_\lambda) = \frac{\log 2}{\log \frac{1}{\lambda}} \quad ;$$

note that this gives a monotonically, strictly increasing map $\lambda \in (0, \frac{1}{2}) \to \dim(C_\lambda) \in (0, 1)$. The iceberg, of which the above comment is only a faint outline, is made immediately approachable in [Falconer, 1990] and [Mattila, 1995].

The case $\lambda = \frac{1}{2}$ is also of special interest. Indeed, the map $\pi_{\frac{1}{2}}$ sends each cylinder set onto a closed interval:

$$\pi_{\frac{1}{2}}(A(x)) = [\pi_{\frac{1}{2}}(x), \pi_{\frac{1}{2}}(x) + 2^{-l(x)}]$$

so that

$$P_p(\pi_{\frac{1}{2}}(A(x))) = P_L(\pi_{\frac{1}{2}}(A(x))) \quad \forall \quad x \in \Xi_o \qquad \text{if and only if} \quad p = \frac{1}{2} .$$

So, the special case $(\lambda = \frac{1}{2}; p = \frac{1}{2})$ of model 4 is isomorphic to model 3.

When $p \neq \frac{1}{2}$ the above situation changes drastically. Consider, for instance, the case $p = 1$; P_1 induces, through $\pi_{\frac{1}{2}}$, a measure $\tilde{P}_1$ that is concentrated on the point $\{1\}$, i.e. for all $A \in \mathcal{F}$ with $\{1\} \notin A : \tilde{P}_1(A) = 0$; so that, in particular

$$\tilde{P}_1(\neg\{1\}) = 0 \quad \text{whereas} \quad P_L(\{1\}) = 0 \quad .$$

A similarly *singular* behavior happens for $p = 0$ and, in fact, for every $p \neq \frac{1}{2}$, namely. there exists some $A \in \mathcal{F}$ (possibly depending on p) such that $\tilde{P}_p(\neg A) = 0$ and $P_L(A) = 0$.

Similar singularities are to be expected with the measures $P_{\lambda,p}$ induced on the Cantor set C_λ by the map π_λ.

In this and the next comment, we indicate some of the reasons why model 2 can be considered typical. The present remark focuses on the Borel structure.

Definition 5.2.1. *Let $(\Omega_1, \mathcal{F}_1)$ and $(\Omega_2, \mathcal{F}_2)$ be two $\sigma-$algebras; a map $f : \Omega_1 \to \Omega_2$ is said to be* measurable *whenever $A \in \mathcal{F}_2 \models f^{-1}(A) \in \mathcal{F}_1$ i.e. $\{\omega \in \Omega_1 \mid f(\omega) \in A\} \in \mathcal{F}_1$. A map $f : \Omega_1 \to \Omega_2$ is said to be an* isomorphism *whenever it is bijective and both f and f^{-1} are measurable. When such a f exists the two spaces $(\Omega_1, \mathcal{F}_1)$ and $(\Omega_2, \mathcal{F}_2)$ are said to be* Borel isomorphic.

We present below a few important families of isomorphic Borel structures. For every separable metric space (Ω, d), let $\mathcal{F}_\Omega$ be the smallest σ-algebra containing all the subsets of Ω that are open in the topology defined by the metric d. (See Appendix D for elementary definitions.) Then two complete separable metric spaces, equipped with the above Borel structure, are Borel isomorphic if and only if they have the same cardinality. Consequently a complete separable metric space must be Borel isomorphic to either of the following three spaces: $\{1, 2, \cdots, n\}$, $\{1, 2, \cdots\} = \mathsf{Z}^+$, or $[0, 1]$. This result can be considerably generalized: one can replace the condition that (Ω, d) be a complete separable metric space, by the weaker condition that Ω be a Polish space – i.e. that as a topological space Ω be homeomorphic to a complete separable metric space – or even be a Lusin space (note, incidentally, that Lusin was Kolmogorov's first advisor!) – i.e. Ω be a Hausdorff space such that there exists a Polish space S and a continuous bijective map $\phi : S \to \Omega$. This family of results goes back at least to the work of Kuratowski [1933] on general topology. As many aspects of measure theory, and allied fields, depend only on Ω being a Lusin space, these isomorphism results have reached the currency of folk-theorems; for modern proofs, see e.g. [Parthasarathy, 1967, Thms.2.12 & 3.9], [Ash, 1972, Thm.6.6.6], [Cohen, 1980, Thm.8.3.6].

Let us now complete the above comment by considering the Lebesgue measure P_L we defined on $[0, 1]$.

We first note that $([0, 1], \mathcal{F}, P_L)$ is non-atomic in the following sense; a probability space $(\Omega, \mathcal{F}, P)$ is said to be non-atomic whenever, for every $\omega \in \Omega$, $P(\{\omega\}) = 0$; we saw that model 2 is non-atomic. The main isomorphism theorem, see e.g. [Parthasarathy, 1977, Prop. 26.6], now states that if Ω is a complete separable metric space, with canonical Borel structure $\mathcal{F}$, and if P is a non-atomic probability measure on $(\Omega, \mathcal{F})$, then $(\Omega, \mathcal{F}, P)$ is isomorphic to our model 2, where we say that two probability spaces $(\Omega_1, \mathcal{F}_1, P_1)$ and $(\Omega_2, \mathcal{F}_2, P_2)$ are isomorphic if there exists a map $f : \Omega_1 \to \Omega_2$ and $N_k \in \mathcal{F}_k$ ($k = 1, 2$) such that: (i) $P_1(N_1) = 0 = P_2(N_2)$; (ii) f is a Borel isomorphism from $(\Omega_1 - N_1, \mathcal{F}_1 \cap (\Omega_1 - N_1))$ onto $(\Omega_2 - N_2, \mathcal{F}_2 \cap (\Omega_2 - N_2))$; and (iii) $P_1 \circ f^{-1} = P_2$.

Note that model 2 is also remarkable by itself for the following reason. The map $\exp : \omega \in [0, 1] \to \exp(2\pi i \omega) \in S^1$ gives a map, which except for 0 and 1 establishes a bijective correspondence between [0,1] and the unit circle S^1. This maps carries with it the measure P_L which now is *uniform* in the sense that it is invariant under the group of all rotations $\phi \in S^1 \to \phi + \phi_o \in S^1$: i.e. P_L is the *Haar measure* of the group S^1.

As already mentioned, the name Lebesgue measure used for the measure introduced in model 2 should properly be reserved to another measure; that measure is uniquely determined by P_L, but defined on a larger collection $\overline{\mathcal{F}}$ of subsets of $[0, 1]$, namely the smallest σ-field which, in addition to the Borel subsets belonging to $\mathcal{F}$, also contains all subsets of any Borel set with zero Lebesgue measure, i.e all subsets of the form $N \subseteq B$ with $B \in \mathcal{F}$ and

$P_L(B) = 0$. The elements of $\overline{\mathcal{F}}$ are said to be the *Lebesgue measurable* subsets of $[0, 1]$. P_L on $\mathcal{F}$ is extended to a countably additive probability measure on $\overline{\mathcal{F}}$ – denoted again by P_L – by imposing $P_L(A \cup N)) = P_L(A)$ for all $A \in \mathcal{F}$ and every subset $N \subseteq B$ with $B \in \mathcal{F}$ and $P_L(B) = 0$.

All these models not only supply the semantics for Kolmogorov's theory, but also show that it is consistent, namely that it admits models (cf. Sect. 1.3 and Appendix E).

Lebesgue integral and random variables. Lebesgue defined originally the σ–algebra of Lebesgue measurable sets of R. His approach is the template we transcribed for [0,1] in the above comment about model 2, namely to start with the semi-closed intervals, to require the σ–algebra property, and then to extend by the inclusion of all subsets of Borel sets of zero measure. Lebesgue obtained already a countably additive measure m, the original Lebesgue measure, defined on all Lebesgue measurable sets, invariant under the translation group acting on R, and such that the probability measure P_L of model 2 is the restriction of m from R to $[0, 1]$. Furthermore, Lebesgue's approach generalizes naturally to R^2, or R^n, replacing intervals by rectangles, or n–dimensional bricks.

The existence of subsets of R which are *not* Lebesgue measurable can be read already from [Vitali, 1905], and the following counterexample is standard [Ash, 1972, Ex.6 of Sect. 1.4], [Cohen, 1980, Thm.1.4.7]. Let Λ be the collections of all cosets $\{S_\lambda\}_{\lambda \in \Lambda}$ of the subgroup Q of the rationals in R, the additive group of reals. Note that for each $\lambda \in \Lambda$, $S_\lambda \cap (0, 1)$ is not empty. Since the cosets – as equivalence classes – are disjoint by definition, the axiom of choice asserts the existence of a subset $B \subset (0, 1)$ such that, for each $\lambda \in \Lambda$, $B \cap S_\lambda$ contains exactly one element. Let now $Q_1 = Q \cap (-1, 1)$ and form the infinite, but countable disjoint union $A = \cup_{x \in Q_1} (B + x)$. If B were Lebesgue measurable, the countable additivity and translation-invariance of the Lebesgue measure would imply, depending on whether $m(B) = 0$ or $m(B) > 0$, that $m(A) = 0$ or ∞, which contradicts $(0, 1) \subseteq A \subseteq (-1, 2)$. Hence B cannot be Lebesgue measurable. *The above proof depends on the axiom of choice.* This is indeed necessary, as Solovay [1970] produced a model of set theory in which every set of reals is Lebesgue measurable – for a didactic account, see also [Jech, 1978, Sect. 42]: the model is of ZF + DC [= Zermelo–Fraenkel + Dependent Choice (which requires less than the full-strength axiom of choice)].

The prime reason why Lebesgue developed his notion of measurability was to alleviate the difficulties encountered with the usual (Riemann) integration theory, e.g. to obtain dispensation from the quasi-Mosaic dictum "Thou shall not interchange limits". In particular let f be the limit of a sequence of Riemann-integrable functions f_n that are uniformly bounded, and converge point-wise on a bounded interval $I \subset \mathsf{R}$; yet, it is possible that

$$\int_I \mathrm{d}x \, (\lim_{n \to \infty} f_n(x)) \neq \lim_{n \to \infty} \int_I \mathrm{d}x \, f_n(x)$$

and that the LHS does not even make sense: f may fail to be Riemann integrable.

Lebesgue's definition of integrability starts by defining a function $f : A \subset \mathsf{R} \to [0, \infty]$ with $A \subset \mathsf{R}$ Lebesgue measurable, to be integrable (we say *Lebesgue integrable*) if and only if $H_f \equiv \{(x, y) \in \mathsf{R}^2 \mid x \in A \, ; \, 0 \leq y \leq f(x)\}$ is Lebesgue measurable in R^2 and of finite measure. The integral of f over A with respect to m is then defined as:

$$\int_A \mathrm{d}m \; f = m(H_f) \quad ; \tag{5.2.5}$$

when no confusion with the Riemann-integral is likely to occur, we use the traditional notation $\int_A \mathrm{d}x \, f(x)$. The extension from positive- to real-valued functions is trivial. This integral shares with the Riemann integral the property to be linear:

$$\int_A \mathrm{d}m \; (a \, f + b \, g) = a \int_A \mathrm{d}m \; f + b \int_A \mathrm{d}m \; g \quad . \tag{5.2.6}$$

In addition it satisfies a precious *continuity* condition, known as the *Lebesgue dominated convergence theorem;* let indeed $\{f_n\}$ be a sequence of Lebesgue-integrable functions, converging point-wise to a function f; if there exists a Lebesgue-integrable function g such that $|f_n| \leq g \; \forall \; n$, then f is Lebesgue-integrable, and

$$\int \mathrm{d}m \; f = \int \mathrm{d}m \; (\lim_{n \to \infty} f_n) = \lim_{n \to \infty} \int \mathrm{d}m \; f_n \quad . \tag{5.2.7}$$

Much of Lebesgue's work on the definition of the integral, including the most important clue for the above therorem, was done in the years 1900–1903; the first edition (1904) of [Lebesgue, 1928] gives a fair view of the program's success; most of the principal results were indeed obtained in the first ten years following the inception of the theory, either by Lebesgue or by his disciples. [Fréchet, 1915] further extended these ideas to produce a theory of integration on abstract spaces. The tools were then all available for Kolmogorov's definition of a random variable and its expectation .

Indeed, it becomes natural in the context on which Kolmogorov focused our attention – namely when $(\Omega, \mathcal{F}, P)$ is singled out as the primary object of the theory – to introduce the following "observable" objects.

Definition 5.2.2. *A random variable is a real-valued measurable function, i.e. – see Def. 5.2.1 – a function*

$$X : \omega \in \Omega \mapsto X(\omega) \in \mathsf{R} \quad \text{such that} \quad \forall \, a \in \mathsf{R} : \{\omega \in \Omega \mid X(\omega) < a\} \in \mathcal{F}(\mathsf{R}) .$$

The probability distribution *of the random variable X is the function*

$$F_X : a \subset \mathsf{R} \mapsto P(X < a)$$

where P_X is the probability measure

$$P_X : A \in \mathcal{F}(\mathsf{R}) \mapsto P(X^{-1}(A)) \in [0,1] \quad ;$$

and $P(X < a)$ is commonly used as an abbreviation for $P(\{\omega \in \Omega \mid X(\omega) < a\})$. Two random variables X and Y are said to be equivalent whenever $P(X \neq Y) = 0$, i.e. whenever they differ at most on a set of measure zero. Provided that X be integrable (in the sense of Lebesgue–Fréchet), the expectation $E(X)$ of a random variable X is the integral

$$E(X) = \int_\Omega \mathrm{d}P \, X = \int_\mathsf{R} \mathrm{d}P_X \, a \quad .$$

The expectation value E inherits directly the linearity, continuity and other properties to be expected from a Lebesgue–Fréchet integral:

$$\left.\begin{array}{l} E(aX + bY) = a \, E(X) + bE(Y) \\ |E(X)| \leq E(|X|) \\ E(X) = E(Y) \quad \text{whenever} P(X \neq Y) = 0 \\ \inf X \leq E(X) \leq \sup X \\ X \leq a < \infty \models E(X)\text{exists} \\ \sum_n E(|X_n|) \text{ converges} \models E(\sum_n X_n) = \sum_n E(X_n) \end{array}\right\} \quad (5.2.8)$$

As Kolmogorov [1933] defined similarly conditional expectations, the materials just presented constitute the syntactic answer to [Hilbert, 1902] question about "mean values"; they provide the key – but only a key – to the semantic building in which a "method in mathematical physics" can function. That, however, was not one of the driving concerns expressed in [Kolmogorov, 1933]: his interest was more in discussing applications such as stochastic independence and the law of large numbers, some aspects of which we address towards the end of this section.

Uniformly distributed sequences modulo 1. This is one of the nicest classes of models that illustrate how the frequentist interpretation of probability can be made to fit into the Kolmogorov mold. It also allows to see how the Boltzmann view of ensembles as time averages naturally appears in the picture. Informally speaking, sequences modulo 1 are obtained from ordinary sequence of real numbers by wrapping the real line on a circle of circumference equal to 1. More formally, the *integral part* of a real number x is the largest integer $[x]$ smaller or equal to x; and the *fractional part* $\dot{x}$ of x is defined as $\dot{x} = x - [x]$; for simple typographical reasons, we chose the notation $\dot{x}$ rather the more familiar symbols $(x \bmod 1)$, or $\{x\}$, which we prefer to keep for other purposes.

   Problem: given an infinite sequence of real numbers $\xi : n \in \mathsf{Z}^+ \mapsto \xi_n \in \mathsf{R}$, determine the distribution of the derived sequence $\dot{\xi} : n \in \mathsf{Z}^+ \mapsto \dot{\xi}_n = \xi_n - [\xi_n] \in [0,1)$. This problem can be traced back to Kronecker; and the decisive impetus in this study is to be attributed to [Weyl, 1916], although Weyl mentions that one of the main theorems appeared in each of three, almost simultaneous, papers: [Bohl, 1909] [Sierpinsky, 1910] and [Weyl, 1910]. We first establish some notation, and state the fundamental result of the theory – see, e.g. [Kuipers and Niederretter, 1974] – followed immediately with an explanation of its significance.

For any subset $A \subseteq [0, 1)$, let χ_A be the characteristic – or indicator – function of A: $\chi_A(x) = 1$ or 0 depending on whether x belongs to A or not. Let further $\operatorname{card}(A, N; \xi)$ denote the number of entries ξ_n of the sequence ξ that satisfy the condition $\dot\xi_n \in A$ for all n with $1 \leq n \leq N$.

Theorem 5.2.2. *On any sequence* $\xi : n \in \mathsf{Z} \mapsto \xi_n \in \mathsf{R}$, *the following six conditions are equivalent:*

1. *for all pairs a, b of real numbers with $0 \leq a < b \leq 1$:*

$$\lim_{N\to\infty} \frac{1}{N} \operatorname{card}([a, b), N; \xi) = b - a \quad ;$$

2. *for all pairs a, b of real numbers with $0 \leq a < b \leq 1$:*

$$\lim_{N\to\infty} \frac{1}{N} \sum_{n=1}^{N} \chi_{[a,b)}(\dot\xi_n) = \int_0^1 dx\, \chi_{[a,b)}(x) \quad ;$$

3. *for all real-valued, continuous functions f on $[0, 1]$:*

$$\lim_{N\to\infty} \frac{1}{N} \sum_{n=1}^{N} f(\dot\xi_n) = \int_0^1 dx\, f(x) \quad ;$$

4. *for all real-valued, Riemann-integrable functions f on $[0, 1]$:*

$$\lim_{N\to\infty} \frac{1}{N} \sum_{n=1}^{N} f(\dot\xi_n) = \int_0^1 dx\, f(x) \quad ;$$

5. *for all complex-valued, continuous functions f on R, of period 1:*

$$\lim_{N\to\infty} \frac{1}{N} \sum_{n=1}^{N} f(\xi_n) = \int_0^1 dx\, f(x) \quad ;$$

6. *for all integers $k \neq 0$:*

$$\lim_{N\to\infty} \frac{1}{N} \sum_{n=1}^{N} \exp\left(2\pi i k \xi_n\right) = 0 \quad .$$

Definition 5.2.3. *A sequence that satisfies one, and therefore all, of the above equivalent conditions is said to be* uniformly distributed modulo 1.

Comments on the definition and its underlying theorem. The LHS of the first condition of the theorem is the limiting relative frequency for the event $\dot{\xi}_n \in [a, b)$, so that the sequence ξ is uniformly distributed modulo 1, exactly when, for all $0 \leq a < b \leq 1$, the limiting relative frequency of the derived sequence $\dot{\xi}$ is equal to the length of the interval $[a, b)$.

The second condition of the theorem is just a rephrasing of the first.

The third, fourth and fifth conditions can be viewed as statements of *ergodicity*: with $n \in \mathbf{Z}^+$ and $x \in [0, 1)$ respectively playing here the roles of time and space, these conditions require the equality of the time-average and the space-average of functions belonging to a specified class.

The sixth condition is known as the *Weyl criterion*. Formally, it is a restriction of the fifth condition; yet the theorem asserts that, in fact, these two conditions are equivalent.

For any $a \in (0, 1)$, let $\xi^{(a)}$ be the sequence $\xi^{(a)} = \{na \,|\, n \in \mathbf{Z}^+\}$. Then $\xi^{(a)}$ is uniformly distributed modulo 1 if and only if a is irrational. To see this, note that if a is rational, then $\xi^{(a)}$ takes only finitely many distinct values, and so does the derived sequence $\dot{\xi}^{(a)}$ which can therefore not be uniformly distributed on $[0, 1)$; for a irrational, we use the Weyl criterion:

$$|\frac{1}{N} \sum_{n=1}^{N} \exp(2\pi i k n a)| = \frac{1}{N} \frac{|\exp(2\pi i k N a) - 1|}{|\exp(2\pi i k a) - 1|} \leq \frac{1}{N} \frac{1}{|\sin(\pi k a)|} \quad .$$

This provides an example of a ergodic property that *fails* on a set of measure zero: the rationals in $(0, 1)$.

The above example is a special case of a more general result: let $\kappa : n \in \mathbf{Z}^+ \mapsto \mathbf{Z}^+$ be a given, strictly increasing, but otherwise arbitrary, sequence of integers; then the set Z_κ of real numbers a for which the sequence $\{\kappa_n a\}$ is *not* uniformly distributed mod 1, is of Lebesgue measure 0. However, it is now harder to determine which a belongs to Z_κ.

It is *not* true that every subsequence of a sequence, which is uniformly distributed mod 1 , inherits this property. For instance, since e is irrational, we know already that $\xi^{(e)} = \{ne \,|\, n \in \mathbf{Z}^+\}$ is *u.d.mod.1*. To see that its subsequence $\{n! \, e \,|\, n \in \mathbf{Z}^+\}$ is not *u.d.mod.1*, consider the Taylor expansion of the exponential function at $x = 1$: the following equality

$$e = 1 + \frac{1}{1!} + \frac{1}{2!} + \cdots + \frac{1}{n!} + \frac{e^\theta}{(n+1)!}$$

holds for some $0 < \theta < 1$; hence for all $n > 2$: the fractional part of $\{n! \, e\}$ is $e^\theta/(n+1)$ which is dominated by $e/(n+1)$. Thus $(n! \, e) \, mod \, 1$ goes to 0 as n goes to infinity; its values are concentrated at the point 0 and thus the sequence $\{n! \, e \,|\, n \in \mathbf{Z}^+\}$ cannot be uniformly distributed modulo 1.

We shall encounter in Sect. 5.4 the allied notion of *normal numbers*. Let it suffice to mention here that a number x is normal in base b, if and only if the sequence $\{b^n x \,|\, n \in \mathbf{Z}^+\}$ is uniformly distributed mod 1.

Sets of measure zero and the laws of large numbers. Besides the intrinsic mathematical interest of the questions they raise, the above models were inserted to entice the Reader to take a dim view of statements to the effect that in the natural sciences one can safely exclude sets of measure zero. Cournot [1843] had captured some of the uneasiness, that had been recognized by the mid-19th century, when he proposed to distinguish between *'physical'* ('real' or 'factual') impossibilities and *'mathematical'* ('logical' or 'metaphysical') impossibilities. The problem persisted well into the 20th century; in this connection, Von Plato [1994] discusses the position of writers such as Bernstein [1912], whom he quotes thus:

> When one relates the values of an experimentally measured quantity to the scales of the reals, one can exclude from the latter in advance any set of measure zero.

This is too brief to do justice to Bernstein; however, see von Plato. As for our own purposes in this work, we feel much closer to the admonition in [Weyl, 1916]:

> one should not evaluate highly the value of theorems in which an unspecified set of exceptions of measure zero appears.

The point came to the forefront with the advent of a theory which went beyond strings of finite length, and handle uncountable sets of infinite sequences (recall Model 3 above). This is true in particular for the revision of the quintessential result of classical probability, namely the laws of large numbers which, as we saw in Chap. 4, give meaning to the intuitive feeling that *if* S_n denotes the number of "heads" in the first n tosses of an uninterrupted sequence of tossings of a coin, *then* there ought to be a number $p \in [0, 1]$ (with $p = \frac{1}{2}$ if the coin is fair) such that the average $\frac{1}{n} S_n$ gets close to p when n becomes large enough. In fact, the matter is not so simple, and it is settled only superficially by Bernoulli's "law of large numbers" (4.3.12). We are now in position to discriminate between two laws of large numbers: the *weak law* and the *strong law*.

The weak law is the one we know already. Let $(\xi_1, \xi_2, ...)$ denote an infinite sequence of independent Bernoulli trials, each with $\xi_k = 0$ or 1, and $< \xi_k >= p$. The laws of large numbers discuss the behavior of:

$$S_n = \sum_{k=1}^{n} \xi_k \quad . \tag{5.2.9}$$

Scholium 5.2.1 (The weak law of large numbers).

$$\lim_{n \to \infty} \text{Prob} \left\{ |\frac{1}{n} S_n - p| \leq \varepsilon \right\} = 1 \quad .$$

This is still the only version of the law that appears in [Borel, 1909a] text, although one finds in his paper of the same year, [Borel, 1909b], the original version of the *strong law* stated below; see also the contribution of Cantelli [1916]. As discussed in Chap. 4, for n large enough, the average $\frac{1}{n}S_n$ is likely to be near p (gambling-houses are built on such an expectation). *Yet*, the above result could not ensure that $\frac{1}{n}S_n$ stays near p as the number of trials increases. This further expectation is supported by the strong law of large numbers, which is indeed much stronger; as it involves infinite sequences rather than just finite strings, we should not be surprised that its proof requires countable additivity.

Theorem 5.2.3 (The strong law of large numbers).

$$\text{Prob}\left\{ \lim_{n\to\infty} (\frac{1}{n}S_n) = p \right\} = 1 \quad .$$

Hence, with probability 1, $(\frac{1}{n}S_n - p)$ not only becomes small, but it also *remains* small. The strength of the "strong" law is well illustrated by the amount of controversy it initially generated. Referred to for a while as *Borel's paradox* by mathematicians with the eminence of a Steinhaus [1923] or a Hausdorff [1927], it does indeed go against the grain: to require that a sequence converges is a severe demand; yet the strong law asserts that *for almost all* sequences, the frequency $\frac{1}{n}S_n$ admits a limit as $n \to \infty$, and that this limit has the right value. To say that "almost all sequences" have a certain property means that all sequences, but a (unspecified) set of measure zero, have this property. It is essential that the "Prob" of the Theorem be the countably additive, p–dependent probability measure P_p – defined in our Model 3 for the Kolmogorov axioms. The situation was made even more puzzling by the law of iterated logarithms, which could seem to support some sentiments opposite to that suggested by the strong law of large numbers. Indeed, let us take a step back and consider the deviation of S_n from its expected value np, and let us measure this deviation $S_n - np$ in terms of its *standard deviation* $\sqrt{npq}$ (where $p + q = 1$) :

$$S_n^* = \frac{S_n - np}{\sqrt{npq}} \quad . \tag{5.2.10}$$

For comparison purpose, recall (4.3.24) which is a result of the "weak" type:

Scholium 5.2.2 (de Moivre–Laplace).

$$\lim_{n\to\infty} \text{Prob}\{a \le S_n^* \le b\} = \frac{1}{\sqrt{2\pi}} \int_a^b \mathrm{d}u \; \mathrm{e}^{-\frac{1}{2}u^2} \quad .$$

The de Moivre–Laplace theorem confirms the intuitive expectation that, for every *fixed* n, it is very improbable that S_n^* be large. *Yet*, we have learned from everyday experience, that *in the long run*, even the most improbable things are bound to happen; we may therefore be intuitively prepared for the fact that S_n^* could pass occasionally through extraordinay high values; this is indeed the case:

Theorem 5.2.4 (The law of iterated logarithms).

$$\mathrm{Prob}\{\limsup\nolimits_{n\to\infty} \frac{S_n^*}{\sqrt{2\log\log n}} = 1\} = 1 \quad,$$

or equivalently

$$\mathrm{Prob}\{\liminf\nolimits_{n\to\infty} \frac{S_n^*}{\sqrt{2\log\log n}} = -1\} = 1 \quad.$$

This means that: (i) for $\lambda > 1$, with probability 1, only *finitely many* of the events

$$S_n > np + \lambda\sqrt{2npq\log\log n} \tag{5.2.11}$$

occur; and (ii) for $\lambda < 1$, with probability 1, (5.2.11) holds for *infinitely many* n's. This theorem, due to Khinchin [1924] and Kolmogorov [1929], fixes mathematical limitations on large oscillations away from the expected value; and yet, we should note that the law of iterated logarithms, and the strong law of large numbers as well, are statements that hold only "in probability", a situation that can make one feel uncomfortable when one recognizes that sets of measure zero do present their own semantic problems.

We presented the above results in the elementary case of Bernoulli trials, i.e. when ξ is restricted to taking only the values 1 or 0 with probability p and $q = (1-p)$. Generalizations to sums of more general random variables have been made available, first under the impetus that came from the St.Petersburg school, with the work of Chebyshev (1821–1894), Markov (1856–1922) and Lyapunov (1857–1918); the history of this mathematical enterprise is informatively recounted in [Maistrov, 1974], albeit in a then "politically correct" rhetoric. Nowadays, with the notion of random variable appearing quite natural, the form to be expected by these generalizations is rather straightforward. Whatever the motivation the difficulty was still, nevertheless, to delineate precisely the necessary and sufficient conditions under which the theorems could be proven; this development took more than half-a-century: from Chebyshev's announcement of his inequality in 1866–67 to the 1935 proof by Feller that the 1922 Lindeberg condition was not only sufficient but also necessary. As we also need the original inequality later, we mention it briefly here.

Scholium 5.2.3 (Chebyshev inequality). *Let X be a random variable such that both its mean $\langle X \rangle = \mu$ and its variance $\langle (X - \mu)^2 \rangle = \sigma^2$ exist. Then for all $a > 0$:*

$$Prob\{|X - m| \geq a\} \leq \frac{1}{a^2}\sigma^2 \quad.$$

In particular, this inequality asserts that if the variance is small, then large deviations from the mean are improbable, a result that confirms the interpretation of the variance. This inequality is often much too conservative an estimate: considerable improvements on the RHS can be obtained upon using more specific information on the random variable considered. We shall nevertheless see an application of this inequality in Sect. 9.5, when we review Khinchin's alternative to the requirement of ergodicity.

The above results generated in turn new questions, as [Petrov, 1995] demonstrates vividly. The mathematically inclined Reader will find there up-to-date formulations, while particular cases that link effectively with the classical record will orient the less technical onlooker. Since however [Feller, 1968, 1971] is more widely available, we refer to the two volumes [F-I and F-II] of the latter text in our all-too-brief mention of the results that were available – or at least in the air – at the time of [Kolmogorov, 1933].

The first stone, laid by the Chebyshev inequality (see Scholium 5.2.3 above) – [see also F-I : Thm. IX.6, p. 233] – was generalized much later into the Kolmogorov inequality [F-I : Thm. IX.7, p. 234], but even without this generalization, it immediately proved to be the key to a proof of a random variable version of the weak law of large numbers [F-I : Sect. X.2] providing the ε^{-2} estimate that evaluates how reliable the limit is; see also [F-II : Thm. VII.7.1, p. 235]. For the strong law of large numbers, cf. [F-II : Thm. VII.8.1 p. 238]. As its name indicates, zero-one law [Kolmogorov, 1933] for tail events [F-II : Thm. IV.6.2, p. 124] asserts that the *tail events* for any sequence $\{X_k\}$ of (independent, identically distributed) random variables, occur with probability 0 or 1 ; loosely speaking a tail event is independent of how the sequence begins, however long that beginning may be; for instance the convergence of the sum $\sum_{k=1}^{n} X_k$ when $n \to \infty$. Any of the many results that generalize to random variables the de Moivre–Laplace theorem, and thus guarantee convergence to a normal distribution, is usually referred to as a *central limit theorem* – this apt name apparently originated in a 1920 paper by Polya – an archetype is given in [F-II : Thm. VIII.4.3, p. 259].

Scholium 5.2.4 (Central limit theorem). *Let $\{X_k \mid k = 1, 2, \cdots\}$ be independent identically distributed random variables, with $E(X_k) = 0$ and $E(X_k^2) = 1$. Then, as $n \to \infty$, the distribution of the normalized sums*

$$S_n{}^* \equiv n^{-\frac{1}{2}} \sum_{k=1}^{n} X_k$$

tends to the normal distribution with density $\frac{1}{2\pi} e^{-\frac{1}{2}x^2}$.

The hypothesis that the random variables be identically distributed is not essential [F-II: Thm.VIII.4.3, p. 262]; but their mutual independence is. As for the law of iterated logarithms, it was already given in quite general form in [Kolmogorov, 1929, 1933]; for further improvements, see [F-I : Fns. p. 211] and [Petrov, 1995, Chap. 7].

In the widening perspective of the above paragraph, the appearance of [Kolmogorov, 1933] might provide a case from which the interplay between the "context of discovery" and the "context of justification" could be studied.

While there can be little doubt that these results supported Kolmogorov's choices in 1933, it is equally true that the modern proofs – and the pertinence – of the theorems owe much of their transparency to the focus gained from Kolmogorov's concise formulation of his axioms. In this study, one should also keep in mind that Kolmogorov's own purposes evolved over a thirty-year period, with shifts in his position on the spectrum extending from intuitionism to formalism (as portrayed in the beginning of this section) to constructivism again; see Chap. 6.

5.3 Shannon's Entropy

The purpose of this section is to examine the justification for the Boltzmann H-function

$$H_\varphi : t \in \mathsf{R}^+ \mapsto \int d\boldsymbol{v}\, \varphi(\boldsymbol{v};t)\, \ln \varphi(\boldsymbol{v};t) \in \mathsf{R} \qquad (5.3.1)$$

which we encountered in the kinetic theory of gases (Chap. 3), where the study of this function was motivated by its apparent link to the thermodynamical entropy. Recall in particular from Theorem 3.3.1 that for every space-homogeneous solution of the Boltzmann equation the H-function approaches monotonically an asymptotic value given by the Maxwell equilibrium distribution (3.2.10), which we denote here φ_{MB}. We further pointed out that the asymptotic value obtained upon inserting φ_{MB} in H is related to the entropy of the ideal gas: see (3.3.26). The reasoning leading to (3.3.26) suggested that we consider, for an arbitrary distribution function f on the phase space $V \times \mathsf{R}^3$, the functional

$$S[f] = -k \int_\Lambda d\boldsymbol{x} \int_{\mathsf{R}^3} d\boldsymbol{v}\, f(\boldsymbol{x}, \boldsymbol{v}) \ln f(\boldsymbol{x}, \boldsymbol{v}) \qquad (5.3.2)$$

as a possible non-equilibrium generalization of the equilibrium thermodynamical entropy; and since we use the phase-space distribution f instead of the velocity distribution φ, we normalize now the function f by

$$\int_\Lambda d\boldsymbol{x} \int_{\mathsf{R}^3} d\boldsymbol{v}\, f(\boldsymbol{x}.\boldsymbol{v}) = N \qquad (5.3.3)$$

N is the total number of particles of a gas enclosed in a box Λ of volume V.

When we look for Boltzmann's own motivation, we are confronted with the problem that Boltzmann – here, but not exclusively here – *"changed his point of view without informing his readers"* [Klein, 1973] Yet, any reconstruction of the "context of discovery" for the Shannon entropy must start in the period 1877–1898, i.e. with Boltzmann's repeated attempts to give a statistical interpretation of the thermodynamical entropy; see our Chap. 3. This reconstruction should also take note that Maxwell had proposed, on various occasions, interpretations of the second law of thermodynamics in terms of the very incomplete knowledge available on the random agitation of molecules; see e.g. [Maxwell, 1878]. The acceptance of explanations along these lines met with many objections, mostly concerning the delineation of their proper domain. An illustration of the situation is provided by the persistent problems that keep popping out about the Maxwell demon, starting with [Szilard, 1929], and later [Brillouin, 1962]; for modern reviews, see [Bennet, 1982], [Leff and Rex, 1990] and [Earmann and Norton, 1998, 1999].

In this section, we chose to follow the injunction in [Reichenbach, 1938], and focus on the "context of justification". We therefore start with [Shannon, 1948]; didactic accounts of the mathematical results which interest us here can be found in [Shannon and Weaver, 1949], [Khinchin, 1957]. In particular, Thm. 5.3.1 below is stated and proven – with some minor variations in its assumptions – rather early in these presentations: [Shannon and Weaver, 1949, p. 19 and App. 2], [Khinchin, 1957, pp. 1–13].

We proceed now with a sequence of four axioms on a functional S that formalize the intuitive idea that entropy should measure the lack of information pertaining to a probability distribution. For didactic purposes, we first limit ourselves to probability distributions that are defined on a *finite* set $\{\omega_1, \omega_2, \ldots, \omega_\nu\}$ of elementary events:

$$P^{(\nu)} = \{p_i \mid i \in \mathbb{Z}_\nu\} \quad \text{with} \quad p_i \geq 0 \ \forall \ i \in \mathbb{Z}_\nu \quad \text{and} \quad \sum_{i \in \mathbb{Z}_\nu} p_i = 1 \quad (5.3.4)$$

where $\mathbb{Z}_\nu$ stands for the set of ν indices $\{0, 1, \ldots, \nu - 1\}$. To express that $S[P]$ is a property of P alone – and not of the order in which one lists its entries – and that it varies continuously with P, we first impose:

Axiom 5.3.1. *(a) For every permutation* $\pi : i \in \mathbb{Z}_\nu \mapsto \pi(i) \in \mathbb{Z}_\nu$ *and any distribution* $P^{(\nu)} = \{p_i \mid i \in \mathbb{Z}_\nu\}$, *let* $P^{(\nu)}_\pi = \{(p_\pi)_i = p_{\pi(i)} \mid i \in \mathbb{Z}_\nu\}$. *Then*

$$S[P^{(\nu)}] = S[P^{(\nu)}_\pi] \quad .$$

(b) $S[P(\nu)]$ *is continuous in each of the* p_i.

The next axiom expresses that the uniform distribution carries the minimum amounts of information:

Axiom 5.3.2. *For every distribution* $P^{(\nu)} = \{p_i \mid \in \mathbb{Z}_\nu\} : S[P^{(\nu)}] \leq S[U^{(\nu)}]$, *where* $U^{(\nu)} \equiv \{p_i = \frac{1}{\nu} \mid i \in \mathbb{Z}_\nu\}$.

In the third axiom, we eliminate the redundancy of adding to our sample an additional event that has zero probability to occur:

Axiom 5.3.3. *If* $\overline{P}^{(\nu+1)} = \{\overline{p}_i \mid i \in \mathbb{Z}_{\nu+1}\}$, *has a zero entry* $\overline{p}_{i_o} = 0$ *then*

$$S[\overline{P}^{(\nu+1)}] = S[P^{(\nu)}]$$

where $P^{(\nu)} = \{p_i \mid i \in \mathbb{Z}_\nu\}$ *is obtained from* $\overline{P}^{(\nu+1)}$ *by deleting its zero entry.*

The next axiom deals with the entropy of composite distributions. For any distribution $P = \{p_i \mid i \in \mathsf{Z}_\nu\}$ and any collection $\{Q_i\}$ of distributions $Q_i = \{q_{j_i} \mid j_i \in \mathsf{Z}_{\nu_i}\}$ the composite distribution $P \vee Q$ is defined by

$$P \vee Q = \{r_{i,j_i} = p_i q_{j_i} \mid i \in \mathsf{Z}_\nu , \ j_i \in \mathsf{Z}_{\nu_i}\} \ . \tag{5.3.5}$$

We then require

Axiom 5.3.4. *If P, Q and $P \vee Q$ are defined as above, then*

$$S[P \vee Q] = S[P] + S[Q|P] \quad \text{where} \quad S[Q|P] = \sum_i p_i S(Q_i) \ .$$

Note that if all distributions Q_i are identical, which we write $Q_i = Q$, i.e. that P and Q are independent distributions, Axiom 4 demands, in particular, that

$$S[P \vee Q] = S[P] + S[Q] \ .$$

Theorem 5.3.1. *Let $\mathcal{P}$ be the collection of all finite probability distributions. Then any functional*

$$S : P \in \mathcal{P} \mapsto H[P] \in \mathsf{R}$$

satisfying Axioms (1) to (4) must necessarily be of the form

$$S[P^{(\nu)}] = -k \sum_{i \in \mathsf{Z}_\nu} p_i \ln p_i$$

where k is a positive constant.

We immediately note that the entropy of the uniform distribution $U^{(\nu)}$ thus becomes

$$S[U^{(\nu)}] = k \ln \nu \ . \tag{5.3.6}$$

We should remark at this point that the characterization of entropy obtained in Thm. 5.3.1 works independently of any interpretations of probabilities, and thus would work as well in the context of the de Finetti approach as in that of limiting frequencies, both of which we carefully examine in Chap. 6. But just to show how it works with a simple version of the frequentist interpretation, we could imagine having N molecules distributed in ν cells C_i, with N_i molecules in cell C_i, and interpret p_i as N_i/N. We would then obtain for

$$S_N[N_1, \ldots N_\nu] \equiv N S_1[p_i, \ldots p_\nu] \tag{5.3.7}$$

and S_1 computed from Thm. 5.3.1 :

$$S_N[N_1 \ldots N_\nu] = -k \left(\sum_{i \in \mathsf{Z}_\nu} N_i \ln N_i - N \ln N \right) \ . \tag{5.3.8}$$

We are now in the position to establish a **first contact with the work of Boltzmann.** Let us indeed consider all the number of ways W – called "complexions" by Boltzmann, pure or microscopic states by physicists, and elementary events by probabilists – in which the above distribution $\{N_i \mid i \in Z_\nu\}$ can be obtained. Whenever these particles are indistinguishable, we find:

$$W[N_1, \ldots, N_\nu] = \frac{N!}{N_1! \ldots N_\nu!} \quad . \tag{5.3.9}$$

Upon assuming now that all these complexions are equally probable, we obtain that S_B defined, in agreement with (5.3.6), as

$$S_B[N_1, \ldots, N_\nu] \equiv k \ln W \tag{5.3.10}$$

gives

$$S_B[N_1, \ldots, N_\nu] = k \ln \frac{N!}{N_1! \ldots N_\nu!} \quad . \tag{5.3.11}$$

First, a minor point: the tradition wants us to pretend that in the above equations W stands for "Wahrscheinlichkeit", but this is truly not a probability, unless some normalization is made, which turns out to be irrelevant, since the entropy can only be defined up to a constant, at least in the present context.

Second, an essential fact: for large values of the N_i's, and thus of N, we can use the Stirling formula $n! \approx n^n e^{-n}$ – see (4.3.15) for more details – to approximate the factorials. We then obtain:

$$S_B[N_1, \ldots, N_\nu] \approx -k \left(\sum_i N_i \ln N_i - N \ln N \right) \tag{5.3.12}$$

which matches (5.3.8); see also (5.3.7). Hence, if S_B is identified, see later, with the entropy of the gas, S_1 is the average entropy per molecule.

Just as Kolmogorov's theory, which provides the syntax but not the semantics for probability, Shannon's theory of entropy can also be seen as leaving the semantics of entropy open. Is the map, $S : P \to R$ (where P is the space of probability distribution), a measure of the orderliness or the amount of information in a system, or even something else? Shannon's theory – despite of its origin – is not limited to any of these interpretations.

To explore whether the informational entropy defined by the four axioms may have monoticity properties akin to the thermodynamical entropy, we turn again to the Ehrenfest *dog-flea model* described in Sect. 3.4. The number of "complexions" of fleas corresponding to the situation where n of the N fleas are on dog A is a particular case $[\nu = 2]$ of (5.3.9); assuming that these complexions are all equally probable, we obtain the Boltzmann entropy (5.3.11):

$$S(n) = k \ln W(n) \quad \text{with} \quad W(n) = \frac{N!}{n!(N-n)!} \quad . \tag{5.3.13}$$

In Sect. 3.4 we observed that, under the dynamics of the model, the number n of fleas on dog A at first shows a decay towards equalization of the populations of fleas on both dogs, but that fluctuations soon take over. The entropy $S(n)$ in (5.3.13) follows these fluctuations, mitigated only by the fact that $W(n)$, and thus $S(n)$, become sharply peaked around $n = N/2$ when N becomes very large. Upon using the Stirling asymptotic formula we find:

$$\text{for large values of } N \quad : \quad \frac{1}{N}S(n = \frac{N}{2}) \approx k\ln 2 \quad , \qquad (5.3.14)$$

so that the entropy per flea corresponds to the uniform distribution between the two dogs - see (5.3.6).

This is as it should be, *except* that we have no systematic increase of the entropy towards its equilibrium value.

Suppose now that, instead of knowing that there are n fleas on dog A, we only know that this occupation occurs with a probability $p(x_t = n)$ the time evolution of which is governed by the dynamics of the model established in Sect. 3.4, namely

$$p(x_{t+1} = n) = \sum_{m=1}^{N} p(x_t = m)\, P_{m,n} \qquad (5.3.15)$$

where the transition probability $P_{m,n}$ is given by 3.4.1). According to Axiom 4 and Thm. 5.3.1 above, the entropy corresponding to this situation is

$$S(t) = -k\sum_{n=1}^{N} p(x_t = n)\ln p(x_t = n) + \sum_{n=1}^{N} p(x_t = n)\left[k\ln W(n)\right]. \quad (5.3.16)$$

Scholium 5.3.1. *For all $t \geq 0$:*

1. $S(t+1) \geq S(t)$;
2. $S(t) \leq S_{max} = -k\sum_{n=1}^{N} p_{max}(n)\ln\left[W(n)^{-1}p_{max}(n)\right]$
   $\quad\quad\quad\quad\quad\quad$ with $\quad p_{max}(n) = 2^{-N}\,W(n)$
3. $S_{max} = kN\ln 2$.

Remarks: Hence, the dog-flea model delivers what one would like to expect:

1. the entropy is a monotonically non-decreasing function of time;
2. the distribution $p_{max}(n)$ in part (2) of the Scholium coïncides with the unique stationary distribution 3.4.4;
3. the entropy (5.3.16) passes through a maximum exactly where predicted by the heuristic argument leading to (5.3.14);
4. Klein [1956] reminds us that, as one is willing to identify (5.3.13) with the Boltzmann entropy, one should identify (5.3.16) with the Gibbs entropy, interpreting the probabilities $\{p(n) \mid n = 1, 2, \cdots N\}$ as a distribution over an *ensemble* of pairs of dogs, each pair harboring a total of N fleas;
5. the proof of the Scholium, as we presently see, involves a convexity argument, a feature common to many entropy computations.

Proof of the Scholium: Rewrite (5.3.16) as

$$S(t) = k \sum_{n=1}^{N} W(n) \, \varphi(\xi_n(t)) \left.\vphantom{\begin{array}{c} \\ \\ \end{array}}\right\}$$

$$\text{where} \quad \xi_n(t) = W(n)^{-1} p(x_t = n) \quad \text{and} \quad \varphi(\xi) = -\xi \ln \xi \, . \qquad (5.3.17)$$

With the change of variables $p \to \xi$, the evolution equation (5.3.15) reads now

$$\xi_n(t+1) = \sum_{m=1}^{N} \xi_m(t) \Lambda_{m,n} \quad \text{where} \quad \Lambda_{m,n} = \frac{W(m)}{W(n)} P_{m,n} \, . \qquad (5.3.18)$$

From Sect. 3.4, we know that

$$\sum_{m=1}^{N} W(m) P_{m,n} = W(n)$$

so that

$$(a) \quad \Lambda_{m,n} \geq 0 \quad ; \quad \sum_{m=1}^{N} \Lambda_{m,n} = 1$$

$$(b) \quad \sum_{n=1}^{N} \Lambda_{m,n} W(n) = W(m) \qquad \left.\vphantom{\begin{array}{c} \\ \\ \end{array}}\right\} \, . \qquad (5.3.19)$$

Hence (5.3.18) and (5.3.19a) tells us that each $\xi_n(t+1)$ is a convex combination of the $\{\xi_m(t) \mid m = 1, 2, \cdots, N\}$; thus, we can take advantage of the *convexity of the function φ* to obtain the inequality:

$$\varphi(\xi_n(t+1)) \equiv \varphi\left(\sum_{m=1}^{N} \xi_m(t) \Lambda_{m,n} \right) \geq \sum_{m=1}^{N} \varphi(\xi_m(t)) \Lambda_{m,n}. \qquad (5.3.20)$$

This inequality, when used in connection with (5.3.17) and (5.3.18), gives:

$$S(t+1) \geq k \sum_{n=1}^{N} \sum_{m=1}^{N} W(n) \Lambda_{m,n} \, \varphi(\xi_m(t)). \qquad (5.3.21)$$

Upon interchanging the order of the (finite!) sums in (5.3.21), and using (5.3.19b), we see that the RHS of (5.3.21) reduces to $S(t)$, thus proving the inequality (1) of the Scholium. One obtains immediately that the entropy (5.3.16) is maximized by the distribution (3.4.4) thus proving part (2). Part (3) is a trivial consequence of Part (2).   **q.e.d.**

Let us now return to (5.3.6) and note that, if instead of looking for the maximum of the entropy obtained in Thm. 5.3.1 and thus finding the uniform distribution $U^{(v)}$, we ask the question of how to determine the probability distribution that maximizes the entropy subject to some prior information.

A general meaning of the "naturalness" of this question emerges from the de Finetti semantics to be presented in Sect. 6.3 below. As for now, let us explore the answer to this question in the specific context of the search for an interpretation of the Boltzmann H-function; the typical situation obtains when we know in advance the expectation value $\langle E \rangle = \sum_i p_i \varepsilon_i$ of a specific observable. The answer is provided by the method of *Lagrange multipliers*.

The essence of the method can be brought to light by an elementary geometric argument. Let f and $\{g_k \mid k = 1, 2, \ldots, K\}$ be real-valued functions defined on a common convex domain $\mathcal{D} \subseteq \mathsf{R}^\nu$ (with $K < \nu$). *Problem: Find the extremum of f subject to the condition $g_k(x_1, \ldots, x_\nu) = c_k \quad \forall k = 1, \ldots, K$; where the c_k are pre-assigned constants.* Note then that for any unit vector $\boldsymbol{u}$, the scalar product $\boldsymbol{u} \cdot \boldsymbol{\nabla} f$ is the derivative of f in the direction $\boldsymbol{u}$; and that $\boldsymbol{u} \cdot \boldsymbol{\nabla} g_k = 0$ expresses that $\boldsymbol{u}$ is tangent to the level surface $g_k = c_k$.

Hence a necessary condition for f to pass through an extremal value at a point $\boldsymbol{x} \in \mathcal{D}$ is that the vector $\boldsymbol{\nabla} f(\boldsymbol{x})$ belong to the vector-space spanned by the vectors $\{\boldsymbol{\nabla} g_k(\boldsymbol{x}) \mid k = 1, \ldots, K\}$, i.e.

$$\boldsymbol{\nabla} f = \sum_k \lambda_k \boldsymbol{\nabla} g_k \qquad for \; some \qquad (\lambda_1, \ldots \lambda_K) \in \mathsf{R}^K \quad . \tag{5.3.22}$$

One must therefore solve this system of linear equations, and finally adjust the constants λ_k in such a manner that the constraints $g_k = c_k$ be satisfied. This method is due to [Lagrange, 1788]; it should be clear that for the above argument to make sense, we must have $\boldsymbol{\nabla} g_k \neq 0$ in $\mathcal{D}$; in this case, (5.3.22) can be derived in a calculus course (e.g. [Fulks, 1978]) as a consequence of the inverse function theorem (Appendix B); the method and its generalizations – in particular when, following [Fourier, 1798], the constraint equalities $g_k = 0$ are replaced by inequalities $g_k \leq 0$ – are known in the literature on nonlinear programming as the method of [Kuhn and Tucker, 1951]. The history of the latter is traced in [Kjeldsen, 2000]. For an instructive application of this method, see the discussion of the "surprise examination (or unexpected hanging) paradox" in [Bornwein, Bornwein, and Maréchal, 2000]. In the case of interest to us, the variables are $(p_1, \ldots, p_\nu) \in R^\nu$, f is the entropy $S = -k \sum_i p_i \ln p_i$, (or simply - $\sum_i p_i \ln p_i$), $g_1 = c_1$ is the constraint $\sum_i p_i \varepsilon_i = \langle E \rangle$; we should moreover not forget that we are looking for extrema over probability distributions: we should also impose the constraint $\sum_i p_i = 1$. The Lagrange multiplier method requires us therefore to solve

$$\partial_{p_i}[-\sum_i p_i \ln p_i] = \lambda \partial_{p_i}[\sum_i p_i \varepsilon_i] + \mu \partial_{p_i}[\sum_i p_i], \tag{5.3.23}$$

and the solution is

$$p_i = \frac{1}{Z(\lambda)} \, \mathrm{e}^{-\lambda \varepsilon_i} \qquad where \qquad Z(\lambda) = \sum_i \mathrm{e}^{-\lambda \varepsilon_i} \quad . \tag{5.3.24}$$

Note that μ has already been adjusted: the normalization $Z(\lambda)$ is precisely defined in such a manner that $\sum_i p_i = 1$; as for λ, we must still impose

$$\sum_i p_i \varepsilon_i = \langle E \rangle \quad .$$ (5.3.25)

Note also that this condition determines λ uniquely. To see that, it is sufficient to verify that $[\sum_i p_i(\lambda)\varepsilon_i]$ is a monotonic function of λ; a straightforward computation shows that we have indeed:

$$\frac{d}{d\lambda}[-\sum_i p_i(\lambda)\varepsilon_i] = -\sum_i p_i(\varepsilon_i - \langle E \rangle)^2 < 0$$ (5.3.26)

as we can assume without loss of generality that ε_i is not constant in i (otherwise, the constraint $\sum_i p_i \varepsilon_i = \langle E \rangle$ would be inoperative, i.e. equivalent to the constraint $\sum_i p_i = 1$. We would then receive, as prescribed by Axiom 3.2, the uniform distribution).

We wish to make here a **second contact with the work of Boltzmann.** In the case considered specifically by Boltzmann, the observable $E = \frac{1}{2}mv^2$ is the kinetic energy of the particles in the gas, of which a discrete approximation was taken. This suggests considering a continuous limit of (5.3.24) to recover the Maxwell–Boltzmann distribution: up to a multiplicative constant, λ would play the role of the inverse temperature, and thus (5.3.26) would confirm that the internal energy $U = \langle E \rangle$ of a perfect gas is an increasing function of its temperature. Hence this remark would lead to an alternate heuristic justification – see our discussion of Corollary (3.3.1) – for the fact that the Maxwell–Boltzmann distribution maximizes the entropy subject to the constraint that the average energy is known; therefore, the logical status of this consequence of the H-theorem would be quite enhanced: its derivation would avoid using the Boltzmann equation. For this contact with Boltzmann's theory to be satisfactory, two requirements still need to be met. The first, and most basic, demand is to justify the analogy between the entropy of Thm. 5.3.1 – established for discrete probability distributions – and the thermodynamical entropy defined from Boltzmann's H-function – involving continuous distributions. The second requirement is to control the limiting procedures involved in the argument of the previous paragraph; in particular, we must elucidate the sense in which the formula, giving the Shannon entropy and involving only discrete probability distributions, can serve as an approximation to an analogous formula for continuous probability distributions. This turns out to be more subtle than first meets the eye: there is a profound difference between an analogy and an approximation. The following result illustrates what is actually going on.

Theorem 5.3.2. *Let S be a functional defined on the continuous probability distributions $f : x \in X \mapsto f(x) \in \mathsf{R}^+$ on finite intervals $X \subset \mathsf{R}$ and satisfying the following conditions*

1. *$S[U^X] = k \ln|X|$ for $U^X(x) \equiv \frac{1}{|X|} \; \forall \, x \in X$;*
2. *for every partition, $\mathcal{F} = \{X_i \mid i \in \mathsf{Z}_\nu\}$, of X in finitely many "cells"*

$$S[f] = \lim_{|\mathcal{F}| \to 0} S_{\mathcal{F}}[f] \quad \text{with} \quad S_{\mathcal{F}}[f] \equiv \left\{ \, S[P^{\nu}] + S[\mathcal{F} \mid P^{(\nu)}] \, \right\}$$

where:

a. $|\mathcal{F}| \equiv \sup_{i \in \mathsf{Z}_\nu} |X_i|$ *and* $|X_i| = \int_{X_i} dx$ *is the size of the cell* X_i;

b. $S[P^{\nu}]$ *is the entropy [defined in Thm. 5.3.1] of the discrete probability distribution*

$$P^{(\nu)} = \{ p_i \mid i \in \mathsf{Z}_\nu \} \quad \text{with} \quad p_i = \int_{X_i} dx \, f(x) \quad ;$$

c. $S[\mathcal{F} \mid P^{(\nu)}]$ *is the conditional entropy*

$$S[\mathcal{F} \mid P^{(\nu)}] = \sum_{i \in \mathsf{Z}_\nu} p_i \, S[U^{X_i}]$$

and $S[U^{X_i}]$ *is the entropy of the uniform distribution* U^{X_i}.

Then

$$\boxed{ S[f] = -k \int dx \, f(x) \ln f(x) }$$

Proof: It is a simple application of the definition of the integral of a continuous function. Indeed, with $f_i = \frac{1}{|X_i|} \int_{X_i} dx \, f(x)$ we have

$$\lim_{|\mathcal{F}| \to 0} \left\{ \, S[P^{\nu}] + S[\mathcal{F} \mid P^{(\nu)}] \, \right\}$$

$$= \lim_{|\mathcal{F}| \to 0} \left\{ \, -k \sum_{i \in \mathsf{Z}_\nu} p_i \ln p_i + k \sum_{i \in \mathsf{Z}_\nu} p_i \ln |X_i| \, \right\}$$

$$= \lim_{|\mathcal{F}| \to 0} \left\{ \, -k \sum_{i \in \mathsf{Z}_\nu} |X_i| \, f_i \ln f_i \, \right\}$$

$$= -k \int_X dx \, f(x) \ln f(x) \quad . \qquad \textbf{q.e.d.}$$

Note that the conditional entropy $S[\mathcal{F} \mid P^{(\nu)}]$ approximates the loss of information one incurs when one replaces the continuous distribution f by the averaged distribution $P^{(\nu)}$ which ignores what happens in each individual cell X_i. This term *cannot* be neglected in the expression of $S[f]$: when the mesh of the partition becomes finer and finer, it compensates for the increase in $S[P^{(\nu)}]$, and thus ensures that the object one is really interested in, namely $S[f]$, remains finite. It would therefore be a mistake of principle to try and approximate the integral $S[f]$ by the discrete sum $S[P^{(\nu)}]$: the two expressions are analogous, but their proper relation is given by Thm. 5.3.2. Furthermore, keeping the term $S[\mathcal{F} \mid P^{(\nu)}]$ allows to control the limit involved in finding the probability distribution that maximizes the entropy.

Before setting out to find this maximizing distribution, we should remark that the form of the entropy found in Thm. 5.3.2 is indeed suggested in [Shannon and Weaver, 1949, Sect. 20] as the proper analogy with the entropy found in Thm. 5.3.1. It is further suggested there that one could proceed directly from there to solving the constrained maximization problem, via the calculus of variations, although this would involve a few sticky mathematical manœuvers.

The calculus of variations is an established branch of classical mathematics; it was first developed by [Euler, 1744b], who was motivated by geometrical optimization problems; and by [Lagrange, 1788], whose earlier interest was as Euler's, but who later shifted his emphasis to lay the grounds for his epoch-making applications in mechanics; see further [Lagrange, 1806, Leçons XXI–XXII] and [Poisson, 1809], the precursor Johann Bernoulli (1667–1748), the "variation principles" attributed to Maupertuis (1698–1759) and d'Alembert (1717–1783), the work of Legendre (1752–1833), the early objections of Poinsot (1777–1859), and the critical contributions of Hamilton (1805–1865), Jacobi (1804–1851), and Weierstrass (1825–1897). For a historical overview, see [Kline, 1972, Chap. 24 and 30], and for the philosophical underpinnings of the controversies prompted by the use of mathematical variational principles in the natural sciences, see [Pulte, 1998].

A simple version of this can be given in the framework of Thm. 5.3.2. The problem now is to find the maximum of $S[f]$ subject to the constraint

$$\langle E \rangle = \int_X dx\, f(x) \varepsilon(x) \tag{5.3.27}$$

where again we assume for simplicity that the function ε is continuous. We then search for the maximum of the approximating functional appearing in the theorem (see condition 2), namely

$$\left. \begin{aligned} S_{\mathcal{F}}[f] &= S[P^\nu] + S[\mathcal{F} \mid P^{(\nu)}] \\ &= -k \sum_{i \in Z_\nu} p_i \ln p_i + k \sum_{i \in Z_\nu} p_i \ln |X_i| \end{aligned} \right\} \tag{5.3.28}$$

subject to the constraint (5.3.18), now discretized by $\mathcal{F}$, namely

$$\langle E \rangle = \sum_{i \in Z_\nu} p_i\, \varepsilon_i \qquad \text{where} \qquad \varepsilon_i = \frac{1}{|X_i|} \int_{X_i} dx\, \varepsilon(x) \quad . \tag{5.3.29}$$

Again from the *discrete* version of the Lagrange multipliers method, we obtain

$$-\ln p_i + \ln |X_i| = -\lambda \varepsilon_i - \mu \quad . \tag{5.3.30}$$

Upon recalling the definitions of p_i and f_i, and taking the $\lim \mathcal{F} \to 0$ over nested partitions, so that every $x \in X$ can be obtained as $x = \cap_i X_i$, we have

$$f(x) = \lim_{|X_i| \to 0} f_i = \lim_{|X_i| \to 0} \frac{1}{|X_i|} p_i \quad . \tag{5.3.31}$$

Hence (5.3.21) and (5.3.22) together give us

$$f(x) = \frac{1}{Z(\lambda)} e^{-\lambda\,\varepsilon(x)} \quad \text{with} \quad Z(\lambda) = \int_X dx\, e^{-\lambda\,\varepsilon(x)} \quad . \tag{5.3.32}$$

Two side-remarks on the above proof:

1. the counter-term $S[\mathcal{F} \mid P^{(\nu)}]$ in (5.3.28) is responsible for the term $\ln|X_i|$ that appears in (5.3.30), and it is therefore essential for the limit (5.3.31) to make sense;
2. the probability distribution obtained in this manner satisfies an equation of the form

$$\frac{\delta}{\delta f}\left[-f \ln f - \lambda f \varepsilon - \mu f\right] = 0 \tag{5.3.33}$$

where λ and μ are constants, adjustable to meet the constraints; this equation is precisely the equation with which we would have started had we proceeded directly with the general methods provided by the calculus of variations.

The general and modern version of the calculus of variations, to which [Shannon and Weaver, 1949] alluded, allows to firm up the above discussion and relax some of the conditions we imposed to keep it elementary. In particular, the conditions on f and ε can be somewhat relaxed, although these adjustments would be of little interest to us here beyond the fact that, when the next extension is made, one must be aware that restrictions on the growth of this functions must be imposed. More importantly, the modern presentation of the theory is not restricted to finite intervals X and finite partitions $\mathcal{F}$; in particular, measurable subsets of $\mathbb{R}^n$ enter naturally into the picture.

This leads to a **third contact with the work of Boltzmann:** We have now the justifications necessary to write the entropy functional for probability distributions on phase space:

$$S[f] = -k \int_V d\boldsymbol{x} \int_{\mathbb{R}^3} d\boldsymbol{v}\, f(d\boldsymbol{x}, d\boldsymbol{v}) \ln f(d\boldsymbol{x}, d\boldsymbol{v}) \tag{5.3.34}$$

and to prove that, under the constraint that the average kinetic energy of the molecules be fixed, this entropy is maximized by the probability distribution

$$f(\boldsymbol{x}, \boldsymbol{v}) = \frac{1}{N} \varphi(\boldsymbol{v}) \tag{5.3.35}$$

where $\varphi(bfv)$ is the Maxwell–Boltzmann distribution. The comparison between $S[f]$ and the thermodynamical entropy discussed in Chap. 3 is now straightforward, and serves to fix the constant k.

We have therefore established some of the desired relations between:

1. the Shannon entropy discussed in this section;
2. the Boltzmann entropy – or equivalently the Boltzmann H-function – that made their appearance in the kinetic theory of gases of Chap. 3;
3. the equilibrium thermodynamical entropy of Chap. 2.

While the intimate link between the statistical definitions of Shannon and of Boltzmann is established *independently* of whether one considers situations outside equilibrium or not, we should keep in mind, however, that the link between the irreversible behavior of the H-function and the thermodynamical entropy is still at the level of a visionary conjecture.

The interpretation of *entropy as a measure of information* is reinforced by the fact that it measures the minimal expected length of a code. Specifically, let S be a finite collection of objects $s_1, s_2, \cdots, s_n$ each of which occur with a probability $p_i = p(s_i)$. S is called the *source*; for instance S could be the characters in which a natural language, like English, is written; it is known that e is the most frequent character, i.e. the probability $p(e)$ is larger than the probability of all the other characters. Let then X be an *alphabet,* and X^* denote the collection of all finite strings of elements of this alphabet. For instance $X = \{0, 1\}$ or $\{\cdot, -\}$. A *code* is defined as a map

$$\phi : S \rightarrow X^* \tag{5.3.36}$$

that satisfies the following two properties.

1. The code is *decodable*, i.e. its natural extension (under concatenations)

$$\phi^* : (s_{i_1}, s_{i_2}, \cdots, s_{i_k}) \in S^* \mapsto \phi(s_{i_1})\phi(s_{i_2}) \cdots \phi(s_{i_k}) \in X^* \tag{5.3.37}$$

   is injective. Thus any message, sent in the code, must be decipherable.
2. The code is *instantaneous*, i.e. one does not need to wait for the end of the message to start deciphering it; in other words, the code not only is unambiguous in the sense of condition (1), but also, if the transmission of the message ends before the end of the message, whatever part of the message that has been transmitted has unique meaning; or the code is *prefix-free*, i.e. $\phi(S)$ is prefix-free.

We may note already here that computational algorithms play an essential role in the discussion of the nature of randomness in Sect. 6.2. We should also add that for the purpose of our intrusions in communication theory, we can safely limit our attention to communication through noiseless channels. Upon returning to the above conditions (1) and (2) above, the Reader is encouraged to try and see whether the Morse code should be augmented with a 'blank' character to satisfy these conditions; hint: consider that in this code

$$\phi(e) = \cdot \qquad \phi(a) = \cdot - \qquad \phi(n) = \cdot \cdot -$$

Upon denoting by $l(\phi(s))$ the length of the coded object s, we define the *expected length* of the code as

$$L(\phi) = \sum_{s \in S} p(s)\, l(\phi(s)) \quad . \tag{5.3.38}$$

Given $\mathcal{S}$ and X, the questions then arise of whether there exist a code with prescribed length spectrum, and of obtaining an estimate of the minimal expected length of a code. The answer to the first question was given in [Kraft, 1949]; see [Hamming, 1986, Calude, 1994]:

Theorem 5.3.3 (Kraft inequality). *A necessary and sufficient condition for the existence of a code $\phi : \mathcal{S} \to X^*$ with preassigned length spectrum $\{l(\phi(s)) \mid s \in \mathcal{S}\}$ is*

$$\sum_{s \in \mathcal{S}} b^{-l(\phi(s))} \leq 1 \qquad \text{where} \quad b = \operatorname{card} X \quad .$$

The method of Lagrange multipliers, used to optimize the expected length (5.3.29) subject to the Kraft inequality, gives:

Corollary 5.3.1.

$$L(\phi) \geq L_{min} \qquad \text{with} \quad L_{min} = \frac{1}{\ln b} \operatorname{H}(\mathcal{S}, p)$$

$$\text{where} \quad \operatorname{H}(\mathcal{S}, p) = -\sum_{s \in \mathcal{S}} p(s) \ln p(s) \quad .$$

Remarks

1. The lower bound L_{min} is not necessarily reached since the Lagrange multipliers method does not take into account that the lengths $l(\phi(s))$ must be integers.
2. The entropy $\operatorname{H}(\mathcal{S}, p)$ depends only on the source, whereas the constant $(\ln b)^{-1}$ depends only of the cardinality of the alphabet X : for a given source $(\mathcal{S}, p)$, the shorter the coding alphabet X, the longer the minimal expected length L_{min} of the code, with the extreme case, $b = 1$, leading – as it should – to the impossibility of coding anything.
3. Another connection between entropy and coding is disccused in Sect. 8.3 when we introduce the notion of dynamical entropy.

One should once again note the similarity between the formal treatments of entropy and of probability. The axiomatic theory or the "syntax" of entropy, as given by Shannon, admits at least two different types of models.

In concluding this section, we mention that the standard *quantum* version of the Shannon entropy stems from the pioneering work of Von Neumann [1927] which assigns to a density matrix ϱ the entropy

$$S[\varrho] = -k \operatorname{Tr} \varrho \ln \varrho \quad . \tag{5.3.39}$$

For a formal definition of quantum states and density matrices, see Def. 6.3.2 and Thm. 6.3.2 below. (5.3.30) means $S[\varrho] = -k \sum_i p_i \ln p_i$ where the density matrix ϱ has been written in diagonal form $\varrho = \sum_i p_i P_i$ in which the P_i are one-dimensional mutually orthogonal projectors. More generally, for any positive operator A on a Hilbert space $\mathcal{H}$, the trace of A is defined as

$\operatorname{Tr} A \equiv \sum_k (A\Psi_k, \Psi_k)$ where $\{\Psi_k\}$ is an orthonormal basis in $\mathcal{H}$; the value $\leq \infty$ of $\operatorname{Tr} A$ is independent of the basis chosen to compute it; this is equivalent to saying that for every unitary [i.e. $UU^* = I = U^*U$] operator U acting on $\mathcal{H}$, we have: $\operatorname{Tr} A = \operatorname{Tr} U^*AU$.

There are several ways to justify this definition. For instance, a quantum equivalent to Thm. 5.3.1 is stated and proven in [Thirring, 1983], namely:

Theorem 5.3.4. $S[\varrho] = -k \operatorname{Tr}\varrho \ln \varrho$ *is the only quantum entropy satisfying the following conditions:*

1. *$S[\varrho]$ is a continuous function of the eigenvalues of ϱ;*
2. *$S[\varrho]$ is normalized by* $\quad S\left[\begin{pmatrix} \frac{1}{2} & 0 \\ 0 & \frac{1}{2} \end{pmatrix}\right] = k \ln 2 \quad$;
3. *$S[\varrho] = S[P^{(\nu)}] + \sum_{i \in \mathsf{Z}_\nu} p_i S[\varrho_i]$*
   whenever
   $P^{(\nu)} = \{p_i \mid i \in \mathsf{Z}_\nu\}$ with $p_i \geq 0$ and $\sum_i p_i = 1$;
   $\{\varrho_i \mid i \in \mathsf{Z}_\nu\}$ are density matrices acting on Hilbert spaces $\mathcal{H}_i$ $(i \in \mathsf{Z}_\nu)$;
   ϱ is the density matrix defined on the Hilbert space $\mathcal{H} = \oplus_{i \in \mathsf{Z}_\nu} \mathcal{H}_i$ by
   $\varrho = \oplus_{i \in \mathsf{Z}_\nu} p_i \varrho_i$.

Under the conditions that the quantum Hamiltonian (i.e. the energy operator) H be positive, and be such that

$$\operatorname{Tr} e^{-\lambda H} < \infty \qquad \text{for all real} \quad \lambda > 0 \quad, \tag{5.3.40}$$

Von Neumann [1927] moreover established that the maximum of the entropy $S[\varrho] = -k\operatorname{Tr}\varrho \ln \varrho$, under the constraint $\operatorname{Tr}\varrho H = \langle E \rangle$, obtains for

$$\varrho = \frac{1}{Z(\beta)} e^{-\beta H} \qquad \text{with} \qquad Z(\beta) = \operatorname{Tr} e^{-\beta H} \quad. \tag{5.3.41}$$

As in the classical situation, the value of β is determined by the value of $\langle E \rangle$ and, in fact, β is to be interpreted as the inverse temperature: $\beta = (kT)^{-1}$.

For modern developments, see [Ohya and Petz, 1993] and Sect. 8.3 below. An original path motivated by Wigner's theory of random matrices – for the latter, see e.g. [Mehta, 1991] – was developped in a series of papers, starting with [Voiculescu, 1993]; see also [Voiculescu, 1997] and, for a new departure, [Voiculescu, 1998].

5.4 The Transcendence of Randomness

The aim of this section is to examine a variety of elementary mathematical objects – prime numbers, irrational and transcendental numbers – to illustrate the expectations one might have, or not have, about what is random. In particular, we would hope to help reconcile an apparent conceptual tension, namely this: *whereas randomness is ubiquitous, it is nevertheless elusive and hard to capture in the mathematical net.* Much of this section is therefore illustrative. However, towards the end of the section, we define an appealing notion of statistical randomness, due to Borel, and we confront it with specific examples such as the Champernowne–Copeland–Erdös numbers, and the decimal expansion of π. In doing so we aim at making more acceptable an appeal to the perhaps less familiar tools involved in recursive function theory; the latter logical aspects of a computationally satisfactory definition of randomness is considered in Chap. 6.

On the one hand, randomness seems to be associated with situations that are partially, or even inherently, unknown and/or non-deterministic. And, one has to take stock of the fact that probability statements have found their way into number theory – the Book of eternal truths, as Erdös put it – as well as mundane questions such as why there seems to be so little order in the decimal expansion of π, or the almost transcendental fact that the quadrature of the circle is a proven impossibility.

On the other hand, even a positivist user [Von Neumann, 1932a] of von Mises' collective had to admit, when later faced with actually producing the kind of randomness necessary for classical Monte-Carlo computations:

> *Anyone who considers arithmetical methods of producing random digits is, of course, in a state of sin ... there is no such thing as a random number – there are only methods to produce random numbers ... and a strict arithmetic procedure ... is not such a method.*
>
> [Von Neumann, 1951]

Mathematically, this might have suggested a definition of randomness along the line that a sequence is to be reputed random if it fails to satisfy regularity (or arithmetic) tests. But a Popperian *falsificationist* approach [Popper, 1935] to a mathematical definition of randomness was apparently not in the cards.

Still, empirical random numbers generators have been proposed that do actually use as input the stochastic behavior encountered in many natural phenomena, prototypes of which are discussed in Chap. 3; in a lighter vein, a device was produced from unorthodox enough a source to warrant notice in *Scientific American* (November 1997, p. 28): it uses digital snapshots of the viscous fluid flow in, of all things, the lava lamps that were of rather moot repute in the mesmerizing, groovy 1960s. We want to retain here the fact that such a simple setting offers already enough randomness to withstand the scrutiny of random-number crunchers. For a colorful palette of mathematically more tractable but "chaotic" dynamical systems, see the marvelous do-it-yourself text by Devaney [1992].

Determinism and randomness; where do the prime numbers stand?
By way of contrast, nothing seems *a priori* more immutable, orderly, and deterministic than the procession of the integers. However, a closer look, focussing on the succession of the prime numbers, reveals intriguing features. For instance, while it is relatively inexpensive to determine whether integers are prime or not, it is extremely expensive to find the prime factorization of

a given integer n, already when the latter happens to be the product of only two large primes; this difficulty is real enough to be exploited commercially in public key cryptography [Stinson, 1995].

To size up this suggestive intrusion of randomness, let us consider the successive sums, as $n \to \infty$, of the reciprocal of the prime numbers not exceeding n; this sequence diverges, at the rate given by the estimate:

$$\sum_{p \leq n} \frac{1}{p} = \log \log n + B + 0(\frac{1}{\log n}) \tag{5.4.1}$$

where B is a constant. This suggests the study of the asymptotic behavior of the number $\nu(m)$ of prime divisors of the integer m.

Remarkably fine results on the asymptotic behavior of their average and variance

$$\frac{1}{n} \sum_{m \leq n} \nu(m) \quad \text{and} \quad \frac{1}{n} \sum_{m \leq n} [\nu(m) - \log \log n]^2$$

were first derived by classical analytic methods, see e.g. [Hardy and Ramanujan, 1917, Narkiewicz, 1977]. Although there had been occasional use of probabilistic *methods* (e.g. the Chebychev inequality, Scholium 5.2.3 above) in Number theory, notably by Turán [1934], it has been argued with some justification that it is the use of the *ideas* of probability theory that should mark the birthdate of the mathematical field now known as *probabilistic number theory*.

The year is 1939, when Erdös met Kac who had been working for some time – largely at the instigation of Steinhaus – on the notion of *statistical independence*. This notion was indeed the key to the result of interest here. Erdös and Kac [1939] noticed that, upon writing

$$\left. \begin{array}{l} \nu(m) = \sum_{p|m} 1 = \sum_{k=1}^{n-1} \delta_k(m) \\ \\ \text{where} \\ \\ \delta_k(m) = \begin{cases} 1 \text{ if } k \text{ is a prime divisor of } m \\ 0 \text{ otherwise} \end{cases} \end{array} \right\} . \tag{5.4.2}$$

one introduces sums of statistically independent random variables, so that *"the central limit theorem of the calculus of probability can be applied."* They actually presented their result in a more general situation; their general theorem reduces here to the following statement: given $-\infty \leq a < b \leq +\infty$, the number

$$K_n(a, b) \equiv \text{card}\{m \in \mathbf{Z}^+ \mid 1 \leq m \leq n, a < \frac{\nu(m) - \log \log m}{\sqrt{\log \log m}} < b\} \tag{5.4.3}$$

obeys the asymptotic law:

$$\lim_{n \to \infty} \frac{1}{n} K_n(a, b) = \frac{1}{\sqrt{2\pi}} \int_a^b dx \, e^{-\frac{1}{2}x^2} . \tag{5.4.4}$$

Thus, they uncovered the presence of de Moivre–Laplace's normal distribution (see 4.3.24) familiar to gamblers, naturalists, demographers or sociologists; and to scientists evaluating the distribution of non-systematic errors in their measurements. But randomness in the orderly field of elementary number theory! How incongruous this intrusion appeared to be for quite some time is preserved in the reminiscences [Kac, 1985, p. 91] of a marvelous *conteur* who was a principal in the unfolding of the original story. Further mathematical developments around the idea of statistical independence in number theory are discussed by Kac [1959c] in his celebrated Carus monograph; for an update, see [Goodman, 1999]. Kac shows, in particular, how these ideas provide a theoretical frame for Borel's notion of normal numbers, which we review later in this section.

Is irrational random enough; what about e and π? We now want to convey a feeling for the sense in which most sequences are random, even though it is very difficult to identify a specific sequence as being definitely random. The full discussion is postponed to Sect. 6.2. Consider first:

$$\left.\begin{array}{rl} \sqrt{2} &= 1.41421356237310\ldots \\ e &= 2.71828182845905\ldots \\ \pi &= 3.14159265358979\ldots \end{array}\right\}. \tag{5.4.5}$$

These three numbers are irrationals, but – from these expansions – we can only say that since we do not detect any periodicity, we cannot infer that they are rationals. To *prove* that they are actually irrationals is another matter. That $\sqrt{2}$ is irrational was known to the Ancients – perhaps not Pythagoras (570–490 BC), but certainly his disciples – and the proof is obtained by a contradiction with the fact that no integer can be both even and odd. For π and e, the answer came much later; the original proofs involved the use of continued fractions, a technique the name of which we owe to Wallis [1655] who used it to study π, or rather its inverse. It is known that every $x \in (0,1)$ can be written, uniquely, as a simple continued fraction, which may or may not stop:

$$x = (b_1 + (b_2 + (b_3 + \cdots)^{-1})^{-1})^{-1} \equiv [b_1, b_2, b_3, \cdots] \tag{5.4.6}$$

where the b's are non-negative integers, and the $\equiv$ sign only indicates that what follows is an abreviated notation. With this notation

$$\left.\begin{array}{rl} \sqrt{2} &= 1 + [2,2,2,2,2,2,2,2, \ldots] \\ e &= 2 + [1,2,1,1,4,1,1,6, \ldots] \\ \pi &= 3 + [7,15,1,292,1,1,1,2, \ldots] \end{array}\right\} \tag{5.4.7}$$

Incidentally, Bombelli [1572] had already derived, somewhat reluctantly, an expression amounting to the simple continued fraction for $\sqrt{2}$; today this particular expansion is immediate: notice that $\sqrt{2} = 1 + y^{-1}$ implies that $y = 2 + y^{-1}$. Bombelli remarked that the expansion does not terminate, thus confirming an earlier and somewhat curious intuition:

an irrational number is not a true number but lies hidden in a kind of cloud
of infinity.

[Stifel, 1544]

Euler [1744a] proved in 1737/44 that the simple continued fraction expansion of every rational number $x = p/q$ terminates, from which he could infer that e is irrational: indeed, he had exhibited a simple continued fraction expansion of e, the first terms of which are reported in (5.4.7). [Liouville, 1840] showed that another classical, more elementary, proof of the irrationality of e could be used to obtain a better – and subsequently important – result, namely that e is not the solution of any equation of the form: $a\,e + b^{-1}e = c$, where a would be a strictly positive integer and b, c integers. The same holds already for e^2. Lambert [1768] established that for any non-zero, but rational number x, $\tan x$ is not rational; hence, since $\tan \frac{\pi}{4} = 1 : \pi$ *is not rational.* He also verified independently the irrationality of e; an interesting story is told in [Laezkovich, 1997].

Thus, it did take some effort to obtain the irrationality of e and π in spite of the fact that most real numbers are irrational: indeed, we proved in Sect. 5.2 that the rational numbers, being countable, form a set of Lebesgue measure zero. For modern mathematical proofs, see [Hardy and Wright, 1979]; and for a gentler view on the historical progression of ideas, [Maor, 1994].

Is transcendental random enough; what about e and π? Compare the three numbers the simple continued fraction of which is given in (5.4.7). Clearly the first is 'more regular' or 'less random' than the other two. This observation can be sharpened as follows.

A number is said to be *algebraic* if it is a root of an algebraic equation with integer coefficients, i.e. if it is the solution of an equation of the form

$$\left.\begin{aligned}
a_0\,x^n + a_1\,x^{n-1} + \cdots + a_{n-1}\,x + a_n = 0 \\
\text{with} \quad a_o \neq 0 \quad \text{and} \quad a_k \in \mathsf{Z}, \ k = 0, 1, \cdots, n
\end{aligned}\right\} . \qquad (5.4.8)$$

Since the union of countable sets is countable, the set A of all algebraic numbers is countable – and hence is a subset of Lebesgue measure zero in R – as was the set of all rationals, which it obviously contains. The inclusion is proper since $\sqrt{2}$ is not rational, but is algebraic: it is a root of $x^2 - 2 = 0$. A theorem by Gauss [1801] asserts that every algebraic integer – i.e. an algebraic number with $a_o = 1$ in (5.4.8) – either is an integer or is irrational.

Lagrange [1769], who had used continued fractions to approximate the irrational roots of algebraic equations, proved the periodicity of the simple continued fraction expansion of any real root of a quadratic equation; [Euler, 1744a] had already seen the converse implication; in particular, Lagrange's result implies directly that the continued fraction expansion of $\sqrt{2}$ must be periodic (as Bombelli had found explicitly: see (5.4.7).

The numbers that are not algebraic are called *transcendental* because, as Euler [1744a] had remarked, *"they transcend the power of algebraic methods."*

The fauna of transcental numbers includes indeed some quite exotic specimen, and all seem hard to catch. Here again, in spite of the fact that most numbers are transcendental, to show that a specific number is transcendental requires some exertion.

Legendre [1794] conjectured that π may be transcendental, adding " *but this seems quite difficult to prove rigorously.*" Difficult it was, indeed. Some fifty years elapsed until Liouville [1844], at long last, exhibited a number that was provably transcendental. He had the insight to show first that irrational algebraic numbers cannot be "too rapidly" approximated by rational numbers; specifically: if x is an algebraic number of degree $n \neq 1$, there exists a constant $C > 0$ (depending on x) such that for all integers p and q (with $q > 0$) the following inequality holds

$$\left| x - \frac{p}{q} \right| > \frac{C}{q^n} \quad . \tag{5.4.9}$$

This suggests the following definition: a *Liouville number* x is a real number such that for every positive integer n the inequality

$$0 < \left| x - \frac{p}{q} \right| < \frac{1}{q^n} \tag{5.4.10}$$

has an infinite set of solutions $(p, q) \in \mathbb{Z} \times \mathbb{Z}^+$. Liouville's main result (5.4.9) thus implies that every Liouville number is transcendental, and

$$x = \sum_{n=1}^{\infty} 2^{-k!}$$

is such a number.

Hermite [1873] proved that e is transcendental, but wrote to a friend: "*I do not dare to attempt to show the transcendance of π.*" Lindemann [1882] established that: if, for $(k = 1, 2, \cdots, n)$, x_k are distinct algebraic numbers, real or complex, and if p_k are positive algebraic numbers, not all zero, then

$$\sum_{k=1}^{n} p_k \, e^{x_k} \neq 0 \quad . \tag{5.4.11}$$

In particular, since $1 + e^{i\pi} = 0$, $i\pi$ is not algebraic; since the product of two algebraic numbers is algebraic, and i is obviously algebraic, π *is not an algebraic number*. This implies *a fortiori* that the simple continued fraction expansion of π is not periodic. Lindemann's result is remarkable in at least two ways:

(1) The method of proof prompted the 7th of the famous problems in [Hilbert, 1902]. Several beautiful results have been secured; for an update, see e.g. [Shidlovskii, 1989]. The hunt is still going on; for instance, Nestrenko [1996] proved that for every positive integer $n : \pi$ and $e^{\pi\sqrt{n}}$ are algebraically

independent, i.e. if a polynomial P in two variables, with coefficients in the integers – or, equivalently, the rationals – satisfies $P(\pi, e^{\pi\sqrt{n}}) = 0$, then P must be identically zero; this implies in particular that $\pi + e^{\pi\sqrt{n}}$ is irrational, and even for $n = 1$ that was new.

(2) Lindemann's result settles – in the negative – the Ancients' problem of the quadrature of the circle: it is not possible *by straightedge-and-compass alone* to construct a square that has the same area as a given circle. Indeed, this would amount to giving a geometric construction of π, and such constructions, using only straightedge and compass, correspond to the solution of (quadratic) algebraic equations.

The latter elementary fact had already been used in [Gauss, 1801] – a monument expounding many pioneering and profound insights in Algebra – to solve another Ancient problem: the criterion for a $n-$sided regular polygon to be contructible with straightedge and compass alone is that

$$n = 2^m \prod_j F_j \qquad (5.4.12)$$

where m is any positive integer, and the F_j are distinct Fermat primes chosen from

$$F_k = 2^{2^k} + 1 \quad k = 0, 1, 2, 3, 4, \cdots [?] \quad . \qquad (5.4.13)$$

The first, geometrically constructible, regular polygon with a prime number of sides that was not known to the Ancients, the heptadecagon $(17 = F_2)$, was exhibited by the teenager Gauss in 1796; as a gauge of the importance of this result, the figure was incised on the tombstone of the *Princeps Mathematicorum* in the Albanifriedhof in Göttingen. For a concise presentation of the mathematics, see [Gallian, 1998, Chap. 23], and for its historical context [Dickson, 1971]. Gauss knew that Euler had proven that F_5 is not a prime; ever since, the existence of a Fermat prime for $k > 5$ is an open problem, hence the question mark in (5.4.13).

It seems reasonable to ask whether every transcendental number must be a Liouville number. This is however not the case [Mahler, 1937]; in fact, for all integers $p, q \geq 2$:

$$|\pi - \frac{p}{q}| > q^{-\beta} \quad ;$$

this was proved by Mahler [1953] with $\beta = 42$ (it does not matter for our immediate purpose whether this value of β is optimal).

Transcendentals are elusive indeed! Having rejected the algebraic numbers as too regular, we still have to ask for a measure allowing to decide whether the ubiquitous but elusive transcendentals provide enough randomness without having to exclude the Liouville numbers.

Is normal (in Borel sense) random enough; what about e and π?
Borel [1909b] introduced the notion of *normal number*, a concept of *statistical randomness* we propose to sketch now.

For any fixed prime number p, let $Z_p = \{0, 1, \cdots, p-1\}$; and let us write $B^{(k)} = (b_1, b_2, \cdots, b_k)$ for the elements of Z_p^k; i.e. $B^{(k)}$ is a string of length k, made of elements of Z_p. For every real number x, consider the sequence $\xi : n \in Z^+ \mapsto \xi_n \in Z_p$ obtained from the p–adic expansion of the fractional part of x :

$$x - [x] = \sum_{n=1}^{\infty} \xi_n p^{-n} \quad \text{with} \quad \xi_n \in Z_p \tag{5.4.14}$$

where $[x]$ denotes the largest integer smaller than, or equal to x. Let us finally denote by $N_n(x, B^{(k)})$ the number of occurences of the string $B^{(k)}$ in the initial string $\xi^{(n)} = (\xi_1, \xi_2, \cdots, \xi_n)$ of the expansion of $x - [x]$.

Definition 5.4.1. *The number x is* normal in base p *whenever, for all $B^{(k)} \in Z_p^k$,*

$$\lim_{n \to \infty} \frac{1}{n} N_n(x, B^{(k)}) = p^{-k} \quad . \tag{5.4.15}$$

Note that

1. this definition requires in particular that the limit in the LHS exists;
2. for the particular case $k = 1$, the above is the requirement that ξ is uniformly distributed;
3. the higher conditions (for $k > 1$) require the disppearance of the corresponding correlations.

It is known that, except for a set of Lebesgue measure zero, almost every real number x is normal in each base p, and thus – since the countable union of null sets is still a null set – for all bases p. In Borel's hands this is tantamount to a strong law of large numbers. Also, it is known that if the ratio of $\log p$ and $\log p'$ is rational then *either* x is normal in the basis of both p and p' *or* x is not normal in the basis of either. [Kac, 1959c, Schmidt, 1960, Mendès-France, 1967, Kamac, 1973, Kuipers and Niederretter, 1974, Goodman, 1999, Nillsen, 2000] establish connections between normality and randomness in a sense akin to von Mises' – see Chap. 6.

Yet again in spite of their theoretical abundance, it is difficult to exhibit *specific* numbers that *are* normal; for instance, it is not known whether e or π, are normal or exceptional. One normal number, constructed by Champernowne [1933], is: .1234567891011121314... ; this number is transcendental. Upon replacing the successive integers by the successive primes, one obtains .1357111317..., which is also normal; this was proven as a particular case by Copeland and Erdös [1946], who showed more generally that, *if $a_1 < a_2 < a_3 < \cdots$ is an increasing sequence of integers such that, for every $\theta < 1$, the number of a's up to N exceeds N^θ, provided N is sufficiently large, then the number $x = .a_1 a_2 a_3 \cdots$ is normal with respect to the basis β* in which the integers a_k are expressed. The condition of the theorem is satisfied for the successive primes since the number of primes smaller than N is larger than $c N / \log N$, for all $c < 1$, an estimate already known to Legendre. Other normal numbers that would be less contrived have not followed suite.

Big-time π. In view of the paucity of theoretical results just mentioned, it is instructive to turn briefly to the empirical evidence. Algorithms to compute π were already known to Wallis [1685] and to Leibniz [1682]; for antecedents, see e.g. [Struik, 1969, p. 351]. Asides from the notorious Gregory–Leibniz expansion of $\pi/4$ as an alternating series of the inverse of the odd integers, a wealth of infinite series and infinite product expansions for π and its powers were obtained by Euler between 1734 and 1755, connecting π with the Bernoulli numbers and the positive integer values of what was recognized later as the Riemann ζ–function; for an account of these developments, see e.g. [Kline, 1972, pp. 439 and 448–449], [Hairer and Wanner, 1996, sections I.4–I.6], or [Lange, 1999].

While these algorithms are very simple to state, their convergence is so slow as to prohibit any serious improvement on Archimedes' approximation $3\frac{10}{71} < \pi < 3\frac{1}{7}$. Advances in both computing theory and computer technology have since then made available plenty of material on the numerical value of π. Easily readable reviews are given in [Wagon, 1985], [Bailey, Borwein, Borwein, and Plouffe, 1997]; see also [Berggren, Bornwein, and Bornwein, 1999]. These allow us to report here only some of the highlights.

About two thousand decimals had been computed in 1949 by ENIAC; more than one million by 1973; some 20 years later, by early 1995, close to 2.5 billions digits in this decimal expansion were known; and in August 1997, the record was 51.5 billions, and, as of October 1999, more than 206 billions! It is difficult to visualize what it really means to know a *natural* object with such precision:

> *four decimals give π with a precision sufficient for pratical applications. From 16 decimals one obtains, within a hair's thickness, the length of the circumference of a circle whose radius would be the average distance from the Earth to the Sun. Now, replace the Sun by the most distant nebula, and the hair by the smallest particle known to the physicists: to reach such a fantastic precision, you would need only 40 decimals.*   [Dubreil, 1948]

Three lessons accrue from the production of long decimal expansions of π.

1. Estimations of randomness require long strings. The persisting news is that no substring has been detected to appear predictably in the decimal expansion of π. Over an initial segment of 10 millon digits, the *observed* relative frequency of a string of five consecutive digits, all different from one another, is $.302488 \cdot 10^{-5}$; under the assumption of independent uniform distribution the *theoretical* value is $.3024 \cdot 10^{-5}$. The error is compatible with the fact that, over strings of length n the observed deviation from the relative frequencies of each digit approaches zero as $1/\sqrt{n}$: see de Moivre–Laplace (4.3.24); specifically, over the first 10 millions digits these relative frequencies are observed to be between $.0999337$ and $.1001093$.

2. This kind of search focuses the development of new algorithms, and makes possible their evaluation. For instance, an algorithm was designed which

computes the digit appearing in any prescribed place without having to compute first the digits preceding it. This is however not faster than the best traditional algorithms; it does not allow to predict directly at which places a prescribed digit will appear – so no strategy is available yet to disprove that the decimal expansion of π is random.

3. Having reliably produced/stored a 'standard model' – i.e. a sufficiently long decimal expansion of π – one can test the hardware or software integrity of other computers by instructing them to compute π and compare their output with the standard model:

> *in 1986, a π calculation program detected some obscure hardware problem in one of the original Cray-2 supercomputer.*   [Bailey, Borwein, Borwein, and Plouffe, 1997]

Be that as it may, we retain from this incursion that empirical evidence indicates the presence of much statistical randomness hiding the *a priori* deterministic nature of such an object as π; this situation begs for a better, more intrinsic, definition of the concept *"randomness"*. In Sects. 6.1 and 6.2 we exploit the diversity of what we have learned in the present section, and delineate a mathematical path – with its definitions and some of its theorems – that has been proposed towards a better understanding of what constitute randomness.

6. Modern Probability: Competing Semantics

Watson won't allow that I know anything of art, but that is mere jealousy, because our views upon the subject differ.
[Doyle, 1902]

The case [Chap. 4] of limiting frequencies *vs.* degrees of belief was opened anew in the 20th century. Two major *semantic* developments occured within ten years around Kolmogorov's decisive contribution [Chap. 5] to the syntax of probability theory: Von Mises [1928] proposed a strictly *frequentist* interpretation, which we introduce in Sect. 6.1; and [De Finetti, 1937] offered a *degree of belief* interpretation [Sect. 6.3]. The semantics of the theory however turned out to be much more controversial than its syntax. Motivated in large part by some of the unresolved problems with von Mises' approach, logicians were led to the clarification of the concept of *randomness*. This task was completed only after the notion of *algorithmic complexity*, pioneered by Kolmogorov [1963], became available and was extended from finite strings to infinite sequences by Martin-Löf [1966] [Sect. 6.2]; in that section, we make use of *recursive function theory*, the mathematical elements of which are summarized in Appendix C.

6.1 Von Mises' Semantics

The project [Von Mises, 1919], followed by [Von Mises, 1928] for which he is most known among today's probabilists, was ambitious and at least in some directions its scope was even more ambitious than could have been anticipated. von Mises wrote in a lively, often polemic style; from the beginnings already, he laid strong claims to originality and, over the years, he alternatively stated that he had been misunderstood, then vindicated, and then again still misunderstood on some essential points. To avoid being bogged down in historical controversy, we therefore decided to quote him extensively from the 1957 English translation of the 'definitive' German 1951 edition (von Mises died in 1953), and also occasionally from his 1940 address to the Institute of Mathematical Statistics [Von Mises, 1941]. We indicate these two sources as [PST] and [IMS]. Although definitely not a formalist, von Mises shared with Hilbert the view that an axiomatization and/or rigorization of probability theory is necessary:

> *The essentially new idea which appeared about 1919 ... was to consider the theory of probability as a science of the same order as geometry or theoretical mechanics.*
>
> [PST vii]

and he made it immediately clear that he had his own view on how to carry this quest:

> *In other words, to maintain that just as the subject matter of geometry is the study of space phenomena, so probability theory deals with mass phenomena and repetitions. By means of the methods of abstraction and idealization ... a system of basic concepts is created upon which the logical structure can then be erected. Owing to the original relation between the basic concepts and the observed primary phenomena, this theoretical structure permits us to draw conclusions concerning the world of reality.*
>
> [PST vii]

This is more than an incidental appeal to positivism: his profound persuasion indeed is firmly on record [Von Mises, 1939]. The same philosophical standpoint may also have contributed to [Von Neumann, 1932a] choice in focusing on von Mises' specific proposal, the *collective*, as guide for his statistical approach to the definition of the state of a quantum system.

In his section on *"The relation to experimental data"*, Kolmogorov [1933] deferred [fn. 4, p. 3] to [Von Mises, 1931] for *"the applicability of the theory of probability to the world of real events."* Compared this to:

> *the basic mathematical investigations were carried out by Kolmogorov. They form an essential part of a complete course on the theory of probability. They do not, however, constitute the foundations of probability ... According to our point of view, such a system of axioms cannot take the place of our attempt to clarify and delimit the concept of probability ... Our presentation of the foundations of probability aims at clarifying precisely that side of the problem which is left aside in the formalist mathematical conception ... [P]robability ... remains a natural science, a theory of certain observable phenomena, which we have idealized in the concept of a collective.*
>
> [PST pp. 99–100]

The natural world on which von Mises wants to make predictions is quite explicitly physical: already in [Von Mises, 1928], the whole Sixth [and last] lecture is on *Statistical Problems in Physics*, among which he singled out specifically: the *Kinetic Theory of Gases* and *Brownian motion*; in particular, we read:

> *We may even expect the application to gases to give especially good results because of the two conditions already mentioned, namely, the immensely large number of elements in the collective, and the extreme disparity between the small causes that influence the molecular collisions and the large effects produced.*
>
> [PST p. 183]

In closing our review of some of the motivations for von Mises' theory, we first mention that from the previous two quotes, it appears that von Mises did not appreciate what Kolmogorov's achievement really amounts to, nor what setting the foundations of probability means. The thought that Kolmogorov's theory may "take the place" of von Mises' own semantic construction shows the confusion. He seemed to be thinking of his work as enriching Kolmogorov's theory and, furthermore, supplying applications for such an enriched theory. Second, we should mention his views on the laws of large numbers, in both their classical and their modern forms. To the first [Bernoulli–Poisson and Bayes], he reproaches that they

> *lose their relation to reality if we do not assume from the beginning the axiom of limiting frequencies*
>
> [PST p. 125]

which we shall see as the first of the defining axioms for von Mises' collectives. As for their modern extensions [Cantelli, Polya, Khinchin and Kolmogorov, all of whom are mentioned in PST p. 127 and p. 231], we read:

> *This formulation of the Strong Law of Large Numbers, and the way we derived it, shows clearly that both the problem and its solution fit into the general scheme of the frequency theory without difficulty. It is a problem of constructing a certain new collective by means of the usual operations. This is all I wanted to show; it is not my purpose to give a discussion of the incorrect expositions which the publication of this proposition provoked.*
>
> [PST p. 129]

Let it be emphasized that nowhere in that discussion is any mention made of the dreaded sets of measure zero.

The foundations of probability in [Von Mises, 1919, 1928] thus are grounded in the necessity to have a cleaner interpretation of the theory directly in terms of frequencies and randomness; for that purpose, he proposed a concept of *collective*. A *collective* is an infinite sequence $\xi : n \in \mathcal{N} \mapsto \xi_n \in \Omega$ that satisfies two axioms the intent of which is stated immediately below. Notice that a collective involve a *single* sequence only, and *not* an ensemble of sequences. For the time-being, we let Ω be unspecified; to avoid all complications, the Reader is advised to first read the following definitions as if Ω were a finite set, e.g. the set $\{0, 1\}$ of two elements.

The **first axiom** requires that, for *every* part $A \subseteq \Omega$, the *relative frequencies*, defined for every $N \in \mathcal{N}$, by $N_A = \mathrm{card}\,\{n \mid 1 \leq n \leq N ,\ \xi_n \in A\}$ stabilize as N becomes very large; specifically, the existence of the following limits is required

$$W_A = \lim_{N \to \infty} \frac{N_A}{N} \quad . \tag{6.1.1}$$

W_A is called the *limiting relative frequency* for the attribute A in the collective; the collection $\{W_A | A \subseteq \Omega\}$ of all limiting frequencies for a collective is called its *distribution*.

The **second axiom** assumes that rules for *place selection* are given, ensuring that every infinite subsequences extracted from the collective according to these rules has again the same distribution, i.e. with ξ' denoting any subsequence extracted from ξ according to the rules of place selection, and with N'_A and W'_A being defined from ξ' as N_A and W_A are from ξ, we must have

$$W'_A = W_A \quad \text{for} \quad \text{all} \quad A \subseteq \Omega \quad . \tag{6.1.2}$$

While the first axiom is explicitly given in terms of a primary notion from classical analysis (the convergence of a sequence), the challenge remains to reduce similarly the second axiom, namely the rule of place selection, to primary concepts already available. There was a proposal for this in [Von Mises, 1928], but it needed some refinements, some of which are included in the definitive 1951 edition. These are discussed at the end of the section, as we need first to complete our survey of the program's scope.

Briefly stated, the rationale for the first axiom is:

> In a certain sense, the kollektiv corresponds to what is called a population in practical statistics. Experience shows that in such sequences the relative frequency of the different results ... varies only slightly, if the number of experiments is large enough. We are therefore prompted to assume that in the kollektiv, i.e. in the theoretical model of the empirical sequence or population, each frequency has a limiting value, if the number of elements increases endlessly. This limiting value is called ... the 'probability of the attibute in question within the kollektiv involved' ... Let me insist on the fact that in no case is a probability value attached to a single event as such, but only to an event inasmuch as it is the element of a well-defined sequence.
>
> [IMS p. 341]

The second axiom is that there ought to be enough *randomness* in the collective to ensure the *impossibility of any gambling strategy:*

> The authors of such systems have all, sooner or later, had the sad experience of finding out that no system is able to improve their chances of winning in the long run, i.e., to affect the relative frequencies with which different colors or numbers appear in a sequence selected from the total sequence of the game. This experience forms the experiemtal basis of our definition of probability.
>
> [PST p. 25]

Hence the famous dictum: *"First the collective – then the probability"* [PST p. 18–20].

With this scheme in hand, von Mises proceeded [PST pp. 38–58] to describe four basic operations for transforming collectives into new ones; they are place selection [see second axiom above], mixing, partitioning and combination, the idea here being that

> by combining these basic processes we can settle all problems in probability theory ... in the clearly defined framework of this theory no space is left for metaphysical speculations ...
>
> [IMS p.343]

Yet, for half a century, several objections were raised against each or both of the basic axioms:

- Ω must be specified, and so should be the subsets $A \subseteq \Omega$ on which the limiting frequency W_A is required to exist.
- While the four operations allow to show that the probability $p : A \mapsto W_A$ satisfies the usual properties expected from the classical theory, only simple (or finite) additivity is derived; can one – or even should one – try and prove countable additivity?
- The rules for the selection of admissible subsequences must be *specified*. That *some* rule must be imposed was realized from the beginning of the investigation. For instance, consider indeed any sequence $\xi : \mathbf{Z}^+ \to \{0, 1\}$ with limiting frequency satisfying $0 < W_{\{1\}} < 1$; then, ξ has infinitely many entries $\xi_{n_k} = 1$, so that it is possible to extract, from ξ, subsequences ξ' with $W'_{\{1\}} = 1$; this is the kind of subsequences one should *not* be allowed to select. Hence the necessity of some selection rule.
- While von Mises claimed that experience allows the idealization consisting in dealing only with infinite sequences that converge, we remember from the law of iterated logarithms – Thm. 5.2.4 – that things are not so simple; von Mises idealizes out streaks of bad luck. Moreover, tail events (such as the convergence of a sequence) are not detectable from finite strings, and only the latter are experimentally available. Since one cannot decide experimentally if actual sequences of real observations are collectives or not, where does this leave the empirical claims of the theory?

The very fact that such objections were raised and regarded as legitimate shows that von Mises was not, or at least did not make it, entirely clear whether he was just supplying *models* for the Kolmogorov syntax, or also providing applications for the theory of probability regarding which possible experiments must be at least conceivable. Some of the objections, especially those regarding infinite sequences, would not be serious objections if what he was doing was to define relational structures which satisfy the Kolmogorov axioms. What Wald and Ville did, as we shall see, are more explicitly the former – supplying models. Wald [1936] succeeded in proving a consistency theorem, asserting the existence of infinitely many collectives $K_p(\mathcal{S}, \mathcal{L})$ for the following given: Ω a space of arbitrary cardinality; $\mathcal{F}$ a countable algebra of subsets of Ω with $\Omega \in \mathcal{F}$; $m : A \in \mathcal{F} \mapsto m(A) \in [0, 1]$ a simply additive set function, with $m(\Omega) = 1$; $\mathcal{L}$ the set of Jordan-measurable subsets of Ω defined from m and $\mathcal{F}$; p the restriction of m to $\mathcal{L}$; $\mathcal{S}$ a countable system of selections.

This theorem calls for a definition of 'selection' and 'collective' which match reasonably closely the intuitive requirements of the [Von Mises, 1928] axioms. We now reproduce these definitions. Throughout, Ω is a set of arbitrary cardinality. For each non-negative integer n, a *selection function* is a function

$$s_n : (\xi_1, \cdots, \xi_n) \in \Omega^n \mapsto s_n(\xi_1, \cdots, \xi_n) \in \{0, 1\} \quad ; \qquad (6.1.3)$$

and we set s_o to be a constant. For every sequence $S = \{s_n \mid n = 0, 1, \cdots\}$ of selection functions, and every sequence $\xi : n \in \mathsf{Z}^+ \mapsto \xi_n \in \Omega$, we define first the auxiliary sequence

$$\zeta : n \in \mathsf{Z}^+ \mapsto \zeta_n \in \{0, 1\} \quad \text{with} \quad \zeta_n = s_{n-1}(\xi_1, \cdots, \xi_{n-1}) \quad . \qquad (6.1.4)$$

We now extract the subsequence $S\xi$ of ξ :

$$S\xi : k \in \mathsf{Z}^+ \mapsto (S\xi)_k = \xi_{n_k} \quad \text{with} \quad n_k \text{ such that} \quad \zeta_{n_k} = 1 \quad . \qquad (6.1.5)$$

This procedure is called a *selection*. Note two things: (i) implicit in the definition of a subsequence is the requirement that the function $k \to n_k$ be strictly increasing; (ii) whether an entry ξ_n is kept for inclusion in the selected subsequence depends only on the preceding entries $\xi_1, \cdots, \xi_{n-1}$ in the original subsequence. This is an important feature in the realization of axiom 2.

For an elementary example of a selection S, consider $\Omega = \{0, 1\}$, and fix an arbitrary finite string $x = (x_1, \cdots, x_m) \in \Omega^m$; a term ξ_n of an infinite sequence ξ is then kept for inclusion in the subsequence $S_x\xi$ if and only if it is immediately preceded in ξ by the string x. Then a sequence $\xi : n \in \mathsf{Z}^+ \mapsto \xi_n \in \{0, 1\}$ – which admits a limiting frequency, say p for $\{1\}$ – is said to be admissible if every subsequence extracted from it by any selection S_x has the same probability p. This selection is of historical interest since it is an essential ingredient in several proposals made independently by various authors: the *indifferent sequences* of [Popper, 1935], the *Nachwirkungsfreie Folgen* of [Reichenbach, 1938], the *admissible numbers* of [Copeland, 1928], and sequences which can serve to define the *collectives* associated to the system of bounded selections of [Ville, 1939].

Let now $\mathcal{S} = \{S\}$ be a collection – one says a system – of selections, with $id \in \mathcal{S}$; and $\mathcal{L} = \{L\}$ be a collection of subsets of Ω. A sequence $\xi : n \in \mathsf{Z}^+ \mapsto \xi_n \in \Omega$ is said to be a *collective* with respect to $\mathcal{S}$ and $\mathcal{L}$ if, there exist a function p on $\mathcal{L}$ such that for every $S \in \mathcal{S}$ and all $L \in \mathcal{L}$, $p(L) \equiv lim_{N\to\infty} \frac{1}{N} \mathrm{card} \{n \mid 1 \leq n \leq N, (S\xi)_n \in L\}$ exists and defines a function $p : L \in \mathcal{L} \to [0, 1]$, that is independent of S. The set of all these collectives is denoted $K(\mathcal{S}, \mathcal{L})$; the set of all the collectives in $K(\mathcal{S}, \mathcal{L})$ which give the same p is denoted $K_p(\mathcal{S}, \mathcal{L})$.

When Ω is not countable, Wald [1936] also showed that, if one were to try to impose that $\mathcal{L}$ consist of the collection of *all* subsets of Ω, there would be no system of selections $\mathcal{S}$ such that there would exist even one collective $K(\mathcal{S}, \mathcal{L})$. Ville [1939] strengthened this result by showing that $\mathcal{L}$ must in general be strictly smaller than the collection of all Lebesgue measurable subsets. Similarly, not all systems of selections are possible: the restriction to countable $\mathcal{S}$ is essential to the proof of consistency. Since, in general, the collection of Jordan subsets is closed only under *finite* unions, it does *not* make sense to ask for p to be *countably* additive, although it is simply additive; in fact, Ville [1939] gave arguments to the effect that the von Mises–Wald formalism is *not* strong enough to allow the proof of any of the usual theorems involving countable additivity.

This is bad enough, but there is something worse: Ville [1939] was able to prove that for any system of selection $\mathcal{S}$ and any $p \in (0,1)$, there exists at least one collective such that the frequency approaches its limit unilaterally, i.e.

$$\forall N \in \mathbb{Z}^+ : \quad \frac{1}{N} \operatorname{card} \{n \mid 1 \leq n \leq N, \, \xi_n = 1\} \geq p \quad .$$

This is a serious objection: a bettor who would consistently bet on 1 at every throw would realize a total gain that is always positive !

Ville [1939] was nevertheless able to find a selection condition stricter than Wald's, involving a notion of *martingale* that allows to bypass this difficulty; in particular, given a martingale σ, and a probability distribution p, the set of new collectives $H(\sigma, p)$ is not only still infinite, but even of full cylindrical measure: $P_p(H(\sigma, p)) = 1$; results were also obtained that cover the law of iterated logarithms.

The work of Wald and Ville shows that while the 'collective' scheme – as originally proposed by von Mises – could not be taken at face-value, it still could be refined – in a manner acceptable to von Mises – and become self-consistent, thus allowing for a mathematical theory. Nevertheless, the question of empirical verification/falsification remains unresolved. Moreover, the extension of von Mises' scheme to the quantum realm presents even greater challenges than its implementation in classical probability theory. Added to the intrinsic difficulties of the original framework, these new problems beg for some drastically different semantic approach; one proposal to this effect is discussed in Sect. 6.3 below.

6.2 Algorithmic Complexity and Randomness

The purpose of this section is to make mathematically precise the following heuristic idea: 'a random string or sequence is one that is its own best description', namely that there should be no sub-pattern, nor sub-structure nor sub-order hidden within that would allow one to reproduce this specific, individual string or sequence, short of copying it *verbatim,* so to speak; this kind of randomness thus is a quality pertaining to the actual, individual outcome; not to the process by which it has been created; for the philosophical underpinnings of this distinction, see [Earman, 1986]. To specify this idea, we deploy two tools.

The first tool is the theory of recursive functions – see e.g. [Hodel, 1995] – the elements of which are summarized in Appendix C; it was introduced in the reformulation of von Mises' collectives by [Church, 1940].

The second tool is the theory of algorithmic complexity – see e.g. [Chaitin, 1987, Calude, 1994, Li and Vitanyi, 1997] – to which we now turn; its beginnings date from the mid 1960s, with three recognized independent sources [Chaitin, 1966, Kolmogorov, 1963, Solomonoff, 1964] – listed in alphabetic order to indicate that we are not establishing priorities, as each of these authors had circulated earlier proposals; see also [Loveland, 1966].

Solomonoff's motivation might be the most illustrative. Imagine a situation where the results of a scientist's observations are collected by her in the form of (finite!) binary strings. She wants to make sense of these, to communicate with her colleagues around the World in a manner that indicates she comprehended the regularities in the data she had gathered, and to make predictions. She aims at reducing her data to a theory, which she expresses also as a finite binary string. This theory *must* reproduce the data faithfully, and *ought to* be as concise as possible. Then, the explanatory power of the theory can be evaluated in terms of the difference between the length of the original string (representing the raw observations) and the length of the final string (the theory).

The first step in the abstraction is to view the theory as a program for a computation that reproduces the original string. We now make this step mathematically precise.

We denote by X^* the (countable) set of all binary strings:

$$\left. \begin{array}{l} X^* = \bigcup_{n \in \mathbb{Z}^+} X^n \\[6pt] \text{with} \\[6pt] X^n = \{(x_1, x_2, \ldots, x_n) \mid x_k \in 0, 1\,;\, k = 1, 2, \ldots, n\} \end{array} \right\} , \qquad (6.2.1)$$

and we denote by $l(x) = n$ the length of a string $(x_1, x_2, \ldots, x_n)$. We reserve the symbol X to denote the set of all (infinite!) binary sequences $\xi : n \in \mathbb{Z}^+ \mapsto \xi_n \in \{0, 1\}$ – see e.g. Thm. 6.2.3. below.

We first need a formal definition of a machine that processes the information contained in strings to produce certain results; the model is a computer that is fed a program, starts processing it without delays, and continues until it has finished printing an output; and then halts. We imagine our computer to be completely reliable, but also without the slightest imagination allowing it to interpret the unformulated expectations of the operator.

Definition 6.2.1. *A computer C is a partial recursive function*

$$C : p \in X^* \mapsto C(p) = s \in X^* \quad ;$$

the finite string p, serving as input (or variable) to C is called a program; *the finite string $C(p) = s$ is called the* output *(or the value) of C.*

We allow C to be only a "partial" function in order to model situations where perfectly good programs p may lead to unending computations on a given computer $C : C(p)$ is not defined for such p. The above definition of a computer is oblivious to the (finite) space a computation may occupy, or the (finite) amount of time it may take: our computer is *ideal* in the sense that it has unlimited storage space and unlimited computational time: we want to be able to focus on the minimal size of the programs, not on the computing resources needed to execute them. Also, in order to avoid technical difficulties, we assume henceforth tacitly that the programs are *self-delimiting*, i.e. are explicitly encoding their length within themselves – a natural assumption that is not as innocuous as it may appears to be [Chaitin, 1987, Calude, 1994].

We now want to approach randomness by formalizing first what it could mean for a string to be hard to describe, i.e. to reproduce by a computation.

Definition 6.2.2. *For any given computer C, the complexity $K_C(x)$ of a binary string x is the length of the shortest program that causes C to output x:*

$$K_C(x) = \min_{\{p|C(p)=x\}} l(p) \quad .$$

If no program makes C compute x, we define $K_C(x) \equiv \infty$.

Complexity, when so defined, obviously depends on the computer we use; this could be annoying, and can be by-passed if we realize that we have no hope to model reasonably randomness with short strings: we are interested only in very long strings, and could therefore tolerate ambiguïties that are small with respect to the complexity we are looking for. The following definition could thus be a hopeful step in this direction.

Definition 6.2.3. *A computer U is said to be* universal *if, for any other computer C, there exists a constant $c(U,C)$ – depending only on C and U – such that*

$$\forall\, x : \quad K_U(x) \leq K_C(x) + c(U,C) \quad .$$

It means that the computer U is such that, for any other computer C, there is a prefix μ such that the program μp makes U do the same computation as C does, by first having μ to cause U to mimick C. That such universal computers do exist – at least in principle – is a consequence of the definition of recursive functions: it amounts to saying that among the partial recursive functions there exist optimal ones. From an asymptotic point of view, what is relevant is that we have a *uniform* bound c :

$$\forall\, x : \quad |K_U(x) - K_C(x)| \leq c(U,C) \quad .$$

We assume henceforth that a choice of a universal computer has been made, and we drop the index U.

From very early on it was realized that randomness is better evidenced when the above 'absolute' complexity K is replaced by the finer notion of relative complexity, namely:

Definition 6.2.4. *Given a computer C, the relative complexity $K_C(x|y)$ of a binary string x is the length of the shortest program that causes C to output* *x under the supplementary condition that the computer has already received, as an additional input, another binary string y* :

$$K_C(x|y) = \min_{\{p|C(p,y)=x\}} l(p) \quad .$$

Clearly the maximal complexity of any finite string x of length $l(x) = n$ is (of the order of) the length of that string; indeed, the program '*print x*' causes C to produce x; except for the instruction '*print*', its length is that of x whatever the length of x may be. This suggests that we measure the departure from randomness as a departure from maximal complexity:

Definition 6.2.5. *With $\varepsilon > 0$, we say that a string x of length n is $\varepsilon -$ random whenever*

$$K(x|l(x)) > n - \varepsilon \quad .$$

In this sense, π is certainly not random, even though its decimal expansion seems to be statistically random: a very short program suffices to produce several billions decimals – see the last subsection of Sect. 5.4. Even without carrying out the computation, we know how to convey what π is, and we saw in elementary calculus some simple rules of formation, namely series expansions. The experimental evidence nevertheless suggests that Def. 6.2.5 opens the door to a concept of randomness that is *stronger* that the Borel normality condition. The Champernowne–Copeland–Erdös numbers, also encountered in Sect. 5.4, provide some support to this conjecture. This conjecture is actually correct: see Scholium 6.2.1.

Kirchherr, Li, and Vitanyi [1997] offered a vivid and very neat scenario to illustrate the meaning of algorithmic complexity. Let us try to stage it in three acts.

Act 1: A Dealer offers odds $2:1$ for bets on heads in a string of n throws of a coin. The length of the string is supposed to be large, say $n = 1000$. The potential Gambler figures that her expected gain would be

$$\langle G \rangle_n = \sum_{k=1}^{n} \left[p_k \cdot 2 + (1 - p_k) \cdot (-1) \right] \quad .$$

If she were to assume that the coin is fair and the Dealer honest, i.e. that $p_k = \frac{1}{2} \, \forall k$, this gives $\langle G \rangle_n = \$\,500$: too tantalizing for her not to question these assumptions. How could the Dealer make a living from the odds he offered?

Act 2: The Gambler makes a counter-offer. She would accept the odds, provided she be allowed to cover her bet with a side-bet: she requests that the Dealer agree to pay her at the end of the game when x, the string of outcomes, is recorded the sum of $S(x)$ dollars, with

$$S(x) = 2^{[n-K(x)]}$$

where $K(x)$ is the complexity of x. She notes that if the assumptions of Act 1 are satisfied, i.e. if x is essentially random, then $[n - K(x)] \approx 0$, so that $S(x) \approx 1$; moreover she points out that the average of S over the 2^n *a priori* equally probable strings x of length n is

$$\langle S \rangle_n \equiv \sum_{\{x | l(x) = n\}} \frac{1}{2^n} S(x) = \sum_{\{x | l(x) = n\}} 2^{-K(x)}$$

which, she argues from Kraft's inequality, is smaller or equal to $\$1$. She thus tries to entice the Dealer by offering to pay him a one-time premium of a few dollars, outright, at the beginning of the game.

Act 3: The Dealer objects. Even when the game is played, with x fully in view, neither he nor the Gambler is able to compute effectively $K(x)$. The Gambler has to concede; she modifies her requested cover to

$$\tilde{S}(x) = 2^{[n - \pi(x)]}$$

where π is any program she would devise at the end of the game to reproduce x. Although less dramatic than her initial counter-offer, this still leads, under the assumptions of Act 1, to $[n - \pi(x)] \approx 0$, and thus $\tilde{S}(x) \approx 1$. The Dealer has been exposed: to earn a living with the odds $2 : 1$ he has to have a system π' that arranges for the outcome x not to present constant $p_k = \frac{1}{2}$; and $l(\pi'(x))$ must be significantly shorter than n; were he to accept the counter-offer, the exponential in $\tilde{S}$ would wipe out his operating capital. A morality tale to enchant Pascal and confound the Chevalier de Méré. It also uncovers a nascent concept of *algorithmic probability*. To a "regular" string x of length $l(x) = n$ and complexity $K(x) \ll n$ one may want to attribute a cause "simpler" than a Bernoulli trial (with $p = \frac{1}{2}$) that results in strings, each and every one of which has probability 2^{-n}; this extension of the Laplace–Occam rule could lead one to say that x is $2^{n-K(x)}$ more likely to have been produced by a cause simpler than a random Bernoulli trial, thus pointing to a distribution $\{2^{-K(x)} \mid l(x) = n\}$. For a detailed discussion of this suggestive idea, see [Li and Vitanyi, 1997, Chap. 4].

The concepts just reviewed for the complexity of finite strings and their approximate randomness provide the springboard for a successions of mathematically more sophisticated papers, the first of which is [Martin-Löf, 1966], allowing to define an infinite sequence to be random if it passes a funnel of non-randomness tests – 'filter' or 'sieve' are also used instead of 'funnel'.

Definition 6.2.6. *A test is a subset $V \subset \mathbb{Z}^+ \times X^*$ such that*

(1) V is recursively enumerable ;
(2) $V_m \supseteq V_{m+1}$, where $V_m = \{x \in \mathbb{Z}^+ \mid (m, x) \in V\}$;
(3) $\text{card}\{X^n \cap V_m\} \leq 2^{n-m}$.

The index $m_V(x)$ *of a string x of length $l(x)$. is $m_V(x) = max\{m \in \mathbb{Z}^+ \mid x \subset V_m\}$.*

Note that (3) gives the key to the interpretation of Def. 6.2.6: there are 2^n strings in X^n and at most a fraction 2^{-m} of those have the regularity necessary to pass the filter V_m; this gives a meaning to the fact that the successive filters test for more and more regularity. $m_V(x)$ is thus the maximal level of regularity exhibited by the string, in the test V; and one has $m_V(x) \leq l(x)$.

Note further that this condition (3) can be rewritten as:

$$2^{-n} \text{card}\{X^n \cap V_m\} \leq 2^{-m} \qquad \text{i.e.} \qquad \sum_{x \in X^n \cap V_m} \left(\frac{1}{2}\right)^n \leq 2^{-m} \quad .$$

Hence Def. 6.2.6 can be generalized to cover tests for distributions other than the fair coin; one may indeed replace (3) by

$$(3') \qquad \qquad \sum_{x \in X^n \cap V_m} p(x) \leq 2^{-m}$$

which Martin-Löf also discusses (essentially in that form), even with special concern for the positivist scruples that one may involve only "computable distributions".

In summary, the choice of (3), here or in Def. 6.2.8 below, prescribes what distribution one decides to test (and it thus dictates, in particular, the index of the string in question). Hence, the definition is, as it should be, for a class of tests. For simple didactic reasons, we pursue here only the special case of a test for a fair coin. The test for biased coin is just slightly more messy – as long as the bias remains the same in the course of time – but it does not appear to contain any principal differences.

Evidently, how regular a string is reputed to be depends in general on the test V; this situation would improve if one could prove that there is at least one test U that satisfies the following condition.

Definition 6.2.7. *A test U is universal if for every test V there is a $c \in \mathsf{Z}^+$ such that*

$$V_{m+c} \subseteq U_m \quad \forall \quad m \in \mathsf{Z}^+ \quad .$$

Theorem 6.2.1.

(a) a universal test U exists;
(b) for every test V, there exists an element $c \in \mathsf{Z}^+$ such that

$$m_V(x) \leq m_U(x) + c \quad \forall \quad x \in X^* \quad .$$

That such a universal test exists is where recursive theory enters.

Proof: Since every test V is recursively enumerable, the set of all these tests is also recursively enumerable, i.e. there exists a subset $T \subset \mathsf{Z}^+ \times \mathsf{Z}^+ \times X^*$ such that to every test V corresponds precisely one $e \in \mathsf{Z}^+$ with

$$(m, x) \in V \quad \text{iff} \quad (e, m, x) \in T \quad . \tag{6.2.2}$$

One then verifies that the test U defined, for fixed $n \in \mathsf{Z}^+$, by

$$(m, x) \in U \quad \text{iff} \quad (n, m+n, x) \in T \qquad (6.2.3)$$

is universal. Part (b) follows immediately from the above.   **q.e.d.**

Theorem 6.2.2. *There is an integer $k \in Z^+$ such that, given K and U*

$$m_U(x) - k \leq l(x) - K(x|l(x)) \leq m_U(x) + k \quad \forall \quad x \in X^* \quad .$$

This provides the beginning of a link between the emerging concepts of complexity (Defs. 6.2.2–5), the lack of regularity in some universal tests (Defs. 6.2.6–7), and randomness. The gist of Martin-Löf's contribution is that his notion of test, and its main properties, can be extended from strings to sequences by the following adjustments.

Definition 6.2.8. *A sequential test is a subset $V \subset Z^+ \times X^*$ such that*

   (1) V is recursively enumerable ;
   (2) $\{x \in V_m , n \leq m , y \geq x\} \Rightarrow y \in V_n$;
   (3) $\operatorname{card} \{X^n \cap V_m\} \leq 2^{n-m}$
        where, as before $V_m = \{x \in Z^+ \mid (m, x) \in V\}$; and now $y \geq x$ means
        $\{ l(y) \geq l(x) \quad$ and $\quad y_k = x_k \quad \forall \quad 1 \leq k \leq l(x) \}$.

Note that a sequential test is indeed a test in the sense of Def. 6.2.6, and that one can again define the index of a string, now with respect to a sequential test. Universal sequential tests exist, again by an appeal to recursive theory. The interest of Def. 6.2.8 is that it allows to consider any sequence $\xi : n \in Z^+ \mapsto \xi_n \in \{0,1\}$ through its initial string $\xi^{(n)} = (\xi_1, \xi_2, \cdots, \xi_n)$ of length n. $\xi_k^{(n)} = \xi_k \ \forall \ 1 \leq k \leq n$. Since, for any sequence ξ and any universal sequential test U, the index $m_U(\xi^{(n)})$ is a non-decreasing function of n, the following limit exists (finite or infinite):

$$m_U(\xi) = \lim_{n \to \infty} m_U(\xi^{(n)}) \quad . \qquad (6.2.4)$$

Hence the index $m_U(\xi)$ of a sequence ξ, with respect to the universal test U, is well-defined.

Definition 6.2.9. *A binary sequence ξ is said to be random in the sense of Martin-Löf iff $m_U(\xi) < \infty$.*

As consequences of Thms. 6.2.1b and 2.2, one obtains the following results.

Theorem 6.2.3.

   (a) The condition that a sequence be random is independent of the universal test used in Def. 6.2.9.
   (b) $m_U(\xi) - k \leq n - K(\xi^{(n)}|n) \leq m_U(\xi) + k \quad \forall \quad \xi \in X$ and $n \in Z^+$.

Note that, in this approach, the notion of "randomness" is defined without any appeal to probabilistic notions, although a connection with the latter is pointed out later in this section. As a rephrasing of part (b) of the above theorem, we see that the complexity of the initial segment $\xi^{(n)}$ of any random sequence ξ must increase sufficiently fast with n to remain within a certain finite range of n. Some converse of this also holds, in the sense of Thm. 6.2.6 below, which also sets us on the road for a link to the version of the law of iterated logarithms proposed in [Martin-Löf, 1971].

Theorem 6.2.4. *Let ξ be any binary sequence, and φ be a recursive function satisfying*

$$\sum_{n=1}^{\infty} 2^{-\varphi(n)} = +\infty \quad .$$

Then

$$\operatorname{card}\{n \in N \mid n - K(\xi^{(n)}|n) > \varphi(n)\} = \infty \quad .$$

Note that: (i) the condition of the theorem is to be interpreted as an upper bound on the increase of φ as n goes to infinity; in particular, $\varphi(n) = c$ satisfies this assumption, and $\varphi(n) = \beta n$, with $0 < \beta < 1$, does not; and (ii) the theorem still allows for the converse inequality,

$$n - K(\xi^{(n)}|n) \leq \varphi(n) \tag{6.2.5}$$

to hold infinitely often.

Theorem 6.2.5. *Let ξ be a binary sequence such that there exists a constant c for which*

$$\operatorname{card}\{n \in N \mid n - K(\xi^{(n)}|n) \leq c\} = \infty \quad .$$

Then ξ is random in the sense of Martin-Löf (Def. 6.2.9).

The next theorem requires us to recall a rather natural definition in recursive function theory.

Definition 6.2.10. $\sum_n f(n)$ *(with $f(n) \geq 0$) is said to be* recursively convergent *if there exists a recursive sequence $\{n_k \mid k \in Z^+\}$ such that*

$$\forall \, k \in Z^+ \quad : \quad \sum_{n=n_k}^{\infty} f(n) < 2^{-k} \quad .$$

With this we can state the next result in the theory of random sequences:

Theorem 6.2.6. *Let ξ be a random sequence, and ψ be a recursive function such that*

$$\sum_n 2^{-\psi(n)}$$

is recursively convergent. Then

$$\operatorname{card}\{n \in N \mid n - K(\xi^{(n)}|n) > \psi(n)\} < \infty \quad .$$

These results call for a few remarks.

1. The above theorem asserts that

$$n - K(\xi^{(n)}|n) \le \psi(n) \qquad (6.2.6)$$

holds for all but finitely many n; this assertion is much stronger that the mere condition

$$\text{card}\,\{n \in N \mid n - K(\xi^{(n)}|n) \le f(n)\} = \infty \qquad . \qquad (6.2.7)$$

2. Upon combining Thms. 6.2.4 and 6.2.6, one sees the sense in which, *even for honest random sequences*, there are oscillations of $[n - K(\xi^{(n)}|n)]$. For certain functions φ, one has: $[n - K(\xi^{(n)}|n)] > \varphi(n)$ for *infinitely many n's*; whereas for other functions ψ, one has: $[n - K(\xi^{(n)}|n)] \le \psi(n)$ for *all, but finitely many n's*.

3. To see in which sense the above differs from the result obtained for a sequence that is not necessarily random, compare Thm. 6.2.7 below with Thm. 6.2.5, and Thm. 6.2.8 below with Thm. 6.2.6.

Theorem 6.2.7. *With probability* 1, *there exists a number* c *(which depends on ξ) such that*

$$\text{card}\,\{n \in N \mid n - K(\xi^{(n)}|n) \le c)\} = \infty \qquad .$$

Note that, consequently, Thms. 6.2.5 and 6.2.7 identify exactly the set of random sequences as a set of full measure:

Corollary 6.2.1. *Most sequences are* *random.*

Theorem 6.2.8. *If ψ is a recursive function with*

$$\sum_{n=1}^{\infty} 2^{-\psi(n)} < \infty \qquad ,$$

then, one has with probability 1 *that*

$$n - K(\xi^{(n)}|n) \le \psi(n)$$

holds for all but finitely many n's.

Hence this theorem tells us that the above inequality only holds with probability 1; compare with Thm. 6.2.4 which actually always hold for *all* random sequences.

Finally, recall (see discussion of Def. 6.2.6) that Martin-Löf also succeeded in [Martin-Löf, 1966], by a modification of condition (3) of Def. 6.2.8, to characterize Bernoulli sequences and that *"these are precisely the sequences for which von Mises introduces the term Kollektiv."*

A wealth of alternate, but ultimately either exactly or very closely equivalent, definitions have been proposed – see e.g. [Levin, 1973, Schnorr, 1977, Gács, 1986, Calude and Chitescu, 1988] – and are discussed in [Chaitin, 1987, Sect. 7.2], [Calude, 1994, Sect. 6.3]. Li and Vitanyi [1997, Sects. 3.6, 4.7] also note that complexity oscillations may be used to discriminate between some of these [Van Lambalgen, 1990]. The above discussion applies to infinite sequences; for finite strings, the original Kolmogorov definition – Def. 6.2.6 above – has also been sharpened; see [Calude, 1995].

In conclusion, the term "random sequence" has now been consistently described at the theoretical level. The algorithmic definition of randomness has progressed beyond the explorations of the previous section, and yet incorporates their sharpest notion; indeed, one can prove now – see e.g. [Calude, 1994, Thm 6.60]:

Scholium 6.2.1. *Every random sequence is normal in the sense of Borel.*

Since we had already noted that there are Borel normal sequences that are *not* random in the sense of this section, we conclude that the new definition of randomness is *strictly* more stringent than the randomness conditions discussed in Sect. 5.3, including the condition of statistical randomness, or Borel normality. Cor. 6.2.1 presents some evidence that the baby was not thrown out with the bathwater. Life however has not become more tractable for all of that, as we should point out that randomness is now *provably* elusive [Chaitin, 1974, 1987, Calude, 1994]:

Scholium 6.2.2. *Let B be the partial function with $B(n)$ the largest natural number of complexity not larger than n; and f be any partial recursive function $f : Z^+ \to Z^+$. Then $B(n) \geq f(n)$ for sufficiently large n, with strict inequality necessarily happening from place to place.*

This is quite a remarkable result, which Chaitin points out to be a sort of algorithmic incompleteness theorem: B is not computable. This is in consonance with the expectation that randomness should be a fugitive quality: we cannot determine empirically that a sequence is random; but if it is not, this will show up after finitely many steps in a refining funnel of tests – *how many* is what we cannot foretell.

Some of the connections that have been recognized between the algorithmic approach to randomness and Shannon's entropy, as well as with Kolmogorov's dynamical entropy, is pointed out in Chap. 8. Nevertheless, one must admit that the definitions of Martin-Löf *et al.* were met with some reticence, some comparatively mild [Fine, 1973], and some more impatient:

> *From the standpoint of practical random number generation ... this is ... the worst definition of randomness that can be imagined.*
>
> [Knuth, 1981]

It is indeed somewhat doubtful that von Neumann (see the discussion after Thm. 6.3.2 below) would have accepted lightly such a long detour to justify in elementary terms what he meant by an ensemble of identically prepared systems. Hence the urge to turn to some more simple-minded semantics, one that would put the emphasis on the information one has on the manner the system has been prepared.

6.3 De Finetti's Semantics

In an attempt to winnow the threshing floor of probability theory, another stake was developed in [De Finetti, 1937]; it is predicated on the following two premises.

First, he had no – or little – quarrel with the syntax of probability theory as it had emerged in the early 1930s:

> *formally, the theory of probability is the theory of additive and non-negative functions of events; opinions diverge only in one point, the question of whether these functions need to be ... additive only on finite sets, or also on denumerable sets*

specifically, [De Finetti, 1937] canonical references to the formal theory were [Lomnicki, 1923, Cantelli, 1932, Kolmogorov, 1933].

Second, however, he refused to be satisfied with either of the two lines in the interpretation of probability (cf. Chap. 4): on the one hand, the purportedly objective view, based on limiting relative frequencies; and on the other hand, the so-called subjective view, articulated around some formulation of degree of belief. Ramsey [1931] gave a similar semantics for degrees of belief earlier than de Finetti, but all evidence indicates that de Finetti was not aware of Ramsey's attempt.

To de Finetti, neither seemed to be fully adequate as a legitimate starting base for the formulation of a probability theory, nor for *"the explanation of its usefulness."* His reproach to the first view was that it was not operationally implementable and, to the second, that it was too vague.

In the latter, however, he found an essential element that needed to be brought into focus: he insisted that one should recognize that a subjective judgement can be legitimate on its own merits, and warned that *"if one seeks to replace it afterwards by something objective, one does not make progress, but merely an error."* Think, for instance, of the type of judgement – subjective vs. objective – that induces one to take an insurance on such a singular event as one's own life. He was aware also that he was trodding there on philosophical territory:

> *The question we pose ourselves now includes in reality the problem of reasoning by induction,* [a question which he emphasized was intimately linked to the notion of] *cause, advanced by David Hume, which I consider the highest peak that has been reached by philosophy.*

From these two premises, de Finetti proposed to develop a mathematically and semantically rigorous approach to probability, which he called subjective.

Our presentation starts with a brief review of the lattice theory relevant to the prepositional calculus of "events". [Readers solely interested in the classical aspects of the theory may skip this portion and go directly to the paragraph preceding Def. 6.3.1.]

We then present de Finetti's doctrine, mainly from its exposition in [De Finetti, 1937], but also with occasional confirmation drawn from the collection of reprints assembled by De Finetti [1972]. We state the main results in the form of three theorems that recover the usual (i.e. Kolmogorov) formulations of probability theory from de Finetti's subjective interpretation: Thms. 6.3.1 and 6.3.3 show how the syntax of probability is recovered under that interpretation and how the conditional probability assignments can be consistently defined as well, while Thm. 6.3.4 establishes a link with the frequency interpretation and Bayes theorem. The section ends with some bibliographical indications for further enquiries about the subjective approach to the theory of probability. In particular, the empirical scope of [De Finetti, 1937] program was later paralleled by Carnap's investigations of the logical relationship between hypothesis and evidence. To facilitate the transition we use, throughout this section, a notation familiar to logicians (see Appendix B), although the motivations and concepts – with which we begin – focus on a more restrictive scope, that we deem closer to de Finetti's original purpose. In particular, we focus our attention on classical systems, so that the universe of discourse can be anchored in the theory of Boolean lattices $\{\mathcal{P}, \models, \wedge, \vee, \neg\}$. Nevertheless, we indicate how this classical scheme can be modified to accommodate the wider universe of quantum theory.

One should note from the outset that the lattice theory of propositions we review below provides, very much like Kolmogorov's theory of probability, the syntax of proposition calculus (classical and quantum). De Finetti's semantics comes in only when the p-functions assigned to such a calculus are given the meaning of being degrees of rational beliefs – i.e. via the three theorems. It is of course not the only possible semantics to be given to such a calculus.

To a logician, $\mathcal{P}$ is a collection of special assertions, satisfying conditions which we delineate presently. With A and $B \in \mathcal{P}$, we say that A *entails* B (which we write $A \models B$) when B is true whenever A is true. We assume that $\models$ is a partial order relation : it is *reflexive* ($A \models A$), *transitive* ($A \models C$ whenever $A \models B$ *and* $B \models C$), and *antisymmetric* ($A = B$ whenever $A \models B$ *and* $B \models A$) ; the latter condition means that we identify assertions that are equivalent. Except towards the end of this section, the elements of $\mathcal{P}$ are specified to be *propositions* and are also referred to as *events*; thus A is the proposition "the event A occurs"; this context demands that we read $\models$ as an empirical or pragmatic relation: $A \models B$ is interpreted as meaning that the event B must occur whenever the event A does occur.

For every pair $(A, B) \in \mathcal{P} \times \mathcal{P}$, we assume that there exist two elements $A \wedge B$ and $A \vee B$ in $\mathcal{P}$, where $A \wedge B$ satisfies the two defining conditions

$$
\begin{array}{lll}
a. & A \wedge B \models A & \text{and} & A \wedge B \models B \\
b. & C \models A \wedge B & \text{whenever} & C \models A \quad and \quad C \models B
\end{array} \tag{6.3.1}
$$

and similarly $A \vee B$ satisfies the two defining conditions

$$
\begin{array}{lll}
c. & A \models A \vee B & \text{and} & B \models A \vee B \\
d. & A \vee B \models D & \text{whenever} & A \models D \quad and \quad B \models D \quad .
\end{array} \tag{6.3.2}
$$

Formally, these relations must satisfy:

$$
\begin{array}{lll}
A \wedge B = B \wedge A & \text{and} & (A \wedge B) \wedge C = A \wedge (B \wedge C) \\
A \vee B = B \vee A & \text{and} & (A \vee B) \vee C = A \vee (B \vee C)
\end{array} \tag{6.3.3}
$$

$$
A \wedge (B \vee A) = A = A \vee (A \wedge B) \quad . \tag{6.3.4}
$$

With the above definitions $\{\mathcal{P}, \models, \wedge, \vee\}$ is now a *lattice* [Birkhoff, 1940, Dubreil and Dubreil-Jacotin, 1961, Kurosh, 1963, Halmos, 1963]. There is a little bit of redundancy in these definitions. In particular, note that the following three conditions are equivalent

$$
B \models A \quad ; \quad A \wedge B = B \quad ; \quad A \vee B = A \quad . \tag{6.3.5}
$$

Hence, conversely, if two binary relations $\wedge$ and $\vee$ on a set $\mathcal{P}$ satisfy conditions (6.3.3–4), any two of the last three conditions (6.3.5) can be used to define the first, i.e. the partial ordering $\models$, thus inducing on $\{\mathcal{P}, \wedge, \vee\}$ the structure of a lattice.

Moreover, our set $\mathcal{P}$ comes naturally equipped with two special elements: the *absurd* proposition $(\emptyset)$ and the *tautology* (I), satisfying for all $A \in \mathcal{P}$: $\emptyset \models A$ and $A \models I$.

Finally, we assume that $\mathcal{P}$ is complemented, i.e. that one is given a bijective map $\neg : A \in \mathcal{P} \mapsto \neg A \in \mathcal{P}$ such that for all $A \in \mathcal{P}$:

$$
\neg(\neg A) = A \quad ; \quad A \wedge \neg A = \emptyset \quad ; \quad A \vee \neg A = I \quad . \tag{6.3.6}
$$

$\neg A$ is interpreted as the *negation* of A. Note also that

$$
A \vee B = \neg\{(\neg A) \wedge (\neg B)\} \quad \text{and} \quad A \wedge B = \neg\{(\neg A) \vee (\neg B)\} \quad . \tag{6.3.7}
$$

Here again, we have a little redundancy: the negation can be used, via (6.3.7), to define $\wedge$ from $\vee$, or $\vee$ from $\wedge$.

If one needs – and is willing – to go beyond finitary mathematics, some completeness condition must be assumed. The most demanding is that for every index set $\mathcal{J}$ and for every subset $\{A_j \mid j \in \mathcal{J}\} \subseteq \mathcal{P}$: there exist two propositions $\wedge_{j \in \mathcal{J}} A_j$ and $\vee_{j \in \mathcal{J}} A_j$ in $\mathcal{P}$ satisfying the straightforward generalizations of (6.3.1–2), namely

$$
\left.
\begin{array}{l}
\wedge_{j \in \mathcal{J}} A_j \models A_i \quad \text{for all} \quad i \in \mathcal{J} \quad , \text{and} \\
C \models \wedge_{j \in \mathcal{J}} A_j \quad \text{whenever} \quad C \models A_i \quad \forall \quad i \in \mathcal{J}
\end{array}
\right\} \tag{6.3.8}
$$

$$
\left.
\begin{array}{l}
A_i \models \vee_{j \in \mathcal{J}} A_j \quad \text{for all} \quad i \in \mathcal{J} \quad , \text{and} \\
\vee_{j \in \mathcal{J}} A_j \models D \quad \text{whenever} \quad A_i \models D \quad \forall \quad i \in \mathcal{J}
\end{array}
\right\} .
\qquad (6.3.9)
$$

When these conditions are satisfied, $\mathcal{P}$ is said to be *complete*. If one demands the above conditions to hold only for countable $\mathcal{J}$, one says that $\mathcal{P}$ is $\sigma-complete$.

We say that A and B are *mutually exclusive* whenever

$$
A \models \neg B \quad \text{or} \quad \text{equivalently} \quad B \models \neg A \; ; \qquad (6.3.10)
$$

and a collection $\{ A_j \mid j \in \mathcal{J} \}$ of mutually exclusive elements of $\mathcal{P}$ is said to be *exhaustive* whenever

$$
\vee_{j \in \mathcal{J}} A_j = I . \qquad (6.3.11)
$$

We now come to the restrictive assumption which focuses the discussion on *classical* systems. Henceforth, we assume that $\{ \mathcal{P}, \models, \wedge, \vee \}$ satisfies the (equivalent) *distributive* laws

$$
\left.
\begin{array}{r}
A \vee (B \wedge C) = (A \vee B) \wedge (A \vee C) \\
A \wedge (B \vee C) = (A \wedge B) \vee (A \wedge C) \\
(A \vee B) \wedge (B \vee C) \wedge (C \vee A) = (A \wedge B) \vee (B \wedge C) \vee (C \wedge A)
\end{array}
\right\} . \qquad (6.3.12)
$$

A *Boolean lattice* is defined as a complemented, distributive lattice. Typically, the set $\mathcal{P}(\Omega)$ of the subsets of a set Ω, equiped with the usual relations $\{ \subseteq, \cap, \cup, c \}$ is a Boolean lattice. In fact, a theorem due to [Stone, 1940] asserts that for every Boolean lattice $\{ \mathcal{P}, \models, \wedge, \vee, \neg \}$ one can construct canonically a set Ω and an injective homomorphism $\varphi : \mathcal{P} \mapsto \mathcal{P}(\Omega)$ that sends $\{ \mathcal{P}, \models, \wedge, \vee, \neg \}$ into $\{ \mathcal{P}(\Omega), \subseteq, \cap, \cup, c \}$. If we wish, we can refer to this Ω (uniquely determined by $\mathcal{P}$) as the phase-space of the classical system described by the Boolean lattice $\mathcal{P}$ of all events relative to the system considered.

Note in particular that, in a Boolean lattice, a necessary and sufficient condition for A and B to be mutually exclusive is that $A \wedge B = \emptyset$; this condition however is not sufficient for more general lattices – for a generic example see below the standard quantum lattice $\mathcal{P}(\mathcal{H})$.

We said earlier that the condition that the lattice of propositions be Boolean restricts the scope to classical physics. Birkhoff and Neumann [1936] proposed a program to define a proposition calculus that would encompass the description of quantum mechanical systems. Their first step was to point out that, in the traditional formalism of quantum theory [Von Neumann, 1932a], we have a natural lattice of special "observables" – those with eigenvalues 0 and 1 – that should qualify as the quantum analog to the classical propositions – with their "yes" or "no" values. These are the projectors on the closed subspaces of the separable Hilbert space $\mathcal{H}$ used to describe what we now call an irreducible quantum mechanical system; the lattice operations are defined there by

$$\left.\begin{array}{l} \neg A = I - A \\ A \wedge B = \lim_{n \to \infty} (AB)^n \\ A \vee B = \neg\{(\neg A) \wedge (\neg B)\} \\ A \models B \text{ whenever } A \wedge B = A \end{array}\right\} \quad . \tag{6.3.13}$$

This lattice – which we denote $\mathcal{P}(\mathcal{H})$, and refer to as the *standard quantum lattice* – is *not* Boolean; nor is, *a fortiori,* the lattice of the projectors of non-abelian von Neumann algebras $\mathcal{N}$; the latter were brought to life in [Murray and Neumann, 1936]; for a systematic updated presentation, see [Kadison and Ringrose, 1983, 1986].

Nevertheless, in case the Hilbert space $\mathcal{H}$ is finite-dimensional, or more generally, when the von Neumann algebra $\mathcal{N}$ is a $type - II_1$ factor, the lattice of projectors satisfies a weaker property: it is *modular,* i.e.

$$\text{if} \quad A \models C \quad \text{then} \quad \forall \ B \in \mathcal{P} \quad : \quad A \vee (B \wedge C) = (A \vee B) \wedge C \quad . \tag{6.3.14}$$

Note that, when $A \models C$ we have $A \vee C = C$. Hence if $\mathcal{P}$ were Boolean, (6.3.14) would result from the first of the relations (6.3.10); i.e. every Boolean lattice is modular. Therefore, if we were to take as an axiom that the lattice of propositions on a physical system is modular, all classical systems would still be covered. The modularity condition is deeply rooted in abstract algebra; for a milestone, see Dedekind [1900].

This condition, however, does not go far enough for the purposes of quantum theory, since modularity is violated when the Hilbert space $\mathcal{H}$ is infinite-dimensional. The search for a further weakening of the distributive condition, one that would allow to address the empirical formulation of quantum theory, was undertaken by Jauch and his students. See [Emch and Piron, 1963, Piron, 1964, Emch and Jauch, 1965, Jauch, 1968, Piron, 1976]; and, for an update, [Moore, 1999].

This program is achieved in two steps. The first requires that $\neg$ be *hereditary,* and the second that if $A \models C$ then A is *compatible* with C, where both italicized terms are defined empirically as follows.

First, whenever $A \models C$, we use the notation: $\mathcal{P}_{A,C} = \{X \in \mathcal{P} \mid A \models X \models C\}$. $\mathcal{P}_{A,C}$ inherits trivially the $\vee$ and $\wedge$ operations defined on $\mathcal{P}$. We now require explicitly that $\mathcal{P}_{A,C}$ become a complemented lattice under the *relative complementation* defined by

$$\neg_{A,C} : X \in \mathcal{P}_{A,C} \mapsto (A \vee \neg X) \wedge C \in \mathcal{P}_{A,C} \quad . \tag{6.3.15}$$

This is therefore a condition of consistency between the description of a system and any reduced description. It is in this sense that we want the negation to be *hereditary*; Piron coined the acronym *CROC* [= *canonically relatively orthocomplemented*] to describe lattices that satisfy this requirement.

Next, we say that two propositions A and C are *compatible,* whenever the sublattice of $\mathcal{P}$ generated from A and C – by repeated applications of the operations $\wedge$ and $\neg$ – is Boolean. We denote this relation by $A \rightleftharpoons C$ (lest it be confused with the usual notations $\leftrightarrow$ or $\Leftrightarrow$ of classical logic). Note that

$$A \rightleftharpoons C \quad \text{if and only if} \quad A \vee \{ \neg A \wedge C \} = C \vee \{ \neg C \wedge A \} \qquad (6.3.16)$$

which emphasizes both the symmetry of this relation, and its nature as a restricted form of distributivity. In the simplest non-classical case, the standard quantum lattice $\mathcal{P}(\mathcal{H})$ of the projectors on closed subspaces of a separable Hilbert space $\mathcal{H}$, this condition is equivalent to the familiar condition that A and C commute with one another: $AC = CA$.

Now comes the last of the empirical axioms defining the quantum proposition calculus:

$$\text{if} \quad A \models C \quad \text{then} \quad A \rightleftharpoons C$$

$i.e.$ $\qquad\qquad\qquad\qquad\qquad\qquad\qquad\qquad\qquad\qquad (6.3.17)$

$$\text{if} \quad A \models C \quad \text{then} \quad A \vee \{ (\neg A) \wedge C \} = C .$$

This relation is known as *orthomodularity* in the lattice-theory community, cf. e.g. [Holland, 1970].

Note that a modular lattice necessarily satisfies (6.3.17), but that the converse is not true: the standard quantum lattice $\mathcal{P}(\mathcal{H})$ is a CROC lattice that satisfies (6.3.17), but it is not modular when $\mathcal{H}$ is infinite-dimensional. In this sense, Jauch's project was successful. Note that the notion of quantum state was an intrinsic part of the program, although the flow of the argument here led us to postpone its introduction – see Def. 6.3.2 and Thm. 6.3.2 below.

Nevertheless, some serious limitations persist. For discussions of the philosophical status of this *quantum proposition calculus* – sometimes, but not universally, viewed as (the germ of) a "quantum logic" – see, e.g. [Hughes, 1989b, Van Fraassen, 1991, Pavicic, 1992, Rédei, 1998, Svozil, 1998]. However, while we mentioned already that the lattice of projectors of a von Neumann algebra satisfies all the conditions we wanted to impose on a (classical or quantum) proposition calculus, it is not known whether there are more esoteric models. The problem of an abstract characterization of $\mathcal{P}(\mathcal{H})$, is discussed carefully in [Varadarajan, 1985]; see already [Piron, 1964, Amemiya and Araki, 1967]; an unexpected twist has been uncovered in [Goldblatt, 1984].

Our Reader will notice that we do not want to commit to an empirical axiom which takes $\mathcal{P}$ to be atomic; an element $P \in \mathcal{P}$ with $P \neq \emptyset$ is said to be an *atom* (or a point) if $A \models P$ implies either $A = \emptyset$ or $A = P$. For instance, the one-dimensional projectors in $\mathcal{P}(\mathcal{H})$ are atoms; moreover this example satisfies the following two properties, essential to its unique characterization: (i) if $A \models B$ with $A \neq B$, then there exists an atom P such that $P \models B$ but not $P \models A$; (ii) for every $A \in \mathcal{P}(\mathcal{H})$ and every atom $P \in \mathcal{P}(\mathcal{H})$, we can only have $A \models C \models A \vee P$ when either $C = A$ or $C = A \vee P$. By contrast the lattice of projectors of a factor of type II_1 – Def. D.2.14 – has no atoms; nor does the (classical) von Neumann algebra $\mathcal{L}^{\infty}([0, 1], \mathrm{d}x)$ of the functions, on the interval $[0, 1]$, which are essentially bounded w.r.t. Lebesgue measure. This shared property of two models – namely to have no atoms – suggested to von Neumann that factors of type II_1 were the direction in which to look for a natural extension to the quantum realm of the classical notions of probability theory and statistical mechanics; see [Von Neumann, 1981]. For recent accounts, supplemented by unpublished documents, see [Rédei, 1999].

Beyond the elaboration of a "quantum logic", the approach involving in particular the lattice of projectors of von Neumann algebras – Def. D.2.11 – has been developed far enough to address issues in the study of systems with infinitely many degrees of freedom and the quantum measuring process, see e.g. [Emch, 1972a, 1984, Whitten-Wolfe and Emch, 1976, Busch, Lahti, and Mittelstaedt, 1991, Rédei, 1998] and references quoted therein.

Yet, in spite of its empirical successes and broad scope, quantum theory is still marred by conceptual problems having to do with the question of whether it is genuinely complete, although many among contemporary mainstream physicists would deny there is a problem [Fuchs and Peres, 2000]. The seeds of doubt were planted by [Einstein, Podolsky, and Rosen, 1935] who devised a *gedanken experiment* to demonstrate that some desirable *element of reality* is lacking; in particular, a pure state of a composite system cannot always be written as a tensor product of pure states of the component systems: it is *"entangled"* in a manner that the description of classical composite systems never has to face.

To remedy this situation [Bohm, 1952] proposed a *hidden variable* interpretation of quantum mechanics, but [Bell, 1964] showed that this formulation of quantum theory could satisfy locality only if some correlation inequalities were satisfied. Experiments established however that these Bell inequalities are actually violated [Aspect, Dalibard, and Roger, 1982]. For assessments, see [Mermin, 1993] and the collection of papers assembled in [Shimony, 1993], in particular No.3 in volume I and Nos. 7–12 in volume II; for extensions, see [Greenberger and Zeilinger, 1995, Summers, 1997]. Along different lines, [Griffiths, 1984, Omnes, 1988, Gell-Mann and Hartle, 1991,] proposed alternate interpretations. Here, unfortunately, is not the place to discuss these.

We made this detour in part to indicate that certain of the ideas in [De Finetti, 1937], although strictly tailored on a classical pattern, are naturally adaptable to fit quantum activities. For simplicity, we describe now the [De Finetti, 1937] semantic approach to probability in a finitary setting, as he did. To situate himself as close as he could to one's primitive numerical intuition of classical probability theory, [De Finetti, 1937] chose to analyze the placing of bets. The scenario is as follows. $\{\mathcal{P}, \wedge, \vee, \neg\}$ is a Boolean lattice. The setting is given by a finite, exhaustive collection $\{A_i \mid i = 1, 2, ...n\}$ of mutually exclusive events $A_i \in \mathcal{P}$. There are two actors: the Bookie and the Bettor. The action is governed only by the requirements that for each A_i, the Bookie chooses the "odds" p_i, which he announces. He is then obligated to accept bets of any sum (or "stake") S_i offered by the Bettor, subject to the agreement that when the event A_i is tested, the Bettor's net gain will be:

$$G_i = \begin{cases} (1 - p_i)\, S_i & \text{if} \quad A_i \quad \text{occurs} \\ -p_i\, S_i & \text{if} \quad A_i \quad \text{does not occur} \end{cases} \quad ; \tag{6.3.18}$$

the possibility $S_i = 0$ is allowed, which amounts to say that the Bettor can decline to bet on any particular event. When A_k occurs – and thus none of the A_j with $j \neq k$ does occur, since the events are mutually exclusive – the Bettor has a total net gain:

$$\mathcal{G}_k = (1 - p_k)\, S_k \; - \; \sum_{j \neq k} p_j\, S_j \quad . \tag{6.3.19}$$

Definition 6.3.1. *A Bookie's assignment $\{p_i\}$ is said to be* coherent, *whenever the Bettor cannot arrange to put his stakes $\{S_i\}$ in such a manner that*

$$\mathcal{G}_k > 0 \quad \forall \quad k = 1, 2, \ldots, n$$

i.e. his total net gain be always strictly positive, independently of the outcomes of the test of the events.

The beauty of this condition, which is called *coherence* in [De Finetti, 1937], and which is also referred to as *fairness*, resides in its semantic simplicity and its syntactic power. We have indeed:

Theorem 6.3.1. *A Bookie's assignment $\{p_i\}$ is coherent if and only if*

(a) $0 \leq p_i \leq 1$

(b) $p_i = 0 \;(\text{resp. } 1) \text{ if } A_i = \emptyset \;(\text{resp. } I)$

(c) $\sum_{i=1}^{n} p_i = 1$.

Hence the coherence condition (Def. 6.3.1) does state that a Bookie makes coherent assignments *if and only if* his assignments conform to the measure-theoretical syntax of [Kolmogorov, 1933].

Proof of the Theorem: To prove that de Finetti's coherence condition implies the Kolmogorov syntax, note the equivalence of (c) above with

$$\det \mathcal{C} = 0 \tag{6.3.20}$$

where $\det \mathcal{C}$ is the determinant of the matrix $\mathcal{C}$:

$$\mathcal{C} = \begin{pmatrix} (1 - p_1) & -p_2 & \cdots & -p_n \\ -p_1 & (1 - p_2) & \cdots & -p_n \\ \cdots & \cdots & \cdots & \cdots \\ -p_1 & -p_2 & \cdots & (1 - p_n) \end{pmatrix} \tag{6.3.21}$$

which is the matrix corresponding to the system of linear equations (6.3.19):

$$\mathcal{G} = \mathcal{C}\, \mathcal{S} \tag{6.3.22}$$

where $\mathcal{G}$ (resp. $\mathcal{S}$) is the column vector with entries $\mathcal{G}_k$ (resp. $\mathcal{S}_k$). The negation of condition (6.3.20) is precisely the condition under which one can solve (6.3.20) for $\mathcal{S}$:

$$S = C^{-1} \mathcal{G} \qquad (6.3.23)$$

i.e. the condition under which the Bettor can choose in advance the gains $\mathcal{G}_k$ he wants to make and compute – see (6.3.23) – the stakes he has to put to ensure the desired outcome. Hence (6.3.20), and thus (c), is necessary for the Bookie's assignments to be coherent. To derive (a), note that $\neg A_i = \vee_{j \neq i} A_j$. The coherence of a betting on $(A, \neg A)$ requires then that $\mathcal{G}_A$ and $\mathcal{G}_{\neg A}$ can never be of the same sign, i.e. that $\mathcal{G}_A \cdot \mathcal{G}_{\neg A} \leq 0$; together with $p_{\neg A} = 1 - p_A$ [a consequence of (c)], this is equivalent to $p_A (1 - p_A) \geq 0$, i.e. $0 \leq p_A \leq 1$ which is (a). $p_I = 1$ is obvious, and so is then $p_\emptyset = 0$.   **q.e.d.**

Moreover, as an immediate consequence of Thm. 6.3.1, we have:

Corollary 6.3.1. *A coherent assignment*

$$p : A \in \mathcal{P} \mapsto p(A) \in [0, 1]$$

is necessarily additive, i.e. satisfies, for all pairs $(A, B) \in \mathcal{P} \times \mathcal{P}$:

$$p(A) + p(B) = p(A \vee B) + p(A \wedge B) \quad .$$

Proof: For any pair (C, D) of mutually exclusive propositions, the coherent Bookie assigns odds that satisfy:

$$p(C) + p(D) + p(\neg C \wedge \neg D) = 1 \qquad (6.3.24)$$

and thus

$$p(C) + p(D) = p(C \vee D) \quad . \qquad (6.3.25)$$

Note now that for any pair (A, B) of not necessarily mutually exclusive propositions, $C = A \wedge B$ and $D = \neg A \wedge B$ are mutually exclusive, with $C \vee D = B$; from (6.3.25), we thus obtain:

$$p(A \wedge B) + p(\neg A \wedge B) = p(B) \quad ; \qquad (6.3.26)$$

since moreover A and $\neg A \wedge B$ are also mutually exclusive, one obtains similarly

$$p(A) + p(\neg A \wedge B) = p(A \vee B) \quad . \qquad (6.3.27)$$

The corollary then follows from substracting (6.3.27) from (6.3.26).   **q.e.d.**

Note that, for any (i.e. not necessarily exhaustive) *finite* collection $\{A_j \mid j \in \mathcal{J}\}$ of mutually exclusive propositions, we have, as a consequence of (6.3.25) :

$$p\left(\bigvee_{j \in \mathcal{J}} A_j \right) = \sum_{j \in \mathcal{J}} p(A_j) \quad . \qquad (6.3.28)$$

However, in case the index set $\mathcal{J}$ is *countably infinite*, the generalization of this relation becomes a separate assumption, specifically referred to as *countable additivity*; the non-finitary nature of this assumption is of marginal interest in an elementary presentation of de Finetti's semantic program for the recovery of Kolmogorov syntax, where countable additivity also appears as a separate assumption. It is also of some semantic interest to note that, already as consequence of the finite additivity proven in Cor. 6.3.1 :

$$p(A \wedge B) = 1 \quad \text{whenever} \quad p(A) = 1 \quad and \quad p(B) = 1 \quad . \tag{6.3.29}$$

Evidently, one can superimpose a statistical interpretation of Def. 6.3.1 in terms of the expectation value $\langle \mathcal{G} \rangle = \sum_{k=1}^{n} p_k \mathcal{G}_k$, namely $\langle \mathcal{G} \rangle = 0$. It is however essential to the [De Finetti, 1937] doctrine that *no* such interpretation is necessary to make sense of Def. 6.3.1. The latter is only a statement to the effect that the Bookie has checked his information and that it is consistent.

It makes no statement on which specific values the Bookie should attribute to the particular "odds" he offers for each event. Any such attribution would require a specification of the information content of the game that goes beyond coherence. We have seen in Sect. 5.2 that the Shannon entropy – in connection with the method of Lagrange multipliers – allows one to optimize this information content, given some supplementary knowledge. In this manner, the classical physicist computes such things as canonical equilibrium states. Incidentally, *so does his quantum colleague* [Von Neumann, 1932a]. [The Reader who is solely interested in the classical version of de Finetti's theory may skip again to the paragraph preceding Def. 6.3.3.] Here again, it is important to distinguish between syntax and semantics.

We deal first with the syntax. In the quantum prepositional calculus a *state* is defined as a $[0, 1]$−valued function p on $\mathcal{P}$, as in Cor. 6.3.1., *except* that the additivity condition is required only for pairs (A, B) of propositions that are compatible in the sense of (6.3.16); in this connection, recall that if A and B are mutually exclusive then A and B are compatible. Formally, we register as follows the generalization to which we just arrived .

Definition 6.3.2. *Let $\mathcal{P}(\mathcal{H})$ be the standard quantum lattice, the elements of which are the projectors on the closed subspaces of a separable Hilbert space $\mathcal{H}$. A (normal) state on $\mathcal{P}(\mathcal{H})$ is a countably additive measure on $\mathcal{P}(\mathcal{H})$, i.e. it is a map*

$$p : A \in \mathcal{P}(\mathcal{H}) \mapsto p(A) \in [0, 1]$$

satisfying

$$p(\emptyset) = 0 \qquad p(I) = 1$$

and for every sequence $\{A_j \mid j \in \mathcal{J}\}$ of mutually exclusive elements $A_j \in \mathcal{P}(\mathcal{H})$ (i.e $A_j \models \neg(A_{j'}) \ \forall \ j, j' \in \mathcal{J}$ with $j \neq j'$) :

$$p\left(\bigvee_{j \in \mathcal{J}} A_j \right) = \sum_{j \in \mathcal{J}} p\left(A_j \right) \quad .$$

We denote by $\mathcal{S}(\mathcal{H})$ the convex set of all states on $\mathcal{P}(\mathcal{H})$, and by $\mathcal{S}_e(\mathcal{H})$ the set of its extreme points, the pure (normal) states.

In the above definition the index set $\mathcal{J}$ is allowed to be countably infinite, and thus the qualifier *"normal"* is to emphasize that the measures we are now prepared to deal with are assumed to be "countably additive", rather than finitely additive as would be in line with [De Finetti, 1937] original proposal; a commitment to finitary mathematics would lead us here too far away from the praxis of standard quantum mechanics. We should nevertheless point out that there are states on quantum systems (in the thermodynamical limit) which are not simulatenously normal, e.g. states that correspond to different temperatures.

In contrast to Def. 6.3.2, Von Neumann [1932a] – who focused on systems with finitely many degrees of freedom – proposed to identify the states, on the quantum system described by $\mathcal{H}$, as the density matrices ϱ on $\mathcal{H}$, i.e. the positive trace-class operators on $\mathcal{H}$ with $\mathrm{Tr}\,\varrho = 1$, or equivalently as those operators ϱ on $\mathcal{H}$ for which there exist:

- a (possibly infinite, but countable) set $\mathcal{J}$;
- a sequence of non-negative numbers, indexed by $\mathcal{J}$, $\{\lambda_j \geq 0 \mid \sum_{j \in \mathcal{J}} \lambda_j = 1\}$;
- a sequence of vectors, indexed by $\mathcal{J}$, $\{\Psi_j \in \mathcal{H} \mid (\Psi_j, \Psi_{j'}) = \delta_{jj'} \; \forall \, j, j' \in \mathcal{J}, \}$;

such that

$$\varrho = \sum_{j \in \mathcal{J}} \lambda_j \, \varrho_j \quad \text{where} \quad \varrho_j : \Psi \in \mathcal{H} \mapsto (\Psi, \Psi_j)\,\Psi_j \in \mathbb{C}\Psi_j \quad . \tag{6.3.30}$$

The latter notation has the didactic advantage of emphasizing that the set $\mathcal{R}(\mathcal{H})$ of all *density matrices* ϱ on $\mathcal{H}$ is a convex set, the extremal points of which are the one-dimensional projectors ϱ_j. These extreme states are referred to as *pure states*, since they are not mixtures of other states. To every pure state ψ corresponds, not simply a vector Ψ_o with $\|\Psi_o\| = 1$, but a *ray* $\{\Psi = \omega\Psi_o | \omega \in \mathbb{C} \text{ with } |\omega| = 1\}$ of normalized vectors. We have thus for every $A \in \mathcal{P}(\mathcal{H}) : \psi(A) = \mathrm{Tr}\varrho_\psi A = (A\Psi, \Psi)$, where ϱ_ψ is the projector on the one-dimensional subspace spanned by Ψ.

Gleason [1957] established the following representation theorem to the effect that von Neumann's definition of state can be faithfully interpreted in terms of de Finetti's semantics; for a didactic proof, see [Varadarajan, 1985].

Theorem 6.3.2. (Gleason) *With $\mathcal{P}(\mathcal{H})$, where $\dim(\mathcal{H}) \geq 3$, $\mathcal{R}(\mathcal{H})$ and $\mathcal{S}(\mathcal{H})$ as above, the map $\varrho \in \mathcal{R}(\mathcal{H}) \mapsto p_\varrho \in \mathcal{S}(\mathcal{H})$ defined by*

$$p_\varrho : A \in \mathcal{P}(\mathcal{H}) \mapsto p_\varrho(A) = \mathrm{Tr}\varrho A \in [0, 1]$$

is bijective and convex.

Indeed, it is easy to verify that for every $\varrho \in \mathcal{R}(\mathcal{H})$, p_ϱ is a countably additive measure on $\mathcal{P}(\mathcal{H})$; and that the map $\varrho \in \mathcal{R}(\mathcal{H}) \to p_\varrho \in \mathcal{S}(\mathcal{H})$ is injective, i.e. if $p_\varrho = p_{\varrho'}$ then $\varrho = \varrho'$; it is also trivial to verify that this map is convex, i.e. that for every $\lambda \in [0, 1]$ and every pair of elements ϱ_o and ϱ_1 in $\mathcal{R}(\mathcal{H})$: $\varrho = (1 - \lambda)\varrho_o + \lambda\varrho_1$ implies $p_\varrho = (1 - \lambda)p_{\varrho_o} + \lambda p_{\varrho_1}$. The hard part was thus to prove the map is surjective, i.e. that for every normal state $p \in \mathcal{S}(\mathcal{H})$ there is a density matrix $\varrho \in \mathcal{R}(\mathcal{H})$ such that $p = p_\varrho$; and that is precisely what Gleason achieved.

One may note at this juncture that the interpretation of the above non-Boolean lattice theory is quite open. Just as we remarked earlier regarding the Boolean calculus and de Finetti's semantics, the above theory only provides the syntax for quantum probability while several viable semantics compete for the best interpretation. The first systematic attempt in this direction [Von Neumann, 1927] is consolidated in [Von Neumann, 1932a].

Recognizing that a consistent formulation of probabilities for observables that are not mutually compatible – i.e. cannot be measured *simultaneously* with arbitrarily small uncertainties – requires some careful attention, von Neumann proposes that we consider *ensembles* of mutually independent systems that have been prepared according to identical protocols. Such ensembles are construed to admit sub-ensembles – with identical statistical properties – on which relative frequencies are established separately for the various observables, say "position" on one sub-ensemble and "momentum" on another. In this connection [see the famous footnote No.156 in [Von Neumann, 1932a]], von Neumann uses the term "collective" [discussed in our Sect. 6.1 above] for which he cites von Mises approvingly. While von Neumann's adhesion to von Mises' semantic often seems to be taken as the authoritative canon of quantum probability theory, footnote No.156 should be contrasted [Rédei, 1999] with the following statement:

> This view, the so-called 'frequency theory of probability' has been very brilliantly upheld and expounded by R. von Mises. This view, however, is not acceptable to us, at least not in the present 'logical' context.
> [Von Neumann, ca 1937]

The 'context' in the above quote is that of [Birkhoff and Neumann, 1936] and of [Murray and Neumann, 1936]; see also the posthumous [Von Neumann, 1981]: the existence of an uniform a priori probability – a countably additive state, invariant under all symmetries, i.e. a finite normal trace – restricts quantum logic to the study of the lattice of projections of von Neumann factors – of type I_n or II_1, and thus excludes the usual formalism for the quantum description of mechanical systems with finitely many degrees of freedom.

Among other interpretative schemes, the potentiality or propensity interpretation has been drawing a great deal of attention; cf. [Heisenberg, 1930, Popper, 1935, 1990, Shimony, 1986].

Let A be an observable with discrete spectrum, and let $\{\Phi_i | i = 1, 2, \cdots\}$ be an orthonormal basis of eigenvectors of A, i.e. $A\Phi_i = a_i\Phi_i$ with $(\Phi_i, \Phi_j) = \delta_{i,j}$. Let further Ψ be any vector with $\|\Psi\| = 1$, and let $\Psi = \sum_i c_i\Phi_i$ be its expansion in the basis $\{\Phi_i\}$. Then, according to the above interpretation, the pure state ψ, corresponding to Ψ, possesses the *potential,* or *propensity,* of "ending up" in the pure state ϕ_i, corresponding to Φ_i, with probability $p(\psi, \phi_i) \equiv |(\Psi, \Phi_i)|^2 = |c_i|^2$, in the course of a measurement of A. Hence, $p(\psi, \phi_i)$ is usually called the *transition probability* between the pure states ψ and ϕ_i. And the result of such "ending up" is called *actualization* [Shimony, 1986]. However, for such an interpretation to work, one must show how any potentiality actualizes; one has to provide the physical mechanism for the process of actualization. This turns out to be one of the toughest problems in the foundations of quantum physics. Even if one constructs an adequate theory for quantum measurement, one is still far from resolving the problem, because quantum measurements – which essentially involve observers – are only a small proper subset of all the processes of actualization – which include all the results (i.e. events) of the interactions between the micro- and macro-systems. If quantum theory is to give us an objective and non-anthropomorphic description of the quantum world, it has to go beyond observations [Shimony, 1986].

Instead of aiming at an explanation of the measuring process – the actualization just discussed – the standard interpretation considers the pure state ψ as the summary of the information one has on the system considered, and notices that to any arbitrary orthonormal basis $\{\Phi_i | i = 1, 2, \cdots\}$ is associated a "complete system of commuting observables" [$\equiv$ csco], i.e. the maximal abelian algebra $\mathcal{A}$ generated by the observables that are diagonal in that basis; these observables are of the form $A\Phi_i = a_i\Phi_i$ where $\{a_i \in \mathsf{R} | i = 1, 2, \cdots\}$. The following straightforward computation gives then $\forall\, A \in \mathcal{A}$:

$$\psi(A) = (A\Psi, \Psi) = \sum_{i,j} c_i c_j^{\,*}(A\Phi_i, \Phi_j) = \sum_i |c_i|^2 (A\Phi_i, \Phi_i) = \sum_i |c_i|^2 \phi_i(A),$$

i.e.

$$\psi \equiv \sum_i |c_i|^2 \phi_i.$$

Hence, restricted to $\mathcal{A}$, the state ψ – which was pure on $\mathcal{B}(\mathcal{H})$ the algebra of all observables on the system – is a convex combination – i.e. a mixture – with coefficients $|c_i|^2$ of states ϕ_i which are pure on $\mathcal{A}$, as well as on $\mathcal{B}(\mathcal{H})$.

Once Def. 6.3.2 is understood, the above result requires no further interpretation: it is a mathematical corollary of Thm. 6.3.2. Naturally, this is not to say that this result is not part of the semantics to be associated with Def. 6.3.2: it is best considered as a consequential part of its interpretation. And it is a necessary commentary, pointing out one important sense in which quantum theory is a genuine generalization of classical theory.

Thus, the notion of state does not presuppose any commitment to some philosophical view that the state of a quantum system depends on the observer, only that the observer makes predictions on the basis of the information he has on the preparation of the system; for a pursuit of this Bayesian line in physics, see [Jaynes, 1967] and other papers in [Jaynes, 1989]; for criticisms and defenses of this point of view, see [Friedman and Shimony, 1971, Seidenfeld, 1979, Uffink, 1995]. Even more urgently, a view such as de Finetti's seems to be necessary if one is to speak of the behavior of an "isolated" system, a task which is demanded today by actual laboratory situations where individual quantum systems (rather than ensembles) are claimed to be prepared and sustained [Zeilinger, 1999].

Let us now return to the presentation of the basic tenets of [De Finetti, 1937] classical approach; as he was prompt to point out, assignments of probability must be open to further tuning, or updating:

> observation cannot confirm or refute an opinion, which is and cannot be other than an opinion and thus neither true nor false; observation can only give us information which is capable of influencing our opinion. The meaning of this statement is very precise: it means that to the probability of a fact conditioned on this information – a probability very distinct from that of the same fact not conditioned on anything else – we can indeed attribute a different value.

It is therefore necessary to produce a pragmatic definition of conditional probabilities in terms of coherent betting schemes. De Finetti [1937] did this along the following lines. For any pair of events $(A, B) \in \mathcal{P} \times \mathcal{P}$, one introduces a happening of a new kind, which de Finetti called a *3-event*, denoted $B|A$, and defined operationally:

Definition 6.3.3.

- $B|A$ occurs *whenever A occurs and B occurs*
- $B|A$ does not occur *whenever A occurs but B does not occur*
- $B|A$ is void *whenever A does not occur.*

A conditional bet involving $B|A$ is declared to be

- won *whenever $B|A$ occurs*
- lost *whenever $B|A$ does not occur*
- called off *whenever $B|A$ is void.*

For X standing for either of A, B or $B|A$, we denote by p_X, S_X and $\mathcal{G}_X$ the corresponding odds assignment, stakes and gains, and write

$$\mathcal{G} = \begin{pmatrix} G_B \\ G_A \\ G_{B|A} \end{pmatrix} \quad and \quad \mathcal{S} = \begin{pmatrix} S_B \\ S_A \\ S_{B|A} \end{pmatrix} \quad .$$

In case $B \models A$, we have

$$\mathcal{G} = \mathcal{C}\mathcal{S} \quad \text{with} \quad \mathcal{C} = \begin{pmatrix} (1-p_B) & (1-p_A) & (1-p_{B|A}) \\ -p_B & (1-p_A) & -p_{B|A} \\ -p_B & -p_A & 0 \end{pmatrix} \tag{6.3.31}$$

and the coherence condition now reads

$$\det \mathcal{C} = 0 \qquad i.e. \qquad p_B - p_{B|A}\, p_A = 0 \quad . \tag{6.3.32}$$

In general, to remove the assumption $B \models A$, it is sufficient to replace in the above B by $A \wedge B$ which obviously satisfies $A \wedge B \models A$ and $(A \wedge B)|A = B|A$. Upon substituting this in (6.3.32) we thus obtain:

Theorem 6.3.3. *For any pair $A, B \in \mathcal{P} \times \mathcal{P}$, a Bookie's assignment on a conditional bet involving $B|A$ is coherent if and only if:*

$$p(A \wedge B) = p(B|A)\, p(A) \quad .$$

While this condition coïncides precisely with the Bayes definition of the *conditional probability* $p(B|A)$, its interpretation is now made directly in terms of a condition on the coherence of subjective assignments of odds. The following result then follows straightforwardly:

Corollary 6.3.2. *The conditional probabilities satisfy*

(1) $0 \leq p(B|A) \leq 1$
(2)   If $\quad A \models B \quad$, then $\quad p(B|A) = 1$
(3)   If $\quad A \models \neg(B \wedge C) \quad$, then $\quad p(B \vee C|A) = p(B|A) + p(C|A)$
(4) $p(B \wedge C|A) = p(B|A \wedge C)\, p(C|A)$

Note that if the conditional probability were taken as the primary object, then one could define the ordinary probability by letting, for every $B \in \mathcal{P} : p(B) = p(B|I)$; in this connection, see the Carnap approach briefly mentioned at the end of this section.

Finally, we must address the question of whether and how the subjective semantic approach to probability theory ought to be brought into contact with the relative frequency semantics.

> *That this principle can only be justified in particular cases is not due to an insufficiency of the method followed, but corresponds logically and necessarily to the essential demands of our point of view. Indeed, probability being subjective, nothing obliges us to choose it close to the frequency; all that can be shown is that such an evaluation follows in a coherent manner from our initial judgement when the latter satisfies certain perfectly clear and natural conditions.*
>
> [De Finetti, 1937]

The central pieces in the forthcoming argument are the definition of exchangeable events and the de Finetti exchangeability theorem. Consider first a finite collection of events $\{E_1, E_2, \ldots, E_n\}$, and the 2^n events: $E_1 \wedge E_2 \wedge \ldots \wedge E_n$, $E_1 \wedge \neg E_2 \wedge \ldots \wedge E_n$, $\neg E_1 \wedge E_2 \wedge \ldots \wedge E_n$, $\neg E_1 \wedge \neg E_2 \wedge \ldots \wedge E_n$, $\ldots$,

$\neg E_1 \wedge \neg E_2 \wedge \ldots \wedge \neg E_n$. These events form an exhaustive collection of mutually exclusive events, and thus assignments of odds can be made in accordance with Thm. 6.3.1. To facilitate the connection with the usual presentation, see e.g. [Feller, 1968, 1971], let us denote by Prob such an assignment, and let us write $E_k = x_k$ with $x_k = 1$ to denote that the event E_k occurs, and with $x_k = 0$ to denote that it does not occur. With these notations, let us define the function:

$$\left. \begin{array}{l} P(E_1, E_2, \ldots, E_n) : (x_1, x_2, \ldots, x_n) \in \{0,1\}^n \mapsto \\ \\ \mathrm{Prob}(E_1 = x_1, E_2 = x_2, \ldots, E_n = x_n) \in [0,1] \end{array} \right\} \qquad (6.3.33)$$

which we call the joint distribution, relative to a chosen assignment of odds for the 2^n joint events just listed. Note that, as a consequence of Thm. 6.3.1, the assignments are consistent over n :

$$\left. \begin{array}{l} \mathrm{Prob}(E_1 = x_1, \ldots, E_{n-1} = x_{n-1}) = \\ \quad \mathrm{Prob}(E_1 = x_1, \ldots, E_{n-1} = x_{n-1}, E_n = x_n) + \\ \quad \mathrm{Prob}(E_1 = x_1, E_{n-1} = x_{n-1}, E_n = 1 - x_n) \end{array} \right\} \quad . \qquad (6.3.34)$$

As usual, we would say that the events in a string $\{E_1, E_2, \ldots, E_n\}$ are independent, whenever

$$\mathrm{Prob}(E_1 = x_1, E_2 = x_2, \ldots, E_n = x_n) = \prod_{k=1}^{n} \mathrm{Prob}(E_k = x_k) \quad . \qquad (6.3.35)$$

In addition, we would say that these events were identically distributed if there existed a single distribution $p_o : x \in \{0,1\} \mapsto p_o(x) \in [0,1]$ such that

$$\mathrm{Prob}(E_k = x_k) = p_o(x_k) \quad \forall \quad k = 1, 2, \ldots, n \quad . \qquad (6.3.36)$$

Note that all these notions make sense in the subjective framework proposed by de Finetti, where what one usually refers to as two trials of the same events are in fact considered as different events. De Finetti [1937] emphasized the propriety of introducing still a different notion:

Definition 6.3.4. *The events pertaining to a finite string are said to be exchangeable whenever the distribution function (6.3.33) is symmetric; i.e. for every permutation π of the indices k :*

$$\begin{array}{l} \mathrm{Prob}(E_1 = x_{\pi(1)}, E_2 = x_{\pi(2)}, \ldots, E_n = x_{\pi(n)}) = \\ \mathrm{Prob}(E_1 = x_1, E_2 = x_2, \ldots, E_n = x_n) = \\ \mathrm{Prob}(E_{\pi^{-1}(1)} = x_1, E_{\pi^{-1}(2)} = x_2, \ldots, E_{\pi^{-1}(n)} = x_n) \quad . \end{array}$$

This is naturally extended from finite strings to infinite sequences: since the assignments of odds are consistent over n , we can view an assignment of odds for an infinite sequence as an inductive limit; and similarly for the corresponding distribution, i.e.

$$P(E_1, E_2, \ldots) = \lim_{n \in \mathbb{N}} P(E_1, E_2, \ldots, E_n) \quad . \qquad (6.3.37)$$

Definition 6.3.5. *The events pertaining to an infinite sequence are said to be exchangeable, with respect to P, whenever the restriction P_n of P to any finite substring of length n makes the events of that substring exchangeable in the sense of Def. 6.3.4.*

The following profound result is due to de Finetti:

Theorem 6.3.4 (de Finetti's Exchangeability Theorem). *To every infinite sequence of exchangeable events corresponds a unique measure μ on $[0,1]$ such that for all pairs of integers (n,k) with $k \leq n$:*

$$\text{Prob}(E_1 = 1, \ldots, E_k = 1, E_{k+1} = 0, \ldots, E_n = 0) = \int_{[0,1]} \mu(\mathrm{d}p)\, p^k\, (1-p)^{n-k} .$$

The following is then an immediate consequence of exchangeability:

$$\text{Prob}\left\{ \sum_{i=1}^{n} x_i = k \right\} = \int_{[0,1]} \mu(\mathrm{d}p) \binom{n}{k} p^k (1-p)^{n-k} \qquad (6.3.38)$$

i.e. the probability that one obtains k successes in n trials is a mixture of Bernoulli trials; compare with (4.2.14). Since the measure μ depends only on P, and in particular is independent of (n,k), the theorem (or 6.3.38) can be viewed as a derivation of Bayes' theorem from first principles. Therefore, for exchangeable events (and thus *a fortiori* for independent identically distributed [$\equiv$ i.i.d.] events), this provides a link between de Finetti's subjective semantics and the relative frequency semantics that is usually associated with the law of large numbers.

It is essential for the unrestricted validity of the theorem, namely for the existence of a unique μ working for all (n,k), that the theorem be formulated for infinite sequences; a positivist mind would certainly frown upon the limit (6.3.37); it is therefore interesting to know that the situation for finite and/or partial exchangeable sequences has been investigated [De Finetti, 1972, Chap. 9]; see moreover [Diaconis and Freedman, 1980b]. With the Exchangeability Theorem in hand, one may argue that de Finetti's semantics provides a *"partial resolution to Hume's problem of induction"* [Zabell, 1988]. For a review of applications, see [Aldous, 1985].

For a discussion of the inverse problem of sufficiency, see [Accardi and Pistone, 1982], where the Reader will also find an exposition of the mathematical structure behind de Finetti's theorem. This structure was clarified first by [Hewitt and Savage, 1955] in the classical framework of symmetric measures on an infinite Cartesian product of a compact Hausdorff space with itself. Their formulation was subsequently generalized to a non-commutative setting in [Størmer, 1969]: the set $\mathcal{S}$ of the symmetric states on an infinite tensor product $\otimes_{k\in\mathbb{Z}^+}\mathcal{A}_k$ of identical copies $\mathcal{A}_k$ of a C^*-algebra $\mathcal{A}$ form a Choquet simplex – i.e. every symmetric state can be decomposed in a unique way as an integral over extremal symmetric states – and the extreme points of $\mathcal{S}$

are product states $\otimes \varrho$, with ϱ a state on $\mathcal{A}$. This quantum de Finetti theorem was extended to apply to the quantum many-body systems of identical bosons [Hudson, 1981, Hudson and Moody, 1976]. For recent discussions of de Finetti's work and related topics, see [Costantini and Galavotti, 1997].

In closing, we make three comments: the first on precedents; the other two on lines of research which de Finetti may not have anticipated, but are somehow in line with his semantic approach to the theory of probability.

Firstly, we should mention that one can find in [Ramsey, 1931] some interesting antecedents (in fact dating back to the mid 1920s) for both the modern emphasis on the subjective aspects of probability, and the Dutch-book approach. Ramsey, however, appears not to have been granted de Finetti's sustained mathematical vigor and missionary zeal. This paper and [De Finetti, 1937], as well as a wealth of earlier papers and a summation by L.S. Savage, have been reprinted with an instructive introduction in [Kyburg and Smokler, 1964]. Some further affinities between the approach of Ramsey and de Finetti can be found in [Keynes, 1921, Jeffreys, 1939, Savage, 1954]. For updated and constructive philosophical criticisms, see [Kyburg, 1974, Skyrms and Harper, 1988].

Secondly, we should point out the intimate connection between de Finetti's approach and Carnap's *logical* approach to probability – also known as an interpretation of probability as the *degree of confirmation;* [Carnap, 1950, 1952]. To appreciate the difference of the two approaches, however, one should see Carnap's in the wider philosophical context of logical positivism/empiricism; for more details, see [Giere and Richardson, 1996]. Now, instead of taking $p(B|A)$ as the degree of belief of B given A, Carnap, who used the notion, $c(h, e)$, took c to be the degree of inductive support that a piece of evidence, e, gives to the hypothesis, h. Both approaches are (different) models of probability calculus in that they are constructed so that they satisfy the axioms of Kolmogorov's syntax – or some slight variations of it. Carnap's approach is a purely logical one because the probability is interpreted as a logical relation between two statements – e.g. a hypothesis and an evidential statement – which quantifies the objective epistemic relation between these statements. In other words, $c(h, e)$ is the counterpart of entailment – $e \models h$ – in inductive logic. In some later presentations of this approach [Kemeny, 1963, Carnap, 1963], $m(.) = c(., t)$ (where t is a tautology) is interpreted as defining a measure m on the set of models, i.e. on the set of states of affairs where every assertion either is true or is false. Each state of affairs then corresponds, in the context of this section, to a particular assignment of probability where the first property in Cor. 6.3.1 is replaced by the stronger requirement:

$$p : A \in \mathcal{P} \mapsto p(A) \in \{0, 1\} \quad . \tag{6.3.39}$$

A classical physicist would call these "models" points in the phase-space, or pure states, for the physical system considered; e.g. for a coin, the two "models" are (head-up, tail-down) and (head-down, tail-up). A notion of coherence stronger than de Finetti's is presented and defended in [Shimony, 1955], which leads to a stronger version of the axiom corresponding to (2) in Cor. 6.3.2, namely $c(h,e) = 1$ if *and only if* $e \models h$. This paper also envisages, along with [Koopman, 1940], the possibility of replacing a quantitative theory, such as [De Finetti, 1937], by a qualitative, or comparative, approach.

This brings us to the third of our closing comments. De Finetti [1937] makes the following two tantalizing remarks:

> *It can be seen that one could ... eliminate everything quantitative, whether in the condition of coherence or in the definition of probability, in order to keep only the purely qualitative aspect of the definition (inequality between probabilities) and the condition of coherence (a small number of very simple axioms).*

> *another definition, equally subjective and very similar, that it may perhaps be useful to compare with the subjective definition of probability, is the utility of Vilfredo Pareto.*

here, de Finetti refers to [Pareto, 1909]; compare this perhaps with [Pareto, 1919]. The concept of utility function, later to be used routinely by the economists, also figures preeminently in the motivations Arrow [1951,1963] offers for his discussion of social choices, where he submits to a rigorous critique the ambiguity in the definition of this function (which could be replaced by its composition with any monotonically increasing function), and proves a general theorem to the effect that this ambiguity, while affordable in individual predictions, can lead to logical inconsistencies when used in aggregate decisions. In closing this section, we want to illustrate this point by adapting to this new purpose the scenario set up in [De Finetti, 1937]. Suppose three horses H_1, H_2 and H_3 are entered in a race; each of three independent Bookies $B^{(1)}, B^{(2)}$ and $B^{(3)}$ chooses the odds: $p^{(j)}{}_i$ denoting the odds placed on the horse H_i by Bookie $B^{(j)}$. These come out:

$$
\begin{array}{llll}
\textit{Bookie No.1} & : & p^{(1)}{}_1 > p^{(1)}{}_2 > p^{(1)}{}_3 & \\
\textit{Bookie No.2} & : & p^{(2)}{}_2 > p^{(2)}{}_3 > p^{(2)}{}_1 & \text{with } \sum_i p^{(j)}{}_i = 1 \text{ for } j = 1,2,3 \,. \\
\textit{Bookie No.3} & : & p^{(3)}{}_3 > p^{(3)}{}_1 > p^{(3)}{}_2 &
\end{array}
$$

$$\tag{6.3.40}$$

Note that each Bookie offers coherent bets. When the specific numerical values of the $p^{(j)}{}_i$ are ignored beyond the fact that they satisfy (6.3.40), the Bookies have in effect produced three orderings $O^{(j)}$ for the issue of the race:

$$
\left.
\begin{array}{llll}
\textit{Ordering No,1} & : & H_1 > H_2 > H_3 \\
\textit{Ordering No.2} & : & H_2 > H_3 > H_1 \\
\textit{Ordering No.3} & : & H_3 > H_1 > H_2
\end{array}
\right\} \,.
\tag{6.3.41}
$$

Now, suppose that N independent bettors come along, ignorant of (6.3.40) and focusing solely on (6.3.41): $N^{(j)}$ of them turn out to believe that $O^{(j)}$ is the right ordering. With

$$\nu_j = \frac{N^{(j)}}{\sum_{i=1}^{3} N^{(i)}} > 0 \qquad \sum_{i=1}^{3} \nu_j = 1 \qquad (6.3.42)$$

denoting the distribution of the bettors population, we describe what happens when the ν_j are among themselves as are the length of the sides of a genuine triangle, namely when

$$\nu_1 + \nu_2 > \nu_3 \quad , \quad \nu_2 + \nu_3 > \nu_1 \quad , \quad \nu_3 + \nu_1 > \nu_2 \quad . \qquad (6.3.43)$$

Upon analysing the meaning of this democratic aggregate of preferences, we find that $N^{(1)} + N^{(3)}$ people believe $H_1 > H_2$, whereas $N^{(2)}$ people believe that $H_2 > H_1$; this gives a majority to the preference $H_1 > H_2$. A similar analysis gives also a majority to the preference $H_2 > H_3$. Upon compounding these two findings, we obtain $H_1 > H_2 > H_3$, and hence by transitivity $\mathbf{H_1 > H_3}$. However, if we count the preferences further, we see that $N^{(1)}$ people believe that $H_1 > H_3$, whereas $N^{(2)} + N^{(3)}$ believe that $H_3 > H_1$, thus giving a majority to the preference $\mathbf{H_3 > H_1}$. Note that all these inequalities are strict (we wrote $>$, and not $\geq$, so that we explicitly excluded $=$). Hence the majority belief pattern is inconsistent with a transitive, and strict ranking ; one would like to avoid having to accept that the aggregate of unambigous individual rankings can lead to a tie between two (and in fact all three) horses: this would be contrary to race-track rules and practice, as well as to the expected transitivity of preference orderings. It should be noted that the condition (6.3.43) is generic (i.e. non-accidental) in the following sense; let Δ_o denote the set of all ternary probability distributions $\nu = (\nu_1, \nu_2, \nu_3)$ with therefore $\nu_i \geq 0$ and $\sum_i \nu_i = 1$. These can be represented graphically as the points of an equilateral triangle, which we also denote by Δ_o since the representation is bijective. The extreme points of this simplex are the distributions $(1,0,0)$, $(0,1,0)$, $(0,0,1)$. Then, the domain covered by the distributions satisfying (5.3.43) is the two-dimensional domain represented graphically as the interior points of the equilateral triangle Δ_1 the vertices of which are the distributions $(\frac{1}{2}, \frac{1}{2}, 0)$, $(0, \frac{1}{2}, \frac{1}{2})$, $(\frac{1}{2}, 0, \frac{1}{2})$; the Lesbesgue measure of Δ_1 is a finite fraction (in fact $\frac{1}{4}$) of that of Δ_o, the space of all ternary distributions. The presence of a generic domain leading to indecisive – or inconsistent – outcomes is one aspect of the phenomena known as the *Condorcet voting paradox* in recognition of the fact that he detected it while trying to establish unambigous balloting procedures for the 18th-century Royal Academy of Sciences in Paris.

A form of this "paradox" also figures as a motivating example in the first edition of [Arrow, 1951,1963]; the second edition also contains a brief historical survey. Arrow's general theorem states that, as soon as more than two candidates are vying for a position, preference rankings by majority rule are undecisive; see also [Arrow, 1972]. This discussion of aggregate rankings does not appear to have been contemplated by de Finetti.

It is interesting for us to notice here that the situation improves considerably when we go beyond the mere preference ranking (6.3.41), and we take into account the specific values of the odds. Consider now the aggregate probability distribution

$$\mu = (\mu_1, \mu_2, \mu_3) \quad \text{with} \quad \mu_i = \sum_j \nu_j \, p^{(j)}{}_i \quad . \tag{6.3.44}$$

It is then easy to prove that the inequalities (6.3.40) imply that the determinant of the matrix $\boldsymbol{P} = \{p^{(j)}{}_i\}$ is different from zero, so that the map

$$\nu \in \Delta_o \mapsto \mu = \nu \boldsymbol{P} \in \Delta_p \tag{6.3.45}$$

is bijective; here Δ_p is the triangle with vertices $p^{(j)} = (p^{(j)}{}_1, p^{(j)}{}_2, p^{(j)}{}_3)$; note also – again as a consequence of the inequalities (6.3.40) – that the uniform distribution $\mu_o = (\frac{1}{3}, \frac{1}{3}, \frac{1}{3})$ belongs to Δ_p, so that there exists always exactly one distribution ν leading to a three-fold tie, i.e. to $\mu = \mu_o$; even a two-fold tie can be seen to obtain only from a straightline segment in Δ_o. Hence the space of indecisive bettor distributions is of measure zero in all cases of ties; this is therefore a considerable improvement over (6.3.43), demonstrating how preference orderings are intrinsically less decisive than probabilistic orderings. Henceforth, the tentative qualitative version of the subjective approach proposed in [De Finetti, 1937] is to be disregarded in favor of his original quantitative version, namely the version that requires bookies to give numerical values to the odds offered to bettors. The problem is therefore reduced to assessing these numerical values; de Finetti's exchangeability theorem is one way to achieve that; another way is discussed in Sect. 5.3 on Shannon's entropy.

7. Setting-up the Ergodic Problem

> *Yet, what is the law which brings this infinity to a fixed measure, and imposes order upon it, this is what I am eager to learn.*
>
> [Augustine of Hippo, 389]

Boltzmann's proposals are discussed in Sect. 7.1, leading to the concept of dynamical systems defined in Sect. 7.2. The idealization provided by the latter allowed the first formal responses to Boltzmann's conjectures: the ergodic theorems of Birkhoff and von Neumann, reviewed in Sect. 7.2.

7.1 Boltzmann's Heuristics

There seem to be little dispute that ergodic hypotheses are to be traced back to [Boltzmann, 1871a]. More than a century later, however, their purpose, reasonableness and success are matters of some continuing controversies, see e.g. [Brush, 1967, Lebowitz and Penrose, 1973, Sklar, 1973, Quay, 1978, Wightman, 1985, Earman and Rédei, 1996, Szasz, 1996]. Difficulties appeared indeed on several fronts. First, ergodic theory involves some drastic idealizations, some of which were irrelevant in the finitist views that presided over Boltzmann's physical speculations on his kinetic theory of gases; these often implicit idealizations cloud a clear delineation of the approximations that must be involved. Second, much of the mathematical framework – measure theory and functional analysis – which we now regard as fundamental was not available to Boltzmann; these tools are necessary for understanding ergodic theory in terms that are unambiguous to a modern reader.

Rather than plunging directly into the question of whether ergodic theory "justifies" statistical mechanics, be it through an ergodic theorem asserting the equality of phase-space and time-averages, or via more stringent conditions higher up in the ergodic hierarchy, we read Boltzmann's theory partly as a proposal to introduce probability and measure theory into mechanics. Some 30 years before Hilbert's quest, and 60 years before Kolmogorov's syntax – see Chap. 5 – Boltzmann's originality resides in the semantics he offered for the notion of ensemble; in Sect. 3.3, we reported that Boltzmann's postulates are of a different kind than those proposed by Maxwell; Gibbs' limiting relative frequency interpretation will be seen to be conceptually closer initially to Boltzmann's setting, although in effect Gibbs' focus on the consequences is consonant with Maxwell's views. We focus on Boltzmann's basic questions: first to justify the use of statistical ensembles in mechanics, and then to justify the choice of a particular ensemble. Once again, we will aim at what Reichenbach [1938] calls the context of justification, while the context of discovery will take a subordinate place, but a place one cannot afford to ignore. In particular, we formulate the problem in modern language, in order to draw attention to some of the original, but often unformulated, presuppositions.

Let us first sketch the essentials of the *mechanical* in statistical mechanics; the review of the Liouville and Poincaré theorems is postponed however to Sect. 7.2. Boltzmann knew that the objects of interest are the *observables* – later on, Kolmogorov was to call them *random variables* (which we have seen in Chap. 5). These observables are functions on the space of the "dynamical states" of the system under consideration: classically, these are the points of a *phase-space* Ω, e.g. :

$$\Omega \subset \mathsf{R}^{2n} = \{x = (p, q) \mid p \in \mathsf{R}^n , q \in \mathsf{R}^n\} \quad . \tag{7.1.1}$$

At the time of Boltzmann, the functions usually encountered in mathematical physics were *continuous* functions; we will tentatively accept this in the main body of this section.

The evolution is governed by the laws of *Hamiltonian* mechanics, with the Hamiltonian H being time-independent, as we will focus on isolated systems; for each value E of the Hamiltonian,

$$\Omega_E \equiv \{x = (p, q) \in \mathsf{R}^{2n} \mid H(p, q) = E\} \tag{7.1.2}$$

is called the *energy surface* corresponding to E. The evolution is assumed to satisfy the Hamilton equations:

$$\dot{p}_k = -(\partial_{q^k})H(p, q) \quad ; \quad \dot{q}^k = (\partial_{p_k})H(p, q) \quad ; \quad k = 1, 2. \cdots, n \tag{7.1.3}$$

where $\dot{x}$ denotes the time derivative of $x : t \in \mathcal{T} \mapsto x_t \in \mathsf{R}^{2n}$. Although $\mathcal{T}$ is here the real line R, we preserve the typographical distinction for later purposes. These equations integrate to a flow on R^{2n}; since (7.1.3) implies

$$\frac{\mathrm{d}H}{\mathrm{d}t} = \sum_k [\partial_{p_k} H \, \dot{p}_k + \partial_{q^k} H \, \dot{q}^k] = 0 \quad ,$$

each Ω_E is stable under the flow; thus, we will henceforth only be concerned with the restriction of the flow to Ω_E :

$$(t, x) \in \mathcal{T} \times \Omega_E \mapsto x_t \in \Omega_E \quad ; \tag{7.1.4}$$

and, as the index E is now superfluous, it will be dropped. Transcribed to the observables, this flow is described by:

$$(t, f) \in \mathcal{T} \times \mathcal{C}(\Omega) \mapsto \tau_t[f] \in \mathcal{C}(\Omega) \quad ; \quad \tau_t[f] : x \in \Omega \mapsto f(x_{-t}) \in \mathsf{R}. \quad (7.1.5)$$

The continuity of the flow entails, for each f and each x, that

$$f_x : t \in \mathcal{T} \mapsto f_x(t) = f_t(x) \in \mathsf{R} \qquad (7.1.6)$$

is a continuous function of the time t.

Boltzmann's argument for the introduction of *ensemble* in *statistical* mechanics can be broken down in two steps which were unfortunately somewhat entangled. This entanglement was due in part to some undeclared assumptions, but also to the fact that Boltzmann's physical intuition was ahead of his times in terms of the mathematics he would have needed for a formulation free from ambiguities. In a nutshell, Boltzmann was trying not only to show that time-averages and phase-space coïncide in the microcanonical ensemble, but he had first to arrive at the very notion of ensemble. We find it useful to separate these two steps.

Step 1: *To find a convenient choice* $\mathcal{C}$ *of time-dependent functions* φ *which admit an invariant mean* $\eta[\varphi]$.

Step 2: *To find a reasonable set of conditions on a class* $\mathcal{O}$ *of observables which would ensure that their time-average over an orbit in phase-space is independent of the orbit.*

First of all, we should recall that Boltzmann had a "resolutely finitist" outlook, documented in [Dugas, 1959] where this contention is supported by numerous quotes from Boltzmann's writings and lecture notes; here is one of them:

> *The concepts of the integral and differential calculus, cut loose from any atomic representation* [i.e. from the finite collections of objects from which they were generated] *are purely metaphysical if, following the famous definition of Mach, we understand this to designate the things* [sic] *for which we forgot how we arrived at them.*
> [Boltzmann, 1896–1898].

Remark A:

1. In Step 1, Boltzmann's paradigm was:

$$\eta[\varphi] \equiv \lim_{T \to \infty} \eta^T[\varphi] \quad \text{where} \quad \eta^T[\varphi] \equiv \frac{1}{2T} \int_{-T}^{T} dt \; \varphi(t) \qquad (7.1.7)$$

   although he did not insist on actually carrying out the limit.
2. Boltzmann seems to have been content with an understanding that the averages were to be taken over "very long times", i.e. times that are sufficiently large to wipe out rapid fluctuations. In their commentary of the work of Boltzmann, Ehrenfest and Ehrenfest [1911] introduced the dog-flea model – which we studied in Sect. 3.4 – warning us of the necessity

to adapt the time-scale of the analysis to the phenomenon we want to describe. In Chaps. 10–15 of this book, we will see explicitly how the matching of the scale to the description of the phenomena has turned out to be one of the major ingredients in the formulation of statistical mechanics; in Chap. 10, in order to focus on macroscopic thermodynamical variables; in Chap. 15, for the description of the approach to equilibrium; and in Chaps. 11–14, to capture phase transitions, where the proper choice of scale – this one in space – will be essential in the renormalization program developed to focus on critical phenomena.

3. The finitism of Boltzmann will also appear later when we discuss the relevance of the distinction between the *strict ergodic hypothesis* and the ergodic hypothesis proper (also called the *quasi-ergodic hypothesis*). See Rems.. C.3, C.4 and D.1. in this section.

For the time being, we will ignore the finitist semantic scruples and read *mathematical limit* where *lim* is written; to avoid Mach's criticism, we shall not forget that the mathematical definition of limit – with its ε and δ_ε, or T_ε – specifically indicates *"how we go"* there; see Def. D.1.10. Since, however, the mathematical limit (7.1.7) does not exist, in general, for all continuous functions, some precautions must be taken. Either one restricts one's attention to some special class of functions for which the limit exists, or one extends the definition of an invariant mean. We do a little of both. Amongst the possible restrictions, the one that seems to be the most innocuous – because it is the one that can be most easily disposed of later – is to assume that our functions are not only continuous but also bounded. We can then simplify the presentation of the main ideas by taking advantage of the following mathematical results:

Theorem 7.1.1. *With C a linear space over R, let $p : \varphi \in C \mapsto p(\varphi) \in R^+$ be subadditive and homogeneous, i.e. $\forall \, \varphi, \psi \in C$ and $\forall \, a \in R^+ : p(\varphi + \psi) \leq p(\varphi) + p(\psi)$ and $p(a\varphi) = a \, p(\varphi)$. Further, let M be a linear subspace of C and $\eta_o : \varphi \in M \mapsto \eta_o(\varphi) \in R$ such that (i) η_o is linear, (ii) $\forall \, \varphi \in M :$ $\eta_o(\varphi) \leq p(\varphi)$.*

Then, there exists an extension $\eta : \varphi \in C \mapsto \eta(\varphi) \in R$ such that (i) η is linear, (ii) $\forall \, \varphi \in C : \eta(\varphi) \leq p(\varphi)$, (iii) $\forall \, \varphi \in M : \eta(\varphi) = \eta_o(\varphi)$.

Corollary 7.1.1. *With C a normed linear space over R and M a linear subspace of C, let $\eta_o : \varphi \in M \mapsto \eta_o(\varphi) \in R$ such that (i) η_o is linear, and (ii) $\forall \, \varphi \in M : \eta_o(\varphi) \leq \|\varphi\|$; let finally $\|\eta_o\| \equiv \sup_{\varphi \in M} |\eta_o(\varphi)|$.*

Then, there exists an extension $\eta : \varphi \in C \mapsto \eta(\varphi) \in R$ such that (i) η is linear, (ii) $\|\eta\| = \|\eta_o\|$ with $\|\eta\| \equiv \sup_{\varphi \in C} |\eta(\varphi)|$, and (iii) $\forall \, \varphi \in M : \eta(\varphi) = \eta_o(\varphi)$.

This theorem and its corollary are both called *the* Hahn–Banach theorem after Hahn [1927] and Banach [1929]; note that they were formulated and proven more than half a century after Boltzmann had started introducing the problems under discussion here.

Example: Let $\mathcal{C} = \mathcal{C}(\mathcal{T})$ be the set of all bounded continuous functions on the real line $\mathcal{T}$; equip $\mathcal{C}(\mathcal{T})$ with the norm $\|\varphi\| = \sup_{t \in \mathcal{T}} |f(t)|$; and let

$$
\mathcal{M} = \left\{ \varphi \in \mathcal{C}(\mathcal{T}) \mid \eta_o[\varphi] \equiv \lim_{T \to \infty} \frac{1}{2T} \int_{-T}^{T} dt\, \varphi(t) \text{ exists} \right\} .
$$

Note that in this example the functional η_o is furthermore positive, and invariant under the time translations $\varphi \to \varphi_s$ with $\varphi_s(t) = \varphi(t - s)$. These properties can be carried over to the extension η.

Corollary 7.1.2. *There exists invariant means on $\mathcal{T}$, i.e. positive linear functionals*

$$
\eta : \varphi \in \mathcal{C}(\mathcal{T}) \mapsto \eta(\varphi) \in \mathsf{R}
$$

normalized to 1, and invariant under translations in $\mathcal{T}$; i.e.

(i) $\eta[\varphi] \geq 0 \ \forall\, \varphi \geq 0$ where $\varphi \geq 0$ means $\varphi(t) \geq 0 \ \forall\, t \in \mathsf{R}$
(ii) η is linear   i.e. $\eta[af + bg] = a\eta[f] + b\eta[g]$
(iii) $\eta[u] = 1$ where $u(x) \equiv 1 \ \forall\, x \in \Omega$
(iv) $\eta[\varphi_s] = \eta[\varphi] \ \forall\, s \in \mathsf{R}$ where $\varphi_s(t) \equiv \varphi(t - s) \ \forall\, s, t \in \mathsf{R}$

such that

$$
\eta(\varphi) = \lim_{T \to \infty} \frac{1}{2T} \int_{-T}^{T} dt\, \varphi(t)
$$

for all φ where the limit in the RHS exists.

Finally, since the time-evolution (7.1.4) is continuous, we can substitute f_x in (7.1.6) for φ in the above results; thus these results allow to define averages on observables as follows:

Corollary 7.1.3. *Let $\mathcal{C}(\Omega)$ denote the set of all bounded continuous functions on Ω. Then for every $f \in \mathcal{C}(\Omega)$ and every $x \in \Omega$, there exists a time-invariant mean given by:*

$$
\eta_x : f \in \mathcal{C}(\Omega) \mapsto \eta_x[f] \equiv \eta[f_x] \in \mathsf{R}
$$

with $\eta_x[f]$ depending only on the orbit $x_{\mathcal{T}} = \{x_t \mid t \in \mathcal{T}\}$, rather than on any particular x belonging to this orbit.

Remark B:

1. The invariant mean(s) η obtained in Cor. 7.1.2, for $\mathcal{T} = \mathsf{R}$, are not to be confused with the Haar measure, i.e. here the Lebesgue measure μ. Both η and μ are translation invariant, but μ cannot be normalized, whereas the means η are [see property (iii) in Cor. 7.1.2]. The definition of the Haar measure requires that the continuous functions φ be of compact support; they need only to be bounded for the definition of η; and in fact the means η of the Corollary are identically zero on functions of compact support. The Haar measure is unique, the means η are not; we shall come back to this in Rem. 4 below.

2. The Haar measure was developed [Haar, 1933, Von Neumann, 1936] almost at the same time as the centrality of measure theory was universally recognized by probabilists after the work of Kolmogorov. The theory of invariant means on topological groups, which got a start in [Agnew and Morse, 1938], came of age in the 1950s and 1960s; for a concise, mathematically cogent account, see [Greenleaf, 1969]; and for a brief review geared to applications in mathematical physics akin to those discussed in the present chapter, [Emch, 1972a, pp. 164–188]; such applications – involving also groups of symmetries beyond the time-evolution – to the theory of phase transitions will be seen in Chaps. 11–14.

3. We focused here on extensions of (7.1.7) to $\mathcal{C}(\mathcal{T})$, but other spaces of functions can be considered (as suggested by the theorem): see [Greenleaf, 1969], and also the theorems of Birkhoff and of von Neumann which will be discussed in Sect. 7.2.

4. The didactic proof of the Hahn–Banach Theorem given in [Dunford and Schwartz, 1964] consists of two parts: first, it establishes the existence of a maximal extension; and then, it shows that the domain of this maximal extension is the whole space $\mathcal{C}$. Note that the first part of the proof uses the *Zorn maximal principle*, an equivalent form of the *axiom of choice*, while the second part is by *reductio ad absurdum*. Hence this central theorem of analysis is *not constructive*. No uniqueness of the extension is claimed, and there are in fact many.

In view of Rem. B.4, we must simply assume henceforth that a choice of some mean η has been made, and proceed with the presentation of the program sketched in [Boltzmann, 1871a]; fortunately, some of the most basic results do not depend on such a choice. What we have gained so far is that we know mathematically what we say when we speak of a mean η; we have also learned that the actual construction of a specific mean is quite another matter: it demands too much, both mathematically as just seen, and also epistemologically when we recall the finitist views Boltzmann harbored when he considered (7.1.7).

We now turn to Step 2, which would be trivial if the following could be sustained:

> ### The (strict) ergodic hypothesis:
>
> *The energy surface contains exactly one orbit,*
> *i.e. for every $x \in \Omega$: $\{x_t \mid t \in \mathcal{T}\} = \Omega$* .

The papers [Boltzmann, 1869, 1871a,b, 1881, 1884/5, 1887, 1892] contain several specific sentences to the effect that Boltzmann believes this hypothesis is justified for systems of interest in kinetic theory, even though Brush [1967] warns, correctly indeed, that such a strict reading of Boltzmann hypotheses is *too literal* in regard to the analogies Boltzmann himself brings to support his case; a proper consideration to the latter is actually conducive to less demanding assumptions, as we shall presently see. Also, note that the above hypothesis was not unreasonable when set against its contemporary mathematical landscape, although later mathematical developments led to the conclusion that the hypothesis – at least in this strict form – is untenable.

Mathematically, the "strict" ergodic hypothesis is requiring that a one-dimensional object – the trajectory $\{ x_t|\ t \in \mathcal{T}\}$ – covers a multidimensional manifold – the energy surface Ω. In a perhaps naïve sense, the above demand appears counter-intuitive and somewhat far-fetched. Yet, to the surprise of most of the contemporary mathematicians, Cantor [1878] established the existence of a *bijective* map between the points of an interval and those of a multidimensional "manifold" such as a square; however, the map was *not continuous.* Upon noting that Netto [1879] and others had attempted to prove on general grounds that this incompatibility was not an accident, Peano [1890] gave an explicit construction of a *continuous* curve that fills a whole square; however, that curve has self-intersections, i.e. does not define a bijective map; moreover it is nowhere differentiable. Other examples were found later, e.g. [Hilbert, 1891], or see [Hairer and Wanner, 1996]. The definitive proof that dimension is a topological invariant of manifolds was given in [Brouwer, 1910]. The relevance of these results for the dismissal of the strict ergodic hypothesis was forthcoming:

> ... *these doubts* [expressed by Ehrenfest and Ehrenfest [1911]] *were correct,* ... *not only is not every gas an ergodic system, but such ergodic gas-systems can not at all exist.*
>
> [Rosenthal, 1913]

Plancherel arrived at the same negative conclusion, first in [Plancherel, 1912] which is also based on [Brouwer, 1910], and second in [Plancherel, 1913], through the use of Lebesgue measure theory, which was then of recent vintage: indeed, Plancherel [1913] cites the 1904 first edition of [Lebesgue, 1928]. The reception of these results is perhaps best described by a contemporary:

> *it does not seem that this abstract hypothesis had ever been really contemplated by the physicists; they probably never believed in systems that would be rigorously ergodic in the mathematical sense of the word, i.e. ... in a trajectory that would rigorously go through every interior point of a square ... ; curves such as Peano's were known only by those who were interested in set theory, and nobody ever thought that such curves could by dynamical trajectories. For the physicist, a trajectory is ergodic if it approaches as close as one wishes to any position compatible with the value of the energy.*
>
> [Borel, 1914a]

Remark C:

1. The question now is whether Boltzmann – or perhaps some of his readers and commentators – had blundered, and if so, how and why. The question was raised almost simultaneously by Hertz [1910] and Ehrenfest and Ehrenfest [1911].

2. It may seem – see e.g. [Brush, 1967] – that some ambiguity on whether or not this question was going against some long-established wisdom persisted for some time, even in the writings of such mathematicians as Poincaré [1894]. The latter, however, when commenting on the work of Maxwell – whom he apparently takes as some sort of surrogate for Boltzmann – states [our emphasis]:

> *he admits that, whatever the initial state of the system, it always passes an infinite number of times, I would not say through every state compatible with the existence of the integrals, but as close as desired to every one of these states*
>
> [Poincaré, 1894]

Prophetic? Poincaré does not play it up; the style indicates that he considers this precaution to be a routine exercise in mathematical rigor, rather than a matter of physical principle.

3. In the beginning of the 20th century, Poincaré – although not specifically nor explicitly motivated by the work of Boltzmann – proposed an interesting concept: the *physical continuum* as distinct from the *mathematical continuum* [Poincaré, 1903, 1904]. The latter is the continuum of Dedekind and Kronecker, in which for the points A, B, C the transitivity property

$$A = B \text{ and } B = C \quad \models \quad A = C \tag{7.1.8}$$

holds. For the physical continuum, Poincaré reinterprets the equality sign " $=$ " to mean "not distinguishable from", thus leading to a situation where (7.1.8) must be rejected, while keeping the symmetry and reflexivity of the relation " $=$ "; Poincaré writes this rejection of transitivity provocatively as the compatibility of the three relations

$$A = B \ , \quad B = C \ , \quad A < C \tag{7.1.9}$$

or more generally

$$A_k = A_{k+1} \text{ for } k = 1, 2, \cdots, n - 1 \text{ and yet } A_1 \neq A_n$$

> *which can be regarded as the formulation of the physical continuum ... there is here, with the principle of contradiction, an intolerable disagreement, and it is the necessity to put an end to this situation that constrained us to invent the mathematical continuum ... which is only a particular system of symbols.*
>
> [Poincaré, 1903] [pp. 36–41]

4. Poincaré's proposal did not seem to have raised great interest at the time. Yet, his idea – that a proper account of physical measurements would require avoiding dubious idealizations from the physical continuum to the mathematical continuum – reappears in the work of Zeeman [1961] and some of his disciples, e.g. Poston [1971a], under the name of "tolerance spaces" and "fuzzy geometry"; for a tantalizing introduction, see: [Poston, 1971b]. In a different vein, see also [Gallavotti, 1995], and especially, his

section 2, on the place of the ergodic theory in continuous versus discrete phase-space.

Boltzmann intuited his ergodic hypothesis from an analogy with the Lissajous figures Lissajous [1857]. He considered a two-dimensional harmonic oscillator, with Hamiltonian

$$H = H_1 + H_2 \text{ where } H_i = \frac{1}{2} p_i{}^2 + \frac{1}{2} w_i{}^2 q_i{}^2 \ (i = 1, 2) \text{ and } \frac{\omega_1}{\omega_2} \notin \mathsf{Q} \quad (7.1.10)$$

where Q denotes the set of rational numbers; recall that, although Q is dense in R, it is countable and thus is a subset of Lebesgue measure zero.

Since H_1 and H_2 are constants of the motion, for every pair $a = (a_1, a_2)$ of positive reals, the two conditions $H_i = E_i \equiv \frac{1}{2} w_i{}^2 a_i$, $i = 1, 2$ determine a $2(= 4 - 2)$–dimensional submanifold $\Omega^{(a)} \subset \Omega_E$ – where Ω_E is the 3–dimensional energy surface corresponding to the total energy $E = E_1 + E_2$; $\Omega^{(a)}$ is stable under the time evolution: the orbits are the curves

$$t \in \mathcal{T} \mapsto q(t) \in \Omega^{(a)} \text{ with components } q_i(t) = a_i \sin[\omega_i t + 2\pi \psi_i)] \ . \quad (7.1.11)$$

Boltzmann describes this situation by stating that each orbit *"covers"* $\Omega^{(a)}$. This cannot be interpreted literally to mean that, given a single orbit, it passes through every point of $\Omega^{(a)}$: it is simply *not true* that for any initial condition $q(0) \in \Omega^{(a)}$

$$q(0) \in \Omega^{(a)} \quad \models \quad \{q(t) \mid t \in \mathcal{T}\} = \Omega^{(a)} \quad . \quad (7.1.12)$$

Nevertheless it is true that any single orbit starting on the torus $\Omega^{(a)}$ passes arbitrarily close to any given point of $\Omega^{(a)}$, i.e. *it is true* that for each initial condition $q_o \in \Omega^{(a)}$:

$$q(0) \in \Omega^{(a)} \models \{q(t) \mid t \in \mathcal{T}\} \text{ is a dense proper subset of } \Omega^{(a)}. \quad (7.1.13)$$

Since this statement is an anchor for the argument, a proof will be given that sheds light on what is going on. Consider to this effect the bijective, bi-continuous map

$$(q_1, q_2) \in \Omega^{(a)} \mapsto (x_1, x_2) \in \mathsf{T}^2 \equiv \mathsf{R}^2/\mathsf{Z}^2 \text{ with } q_i = a_i \sin[2\pi x_i]. \quad (7.1.14)$$

This map straightens the orbit (7.1.11) to:

$$t \in \mathcal{T} \mapsto x(t) \in \mathsf{T}^2 \text{ with components } x_i(t) \equiv (\frac{1}{2\pi} \omega_i t + \psi_i) \bmod 1 \ . \quad (7.1.15)$$

The successive intersections of this curve with the horizontal line through $x(0) = \psi$ is the set

$$S_\psi = \{x(t_n) = (\xi_n + \psi_1) \bmod 1 \mid n \in \mathsf{Z}\} \quad \text{with} \quad \xi_n = \frac{\omega_1}{\omega_2} n \quad . \quad (7.1.16)$$

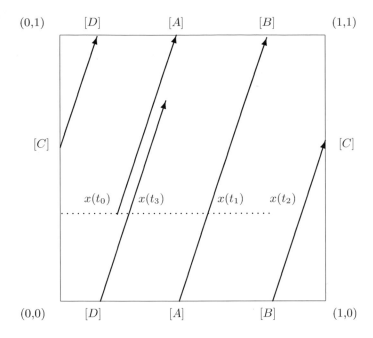

Fig. 7.1. An orbit of the 2-dim. harmonic oscillator

Since ω_1/ω_2 is irrational, $\{\xi_n \mid n \in \mathsf{Z}\}$ is one of Weyl's "uniformly distributed sequences modulo one" [Weyl, 1916] which we studied in Sect. 5.3 : we know that S_ψ is a dense proper subset of $[0,1]$. Consequently

$$x(0) \in \mathsf{T}^2 \quad \models \quad \{x(t) \mid t \in \mathcal{T}\} \text{ is a dense proper subset of } \mathsf{T}^2 \quad . \quad (7.1.17)$$

Now, since (7.1.14) is a bijective, bi-continuous map, (7.1.17) implies (7.1.13), so that (7.1.12) is certainly an overstatement that cannot sustain scrutiny.

Three questions are proper here:

1. what is (are) the physical reason(s) why one would not want to care and distinguish between the two assertions (7.1.12) and (7.1.13)?
2. what is (are) the physical reason(s) one could bring to support either of these two assertions as "generic" in the kinetic theory of gases?
3. even if (7.1.13) were "generic", what would one gain from knowing that it holds for a given system of interest?

Remark D:

1. To the first question, we would simply answer that to a committed finitist – like Boltzmann seems to have been – it does not make sense to argue on whether something would happen in a very long time, or only in an infinite amount of time; the two statements (7.1.12) and (7.1.13) cannot be

distinguished in any finite amount of time. Recall also in this connection Rems. C.3 and C.4 above. We were faced with a similar difficulty when we studied the notion of randomness in Sect. 6.2; even closer to the above model, a pragmatic distinction between (7.1.12) and (7.1.13) would be tantamount to a pragmatic distinction between replacing the irrational quotient ω_1/ω_2 with an arbitrarily close, but rational, approximation. Note nevertheless that Q^c rather than Q would have more of a claim at being "generic" in R, since, w.r.t. the Lebesgue measure $\mu(\mathsf{Q}^c) = 1$ while $\mu(\mathsf{Q}) = 0$.

2. An answer to the second question takes on where the first leaves off: so far, the word "generic" has been used in a loose sense to try and convey the idea that one is after a property which "is to be expected in usual circumstances"; we will specify a precise meaning to the word "generic" in Sect. 9.1. For the time-being, one could argue for instance that the interactions between many degrees of freedom would destroy the accidental constants of the motion, in the same way that the close, periodic orbits one receives for the harmonic oscillator with $\omega_1/\omega_2 \in \mathsf{Q}$ (which is accidental) get stretched into dense orbits when this ratio becomes irrational, i.e. when $\omega_1/\omega_2 \in \mathsf{Q}^c$ (which is "generic"). To buttress this line of speculation, Boltzmann was calling upon small stochastic perturbations, while Maxwell expected the influence of the walls to provide such a mechanism. Nowadays, one has to cope with precise mathematical arguments to the effect that an extension to the whole energy surface of even (7.1.13) would not be generic for finite Hamiltonian systems; see Sect. 9.1.

3. *A posteriori*, the third question is the easiest to answer, when we consider again the Boltzmann 2–dimensional harmonic oscillator analogy just discussed. As ξ_n is uniformly distributed modulo, we have for every Lebesgue measurable $\Delta_1 \subseteq [0,1] \times \{x(0)\}$:

$$\lim_{T \to \infty} \frac{1}{2T} \operatorname{card}\{t \in [-T,T] \mid x(t) \in \Delta_1\} = \mu_1(\Delta_1) \quad . \qquad (7.1.18)$$

The limiting frequency thus obtained, for the successive intersections of the orbit with the horizontal line passing through $x(0)$, leads to the following evaluation of the average time the orbit spends in a measurable two-dimensional subset $\Delta \subseteq \mathsf{T}^2$:

$$\lim_{T \to \infty} \frac{1}{2T} \int_{-T}^{T} dt\, \chi_\Delta(x_t) = \mu(\Delta) \qquad (7.1.19)$$

where, as usual, χ_Δ is the indicator function of the set Δ, i.e.

$$\chi_\Delta(x_t) = \begin{cases} 1 & \text{if } x_t \in \Delta \\ 0 & \text{if } x_t \notin \Delta \end{cases} ,$$

and μ is the restriction of the Lebesgue measure – and hence the Haar measure – on the torus.

The formula (7.1.19) appears thus in our context as the two-dimensional continuum analog of Weyl's formula (7.1.18); as a matter of fact, Weyl [1916] also considers the situation addressed here.

The pioneering originality of Boltzmann was to view (7.1.19), not as an equality to be proven, but as a *pragmatic definition* of the measure μ, which he termed *ensemble*, and thus obtaining these measure(s) together with their limiting-frequency interpretation.

The above analysis of the 2−dimensional oscillator analogy, which was essential to Boltzmann argument, suggests a form of the ergodic hypothesis that is *weaker* that its strict form, namely the modified form:

The (modified) ergodic hypothesis:

Each orbit is dense on its energy surface, i.e.
for each $x_o \in \Omega$:
$\{x_t \mid t \in \mathcal{T}\}$ *is a dense proper subset of* Ω.

This is the form of the ergodic hypothesis that was introduced almost simultaneously by Hertz [1910] and Ehrenfest and Ehrenfest [1911], the latter dubbing it the *quasi-ergodic* hypothesis. For a precedent – genuine, albeit without much fanfare – see [Poincaré, 1894], as reported in Rem. C.2 above. This ergodic hypothesis reduces then Step 2 to the assumption that for every $f \in \mathcal{C}[\Omega]$, $\eta_x[f]$ – see in Cor. 7.1.3 above – be continuous. Is it ?

The exactly solvable model invoked by Boltzmann and discussed above – the 2−dimensional harmonic oscillator with ω_1/ω_2 irrational – is also instructive in this respect, although the lesson has to be taken *cum grano salis*; the uniform continuity argument leading to (7.1.24) shows how this works in the context developed so far.

On the one hand, since η is positive, linear and normalized, we have:

$$|\eta_y[f] - \eta_x[f]| = |\eta[f_y - f_x]| \leq \|f_y - f_x\| \equiv \sup_{t \in \mathcal{T}} |f(y_t) - f(x_t)| \quad . \quad (7.1.20)$$

On the other hand, since T^2 is compact, f is not only continuous, but it is also uniformly continuous, i.e. for all $\varepsilon > 0$, there exist a $\delta_{(f,\varepsilon)} > 0$ such that

$$\xi, \zeta \in \mathsf{T}^2 \text{ and } d(\xi, \zeta) < \delta_{(f,\varepsilon)} \quad \models \quad |f(\xi) - f(\zeta)| \leq \varepsilon \quad , \quad (7.1.21)$$

where d denotes the Euclidean distance on the flat torus $\mathsf{T}^2 \equiv \mathsf{R}^2/\mathsf{Z}^2$. Finally, the Boltzmann model enjoys a peculiarly gentle – but otherwise damning – property, namely:

$$x, y \in \mathsf{T}^2 \text{ and } t \in \mathcal{T} \quad \models \quad d(y_t, x_t) = d(x, y) \quad ; \quad (7.1.22)$$

this is evidently much stronger than continuous dependence on initial conditions.

Now, (7.1.20–22) together imply that for every $f \in \mathsf{T}^2$ and every $\varepsilon > 0$, there exists a $\delta_{(f,\varepsilon)}$ such that:

$$d(y, x) < \delta_{(f,\varepsilon)} \quad \models \quad |\eta_y[f] - \eta_x[f]| < \varepsilon \quad . \tag{7.1.23}$$

This continuity property allows to complete Boltzmann's argument. Since the orbits are dense in T^2, (7.1.21) implies that $\eta_x[f]$ is independent of $x \in \mathsf{T}^2$, and thus defines a map

$$\tilde{\eta} : f \in \mathcal{C}(\mathsf{T}^2) \mapsto \tilde{\eta}[f] \in \mathsf{R} \quad \text{with} \quad \tilde{\eta}[f] = \eta_x[f] \quad \forall x \in \mathsf{T}^2 \quad . \tag{7.1.24}$$

Since η_x is continuous in $f \in \mathcal{C}(\mathsf{T}^2)$, so is $\tilde{\eta}$.

This is what Boltzmann calls an ensemble or a representative ensemble. In a semantic sweep by which he identifies the primary concept – the time-average – with a subordinate notion – its representation by a probability distribution over the part of the phase-space available to the system:

$$\tilde{\eta}[f] \equiv \langle f \rangle \tag{7.1.25}$$

where $\tilde{\eta}[f]$ is the time-average, and $< f >$ is the space-average with respect to a distribution which the above equality is meant to *define*.

Boltzmann anticipates here the modern definition of an integral or a measure – see e.g. [Schwartz, 1957] who naturally cites [Bourbaki, 1952] – as a positive linear functional on a properly chosen linear space of test functions, hypothetically denoted here by $\mathcal{C}$:

$$\tilde{\eta} : f \in \mathcal{C} \mapsto \eta[f] \in \mathsf{R} \tag{7.1.26}$$

defines the measure μ with respect to which it is represented by the integral

$$\tilde{\eta}[f] \equiv \int \mathrm{d}\mu(x) f(x) \quad . \tag{7.1.27}$$

To be a proper mathematical definition, where μ becomes a set function satisfying the conditions discussed in Sect. 5.2, the above would require some additional properties of the functional $\tilde{\eta}$ and the space $\mathcal{C}$; under the further assumption that there exists on $\mathcal{C}$ a standard measure dx with respect to which $\mathrm{d}\mu$ is absolutely continuous, one has

$$\mathrm{d}\mu(x) = \varrho(x)\,\mathrm{d}x \quad , \tag{7.1.28}$$

where ϱ is a probability distribution when η is normalized, say by $\eta[1] = 1$.

What is remarkable in the specific oscillator model chosen by Boltzmann is that all of the above conditions happen to be satisfied. Indeed, since the functional $\tilde{\eta}$ obtained there is invariant under the translations on the – compact – torus T^2, the corresponding measure in (7.1.27) coïncides with the Haar measure:

$$\tilde{\eta}[f] = \int_{\mathsf{T}^2} \mathrm{d}\mu(x)\, f(x) \quad \text{with} \quad \mathrm{d}\mu(x) = \mathrm{d}x_1\, \mathrm{d}x_2 \quad . \tag{7.1.29}$$

As we noticed following (7.1.17), the results of [Weyl, 1916] show that this model is particular in the sense that, for the space $\mathcal{C}$ of continuous functions, one can replace $\tilde{\eta}$ by the actual ergodic limit, i.e. with $\mathrm{d}\mu(x) = \mathrm{d}x_1\, \mathrm{d}x_2$:

$$\tilde{\eta}[f] = \lim_{T \to \infty} \frac{1}{2T} \int_{-T}^{T} \mathrm{d}t f(x_t) = \int_{\mathsf{T}^2} \mathrm{d}\mu(x)\, f(x) \,. \tag{7.1.30}$$

Intuitively, and in line with Rem. D.3 above, the connection between the definition of a measure as a set function, and its definition via a linear functional, is based on the idea that we would want to call a set $\Delta \in \mathsf{T}^2$ measurable whenever $\tilde{\eta}$ can be extended to its indicator function χ_Δ. In an empirical approach – such as Boltzmann was advocating – it is natural to prefer a definition that emphasizes the expectation values of the observables; that is what the modern definition does, and that happens to be precisely how Boltzmann introduced the notion of ensembles, much before the modern formalization was cast.

Note the essential programmatic difference between having the equality of time-average and space-average appear as a *theorem* – as in (7.1.30), [Weyl, 1916] – and viewing it as an *empirical definition* of a measure – as in (7.1.25), [Boltzmann, 1871a]. For such a definition to be possible in general would involve mathematical subtleties, such as convergence, a proper choice of the topology with respect to which limits are taken, and/or issues of constructibility, which were beyond Boltzmann's interests, as well as beyond the reach of the mathematics available in his times. Moreover, it appeared later that, in general, these conditions could not be all satisfied; whence the "ergodic problems" that were to come.

Boltzmann's model nevertheless served the pioneering purpose of urging attention to some of the idealizations involved in the exploration of systems one would expect to encounter in the kinetic theory of gases. One specific such idealization is the notion of dynamical system which we now turn to introduce.

7.2 Formal Responses: Birkhoff/von Neumann

One of the major conceptual developments that allowed the notion of dynamical system to emerge from the early ergodic theory was the geometrization of Hamilton's mechanics in the hands of Poincaré; see for instance, his treatment of integral invariants in Tome III of [Poincaré, 1892–1907]. Birkhoff, in a series of five papers, starting in 1912 and leading to [Birkhoff and Smith, 1928], mentions this book amongst several other contributions by Poincaré dated from 1881 to 1907.

In conjunction with this shift in the approach – the authors of the 1930's could avail themselves of the newly developed techniques in topological vector spaces [Banach, 1932]. How new these methods were then can be appreciated from Banach's short preface, where he draws his reader's attention to the fact that the periodical *Studia Mathematica* had just been created for the primary purpose of collecting researches on functional analysis and its applications; the first volume appeared in 1929. Here again, there were precedents; a crucial one of these was Lebesgue's integration theory [Lebesgue, 1928], which is precisely where the presentation in [Banach, 1932] starts. We briefly mention here two results in *mechanics* that belong to the transition period.

Theorem 7.2.1 (Liouville). *The flow $\{\phi_t | t \in \mathcal{T}\}$ on R^{2n} generated by the Hamiltonian equations (7.1.3) preserves the volume V in phase-space.*

Theorem 7.2.2 (Poincaré). *Let $\Omega \subset \mathsf{R}^{2n}$ be closed, bounded and stable under the flow ϕ of Theorem 2. Then for every measurable $A \subseteq \Omega$ and almost every $x \in \Omega$, there exists an infinite subsequence $\{t_n \in \mathcal{Z} \mid n \in \mathsf{Z}\}$ such that $x_{t_n} \in A$.*

The proofs of both of these theorems are instructive in that they are a guide to the forthcoming developments.

Sketch for the proof of Thm. 7.2.1. Let A be any measurable subset of R^{2n} , and let us denote by A_s the image of A through the flow at time s. From the Gauss theorem [Sect. B.2] applied to the flow cylinder

$$M \equiv \{\, x = (\xi_s, s) \in \mathsf{R}^{2n+1} \mid \xi_o \in A \,; s \in [0,t] \,\}$$

we obtain:

$$\left.\begin{array}{l} V(A_t) - V(A) \equiv \\[1ex] \int_{A_{s=t}} dp_1 dq_1 \cdots dp_n dq_n \; - \; \int_{A_{s=0}} dp_1 dq_1 \cdots dp_n dq_n = \\[1ex] \int_M dp_1 dq_1 \cdots dp_n dq_n \, ds \;\; \mathrm{div}\,(\dot{p}_1, \dot{q}_1, \cdots, \dot{p}_n, \dot{q}_n, 1) \end{array}\right\} . \qquad (7.2.1)$$

It is therefore sufficient to show that a Hamiltonian flow is always divergence-free, i.e. that the divergence

$$\mathrm{div}\,(\dot{p}, \dot{q}, 1) \;=\; \mathrm{div}\,(\dot{p}, \dot{q}) \equiv \sum_{k=1}^{n} \frac{\partial \dot{p}_k}{\partial p_k} + \frac{\partial \dot{q}_k}{\partial q_k} \qquad (7.2.2)$$

of its velocity field vanishes everywhere. The Hamiltonian equations (7.1.3) give in the RHS of (7.2.2):

$$\frac{\partial \dot{p}_k}{\partial p_k} + \frac{\partial \dot{q}_k}{\partial q_k} \;=\; -\frac{\partial^2 H}{\partial p_k \partial q_k} + \frac{\partial^2 H}{\partial q_k \partial p_k} \qquad (7.2.3)$$

which, together with the differentiability of H , gives in (7.2.2), $\mathrm{div}(\dot{p}, \dot{q}) = 0$, and thus proves Thm. 7.2.1. **q.e.d.**

Remark A:

1. The Liouville theorem thus associates to every isolated Hamiltonian system in R^{2n}, a canonical invariant volume element, or measure.
2. This theorem appears today as an elementary result in symplectic geometry; see for instance Proposition 3.3.4 in [Abraham and Marsden, 1978].

3. From R^{2n} the existence of an invariant measure carries over directly to any submanifold that is stable under the flow, and so to the energy surface Ω_E.
4. The Liouville measure, *qua* measure, coïncides with the equilibrium ensemble of Boltzmann, which Gibbs called *microcanonical*. However, we emphasize again the profound differences in the semantic understandings that presided over the derivation of the same syntactic structure.

Sketch for the proof of Thm. 7.2.2 Let $\phi_t[x] = x_t$ denote the image of x under the Hamiltonian flow at time t. For $s \geq 0$ let $U_s \equiv \cup_{t \geq s} \phi_t[A]$. Since Ω is closed and bounded, $V(\Omega) < \infty$; furthermore, since Ω is stable under the flow, $U_s \subset \Omega$ and thus $V(U_s) < \infty$. Furthermore, since $\phi_{-s}[U_s] = U_0$ the Liouville theorem implies $V(U_s) = V(U_0)$. And yet, the definition of U_s implies that $U_s \subset U_0$. Hence:

$$V(U_0 \backslash U_s) = 0 \quad \text{where} \quad U_0 \backslash U_s \equiv \{x \in U_o \mid x \notin U_s\} \quad . \tag{7.2.4}$$

But $A \subset U_0$; hence (7.2.4) implies in particular that

$$\forall s > 0 \; : \; V(\{x \notin U_s \mid x \in A\}) = 0 \tag{7.2.5}$$

which says that the set of points of A which do not return to A at any time $t \geq s$ is of measure zero. Since this holds for all s, we can choose them to satisfy the conclusion of Thm. 7.2.2.   **q.e.d.**

Remark B:

1. In the literature on statistical mechanics, and more particularly on the kinetic theory of gases – see Sects. 3.3–4 – this result is known as the Poincaré recurrence theorem, or the Zermelo paradox, or a mixture of the two attributions; see e.g. [Carathéodory, 1919]. As noted in [Wightman, 1985], Poincaré himself, referred to this result as *"stabilité à la Poisson."*
2. The proof, given in [Poincaré, 1890], holds in a context more general than the one in which the theorem was stated above, While the original proof was phrased in terms of the hydrodynamics of incompressible fluids, the proof itself – as indicated by the sketch given above – shows how the assumptions that are essential to the conclusion are exactly those embodied in Def. 7.2.1 below; in particular, one does not need to assume that the evolution is Hamiltonian, as long as one makes some other postulate to the effect that there is a finite measure, invariant under the flow, thus bypassing Liouville's theorem by *fiat*.

Definition 7.2.1. *A classical dynamical system* $\{\Omega, \mathcal{F}, \mu; \phi\}$ *consists of*

1. $\{\Omega, \mathcal{F}, \mu\}$, *a probability space in the sense of Kolmogorov – see Sect. 5.1;*
2. $\mathcal{T} = \mathsf{Z}$ *or* R, *a group in one parameter called "time";*
3. ϕ, *a group action of* $\mathcal{T}$ *on* Ω, *i.e.*

$$\phi : (t, x) \in \mathcal{T} \times \Omega \mapsto \phi_t[x] \in \Omega$$

satisfying the following conditions:

$$
\begin{aligned}
(a) \quad & f : x \in \Omega \mapsto f(x) \in \mathsf{R} \text{ measurable} \quad \models \\
& f : (t, x) \in \mathsf{R} \times \Omega \mapsto f(\phi_t[x]) \in \mathsf{R} \text{ measurable} \quad ; \\
(b) \quad & t, s \in \mathcal{T} \models \phi_t \circ \phi_s = \phi_{t+s} \quad ; \\
(c) \quad & t \in \mathcal{T} \text{ and } A \in \mathcal{F} \models \mu(\phi_t(A)) = \mu(A) \quad .
\end{aligned}
$$

Remark C:

1. In conformity with the literature in dynamical system, we wrote here μ for the probability measure P introduced in our discussion of the Kolmogorov axioms; in particular, we tacitly assume $\mu(\Omega) = 1$, although what is important is that $\mu(\Omega)$ be finite.
2. In case $\mathcal{T} = \mathsf{R}$, ϕ is referred to as a *flow*; if $\mathcal{T} = \mathsf{Z}$, ϕ is said to be a *cascade*.
3. For condition (a), we recall that a function $f : x \in \Omega \mapsto f(x) \in \mathsf{R}$ (with the values $\pm\infty$ allowed) is said to be measurable iff

$$\forall \, a \subset \mathsf{R} \; : \; \{x \in \Omega \mid f(x) > a\} \in \mathcal{F} \quad ;$$

   equivalently, the condition $f(x) > a$ may be replaced in this definition by any of the conditions: $f(x) \geq a$, $f(x) \leq a$, or $f(x) < a$.
4. In case of a cascade, condition (a) is often written in the form:

$$\forall \, A \in \mathcal{F} \; : \; \{x \in \Omega \mid \phi_t(x) \in A\} \in \mathcal{F} \quad .$$

5. We have adopted here the definition of a dynamical system according to [Sinai, 1976]. Other points of view can be taken: Arnold and Avez [1968] assume that Ω is a smooth differentiable manifold; and Ollagnier [1985] focuses on topological dynamics, where Ω is a compact Hausdorff space. The definition of $\phi(t)$ is then adapted accordingly: it is assumed to be smooth in case Ω is taken to be a differentiable manifold, or to be a homomorphism when Ω is assumed to be a topological space.
6. The Boltzmann harmonic oscillator model is a dynamical system in the sense of Def. 7.2.1, but also in any of the senses of Rem. 5, with $\mathcal{T} = \mathsf{R}$ parametrizing the Hamiltonian evolution group action on the torus $\Omega^{(a)} \approx \mathsf{T}^2$.
7. We will have occasions, particularly in Chaps. 11–14, to consider situations where $\mathcal{T} \, (= \mathsf{Z}$ or $\mathsf{R})$ is replaced by more general topological groups, the main examples being provided by translation-groups.

Amongst the first successes of the nascent ergodic theory are the mathematical results known as the ergodic theorems of Birkhoff [1931] and Von Neumann [1932b], although it would be fair to add references to [Khinchin, 1932, 1933, Hopf, 1932]; see also [Hopf, 1937].

Theorem 7.2.3 (Birkhoff). *Let $\{\Omega, \mathcal{F}, \mu; \phi\}$ be a classical dynamical system in the sense of Def. 7.2.1. For every μ−integrable function $f \in \mathcal{L}^1(\Omega, \mathcal{F}; \mu)$, let*

$$\eta_x{}^T[f] \equiv \frac{1}{2T} \int_{-T}^{T} \mathrm{d}t \; f(\phi_t[x]) \quad .$$

Then, there exists $A_f \in \mathcal{F}$ with $\mu(A_f) = 1$ such that

1. *$\eta_x[f] = \lim_{T \to \infty} \eta_x{}^T[f]$ exists for all $x \in A_f$;*
2. *$\forall (t, x) \in \mathcal{T} \times A_f : \eta_{\phi_t[x]}[f] = \eta_x[f]$;*
3. *$\int_\Omega \mathrm{d}\mu(x) \; \eta_x[f] = \int_\Omega \mathrm{d}\mu(x) \; f(x)$.*

Theorem 7.2.4 (von Neumann). *Let $\{\Omega, \mathcal{F}, \mu; \phi\}$ be a classical dynamical system. For every μ−square-integrable function $f \in \mathcal{L}^2(\Omega, \mathcal{F}; \mu)$, let*

$$\eta_x{}^T[f] \equiv \frac{1}{2T} \int_{-T}^{T} \mathrm{d}t f(\phi_t[x]) \quad .$$

Then, there exists a μ−square-integrable function $\overline{f} \in \mathcal{L}^2(\Omega, \mathcal{F}; \mu)$, such that

$$\lim_{T \to \infty} \int_\Omega \mathrm{d}\mu(x) \left| \eta_x{}^T[f] - \overline{f}(x) \right|^2 = 0$$

Remark D:

1. These theorems take to heart Boltzmann's long-time averages, but the emphasis has shifted: where Boltzmann strives to justify the use of *smooth* measures – i.e. his ensembles – as time-averages of *singular* measures – i.e. δ_{x_t} – Birkhoff and von Neumann take for granted the existence of a canonical measure μ, and they then concentrate on finding mathematically sound theorems that ensure the existence of the limit. Since the limit in general does not exist for continuous functions, they change the class of admissible functions, shooting for consistency: one should not require more from the original functions than what one can require from their average. If one resolves to deal only with integrals – in particular so as to avoid Weyl's objections to the indiscriminate use of sets of measure zero [see Sect. 5.2] – then one should be ready not to want to distinguish between functions that differ only on sets of measure zero. The vector spaces $\mathcal{L}^1(\Omega, \mathcal{F}, \mu)$, $\mathcal{L}^2(\Omega, \mathcal{F}, \mu)$ or $\mathcal{L}^\infty(\Omega, \mathcal{F}, \mu)$ fit that bill: their elements are indeed classes of equivalence of functions defined up to sets of measure zero. Hence, the fact that these theorems hold only up to set of measure zero is consistent with the assumptions made on admissible functions.

2. For the definition of the Banach spaces $\mathcal{L}^p$ with $1 \leq p \leq \infty$, and their dual properties, see in Appendix D, Exemples 2.3–4 and Defs. 7.2.4–5.

3. von Neumann's theorem presents the advantage to give a *convergence in mean-square deviation,* while Birkhoff's theorem conveys a *convergence in probability.*

4. von Neumann's theorem is one of the first illustrations of the power of the Hilbert space methods proposed for classical mechanics by Koopman [1931].

We pause briefly to outline the Koopman formalism, collecting its main mathematical properties in the form of elementary Scholia. From the start however, we want to point out that the formalism is more than just a collection of mathematical niceties; it involves a conceptual shift in emphasis: away from the points – or subsets of points – in phase-space, towards a focus on the algebra of observables.

In the Koopman formalism, the bounded observables are chosen to be $\mathcal{L}^\infty$–functions on (Ω, μ); and their expectation value for a distribution with μ–density $\varrho \in \mathcal{L}^1$ are given by

$$f \in \mathcal{L}^\infty \mapsto \langle \varrho, f \rangle \in \mathsf{R} \quad \text{with} \quad \langle \varrho, f \rangle_\mu \equiv \int_\Omega d\mu(x)\, \varrho(x)\, f(x) \quad . \qquad (7.2.6)$$

The Koopman formalism begins with the construction of a representation – see Scholium 7.2.1 below – of the algebra $\mathcal{L}^\infty$ of observables as a sub-algebra of the algebra $\mathcal{B}(\mathcal{H})$ of all bounded linear operators acting on the complex Hilbert space $\mathcal{H} \equiv \mathcal{L}^2$:

$$\pi_\mu : f \in L^\infty \mapsto \pi_\mu(f) \in \mathcal{B}(\mathcal{H}) \quad . \qquad (7.2.7)$$

It is defined by

$$\pi_\mu(f) : \Psi \in \mathcal{H} \mapsto \pi_\mu(f)\Psi \in \mathcal{H} \quad \text{with} \quad [\pi_\mu(f)\Psi](x) \equiv f(x)\Psi(x) \quad \text{a.e.} \qquad (7.2.8)$$

Scholium 7.2.1. *The map (7.2.7) is a* representation; *i.e. it preserves the algebraic structures, namely:*

1. *in $\mathcal{L}^\infty$: point-wise addition and multiplication of functions, complex conjugation;*
2. *in $\mathcal{B}(\mathcal{H})$: addition and composition of operators, hermitian conjugation;*
3. $\|\pi_\mu(f)\|_{\mathcal{B}(\mathcal{H})} \leq \|f\|_\infty$.

Remark E: (1) and (2) are straightforward; and (3) follows directly from (1) and (2).

Scholium 7.2.2. *The representation (7.2.7) is* faithful *and* normal; *i.e.*

1. $\pi_\mu(f) = 0 \quad \models \quad f = 0$;

2. *For every bounded increasing filter $F \subseteq \{\mathcal{L}^\infty\}^+$: $\sup_{f \in F}[\pi_\mu(f)] = \pi_\mu([\sup_{f \in F} f])$.*

The second of the properties listed in Scholium 7.2.2 is a continuity condition. For completeness, we note that the relation $f \leq g$ a.e. defines a partial ordering on $\mathcal{L}^{\infty+}$, the cone of all elements of $\mathcal{L}^\infty$ that are positive almost everywhere. An increasing filter is a partially ordered subset $F \subseteq \mathcal{L}^{\infty+}$ such that for each pair f, g of elements of F, there is at least one element $h \in F$ – depending in general on f and g – for which both $f \leq h$ and $g \leq h$. A filter is said to be bounded if there is an element $f_o \in \mathcal{L}^{\infty+}$ such that $\forall f \in F$: $f \leq f_o$.

We have now completely specified the sense in which $\mathcal{L}^\infty$ and its image through π_μ are the same mathematical objects; consequently, which of these two realizations one chooses will be immaterial to the physics and thus only a matter of mathematical expediency.

The above result entails a trivial corollary that we mention here as it will carry over to the quantum formulation of dynamical systems, and will be of great importance there.

Scholium 7.2.3. *The vector $\Phi_o \in \mathcal{H}$, defined by $\Phi_o(x) = 1$ for all $x \in \Omega$, is cyclic and separating for the representation (7.2.7), i.e.*

1. *the vector space $\{\pi_\mu(f)\Phi_o \mid f \in \mathcal{L}^\infty\}$ is dense in $\mathcal{H}$;*
2. *the action of $\pi_\mu(f)$ on the vector Φ_o alone determines uniquely $f \in \mathcal{L}^\infty$, since*

$$f \in \mathcal{L}^\infty \text{ and } \pi_\mu(f)\Phi_o = 0 \quad \models \quad f \equiv 0 \ a.e.$$

Remark F: The general mathematical theory lurking behind the formalism was yet to be brought to light when Koopman proposed his particular realization; let it suffice for the moment to drop a few anchors [for more rope, see e.g. [Emch, 1972a, Landsman, 1998]]:

1. The theory of operator algebras, and in particular of von Neumann algebras, proved particularly adapted to the axiomatization and further developments of the statistical mechanics necessary to describe – quantum as well as classical – systems in the thermodynamical limit.
2. Our first contact with these powerful mathematical structures occurred already in Sect. 6.3; we explained elsewhere [Emch, 1972a, 1984] how the basic tenets of classical ergodic theory can be extended to a non-commutative context; for refs. to quantum Kolmogorov and Anosov flows, see Rems. F.3 in Sect. 8.2, H.3 in Sect. 8.4, and Model 3 in Sect. 15.3.
3. In particular, Scholium 7.2.1 and Scholium 7.2.2 spell out the fact that $\mathcal{L}^\infty$ and $\pi_\mu(\mathcal{L}^\infty)$ are isomorphic as von Neumann algebras.
4. One of the tasks of the present section is therefore to show what insight the Koopman formalism brings to ergodic theory: the opening key obtains in the following result.

Scholium 7.2.4. *Let α_μ be the group action*

$$\alpha_\mu : (t, \pi_\mu(f)) \in \mathcal{T} \times \pi_\mu(\mathcal{L}^\infty) \mapsto \alpha_\mu(t)[\pi_\mu(f)] \in \pi_\mu(\mathcal{L}^\infty)$$

defined by

$$\alpha_\mu(t)[\pi_\mu(f)] = \pi_\mu(f \circ \phi_{-t})$$

where ϕ denotes the evolution in the Def.7.2.1 of a dynamical system; and for almost all $x \in \Omega$: $[f \circ \phi_{-t}](x) \equiv f(\phi_{-t}[x]) \equiv f_t(x)$.

Then α_μ is unitarily implemented, i.e.

$$\alpha_\mu(t)[\pi_\mu(f)] = U_\mu(t)\pi_\mu(f)U_\mu(-t)$$

by the continuous, one parameter group U_μ of unitary operators defined on $\mathcal{H}$ by:

$$U_\mu : (t, \Psi) \in \mathcal{T} \times \mathcal{H} \mapsto U_\mu(t)\Psi \in \mathcal{H}$$

with

$$[U_\mu(t)\Psi](x) = \Psi(\phi_{-t}[x]) \quad \text{a.e.} \quad .$$

Now we can see the mathematical advantage of the Koopman formalism: it transfers the group action

$$\phi : \mathcal{T} \times \Omega \;\to\; \Omega$$

of a dynamical system, first to an algebraic group action, namely

$$\alpha_\mu : \mathcal{T} \times \pi_\mu(\mathcal{L}^\infty) \;\to\; \pi_\mu(\mathcal{L}^\infty)$$

– which gives the evolution in terms of its action on observables – and then transfers further this group action to a unitary group action on the Hilbert space $\mathcal{H} = \mathcal{L}^2$:

$$U_\mu : \mathcal{T} \times \mathcal{H} \;\to\; \mathcal{H}.$$

The latter is one of the best studied categories in Hilbert space, and one of the most directly interpretable; see e.g. [Stone, 1932, Kato, 1966, Reed and Simon, 1972–1980].

For instance, in this context, the Fourier spectral analysis takes the form

$$U(t) = \int_{\mathsf{R}} dE_\lambda \, e^{-i\lambda t} = e^{-iLt} \tag{7.2.9}$$

where E is an increasing family of projectors, the spectral family of the self-adjoint generator L of U :

$$L = \int_{\mathsf{R}} dE_\lambda \, \lambda \quad . \tag{7.2.10}$$

In our context, L is the Liouville operator generating the evolution. The discrete part of its spectrum is defined to be the set of the eigenvalues of L :

$$\left.\begin{array}{l} \mathrm{Sp}_d(U) \equiv \mathrm{Sp}_d(L) = \\[2ex] \{\lambda \in \mathsf{R} \mid \exists \Psi \in \mathcal{H} \text{ with } \forall t \in \mathsf{R} \; : \; U(t)\Psi = \mathrm{e}^{-\mathrm{i}\lambda t}\Psi\} \end{array}\right\}. \qquad (7.2.11)$$

We will denote by E_o the invariant eigensubspace of U :

$$E_o \equiv \{\Psi \in \mathcal{H} \mid \forall \, t \in \mathsf{R} \; : \; U(t)\Psi = \Psi\} \quad . \qquad (7.2.12)$$

This closes our review of the Koopman formalism, and we can now return to our study of classical ergodic theory proper. In the frame of the idealization provided by the concepts of classical dynamical system, the theorems of Birkhoff and von Neumann make the following definition a natural formalization of Boltzmann ergodic hypothesis.

Definition 7.2.2. *A classical dynamical system is said to be* ergodic *if for every $\mu-$summable function $f \in \mathcal{L}^1(\Omega, \mathcal{F}; \mu)$ there exists $A_f \in \mathcal{F}$ with $\mu(A_f) = 1$ such that*

$$\forall \, x \in A_f \; : \; \eta_x[f] = \int_\Omega \mathrm{d}\mu(x) f(x) \quad .$$

The physical relevance of this definition will be explored in Chap. 9; in particular, no point is made here as to whether finite Hamiltonian systems are generically ergodic – they are not, and further limiting procedures will need to be considered in order to carry forward the good properties that ergodic dynamical systems do bring forth.

The following property, which was introduced in [Birkhoff and Smith, 1928], sheds light on the nature of the ergodicity required in the above definition and on its relation with Boltzmann's idea – namely that the evolution brings any point arbitrarily close to any other.

Definition 7.2.3. *A classical dynamical system $\{\Omega, \mathcal{F}, \mu; \phi\}$ is said to be* metrically transitive *if there exists no subset $A \subset \Omega$ which is invariant and measurable with $0 < \mu(A) < 1$; i.e. : $A \in \mathcal{F}$ and $\phi_t[A] = A \; \forall \, t \in \mathcal{T} \models \mu(A) = 0$ or 1.*

We now have:

Theorem 7.2.5. *Let $\{\Omega, \mathcal{F}, \mu; \phi\}$ be a classical dynamical system. Then the following three conditions are equivalent:*

1. *$\{\Omega, \mathcal{F}, \mu; \phi\}$ is ergodic in the sense of Def. 7.2.2;*
2. *$\{\Omega, \mathcal{F}, \mu; \phi\}$ is metrically transitive in the sense of Def. 7.2.3;*
3. *For any distribution $\varrho \in \mathcal{L}^1$ with $\varrho \geq 0$ and $\|\varrho\|_1 = 1$, the linear functional*

$$\tilde{\varrho} : f \in \mathcal{L}^\infty \mapsto \langle \tilde{\varrho}; f \rangle \equiv \int_\Omega \mathrm{d}\mu(x)\, \varrho(x)\, f(x) \in \mathsf{R}$$

   is invariant under the evolution, i.e.

$$\forall t \in \mathcal{T} : \quad \langle \varrho; \phi_t[f] \rangle = \langle \varrho; f \rangle$$

iff $\tilde{\varrho} = \tilde{\mu}$, *i.e.* iff

$$\forall f \in \mathcal{L}^\infty : \quad \langle \tilde{\varrho}; f \rangle = \int_\Omega \mathrm{d}\mu(x)\, f(x) \,,$$

i.e. iff

$$\mu(\{\, x \in \Omega \mid \varrho(x) = 1 \,\}) = 1 \quad;$$

4. *Any observable $f \in \mathcal{L}^\infty$, invariant under the time evolution, i.e. such that*

$$\forall\, t \in \mathcal{T} : \quad f \circ \phi_{-t} = f \quad,$$

is a constant (almost everywhere), i.e.

$$f(x) = c \quad \text{a.e.} \quad;$$

5. *With E_o , the subspace defined by (7.2.12):*

$$\dim E_o = 1 \quad.$$

Remark G: The equivalence of these five conditions is straightforward; see e.g. [Arnold and Avez, 1968, Brown, 1976]. The interpretation of the five conditions indicates that the early authors had captured some of the essential structure.

1. Once the existence of the limit η_x is established – see Thms. 7.2.3–4 – ergodicity is defined as the equality between time and phase-space average. Thus the equivalence between conditions (1) and (2) in Thm. 7.2.5 is referred to, sometimes, as "the" ergodic theorem, or even Birkhoff's ergodic theorem. In our presentation, we preferred to separate out the *existence* of the time-average, which is Birkhoff's breakthrough in establishing Thm. 7.2.3, and its *equality* with the phase-space average, which only results when the metric transitivity condition is satisfied.
2. A measurable subset of phase-space is stable under the flow iff its complement is; thus metric transitivity asserts the absence of measurable subsets that would be stable under the flow and of measure non-zero.
3. No time-invariant, smooth distribution exists beyond the invariant measure which enters the definition of the dynamical system considered.
4. There are no non-trivial constants of the motion.
5. When the evolution is viewed as a flow in the Hilbert space $\mathcal{H} = \mathcal{L}^2$, it leaves no vector invariant, except the trivially-invariant vectors Φ_o that are constant a.e.; recall – Scholium. 7.2.3 – that these vectors are cyclic and separating; therefore, in that sense also, ergodic dynamical systems are indecomposable.

8. Models and Ergodic Hierarchy

One could say here that, as is often the case, the mathematicians know a great deal about very little and the physicists very little about a great deal.

[Ulam, 1955]

Before we enter in the present chapter, let us consider where we stand and where we are going. In the preceding chapter, we saw how Boltzmann's quest for ergodicity led to an idealization – namely the notion of dynamical system – with precise enough a mathematical structure to support an unambiguous formulation of what is meant by an ergodic system. The next question should be whether such conditions are realistic enough to provide an adequate foundation to statistical mechanics. While we present some positive evidences in this chapter, we must preface this presentation by a warning. In the next chapter, we argue indeed that some doubts must be raised against unrestricted claims: it turns out that the condition of ergodicity – even in the version we called the "modified" ergodic hypothesis in Sect. 7.1 or its mathematical version in the equivalent defintions 7.2.2 and 7.2.3 – is too drastic to be satisfied by most finite Hamiltonian systems. The problem of how to save ergodic theory from this disaster is postponed to in Chap. 9.

In the present chapter we evaluate some of the aspects of ergodic theory that are worth saving in the light gained in an exploration of models; these simple models were chosen because their very simplicity allows rigorous and complete descriptions of their most intimate details. Our exploration through these models leads to a hierarchy of rich properties that seem to capture in a finer and finer net some conditions akin to those one would wish more realistic models to enjoy. For the purposes of such an exploration, it seemed unwise to focus immediately and exclusively on Hamiltonian systems, although some of the models we present are Hamiltonian.

The lessons to be drawn from the models discussed in this chapter can be made manifest only through some detailed analysis; several of the mathematical tools needed to complete this analysis are surveyed in Appendices B and D. Yet, lest the Reader gets lost in a maze of abstract forays, we wish to indicate from the onset some physical highlights to be obtained as we proceed: even some of the apparently most artificial mathematical models do have counterparts in a more mundane world and elicit some insights into the expected randomness of statistical mechanics; among the latter, the notion of dynamical entropy.

Models 1 and 2 illustrate the relation between ergodicity and mixing. Model 3 re-examine Bernoulli's coin tossing trials in the light just gained. Model 4 presents the Markov structure of the Brownian motion, an idealization that led to profound insights in mathematics as well as in physics.

As part of a preliminary discussion of the problem of reducing the description of a macroscopic dissipative system by embedding it in a conservative microscopic system, Rem. F.2 offers the Bloch equation for spin relaxation to showcase that the very existence of such a reduction imposes empirically verifiable restrictions on the dissipative dynamics. Rem. H.4 points out that the two-dimensional structure of Model 5 – the Arnold Cat map – does not preclude its quantization to a description that shares some structural aspects with the kinematics of the quantum Hall effect. Rem. I.5 points out that the hyperbolic structure of the geodesic flow on a Riemann surface of constant negative curvature extends to situations encountered in general relativity.

Closer to earth, the relation between the mechanism responsible for the ergodic behavior of the geodesic flow and the more immediately relevant behavior of billiard systems is pointed out in Model 7 and Rem. J.

8.1 Mixing Properties

We start with one of the simplest – and oldest – model for ergodicity.

Model 1. This model is suggested by Boltzmann's investigation of the $2-$dimensional harmonic oscillator – see Chap. 7, and in particular Fig. 7.1. The phase-space for this model is the flat $2-$torus $\Omega = T^2 = \mathsf{R}^2/\mathsf{Z}^2$; $\mathcal{F}$ is the $\sigma-$ring of its Borel sets; and μ is the restriction to T^2 of the Lebesgue measure on R^2. The evolution is:

$$\phi : (t; x_1, x_2) \in \mathsf{R} \times T^2 \mapsto (x_1(t), x_2(t)) \in T^2$$

with $\qquad\qquad\qquad\qquad\qquad\qquad\qquad\qquad\qquad\qquad\qquad$ (8.1.1)

$$x_k(t) \equiv x_k + \omega_k t/2\pi \ \text{mod} \ 1 \quad .$$

This system is ergodic iff ω_1/ω_2 is irrational – as illustrated in Sect. 7.1. Note nevertheless that, even when this system is ergodic, the distance between any two points x and y is constant over time; hence all the points of any region in phase-space move in concert and the shape of the region remains the same in the course of time: the evolution is rigid and thus too coherent to accommodate the kind of randomness we are looking for. The following model – although somewhat contrived – is better in this regard.

Model 2. This model is known as the *kneading system*, or the *baker's transform*. It draws its name from what a baker does – or rather could do – when he kneads his dough. The dough is the phase-space of the system: we take $(\Omega, \mathcal{F}, \mu)$ as in Model 1 above. The evolution is discrete

$$\phi : (n; x_1, x_2) \in \mathbb{Z} \times T^2 \mapsto (x_1(n), x_2(n)) \in T^2 \qquad (8.1.2)$$

where $\phi_n = \psi \circ \psi \circ \ldots \circ \psi$ is the n-th iterate of a transformation – the *baker's transform*:

$$\phi_1 \equiv \psi : (x_1, x_2) \in T^2 \mapsto (\overline{x}_1, \overline{x}_2) \in T^2 \qquad (8.1.3)$$

which we now describe, and illustrate in Fig. 8.1.

The baker first "pulls" on his dough and flattens it, to twice its length and half its height:

$$\pi : (x_1, x_2) \in T^2 \equiv [0, 1) \times [0, 1) \mapsto$$
$$(\tilde{x}_1, \tilde{x}_2) = (2x_1, \tfrac{1}{2}x_2) \in \tilde{\Omega} \equiv [0, 2) \times [0, \tfrac{1}{2}) \qquad (8.1.4)$$

following which he "cuts" the right half, and then places it on top of the left half:

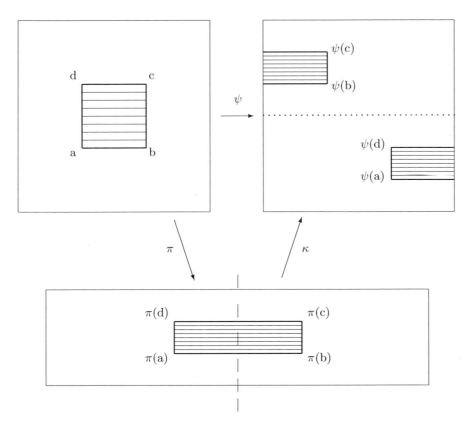

Fig. 8.1. The baker's transform: definition of $\psi = \kappa \circ \pi$

$$\kappa : (\tilde{x}_1, \tilde{x}_2) \in \tilde{\Omega} \mapsto (\overline{x}_1, \overline{x}_2) \in T^2 \tag{8.1.5}$$

with

$$(\overline{x}_1, \overline{x}_2) = \begin{cases} (\tilde{x}_1, \tilde{x}_2) & \text{if} \quad 0 \le \tilde{x}_1 < 1 \\ (\tilde{x}_1 - 1, \tilde{x}_2 + \frac{1}{2}) & \text{if} \quad 1 \le \tilde{x}_1 < 2 \end{cases} . \tag{8.1.6}$$

The baker's transform in (8.1.3) $\psi = \kappa \circ \pi$, which composes the two elementary mappings π followed by κ, is then given by

$$(\overline{x}_1, \overline{x}_2) = \begin{cases} (2x_1, \frac{1}{2}x_2) \bmod 1 & \text{if} \quad 0 \le x_1 < \frac{1}{2} \\ (2x_1, \frac{1}{2}(x_2 + 1)) \bmod 1 & \text{if} \quad \frac{1}{2} \le x_1 < 1 \end{cases} . \tag{8.1.7}$$

In Fig. 8.1, the square (a, b, c, d) is evidently not part of the definition of ψ : it is included to serve as a marker. One infers immediately from the picture that the kneading action, obtained by iterating the baker's transform ψ, results in spreading this marker into thinner and thinner disconnected slices more and more uniformly distributed over the height of the dough, thus producing a structure reminiscent of the pastry known as a "napoleon" or a "millefeuille". To make manifest this spreading, we look at the successive iterations from a measurable fixed partition of the phase-space Ω. Fig. 8.2 illustrates such a partition, here in nine regions A_k with $k = 1, 2, \dots, 9$.

We measure then at each iterate $n = 0, 1, 2, \dots$ how much of an initially marked region B spreads through each region A_k of the partition. Fig. 8.3 reports the ratio

$$\mu_k = \frac{\mu(A_k \cap \phi_n[B])}{\mu(B)} . \tag{8.1.8}$$

for the first four iterations $0 \le n \le 3$ with $B = A_5$. What emerges suggests indeed a tendency towards uniformization: the aim of kneading – i.e. of iterated repetitions of the baker's transform – is surely to produce a uniform dough.

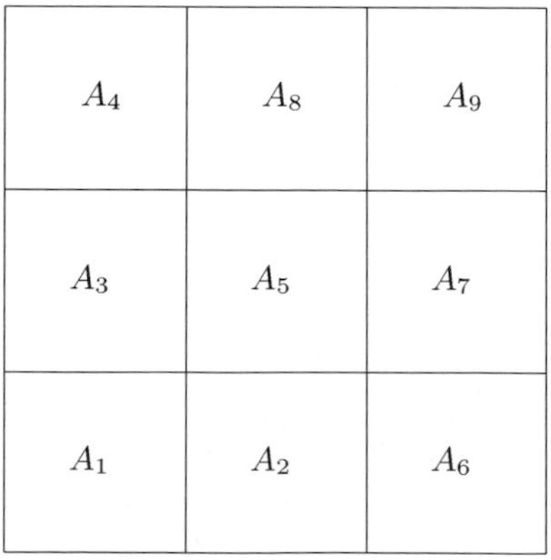

Fig. 8.2. A partition $\{A_k \mid k = 1, 2, \cdots, 9\}$

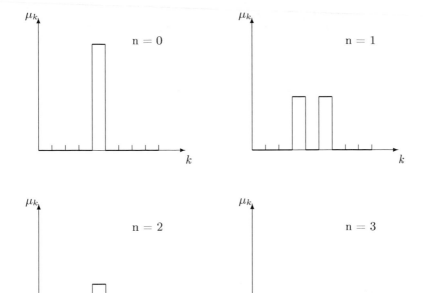

Fig. 8.3. The effect of kneading: $\mu_k = \mu(A_k \cap \phi_n[A_5]) \, / \, \mu(A_5)$

Remark A:

1. The kneading action just depicted indicates that the tendency to uniformization is compatible with the fact that the baker's transform is measure-preserving: $\mu \circ \psi = \mu$.

2. This tendency to uniformization, as seen from a measurable partition, is also compatible with the existence of recurrences, even the existence of periodic points; for instance,

$$\psi[b] = d \text{ and } \psi[d] = b \quad \text{for} \quad b = (\tfrac{2}{3}, \tfrac{1}{3}) \text{ and } d = (\tfrac{1}{3}, \tfrac{2}{3}) \qquad (8.1.9)$$

so that $\{b, d\}$ is an orbit of period 2.

3. In statistical mechanics, the importance of such a tendency to uniformization, was stressed by Gibbs who proposed the following analogy:

> *Let us suppose the liquid* [an incompressible fluid] *to contain a certain amount of coloring matter which ... is distributed with variable density. If we give the liquid any motion whatever, subject only to the hydrodynamic law of incompressibility ... the density of the coloring matter at the same point of the liquid will be unchanged ... Yet no fact is more familiar to us than that stirring tends to bring a liquid to a state of uniform mixture, or uniform densities of its components*

> *... The contradiction is to be traced to the notion of the* density *of the*
> *coloring matter ... in an element of space ...*
>
> [Gibbs, 1902][pp. 144–145]

The baker's transform suggests therefore the following:

Definition 8.1.1. *A classical dynamical system* $(\Omega, \mathcal{F}, \mu; \phi)$ *is said to be*
mixing whenever, for every pair of measurable subsets A, B :

$$\lim_{t \to \infty} \frac{\mu(A \cap \phi_t[B])}{\mu(B)} = \mu(A) \quad .$$

Remark B:

1. The kneading system – i.e. the iterated baker's transform of Model 2
   – is mixing in the sense of the above definition; this can be seen most
   readily from Scholium 8.2.1 and Rem. D.1 below. The dynamical system described in Model 1, however, is *not* mixing although it is ergodic.
   Hence, ergodicity does not entail mixing, but not conversely; see Thm.
   8.1.1 below.
2. This definition says that the *fraction* of B that is to be found in A after
   a long time t tends to be equal to the "volume" of A.
3. This property is also called *strong mixing*, to distinguish it from the
   property of *weak mixing*, namely:

$$\lim_{T \to \infty} \frac{1}{T} \int_0^T dt \; |\frac{\mu(A \cap \phi_t[B])}{\mu(B)} - \mu(A)| = 0 \quad .$$

   Clearly, strong mixing entails weak mixing.
4. When written in the equivalent form

$$\lim_{t \to \infty} \mu(A \cap \phi_t[B]) = \mu(A)\,\mu(B) \quad ,$$

   the mixing property is also known as the *clustering* property; the property
   of *weak clustering* is defined similarly, as a rewriting of the weak mixing property. The interest of multi-time clustering properties has been
   recognized, although we do not need this extension here.
5. In the above, we formulated the mixing conditions with respect to the
   very distant future; analogous conditions can be formulated with respect
   to the very remote past; note, for instance, that the next theorem holds
   for both.

Theorem 8.1.1. *Every mixing dynamical system is ergodic, but the converse*
is not true.

Proof. We give the proof for weak mixing; since strong mixing entails
weak mixing, the theorem holds for both. Let A be an invariant measurable subset. We have for all t : $A = A \cap \phi_t[A]$ and thus $\mu(A) =$

$\lim_{T\to\infty} \frac{1}{T}\int_0^T dt\ \mu(A\cap\phi_t[A])$. The weak mixing property applied to the RHS entails $\mu(A) = \mu(A)\,\mu(A)$ and thus either $\mu(A) = 0$ or $\mu(A) = 1$. Conversely, we have already noticed – Rem. B.1 – that Model 1 is ergodic but not mixing.

q.e.d.

Here again, mixing can be characterized completely within the Koopman formalism.

Theorem 8.1.2. *Let* $(\Omega, \mathcal{F}, \mu; \phi)$ *be a dynamical system. Then the following conditions are equivalent*

1. $(\Omega, \mathcal{F}, \mu; \phi)$ *is mixing, i.e. for all* $A_1, A_2 \in \mathcal{F}$ *:*

$$\lim_{t\to\infty} \mu(\phi_t[A_1]\cap A_2) = \mu(A_1)\mu(A_2)\,;$$

2. for all $\Psi_1, \Psi_2 \in \mathcal{L}^2$ *:*

$$\lim_{t\to\infty} (U(t)\Psi_1, \Psi_2) = (\Psi_1, \Phi_o)(\Phi_o, \Psi_2)$$

where $U(t)$ *is the Koopman unitary operator of Scholium 7.2.4, and* $\Phi_o(x) = 1$ *for all* $x \in \Omega$ *.*

Proof. For $k = 1, 2$, let $\Psi_k = \chi_{A_k}$ be the indicator function of the set $A_k \in \mathcal{F}$; the second condition of the theorem reduces then precisely to the first condition, thus proving $(2) \models (1)$. The proof in the other direction follows then from the fact that the set of all simple functions – i.e. the finite linear combinations of indicator functions – is dense in $\mathcal{L}^2$. **q.e.d.**

Remark C:

1. Condition (2) of the theorem implies that the projector E_o – see (7.2.12) – is one–dimensional; hence, upon using the characterization (5) of ergodicity given in Thm. 7.2.5, we can derive from Thm. 8.1.2 above an alternate proof of Thm. 8.1.1 – namely that mixing entails ergodicity.
2. Condition (2) implies, by straightforward contradiction, that the evolution has *properly continuous spectrum;* this requires two things: (i) the spectrum admits no eigenvalue besides $\lambda = 0$ – i.e., see (7.2.11), $\mathrm{Sp}_d(U) = \{0\}$ – and (ii) $\lambda = 0$ is non-degenerate. Note that this terminology, used in much of ergodic theory, is not standard in other fields of analysis.
3. However, to ensure that the dynamical system be strongly mixing it is not sufficient to assume that the evolution has properly continuous spectrum: the latter condition is equivalent [Halmos, 1958] to the condition that the dynamical system be weakly mixing. Thus, the Koopman formalism provides a spectral characterization for both strong and weak mixing.

8.2 K-Systems

The baker's transform of Model 2, with its mixing property, seems – as Gibbs suggested in the above quote – to bring us in contact with structures that are suggestive of, or pertinent to, statistical mechanics. In particular mixing is routinely associated with the production of randomness: think of what you expect from a honest card dealer upon requesting him to mix a deck of cards. We reinforce this idea with yet another (class of) model(s).

Model 3. This model is referred to as the *Bernoulli shift* $B(p,q)$, with $0 < p < 1$ kept fixed throughout, and $q = 1-p$. It describes time–translations in the space of doubly–infinite binary sequences. They derive their name from the Bernoulli trials – sequences of coin tossings – which we began studying in Chap. 4.

The probability space $(\Omega, \mathcal{F}, \mu)$ of the present model is defined in complete analogy with Model 3 of Sect. 5.2 – except we have now Z replacing Z^+ – to obtain:

$$\Omega \equiv \Xi \equiv \{\, \xi : k \in \mathsf{Z} \to \xi_k \in \{0,1\} \,\} \quad . \tag{8.2.1}$$

$\mathcal{F}$ is the $\sigma-$ring generated by the cylinder sets of the following form. For each pair $(k,i) \in \mathsf{Z} \times \{0,1\}$ the elementary cylinder set A_k^i is the collection of binary sequences, the $k-$th entry of which is 0 or 1, according to the value of i:

$$A_k^i = \{\, \xi \in \Xi \mid \xi_k = i \,\} \quad . \tag{8.2.2}$$

The measure μ is determined by its value on cylinder sets; for arbitrary n, $-\infty < k_1 < k_2 < \cdots < k_n < \infty$ and $i_\nu \in \{0,1\}$:

$$\mu(A_{k_1}^{i_1} \cap A_{k_2}^{i_2} \cap \cdots \cap A_{k_\nu}^{i_\nu} \cdots \cap A_{k_n}^{i_n}) = \prod_{\nu=1}^n p(i_\nu)$$

$$\text{with} \quad p(1) = p \quad \text{and} \quad p(0) = q \quad . \tag{8.2.3}$$

Finally, the evolution is given by the shifts along the sequences:

$$\phi : (n, \xi) \in \mathsf{Z} \times \Xi \mapsto \xi(n) \in \Xi \quad \text{with} \quad \{\xi(n)\}_m = \xi_{m-n} \quad . \tag{8.2.4}$$

This completes the description of Model 3. We examine now whether this dynamical system enriches our panoply of ergodic properties.

Scholium 8.2.1. *The map*

$$\chi : \xi \in \Xi \mapsto (x_1, x_2) \in T^2$$

$$\text{with} \quad x_1 = \sum_{n=0}^{\infty} \frac{1}{2^{n+1}} \xi_{-n} \quad \text{and} \quad x_2 = \sum_{n=1}^{\infty} \frac{1}{2^n} \xi_n$$

establishes an isomorphism between the particular Bernoulli shift $B(\frac{1}{2}, \frac{1}{2})$ in Model 3, and the kneading system – or iterated baker's transform – described in Model 2.

Sketch of the proof. This proceeds in the following three steps, for which it suffices here to indicate what is to be shown and what precautions are called for.

1. χ is a bijection, except on those points $(x_1, x_2) \in T^2$ for which x_1 or x_2 is a dyadic fraction. These exceptions do not matter as these points are countable, and thus form a set of measure zero.
2. χ is measure preserving; it is here that $p = q = \frac{1}{2}$ enters.
3. $\chi \circ \tau = \psi \circ \chi$ where τ denotes the unit translation ϕ_1 in (8.2.4) and ψ is the baker's transform (8.1.3). This is to say that the following diagram is commutative:

$$
\begin{array}{ccc}
\Xi & \xrightarrow{\ \tau\ } & \Xi \\
\chi \downarrow & & \downarrow \chi \\
T^2 & \xrightarrow{\ \psi\ } & T^2
\end{array}
$$

q.e.d.

Remark D:

1. $B(p, q)$ is mixing. To see that, consider any two cylinder sets as in (8.2.2), say A and A' where:

$$A = A_{k_1}^{i_1} \cap \cdots \cap A_{k_m}^{i_m} \qquad \text{with} \quad -\infty < k_1 < \cdots < k_m < \infty$$

$$A' = A_{l_1}^{j_1} \cap \cdots \cap A_{l_{m'}}^{j_{m'}} \qquad \text{with} \quad -\infty < l_1 < \cdots < l_{m'} < \infty \quad .$$

It is then always possible to find N finite – depending on A and A' – such that

$$-\infty < k_1 < \cdots < k_m < l_{1+N} < \cdots < l_{m'+N} < \infty$$

and thus,

$$\forall \, n \geq N \quad : \quad \mu(A \cap \phi_n[A']) = \mu(A)\,\mu(A') \quad .$$

2. A more general class of Bernoulli shifts can be defined, where the entries of the sequences take any of n distinct values, instead of just 0 or 1, i.e. where $\{0, 1\} = \mathsf{Z}_2$ is replaced by $\{0, 1, \ldots, n-1\} = \mathsf{Z}_n$. The measure again is defined from its value on cylinder sets, and it is determined by the probability distribution

$$p : i \in \mathsf{Z}_n \mapsto p_i \in (0, 1) \quad \text{with} \quad \sum_{i=0}^{n-1} p_i = 1 \,. \tag{8.2.5}$$

The dynamical map ϕ again describes the translations along the sequences. Later in this section – see Def. 8.3.2 – we review the concept of dynamical entropy; let it suffice here to say that it attributes to dynamical systems a number that is invariant under isomorphisms, i.e. it is the same for two dynamical systems that are isomorphic. For any Bernoulli shift $B(p_o, \ldots, p_{n-1})$ this dynamical entropy happens to coïncide – see Rem. G.2 below – with the entropy

$$h(p) = -\sum_{i=0}^{n-1} p_i \ln p_i \qquad (8.2.6)$$

of the distribution p in (8.2.5).

3. Since the dynamical entropies of two Bernoulli shifts $B(p, q)$ and $B(p', q')$ with p' different from p or q are easily verified to be different, these systems cannot be isomorphic. In case $(p', q') = (q, p)$ the two systems are isomorphic, and the isomorphism is in fact simply obtained by exchanging the roles of 0 and 1. We therefore have a continuum $0 < p \leq \frac{1}{2}$ of non-isomorphic dynamical systems, each one of which is mixing.

4. It must be noticed that in the class of the general Bernoulli shifts of Rem. D.2, the dynamical entropy is a *complete* invariant i.e. two Bernoulli shifts are isomorphic if and only if they have the same dynamical entropy [Ornstein, 1970]. A didactic version of Ornstein's proof can be found in [Shields, 1973]; for this result, and more, see also [Ornstein, 1974]. Partial results had been known for some time [Meshalkin, 1959, Blum and Hanson, 1963]. For instance, $B(\frac{1}{2}, \frac{1}{2})$ and $B(\frac{1}{3}, \frac{1}{3}, \frac{1}{3})$ are not isomorphic; but $B(\frac{1}{2}, \frac{1}{8}, \frac{1}{8}, \frac{1}{8}, \frac{1}{8})$ and $B(\frac{1}{4}, \frac{1}{4}, \frac{1}{4}, \frac{1}{4})$ are isomorphic.

The recognition that the Bernoulli systems are mixing falls still very short from telling us how stochastic and unpredictable these dynamical systems really are. They indeed enjoy two properties that we should discuss here; see Def. 8.2.1 below, and Def. 8.3.2 in the next section. Both of these properties were introduced by Kolmogorov [Kolmogorov, 1958, 1959]; see also [Sinai, 1959]. The equivalence of these two properties was proven by Rohlin and Sinai [Rohlin and Sinai, 1961]; see Thm. 8.3.1.

Definition 8.2.1. *[Kolmogorov, 1958] A classical dynamical system $(\Omega, \mathcal{F}, \mu; \phi)$ is said to be a K–system if there exists a σ–subring $\mathcal{A}$ of $\mathcal{F}$ such that:*

1. $\forall \, t > 0 \; : \; \mathcal{A} \subset \phi_t[\mathcal{A}]$
2. $\bigwedge_{t \in T} \phi_t[\mathcal{A}] \equiv \{\emptyset, \Omega\}$
3. $\bigvee_{t \in T} \phi_t[\mathcal{A}] \equiv \mathcal{F}$.

Remark E: This lapidary definition requires a few explanations, which we give together with an illustration to the effect that Bernoulli shifts are K-systems (since the latter are discrete, we write again n or k, instead of t, for time).

a. For any Bernoulli shift $B(p, q)$, we choose for $\mathcal{A}$ the σ–ring generated by all cylinder sets A_k^i with $k \leq 0$ and $i \in \{0, 1\}$. Recall that $\mathcal{F}$ is similarly generated, but with no restriction on k to exclude the future.

b. With $\mathcal{A}$ as in (a) above, the σ−ring $\phi_n[\mathcal{A}]$ is generated now by all cylinder sets A_k^i with $k \leq n$ and $i \in \{0, 1\}$. When $n > 0$, this new σ−ring is a "refinement" of $\mathcal{A}$: it makes distinctions based on the sequences' entries at places $k = 1, \ldots, n$, and therefore it contains more and finer measurable sets; this is the fact we denote by $\mathcal{A} \subset \phi_n[\mathcal{A}]$; see condition (1) of the above Definition.

c. The second condition in the Definition states that the only measurable subsets that are contained in all $\phi_n[\mathcal{A}]$ are of measure 0 or 1. To show that this is the case for our illustration, recall – Rem. D.1 – the argument we used in order to show that Bernoulli systems are mixing, and denote by $\mathcal{B}$ the ring finitely generated by the cylinder sets A_k^i. For every $B \in \mathcal{B}$ – the definition of which involves thus finitely many A_k^i – we can find therefore a N such that for this B and for all $n \geq N$ and $A \in \mathcal{A}$: $\mu(A \cap \phi_n[B]) = \mu(A)\,\mu(B)$, i.e. since $\mu \circ \phi_{-n} = \mu$: $\mu(\phi_{-n}[A] \cap B) = \mu(A)\,\mu(B)$. Because of (b), this implies that for all $A \in \bigwedge_{n=0}^{\infty} \phi_{-n}[\mathcal{A}]$ and all $B \in \mathcal{B}$: $\mu(A \cap B) = \mu(A)\,\mu(B)$. Since the elements in $\mathcal{B}$ do in fact generate $\mathcal{F}$, the above relation still holds for any $B \in \mathcal{F}$ and thus in particular for $B = A$; we thus have $\mu(A) = \mu(A \cap A) = \mu(A)^2$, i.e. $\mu(A) = 0$ or 1 for all $A \in \bigwedge_{n=0}^{\infty} \phi_{-n}[\mathcal{A}] = \bigwedge_{n=-\infty}^{0} \phi_n[\mathcal{A}] = \bigwedge_{n=-\infty}^{\infty} \phi_n[\mathcal{A}]$ which is the content of condition 2 of the definition of a K-system.

d. The third condition in the above Definition states that $\mathcal{F}$ is the smallest of its σ−subrings that contains all $\phi_n[\mathcal{A}]$. For $B(p, q)$ the very definition of $\mathcal{A}$ and $\mathcal{F}$ shows that this is indeed the case.

Hence Bernoulli shifts are K-systems with discrete time parameter, i.e. cascades. K-flows – i.e. K-systems with $\mathcal{T} = \mathsf{R}$ – also exist, and we give an example below.

Model 4. This is the so-called *flow of Brownian motion*; see [Hida, 1970]. Its interest for us is two-fold. First, it establishes the existence of K-systems with continuous time; and second, this dynamical system derives its name from the fact that it is the natural conservative extension of the dissipative system described by a diffusion equation; see Rem. F below. Model 4 is more technical and specialized than the previous models; lest the Reader gets lost in technicalities, we mark the beginning of each of the successive steps in the construction of this model.

• *The phase-space $\Omega = \mathcal{S}(\mathsf{R})^*$ is the space dual to the Schwartz space $\mathcal{S}(\mathsf{R})$ of real-valued, rapidly decreasing functions on R, with $\mathcal{S}(\mathsf{R})$ equipped with the topology provided by the family of semi-norms

$$p_{m,n} = \sup_{x \in \mathsf{R}} \left| x^m \frac{\mathrm{d}^n f}{\mathrm{d}x^n}(x) \right| < \infty \quad .$$

For background on the Schwartz space and its dual, see in Appendix D: Def. D.20, Thms. D.5-6, and Example 12.

- *The $\sigma-$algebra $\mathcal{F}$ of subsets of Ω. $\mathcal{F}$ is generated by the cylinder sets*

$$A(f_1, \cdots, f_n; B) = \{\omega \in \Omega \mid (\langle \omega; f_1 \rangle, \cdots, \langle \omega; f_n \rangle) \in B\}$$

where n runs over Z^+, $\{f_1, \cdots, f_n\}$ runs over the collection of $n-$tuples of elements $f_k \in \mathcal{S}(\mathsf{R})$, and B runs over the Borel subsets of R^n.

- *The measure μ on the phase space Ω.* To define a measure on the infinite-dimensional vector space $\Omega = \mathcal{S}(\mathsf{R})^*$ we first recall the classical Bochner theorem – see e.g. [Yosida, 1974][Sect. XI.13] or [Gel'fand and Vilenkin, 1964][Sect.II.3.2] – which characterizes the positive measures on R^n by their Fourier transform:

$$\chi(k) = \int d\mu(x) \exp\{-i\, k \cdot x\}$$

e.g. $\chi(k) = \exp\{-\frac{1}{2}\|k\|^2\}$ $d\mu(x) = (2\pi)^{-n/2} \exp\{-\frac{1}{2}\|x\|^2\}\, dx$

where χ satisfies in general the necessary *and sufficient* conditions that it is continuous, and is of positive type, i.e. that for all finite choices of $\lambda_k \in \mathsf{C}$ and $x_k \in \mathsf{R}^n$:

$$\sum_{j,k} \lambda_j^* \lambda_k \, \chi(x_j - x_k) \geq 0\,.$$

The extension from R^n to $\mathcal{S}(\mathsf{R})^*$ is the Bochner–Minlos theorem – see e.g. [Glimm and Jaffe, 1981][Thm.3.4.2 and Remark] or [Gel'fand and Vilenkin, 1964][Thm.II.3.3.3 or IV.4.2.2].

We use this theorem to assert that

$$\chi(f) = \int_\Omega d\mu(\omega) \exp\{-\mathrm{i}\langle \omega; f \rangle\}$$

with $\chi : f \in \mathcal{S}(\mathsf{R}) \mapsto \exp\{-\frac{1}{4}\Theta\|f\|^2\}$

where $\|\ldots\|$ denotes the $\mathcal{L}^2-$norm on $\mathcal{S}(\mathsf{R})$, and $\Theta \in \mathsf{R}$ with $\Theta \geq 1$, defines a unique, normalized ("Gaussian") measure μ on the measurable space $(\Omega, \mathcal{F})$: $(\Omega, \mathcal{F}, \mu)$ is the probability space of our Model 4.

- *The time-evolution $\phi : (t, \omega) \in \mathsf{R} \times \Omega \mapsto \phi_t[\omega] \in \Omega$ is defined by:*

$$\phi_t[\omega] : f \in \mathcal{S}(\mathsf{R}) \mapsto \langle \phi_t[\omega]; f \rangle \equiv \langle \omega; \phi_t^\#[f] \rangle \in \mathsf{R}$$

with $\phi_t^\#[f](x) \equiv f(x - t)$. Since, for all $t \in \mathsf{R}$, $\|\phi_t^\#[f]\| = \|f\| : \chi \circ \phi_t^\# = \chi$, and thus $\mu \circ \phi_t = \mu$, i.e. the measure μ is invariant under the time-evolution. This completes the description of our dynamical system.

- *The K-flow property.* To show that we constructed a K-flow, it is sufficient to show that the three conditions of Def. 8.2.1 above are satisfied by the $\sigma-$ring $\mathcal{A}$ generated by the cylinder sets of the form $A(f_1, \cdots, f_n; B)$ with now the f_k restricted by the condition that their support be contained in $(-\infty, 0]$: as time proceeds, the f_k are restricted only by the condition that their support be contained in $(-\infty, t]$; from this we can infer that $\phi_t[\mathcal{A}]$ satisfy indeed the conditions of Def. 8.2.1.

Remark F:

1. The physical interest of this model derives from the fact that it obtains canonically through the classical Kolmogorov–Daniell construction – see for instance [Tucker, 1967][Thm. 2.3.2] – as the minimal Markov process associated to a Markov semi-group that, describes here a diffusion equation. From a physical point of view, this result gives a possible handle towards the solution to the following problem of central importance in non-equilibrium statistical mechanics: given a dissipative system Σ_S with dissipative evolution $\{\gamma(t) \mid t \in \mathsf{R}^+\}$, find some system Σ_R, and an interaction between these two systems, such that the composite system $\Sigma = \Sigma_S \otimes \Sigma_R$ be a dynamical system with a (conservative) evolution $\{\phi(t) \mid t \in \mathsf{R}\}$ that, for $t \geq 0$, reduces to $\gamma(t)$ when observed from Σ_S alone.

   The commutative diagram of Fig. 8.4a illustrates this situation. i denotes the embedding of the observed system Σ_S into the composite – or larger – system Σ; and $\mathcal{E}$ denotes the projection – or, in probabilistic language, the conditional expectation – of Σ onto Σ_S. The commutativity of the diagram expresses that the reduced evolution is $\gamma(t) = \mathcal{E} \circ \phi(t) \circ i$.

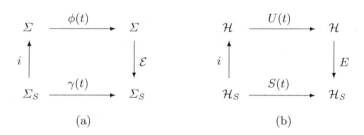

(a)   (b)

Fig. 8.4a,b. Dilations of semigroups

2. Consider for Σ_S the system consisting of a single spin $1/2$ particle evolving according to some dynamics

$$\gamma(t, A) \in \mathsf{R}^+ \times \mathcal{A} \mapsto \gamma(t) \in \mathcal{A}$$

   where $\mathcal{A}$ is the algebra of 2×2 matrices with complex entries, generated by the *Pauli matrices*

$$\sigma_x = \begin{pmatrix} 0 & 1 \\ 1 & 0 \end{pmatrix} \quad \sigma_y = \begin{pmatrix} 0 & -i \\ i & 0 \end{pmatrix} \quad \sigma_z = \begin{pmatrix} 1 & 0 \\ 0 & -1 \end{pmatrix}$$

   and suppose that the dynamics satisfies the Markovian condition $\gamma(t_1) \circ \gamma(t_2) = \gamma(t_1 + t_2)$ for all $t_1, t_2 \geq 0$.

   The condition that such a dynamics obtains as the reduced description – in the sense of Fig. 8.4a – of a conservative system at finite temperature implies that one can find axes x, y, z such that the evolution $\sigma_k(t) = \gamma(t)[\sigma_k]$ satisfies the differential equation

$$\begin{cases} \frac{\mathrm{d}}{\mathrm{d}t}\,\sigma_x(t) & = -\lambda_\perp \sigma_x(t) - \omega \sigma_y(t) \\[2mm] \frac{\mathrm{d}}{\mathrm{d}t}\,\sigma_y(t) & = -\lambda_\perp \sigma_y(t) + \omega \sigma_x(t) \\[2mm] \frac{\mathrm{d}}{\mathrm{d}t}\,[\sigma_z(t) - \varepsilon I] = -\lambda_\| \,[\sigma_z(t) - \varepsilon I] \end{cases}$$

where ε (with $|\varepsilon| < 1$) is a constant, the equilibrium value of σ_z; ω describes a periodic precession around the $z-$axis; and the dissipative part of the evolution is described by $0 \le \lambda_\| \le 2\lambda_\perp$. This translates to the relation $T_\| \ge \frac{1}{2}T_\perp$ between the parallel and transverse relaxation times, a prediction that appears to be empirically verified. This very special form of the *Bloch equation* obtains without assuming any specific coupling to a thermal bath, but only that such a coupling exists; see [Emch and Varilly, 1979], and references quoted therein for earlier evidences. Quantum models for the approach to equilibrium are discussed in Chap. 15.

3. The project sketched in (1) is still incomplete in the sense that one still needs to address a major question, namely to find an interpretation of the systems Σ as a specific physical system. To ask that the system Σ be mechanical, with realistic Hamiltonian, would be asking too much in general. What may be reasonable is that Σ be obtained as a result of further idealizations, such as the *thermodynamical limit* and a weak-coupling limit *à la van Hove*. Such a program is summarized in [Emch, 1982] inspired by the model of Ford, Kac, and Mazur [1965], a rigorous treatment of which is given in [Davies, 1972].

   Note that while the original model was first presented in the framework of classical mechanics, its extension to the quantum realm was immediately recognized; the latter is discussed as see Model 3, Sect. 15.3. This model, in turn, suggested the quantum extension of the notion of Kolmogorov system [Emch, 1976a]; see also [Emch, Albeverio, and Eckmann, 1978]; and for a didactic and detailed account, see [Emch, 1976b]. A discussion of the extension to the quantum realm of the notion of Markov processes and dilations beyond the Kolmogorov–Daniell reconstruction can be found in [Kümmerer, 1985a,b, 1988]; for general formulations of quantum stochastic processes, see [Accardi, Frigerio, and Lewis, 1982, Hudson and Parthasarathy, 1984, Lindblad, 1993, Belavkin, 1998, Goswami and Sinha, 1999].

4. With a very different motivation, the classical reconstruction theorem for Markov processes plays a central role in the mathematical implementation of the Symanzik Euclidean program in quantum field theory; see [Guerra, Rosen, and Simon, 1975].

5. Let us finally mention that the idea of extending a contractive semi-group to a unitary group acting in a bigger space was pioneered in the framework of Hilbert space theory by Sz.-Nagy in the Appendix to [Riesz and Sz.-Nagy, 1955]; see Fig. 8.4b. Note nevertheless that his condition of minimality must be modified to allow for the construction of Markov processes from Markov semi-groups; suffice it to say here that this also has been understood.

Model 4'. This is the simplest possible model for the Sz.-Nagy extension sketched in Fig. 8.4b. Take $\mathcal{H}_S = \mathsf{C}$ and $\mathcal{H} = \mathcal{L}^2_\mathsf{C}(\mathsf{R}, dx)$. Let $\lambda > 0$ fixed, and

$$S : (t, z) \in \mathsf{R}^+ \times \mathcal{H}_S \mapsto S(t)z \in \mathcal{H}_S \quad \text{with} \quad S(t)z = e^{-\lambda t}z \quad ;$$

$$U : (t, \Psi) \in \mathsf{R} \times \mathcal{H} \mapsto U(t)\Psi \in \mathsf{H} \quad \text{with} \quad [U(t)\Psi](x) = e^{-ixt}\Psi(x) .$$

Let further

$$\left.\begin{array}{l} E : \Psi \in \mathcal{H} \mapsto (\Psi, \Phi_o) \in \mathcal{H}_S \\[2mm] i : z \in \mathcal{H}_S \mapsto z\Phi_o \end{array}\right\} \quad \text{with} \quad \Phi_o(x) - [\frac{\lambda}{\pi(x^2 + \lambda^2)}]^{\frac{1}{2}} .$$

Since

$$(U(t)\Phi_o, \Phi_o) = \int_\mathsf{R} dx\, e^{-ixt} \frac{\lambda}{\pi(x^2 + \lambda^2)} = e^{-\lambda|t|} ,$$

we verify immediately that we have indeed the desired result, namely:

$$E \circ U(t) \circ i = S(|t|) .$$

The physical interest of this model is that its second quantization provides precisely the quantum version of the flow of Brownian motion, and thus the embedding demanded by the quantum version of the Ford–Kac–Mazur oscillator model; see [Emch, 1976a] and Rem. F.3 above for further references. K-systems satisfy properties that are much stronger than simple ergodicity. To see this we review an intermediary notion, and three related powerful results.

Recall first – see Thm. 8.1.1 – that mixing implies ergodicity, but not the converse.

Definition 8.2.2. *A cascade, i.e. a dynamical system with discrete time parameter $n \in \mathsf{Z}$, is said to have* Lebesgue spectrum *whenever, in the Koopman formalism, there exists an orthonormal basis $\{\Psi_{(j,n)} \mid (j, n) \in J \times \mathsf{Z}\}$ in the space $\mathcal{H}^\perp$ orthogonal to the a.e. constant functions, such that*

$$\forall\, (j, n) \in J \times \mathsf{Z} : U\Psi_{(j,n)} = \Psi_{(j,n+1)} .$$

The cardinality of the set J is called the multiplicity *of the Lebesgue spectrum. A flow, i.e. a dynamical system with continuous time parameter $t \in \mathsf{R}$, is said to have Lebesgue spectrum whenever, for every $\alpha > 0$ the discrete dynamical system obtained by restricting t to $\alpha\mathsf{Z}$ has Lebesgue spectrum.*

It can be shown that a dynamical system $\{\Omega, \mathcal{F}, \mu, \phi\}$ with a continuous time parameter has Lebesgue spectrum in the sense of the above definition if and only if the spectral measure

$$\mu_{\Psi}((-\infty, \lambda]) = (E^{\perp}(\lambda)\Psi, \Psi)$$

is absolutely continuous with respect to Lebesgue measure for all $\Psi \in \mathcal{H}^{\perp}$. This spectral measure is defined as the spectral resolution of the unitary group

$$U^{\perp}(t) = \int_{R} dE^{\perp}(\lambda) \, \exp\{-i\lambda t\}$$

where $\{U^{\perp}(t) \mid t \in R\}$ is the restriction of $\{U(t) \mid t \in R\}$ to $\mathcal{H}^{\perp}$. This explain the association of the name of Lebesgue to the spectral property described by Def. 8.2.2.

The value of this intermediary notion resides in the following results, the proofs of which we sketch.

Lemma 8.2.1. *Lebesgue spectrum implies mixing, but the converse is not true.*

Theorem 8.2.1. *Every K-system has Lebesgue spectrum with countable multiplicity.*

Corollary 8.2.1. *Every K-system is mixing, but the converse is not true.*

Sketch of the proof of Lemma 8.2.1. For every pair of vectors in $\mathcal{H}$:

$$(U(t)\Psi_1, \Psi_2) = (\Psi_1, 1)(1, \Psi_2) + (U^{\perp}(t)\Psi_1^{\perp}, \Psi_2^{\perp}) \quad .$$

Hence, by Thm. 8.1.2, it is sufficient to prove

$$\lim_{t \to \infty} (U^{\perp}(t)\Psi_1^{\perp}, \Psi_2^{\perp}) = 0 \quad . \tag{8.2.7}$$

For continuous time, by definition, the spectrum of $U^{\perp}$ is absolutely continuous with respect to Lebesgue measure. (8.2.7) reduces to the classical Riemann–Lebesgue lemma which asserts that the Fourier transform maps $\mathcal{L}^1(R)$ into $\mathcal{C}_{\infty}(R)$, the space of all continuous functions that vanish at infinity. For discrete time, the expansion of $\Psi_k^{\perp}$ ($k = 1, 2$) in the orthonormal basis $\{\Psi_{(j,n)}\}$ of Def. 8.2.2, gives:

$$(U^{\perp}(N)(\Psi_1^{\perp}, \Psi_2^{\perp}) = \sum_{j,m} \sum_{k,n} (\Psi_1, \Psi_{(j,m)})(U^N\Psi_{(j,m)}, \Psi_{(k,n)})(\Psi_{(k,n)}, \Psi_2)$$

which can be approximated by a finite sum. Since

$$(U^N\Psi_{(j,m)}, \Psi_{(k,n)}) = (\Psi_{(j,m+N)}, \Psi_{(k,n)}) = \delta_{j,k} \, \delta_{m+N,n}$$

all these terms vanish as N becomes sufficiently large, thus proving (8.2.7). For the converse part of the Lemma, notice simply that having Lebesgue spectrum is evidently a stronger condition than the spectral properties required by mixing alone.   **q.e.d.**

The role of spectral conditions in the occurrence of mixing in ergodic theory was first recognized by Koopman and Neumann [1932].

Since Cor. 8.2.1 follows directly from the combination of Lemma 8.2.1 – the proof of which we just sketched – and of Thm. 8.2.1, we only have now to prove the latter.

Sketch of the proof of Thm. 8.2.1. We first indicate how the argument goes in the case of K-flows, i.e. for continuous time $\mathcal{T} = \mathsf{R}$. Let $\mathcal{A}$ be the refining subalgebra of Def. 8.2.1, and $\mathcal{A}_t = \phi_t \mathcal{A}$. We denote by E_t the closed subspace of $L^2(\Omega, \mathcal{F}, \mu)$ generated by the characteristic functions χ_{A_t} of the elements $A_t \in \mathcal{A}_t$. With $U(t)$ denoting as usual the unitary operator associated to $\phi(t)$ by the Koopman formalism, we have:

$$\forall \ s,t \in \mathsf{R} \ : \ U(t)\,E(s)\,U(-t) = E(s+t) \tag{8.2.8}$$

which expresses Property (1) of the definition of a K-system. Properties (2) and (3) imply that

$$\lim_{s \to -\infty} E(s) = E_{\Phi_o} \quad \text{and} \quad \lim_{s \to \infty} E(s) = I \tag{8.2.9}$$

where E_{Φ_o} is the projector on the one-dimensional space of the functions that are constant a.e.; and I is the identity projector; we use the usual notation:

$$\forall \ \Psi \in \mathcal{L}^2 \ : \ \begin{cases} E_{\Phi_o}\Psi = (\Psi, \Phi_o)\Phi_o \quad \text{with} \quad \Phi_o(x) = 1 \ \forall x \in \Omega \\[2mm] I\Psi = \Psi \end{cases}.$$

Since E_{Φ_o} commutes with all $U(t)$, we can rewrite (8.2.8) and (8.2.9) in the form

$$\forall s,t \in \mathsf{R} \ : \ U^{\perp}(t)E^{\perp}(s)U^{\perp}(-t) = E^{\perp}(s+t) \quad \text{with}$$

$$\lim_{s \to -\infty} E^{\perp}(s) = 0^{\perp} \quad \text{and} \quad \lim_{s \to \infty} E^{\perp}(s) = I^{\perp} \tag{8.2.10}$$

where $^{\perp}$ denotes the restriction of the corresponding operator to the subspace orthogonal to the one-dimensional subspace of the constant functions that are a.e. constant. The relation (8.2.10) happens to be the well-known reformulation – in the language of the Mackey systems of imprimitivity; see e.g. [Emch, 1984][Sect. 8.3.f; in particular eq.(181)] – of the Heisenberg canonical commutation relation. One of the most basic theorems of quantum mechanics – due to [Stone, 1930, Von Neumann, 1931] and, in the form used here, to [Mackey, 1949] – asserts that every representation of (8.2.10) as operators on a [separable] Hilbert space is the direct sum of copies of the Schrödinger representation. This implies in particular that

$$U(t) = \oplus_n U_n(t) \quad \text{with} \quad U_n(t) \simeq U_S(t) \tag{8.2.11}$$

where

$$\forall \Psi \in \mathcal{L}^2(R, dx) \; : \; [U_S(t)\Psi](x) = e^{-ixt}\Psi(x) \tag{8.2.12}$$

which is to say that the spectrum of $U^\perp$ is absolutely continuous with respect to Lebesgue measure. In fact, more can be shown: the multiplicity of the spectrum is countably infinite [Sinai, 1961], namely $U^\perp$ is the direct sum of infinitely many copies of the Schrödinger U_S.

For K-cascade, i.e. K-systems for which times runs over $\mathcal{T} = \mathsf{Z}$, the argument is as follows [Kolmogorov, 1958]. Replace in the above $\{U(t) \mid t \in \mathsf{R}\}$ and $\{E(s) \mid s \in \mathsf{R}\}$ by $\{U^n) \mid n \in \mathsf{Z}\}$ and $\{E_m \mid m \in \mathsf{Z}\}$. With $\mathcal{H}_m = E_m L^2$, let then $\{\Psi_j \mid j \in J\}$ be an orthonormal basis in $\mathcal{H}_1 - \mathcal{H}_o$; we have then that $\{\Psi_j^n = U^n\Psi_j \mid (n,j) \in \mathsf{Z} \times J\}$ is an orthonormal basis in $\mathcal{H}^\perp = \{\Psi \in L^2 \mid (\Psi, \Phi_o) = 0\}$, with $U\Psi_j^n = \Psi_j^{n+1}$. Kolmogorov [1958] also proved – see e.g. [Arnold and Avez, 1968][App. 17]. or [Walters, 1982][Thm 4.33] – that the space $\mathcal{H}_1 - \mathcal{H}_o$ is infinite-dimensional, i.e. that the index set $J = \mathsf{Z}^+$.   q.e.d.

8.3 Dynamical Entropy

From the definition of a K-system, one can infer that a knowledge of the past – as given by all observables measurable with respect to the σ−ring $\mathcal{A}$ which contains all $\mathcal{A}_{t\leq 0}$ – is not sufficient to predict the future – as described by all observables measurable with respect to any of the σ−ring $\mathcal{A}_t$ with $t > 0$. This qualitative remark can be sharpened to a quantitative one, thanks to the notion of dynamical entropy, introduced in [Kolmogorov, 1959, Sinai, 1959], which we now review; for more detail and/or background, see any of the following classic texts, see e.g. [Billingsley, 1965, Arnold and Avez, 1968, Parry, 1969, Walters, 1982].

We first discuss the case of the dynamical systems $(\Omega, \mathcal{F}, \mu, \phi)$ with discrete time, i.e. $\mathcal{T} = \mathsf{Z}$. Let then P be any finite measurable partition of Ω, i.e. $P = \{A_k \in \mathcal{F} \mid k = 1, \cdots, n\}$ with n finite, $\mu(\Omega - \cup_{k=1}^n A_k) = 0$, $\mu(A_k \cap A_l) = 0$ whenever $k \neq l$; as we systematically neglect sets of measure zero in the sequel, the last two conditions are simply written $\cup_{k=1}^n A_k = \Omega$ and $A_k \cap A_l = \emptyset$ whenever $k \neq l$.

In line with our discussion in Sect. 5.3, we define the entropy of a finite measurable partition P by

$$\left. \begin{array}{l} S(P) = \sum_{k=1}^n h[\mu(A_k)] \quad \text{with} \\[2ex] h : x \in (0, 1) \mapsto -x \ln x \quad \text{and} \quad h(0) = 0 = h(1) \end{array} \right\}. \tag{8.3.1}$$

We assume that a base for the logarithm has been chosen and absorbed in the constant k that appears in Thm. 5.3.1.

Recall that for two finite measurable partitions $P = \{A_k \in \mathcal{F} \mid k = 1, \cdots, m\}$ and $Q = \{B_l \in \mathcal{F} \mid l = 1, \cdots, n\}$, the join $P \vee Q$ is the finite measurable partition defined by $P \vee Q \equiv \{A_k \cap B_l \in \mathcal{F} \mid k = 1, \cdots, m \,; l = 1, \cdots, n\}$. Recall also that the conditional entropy $S(P \mid Q)$ of P given Q is defined as:

$$S(P \mid Q) = \sum_{l=1}^{n} \mu(B_l) \sum_{k=1}^{m} h\left[\frac{\mu(A_k \cap B_l)}{\mu(B_l)}\right] \tag{8.3.2}$$

so that

$$S(P \vee Q) = S(Q) + S(P \mid Q) \quad . \tag{8.3.3}$$

Note finally that $\phi[P] \equiv \{\phi[A_k] \mid A_k \in P\}$ is again a finite measurable partition.

Lemma 8.3.1. *For any finite measurable partition P and all positive integers N, let*

$$H_N \equiv S(\bigvee_{\nu=0}^{N-1} \phi^{\nu}[P]) \quad \text{and} \quad s_N \equiv H_{N+1} - H_N \quad .$$

Then the following limits exist and are equal:

$$\lim_{N \to \infty} s_N \quad , \quad \lim_{N \to \infty} S(P \mid \bigvee_{\nu=1}^{N} \phi^{-\nu}[P]) \quad , \quad \lim_{N \to \infty} \frac{1}{N} H_N \quad .$$

The purpose of the above Lemma – for its proof, see below – is to take the first step necessary to making the following definition meaningful.

Definition 8.3.1. *The quantity defined by*

$$H(\phi, P) \equiv \lim_{N \to \infty} \frac{1}{N} S(\bigvee_{\nu=0}^{N-1} \phi^{\nu}[P]) = \lim_{N \to \infty} S(P \mid \bigvee_{\nu=1}^{N} \phi^{-\nu}[P])$$

is called the entropy of the partition P with respect to the evolution ϕ.

The interpretation of the entropy just defined follows from the interpretation of the ordinary entropy (8.3.1) as the information gained from a measurement of P. Equivalently, $H(\phi, P)$ can be interpreted as the limit, as $N \to \infty$, of the average, per measurement, of the information gained during the sequence of measurements $\{P, \phi[P], \cdots, \phi^{N-1}[P]\}$; or as the information gained by measuring P conditioned by its complete past. Note that if the action of ϕ were to be periodic on P, say of period k, we would have $H(\phi, P) = 0$: one would indeed not gain any new information by repeating the measurement more than k times. When $H(\phi, P) > 0$, both interpretations assert that the dynamical system keeps producing randomness: the behavior of the system cannot be predicted from its past history, however far back the latter goes.

We refine this notion in Def. 8.3.2 below. First, however, we must convey an idea of the proof of the above Lemma.

Proof of Lemma 8.3.1 : From (8.3.3), and $\mu \circ \phi = \mu$, we have:

$$H_{N+1} \equiv S(\bigvee_{\nu=0}^{N} \phi^{\nu}[P]) = S(\{\bigvee_{\nu=0}^{N-1} \phi^{\nu}[P]\} \vee \phi^{N}[P])$$

$$= S(\bigvee_{\nu=0}^{N-1} \phi^{\nu}[P]) + S(\phi^{N}[P] \mid \bigvee_{\nu=0}^{N-1} \phi^{\nu}[P]) \tag{8.3.4}$$

$$= H_N + S(P \mid \bigvee_{\nu=1}^{N} \phi^{-\nu}[P])$$

and thus, since $\bigvee_{\nu=1}^{N} \phi^{-\nu}[P] \supseteq \bigvee_{\nu=1}^{N-1} \phi^{-\nu}[P]$:

$$s_N = S(P \mid \bigvee_{\nu=1}^{N} \phi^{-\nu}[P]) \leq S(P \mid \bigvee_{\nu=1}^{N-1} \phi^{-\nu}[P]) = s_{N-1} \quad . \tag{8.3.5}$$

Hence $\{s_N \mid N \in \mathbf{Z}^+\}$ is a non-increasing sequence of non-negative real numbers. Consequently it admits a limit, i.e.:

$$\lim_{N \to \infty} s_N \quad \text{exists} \quad . \tag{8.3.6}$$

Upon dividing the following expression by N

$$S(P) + \sum_{k=1}^{N} s_N = S(P) + \sum_{k=1}^{N} [H_{k+1} - H_k] = S(P) + H_{N+1} - H_1 = H_{N+1}$$
$$\tag{8.3.7}$$

(8.3.6) implies – since P thus $S(P)$ are finite – that the following limits exist and are equal:

$$\lim_{N \to \infty} \frac{1}{N} H_N = \lim_{N \to \infty} s_N \quad . \tag{8.3.8}$$

The theorem now follows from (8.3.8) and the first equality in (8.3.5). **q.e.d.**

In order to convey that the history of a system behaves randomly, we may want to distinguish whether the above behavior happens just for some finite measurable partitions or for all of them. Hence the following definition.

Definition 8.3.2. *The* dynamical entropy *of a dynamical system is* $H(\phi) = \sup_P H(\phi, P)$ *where the supremum is taken over all finite measurable partitions. The system is said to have* completely positive *dynamical entropy if* $H(\phi, P) > 0$ *for all finite measurable partitions – except the trivial partition* $\{\emptyset, \Omega\}$.

The following result justifies the name "entropy per unit time" which Kolmogorov uses in the title of his paper [Kolmogorov, 1959].

Scholium 8.3.1. *For every integer $k > 0$:*

$$\frac{1}{k}H(\phi^k) = H(\phi).$$

This result also makes it natural to define the entropy of a dynamical system with continuous time-parameter $\{\phi_t \mid t \in \mathsf{R}\}$ as $H(\phi_1)$. The main result connecting K-systems and entropy is due to [Rohlin and Sinai, 1961]; see also [Rohlin, 1967].

Theorem 8.3.1. *A dynamical system is a K-system iff it has completely positive dynamical entropy.*

Remark G:

1. Theorem 8.3.1 gives the mathematical confirmation supporting the intuitive impression gained from the above discussion, namely that the K-property is a much stronger form of ergodicity than mixing.
   In fact, this property is even stronger than the Lebesgue spectrum property – see Thm. 8.2.1 and its proof. One knows indeed of dynamical systems with countably infinite Lebesgue spectrum and yet zero dynamical entropy: an example of such a system is provided by the *horocycle* flow on a compact surface of constant negative curvature [Gurevitch, 1961]; for a definition of horocycles, see e.g. [Arnold and Avez, 1968] [App. 29]. The matter is fortunately quite different for the associated *geodesic* flow; see Model 6 below.
2. It could be quite difficult to compute explicitly the sup involved in Def. 8.3.2 of the dynamical entropy. Fortunately, one knows that *if* a finite partition P is a *generator* – i.e. is such that the collection $\{\phi_n[P] \mid n \in \mathsf{Z}\}$ generates the whole $\sigma-$ring $\mathcal{F}$ – *then* $H(\phi, P) = H(\phi)$; for a detailed study of this aspect of the theory, see [Parry, 1969]. In particular for any Bernoulli shift $B(p_o, \cdots, p_{n-1})$ the partition P_o, the n elements of which are the cylinder sets $A_o^i = \{\xi \in \varXi \mid \xi_o = i\}$, is a generator and one obtains readily – see e.g. [Arnold and Avez, 1968] – that

$$H(\phi, P_o) = -\sum_{i=o}^{n-1} p_i \ln p_i$$

   as anticipated in (8.2.6).
3. The dynamical entropy is preserved under isomorphisms; as we pointed out already in Rem. D.4; moreover, the dynamical entropy is a *complete invariant in the class of Bernoulli shifts,* i.e. two Bernoulli shifts are isomorphic dynamical systems iff they have the same entropy [Ornstein, 1970].
4. The existence of a connection between (dynamical) entropy and the complexity of message coding could be read in the watermark of the papers by Shannon which we reviewed in Sect. 5.3. How this reading can be

made precise in terms of the complexity of trajectories of dynamical systems was announced in [Brudno, 1978], with the details given in [Brudno, 1983]. Three results must be reported here.

Let $\{\Omega, \mathcal{F}, \mu, \phi\}$ be a classical dynamical system with discrete time parameter; assume further that $\Omega = X$ is a compact metric space, and $\{X, \mathcal{F}, \mu, \phi\}$ is ergodic. The *first result* is that, for $\mu-$almost all $x \in X$, the complexity $K(x, \phi)$ of the forward trajectory $\{\phi^n[x] \mid n \in Z^+\}$ is precisely equal to the dynamical entropy $H_\mu(\phi)$ of the system. The essence of the paper is to find a natural definition of the complexity of a trajectory; Brudno proposes:

$$K(x, \phi) = \sup_{P \in \mathcal{P}} K(x, \phi | P)$$

where $\mathcal{P}$ is the collection of all finite measurable partitions of X. In the above expression, the conditional complexity $K(x, \phi | \mathcal{U})$ of a trajectory with respect to a finite cover $\mathcal{U} = \{U_i | i \in Z_n, U_i \in \mathcal{F}, \cup_{i \in Z_n} U_i = X\}$ is defined as:

$$K(x, \phi | \mathcal{U}) = \lim \sup_{n \to \infty} \frac{1}{n} \min_{\xi \in \psi(x | \mathcal{U})} K(\xi^n)$$

where $K(\xi^n)$ is the Kolmogorov complexity – see Sect. 6.2 – of the finite string ξ^n formed by the first n entries of an infinite sequence ξ; and

$$\psi(x | \mathcal{U}) = \{\xi : k \in Z^+ \mapsto Z_n \mid \forall k \in Z^+ : \phi^k[x] \in U_{\xi_k}\} \quad .$$

The Reader will note that this definition of the complexity of a trajectory involves modeling a dynamical system by Bernoulli shifts with increasingly large alphabets Z_n; indeed the work of Brudno belongs to the program on symbolic dynamics conducted by Alekseev; see e.g. the review [Alekseev and Yakobson, 1981].

The *second result* is that an overall upper bound of the complexity of individual trajectories is given by the topological entropy, i.e. the supremum of all $H_\mu(\phi)$ when μ runs over all probability measures on X that are invariant under the evolution ϕ. Finally, Brudno shows a *third result*, namely that if a sequence $\xi \in Z_n^{Z^+}$ is random in the sense of Martin-Löf – see Sect. 6.2 – then its complexity $K(\xi) = \lim \sup_n K(\xi^n)/n$ is equal to the dynamical entropy of the Bernoulli shift obtained from the uniform measure on Z_n. Thus, these three results support our interpretation of entropy as a measure of complexity.

5. The present remark, and the one following it, may be omitted in a first reading: they are inserted here only for the purpose of indicating a concern, namely that a classical notion as fundamental as the dynamical entropy ought to have an analog in the quantum realm.

Two main difficulties have to be mastered in order to arrive at a *quantum generalization* of the notion of dynamical entropy. First, the quantum measuring process described by von Neumann – which can be viewed as a model that forces the classical structure of the measuring apparatus on the quantum system – must be tamed so that the instantaneous entropy increase due to the measurement can be separated from the entropy increase due to the long-term evolution. Second, while the time-image of a finite partition – into mutually orthogonal projectors adding up to the identity – is a new finite partition, these two partitions may not commute, and thus there may not exist any finite partition that refines them.

A solution to the first problem was proposed in [Emch, 1974, 1976a], and its conditions are satisfied in the particular case where the basic quantum invariant state ϕ replacing the classical invariant measure μ is tracial, i.e. when ϕ is such that the expectation values $\langle \phi; AB \rangle$ and $\langle \phi; BA \rangle$ are equal even though the observables A and B may not commute. This situation is realized in the quantum version of the Arnold cat; see [Benatti, Narnhofer, and Sewell, 1991].

A solution to the second problem was proposed by Connes and Størmer [1975] for the case where the state is a tracial state; for a recent contribution that illustrates how close the non-commutative results can be formulated in parallel to their classical analog – although the mathematical formalism is certainly not an immediate translation – see [Størmer, 2000] and references cited therein. The tracial state assumption was removed in [Connes, 1985, Narnhofer and Thirring, 1985, Connes, Narnhofer, and Thirring, 1987]; we refer to the resulting dynamical entropy as H_{CNT}.

For a reformulation bringing in the notion of coupling to another system (taken to be classical), see [Sauvageot and Thouvenot, 1992]; and for a didactic presentation adapted to the language of quantum information theory, see [Ohya and Petz, 1993][Chap. 10].

While the CNT approach still involves some classical modeling, a new approach based on the quantum modeling provided by the shift on a spin chain is proposed in [Alicki and Fannes, 1994, Andries, Fannes, Tuyls, and Alicki, 1996].

This dynamical entropy, which we refer to as H_{LAF}, is constructed from ideas dating back to [Lindblad, 1979, 1988]. For a perspective on the latter papers, see [Lindblad, 1993]; and for a review of the dynamical entropy H_{LAF}, as compared to H_{CNT}, see [Tuyls, 1998], where evidence for $H_{CNT} \leq H_{LAF}$ is presented; see already Alicki and Narnhofer [1995] who had proposed the following tentative interpretation: H_{LAF} would seem to be "related to how quickly, by repeated measurements, the information on a given system grows", whereas H_{CNT} "controls how quickly operators become independent from one another"; note that this distinction would be a purely quantum effect since the two entropies coïncide in the classical case. For still other quantum dynamical entropy proposals, see [Voiculescu, 1995, Hudetz, 1998].

6. Whereas these dynamical entropies – quantum as well as classical – do model certain aspects of the thermodynamical entropy production encountered in the non-equilibrium thermodynamics, the analogy is still a far cry from a formal relation.

8.4 Anosov Property

Going back to classical dynamical systems, recall that we saw several ergodic properties which form a strict hierarchy where each is strictly stronger than the following one: K-systems $\Rightarrow$ Lebesgue spectrum $\Rightarrow$ Strong mixing $\Rightarrow$ Weak mixing $\Rightarrow$ Ergodicity. The next question we must consider now is whether (any of) these nice properties are compatible with an additional structure that the idealization to dynamical systems considered up to here has ignored, namely the Hamiltonian aspect of classical mechanics. The answer is positive: billiard systems that *are* K-systems have been specified; a brief introduction to these systems is given at the end of this section. We first present two models that help understand what happens in more difficult circumstances. Nevertheless, we must warn our Reader that we attach a disclaimer in the next chapter to the effect that the finite Hamiltonian systems exhibiting these properties are exceptional, in a sense we make precise there.

Model 5. This model is affectionately known as the Arnold cat; while the sympathy may come from the picture of the very mixed-up cat used in [Arnold and Avez, 1968] to illustrate the behavior of the model, **CAT** stands coldly for **A**utomorphism of the **T**orus satisfying condition **C** [in this nomenclature, we follow [Arnold and Avez, 1968]; others, following the English translation of the original paper [Anosov, 1967], call this "condition U ", which Anosov used to denote the condition characterizing what we now call the Anosov property; this "U" is a transliteration of у, the initial letter of the Russian word for "condition"; and it is pronounced as the *u* in *rule*.]

We first construct the dynamical system $(\Omega, \mathcal{F}, \mu; \phi)$ and then show that, in addition to the dynamics ϕ, the Arnold cat model admits two flows – the expanding flow ψ_+ and the contracting flow ψ_- – that satisfy the relation (8.4.5) below.

The *phase-space* $(\Omega, \mathcal{F}, \mu)$ is the torus $\Omega = \mathsf{T}^2 = \mathsf{R}^2/\mathsf{Z}^2$ equipped with the Borel structure $\mathcal{F}$ and Lebesgue measure μ it inherits from R^2. It is also important to note that T^2 is locally "like" R^2, i.e. inherits from it the structure of a differentiable manifold; and that the identification of points of R^2 whose coordinates differ by integers makes this manifold compact. Finally, the 2-form $\omega = dx_1 \wedge dx_2$ being closed, it equips Ω with the structure of a symplectic manifold.

The *evolution* $\phi : (n; x_1, x_2) \in \mathsf{Z} \times \mathsf{T}^2 \mapsto \phi_n[(x_1, x_2)] \in \mathsf{T}^2$ is given by the iterations $\phi_n = \phi^n$ of the quotient over Z^2 of the map

$$\tilde{\phi} : (x_1, x_2) \in \mathsf{R}^2 \mapsto \tilde{\phi}[(x_1, x_2)] = T \begin{pmatrix} x_1 \\ x_2 \end{pmatrix} \in \mathsf{R}^2 \qquad (8.4.1)$$

where $T \in SL(2, \mathsf{Z})$ with $\operatorname{Tr} T > 2$, i.e.

$$T = \begin{pmatrix} a & b \\ c & d \end{pmatrix} \quad \text{with} \quad a, b, c, d \in \mathsf{Z}, \ ad - bc = 1 \ \text{and} \ a + d > 2 \quad . \quad (8.4.2)$$

We first explain the rationale for the three conditions imposed on T:

1. $a, b, c, d \in \mathsf{Z}$ ensures $\tilde{\phi}(\mathsf{Z}^2) \subseteq \mathsf{Z}^2$, so that the quotient map $\phi : (x_1, x_2) \in \mathsf{T}^2 \mapsto \phi[(x_1, x_2)] \in \mathsf{T}^2$ be properly defined as

$$\phi[(x_1, x_2)] \equiv T \begin{pmatrix} x_1 \\ x_2 \end{pmatrix} \ \operatorname{mod} 1 \quad . \quad (8.4.3)$$

2. $ad - bc = 1$, i.e. $\det T = 1$, ensures that T preserves the volume, i.e. the Lebesgue measure, and in fact the symplectic form $dx_1 \wedge dx_2$. Note also that since $\det T \neq 0$, the inverse T^{-1} *exists*.

3. The additional condition $a + d > 2$, i.e. $\operatorname{Tr} T > 2$, when taken together with the other two conditions, ensures that T has two distinct eigenvalues

$$0 < \Lambda_- < 1 < \Lambda_+ \ \text{with} \ \Lambda_- \cdot \Lambda_+ = 1 \ \text{and} \ \Lambda_-, \ \Lambda_+ \ \text{irrational}. \quad (8.4.4)$$

This implies in particular that the corresponding principal directions (or eigenvectors) $X_\pm$ of T have irrational slopes, and thus each of their integral curves is a dense helix in T^2. These two directions are not necessarily orthogonal, unless the matrix T is symmetric, which is not essential to the ergodic properties of the model.

4. The algebraic consequences of the three conditions imposed on T are easily verified in the simplest example:

$$T = \begin{pmatrix} 1 & 1 \\ 1 & 2 \end{pmatrix} \quad \text{with} \quad T^{-1} = \begin{pmatrix} 2 & -1 \\ -1 & 1 \end{pmatrix} \quad .$$

From the point of view of ergodic theory, the principal property of the model is that the image $\phi^n[A]$ of a measurable region $A \subset \Omega$ in the remote past ($n \ll 0$) or the distant further ($n \gg 0$) becomes a long ribbon asymptotically winding along a dense helix, given by an integral curve of X_- for $n \ll 0$ or of X_+ for $n \gg 0$. This consequence of the hyperbolic character of the map $\tilde{\phi}$ on R^2, with its exponentially *contracting* direction X_- and exponentially *expanding* direction X_+ is captured by the following hyperbolicity relations:

$$\phi^n \circ \psi_\pm(\tau) \circ \phi^{-n} \equiv \psi_\pm(\Lambda_\pm{}^n \tau) \qquad (8.4.5)$$

where $\psi_\pm$ denotes the translation along the direction $X_\pm : \psi_\pm(\tau)[x] \equiv x + \tau X_\pm \ \operatorname{mod} 1$. The real numbers $\lambda_\pm = \ln(\Lambda_\pm)$ are the *Liapunov coefficients* of this system. Note that here $\lambda_\pm = \pm\lambda$ with $\lambda > 0$, which are equivalent to the first two properties in (8.4.4).

Since (8.4.5) plays a archetypical role in the forthcoming discussion, we rephrase it in an equivalent form. Let indeed x and y be two neighboring points on $\mathsf{T}^2 = \mathsf{R}^2/\mathsf{Z}^2$, with $\eta = x - y$ sufficiently small. Upon writing η as a linear combination $\eta = c_- X_- + c_+ X_+$ of the two eigenvectors of the matrix T in (8.4.1), we have:

$$\phi^n[\eta] = c_-\, e^{-\lambda n}\, X_- + c_+\, e^{+\lambda n}\, X_+ \tag{8.4.6}$$

Hence, in the future ($n > 0$), the evolution produces an exponential expansion along the direction X_+ (and an exponential contraction in the direction X_-), whereas the roles of X_+ and X_- are interchanged in the past ($n < 0$), i.e. exponential contraction along X_+ and exponential expansion along X_-. This is the exponential sensitivity to initial conditions, which Hadamard [1898] noticed in Model 6 below – in particular Rem. I.3 – and which Poincaré [1908a] – also [Poincaré, 1912] – perceived as central to the justification of probabilistic methods in the kinetic theory of gases, see Sect. 5.1. We come back to this later on in this section and in Chap. 9.

When expressed in differential form, the relations (8.4.5) are a stronger form – strict equalities replacing inequalities, and one-dimensional instead of finitely-dimensional contracting and dilating subspaces – of *Anosov's condition C,* namely for ξ_k in the vector space spanned by X_k :

$$\left.\begin{aligned} \|\phi^*[\xi]_k\| \leq \Lambda_k\, \|\xi_k\| \quad \text{if} \quad 0 < \Lambda_k < 1 \\[2mm] \|\phi^*[\xi]_k\| \geq \Lambda_k\, \|\xi_k\| \quad \text{if} \quad 1 < \Lambda_k \end{aligned}\right\} . \tag{8.4.7}$$

Remark H:

1. For a general definition of Anosov systems, see e.g. [Arnold and Avez, 1968]. The Reader will note that for these dynamical systems, the space Ω must be a differentiable manifold – for definitions, see Sect. B.3 – so that the relations (8.4.7), which involve the tangent vectors to the manifold, are well defined; accordingly, the evolution ϕ must be required to be smooth enough to preserve the differentiable structures.

2. Very strong ergodic conditions are satisfied by Anosov systems, of which the Arnold cat is perhaps the simplest prototype. First of all these systems are ergodic, see e.g. [Arnold and Avez, 1968][Thm. 17.9]. Moreover, every Anosov cascade satisfies the K-property, see e.g. [Arnold and Avez, 1968][Thm. 17.10]; for Anosov flows – i.e. when the time-parameter is continuous – this conclusion persists provided – see e.g. [Arnold and Avez, 1968][Thm 17.11] – one imposes in addition that the functions Ψ such that $\phi_t \Psi = \Psi \ \forall\, t \in \mathsf{R}$ must be constant.

3. Recall from Rem. F.3 in Sect. 8.2 above, that the classical notion of K-system admits a quantum generalization [Emch, 1976a]. A similar test of the content of the theory is that the Anosov property also seems to have an analog in the study of quantum dynamical systems; for exploratory models, see [Benatti, Narnhofer, and Sewell, 1991, Narnhofer,

1992, Emch, Narnhofer, Thirring, and Sewell, 1994, Narnhofer, Peter, and Thirring, 1996, Peter and Emch, 1998]. Whereas the quantization of the K-property involves carrying over probabilistic notions to the quantum realm, the quantization of the Anosov property requires a transposition of differential geometric concepts to the context of the non-commutative geometry of [Connes, 1990]; see already [Connes and Rieffel, 1987]. The connection between quantum Anosov systems and quantum K-properties has recently been clarified in [Narnhofer and Thirring, 2000].

4. While the value of the Arnold cat has been primarily to provide a stimulating mathematical model for ergodic theory, it could be argued that its 2-dimensional character precludes any real relevance to physics. This argument misses some exciting modern technological and scientific developments: 2−d physics is now an experimental reality; for a glimpse of this, see e.g. [Levi, 1998b]. As a matter of fact, it was noticed in [Benatti, Narnhofer, and Sewell, 1991, Narnhofer, 1992] that the structure of the quantum version of the Arnold Cat appears in the description of the kinematics in the quantum Hall effect, the mathematical, physical and historical aspects of which are summarized in [Bellissard, 1986][Sect. 7]. Let it suffice here to say that the effect is a consequence of the Lorentz force $F = q\,v \times B/c$ exercised by a magnetic field B on a charge q moving with velocity v (c is the speed of light). It is important to us that F and v span a fixed plane, perpendicular to B, so that the motion of the charges can be described in a two-dimensional space. The classical effect, discovered by E. Hall in 1879, immediately captured the attention of his contemporaries who saw it as a means to measure the velocity of "electricity" – we use today the atomic picture conveyed by the term "charge carriers"– in solids; see in particular [Boltzmann, 1880]. The quantum effect, discovered a century later by von Klitzing's group, manifests itself at very low temperature and under very high magnetic field: the electrical resistance associated to the effect occurs so precisely in *integer* multiples of the combination e^2/h of the fundamental constants e [charge of the electron] and h [Planck constant] that it is now used to define the unit standard for the Ohm; this discovery was marked by the 1985 Nobel prize. Some years later, in addition to this integral quantum Hall effect, a fractional quantum Hall effect was discovered, marked by yet another Nobel prize (1998); for a brief introduction to both of these quantum Hall effects, as well as to the classical Hall effect, see [Schwarzschild, 1998].

Model 6. The construction of this model involves the following eight principal steps: (1) we introduce the Poincaré half-plane $\tilde{M}$ as a simply connected Riemann manifold of constant negative curvature; (2) we show how the Poincaré half-plane $\tilde{M}$ is a homogeneous space, namely $SL(2,\mathsf{R})/SO(2)$; (3) we identify the Poincaré half-plane with a subgoup of $SL(2,\mathsf{R})$; (4) we use the action of the symmetries to construct the geodesics; (5) we identify $SL(2,\mathsf{R})$ itself with the phase-space obtained as the unit tangent space of the

Poincaré half-plane; (6) we implement the geodesic flow and show that this space is hyperbolic; (7) we recall that every compact manifold M of constant negative curvature is obtained by identifying points of $\tilde{M}$ that are linked by an element of a discrete subgroup $\Gamma \subset SL(2, \mathsf{R})$, very much like the flat torus – $T^2 = Z^2 \backslash R^2$ – is obtained by identifying points that differ by an element of $Z^2 \subset R^2$. Here $M = \Gamma \backslash \tilde{M} = \Gamma \backslash SL(2, \mathsf{R})/SO(2)$ so that its unit tangent plane – the energy shell of our model – is $T_1 M = \Gamma \backslash SL(2, \mathsf{R})$; (8) we finally indicate how this allows to obtain the Anosov structure as a projection, via Γ, of the hyperbolic structure of the unit tangent bundle of the Poincaré half-plane.

(1) The Poincaré half-plane is the 2-dimensional Riemann manifold

$$\tilde{M} \equiv \{z = x + iy \mid (x, y) \in \mathsf{R}^2 \,,\, y > 0\} \tag{8.4.8}$$

equipped with the differential manifold structure it inherits from its being embedded in R^2; and the metric

$$g \equiv \frac{1}{y^2}(dx^2 + dy^2) = -4\frac{1}{(z^* - z)^2}\, dz^* dz \tag{8.4.9}$$

from which one computes that $\tilde{M}$ has constant negative curvature $\kappa = -1$.

(2) The symmetry group S of $\tilde{M}$; its subgroup K; and $\tilde{M} \approx S/K$. The group of isometries of $\tilde{M}$ is

$$S \equiv SL(2, \mathsf{R}) = \left\{ s = \begin{pmatrix} a & b \\ c & d \end{pmatrix} \;\middle|\; a, b, c, d \in \mathsf{R}\,,\, ad - bc = 1 \right\} \tag{8.4.10}$$

with the Möbius action on $\tilde{M}$ defined by

$$\pi : (s, z) \in S \times \tilde{M} \mapsto s[z] \in \tilde{M} \quad with \quad s[z] \equiv \frac{az + b}{cz + d} \quad . \tag{8.4.11}$$

This action is indeed isometric, i.e. preserves the metric; and it is transitive, i.e. given any $z_1, z_2 \in \tilde{M}$ there exists a $s \in S$ such that $s[z_1] = z_2$. This element $s \in S$ however – and fortunately, as we shall see – cannot be unique: $\dim(\tilde{M}) = 2 < \dim(S) = 3$; specifically, each point $z \in \tilde{M}$ is left invariant by a subgroup $K_z \subset S$ conjugate to:

$$K \equiv \{k \in S \mid k[i] = i\} = \left\{ \begin{pmatrix} \cos(\theta/2) & \sin(\theta/2) \\ -\sin(\theta/2) & \cos(\theta/2) \end{pmatrix} \;\middle|\; \theta \in [0, 4\pi) \right\} . \tag{8.4.12}$$

It is in this sense that the Poincaré half-plane $\tilde{M}$ is a homogeneous manifold:

$$\tilde{M} = S/K = SL(2, \mathsf{R})/SO(2) \quad . \tag{8.4.13}$$

(3) $\tilde{M}$ *identified as the subgroup* $H \subset S$. The above structure suggests that we consider the subgroup $H \subset S$ given by

$$H = \left\{ h = \begin{pmatrix} a & b \\ 0 & a^{-1} \end{pmatrix} \mid a > 0,\, b \in \mathsf{R} \right\} \tag{8.4.14}$$

which enjoys two remarkable properties in our context.

First, it lifts the ambiguity in the action of S on $\tilde{M}$: to every pair (z_1, z_2) of elements of $\tilde{M}$ corresponds *exactly one* element $h \in H$ such that $h[z_1] = z_2$; to verify this, solve $h_k[i] = z_k$, which is immediate, and use the fact that H is a group. Hence, the bijection

$$\pi^{-1} : \begin{pmatrix} a & b \\ 0 & a^{-1} \end{pmatrix} \in H \mapsto ab + ia^2 \in \tilde{M} \tag{8.4.15}$$

allows to *identify* the manifold $\tilde{M}$ and the manifold of the Lie group H; this refinement of (8.4.13) will be useful in the sequel.

Second, one verifies – by straightforward computation here, although it is the expression of a general theorem known as the Iwasawa decomposition – that every element $s \in S$ can be written *in a unique way* in the form

$$S \ni s = h\,k \quad \text{with } h \in H \text{ and } k \in K \qquad \text{i.e.} \qquad S = HK \quad . \tag{8.4.16}$$

This unique decomposition provides the key to taking advantage of the ambiguity (8.4.12) and identifying our isometry group $S = SL(2, R)$ with the unit tangent bundle $T_1 \tilde{M}$ of our space $\tilde{M}$, i.e. with the typical energy surface for the geodesic flow on the Riemann manifold $(\tilde{M}, g)$ defined by (8.4.8–9). Note that there is still a residual ambiguity, as $Z_2 = \{I, -I\} \subset S$ acts trivially on $\tilde{M}$; we should therefore consider only S/Z_2, instead of S. We forgo this distinction, as this legerdemain is innocuous in our context: the elements of Z_2 commute with all elements of S.

(4) The geodesics on $\tilde{M}$. Recall that *geodesics* are defined as lines that, locally, minimize arc-length. Solving this variational problem here is identical to finding the trajectories of the Hamiltonian system, defined on the cotangent bundle $T^*\tilde{M}$ of our manifold $\tilde{M}$ by the Hamiltonian function $H = y^2(p_x{}^2 + p_y{}^2)$. The answer is that a geodesic is either (1) any straight Euclidean half-line parallel to the axis $x = 0$; or (2) any Euclidean semi-circle with center on the axis $y = 0$ and arbitrary finite radius. To see this, it is sufficient to find one particular geodesic γ_o, and to use the fact that S is a group of isometries – thus sends geodesics to geodesics – with H acting transitively on $\tilde{M}$, and K describing the rotations by an angle θ around $i \in \tilde{M}$. The special geodesic that is easy to identify as such is the straight half-line

$$\gamma_o : t \in \mathsf{R} \mapsto \gamma_o(t) \in \tilde{M} \quad \text{with} \quad \gamma_o(t) = (0, e^{-t})$$

i.e. with the identification (8.4.15) of H and $\tilde{M}$:

$$\gamma_o : t \in \mathsf{R} \mapsto \gamma_o(t) \in H \quad \text{with} \quad \gamma_o(t) = \begin{pmatrix} e^{-t/2} & 0 \\ 0 & e^{t/2} \end{pmatrix} \quad . \tag{8.4.17}$$

We then obtain, by action *from the left* in S:

(i) All the geodesics through i, namely $\gamma(t) = k[\gamma_o(t)] = k\gamma_o(t)$ with k as in (8.4.12) and $\theta \in [0, 2\pi)$; it is not hard to verify that, these correspond to all Euclidean circles, with center on $y = 0$ and passing through i. For instance

$$\begin{pmatrix} \cos(\pi/4) & \sin(\pi/4) \\ -\sin(\pi/4) & \cos(\pi/4) \end{pmatrix} \begin{pmatrix} e^{-t/2} & 0 \\ 0 & e^{t/2} \end{pmatrix}$$

gives, via (8.4.11), the Euclidean semi-circle of radius 1 centered at $(0,0)$:

$$\gamma = \{\, (x(t), y(t)) = (\frac{\sinh t}{\cosh t}, \frac{1}{\cosh t}) \mid t \in \mathsf{R} \,\} \quad .$$

(ii) All horizontal translates of our special geodesic, namely $\gamma(t) = h[\gamma_o(t)] = h\gamma_o(t)$ with h as in (8.4.14) and $a = 1$ and b arbitrary; these correspond to the Euclidean straight half-lines parallel to the y-axis:

$$\begin{pmatrix} 1 & s \\ 0 & 1 \end{pmatrix} \begin{pmatrix} e^{-t/2} & 0 \\ 0 & e^{t/2} \end{pmatrix}$$

gives, via (8.4.11), the Euclidean straight half-line

$$\gamma = \{\, (x(t), y(t)) = (s, e^{-t}) \mid t \in \mathsf{R} \,\} \quad .$$

(iii) All geodesics, upon combining (ii) with (i).

(5) The energy shell $T_1 \tilde{M}$. Now, we can take advantage of the unique decomposition (8.4.16) to describe explicitly the identification of the unit tangent bundle $T_1 \tilde{M}$ over $\tilde{M}$; it is given by the extension of π in (8.4.15) to the bijective map

$$\pi^* \; : \; X_z \in T_1 \tilde{M} \; \longrightarrow \; s(X_z) = h(z)\, k(X_z) \in S = H\,K \tag{8.4.18}$$

constructed as follows. For any unit vector X_z attached to a point z, let $h \in H$ be the unique element such that $h^{-1}[z] = i$; let then γ_{X_z} be the unique geodesic γ_z tangent to X_z at $t = 0$ i.e. such that $\gamma_z(0) = z$ and $\dot{\gamma}(0) = X_z$; its image $h^{-1}[\gamma_{X_z}]$ is a new geodesic, passing through i at time $t = 0$; let

$$Y_i = \frac{d}{dt} h^{-1}[\gamma_{X_z}(t)] \Big|_{t=0}$$

be the tangent vector to the new geodesic at i; finally k^{-1} is the rotation that brings the tangent vector Y_i to coincide with the tangent vector $\dot{\gamma}_o(0)$ to the special geodesic γ_o – see (8.4.17). In summary, we have, for all $t \in \mathsf{R}$, $k^{-1}h^{-1}[\gamma_z(t)] = \gamma_o(t)$, and in particular: $k^{-1}h^{-1}[X_z] = \dot{\gamma}_o(0)$, which specify $h \in H$ and $k \in K$ uniquely.

(6) The hyperbolicity of the geodesic flow on $T_1\tilde{M}$. The identification (8.4.18) just described allows to demonstrate immediately the hyperbolic structure of the unit tangent bundle of the Poincaré half-plane; we have indeed the following three right-actions (i.e. actions operating *from the right*):

1. the geodesic flow

$$\phi : (t, s) \in \mathsf{R} \times S \mapsto \phi_t[s] \in S \quad \text{where}$$

$$\phi_t[\begin{pmatrix} a & b \\ c & d \end{pmatrix}] = \begin{pmatrix} a & b \\ c & d \end{pmatrix} \begin{pmatrix} e^{-t/2} & 0 \\ 0 & e^{t/2} \end{pmatrix} \tag{8.4.19}$$

2. the two horocyclic flows

$$\psi_+ : (\tau, s) \in \mathsf{R} \times S \mapsto \psi_+(\tau)[s] \in S \quad \text{where}$$

$$\psi_+(\tau)[\begin{pmatrix} a & b \\ c & d \end{pmatrix}] = \begin{pmatrix} a & b \\ c & d \end{pmatrix} \begin{pmatrix} 1 & \tau \\ 0 & 1 \end{pmatrix} \tag{8.4.20}$$

$$\psi_- : (t, s) \in \mathsf{R} \times S \mapsto \psi_-(\tau)[s] \in S \quad \text{where}$$

$$\psi_-(\tau)[\begin{pmatrix} a & b \\ c & d \end{pmatrix}] = \begin{pmatrix} a & b \\ c & d \end{pmatrix} \begin{pmatrix} 1 & 0 \\ \tau & 1 \end{pmatrix} \quad . \tag{8.4.21}$$

One verifies readily that these actions satisfy the hyperbolic relations:

$$\phi_t \circ \psi_+(\tau) \circ \phi_{-t} = \psi_\pm(e^{\pm t}\tau) \tag{8.4.22}$$

which are the continuous time analog – $t \in \mathsf{R}$ instead of $n \in \mathsf{Z}$ – of property (8.4.5) in Model 5, with here $\Lambda_\pm = e^{\pm 1}$.

(7) The compact configuration space $M = \Gamma \backslash \tilde{M}$. The final step in the construction of Model 6 is to notice that every two-dimensional Riemann manifold that is complete, connected and of constant negative curvature $\kappa = -1$ is – see, e.g. [Wolf, 1967] – of the form $\Gamma \backslash \tilde{M}$, i.e. is obtained from the Poincaré half-plane $\tilde{M}$ by identifying points that differ at most by an element of a subgroup Γ of isometries – i.e. $\Gamma \subset S$ – that satisfy two properties:

1. Γ acts freely on $\tilde{M}$, i.e. for every $z \in \tilde{M}$, $\gamma[z] = z$ for $\gamma \in \Gamma$ implies γ is the identity in S;
2. Γ acts properly discontinuously, i.e. for every $z \in \tilde{M}$ there exists a neighborhood U of z such that $\Gamma_U \equiv \{\gamma \in \Gamma \mid \gamma[U] \cap U \neq \emptyset\}$ is finite; this ensures that the orbit space is a Hausdorff topological space.

However, this allows still some spectral pathologies of the type exemplified by the Selberg trace formula [Heijhal, 1976] pertaining to $\Gamma_\mathsf{Z} = SL(2, \mathsf{Z})$ which satisfies both of the above properties; but, nevertheless, $\Gamma_\mathsf{Z} \backslash \tilde{M}$ is not compact, even though it is of finite volume. To obtain good ergodic behavior based on a natural invariant probability measure, we assume specifically that Γ is co-compact, i.e. that $\Gamma\backslash\tilde{M}$ is compact. Let it suffice to say here that there are many such subgroups $\Gamma \subset S$; their existence is known since a series of papers by Poincaré that occupy much of Tome II of his *Oeuvres*; we mention here only [Poincaré, 1882]; for modern presentations, see [Beardon, 1983, Iversen, 1992].

(8) The Anosov property on $T_1 M$. From $M = \Gamma\backslash\tilde{M} = \Gamma\backslash S/K$, and the construction of (8.4.18), we conclude that there is a bijection $T_1 M \longrightarrow \Gamma\backslash S$ that allows us to identify the unit tangent bundle $T_1 M$ of our compact manifold M – i.e. the energy surface of our model – on which the three one-parameter group actions ϕ and $\psi_\pm$ – see (8.4.19) and (8.4.20–21) – can be projected without further ado, since $M = \Gamma\backslash S$ is a *left*-quotient and these group actions act *from the right*. Note that the geodesic flow preserves the probability measure defined on $T_1 M$ by $d\mu = y^{-2}dxdyd\theta$. These facts establish the main property of the model $(T_1 M, \mu, \phi)$, namely that the geodesic flow on a compact manifold of constant negative curvature $\kappa = -1$ is a Anosov flow. It is also a K-system; see e.g. [Arnold and Avez, 1968][Cor.17.12].

Remark I:

1. The geodesic flow on compact manifolds of constant negative curvature is the first Hamiltonian system ever proven to be ergodic, and even strongly mixing [Hedlund, 1935, Hopf, 1939]; in fact, they satisfy the Anosov property, as we just saw. This model therefore establishes firmly that there are *no intrinsic incompatibility between a strong ergodic behavior – here the Anosov- and K- properties – and the Hamiltonian tenets of classical mechanics.*

2. Our presentation of the model was based on the elegant group-theoretical proof of ergodicity – actually even Lebesgue spectrum – given by [Gel'-fand and Fomin, 1952]; see also [Arnold and Avez, 1968].

3. The basic feature behind the behavior of the model – namely the hyperbolicity (8.4.22) – was already recognized by [Hadamard, 1898], apparently following an even earlier remark by Lobatchevski; it was emphasized, very early on, that the effect of negative curvature is to make this dynamical system extremely sensitive to small changes in its initial conditions since neighboring geodesics diverge *exponentially* from one another; [contrast this with the flat, or Euclidean case, where the divergence is only linear; or even worse, the positive curvature of an ordinary sphere where geodesics initiating at a pole meet again at the opposite pole].

4. The instability of the geodesics on a surface of *negative* curvature is reminiscent of the instability Boltzmann was looking for in his considerations on the effect of collisions in a gas, an idea which Poincaré took over, as we saw in Sect. 5.1. Following the version proposed by Krylov [1979], we sketch in Model 7 and Rem. J below some of the reasons that can be brought in support of this conjecture.

5. Although we presented the model here for *constant* curvature, much of its properties extend to variable curvature: only the negativity of the curvature is essential. Hyperbolicity of the time-like geodesic flow on Lorentz – rather than Riemann – manifolds has been explored in [Emch and Hong, 1987, Ellis and Tavakol, 1994].

6. While the geodesic flow on a surface of constant curvature appears as an over-restriction to free Hamiltonian flows, this reflects only that the underlying metric is very simple; in fact, the Hamilton–Jacobi method in classical mechanics shows the conceptual centrality of geodesic flows with respect to a suitable metric obtained canonically from the original Hamiltonian.

Model 7. Consider two circular disks of radius r, moving on a billiard table in the form of a flat torus with periodic boundary conditions.

Equivalently, we can regard this as the motion of a point particle with a fixed circular obstacle of radius $2r$. This obstacle acts as a tremendous amplifier of the effect of the initial condition: the point just squirting the obstacle, or hitting it, even marginally; a collision of the latter type is depicted in Fig. 8.5, with the additional mental picture – which we viewed first in [Arnold, 1978] – that we have indeed two toral billiards attached to one another along the circle delimiting the obstacle: at each collision, the representing point changes from one to the other torus.

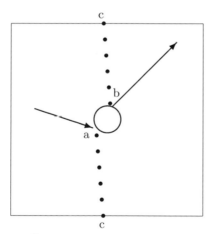

Fig. 8.5. T^2 billiard table with circular obstacle

In line with Rems. I.4–5 above, the connection is made with Model 6, by first smoothing-out this sharp edge: it becomes a surface with opposite sectional curvatures, the kind of model Hadamard had in mind. When the point is far from the "collar", it moves along a Euclidean straight-line, but the effect of the negative curvature at the smooth edge is to make neighboring trajectories diverge from one another all the more so when the curvature is made sharper. Upon taking the sharp limit of a collar of infinite curvature, one may hope to recover the original billiard problem. The proof that this is indeed the case is the spectacular result of [Sinai, 1963, 1970].

Remark J: It was at first believed that the generalization to an ideal gas of N hard spheres would be straightforward. After another twenty years of imaginative and coldly scrutinized work, it appeared that *"The announcement made in [Sinai, 1963] for the general situation must be regarded as premature"* as stated in [Sinai and Chernov, 1987], where the ergodic properties of systems of $N > 2$ hard discs *are* established. In this case, the 2–dimensional torus T^2 is replaced by the $2N$–dimensional torus $\mathsf{T}^2 \times \mathsf{T}^2 \times \cdots \times \mathsf{T}^2$; the removal of the circular obstacle of Fig. 8.5 is replaced by the removal of the union of the interiors of the $\frac{1}{2}N(N-1)$ cylinders given by

$$Q_{ij} \equiv \{(q^i{}_1 - q^j{}_1)^2 + (q^i{}_2 - q^j{}_2)^2 = (2r)^2 \bmod 1\}$$

with $1 \leq i < j \leq N$. This paper also indicates the adaptations that are necessary to cover systems of 3–dimensional hard balls.

For the modern developments, see e.g. [Szasz, 1996]; and for specific results on ergodicity and K-properties of billiard systems, see [Szasz, 1992, 1993] and references quoted therein.

9. Ergodicity vs. Integrability

Les hommes pleurent parce que les choses ne sont pas ce qu'elles devraient être.

[Camus, 1947]

9.1 Is Ergodicity Generic?

The description of ergodic theory – its main theorems, its hierarchy and some of its explicit models – given in Sect. 7.2 and Chap. 8 was intended to show that a solid mathematical edifice has been erected. Yet, one may be reminded of a whimsical remark Einstein once made, albeit about another area of mathematical physics:

> *For the moment, my feeling for this theory is like that of a starved ape, who after a long search, has found an amazing coconut, but cannot open it; so he doesn't even know whether there is anything inside.*
> [Cartan and Einstein, 1979].

In our case the search had been motivated by the need to provide a perspective deeper than that available to Boltzmann or Maxwell – Sect. 7.1 – towards the problem of the putative usefulness of the concepts of ergodic theory for the foundations of statistical mechanics. We saw already that *there do exist Hamiltonian systems that are ergodic*. In particular, Models 1 and 6 of Chap. 8 were already well assimilated by the end of the 1930s, and a wide domain of applications for the general theory was expected by the best authorities of the time, see e.g. [Wiener and Wintner, 1941, Birkhoff, 1942]. This optimism was reinforced by a theorem [Oxtoby and Ulam, 1941] to the effect that ergodicity *is* generic for a specified class of *topological* dynamical systems. The Reader is referred to Appendix D for the topological notions assumed in the account presented below.

To understand what is going on, let us start with the simplest case, namely where X is a n–simplex in R^N : $(n+1)$ points $\{a_k \in \mathsf{R}^N \mid 0 \leq k \leq n\}$ are chosen, subject to the condition that $\{\overline{a_o a_n}\}$ are n linearly independent vectors; we assume throughout that $n \geq 2$. By definition,

$$X = \{x = \sum_{k=0}^{n} \lambda_k a_k \mid \lambda_k \in \mathsf{R}, \, \lambda_k \geq 0, \, \sum_{k=0}^{n} \lambda_k = 1\} \quad .$$

The space X inherits a metric d and a measure μ from the Euclidean distance and the Lebesgue measure in $\mathsf{R}^n \subseteq \mathsf{R}^N$. Note that: (i) (X, d) is a complete metric space; (ii) X is compact; and (iii) the set X_o of its regular points – i.e. the set of points of X which admit a neighborhood that is homeomorphic to a sphere in R^n – is connected and dense in X; this is to say, by definition, that X is *regularly connected*. For every homeomorphism ϕ of the topological space X, we define a topological dynamical system with evolution given by $n \in \mathsf{Z} \mapsto \phi^n$. The fact that (X, d) is a compact metric space allows to define a metric $\tilde{d}$ on the set $H(X)$ of all the topological dynamical systems so defined:

$$\tilde{d}(\phi_1, \phi_2) = \max\{ \max_{x \in X} d(\phi_1[x], \phi_2[x]) \, , \, \max_{x \in X} d(\phi_1{}^{-1}[x], \phi_2{}^{-1}[x]) \} \quad .$$

The resulting metric space, denoted $H(X, \tilde{d})$, is complete. Moreover the composition of homeomorphisms $\phi \circ \psi$ and the inverse operation $\phi \mapsto \phi^{-1}$ are continuous in this metric, so that $H(X, \tilde{d})$ is a topological group. Finally, we introduce the additional requirement that the evolution preserves the measure μ inherited by X from the Lebesgue measure on R^n, and thus define the subgroup

$$H_\mu(X, \tilde{d}) = \{\phi \in H(X, \tilde{d}) \mid \mu \circ \phi = \mu\}$$

which is closed. This is the object of interest: the set of measure preserving homeomorphisms is a complete metric space. It is therefore a Baire space, and we know what it means for a property to be generic in this space; see Def. D.4.3. The following is then a very particular case of the result proven by [Oxtoby and Ulam, 1941].

Scholium 9.1.1. *For $n \geq 2$, metric transitivity is generic in $H_\mu(X, \tilde{d})$.*

Indeed, the construction we just presented extends naturally to more general dynamical systems. We can first construct $n-$dimensional Euclidean polyhedra by putting together collections of simplices satisfying simple consistency conditions. One needs, however to impose separately that the resulting object is compact; and that it is regularly connected, i.e. that the set of its regular points is connected and dense. The class of topological dynamical systems we defined above on simplices can then be extended to topological, measure preserving dynamical systems that are constructed similarly on Euclidean polyhedra; for every Euclidean polyhedron, the collection of these systems again forms a complete metric space. This still obtains at the next level of generalization [Oxtoby and Ulam, 1941].

In this paper the construction, sketched above for the simplest particular case, is presented in its genuine context, namely for a much wider class of measure-preserving topological dynamical systems which are defined on topological objects that are compact polyhedra in the sense of [Alexandroff and Hopf, 1935], equipped with a Stieljes–Lebesgue measure invariant under the evolution. *The result that metric transitivity is generic still holds* *under these general circumstances, provided the polyhedron is assumed to be regularly connected, and of dimension $n \geq 2$* – all conditions we saw to be evidently satisfied by the simplices we considered in our simplified presentation.

So far, so good: the apparently so reassuring achievement of [Oxtoby and Ulam, 1941] thus remained unchallenged for a generation. Yet, the classical texts on mechanics leave us with a paucity of ergodic systems, while they present us with an apparently endless variety of systems that are integrable, i.e. the opposite of ergodic. The general relevance of ergodic theory for Hamiltonian systems thus remains a conjecture. A benchmark in the classical physicist's view of the ergodic hypothesis is registered in Uhlenbeck's summarizing remarks at the conclusion of the 1966 IUPAP Conference on Statistical Mechanics:

> *You heard the great result of Sinai: the motion of N particles ($N \geq 3$) with short range repulsive forces (for instance hard spheres) is metrically transitive. When this is right (I certainly have not yet digested the proof!), and when it can be generalized to the case when also attractive forces are present, then I think one can say that the old problem of Boltzmann of explaining thermodynamics from the laws of mechanics has been finally solved.*

[Uhlenbeck, 1967]

The reactions to these remarks were mixed. *First of all,* everybody at the conference was in awe before the technical achievement, even though few of us had been able to go through the proof, as Uhlenbeck rightly hinted; see indeed Rem. J in Sect. 8.4. *Second,* somewhat vague doubts were voiced as to whether a result that is already valid when only a few particles were present could already incorporate the many-body structures perceived as inherent to statistical mechanics. *Third,* no mention seems to have been heard – neither in Uhlenbeck's lecture, nor in the conference hallways – against the tacit paradigm according to which a physical property should be robust against arbitrarily small perturbations ... unless there are paramount principles to the contrary: if ergodicity was to be a general physical property of statistical mechanical systems, surely it would survive the addition of small attractive forces, or even some softening of the hard core repulsive forces. In other words, it was widely expected that ergodicity, if it was to obtain at all, would be generic for Hamiltonian dynamical systems .. as it was proven to be for the topological dynamical systems of [Oxtoby and Ulam, 1941]; see the discussion of Scholium 9.1.1 above.

Our task is thus to incorporate the tenets of Hamiltonian dynamics explicitly into the framework of ergodic theory. From the onset, we should warn the Reader that the enthusiasm of the 1960s turned out to require severe dampening for reasons which we discuss in the next section.

We begin by stating the main no-go theorem [Markus and Meyer, 1974], and we proceed to explain its content; we review as we go the necessary auxiliary definitions that are needed beyond those given in Appendix D.

Theorem 9.1.1. *Let the configuration space* (M^n, g) *be a fixed* $n-$*dimensional compact Riemann manifold; and let* $(\mathcal{H}^g, \mathcal{C}^\infty)$ *be the topological space of all Hamiltonian flows on the phase-space* T^*M^n *with Hamiltonians of the form* $H^g = T^g + V$, *where* T^g *is the kinetic energy corresponding to the metric* g, *and the potential energy* V *is a smooth function* $V : M^n \to \mathsf{R}$. *Then the subset* $\mathcal{H}^g_E \subset \mathcal{H}^g$ *of such Hamiltonian systems that are ergodic is nowhere dense in* $\mathcal{H}^g$.

Following [Markus and Meyer, 1974] we assume that everything in sight is $\mathcal{C}^\infty$; in particular all functions to be considered are continuously differentiable to all orders. The mathematical formalism of Hamiltonian mechanics – along with the allied notions of differential and symplectic geometry, or symplectic manifold theory – is sketched in Sect. B.3. Recall that given a Hamiltonian $H : (p, q) \in T^*M^n \mapsto H(p, q) \in \mathsf{R}$, the dynamics is given by the Hamilton equations

$$\dot{p}_k = -\partial_{q^k} H \quad \text{and} \quad \dot{q}^k = \partial_{p_k} H \tag{9.1.1}$$

so that for any observable $f : (p, q) \in T^*M^n \mapsto f(p, q) \in \mathsf{R}$, we have, upon defining $f_t(p, q) = f(p(t), q(t))$:

$$\frac{d}{dt} f_t = \xi_H f_t \quad \text{with} \quad \xi_H = \sum_{k=1}^n \left[-\partial_{q^k} H \, \partial_{p_k} + \partial_{p_k} H \, \partial_{q^k} \right] \quad . \tag{9.1.2}$$

In [Markus and Meyer, 1974] a Hamiltonian H on a symplectic manifold M^{2n} – here $M^{2n} = T^*M^n$ – is said to be *ergodic* whenever, for a dense set of values $E \in \mathsf{R}$, the energy surface

$$\Omega^H{}_E = H^{-1}(E) = \{x \in M^{2n} \mid H(x) = E\} \tag{9.1.3}$$

satisfies the following two conditions:

(i)   if $\Omega^H{}_E \neq \emptyset$, then for every $x \in \Omega^H{}_E$: $dH|_x \neq 0$;   (9.1.4)

(ii)   the flow of ξ_H is metrically transitive on at least
one component $M^H{}_E$ of $\Omega^H{}_E$.   (9.1.5)

Condition (9.1.4) is referred to by saying that E is a *regular value* of H; when this condition is satisfied, $\Omega^H{}_E$ consists of disjoint components each of which is a differentiable manifold embedded in M^{2n}. When condition (9.1.4) is not satisfied – i.e. when *either* $\Omega^H{}_E = \emptyset$ *or* dH vanishes at some $x \in \Omega^H{}_E$, – we say that E is a *critical value* of H. Sard's theorem ensures that the set of critical values of our $\mathcal{C}^\infty$ Hamiltonian H is of measure zero; see e.g. [Sard, 1942], [Sternberg, 1983][Theorem II.3.1]; or [Abraham and Marsden, 1978][Thm 1.6.19] which refers to [Abraham and Robbin, 1967][Chap. III]. Hence the crux is in condition (8.1.5). As M^{2n} is a symplectic manifold, its symplectic form ω defines a measure $\mu \equiv \omega^n \equiv \omega \wedge \omega \wedge \cdots \wedge \omega$. Liouville's theorem – stated in case $M^{2n} = R^{2n}$ as Thm. 7.2.1 – ensures in general that μ is invariant under any Hamiltonian flow.

Moreover, this measure induces on $\Omega^H{}_E$ a measure $\mu^H{}_E$ to be visualized as the infinitely thin limit $\delta E \to 0$ of the restriction to the shell $\Omega^H{}_E \times \delta E$ of $\|\nabla H\|^{-1}\mu$, where ∇H is the gradient of H, namely the vector $(\partial_{p_1} H, \cdots, \partial_{p_n} H, \partial_{q_1} H, \cdots, \partial_{q_n} H)$. The Hamiltonian flow, restricted to any connected component $M^H{}_E$ of the energy surface $\Omega^H{}_E$, equipped with the invariant measure $\mu^H{}_E$, is a dynamical system, which Markus and Meyer [1974] define to be *metrically transitive* whenever for all measurable subsets A and B of $M^H{}_E$ with $\mu^H{}_E(A) > 0$ and $\mu^H{}_E(B) > 0$, there is some $t \in \mathsf{R}$ such that $\mu^H{}_E(\phi_t(A) \cap B) > 0$. Hence, upon comparison with Defs. 7.2.2 & 7.2.3, and Thm. 7.2.5, we see that the definition of ergodicity that is used by Markus and Meyer [1974] is general enough to cover our purposes.

The next step in our discussion of the meaning of the theorem, is to agree on a definition of the topology with which $\mathcal{H}^g$ is to be equipped; i.e. we still have to specify what is involved in viewing two of our Hamiltonian dynamical systems as reasonably close to one another. Markus and Meyer [1974] chose the C^∞ − *Whitney topology,* which expresses that two such systems are close whenever the difference $\xi_H - \xi_{H_o}$ between their Hamiltonian vector fields is small, as well as are its derivatives to all orders.

We first approach this notion in the simple case where the configuration manifold is flat, specifically $M^n \subseteq \mathsf{R}^n$, so that the phase-space $M^{2n} \equiv T^*M^n \subseteq \mathsf{R}^{2n}$. We denote by q the points of M^n and by $x = (p, q)$ the points of M^{2n}. We view the Hamiltonian vector field (9.1.2) as a function

$$\xi_H : x \in M^{2n} \mapsto \xi_H(x) \in \mathsf{R}^{2n} \quad . \tag{9.1.6}$$

In Appendix B, the derivative $d\xi_H|_x$ of the function ξ_H at x is defined as the best linear approximation of ξ_H around x; hence $d\xi_H|_x$ is an element of the linear space $L(\mathsf{R}^{2n}, \mathsf{R}^{2n})$. We introduce the notation

$$D\xi_H : x \in M^{2n} \mapsto D\xi_H(x) \equiv d\xi_H|_x \in L(\mathsf{R}^{2n}, \mathsf{R}^{2n}) \quad . \tag{9.1.7}$$

As a function on M^{2n}, $D\xi_H$ in turn admits a derivative and we define by recursion:

$$D^k\xi_H : x \in M^{2n} \mapsto D^k\xi_H(x) \equiv D(D^{k-1}\xi_H)(x) \in L^k(\mathsf{R}^{2n}, \mathsf{R}^{2n})$$
$$\text{with} \quad L^k(\mathsf{R}^{2n}, \mathsf{R}^{2n}) \equiv L(\mathsf{R}^{2n}, L^{k-1}(\mathsf{R}^{2n}, \mathsf{R}^{2n})) \tag{9.1.8}$$

For each $x \in M^{2n}$ separately, we define the norm of the linear operator

$$D^k\xi(x) : h \in \mathsf{R}^{2n} \mapsto D^k\xi_H(x)\,h \in L^k(\mathsf{R}^{2n}, \mathsf{R}^{2n}) \tag{9.1.9}$$

as usual by:

$$\|D^k\xi_H(x)\| = \sup_{h \in \mathsf{R}^{2n}, \|h\| \leq 1} \|D^k\xi_H(x)\,h\| \quad . \tag{9.1.10}$$

The neighborhoods of a Hamiltonian vector field ξ_{H_o} are indexed by finite collections of strictly positive functions $\{\varepsilon_k : M^{2n} \to \mathsf{R}^+ \mid k = 1, 2, \cdots, r\}$ with $r < \infty$. The neighborhoods are then defined by

$$N_{\varepsilon_1, \cdots, \varepsilon_r}(\xi_{H_o}) = \left.\begin{array}{l} \\ \{\xi_H \mid \forall x \in M^{2n} \text{ and } 1 \le k \le r : \|D^k[\xi_H - \xi_{H_o}](x)\| < \varepsilon_k(x)\} \end{array}\right\} . \quad (9.1.11)$$

The collection of all these neighborhoods, with r arbitrary, satisfies the axioms of Def. D.1.7, and therefore specifies a topology on $\mathcal{H}^g$: the $\mathcal{C}^\infty - Whitney$ *topology.*

For the particular needs of Thm. 9.1.1 above, we note that two simplifications appear, but that one technical extension is necessary.

The first simplification is that due to the separation of the Hamiltonian into a fixed kinetic energy and a smooth potential energy, $H - H_o = V - V_o$ so that the whole study can be reduced from the $2n$−dimensional phase-space $M^{2n} = T^*M^n$ to the n−dimensional configuration space M^n. The second simplification is that, since M^n is assumed to be compact, we can appeal to Thm. D.3.7 to introduce the sup-norms in (9.1.11):

$$\|D^k[\xi_H - \xi_{H_o}]\| \equiv \sup_{q \in M^n} \|D^k[\xi_{V - V_o}](q)\| \quad . \quad (9.1.12)$$

This has the advantage that one can now index the neighborhoods with finite collections of constants $\varepsilon_k > 0$ rather than with functions.

The only generalization that is necessary is that in case M^n is not flat, one should replace the ordinary derivative D in (9.1.7–12) by the covariant derivative canonically associated – see e.g. [Abraham and Marsden, 1978] – to the Riemann metric g; in the present context, this is a technicality the details of which are not likely to contribute to the basic understanding of the theorem; therefore, we do not discuss it here, but we subsume it in our keeping the notation D unchanged.

When all the above is done, the space $\mathcal{H}^g$ of the theorem becomes a complete metric space, with distance:

$$d(\xi_H, \xi_{H_o}) = \sum_{k=1}^{\infty} 2^{-k} \frac{\|D^k[\xi_H - \xi_{H_o}]\|}{1 + \|D^k[\xi_H - \xi_{H_o}]\|} \quad . \quad (9.1.13)$$

By the Baire–Hausdorff theorem – see Thm. D.4.3 – one obtains immediately that the space $\mathcal{H}^g$, equipped with the Whitney $\mathcal{C}^\infty$−topology just reviewed, is a Baire space, i.e. – see Def. D.4.2 – a space in which every meager set has empty interior; for some equivalent definitions, see also Thm. D.4.1. This is precisely what we need in order to discuss whether ergodicity is generic. Indeed, the theorem asserts that the subset $\mathcal{H}^g_E \subset \mathcal{H}^g$ of dynamical systems that are ergodic is nowhere dense; hence it is trivially the countable intersection of nowhere dense subsets, i.e. $\mathcal{H}^g_E$ *is a meager subset of the Baire space* $\mathcal{H}^g$. This is to say that ergodicity is *not* generic in the sense of Def. D.4.3.

Markus and Meyer [1974] also obtain a generalization of this result for Hamiltonian systems for which the Hamiltonians do not necessarily split into a sum of kinetic and potential energies, for which the phase-space is a compact $2n$−dimensional manifold that is not necessarily of the form T^*M^n. In these circumstances, the subset of ergodic systems is again meager, so that ergodicity is not generic in that framework as well.

In connection with the paradigm mentioned in our discussion of Uhlenbeck's remarks quoted earlier in this section, recall that every meager subset of a Baire space has empty interior; here this means that

- *if* a dynamical system in $\mathcal{H}^g$, with Hamiltonian H_o, is ergodic – i.e. the Hamiltonian flow corresponding to H_o belongs to $\mathcal{H}^g{}_E$
- *then* it is nevertheless impossible to find an open neighborhood $N(\xi_{H_o})$ of this system that would still be contained in $\mathcal{H}^g{}_E$, i.e. a neighborhood such that *all* corresponding Hamiltonian flows with $\xi_H \in N(\xi_{H_o})$ would also be ergodic.

Hence ergodicity is *not* the robust notion it was expected to be. If the paradigm is to be sustained – even in some weakened form – this requires an explanation. In a nutshell, the explanation is the persistence of KAM tori, a notion that we review in Sect. 9.4. Yet, we should not want to leave this section without a mention that *some things at least* are robust in the world of ergodic theory; for instance, Anosov dynamical systems – see (8.4.7) or models 5 and 6 in Sect. 8.4 – are structurally stable in a sense made precise by the Anosov Theorem: see [Arnold and Avez, 1968][Thm. 16.5 and App. 25].

9.2 Integrable Systems

Among the several surveys of the theory of integrable systems, [Arnold, 1978][Chap. 10 and App.8] may still be the best suited to a pedagogical mathematical introduction; presentations directed to the explanation of chaos can also be found in [Gutzwiller, 1990] and [Lichtenberg and Lieberman, 1992].

Throughout this section we continue to assume that everything in sight is smooth and is defined on a phase-space M^{2n}, i.e. a $2n$–dimensional differentiable manifold equipped with a symplectic form ω. By Darboux theorem B.3.1, around every $x \in M^{2n}$ there exist local coordinates $(p,q) = (p_1, \cdots, p_n, q^1, \cdots, q^n)$ such that $\omega = \sum_{k=1}^n \mathrm{d}p_k \wedge \mathrm{d}q^k$. The Hamilton equations (9.1.1) are supposed to be read in terms of these (local) coordinates. These are the $2n$ coupled differential equations that must be integrated simultaneously to give the flow generated by the Hamiltonian vector field ξ_H. This task would be much easier if only it were possible to decouple these equations into n pairs of independent equations, one for each degree of freedom. As this would require the existence of n independent constants of the motion, this separation is not possible in general. The special cases for which it *is* possible are collected under the name of completely integrable systems, which we now make mathematically precise.

Recall from Sect. B.3:

Definition 9.2.1. *Let (M^{2n}, ω) be a symplectic manifold. The Poisson bracket between two smooth functions f and g from M^{2n} to* R *is the function $\{f, g\} = \omega(\xi_g, \xi_f)$ where ξ_f (resp. ξ_g) is the Hamiltonian vector field defined by $\xi_f \rfloor \omega = df$ (resp. $\xi_g \rfloor \omega = dg$).*

We saw that if ω is written locally in its canonical form $\omega = \sum_{k=1}^{n} dp_k \wedge dq^k$, then ξ_f is given by (9.1.2) and

$$\{f, g\} = \sum_{k=1}^{n} [\partial_{q^k} f \cdot \partial_{p_k} g - \partial_{p_k} f \cdot \partial_{q^k} g] \quad . \tag{9.2.1}$$

The Poisson bracket enjoys the following properties:

1. it is skew-symmetric, i.e. $\{g, f\} = -\{f, g\}$;
2. it is linear, i.e. $\{a\ f + bg, h\} = a\{f, h\} + b\{g, h\}$;
3. it satisfies $\{fg, h\} = \{f, h\}g + f\{g, h\}$, i.e. at fixed h , the operation $f \to \{f, h\}$ is a *derivation*;
4. it satisfies the *Jacobi identity* i.e. $\{f, \{g, h\}\} + \{g, \{h, f\}\} + \{h, \{f, g\}\} = 0$.

Definition 9.2.2. *The Hamiltonian $H : M^{2n} \to$* R *is said to be completely integrable [or simply integrable] whenever there exist $I_k : M^{2n} \to$* R *, ($k = 1, \cdots, n$) such that:*

1. $\forall\, (p, q) \in M^{2n}\ :\ H(p, q) = \tilde{H}(I_1(p, q), \cdots, I_n(p, q)) \in$ R;
2. *for all $1 \leq j, k \leq n\ :\ [I_j, I_k] = 0$;*
3. *for all $x \in M^{2n}\ :\ \{dI_k|_x \mid 1 \leq k \leq n\}$ are linearly independent.*

Note that (1) and (2) imply:

$$\forall\, 1 \leq k \leq n\ :\ [H, I_k] = 0 \quad \text{and thus} \quad \frac{d}{dt}(I_k)_t = 0 \tag{9.2.2}$$

where the second implication follows from (9.2.1); each of the n functions I_k is thus required to be a *constant of the motion*. In geometrical terms, these conditions require that the Hamiltonian vector field ξ_H be tangent to every level set $I_k^{-1}(E_k)$, which is to say that the latter subsets are stable under the Hamiltonian flow corresponding to H , i.e. a trajectory of this flow, passing through any point of $I_k^{-1}(E_k)$, is entirely contained in that set.

Note moreover that the above geometrical remark applies as well to every pair $\{I_j, I_k\}$ by condition (2) of the Definition; the lattercondition is referred to by saying that the functions $\{I_k | 1 \leq k \leq n\}$ are *in involution*.

Condition (3) says in effect that, at every $x \in M^{2n}$ the vectors $\{\xi_{I_k}|_x \mid 1 \leq k \leq n\}$ are linearly independent. Together with the geometric interpretation of condition (2) given above, this implies that these vectors span the tangent plane at x to a $n-$dimensional smooth manifold. Relation (8.2.2) ensures that this manifold is stable under the flow of the Hamiltonian H .

The actual severity of condition (3) needs some commentary, summarized in the comparison between conditions (9.2.6) and (9.2.7) below.

Given

$$\boldsymbol{E} \equiv (E_1, E_2, \cdots, E_n) \in \mathsf{R}^n \tag{9.2.3}$$

and the smooth map

$$\boldsymbol{I} : x \in M^{2n} \mapsto (I_1(x), I_2(x), \cdots, I_n(x)) \in \mathsf{R}^n \quad, \tag{9.2.4}$$

we now review the consequences of conditions (2) and (3) in the Definition on the structure of the level set

$$\boldsymbol{I}^{-1}(\boldsymbol{E}) \equiv \{x \in M^{2n} \mid \forall 1 \leq k \leq n : I_k(x) = E_k\} \quad . \tag{9.2.5}$$

We want to compare condition (3) of Def. 9.2.2 to the result of a version of the Sard Theorem, slightly stronger than the version which we used in our discussion of Thm. 9.1.1; the references we gave then still cover the version we need now.

A point $x \in M^{2n}$ is said to be a critical point of $\boldsymbol{I}$ if the Jacobian matrix

$$J[\boldsymbol{I}(x)] \equiv (\partial_{x^\mu} I_k|_x)$$

has rank strictly smaller than n; and a point $\boldsymbol{E} \in \mathsf{R}^n$ is called a critical value of $\boldsymbol{I}$ if $\boldsymbol{I}^{-1}(\boldsymbol{E}) \subset M^{2n}$ contains at least one critical point of $\boldsymbol{I}$. Now, since we have decided to deal with $\mathcal{C}^\infty$–functions only, the stronger version of the Sard Theorem asserts that the set of all critical values of $\boldsymbol{I}$ is a set of measure zero in R^n. In other words:

for almost all $\boldsymbol{E} \in \mathsf{R}^n : \boldsymbol{I}^{-1}(\boldsymbol{E})$ contains no critical point x of $\boldsymbol{I}$. (9.2.6)

However, one has little control on how many critical points $x \in M^{2n}$ belong to the level set of an exceptional values $\boldsymbol{E} \in \mathsf{R}^n$. In particular, the Sard Theorem does *not* provide an answer to the question of whether the following three equivalent assertions hold true:

$$\text{for almost all } x \in M^{2n} : \begin{cases} (a)\, \text{rank } J[\boldsymbol{I}(x)] = n \\ (b)\, (dI_1 \wedge dI_2 \wedge \cdots \wedge dI_n)|_x \neq 0 \\ (c)\, \{dI_k \mid 1 \leq k \leq n\} \text{ are lin. indep.} \end{cases} \tag{9.2.7}$$

Hence, in spite of (9.2.6) the content of condition (3) in the Def. 9.2.2 of integrable systems is more than just replacing in (9.2.7) "almost all" by "all": (9.2.7) itself needs to be postulated. While it is possible – see [Abraham and Marsden, 1978][5.2.20-24] – to pursue the theory and impose only (9.2.7) as a precondition, the exposition becomes lighter if we impose the slightly stronger condition (3) in the Definition of completely integrable systems; we follow in this the pedagogical route traced in [Arnold, 1978].

We are now ready for the statement of the main result on integrable systems.

Theorem 9.2.1. *Let* (M^{2n}, ω) *be a symplectic manifold;* $H : M^{2n} \to \mathsf{R}$ *be an integrable Hamiltonian as in Def. 9.2.2; and* $\boldsymbol{I}^{-1}(\boldsymbol{E})$ *be a level set as in (9.2.5). Then* $\Omega^n \equiv \boldsymbol{I}^{-1}(\boldsymbol{E})$ *is a smooth* $n-dimensional$ *manifold, stable under the flow of* H. *Let* $\{\phi_t \mid t \in \mathsf{R}\}$ *denote the flow obtained as the restriction to* Ω^n *of the flow on* M^{2n} *determined by the Hamiltonian* H.

If moreover $\boldsymbol{I}^{-1}(\boldsymbol{E})$ *is compact, then there exists a diffeomorphism* Ψ : $M^n \to T^n$ *where* T^n *is a* $n-dimensional$ *torus*

$$T^n \equiv \{\boldsymbol{\varphi} = (\varphi_1, \cdots, \varphi_n) \mid 0 \leq \varphi_k < 2\pi\}$$

such that the flow on T^n, *defined by* $\psi_t = \Psi \circ \phi_t \circ \Psi^{-1}$ *takes the form*

$$\psi_t(\varphi_k) = \varphi_k(t) \equiv \varphi_k + 2\pi \, \nu_k \, t \quad \mathrm{mod} \, 2\pi$$

where $\{\nu_k \mid 1 \leq k \leq n\}$ *are non-negative numbers, depending only on* $\boldsymbol{E}$. *Finally, one can assume without loss of generality that the* n *constants of the motion* $\{I_k \mid 1 \leq k \leq n\}$ *are such that the symplectic form on* M^{2n} *can be written in the diagonal form*

$$\omega = \sum_{k=1}^{n} \mathrm{d}I_k \wedge \mathrm{d}\varphi_k \quad .$$

As a point of nomenclature, the coordinates φ_k are called *angle-variables*, and the functions I_k diagonalizing the symplectic form are called *action-variables*. Several classical examples are presented in [Arnold, 1978], with the comment:

> *In fact, the theorem of Liouville ... covers all the problems of dynamics which have been integrated to the present day.*

Incidentally, Arnold attaches the name of Liouville to Thm. 9.2.1 above, while almost everybody else attributes it to Arnold. A perusal through any of the classical texts on mechanics will convince the Reader that the explicit construction of the action-angle variables is a matter to be studied on a case-by-case basis. It may be no exaggeration to say that this is where classical mechanics becomes an art: the theorem asserts the *existence* of action-angle variables, under the assumptions that the system is integrable, and that the level-set $\boldsymbol{I}^{-1}(\boldsymbol{E})$ is compact. This lifts another question:

> *Are there any general methods to test for the integrability of a given Hamiltonian? The answer, for the moment, is no.*
>
> [Lichtenberg and Lieberman, 1992]

9.3 A Tale of Two Models: Approximation or Error?

These two models help situate the problem and suggest ways to proceed further. The genesis of the models unfolded in three phases. When Toda [1967] proposed the first model, the Toda lattice, he exhibited soliton solutions with the help of an impressive familiarity with elliptic functions.

Yet, the close resemblance [Toda, 1974] of the first model with the second model, a known chaotic system, the Hénon–Heiles system [Hénon and Heiles, 1964], made the Toda lattice appear to be a good candidate for chaotic behavior. The support for this candidacy was then undermined by numerical simulation [Ford, Stoddard, and Turner, 1973]. Soon thereafter, analytical proofs that this system is integrable appeared [Flaschka, 1974, Hénon, 1974].

We now review the Toda model with only the details that are necessary to appreciate the above introductory remarks. We focus on the following version of the model: 3 particles are moving on a ring of radius L, and interact via an exponentially decreasing 2–body potential; the Hamiltonian is:

$$H_\lambda = T + \lambda V \quad \text{with} \quad \begin{cases} T = \frac{1}{2} \sum_{k=1}^{3} p_k{}^2 \\ V = \sum_{k=1}^{3} \exp[-(q_{k+1} - q_k)] \end{cases} \tag{9.3.1}$$

where the periodic boundary condition $q_{3+1} = q_1$ is imposed. The coupling constant $\lambda > 0$ is an arbitrary positive number. Note that the model is usually presented for the case $\lambda = 1$, and the summation in (9.3.1) running over Z_N with the number of particles N fixed but arbitrary; thus with periodic boundary condition $q_{N+1} = q_1$. The above version is general enough to encompass the characteristic features of the original model, and simple enough to make our account elementary throughout.

The change of variables

$$\begin{cases} P_1 = p_1 \\ P_2 = p_2 \\ P = p_1 + p_2 + p_3 \end{cases} \quad \left. \begin{array}{l} Q_1 = q_1 - q_3 \\ Q_2 = q_2 - q_3 \\ Q = q_3 \end{array} \right\} \quad . \tag{9.3.2}$$

is *canonical,* i.e. it keeps the symplectic form in diagonal form:

$$\sum_{k=1}^{3} dP_k \wedge dQ_k = \omega = \sum_{k=1}^{3} dp_k \wedge dq_k \quad . \tag{9.3.3}$$

This choice of variables is motivated by the fact that the potential V in (9.3.1) depends only on the relative position variables Q_1 and Q_2 introduced in (9.3.2):

$$V = \exp[-(Q_2 - Q_1)] + \exp[Q_2] + \exp[-Q_1] \quad . \tag{9.3.4}$$

Since V, and thus H, do not depend on Q, we have that P – which is the canonical variable conjugate to Q – is a constant of the motion:

$$P(t) = P(0) \quad \text{and} \quad Q(t) = P(0) t + Q(0) \mod(2\pi L) \quad , \tag{9.3.5}$$

i.e. $\{P, Q\}$ is a pair of action-angle variables. This reflects the invariance of the model under any global rotation $\{q_k \to q_k + \theta \mid k = 1, 2, 3\}$.

The *first lesson* is thus: when looking for constants of the motion, it is a good strategy to look first for continuous one-parameter groups of symmetries of the Hamiltonian system under consideration; indeed the generator(s) – here the momentum – of a continuous group – here the translations in configuration space – of symmetries of the Hamiltonian are constants of the motion. When rephrased with the proper mathematical care, this is known as the Noether theorem, see e.g. [Abraham and Marsden, 1978][Cor.4.2.14]; for the historical and physical background, see [Kastrup, 1987].

Upon interpreting thus the variables Q_1 and Q_2 as rotating coordinates, we do not loose any generality if we impose $P = 0$ in (9.3.1). Together with (9.3.4), this reduces the problem to studying the $2-$degree of freedom system described by the Hamiltonian $\hat{H}_\lambda = \hat{T} + \lambda \hat{V}$ with:

$$\left. \begin{array}{l} \hat{T} = \frac{1}{2}\left[{P_1}^2 + {P_2}^2 + (P_1 + P_2)^2\right] \\[2ex] \hat{V} = \exp[-(Q_2 - Q_1)] + \exp[Q_2] + \exp[-Q_1] \end{array} \right\} . \tag{9.3.6}$$

It is quite a remarkable fact that

$$\left. \begin{array}{l} \hat{I}_\lambda \equiv (P_1 + P_2)P_1 P_2 + \\[2ex] \lambda\{-(P_1 + P_2)\exp[-(Q_2 - Q_1)] + P_1 \exp[Q_2] + P_2 \exp[-Q_1]\} \end{array} \right\} \tag{9.3.7}$$

is such that the Poisson bracket $[\hat{H}_\lambda, \hat{I}_\lambda]$ vanishes, which is to say that $\hat{I}_\lambda$ is a constant of the motion for the system (9.3.6); since $\hat{I}_\lambda$ and $\hat{H}_\lambda$ are independent, this dynamical system governed by the Hamiltonian $\hat{H}_\lambda$ is indeed integrable.

The *second lesson* we learn from the model is that, as the Hamiltonian (9.3.6) has no obvious symmetry, Noether's theorem could be of no help in determining whether this $2-$degree of freedom system has a constant of the motion, besides $\hat{H}_\lambda$ itself. The complexity of (9.3.7) illustrates our earlier contention – backed by a quote from [Lichtenberg and Lieberman, 1992] – that identifying constants of the motion of a system, and/or determining whether a specific Hamiltonian is completely integrable, are matters where calling upon one's intuition, ingenuity and/or perseverance might be necessary! In the present case, the guessing is helped by the following elementary algebraic trick, a version of which was noticed by Lax [1968].

Scholium 9.3.1. *If two differentiable functions*

$$L : t \in \mathsf{R} \mapsto L(t) \in M(n, \mathsf{R}) \quad and \quad M : t \in \mathsf{R} \mapsto M(t) \in M(n, \mathsf{R})$$

satisfy the commutation relations

$$\forall t \in \mathsf{R} : \quad \frac{\mathrm{d}L}{\mathrm{d}t}(t) = M(t)L(t) - L(t)M(t) \quad ,$$

then

$$\forall\, k \in \mathsf{Z}^{+} \quad and \quad \forall\, t \in \mathsf{R} : \quad \frac{d\left[TrL^{k}\right]}{dt}(t) = 0 \quad .$$

Proof: One first check by induction over k that the time-derivative of the $k-$th power of L is given by

$$\frac{dL^{k}}{dt} = ML^{k} - L^{k}M \,.$$

Since Tr is a linear operation,

$$\frac{dTrL^{k}}{dt} = Tr\frac{dL^{k}}{dt}.$$

The scholium then follows immediately from these two relations, and the fact that for any pair of matrices A and B :

$$Tr(AB - BA) = 0\,.$$

Note that the conditions

$$\forall\, t \in \mathsf{R} : \quad L(t)\,\mathrm{symmetric}\ \ and\ \ M(t)\,\mathrm{antisymmetric}$$

are compatible with the commutation relation assumed in the Scholium.

When the matrices L and M can be written as functions defined on the phase-space of a dynamical system, so that their $t-$dependence can be linked to the evolution of this dynamical system, the pair $\{L, M\}$ is said to be a *Lax pair* for this system; the Scholium gives then a way to construct (at least some of) its constants of the motion, namely the TrL^{k} which are now real-valued functions on the phase-space of the system. The Scholium thus reduces the guesswork to finding whether and how a Lax pair can be constructed for the problem at hand.

A straightforward computation verifies indeed that the following time-dependent matrices constitute a Lax pair for the Hamiltonian (9.3.1).

$$L = \begin{pmatrix} [P - (P_1 + P_2)] & \alpha & \beta \\ \alpha & P_1 & \gamma \\ \beta & \gamma & P_2 \end{pmatrix} \quad and \quad M = \frac{1}{2}\begin{pmatrix} 0 & \alpha & -\beta \\ -\alpha & 0 & \gamma \\ \beta & -\gamma & 0 \end{pmatrix} \tag{9.3.8}$$

with

$$\alpha = \mu\,\exp{-\tfrac{1}{2}Q_1} $$

$$\beta = \mu\,\exp{\ \tfrac{1}{2}Q_2} \tag{9.3.9}$$

$$\gamma = \mu\,\exp{-\tfrac{1}{2}(Q_2 - Q_1)}$$

where $\mu^2 = \lambda$ [the coupling constant in (9.3.1)], and $(P_1, P_2, P; Q_1, Q_2)$ are the dynamical variables defined in (9.3.2).

One further verifies that the first two constants of the motion are indeed recovered from the conclusion of the Scholium with $k = 1, 2$:

$$TrL = P \quad \text{and} \quad \frac{1}{2} TrL^2 = H \quad . \tag{9.3.10}$$

A final, and slightly longer computation, delivers the prize, namely the third constant of the motion for the Hamiltonian (9.3.1); expressed in the variables (9.3.2), this constant is:

$$\left. \begin{array}{l} \mathrm{Tr} L^3 = A + 3\lambda\, B + 6\lambda^{\frac{3}{2}} \quad \text{with} \\[2mm] A = P_1{}^3 + P_2{}^3 + [P - (P_1 + P_2)]^3 \\[2mm] B = [(P_1 + P_2)\exp-(Q_2 - Q_1) + \\ \quad (P - P_1)\exp Q_2 + (P - P_2)\exp-Q_1] \end{array} \right\} . \tag{9.3.11}$$

Remark A:

1. As (9.3.11) is clearly independent of P and H, the Hamiltonian system (9.3.1) – which has 3 degrees of freedom – is completely integrable.
2. The whole discussion above – including the use of Scholium 9.3.1, and a straightforward generalization of the construction (9.3.8–9) – extends from $N = 3$ in (9.3.1) to any arbitrary positive integer $N \geq 3$ and leads to the conclusion that the general Toda lattice is also completely integrable.
3. When restricted to the level set $P = 0$, (9.3.11) gives the constant of the motion for the reduced Hamiltonian (9.3.6); up to irrelevant numerical constants, this reduced constant of the motion coïncides with (9.3.7), namely:

$$I_\lambda = -\frac{1}{3}\,[\,TrL^3|_{P=0} - 6\lambda^{\frac{3}{2}}\,] \quad . \tag{9.3.12}$$

While we presented (9.3.7) as a guess that seemed to come straight out from the wilderness, we see now that it can be derived from Scholium 9.3.1; this derivation however still involves a guess – namely (9.3.8) – which should appear somewhat less bewildering.

This little incursion – started with Scholium 9.3.1 – was taken as a suggestion that the above *second lesson* may be mollified. While there are indeed no straightforward recipe to find out whether a system is integrable, some methods are nevertheless available to guide the search: following in the pioneering steps of [Lax, 1968], variants of Scholium 9.3.1 have been used to exhibit the integrability of several systems governed by non-linear equations of the motion.

The *third lesson* to be drawn from the model is that numerical exploration of the trajectories of a dynamical system – one speaks of 'drawing its phase portrait'; see the discussion at the end of Sect. 9.4 – can be useful as an indication of whether this system is integrable. Historically, it was a decisive step in the analysis of the present model: the phase portrait of the Toda lattice is smooth at all energies.

There is moreover a fourth lesson, which derives from the connection with the Hénon–Heiles system. To make this explicit, consider the further change of variables

$$
\left\{
\begin{array}{ll}
\pi_x = 2\sqrt{3}\,(P_1 + P_2) & x = \frac{1}{4\sqrt{3}}\,(Q_1 + Q_2) \\[2mm]
\pi_y = 2\phantom{\sqrt{3}}\,(P_2 - P_1) & y = \frac{1}{4}\,(Q_2 - Q_1)
\end{array}
\right\}. \tag{9.3.13}
$$

Again this is canonical – i.e. $\mathrm{d}\pi_x \wedge \mathrm{d}x + \mathrm{d}\pi_y \mathrm{d}y = \mathrm{d}P_1 \wedge \mathrm{d}Q_1 + \mathrm{d}P_2 \wedge \mathrm{d}Q_2$ – and it brings the reduced Hamiltonian (9.3.6) to the form

$$
\left.
\begin{array}{l}
\hat{H}_\lambda = \frac{1}{8}\left[\frac{1}{2}[\pi_x{}^2 + \pi_y{}^2]\right. \\[3mm]
\left. \phantom{\hat{H}_\lambda =} + 8\,\lambda\{\exp[-4y] + \exp[2(y + \sqrt{3}x)] + \exp[2(y - \sqrt{3}x)]\}\right]
\end{array}
\right\}. \tag{9.3.14}
$$

It is tempting to expand the potential part of the Hamiltonian in some truncated power series, and try the approximation $e^z \approx 1 + z + \frac{1}{2!}z^2 + \frac{1}{3!}z^3$. Together with the substitution

$$
\pi_k \to p_k = (8\sqrt{3\lambda})^{-1}\,\pi_k \quad \text{with} \quad x, y \quad \text{unchanged} \tag{9.3.15}
$$

one obtains the approximation:

$$
\frac{1}{24\lambda}\hat{H}_\lambda \approx \tilde{H} \equiv \frac{1}{2}\,[p_x{}^2 + p_y{}^2] + \frac{1}{2}\,[x^2 + y^2] + x^2 y - \frac{1}{3}y^3 + const. \tag{9.3.16}
$$

where $\tilde{H}$ is the Hamiltonian of our second model, the Hénon–Heiles system. Note that the factor $[24\lambda]^{-1}$ can be absorbed in a change of the time-scale, and does not affect the integrability.

Numerical computations – see the discussion at the end of Sect. 9.4 below – have shown that the non-linear model (9.3.16) is non-integrable.

As an approximation of the reduced Hamiltonian (9.3.6), the seemingly mild manipulation leading to (9.3.16) from (9.3.14) – the latter being only a canonical transformation of (9.3.6) – turned out to be quite misleading; not so much because (9.3.15) is not canonical – this can be fixed – but for an essential reason: the Hamiltonian (9.3.6) *is* integrable, whereas (9.3.16) *is not* integrable.

The *fourth lesson* is now this: while the integrable Hamiltonian $\hat{H}_o = \hat{T}$ admits a family $\{\hat{H}_\lambda \mid \lambda > 0\}$ of perturbations each of which is integrable, there exist, close to these, other perturbations – see (9.3.16) – that are not integrable. This study of the Toda model suggests therefore that integrability is not a robust notion.

We now return to the general theory. In connection with this fourth lesson drawn from the Toda lattice model, we now know from [Markus and Meyer,

1974] that – very much like ergodicity (see Sect. 9.1 above) – *integrability is not a generic property of Hamiltonian systems:* they form a meager subset, hence no integrable system possesses any neighborhood such that every Hamiltonian system in that neighborhood is also integrable; in other words, there always exist arbitrarily small perturbations that destroy integrability. That being so, we now turn to the question of how the study of integrability could help us delineate better the role of ergodicity in statistical mechanics.

9.4 The KAM Tori

One of the best ways to come to grip with the subject of this section is to go straight to the basic meta-theorem stated in [Arnold, 1978][App. 8, p. 405], which we go over immediately below. After a brief mention of some benchmark papers, we begin our study of this theorem and go over the principal definitions. We finally discuss its connection with the ergodic problem.

Theorem 9.4.1. *If an unperturbed system is non-degenerate, then for sufficiently small conservative Hamiltonian perturbations, most non-resonant invariant tori do not vanish, but are only slightly deformed, so that in the phase-space of the perturbed system, too, there are invariant tori densely filled with phase curves winding around them conditionally-periodically, with a number of independent frequencies equal to the number of degrees of freedom. These invariant tori form a majority in the sense that the measure of the complement of their union is small when the perturbation is small.*

The reason this is a meta-theorem is that it can only be proven under some additional technical assumptions. The theorem was stated by Kolmogorov [1954]; see also [Kolmogorov, 1957]. It was proven under analyticity assumptions by Arnold [1963]; these analyticity requirements were eased by Moser [1962] to some C^r−differentiability (with r quite large, however).

First of all, we should hasten to say that, although the theory admits generalizations to some wider class of dynamical systems, it dealt initially with Hamiltonian dynamics; we limit our discussion to the latter.

Second, the "unperturbed system" of the theorem is assumed to have n degrees of freedom and to be integrable, so that the unperturbed Hamiltonian H_o is a function of n independent action variables, the constants of the motion $\{I_k \mid 1 \leq k \leq n\}$; these are in involution; see condition 2 in Def. 9.2.2. The joint level sets of these action variables are n−dimensional manifolds, invariant under the time-evolution governed by H_o; when compact, these are tori T^n. The theorem is about the persistence of such tori under small perturbations, i.e. when the Hamiltonian H_o is replaced by a Hamiltonian H such that $V \equiv H - H_o$ can be viewed as small. As mentioned at the end of Sect. 9.3, we know that integrability is not a generic property of Hamiltonian systems; hence it would not be reasonable to hope that H be still integrable.

The claim thus is that the departed integrability leaves behind a ghost in the form of deformed tori.

Third, the tori that are "preserved" are special tori, namely the tori that are "non-resonant", a concept which we now review. Recall from Sect. 9.2 that each action variable I_k is the conjugate of an angle variable φ_k and defines a function $\omega_k \equiv \dot{\varphi}_k = \partial_{I_k} H$ which is constant when restricted to any of the level sets T^n. The torus T^n is said to be *non-resonant* when the n numbers $\{\omega_k \mid 1 \leq k \leq n\}$ are *arithmetically independent,* i.e. when

$$\left.\begin{array}{l} \sum_{k=1}^n a_k\omega_k \neq 0 \\[2ex] \forall \ \ \{a_k \mid a_k \in \mathsf{Q}\,,\, 1 \leq k \leq n\,,\, a_k \neq 0 \text{ for at least one } k\} \end{array}\right\} \qquad (9.4.1)$$

where Q denotes the set of all rational numbers. Note that we can replace in the above definition "rational" by "integer" without affecting the definition itself.

The name "non-resonant" comes from the early Laplacian stage of perturbation theory, which Laplace pioneered in his study of the perturbations of the Keplerian orbits of two heavy planets due to their mutual gravitational interaction. Laplace noticed that integer combinations of the frequencies of these orbits appeared in the *denominators* of summands in his expansions. This creates a *"small divisor problem"*: if there are integers a_1 and a_2 such that $a_1\omega_1 + a_2\omega_2$ is very small, the corresponding terms in the expansion blows up: the two orbits are close to resonance. This takes from the phenomena which were first perceived in Pythagoras' musical harmonics, later explained by Galileo in terms of arithmetic combination of frequencies, and to this day rest at the base of the theory of sound production in musical instruments. The Tacoma Narrows bridge provides a more dramatic illustration of the effect of non-linearity in (sub-)harmonic resonant excitations; for an update, see [McKenna, 1999].

Back to astronomy, note with Laplace that the ratio of the Jupiter year and the Saturn year is (11.86 Earth yr/ 29.46 Earth yr), i.e. 0.4026 which is indeed very close to the rational number 2/5. Hence a close resonance appears early in the perturbation expansion, making it unreliable over times of the order (or larger) than a thousand Earth years ... which, however, is very short as celestial things go! This initial impetus of Laplace, cultivated in the course of the 19th century, blossomed into an epoch-making *œuvre* [Poincaré, 1892–1907] which developed, among other things, the theory of perturbations termed secular (= *"séculaires" ≠ séculières.*) As [Poincaré, 1890, 1892–1907] are hard reading indeed, the Reader may want to approach it from Poincaré's own didactic version [Poincaré, 1905–1910]. For a recent historical account of Poincaré's life-long preoccupation with the three-body problem – of which the above is one particular aspect – the Reader may want to consult [Barrow-Green, 1997], an informative review of which can be found in [Diacu, 1999]; see also [Diacu and Holmes, 1996, Goroff, 1993]. These authors emphasize that Poincaré's corpus provides a historical glimpse into the compatibility of mechanics and chaotic behavior, and a first step towards the KAM theorem reported above as Thm. 9.4.1.

The specific attention to the small divisor problem is emphasized in the title of [Poincaré, 1908b]. Any mechanical theory – not just celestial mechanics – which is primarily concerned with the long-term behavior of the systems it wants to study must investigate the presence and persistence of (non-)resonant tori.

To grasp some of the significance of these tori, recall the 2–dimensional torus encountered earlier in our discussions of the Weyl sequences modulo 1 – see Def. 5.2.3 – and of the Boltzmann pair of harmonic oscillators – see Model 1 in Sect. 8.1: when the ratio of the two characteristic frequencies is rational, close orbits do appear, whereas when this ratio is an irrational number, the trajectories are everywhere dense. This is easily seen to carry over to n–dimensional tori: the trajectories are dense iff the n frequencies ω_k are arithmetically independent, i.e. iff the tori are non-resonant. This is in essence what is meant by *"conditionally-periodic phase curves"* in the theorem; the terms *"quasi-periodic"* and *"almost periodic"* are also used. Recall in this context Poincaré's recurrence theorem reported here as Thm. 7.2.2.

Remember now that the rational numbers form a meager set in R, whereas the irrationals are generic. In the present context this transposes to the following central fact: the Lebesgue measure of the union of all invariant resonant tori is equal to zero. This evidently does not deny that the resonant tori form a dense subset; it only asserts that the probability is zero that an invariant torus picked "at random" is resonant. By focusing on non-resonant invariant tori, the theorem disregards a set of measure zero and concentrate instead on a generic feature of integrable Hamiltonian systems.

Finally, an integrable Hamiltonian H_o is said to be *non-degenerate* if the Jacobian

$$det\,[\partial_{I_k}\omega_j] \equiv det\,[\partial_{I_j}\,\partial_{I_k}\,H_o] \neq 0 \quad , \tag{9.4.2}$$

which ensures the functional independence of the frequencies ω_k across the phase-space. Recall indeed that, since H_o is integrable, it is a function only of the n constants of the motion $\{I_k \mid k = 1, \cdots, n\}$, and that, by definition, $\omega_j \equiv \partial_{I_j} H_o$. Hence, the $\{\omega_j \mid j = 1, \cdots, n\}$ are functions of the $\{I_k \mid k = 1, \cdots, n\}$. Hence, again, the non-degeneracy condition (9.4.2) allows – via the inverse function theorem, see Thm. B.1.1 – to view the I_k as functions of the ω_j, i.e. it allows to label (locally at least) the non-resonant tori with the help of their characteristic frequencies.

All the terms appearing in Thm. 9.4.1 have now been reviewed, and we can turn to its conclusions.

We noted at the end of the previous section that since the condition of integrability is sharp, it must either be satisfied or fail entirely; and that, in fact, it does fail in most Hamiltonian systems, because it is not a generic property of Hamiltonian systems.

What we learn from Thm. 9.4.1 is that, when a Hamiltonian system is modified by a perturbation that destroys its integrability, a remnant is nevertheless left behind: the invariant non-resonant tori – defined by the constants of the motion of the unperturbed system – do not disappear, but are instead only deformed as invariant subsets under the new evolution; moreover, while the invariant non-resonant tori of the unperturbed system covered the whole phase-space – up to a set of measure zero – the deformed subsets resulting from the perturbation are invariant under the perturbed evolution; and they still cover a large part of the phase-space, although they may now miss a set of strictly positive measure.

The theorem holds around any integrable Hamiltonian system, independently of its number of degrees of freedom, provided this number be finite. Following Poincaré's investigations half a century earlier, the KAM theorem is one of the first new results in the general theory of non-linear – and thus not necessarily integrable – Hamiltonian systems; both its strengths and its limitations come from the fact that it is a statement about Hamiltonian systems that are close to being integrable; these are loosely referred to as *nearly-integrable* systems. In that sense, ergodic Hamiltonian systems, such as Sinai's hard spheres or his billiards – see end of Sect. 8.4 – are not covered by the umbrella of Thm. 9.4.1: these systems are not to be regarded as being a *"sufficiently small"* perturbation away from some integrable system.

Theoremm 9.4.1 is a *qualitative result:* it is silent on the *quantitative* aspects of the perturbation. Among the questions that the theorem leaves open are:

1. How small is the part of the phase-space that is not covered?
2. How small are the perturbations admitted as "sufficiently small"?
3. How do the answers to the above two questions depend on the number of degrees of freedom?
4. How non-linear should the system be for it to be non-integrable? The Toda model (9.3.1) is badly non-linear – *exponential* interactions – and yet the model is integrable, whereas the Hénon model (9.3.16) – with only *cubic* terms in the interaction – is already not integrable (at least above a known critical total energy).

Nevertheless, the theorem became a new paradigm, invoked in diverse areas of dynamics, from the early motivation in the study of the stability of the solar system to stellar astrophysics [Buchler, Perdang, and Spiegel, 1985], the design of stable high-energy particle accelerators [Dragt, 1985], and the foundations of statistical mechanics; we evidently limit our review to the last of these areas.

Very early in the theory of non-linear systems, it was realized that the traditional perturbation theory was not the most immediate way to approach the above problems: computer simulations are a more effective guide.

Perhaps the first such attempt was [Fermi, Pasta, and Ulam, 1940], indicating that the equipartition of the energy between the normal modes of the unperturbed system may fail, and thus casting doubts on the ergodicity of the perturbed system they considered, namely an assembly of weakly non-linearly coupled oscillators. We shortly come back to this line of investigation.

In our discussion of the Toda model – Sect. 9.3 – we mentioned the pioneering role of the non-integrable system of [Hénon and Heiles, 1964]; see also [Hénon, 1974, 1983]. More non-linear Hamiltonian systems, with a small number of degrees of freedom $(n = 2, 3)$, are also discussed in [Ford, 1970] on the basis of computer calculations, and the results are presented with an interpretation in terms of the KAM theory, i.e. Thm. 9.4.1 above. Note the venue of this paper, which also provides some of the – then recent – references to earlier works.

To outline the strategy followed in these investigations, consider first the case $n = 2$. The phase-space is 4–dimensional, and thus the energy hypersurfaces are 3–dimensional. To visualize what is happening on any of these, one chooses a 2–dimensional plane that intersects transversally the Hamiltonian flow, and one lets the computer plot the intersections of a single trajectory with this plane. When the trajectory moves on an invariant non-resonant torus, long-run computations produce a collection of dots nicely arranged along what appears as a continuous curve. The early investigations [Hénon and Heiles, 1964, Ford, 1970] already showed that there exists a threshold in the values of the energy chosen to define the energy hypersurface; below this threshold the trajectories showed the pattern just described. Going above this threshold, however, revealed an interesting phenomenon: the intersections with the plane consist of a set of randomly scattered points; moreover, the set so produced by a single trajectory seemed to touch on 'every' open subset of the complement of the union of the non-resonant tori. Ford considered also Hamiltonian models with 3 degrees of freedom; as another constant of the motion – besides the energy – happens to be present for these models, the study of their trajectories can use the same strategy as for systems with 2 degrees of freedom, although their dynamics is somewhat richer.

It is important for us here that the models – with $n = 2$ or 3 – considered by Ford [1970] have adjustable coupling constant(s):

> [as the coupling constant(s)] are increased, or as the energy is increased, the stochastic zone increases in size until it almost completely covers the allowed region of the plane ... the stochastic irreversibility discussed here is an inherent property of the mechanical equations of motion ... no additional assumptions of a non-mechanical nature are needed. Thus one has here that beginning understanding of the ultimate source of irreversibility which contributes significantly to statistical mechanics
>
> [Ford, 1970]

This quote was inserted here for two reasons. While one of these is immediately and specifically germane to our current discussion, the other reason involves more general issues of interpretation that may illuminate the course of the investigations that were carried out. We thus examine the background reason first, and do so with the corroborative evidence in [Ford and Lunsford, 1970]; see also [Lunsford and Ford, 1972].

In the vocabulary of Ford and his School, the word *"stochastic"* is used when they see orbits intersecting the reference plane in a manner that appears so random that they feel justified to assert that the orbit covers densely a large portion of the energy surface, a region they deem to be significantly much larger that the portion available if the system were completely integrable. It is also often used when they observe that the distance between two initially very close orbits increases in a manner that is compatible with an exponential growth; this is the experimental manifestation of the *exponential sensitivity to initial conditions* which we discussed earlier, the last time in Rems. I.3–4 in Sect. 8.4. Taken separately, together or in some combination with others, these two phenomena were later generally accepted as the recognizable signature of *chaos*; see e.g. [Devaney, 1989, Holmes, 1989] and also [Eckmann and Ruelle, 1985].

As for the *irreversibility* of statistical mechanics, the one present in transport phenomena such as diffusion or electric resistance, one must admit that the use of the word, at the time at least, was mostly an expression of wishful thinking that went beyond the recognized fact that both of the above phenomena make it illusory to try and trace the initial conditions from the knowledge of the position of an orbit at some distant time in the future or in the past: information is irretrievably lost in the course of the evolution. The theoretical compatibility of the three notions of "stochastic" behavior, of "sensitivity to the initial conditions", and of some "irreversibility" with *conservative laws of motion* is part and parcel of ergodic theory; see Sect. 8.2 (K systems); Sect. 8.3 (Dynamical entropy); and Sect. 8.4 (Anosov property).

The computer investigations were informed by these theoretical constructs, and they do cite indeed the relevant literature. One of the great merits of machine computations of the type we just discussed was to bring up all three of these notions to some level of experimental test, and thus serve as a stimulating guide towards future theories.

The second, but primary, reason for our inserting the above quote from [Ford, 1970] is that the quantitative results reported suggested a smooth transition from integrability, through 'near-integrability', to ergodicity. While these machine calculations could not provide general proofs, it was anticipated that the causes of the phenomenon, which form the skeleton of the KAM theory, could be generic. In particular, arguing from the fact that the habitual abode of statistical mechanics lies within the study of systems with a large number of degrees of freedom, it is of genuine physical interest to know whether the early results just discussed would extend from $n = 2, 3$ to large n, typically $n \approx 10^{23}$ (give-or-take a few orders of magnitude).

This question was approached numerically in [Györgyi, Ling, and Schmidt, 1989, Falcioni, Marconi, and Vulpiani, 1991, Hurd, Grebogi, and Ott, 1994], leading to the conclusion that, for the systems studied, the region of observed stochasticity becomes predominant as the number, n, of degrees of freedom becomes large; specifically that the fraction of the volume in phase-space occupied by the invariant KAM tori *decreases exponentially* as n increases.

To a devotee of ergodic theory, this should sound like good news. Nevertheless, it is clear from [Falcioni, Marconi, and Vulpiani, 1991, Hurd, Grebogi, and Ott, 1994] that this result must be viewed as somewhat of a Pyrrhic victory. Indeed, although reliable computation of the time-evolution on very long time-scales poses genuine technical challenges, these numerical studies indicate that individual trajectories forget their initial conditions and actually invade a reasonable part of the phase-space *only after an extremely long time*: the problem of the time-scale, which had been pushed in the background by the usual statement of the KAM theorem, had re-entered the purview of the study of these dynamical systems.

Some theoretical support for such observed long time-scales may be derived from the work of Nekhoroshev [1977] – see also [Lochak, 1995] and references cited therein – as argued in [Galgani, 1985, 1987, Galgani, Giorgilli, Martinoli, and Vanzini, 1992, Giorgilli, 1995]. These authors reconsidered the negative result of the early numerical experiment on the equipartition of the energy performed in [Fermi, Pasta, and Ulam, 1940] for the model:

$$H_{\alpha,\beta} = H_o + V \quad \text{with}$$

$$H_o = \tfrac{1}{2} \sum_{0 \leq k \leq n} p_k{}^2 + (q_{k+1} - q_k)^2 \tag{9.4.3}$$

$$V = \tfrac{\alpha}{3} \sum_{0 \leq k \leq n} (q_{k+1} - q_k)^3 + \tfrac{\beta}{4} \sum_{0 \leq k \leq n} (q_{k+1} - q_k)^4 \quad .$$

The normal modes of the unperturbed Hamiltonian H_o, when driven by the weak non-linear coupling V in H, were found to have different and long relaxation times: consequently, in the course of time, the individual modes would only start selectively to partake in the equipartition. A computer experiment would then see equipartition only inasmuch as it is run over a very long time.

The general idea behind this interpretation is to look for what happens to the action variables $\boldsymbol{I} = (I_1, \cdots, I_n)$ of the unperturbed Hamiltonian when a weak interaction is switched on, and the evolution is governed by $H(\boldsymbol{I}, \boldsymbol{\varphi}) = H_o(\boldsymbol{I}) + \lambda V(\boldsymbol{I}, \boldsymbol{\varphi})$. As the time-invariance of $\boldsymbol{I}$ is destroyed, the question is to estimate the time T it takes for the $I_k(t)$ to differ perceptibly from their initial value $I_k(0)$. Under rather general circumstances [Nekhoroshev, 1977, Lochak, 1995] it is possible to find strictly positive constants $\lambda_o, \lambda_*, T_o, A, a, b$ such that

$$\left. \begin{array}{ll} \text{for every} & (\lambda, t) \in [0, \lambda_o) \times [0, T(\lambda)) \\ \text{where} & T(\lambda) \approx T_o \, \exp[(\tfrac{\lambda_*}{\lambda})^a] \end{array} \right\} : \|\boldsymbol{I}(t) - \boldsymbol{I}(0)\| \leq A\lambda^b . \tag{9.4.4}$$

Note that this is evidently consistent with the fact that $\boldsymbol{I}$ is a constant when $\lambda = 0$; the additional information contained in this estimate is that *for small*

λ, $T(\lambda)$ *varies exponentially with* λ^{-a}, so that the relaxation time remains – if not infinite, at least – very long, while the tolerated variation of $\boldsymbol{I}(t)$ is of the order of λ^b. So far, so good. The next problem is to determine how reliable this theoretical estimate becomes when the number of degrees of freedom becomes large. A tentative answer, together with an optimistic conjecture, have been proposed, namely:

> *The numerical investigation of systems of the FPU* [= Fermi–Pasta–Ulam] *type suggest that Nekhoroshev's estimates* [i.e. the estimates briefly reviewed above] *concerning the actions of a near-to-integrable Hamiltonian system cannot be extended to the case of systems with a large number of degrees of freedom. This fact seems to support the common belief that for statistical systems there are no adiabatic invariants, and that equipartition of energy should hold.*

> [Giorgilli, 1995]

Evidently, the logic does not pretend to be that of a mathematical derivation. Nonetheless, it is certainly an informed conjecture: circumstances are known [Lochak, 1995] where the constants a and b in (9.4.4) decrease as the inverse power of the number of degrees of freedom.

In any case, the above quote illustrates the remark [or hope?] that the predictive failures of a scientific theory are often the articulation points from which new insights can be gained. In the above account, we started with the historical view that ergodicity should be central to the understanding of statistical mechanics; we pointed out, however, that ergodicity is not a generic property of mechanical systems. This necessitated a change of perspective. Taking then a view from the teachings of classical mechanics, we noticed that integrability was the lamp-post around which textbooks were built. Unfortunately, integrability is not generic either in the space of Hamiltonian dynamical systems. Nevertheless, the KAM theory showed how the lessons one learns from integrability are flexible and can be extended outside to provide information on nearly integrable systems. The residual part of the phase-space, not occupied by the KAM tori, first taken to be negligible for sufficiently small perturbations of systems with a few degrees of freedom, turned out to be where stochasticity – and hopefully ergodicity – are waiting to come to the fore; this drawing out was partially, but quite substantially, achieved by considering the Hamiltonian dynamics restricted to energy shells of higher energy, or systems with a very large number of degrees of freedom.

9.5 Conclusions and Remaining Issues

The purpose of this section is to assess the power and limitations of the idealizations involved in ergodic theory. For a general discussion of idealization, see Chap. 16. Ergodic theory became the object of a renewed interest in the physics community only after a long period during which it had been cultivated mainly by mathematicians [Hopf, 1937, Billingsley, 1965, Arnold and

Avez, 1968, Parry, 1969, Ornstein, 1974, Sinai, 1976, Brown, 1976, Walters, 1982, Cornfeld, Fomin, and Sinai, 1982, Mane, 1983, Ollagnier, 1985, Ellis, 1985].

A different presentation was necessary to broadcast these mathematical results in terms to which physicists would listen and by which they could be prompted to start evaluating how the issues had progressed since Boltzmann, Maxwell and the Ehrenfests; penetrating contributions to this effort can be found in [Jancel, 1969, Farquhar, 1964, Lebowitz and Penrose, 1973, Lanford, 1973, Eckmann and Ruelle, 1985, Ruelle, 1969].

Stimulated in part by this activity, an elaborate critique of the principles – in particular the alleged compatibility of probabilistic concepts with the tenets of a deterministic classical mechanics – was conducted by philosophers; for a sampling see [Sklar, 1973, Friedman, 1976, Lavis, 1977, Quay, 1978, Malament and Zabell, 1980, Leeds, 1989, Batterman, 1990, Rédei, 1992, Sklar, 1993, Earman and Rédei, 1996, Uffink, 1996b, Batterman, 1998, Vranas, 1998].

Each of these three lists is frightfully incomplete; they only indicate the sweep of ideas considered. We propose to begin and summarize where we stand in our own enterprise; then we direct what we have learned towards the sharpening of the issues raised in these papers. In spite of a few happy precedents, we are reminded of the admonition:

> To pursue the discussion at a purely metaphysical and literary level is like driving a car blindfolded: it can only lead to disaster.
>
> [Ruelle, 1991]

We may therefore have to request the Reader to peek with us from under the blindfold, and refer to some of the materials we offered in Chaps. 7, 8 and 9 for the technical background of some of our comments. We started from an idealization, the notion of dynamical system [Def. 7.2.1], on which ergodic theory is built. The idealization was motivated by Boltzmann's inquiries; these resulted in two permanent additions to the language of the kinetic theory of gases, and then of statistical mechanics: (i) the ensembles; and (ii) the ergodic hypothesis [Sect. 7.1]. The notion of dynamical system incorporates the former, and it allows to reformulate the latter as the condition of metric transitivity [Def. 7.2.3]. At that stage no Hamiltonian constraints are imposed on the dynamics, beyond borrowing from Liouville's theorem [Thm. 7.2.1] the idea that the dynamics preserves a measure. The measure itself is interpreted, in line with Boltzmann's approach, as a long-time average. The idealization is to view this simple mathematical structure as a model of what matter in bulk could look like at a microscopic level.

One of the successes of ergodic theory is to have provided a bridge between conservative and dissipative dynamical systems; here conservative systems are represented as measure-preserving flows, and the road to dissipative systems is paved with the conceptualization of some characteristic properties for these systems. Defying Zermelo's objection, ergodic theory provides specific models that help delineate these properties and establish their compatibility with conservative dynamics.

These properties form a hierarchy [Chap. 8]. At the bottom, just above metric transitivity itself, sit mixing properties (Sect. 8.1) that seem to offer a mathematical model responding to some of Gibbs requirements. On top of the hierarchy, stand two notions: dynamical entropy [Sects. 8.2 and 8.3], and sensitive dependence on initial conditions [Sect. 8.4]. The distinctions established between the different levels of the ergodic hierarchy turn out to be important investigative and evaluating assets; see e.g. Sect. 9.3, and the discussions at the end of Sect. 9.4.

Yet, at this mainly mathematical stage of its development, ergodic theory is more concerned with *instantiating* specific mathematical properties, rather than asking which specific physical processes can be modeled by ergodic systems. It does not really address too closely any realistic description of matter in bulk. In particular, one should not look into this kind of ergodic theory for a sufficient justification for the foundations of statistical mechanics.

Rather, it *explores worlds that could be*. And it does it with a purpose: to confront and sharpen as yet poorly formulated ideas. Indeed, serious problems, even at the syntactic level, start when one tries to see whether the various properties of ergodic theory genuinely pertain to Hamiltonian mechanics. While it is true that even the top of the ergodic hierarchy is *compatible* with a Hamiltonian dynamics, this situation is *exceptional* [Sect. 9.1]: even the weakest level of the ergodic hierarchy is not generic among finite Hamiltonian systems. The coat of ergodic theory has become too tight for the body it is asked to cover: the idealizations involved have reached an *impass.*

Hence, to the mythical question of whether the ergodic hypothesis justifies statistical mechanics, the answer is worse than no : it is not the right question. Indeed, the question cannot be anymore whether nature strictly obeys the demands of ergodicity; rather, the question ought to ask how good an idealization the theory really is.

The issues we want to address are divided into three classes.

I. We claim that ergodic theory is a successful physical theory. Its earliest and still one of its greatest successes is that *it helped Boltzmann introduce the concept of ensemble in statistical mechanics.* We argued that, in Boltzmann's view, an ensemble represents the information one has obtained by averaging the results of measurements over the time necessary for the measurement to be completed. At this first level, we isolate three of the issues raised by the proposal to use ergodic theory for pragmatic purposes.

Issue I.1. The most immediate of these issues is the time-scale pertaining to ergodic theory. Note from the onset that speaking of "very-long" time when discussing physical systems is meaningless in the absence of the choice of a time-scale intrinsic to the aspects of the system one wishes to describe. Looking at a 12th-century stained glass window, we are not inclined to think of glass as a liquid ... albeit one with such a huge viscosity that the window in its beauty seems eternal.

In the context of this chapter, the very long time of a description at the macroscopic level needs to appear to be "practically incommensurable" with the scale prevailing at a microscopic level.

Thus as theorists, we must first choose what we want to describe. We typically introduce long-time limits to wash out fluctuations we deem irrelevant. In doing so, we intend to focus on asymptotic behaviors at the detriment of transient phenomena that are purposefully ignored or, at least, pushed to the background. We come back shortly to this general strategy, and offer to support it by numerical estimates of the time ratio to be expected.

But first, we must make a distinction between two different situations. In the non-equilibrium statistical mechanics that studies the approach to equilibrium these long-time considerations involve complex, mixed limits [e.g. Grad, van Hove, Vlasov, etc], the simplest of which is discussed in Chap. 15; let us only say here that the complexity of these limits has to do with the fact that in non-equilibrium statistical mechanics, we have to juggle with three (not just two) time-scales, namely the microscopic time-scale, the time-scale presiding over the duration of the individual measurements, and the time-scale of the transport phenomenon to be considered. In the present section, we limit our scope to equilibrium statistical mechanics.

In order not to have to decide in advance *precisely* how long that time should be, the theorist talks about *arbitrary long times,* or considers *infinite time limits* with the mathematical proviso that the limits exist. The ergodic theorems of Birkhoff and von Neumann show the mathematical sense in which this is realistic; in accordance with these results, there are two pragmatic limitations: sets of measure zero, and $T \to \infty$. Two extreme stances towards these limitations have been taken by the Bold: some reject outright the idealization proposed by ergodic theory; others act as if these limitations did not exist. Both stances amount to saying that ergodic theory cannot serve any purpose in the foundations of statistical mechanics.

A milder attitude is to apply the theory only when one has reasons to expect these limitations not to matter. To be specific, consider, for instance, a normal gas at room temperature and under atmospheric pressure. For a snapshot idea of the microscopic landscape, note that the molecular diameter is of the order of $\approx 10^{-8}$ cm; the average distance between the molecules is $\approx 10^{-7}$ cm, so that the molecules occupy about one thousandth $[(10^{-8}/10^{-7})^3]$ of the total volume of the gas. There is however a lot of action down there. The average velocity of the molecules is $\approx 5 \cdot 10^4$ cm/sec; and their mean free path is $\approx 10^{-5}$ cm, so that the average time between collisions is $\approx 2 \cdot 10^{-10}$ sec. If a measurement of the macroscopic state of the gas were done in the blink of an eye, say 2/10 sec, the ratio between the macroscopic time and the microscopic time would be $\approx 10^9$. To refuse to consider this a "long time" would be tantamount to the situation where a client would request his insurance agent to take into account the age of the universe, something like $\approx 10^{10}$ years, when she writes a policy on his life.

A less elaborate estimate comes from every-morning's life. If we slumber for one hour or so, we find that our cup of coffee has gotten 'cold', i.e. its temperature can be viewed to be close enough to the room temperature that no improvement is to be expected from waiting another hour, several days, or even longer.

There is a hidden rationale behind this mundane observation: deviations from the trend are expected to be rare; in order to avoid the traps of probability theory, such as the law of iterated logarithms [see Thm. 5.2.4], the physicists may argue that in their context, deterministic rather than probabilistic models are of the essence, and that anyhow, over exceedingly long times, the circumstances of the experiment may change, or other phenomena take over, subverting the usefulness of the original model. For instance, within a few hours, the Sun may start shining over the cup of coffee left over from the above discussion, or chemical reactions besiege it, spoiling our very idea of a cup of coffee. We discuss in Issue I.2 below why *this rationale by itself is not sufficient.*

Nevertheless, note that *if and when* a model is constructed in such a way that the long-time limit is proven to exist, the lines of battle are well drawn: by its very definition, the mathematical notion of limit means that you give us the error you want to tolerate (the ε of the mathematician; the only constraint is that $\varepsilon > 0$, otherwise you can choose it arbitrarily); we then tell you how long you should wait (the mathematician's estimates of the $\delta_\varepsilon > 0$; read here $T_\varepsilon = \frac{1}{\delta_\varepsilon} < \infty$) to be sure that no deviations greater than ε will *ever* occur after the *finite* time T_ε has elapsed. The smaller ε is chosen, the larger T_ε will have to be, yet for each separate choice of the tolerance parameters *all is carried out within finite bounds.* Hence a finitist attitude like Boltzmann's is compatible with the mathematical consideration of infinite time limits which are – by their very definition – controllable approximations, see Sect. 16.2. Naturally there is still a proviso: the limit has to be proven to exist for *all* the initial conditions that are relevant; this is our next point.

Issue I.2. If the theorists were required to prove that the above limit exists point-wise for the functions – observables – they are interested in, they would be out of luck: Birkhoff's theorem only asserts that limits exist for all points *but a set of points of measure zero.* Since the theory does not give any hint about where these points are located in the phase-space, this could be a serious objection. It is by-passed by the remark that such sets do *not* contribute to the integrals that are involved in computing expectation values of smooth observables – in fact here 'smooth' is much more than enough, as 'measurable' would suffice; but we promised not to be overly technical. What we want to emphasize is that we touch here the power of Boltzmann's contribution: while the elementary classical mechanics of falling apples and planetary motions is primarily concerned with the trajectories of *material* *points,* statistical mechanics is in essence the study of *ensembles,* or more precisely of *measurable sets.*

The states of simple mechanical systems routinely are identified with points of the phase-space, whereas the concept of state of a system that is relevant to statistical mechanics is that of measure, or density distribution function – via the Radon–Nikodym theorem, see e.g. [Reed and Simon, 1972–1980] – when the situation at hand singles out a class of mutually absolutely continuous measures. We need to explain, however briefly, how and why this change of perspective is hardly an idealization to be held against ergodic theory. In this regard, the classical mechanics of material points is *more* an idealization than is statistical mechanics. In the first expositions of *classical mechanics,* the students are told to think in terms of *material points.* And they quickly forget that Newton's proverbial apple is not a material point; it has a structure of its own. While this structure is certainly essential to people who cultivate apples, eat them, or propagate creation myths, Newton *chose* to compute instead the motion of the center of mass of the falling apple, which is a material point. This is a choice, suggested by what Newton wanted to describe, an idealization tacitly based on the assumption that the mass of the apple is smoothly distributed within its round confines.

We bring up that idealization here for three reasons. First, the usefulness of the idealization evidently extends beyond apples: at a deeper level, Newton's inverse-square law of gravity allows to replace the attraction of a spherical shell with uniform material density – and by an immediate but genuine extension any object with spherical symmetry, like the Earth, or the Sun – by that of a material point of the same total mass placed at its center. Second, as with any idealization, this has a limited domain; here the description of the attraction at a place outside the body: inside a spherical shell, the force of attraction vanishes; this was well–known by the time of Cavendish, but he alone seems to have understood it well enough to use it ... and discover experimentally the Coulomb law of electrostatics, ten years before Coulomb, and with a much better experimental reliability on the (-2) of the "inverse-square' law [Maxwell, 1879a]. And third, this idealization is often too routinely accepted, even though it was also realized early that the idealization of a material point can be misleading when applied to the dynamics of matter in bulk: as even the poet knows, the movement of a cloud in the sky and that of the heavenly bodies require different descriptions; hence the languages of meteorology and of position astronomy differ in essential ways. Why is that so ?

When dealing with a gas, the experimentalist does not *observe* directly the trajectories of the individual molecules, nor can the theorist *compute* them. Sinai credits [Krylov, 1979] for realizing that the reason for these limitations is this: due to repeated collisions, the individual trajectories exhibit an exponentially sensitive dependence on initial conditions; therefore, the boundary of any smooth convex region of the gas looses its smoothness extremely fast to resemble an octopus-with-a-bad-dream more than it does a slightly deformed bubble. Nothing like that happens to the falling apple: it retains its integrity – at least up to its splashing on the ground upon landing, which is beyond the domain of the description. This is what Boltzmann understood when he substituted his ensembles to the individual trajectories ... and it is the idea that was so brilliantly exploited by the Russian school of modern ergodic theory – recall Sect. 8.4.

Issue I.3. We just saw one aspect of the role of initial conditions. This has also been seen from another angle, according to which statistical mechanics would be the mechanics of *incompletely specified systems*.

This view may tend to displace the balance between the role of probability theory and mechanics, and it could lead to a premature emphasis on stochastic processes which naturally appear only later on in the analysis, namely when combined limits come into play to separate different regimes, such as the random walk model for diffusion, the Grad limit in the derivation of the classical Boltzmann equation, or its distant cousin in quantum theory, the van Hove limit introduced in the discussion of the master equation – see Sect. 15.2.

Nevertheless, the pragmatic impossibility to pin down the initial conditions to a point in phase-space did lend additional pragmatic relevance to an important concept, the *coarse-grained observables,* i.e. functions obtained as local averages, ultimately leading from mechanics to some reduced – self-contained or closed – description that would focus on the macroscopic properties of the systems under consideration. We have to postpone further discussion of this issue to our subsequent chapters on statistical mechanics proper. In the meantime, we refer the Reader to Issue II.5 below, where we discuss another incomplete, or reduced, description relevant to the statistical mechanics of large systems, focusing on the special observables introduced by Khinchin under the name of *sum-functions.*

II. The second collection of issues we want to review clusters around one question: Has an essential part of ergodic theory become the exclusive dominion of Mathematics, outside the realm of Physics? There is an undebatable reality that the study of dynamical systems has developed into an almost autonomous mathematical discipline at the crossroads of analysis, geometry, topology, probability and mathematical physics, out of the mainstream of what a large majority of physicists would call real physics. The richness of the mathematical theory is such that it has its own vibrant sub-disciplines; the situation, however, has not gone out-of-hand, as is nicely illustrated by [Young, 1998]; here again note the venue.

Issue II.1. The mathematical theory has discovered an *ergodic hierarchy* that took off where Gibbs had left it, after insisting that statistical mechanics is *not* interested in *ergodicity per se*, but that the property of *mixing* is essential. This is indeed the first step in a strictly increasing hierarchy of conditions: ergodicity, mixing, Lebesgue spectrum, K-systems, Anosov systems, and Bernoulli systems; a net of theorems, examples, and counter-examples has shown that a system satisfies one of these successive conditions only if it satisfies the conditions below it in the hierarchy; and each of these conditions is strictly more stringent than the preceding one. In particular, ergodicity must be satisfied if any of the other conditions is to be satisfied.

Mixing was known beforehand; Lebesgue spectrum gives more information on correlation-decay rates. The concept of dynamical entropy was brought to light by the mathematical theory; and its interpretation is striking, as the dynamical entropy gives a numerical estimate on how much information is lost in the course of the evolution. K-systems are precisely those dynamical systems that have positive dynamical entropy, and the flow of Brownian motion is a K-system; so are certain billiards. The Liapunov coefficients of Anosov systems are measurable quantities.

Issue II.2. Most of the models of mathematical ergodic theory are not Hamiltonian; exceptions are known, but there are physicists who would still say that these are unphysical. The jury is still out [Stoddard and Ford, 1973, Wightman, 1985, Kazumasa and Tomohei, 1996, Vranas, 1998] as to whether some kind of ergodicity is satisfied by a gas with intermolecular interactions governed by a realistic two-body potential, such as a Lennard–Jones potential

$$u(r) = 4\varepsilon \left[(r_o/r)^{12} - (r_o/r)^6 \right] \qquad (9.5.1)$$

characterized by a steep repulsive core for $0 < r < r_o$ followed for $r > r_o$ by a short range attractive well of depth ε; the exponents 12 and 6 are empirical data.

This lack of physical models is a serious objection, and one that mathematics can confirm: among Hamiltonian systems, ergodic systems are *not generic,* which means that the slightest perturbation can kill ergodicity, and with it every other condition higher up in the hierarchy. No physically realistic, *a priori,* condition is known, which when imposed on the Hamiltonian, would ensure the ergodicity of the system.

Issue II.3. While the ergodic hierarchy is mathematically strict, pragmatic evidence is more flexible: a slight failure of one of the ergodic conditions may not be detected immediately from a check of one of the higher conditions. For instance, a mathematician would say that ergodicity is an absolute: either a system is ergodic, or it is not. Some philosophers agree, and some don't. At a pragmatic level, there is a meaning to saying that a system is nearly-ergodic, as there is a meaning to saying that another system is nearly integrable: the set of points where the coveted property is violated is of very small, but non-zero measure. We have seen this happen in Sect. 9.4, and we are going to comment further on this point – see Issues II.4, III.3 and III.4 below.

Issue II.4. The juxtaposition of the following contentions seems to have caused much trouble:

1. the praxis of statistical mechanics seems to focus on systems with a very large number of degrees of freedom;
2. the foundations of statistical mechanics are presented as if they were justified by ergodic theory;
3. within the confines of Hamiltonian dynamics with only a few degrees of freedom (say $n = 2$), specific models have been constructed – Sect. 8.4 – that are not only ergodic, but are also K-systems, and even Anosov systems;
4. ergodicity is not generic among finite Hamiltonian systems – Thm. 9.1.1;
5. at the opposite of ergodicity, complete integrability, although not generic either, seems to endure in the form of the KAM tori, under small perturbations – Thm. 9.2.1.

As (3), (4) and (5) are mathematical facts, it would be hard to square (1) and (2), and even harder to pretend that ergodic theory could, by itself, explain statistical mechanics. Historically, one finds on both sides of the issue eminent authorities giving opposite answers to the question of whether the systems which statistical mechanics should aspire to describe *ought to* have a large number of degrees of freedom.

> *The laws of statistical mechanics apply to conservative systems of any number of degrees of freedom, and are exact. This does not make them more difficult to establish than the approximate laws for systems of a great many degrees of freedom ... The reverse is rather the case ...*
>
> [Gibbs, 1902]

> *The single feature which distinguishes statistical mechanics from ordinary mechanics is the large number of degrees of freedom*
>
> [Grad, 1967]

Not being able to navigate by these conflicting lights, let us first consider some elementary geometric evidence to the effect that large n can help. Suppose that a n-dimensional sphere of radius R contains a uniform distribution of points; consider then the subregion of that sphere consisting of another sphere, concentric to the first, and of radius $R - a$. Then the ratio of the number of points contained in the smaller sphere, over the total number of points is $[(R-a)/R]^n$; this implies that for large n the overwhelming majority of points are in the shell between the two spheres. Physically, this models a situation in which one would impose that the total kinetic energy of an assembly of n particles satisfies $v_1{}^2 + v_2{}^2 + \cdots + v_n{}^2 \leq 2mE$; and, subject only to this condition, we would assume that the n velocities are distributed uniformly; then we just learned that the imposed inequality can be approximated by an equality: $\langle v_1{}^2 + v_2{}^2 + \cdots + v_n{}^2 \rangle \approx 2mE$, provided that n becomes sufficiently large.

Some more sophisticated evidence is presented separately just below – see Issue II.5 – for the role played by the fact that the systems of interest to statistical mechanics involve a *large* number of degrees of freedom.

Issue II.5. In Chap. 5 already, we discussed some of the simplifying features brought about by the law of large numbers in probability theory. Khinchin, who had himself contributed to some of the sharpest results in the subject, also wrote on its application to statistical mechanics [Khinchin, 1949][Sects. 16, 36]; for discussions, see [Jancel, 1969][Sects.I.III.1, I.III.4], [Truesdell, 1961][Sect.5,6] and also [Batterman, 1998].

Khinchin considers a special kind of observables, which he calls *sum functions,* defined on a $n-$dimensional space $R^n = \{x = (x_1, x_2, \cdots, x_n)\}$ and of the form

$$f(x) = \sum_{k=1}^{n} f_k(x_k) \quad \text{with} \quad 0 < a < f_k < b < \infty \quad , \tag{9.5.2}$$

so that their average $\langle f \rangle$ is of the order of n :

$$m \equiv \langle f \rangle = O(n) \quad . \tag{9.5.3}$$

Khinchin's argument comes in two steps. The first to show that the sum functions (9.5.2) enjoy the remarkable property:

$$\sigma^2 \equiv \langle [f - \langle f \rangle]^2 \rangle = O(n) \quad . \tag{9.5.4}$$

Upon inserting this in the Chebyshev inequality – Scholium 5.2.3 – we receive for all $a > 0$:

$$\text{Prob} \, \{ \, | \, f - \langle f \rangle| \geq a \, \} \, \leq \, \frac{1}{a^2} O(n) \quad ,$$

i.e., upon taking (9.5.3) into account:

$$\text{Prob} \left\{ \frac{|f - \langle f \rangle|}{\langle f \rangle} \geq a \right\} \leq \frac{1}{a^2} O(n^{-1}) \quad , \tag{9.5.5}$$

and, upon using the Schwartz inequality on (9.5.4):

$$\langle \frac{|f - \langle f \rangle|}{\langle f \rangle} \rangle \leq O(n^{-\frac{1}{2}}) \quad . \tag{9.5.6}$$

So far, the argument involved only averages over phase-space.

The second part of Khinchin's argument is to notice that under quite general circumstances – they do not even involve the assumption that the functions considered are sum-functions – the following relations obtain:

$$\eta_x^T[f] \equiv \frac{1}{2T} \int_{-T}^{T} dt f(x_t) \quad \Rightarrow \quad \left\{ \begin{array}{l} \langle \eta_x^T[f] \rangle = \langle f \rangle \\ \langle (\eta_x^T[f])^2 \rangle \leq \langle f^2 \rangle \end{array} \right\} \quad . \tag{9.5.7}$$

Since the results of Birkhoff–von Neumann – Thms. 7.2.3 and 4 – ensure convergence, a.e. and in the mean, for

$$\eta_x[f] \equiv \lim_{T\to\infty} \frac{1}{2T} \int_{-T}^{T} dt\, f(x_t) \quad,$$

(9.5.7) implies, *without assuming metric transitivity*, that

$$\langle \eta_x[f]\rangle = \langle f\rangle \quad\text{and}\quad \langle(\eta_x[f])^2\rangle \leq \langle f^2\rangle \quad. \tag{9.5.8}$$

Hence:

$$\langle(\eta_x[f] - \langle f\rangle)^2\rangle \leq \langle(f - \langle f\rangle)^2\rangle \quad. \tag{9.5.9}$$

Upon bringing together the various parts of Khinchin's argument, namely (9.5.3–4) and (9.5.8–9), we obtain as another consequence of the Chebyshev inequality:

$$\text{Prob}\left\{\frac{|\eta_x[f] - \langle f\rangle|}{\langle f\rangle} \geq a\right\} \leq \frac{1}{a^2} O(n^{-1}) \tag{9.5.10}$$

which, as (9.5.5), is thus valid for any sum-function f.

This inequality asserts that the trajectories which result in a large value of the relative difference between the time-average and the space-average of any observable that is a sum-function are very rare when n is very large.

> *This fact establishes the representability of the mean values of the sum functions, and permits us to identify them with the time-average which represent the direct results of any physical measurement.*
>
> [Khinchin, 1949]

The above results did not assume any special property of the measure with respect to which the averages $\langle\cdots\rangle$ and the probabilities $\text{Prob}\{\cdots\}$ are computed, beyond the conditions that it be countably additive and time-invariant. In particular they were obtained without any ergodicity assumption such as metric transitivity. The presentation depended essentially of the fact that the functions satisfy the condition (9.5.2); moreover, Khinchin's approach has been extended by Mazur and Van der Linden [1963].

As with any other probabilistic statement having to do with the law of large numbers, the drawback in (9.5.10) is that the theory does not tell us for which trajectories the inequality is not satisfied; wanting to know that would be hopeless, were it only because the existence $\eta_x[f]$ is asserted only for a.e. x. In this respect, Khinchin's approach does not eliminate the problem of the undetermined small sets on which things go sour. The progress nevertheless was to show how the explicit assumption of metric transitivity can be avoided when the class of observables is restricted in a manner that involves explicitly the fact that statistical mechanics is to be concerned with systems having a large number of degrees of freedom. The estimate (9.5.10) gives that the probability of a relative deviation larger than one part in a million, i.e. $a \approx 10^{-6}$, is of the order of $2 \cdot 10^{-12}$ when $n \approx 6 \cdot 10^{23}$ (the Avogadro number, i.e. the number of molecules in a mole). This *is* indeed an awfully small probability, but this very smallness cannot be ignored without risks, when appraising the pragmatic success of a statistical mechanics that equates space- and time-averages.

III. Four other issues come to mind in connection with some of the above discussions. Since they seem to have a life of their own, we review them in a separate category.

Issue III.1. When ergodic theory studies general dynamical systems, it has no natural way to chose *a priori* the measure that is invariant under the time-evolution, except to impose *a posteriori* that it be extremal invariant (metric indecomposability), or mixing, or that it satisfy any other stronger ergodic conditions; the matter is left to the ingenuity of the model builder. In contrast, for a physical theory that deals with Hamiltonian systems, the choice of the measure is restricted by the physical circumstances under which the system is considered. Much of the ergodic theory considered so far is presented in light of the Liouville theorem – Thm. 7.2.1 – and thus it seems to lead to a uniform distribution on the energy shell, thus placing the microcanonical ensemble at center stage. Physically, this amounts to considering only isolated ensembles. If this were the only possible approach, it would be pretty bad, and this on two accounts, irrespectively of the separate fact – discussed through this chapter – that ergodicity is not generic in the class of Hamiltonian systems.

First, on pragmatic grounds: systems observed in the laboratory are usually not isolated: they interact with the measuring apparatus, and they need to be confined by material or electromagnetic walls.

Second, on theoretical grounds: the assumption that the system is intrinsically isolated is self-defeating since the very definition of temperature depends on the fact that the system considered can exchange energy with its surroundings. An ergodic theory with its primary emphasis on isolated systems would be an idealization from which one must depart. Allowing the system to exchange energy and/or matter would lead not to the microcanonical ensemble, but to the canonical and/or grand canonical ensembles.

One way out would then be to consider the aggregate – system of interest + reservoir – as being isolated. A preliminary proposal for how that could be made to work is presented in Rem. F in Sect. 8.2; the problem is pursued further in Chap. 15. Another way is to abandon ergodicity as the ultimate justification and walk along the steps of de Finetti and Jaynes – see Sect. 6.3 together with Sect. 5.3 – taking stock only of the partial information one actually has on the system. This stand however would not in itself be an entire cop-out from ergodic theory: the interest of idealization offered by ergodic theory is not limited to the issue just addressed.

When we discuss statistical mechanics in the next chapters, we will see that ergodic theory can be used directly also within the framework of the canonical and/or grand canonical ensembles.

Issue III.2. In Chaps. 7–9, we have limited our attention to the time-evolution. While the study of dynamics is certainly of interest, we have ignored so far that, as a mathematical enterprise, ergodic theory is not limited to the 1−dimensional abelian group R. The general ergodic theory can thus be brought to task for the analysis of other symmetry groups of interest to physics. The theory of phase transitions, to be covered in Chaps. 11–14, provides examples.

Issue III.3. Ergodicity, mixing, or any of the subsequent levels of the hierarchy, are properties that are sharply defined within the idealizations offered by ergodic theory. Since, however these properties are not generic in finite Hamiltonian mechanics, the themes they suggest should be read with the idea to transpose them, asking not "whether" but "how much" these properties are satisfied in actual systems of interest to statistical mechanics. This has been attempted with some success in at least two ways; for illustrations, see Sects. 9.3 and 4.

First, one looked for approximate senses in which the properties are nevertheless apparent; this has been done in machine computations which provide evidence of large regions of phase-space dominated by the kind of stochastic behavior one would expect at even the highest level of the ergodic hierarchy.

Once this was elicited, at least in some models, a second level of inquiry was to look for trends by varying the control parameters of the experiment: energy, interaction strength, or number of particles, in order to determine the type of situations that make these regions larger. One of the conjectures such experimental results seem to support is that ergodicity may be less exceptional in the thermodynamical limit where cooperative phenomena have an opportunity to take over, and equilibrium states become indecomposable, with the proviso of further restrictions to the effect that each thermodynamical phase should be considered separately.

Issue III.4. We should finally at least mention the so-called *thermodynamical limit*. Here again, we warn the Reader that we are talking of a limit, in the mathematical sense of the word: nobody is going to pretend that a cup of coffee is infinite, only that a description that sets aside the finiteness of the cup of coffee is a more intelligible description than one that would insist on describing all finite size or surface effects.

Thus, the motivation for considering the thermodynamical limit is to focus on the properties of matter in bulk, which are in the purview of equilibrium and non-equilibrium statistical mechanics. To focus on such properties requires a control of the effects of the boundaries; this means to eliminate those vagaries of the boundaries that are of range much shorter than the size of the system considered, and to understand the role of those that are long-range. This is facilitated by taking a limit where the volume of the sample and the number of particles it contains tend to infinity, while their ratio is kept constant. For instance, upon taking advantage of the formalism of equilibrium statistical mechanics of finite systems, one already learns much of what is to be learned by restricting one's attention to the partition function, or the equilibrium expectation value of observables that one expects to be good candidates for the status of macroscopic observables. To prove the existence of such limits with sufficient generality, however, is everything but easy, but it was done; see the pioneering work of Ruelle [1969], some of which is situated in Chap. 11.

Moreover the ideas first developed for classical systems can be translated to the quantum realm. Our point is then that ergodic theory can be applied to these systems, and not only for the time-evolution, but first and foremost for symmetries such as translations. That extension of the usual ergodic theory allows one to decide whether the equilibrium state is unique, or extremal-invariant; or what kind of decays in the space and/or time correlation functions one might expect; see Chaps. 11–14.

As for the higher ergodic properties of the time evolution in the thermodynamical limit, some results have already been obtained, asserting for instance that the K-property holds for an infinite ideal gas; for a review, see section 11 in [Szasz, 1996] where the Reader will again find that much of the credit must go to the schools of Sinai and of Lebowitz.

From the above review, we conclude that ergodic theory is an idealization. In and by itself, it is not a justification of statistical mechanics. But it provides an imaginative guide for conjectures, and a tool for a penetrating analysis of physical properties of matter in bulk. It is an essential player, one that performs an intrinsic part in the concert of statistical mechanics.

10. The Gibbs Canonical Ensembles

Gibbs simply did not consider ergodicity as relevant to the foundations of the subject. Of course, he was far too polite a man to say so openly; and so he made the point simply by developing his theory without making any use of it. Unfortunately, this tactic was too subtle to be appreciated, and the few who did notice it took it to be a defect in Gibbs' presentation, in need of correction by others.

[Jaynes, 1967]

10.1 Classical Settings

The birth certificate of statistical mechanics is [Gibbs, 1902]. As posted in its subtitle, the work was *"developed with special reference to the rational foundation of thermodynamics"*. Gibbs proposed three "ensembles", which he called microcanonical, canonical, and grand canonical. As we often did in the previous chapters, we defer the discussion of the semantics, and first present the syntax. The mathematical context is that of conservative Hamiltonian mechanics, with phase-space Γ and time-independent Hamiltonian H. Typically in Gibbs, $\Gamma = T^*\Lambda = R^n \times \Lambda = \{(p,q)\}$ where the configuration space $\Lambda \subset R^n$ is a bounded closed region; for the time being n is any integer with $n > 2$. The Hamiltonian is supposed to be a function defined on Γ and possibly depending, in addition, on finitely many external parameters $\{a_i | 1 \leq i \leq I < \infty\}$. It is sometimes useful, but in general not necessary, to consider the tidy circumstance where the Hamiltonian splits nicely in kinetic (T) and potential energy (V) with $H(p,q;a) = K(p) + V(q;a)$.

We now introduce the statistical elements.

Definition 10.1.1. *The* microcanonical *probability distribution is the uniform probability distribution on an energy surface* $\Omega_E = \{(p,q) \in \Gamma \mid H(p,q) = E\}$.

This corresponds to the measure that Boltzmann wanted to be ergodic; see Chap. 7. We come back to this only in connection with its relation with the next definition.

Definition 10.1.2. *The* canonical *probability distribution is* $\varrho \equiv e^{(\Psi - H)/\Theta}$ *where* Θ *is a positive number, and* Ψ *is determined by the normalization* $\int_\Gamma dp \, dq \, \varrho = 1$.

The grand canonical ensemble is defined at the end of this section.

We begin our analysis with the canonical distribution of Def. 10.1.1 and establish the following contact with the syntax of thermodynamics; [Gibbs, 1902][Chap. IV, eqns.(110) and (114)].

Theorem 10.1.1. *Let* $\varrho = e^{(\Psi - H)/\Theta}$ *be a canonical probability distribution as in Def. 10.1.2. Then the quantities*

$$U \equiv \int_\Gamma \varrho H$$

$$S \equiv -k \int_\Gamma \varrho \ln \varrho$$

$$A_i \equiv \int_\Gamma \varrho \left(-\frac{\partial H}{\partial a_i}\right)$$

$$T \equiv \Theta/k$$

with k a positive real number, satisfy the following properties

$$\Psi = U - TS$$

$$T\,dS = dU + \sum_i A_i\,da_i \quad .$$

Proof:

$$S = -k \int_\Gamma \varrho \ln \varrho = -k \int_\Gamma \varrho\,(\Psi - H)/\Theta = -k \left(\Psi - \int_\Gamma \varrho H\right)/\Theta \quad (10.1.1)$$

which is equivalent to the first of the properties claimed by the theorem. From that relation, we have

$$d\Psi = dU - S\,dT - T\,dS \quad . \tag{10.1.2}$$

Moreover, from the normalization $\int \varrho = 1$:

$$e^{-\Psi/\Theta} = \int_\Gamma e^{-H/\Theta} \quad . \tag{10.1.3}$$

Upon taking the differential of both sides, we obtain with Gibbs:

$$\left.\begin{aligned} e^{-\Psi/\Theta}\left[-\tfrac{1}{\Theta}d\Psi + \tfrac{1}{\Theta^2}\Psi d\Theta\right] = \\[4pt] \int_\Gamma e^{-\frac{1}{\Theta}H}\left[\tfrac{1}{\Theta^2} H\,d\Theta + \tfrac{1}{\Theta}\sum_i(-\tfrac{\partial H}{\partial a_i})\,da_i\right] \end{aligned}\right\} \tag{10.1.4}$$

and upon multiplying (10.1.4) by $-\tfrac{1}{\Theta}e^{\Psi/\Theta}$ we obtain

$$d\Psi = \frac{1}{\Theta}(\Psi - U)\,d\Theta - \sum_i A_i\,da_i \quad . \tag{10.1.5}$$

Finally, (10.1.2) and (10.1.5), together with the already established property $\Psi = U - TS$ and the definition of T establish the second of the properties claimed by the theorem.   **q.e.d.**

It seems inescapable to notice the analogy between the second of the relations stated in the theorem and the second law of thermodynamics – see Sect. 2.4 and in particular the differential equations (2.4.5) and (2.4.6); moreover, the relation $\Psi = U - TS$ is analogous to the definition of the thermodynamical potential known as the Helmholtz free energy. Yet, it should be borne in mind that, for the time being, these analogies are purely at the formal syntactic level: the empirical meaning of the probability distribution ϱ is still to be discussed. To approach the latter problem, one has to justify the choice of this ϱ rather than any other one.

Gibbs at first introduced this distribution as a matter of simplicity – although he knew better, as we shall see in Thm. 10.1.2: he wanted ϱ to be an equilibrium distribution, i.e. to be invariant under the time evolution, so the most immediate candidates were to be functions of the energy; moreover, he proposed that *"the most simple case conceivable"* is to take $\ln \varrho$ linear in the energy, a condition he could write without loss of generality, as $\ln \varrho = (\Psi - H)/\Theta$. This is simple enough indeed ... provided one understands why it is $\ln \varrho$ that has to be written that way, and not any other function of ϱ. This choice can be made reasonable – see the discussion following Scholium 10.1.1 below – by considering the aggregation of two systems. But then again, we do not know yet what the empirical meaning of the probability is – and thus what it means to couple individual systems, especially if one takes the view, as Gibbs does in his initial presentation, that probabilities are representing "ensembles" of a large number of structurally identical systems. We are here in danger again of being drawn back to the controversy between the frequentist and the subjective interpretations of probabilities, although von Mises and de Finetti – see Chaps. 4 and 6 – were still a quarter of a century away in Gibbs' times. Let us therefore keep exploring the consequences of the syntax alone, postponing the semantics until after the proof of Thm. 10.1.2.

As a background toward an *a posteriori* justification of his *a priori* linearity argument, Gibbs established the following property of what he calls the *"average of the index of probability"* – [Gibbs, 1902][Chap. XI, Theorem VII] – and which we recognize as the informational entropy systematized half a century later by Shannon; see our Sect. 5.3.

Scholium 10.1.1. *Let ϱ_0 be a probability distribution defined on the aggregate phase-space $\Gamma_0 = \Gamma_1 \times \Gamma_2$; and let ϱ_i, $i = 1, 2$ be the reduced probability distributions*

$$\varrho_1 \equiv \int_{\Gamma_2} \varrho_0 \quad ; \quad \varrho_2 \equiv \int_{\Gamma_1} \varrho_0 \quad .$$

Then the corresponding entropies

$$S_k \equiv -k \int_{\Gamma_k} \varrho_k \ln \varrho_k \qquad \text{with} \qquad k = 0, 1, 2$$

satisfy the inequality
$$S_0 \leq S_1 + S_2$$

with

$$S_0 = S_1 + S_2 \quad \text{iff} \quad \varrho_0 = \varrho_1 \cdot \varrho_2 \quad \text{i.e.} \quad \ln \varrho_0 = \ln \varrho_1 + \ln \varrho_2 \, .$$

Proof: We can assume, without loss of generality, that ϱ_0, and hence ϱ_1 and ϱ_2 are strictly positive. We have

$$S_0 - S_1 - S_2 = -k \left[\int_{\Gamma_0} \varrho_0 \ln \varrho_0 - \int_{\Gamma_1} \varrho_1 \ln \varrho_1 - \int_{\Gamma_2} \varrho_2 \ln \varrho_2 \right]$$

$$= -k \int_{\Gamma_0} \varrho_0 \left[\ln \varrho_0 - \ln \varrho_1 - \ln \varrho_2 \right]$$
(10.1.6)

or equivalently

$$S_0 - S_1 - S_2 = -k \int_{\Gamma_0} \varrho_1 \varrho_2 \left[\frac{\varrho_0}{\varrho_1 \varrho_2} \right] \ln \left[\frac{\varrho_0}{\varrho_1 \varrho_2} \right] \, . \tag{10.1.7}$$

Since ϱ_0, and hence ϱ_1 and ϱ_2, are normalized probability distributions:

$$\int_{\Gamma_0} \varrho_1 \varrho_2 \left(1 - \left[\frac{\varrho_0}{\varrho_1 \varrho_2} \right] \right) = \int_{\Gamma_1} \varrho_1 \int_{\Gamma_2} \varrho_2 - \int_{\Gamma_0} \varrho_0 = 1 \cdot 1 - 1 = 0 \quad , \tag{10.1.8}$$

which can be inserted in (10.1.7) to bring it to the form:

$$S_0 - S_1 - S_2 = -k \int_{\Gamma_0} \varrho_1 \varrho_2 \, f \left(\left[\frac{\varrho_0}{\varrho_1 \varrho_2} \right] \right) \quad \text{where} \quad f(x) \equiv x \ln x + 1 - x \quad . \tag{10.1.9}$$

The proof is completed by noticing that the function f satisfies:

$$f(x) \geq 0 \quad \forall \, x \in \mathbb{R} \quad \text{with} \quad f(x) = 0 \quad \text{iff} \quad x = 1 \quad . \tag{10.1.10}$$

q.e.d.

The Scholium evidently does *not* depend on the interpretation of S as the informational entropy; yet the result is so compelling that it was taken later as one of the axioms for Shannon's informational entropy – see Axiom 5.3.4. Indeed, the aggregate distribution $\varrho = \varrho_1 \cdot \varrho_2$ of the reduced distributions ϱ_1 and ϱ_2 should convey no more information than that conveyed by the original distribution ϱ_0, since the latter may include correlations that are washed out by the reduction; and the entropy of the original distribution is equal to the sum of the entropies of the reduced distributions if and only if the original distribution is a product of the reduced distributions, i.e. iff the reduced distributions are *independent*.

Gibbs specifically recognized the condition for the equality, proven in the second part of the Scholium, as a condition of independence, thus providing us with a conceptual context for his initial motivation to single out the canonical distribution. With the notation of the Scholium, what he was saying [Gibbs, 1902][pp. 33 and 135] is that in the special case where ϱ_0 is the canonical distribution

$$\varrho_0 = e^{(\Psi_0 - H_0)/\Theta_o} \tag{10.1.11}$$

and where, in addition, the Hamiltonian is of the form

$$H_0 = H_1 + H_2 \tag{10.1.12}$$

with H_1 (resp. H_2) independent of Γ_2 (resp. Γ_1), then the reduced distributions ϱ_1 and ϱ_2 are independent, in the sense that $\varrho_0 = \varrho_1 \cdot \varrho_2$; and they are themselves *canonical* distributions again:

$$\varrho_1 = e^{(\Psi_1 - H_1)/\Theta_1} \quad ; \quad \varrho_2 = e^{(\Psi_2 - H_2)/\Theta_2} \tag{10.1.13}$$

> – *a property which enormously simplifies the discussion, and is the foundation of extremely important relations to thermodynamics.*
> [Gibbs, 1902][p. 33]

Indeed the analogy is all the more powerful that in (10.1.13) above:

$$\Theta_1 = \Theta_0 = \Theta_2 \quad ; \tag{10.1.14}$$

note that we have also

$$\Psi_1 + \Psi_2 = \Psi_0 \quad \text{and} \quad S_0 = S_1 + S_2 \quad . \tag{10.1.15}$$

Now, the analogy is crying out for attention: in thermodynamics, two systems in equilibrium at the same temperature are in equilibrium with one another. To make the analogy even more compelling, two further points must be attended to. The first is that the empirical determination that two systems be in equilibrium with one another requires that they interact and do so very weakly. Gibbs renders this by adding a small term H_{int} to (10.1.12) and pretending that this term is negligible when discussing the reduction of the probability distribution. The second point is the perennial question of the sense in which results that are true in probability in the nascent statistical mechanics can tell us anything about the individual systems of thermodynamics. So far in our presentation, we only have analogies, albeit very suggestive ones. Moreover, with the help of the next result below, Gibbs further offers an argument, according to which there are statistical equivalents to the thermodynamical facts that *"when two bodies of different temperatures are brought together, that which has the higher temperature will loose energy"* [Gibbs, 1902][p. 158–160].

The following result is the variational principle that is the first step in the semantics of the canonical distribution; [Gibbs, 1902][Chap. XI, Theorem II].

Theorem 10.1.2. *The canonical distribution*

$$\varrho_0 = e^{(\Psi - H)/\Theta}$$

maximizes

$$S(\varrho) = -k \int_\Gamma \varrho \ln \varrho$$

subject to the constraint

$$U = \int_\Gamma \varrho H \quad .$$

Moreover, the values of Θ and Ψ are determined uniquely from the condition

$$U = -\frac{\partial}{\partial \beta} \ln Z \quad \text{with} \quad \beta \equiv 1/\Theta \quad \text{and} \quad Z \equiv \int_\Gamma e^{-\beta H} \quad .$$

Proof: We already saw a variant of this theorem in our discussion of Thm. 5.3.2, where we mentioned that its general proof belongs to the calculus of variations. Gibbs gave a simple-minded proof, and we reproduce it for what it is worth, i.e. without worrying about conditions of smoothness and behavior at infinity. Let ϱ be a probability distribution on Γ satisfying the conditions:

$$\int_\Gamma \varrho = \int_\Gamma \varrho_0 = 1 \quad \text{and} \quad \int_\Gamma \varrho H = \int_\Gamma \varrho_0 H = U \tag{10.1.16}$$

and let

$$\Delta \equiv \ln \left[\frac{\varrho}{\varrho_0} \right] \quad i.e. \quad \varrho = e^{\beta(\Psi - H) + \Delta} \quad . \tag{10.1.17}$$

Let us now evaluate

$$S(\varrho) - S(\varrho_0) = -k \left[\int_\Gamma \varrho \ln \varrho - \int_\Gamma \varrho_0 \ln \varrho_0 \right]$$

$$= -k \int_\Gamma \left[\varrho \cdot (\beta(\Psi - H) + \Delta) - \varrho_0 \cdot (\beta(\Psi - H)) \right] \tag{10.1.18}$$

$$= -k\beta\Psi \int_\Gamma [\varrho - \varrho_0] + k\beta \int_\Gamma [\varrho - \varrho_0] H - k \int_\Gamma \varrho \Delta .$$

The two conditions (10.1.16) imply that the first two terms in this expression vanish, so that (10.1.18) reduces to:

$$S(\varrho) - S(\varrho_0) = -k \int_\Gamma \varrho \Delta \tag{10.1.19}$$

which we can rewrite as

$$S(\varrho) - S(\varrho_0) = -k \int_\Gamma \varrho_0 \left[\frac{\varrho}{\varrho_0} \right] \ln \left[\frac{\varrho}{\varrho_0} \right]$$

$$= -k \int_\Gamma \varrho_0 \left(\left[\frac{\varrho}{\varrho_0} \right] \ln \left[\frac{\varrho}{\varrho_0} \right] + 1 - \left[\frac{\varrho}{\varrho_0} \right] \right) \tag{10.1.20}$$

$$= -k \int_\Gamma \varrho_0 \, f \left(\left[\frac{\varrho}{\varrho_0} \right] \right)$$

where f is the function encountered already in (10.1.9), and satisfying (10.1.10), so that we obtain indeed

$$S(\varrho) - S(\varrho_0) \leq 0 \qquad \text{with} \qquad S(\varrho) - S(\varrho_0) = 0 \quad \text{iff} \quad \varrho = \varrho_0 \quad . \quad (10.1.21)$$

To prove the second part of the theorem, note that

$$\partial_\beta \ln Z = Z^{-1} \partial_\beta Z = Z^{-1} \int_\Gamma e^{-\beta H} (-H) = -\int_\Gamma \varrho_0 H$$

$$\partial_\beta{}^2 \ln Z = -Z^{-2} (\partial_\beta Z)^2 + Z^{-1} \partial_\beta{}^2 Z = \qquad\qquad (10.1.22)$$

$$- \left(\int_\Gamma \varrho_0 H \right)^2 + \int_\Gamma \varrho_0 (H)^2 = \int_\Gamma \varrho_0 \left(H - \int_\Gamma \varrho_0 H \right)^2 \geq 0 .$$

The first of the equations (10.1.22) establishes the desired relation between the partition function Z and the expectation value of H. The second of these equations shows that the expectation value of H is a monotonically decreasing function of the parameter $\beta = 1/\Theta$, so that there is only one value of Θ for which the constraint is satisfied; this in turn implies that the value of Z and hence Ψ are uniquely determined by the constraint. **q.e.d.**

This result – which Gibbs gives with no more fanfare than he makes over any of the other theorems he proved in his Chap. XI – was forcefully promoted half-a-century later by Jaynes in a series of papers [Jaynes, 1989] as one of Gibbs' major contributions. Indeed, at this point, one can turn the table around: forget about the original justifications Gibbs gave for his canonical ensemble, and take this as an *alternate definition* of the canonical distribution.

While, at the *syntactic* level, the relative emphasis is only a matter of taste, it makes a huge difference at the *semantic* level: the canonical distribution is the one probability distribution that incorporates the information contained in the constraints and no more. Rather than being attached to a frequentist interpretation based on pragmatically ill-defined ensembles, statistical mechanics is now open to an information-theoretic interpretation of probability *à la de Finetti* – see Sect. 6.3: the canonical distribution is the probability distribution that expresses best the available knowledge; thus, in de Finetti's perspective, it would be irrational to bet on anything else, which would amount to betting on the basis of information that is not available.

Objections, however, may be raised against attributing to the constraints the status of empirical data; see e.g. [Uffink, 1995, 1996a] and references given therein. One of the most serious flaws could be that the acceptance of an *average* energy as a datum would require measurements on a collection of identically prepared systems, and thus would seem to appeal again to a frequentist notion of ensemble; how could one, from a measurement carried on a single system, be reasonably sure that the datum is representative of the situation, and not just a fake resulting from a large fluctuation? For such datum to become a reliable constraint one would need to know, *at least,* that the intrinsic fluctuations of the system under consideration are much smaller than the extrinsic experimental accuracy of the measuring process.

A response to this objection can be construed in a syllogism assembled from Gibbs' own writings:

(a) *"the average square of the anomalies of ε,* [Gibbs' notation for the energy] *that is, of the deviation of the individual values from the average, is in general of the same order of magnitude as the reciprocal number of degrees of freedom, and therefore to human observation the individual values are indistinguishable from the average values when the number of degrees of freedom is very great"* [Gibbs, 1902][p. 168];

(b) *"the distribution of energy between the parts of a system, in case of thermal equilibrium, is such that the temperatures of the parts are equal"* [Gibbs, 1902][p. 174];

(c) *"if a system of a great number of degrees of freedom is microscopically distributed in phase, any very small part of it may be regarded as canonically distributed"* [Gibbs, 1902][p. 183].

We propose to examine each of these points separately in order to gauge their generality and the semantics they support.

Before doing so, however, we note that every thing we presented above – up to and including Thm. 10.1.2 – is completely insensitive to the number $n(= 3N)$ of degrees of freedom of the mechanical systems considered. Yet, systems with a large number of degrees of freedom do play a special role in thermodynamics, would it only be so when one introduces in the discussion such things as "heat baths" and "reservoirs": by definition, the latter are supposed to be so large that their properties cannot be observably affected by letting much smaller systems be brought in contact with them. This is going to be taken into account presently, as the major premise (a), and thus the conclusion (c), appeal explicitly to the presence of a large number of degrees of freedom.

For the origin of Premise (a) consider the relative dispersion of the energy of a canonical distribution around its expectation value. From (10.1.22) we read:

$$\frac{\langle (H - \langle H \rangle)^2 \rangle}{\langle H \rangle^2} = \frac{\partial_\beta^2 \ln Z}{(\partial_\beta \ln Z)^2} \tag{10.1.23}$$

the dependence of which against the number n of degrees of freedom has to be evaluated. Evidently, this may be model dependent, which implies that we should explore the n–dependence on a model that stands a chance to be "typical". The archetype of choice is, here again, the ideal gas modeled by a $6N$–dimensional phase-space Γ with configuration space $\Lambda = D^N$ with $D \subset R^3$ and $\int_D d\boldsymbol{q} \equiv V$; and Hamiltonian consisting only of a kinetic term

$$H = \frac{1}{2m} \sum_{k=1}^{N} |\boldsymbol{p}_k|^2 \tag{10.1.24}$$

so that the canonical distribution ϱ and the partition function Z take the form

$$\varrho = Z^{-1} \int_\Gamma d\boldsymbol{p}_1 \cdots d\boldsymbol{p}_N \, d\boldsymbol{q}_1 \cdots d\boldsymbol{q}_N \, e^{-\beta H} \; ; \; Z = \left[(2\pi m \Theta)^{\frac{3}{2}} V \right]^N \tag{10.1.25}$$

which, *up to normalization,* we knew as the Maxwell distribution (3.2.10). Upon inserting in (10.1.23) the result of the computations suggested by (10.1.22), namely:

$$\left.\begin{array}{c} \ln Z = N\left[\ln V + \frac{3}{2}\ln(2\pi m\Theta)\right] \\[2mm] \langle H\rangle \equiv U = -\partial_\beta \ln Z = \frac{3}{2}N\Theta \\[2mm] \langle (H - \langle H\rangle)^2\rangle \equiv {\partial_\beta}^2 \ln Z = \frac{3}{2}N\Theta^2 \end{array}\right\}. \qquad (10.1.26)$$

we obtain

$$\frac{\sqrt{\langle (H - \langle H\rangle)^2\rangle}}{\langle H\rangle} = (\frac{1}{2}3N)^{-\frac{1}{2}} = O(n^{-\frac{1}{2}}) \quad . \qquad (10.1.27)$$

Hence, the average of the standard deviation of the energy, as compared to the energy itself decreases as the inverse of the square-root of $n = 3N$, the number of degrees of freedom. Therefore, as n becomes very large, the energy is very closely equal to its average value, i.e. *the canonical distribution becomes indistinguishable from the microcanonical distribution.* The decrease of relative fluctuations as n increases is part of the argument Gibbs was to develop in favor of considering what was to become the study of the thermodynamical limit:

> *A very little study of the statistical properties of conservative systems of a finite number of degrees of freedom is sufficient to make it appear, more or less distinctly, that the general laws of thermodynamics are the limit toward which the exact laws of such systems approximate, when their number of degrees of freedom is indefinitely increased.*
>
> [Gibbs, 1902][p. 166]

As presented here, the computation is done for the ideal gas model only. A similar computation for an assembly of n harmonic oscillators would give the same result. Upon re-examining the proofs, one notices that the argument centers on

$$\langle (H - \langle H\rangle)^2\rangle = {\partial_\beta}^2 \ln Z = -\partial_\beta U = kT^2\partial_T U = kT^2 C_V \qquad (10.1.28)$$

so that (10.1.27) expresses the pragmatic expectation that the energy U and the specific heat C_V be both proportional to N, the number of molecules. From this, one would surmise that the result (10.1.27) should persist for a large class of models, although Gibbs himself warned, [Gibbs, 1902][p. 75], that this could not be a universal proposition; indeed, we shall see in Chaps. 11–14 how large fluctuations do indeed occur in systems presenting phase transitions. If something go wrong in the argument, we may therefore have to re-examine its dependence on Premise (a). For the time being, we accept this premise on the basis of the evidence presented so far. Nevertheless, we return to this point shortly.

Premise (b) is to be seen in the present argument as a semantic comment on (10.1.11–15), the latter having been derived above as a purely syntactic statement.

Conclusion (c) follows then from (a) and (b). By definition of the microcanonical distribution, the energy of a system (Γ, H) has a sharp value, say U. When this system is very large, this distribution is empirically indistinguishable from a canonical distribution with average energy U. When considering an aggregate system $(\Gamma = \Gamma_S \times \Gamma_R, H = H_S + H_R + H_{int})$, the canonical distribution when reduced to the part S of the system is the canonical distribution for that system, with respect to its own Hamiltonian H_S, provided that the interaction energy H_{int} be negligible; this holds irrespectively of the number of degrees of freedom of the subsystem (Γ_S, H_S), although the first step of the argument is premised on the assumption that the aggregate system (Γ, H) has a large number of degrees of freedom.

This provides a justification for considering as empirical data the constraints which are imposed to derive the canonical distribution from the informational entropy, in accordance with the alternate definition inferred from Thm. 10.1.2.

In reference to our argument in favor of Premise (a), we should mention that Conclusion (c) – and thus Premise (a) – can be viewed, more generally, as a consequence of the central limit theorem; indeed compare relation (10.1.27) with (9.5.6) which is due to Khinchin [1949]; moreover this approach carries over to the quantum domain [Jancel, 1969]. However, we recall Gibbs' warning: since there are systems for which (10.1.27) does not hold, any such proof must contain some explicit limitation on the generality of its assumptions. In the approach just mentioned, a natural assumption seems to be the independence of the stochastic variables considered; this independence can be spoiled by the cooperative phenomena responsible for phase transitions.

In connection with the proposal for a justification of viewing the constraint $\langle H \rangle = U$ in Thm. 10.1.2 as being empirically meaningful, we should mention that Gibbs also pursued two other tracks.

First, he wanted to interpret the microcanonical ensemble as a *time-ensemble,* [Gibbs, 1902][pp. 169 & ff.] a resurgence of the ergodic theory which is however not directly relevant to our purpose here when Premise (a) is accepted. Moreover there are some pragmatic objections to relying too strictly on the microcanonical ensemble; indeed this ensemble corresponds to the physical situation where the system considered is *completely* isolated from its surroundings: in particular, assuming no interaction with the surroundings would preclude making any experiment on the system. Hence the microcanonical ensemble must be regarded as an idealization, away from which accommodations are necessary. The canonical distribution is better suited to the project of modeling thermodynamics, since it allows the system to exchange energy with its surroundings. And, in fact, as one follows the flow of his argument, one sees that Gibbs returned to the canonical distribution as the one that corresponds to the situation *"when the quantity of the bath is indefinitely increased"* [Gibbs, 1902][p. 181] ... which was the point of the above syllogism.

This still leaves a concern one can identify in the critical examinations of almost any subjective semantics, namely to try and force on it an objective reinterpretation. One way to respond is to argue – with [De Finetti, 1937, 1972] and [Jaynes, 1967, 1989] – that a theory should ultimately be judged by the predictions it makes, and then to note that the thermodynamical predictions of statistical mechanics would fare well by this criterion. The other way is to entertain the rhetorical question of how the system "knows" (i.e. reacts to the fact) that we regard it under such and such constraints. The above discussion makes a first step in the direction of an answer: the system is made to "know" by being brought into contact with a thermal bath.

Yet, the discussion so far is silent on *how* the system reacts. In this juncture enters the *second,* and more demanding, track Gibbs wanted to pursue. He offered to replace the static Conclusion (c) by a dynamical one on thermalization, based on the replacement of (b) above by the stronger requirement, namely: (b')*"a body of which the original thermal state may be entirely arbitrary, may be brought approximately into a state of thermal equilibrium with any given temperature by repeated connections with other bodies of that temperature."* [Gibbs, 1902][p. 161]. This is an extension of the already quoted conclusion that energy flows from the hotter to the colder body, itself presented by Gibbs as consequence of Scholium 10.1.1 and an appeal to Thm. 10.1.2 via the analogy with entropy increases in thermodynamics. To hope for this to be true could be construed as tantamount to requiring a definite *approach* towards equipartition of the energy among all degrees of freedom. However, this process can be so extremely slow as to be hard to observe; recall the discussion of the Fermi–Pasta–Ulam model towards the end of Sect. 9.4. For foundational purposes, we prefer to separate the definition of the equilibrium distribution from the harder question of whether and how this distribution can be reached. We consider the first issue to be essentially settled: the information-theoretic definition following from Thm. 10.1.2 is clean and unambiguous; we argued that it is pragmatically acceptable, even in terms of the empirical character of the constraints it imposes. As for the second issue, we must emphasize that a rigorous – and verifiable – formulation of the mechanical conditions, which would entail any general conclusion to the effect that specified Hamiltonian microscopic models would necessarily approach thermodynamical equilibrium, is still a largely open problem in non-equilibrium statistical mechanics. Nevertheless, some encouraging results have been obtained in this direction; see part of the discussion at the end of Sect. 10.2b below, and in Chap. 15.

Once agreed on the definition of the canonical distribution, one can proceed similarly for the so-called grand canonical distribution, where not only the energy but also the number of particles is subject to fluctuations.

Definition 10.1.3. *The* grand canonical distribution *is the probability distribution, defined on* $\Gamma = \cup_{N=0}^{\infty} \Gamma_N$ *by:*

$$\varrho_N(p_1, \cdots, p_N, q_1, \cdots q_N) = \mathcal{Z}^{-1} e^{(\mu N - H_N)/\Theta}$$

with $\mathcal{Z}$ *denoting the grand partition function:*

$$\mathcal{Z} = \sum_{N=0}^{\infty} \int_{\Gamma_N} e^{(\mu N - H_N)/\Theta}$$

and Θ, μ *real.*

Theorem 10.1.3. *The grand canonical distribution maximizes*

$$S(\varrho) = -k \sum_N \int_{\Gamma_N} \varrho_N \ln \varrho_N$$

subject to the constraints

$$U = \sum_N \int_{\Gamma_N} \varrho_N H_N \qquad \text{and} \qquad \langle N \rangle = \sum_N \int_{\Gamma_N} \varrho_N N \quad .$$

Here again, this variational result – which can be formally derived by the method of Lagrange multipliers – can now be taken as an alternate definition of the grand canonical distribution. With $\beta = 1/\Theta$ and $\beta\mu$ considered as independent variables, one has the relations

$$\left. \begin{aligned} U &= -\partial_\beta \ln \mathcal{Z}|_{\beta\mu} \\[2mm] \langle N \rangle &= \partial_{\beta\mu} \ln \mathcal{Z}|_\beta \\[2mm] S &= k\left(\ln \mathcal{Z} + \beta U - \beta\mu\langle N \rangle \right) \end{aligned} \right\} . \tag{10.1.29}$$

Again, the analogy with thermodynamics is manifest, and points to a semantics for the grand canonical distribution: $T = \Theta/k$ is again the temperature, S the thermodynamical entropy, U the internal energy, and now $\langle N \rangle$ is the average number of particles, while μ is the chemical potential. The discussion of this semantics, including the appeal to a large reservoir, parallels the discussion we outlined for the canonical distribution. We should moreover mention that there is an immediate generalization of Def. 10.1.3 to the case of several species of particles, each appearing now with its own chemical potential μ_k. We do not pursue this avenue here, but refer to [Gibbs, 1902][Chap. XV], and to [Gibbs, 1961][papers III and IV, and much of the "unpublished" fragments collected in IX].

Returning for a moment to the canonical distribution, we should emphasize that the last word could not have been said in [Gibbs, 1902]. The Gibbs paradox illustrates that an essential and new avenue was still open, which is the subject of Sect. 10.2 below.

The ideal gas suffices to our purpose here. With the notation of Thm. 10.1.2, we have

$$S = k \left(\ln Z + \frac{1}{\Theta} U \right) \tag{10.1.30}$$

in which we insert the computations (10.1.26) to obtain:

$$S = kN \ln \left(V \Theta^{\frac{3}{2}} \right) + \frac{3}{2} kN \left(1 + \ln(2\pi m) \right) \tag{10.1.31}$$

where $V = |\Lambda|$ is the volume in configuration space.

Imagine now the following gedanken experiment. Two vessels separated by a thin wall are filled with a certain amount of a gas; suppose the molecules of the gas on both sides have all the same mass m; and that both sides are in equilibrium at the same temperature and same pressure. Now imagine that the wall between the two vessels is gently removed. The aggregate vessel is now filled with a gas that is still in equilibrium at the same temperature and pressure. Let us denote by S_1 and S_2 the entropies of the gas on each side of the wall; since the corresponding canonical distributions are independent of one another, the entropy of the aggregate – before the wall is removed – is $S_1 + S_2$. Let further S_{12} be the entropy of the gas after the wall has been removed. From (10.1.31) we read

$$\left. \begin{array}{l} S_{12} - (S_1 + S_2) = \\[2mm] \quad k\left[(N_1 + N_2) \ln(V_1 + V_2) - N_1 \ln V_1 - N_2 \ln V_2 \right] = \\[2mm] \quad -k\left[N_1 \ln \frac{V_1}{V_1 + V_2} + N_2 \ln \frac{V_2}{V_1 + V_2} \right] > 0 \end{array} \right\} \quad . \tag{10.1.32}$$

So, the informational entropy has increased, while we expected the thermodynamical entropy to remain constant, since the wall between the two vessels is immaterial from a thermodynamical point of view.

What has happened to the seemingly perfect analogy between the informational syntax of the canonical distribution and the syntax of thermodynamics? How come that they could lead to two mutually exclusive conclusions? This is the Gibbs paradox.

To see that we have indeed lost information in the course of removing the wall, suppose that before removing the wall we had painted in blue the molecules in one vessel, and in yellow those of the other. As the wall is removed, the molecules of each vessel are allowed to roam the whole vessel so that, after a while, we get an uniform greenish mixture throughout the aggregate vessel. Whereas we knew, with probability one, that a blue molecule was initially in the vessel where we had put it, we only know now that it is in that part of the aggregate vessel with probability $N_1/(N_1 + N_2)$.

The resolution of the paradox is therefore that the classical formalism of the canonical distribution has not so far taken into account that the molecules of a gas are not distinguishable from one another. Note that we did not get into this difficulty when we were presenting kinetic theory; compare indeed the strict inequality (10.1.32) above and the equality (3.3.27). Let us denote by S_B the entropy obtained in (3.3.26), and by S_{can} the entropy obtained now in (10.1.31); we have

$$S_B = S_{can} - N \ln N \quad . \tag{10.1.33}$$

Upon following the computations that lead to each of these two formulas, one sees that the difference did appear because of a difference in normalization, difference between the definitions of the entropy: $S = -k \int \varrho_1 \ln \varrho_1$ vs. $S = -k \int f \ln f$, with $f = N \varrho_o$. However, this formal remark does not address the fundamental issue of the indistinguishability of the molecules. Instead, let us recall that we are exploring the connection with thermodynamics, where large systems are the order of the day. We know from Stirling formula, that for large values of N, $\ln(N!) \approx N \ln N$. Together with (10.1.33), this suggests replacing the canonical partition function by:

$$Z_B = \frac{1}{N!} Z_{can} \quad . \tag{10.1.34}$$

It is easy to verify that this additional factor does not affect any of the thermodynamical relations. And, at least, the extra factor $N!$ has for it that it is the number of ways there are to distribute N particles in the same configuration. Nevertheless, the so-called "correct Boltzmann counting" (10.1.34) *must* appear as an *ad hoc* Ansatz; it is certainly on a par with the marvelous ingenuity manifested in other parts of Gibbs' work; more importantly, it stands by the road towards the final resolution of the paradox. This is to be done in two steps. First, one should come to recognize that quantum mechanics is the place where the indistinguishability of particles enters in a natural and inescapable way; and second, one should bring to bear that one expects the predictions of classical and quantum mechanics to agree in the high temperature limit. Now indeed, this is what happens: the above factor $N!$ appears when one takes the high temperature limit of the quantum canonical partition function.

Gibbs had died in 1903, just as quantum mechanics started to emerge; he did not live long enough to contribute to the forthcoming quantum statistical mechanics. All the same, he did leave behind an impressive bridge connecting ideas from classical mechanics and thermodynamics through the use of statistical assumptions that he rightly called canonical. Nevertheless, the strength of the structure rested on two somewhat ill-defined notions: the ensembles and the appeal to large systems. We discussed above how the subsequent developments in the semantics of probability theory allowed to turn Gibbs' variational theorems – Thms. 10.1.1 and 2 – into definitions with empirical grounding, thus allowing to dispense with the necessity of introducing ensembles. Evidently, this is not to deny the reality that a cup of coffee is a collection of particles, but only to reject as superfluous to the foundation of the theory the notion that one should consider ensembles, the elements of which are copies of a cup of coffee. As for controlling the thermodynamical limit, this has been a major task in mathematical physics, a task that was largely accomplished in [Ruelle, 1969]; we sketch a quantum version of this program towards the end of Sect. 12.1.

The thermodynamical limit formalism allows to place in broader perspective the use of the central limit theorem following [Khinchin, 1949] to derive the canonical distribution from the microcanonical distribution, and thus give a more explicit justification for the conclusion of the Gibbs syllogism discussed earlier in this section; see e.g. [Dobrushin and Tirozzi, 1977]. In modern treatments, see e.g. [Deuschel, Stroock, and Zessin, 1991, Georgii, 1993, Kipnis and Landim, 1999], the reliance on the central limit theorem is superseded by the theory of large deviations [Ellis, 1985, Deuschel and Stroock, 1989] allowing here to deal with deviations of the order of the volume $|\Lambda|$ of the configuration space, rather than $|\Lambda|^{\frac{1}{2}}$.

A sense in which the diverse equilibrium ensembles are related in the thermodynamical limit was established for a class of continuous systems much richer than the ideal gas primarily covered by Khinchin. Indeed, [Ruelle, 1969][see also [Lanford, 1973, Georgii, 1988]] showed how the pressure $p(z,\beta)$ defined from the grand canonical partition function $\mathcal{Z}_\Lambda(z,\beta) = \sum_N z^N Z_\Lambda(N,\beta)$ – where $Z_\Lambda(N,\beta)$ is the configurational canonical partition function – by:

$$\beta\, p(z,\beta) = \lim_{|\Lambda|\to\infty} |\Lambda|^{-1} \mathcal{Z}_\Lambda(z,\beta) \quad ; \tag{10.1.35}$$

the free-energy $f(\varrho,\beta)$ per unit volume defined from the canonical partition function $Z_\Lambda(N,\beta)$ by:

$$-\beta f(\varrho,\beta) = \lim_{|\Lambda|\to\infty} |\Lambda|^{-1} Z_\Lambda(N,\beta) \quad \text{with} \quad \varrho = \frac{N}{|\Lambda|} \quad \text{fixed} \quad, \tag{10.1.36}$$

and the entropy $s(\varrho,\varepsilon)$ per unit volume defined from the microcanonical partition function $\Omega_\Lambda(N,E)$ by:

$$s(\varrho,\varepsilon) = \lim_{|\Lambda|\to\infty} |\Lambda|^{-1} \Omega_\Lambda(N,E) \quad \text{with} \quad \varepsilon = \frac{E}{|\Lambda|} \quad \text{fixed} \quad, \tag{10.1.37}$$

satisfy the relations

$$p(z,\beta) = \sup_{\varrho<\varrho_{cp}} [\beta^{-1}\varrho \ln z - f(\varrho,\beta)] \tag{10.1.38}$$

$$f(\varrho,\beta) = \inf_{\varepsilon>\varepsilon_o(\varrho)} [\varepsilon - \beta^{-1}s(\varrho,\varepsilon)] \quad . \tag{10.1.39}$$

The convexity properties of these functions are essential both to the mathematics of these relations and to the stability of the physical systems considered; the latter aspect is illustrated in Chap. 12. The relations (10.1.38–39) are of the form of the conjugation relation in convexity theory, the Legendre transformation familiar from the passage from Lagrangian mechanics to Hamiltonian mechanics, e.g. in the free case:

$$\frac{1}{2m} p^2 = \max_v [pv - \frac{1}{2}mv^2]$$

or more generally, the Hamiltonain $H(q,p)$ and the Lagrangian $L(q,\dot{q})$ are conjugate in the sense that they are convex functions satisfying

$$H(q,p) = \max_{\dot{q}} [p\dot{q} - L(q,\dot{q})] \quad .$$

For a recent discussion of the link between the above functional equivalence and the original measure theoretical equivalence, see [Lewis, Pfister, and Sullivan, 1994] who use a simpler form – the "principle of the largest term" – of the large deviation principle used by earlier authors to tackle the problem of equivalence of ensembles. For an interesting transposition of the large deviation principle to some dynamical ensembles in non-equilibrium statistical mechanics, see the related papers [Gallavotti and Cohen, 1995a,b, Eckmann, Pillet, and Rey-Bellet, 1999a, Lebowitz and Spohn, 1999, Gallavotti, 2000b,a].

10.2 Quantum Extensions

10.2.1 Traditional Formalism

The first extension of the Gibbs formalism from its initial classical setting to the quantum realm is conceptually straightforward. Recall that in the traditional formulation of quantum mechanics [Von Neumann, 1932a], the observables are represented by self-adjoint operators acting on a Hilbert space $\mathcal{H}$, and the states by density matrices ϱ, i.e. trace-class, positive operators, normalized by $\mathrm{Tr}\varrho = 1$, so that the expected value of an observable A, when the system is in the state ϱ, is given by $\langle A \rangle = \mathrm{Tr}\varrho A$. Recall further, from Thm. 5.4.3, the reasons we have to define the entropy of the quantum state ϱ as $S(\varrho) = -k\mathrm{Tr}\varrho \ln \varrho$.

In the sequel we denote by $\mathcal{B}(\mathcal{H})$ the set of all bounded operators acting on the Hilbert space $\mathcal{H}$, and by $\mathcal{R}(\mathcal{H})$ the set of all density matrices on $\mathcal{H}$. Note that every convex combination of two density matrices is again a density matrix, i.e. ϱ_1, $\varrho_2 \in \mathcal{R}(\mathcal{H})$ and $\lambda_1, \lambda_2 \in \mathsf{R}^+$ with $\lambda_1 + \lambda_2 = 1$ implies that the convex combination $\lambda_1\varrho_1 + \lambda_2\varrho_2$ is also in $\mathcal{R}(\mathcal{H})$; the special density matrices which cannot be written as mixtures (i.e. as convex combinations) of two different density matrices are called *pure states;* they are one-dimensional projectors, i.e. there exists a unit vector $\Psi_\varrho \in \mathcal{H}$ unique up to a phase, such that $\varrho : \Psi \in \mathcal{H} \to (\Psi, \Psi_\varrho)\Psi_\varrho$; and thus for the pure states: $\langle A \rangle = \mathrm{Tr}\varrho A$ takes the form $\langle A \rangle = (A\Psi_\varrho, \Psi_\varrho)$.

Theorem 10.2.1. *Let H be a self-adjoint operator, acting on the Hilbert space $\mathcal{H}$, and satisfying*

$$\forall \, \beta \in (0, \infty) \; : \; \mathrm{Tr}\, e^{-\beta H} < \infty \quad .$$

Then for every $U \in \mathbb{R}$ such that

$$\inf_{\Psi \in \mathcal{H}\,;\,\|\Psi\|=1} (H\Psi, \Psi) \; < \; U \; < \; \sup_{\Psi \in \mathcal{H}\,;\,\|\Psi\|=1} (H\Psi, \Psi)$$

there is a unique $\varrho \in \mathcal{R}(\mathcal{H})$ that maximizes

$$S(\varrho) = -k \mathrm{Tr}\varrho \ln \varrho \,,$$

subject to the constraint

$$\mathrm{Tr}\varrho H = U \,.$$

This density matrix is of the form

$$\varrho_\beta \equiv \frac{1}{\mathrm{Tr}\exp(-\beta H)} \, \exp(-\beta H)$$

with $\beta \in (0, \infty)$ uniquely determined by U.

Remark: The condition that $\mathrm{Tr}\, e^{-\beta H}$ be defined and finite for all $\beta \in (0, \infty)$ implies:

1. H is bounded below.
2. H has discrete spectrum, namely $\mathcal{H}$ admits a basis $\{\Psi_n\} \subset \mathcal{H}$ which is orthonormal – i.e. $\forall \, m, n \; : \; (\Psi_m, \Psi_n) = \delta_{mn}$ – and such that $H\Psi_n = \varepsilon_n \Psi_n$; the Ψ_n are called the eigenvectors of H, and the ε_n its eigenvalues.
3. Each eigenvalue ε_n of H is at most finitely degenerate, i.e. the eigensubspace $\mathcal{H}_n \equiv \{\Psi \in \mathcal{H} \mid H\Psi = \varepsilon_n \Psi\}$ attached to the eigenvalue ε_n of H is finite-dimensional, i.e. $\dim \mathcal{H}_n < \infty$.
4. the spectrum $\sigma(H) \equiv \{\varepsilon_n\} \subset \mathbb{R}$ has no accumulation point; in particular, when $\dim \mathcal{H} = \infty \; : \; \varepsilon_n \to \infty$ as $n \to \infty$.

We show next how the proof of the above *quantum* Thm. 10.2.1, already announced in [Von Neumann, 1927], can be reduced to a *classical* application of the Lagrange multiplier method, by the following Lemma, first established in [Von Neumann, 1932a].

Lemma 10.2.1. *With H as in the Theorem, and ε_n, Ψ_n as in the above Remark, let P_n be the projector*

$$P_n : \Psi \in \mathcal{H} \to (\Psi, \Psi_n)\,\Psi_n \in \mathcal{H} \quad .$$

And let ϱ be an arbitrary density matrix. Then the density matrix

$$\tilde{\varrho} \equiv \sum P_n P_n \quad \text{with} \quad p_n \equiv \mathrm{Tr}\varrho P_n$$

satisfies

$$\mathrm{Tr}\, \tilde{\varrho} H = \mathrm{Tr}\, \varrho H \quad \text{and} \quad S(\tilde{\varrho}) \geq S(\varrho) \quad .$$

Proof of Lemma 2.1 $\tilde{\varrho}$ is clearly positive with $\mathrm{Tr}\tilde{\varrho} = 1$, i.e. is a density matrix. From the Remark above, $H = \sum \varepsilon_n P_n$; upon substituting this in $\mathrm{Tr}\tilde{\varrho}H$, we obtain $\mathrm{Tr}\tilde{\varrho}H = \mathrm{Tr}\varrho H$. Since $\varrho \in \mathcal{R}(\mathcal{H})$: $\mathcal{H}$ admits an orthonomal basis $\{\Phi_m\}$ such that $\varrho\Phi_m = q_m\Phi_m$, with $q_n \geq 0$ and $\sum_m q_m = 1$. Let Q_m denote the projector $Q_m : \Phi \in \mathcal{H} \mapsto (\Phi, \Phi_m)\,\Phi_m \in \mathcal{H}$. With

$$h(x) = \begin{cases} -x\ln x & \text{if} \quad 0 < x \leq 1 \\ 0 & \text{if} \quad 0 = x \end{cases} \tag{10.2.1}$$

we have:

$$\forall\, x \in [0,1] : h(x) \geq 0 \quad and \quad h''(x) < 0 \quad . \tag{10.2.2}$$

Thus:

$$\left.\begin{array}{c} \{\lambda_m \geq 0\,,\ \sum_m \lambda_m = 1\,,\ x_m \in [0,1]\} \\[2mm] \models \quad h(\sum_m \lambda_m\, x_m) \geq \sum_m \lambda_m h(x_m) \end{array}\right\} . \tag{10.2.3}$$

Now, with

$$\lambda_m \equiv \mathrm{Tr}\, Q_m P_n \quad and \quad x_m \equiv q_m \tag{10.2.4}$$

which satisfy the assumptions of (10.2.3), we obtain:

$$\left.\begin{array}{l} S(\tilde{\varrho}) = \\[2mm] k\sum_n h(p_n) = k\sum_n h(\mathrm{Tr}\,[\varrho P_n]) = k\sum_n h(\mathrm{Tr}\,[\sum_m q_m Q_m P_n]) = \\[2mm] k\sum_n h(\sum_m \mathrm{Tr}[Q_m P_n]\, q_m) \geq k\sum_n \sum_m \mathrm{Tr}[Q_m P_n]\, h(q_m) = \\[2mm] k\sum_m h(q_m) = S(\varrho) \end{array}\right\} \tag{10.2.5}$$

where the first equality in (10.2.5) came from the fact that the p_n are the eigenvalues of $\tilde{\varrho}$; the second equality ensued from the definition $p_n = \mathrm{Tr}\,(\varrho P_n)$; the third equality came from $\varrho = \sum_m q_m Q_m$; the fourth equality followed from the linearity of the trace Tr. The inequality in the second line of (10.2.5) followed from (10.2.3) with the substitution (10.2.4). The first equality of the third line in (10.2.5) came from $\sum_m Q_m = I$, which expresses the facts that $\{\Phi_m\}$ is orthonormal basis in $\mathcal{H}$, and $\mathrm{Tr}P_n = 1$; and finally the last equality came from the fact that the q_m are the eigenvalues of ϱ. In summary, (10.2.5) reads $S(\tilde{\varrho}) \geq S(\varrho)$.   **q.e.d.**

Proof of Thm. 10.2.1 From the Lemma, it follows that $S(\varrho)$ reaches it maximum with ϱ of the form $\varrho = \sum_n p_n P_n$. Consequently, we must maximize

$$S(\varrho) \equiv -k\sum_n p_n \ln p_n \tag{10.2.6}$$

subject to the constraints

$$p_n \geq 0 \quad ; \quad \mathrm{Tr}\varrho \equiv \sum_n p_n = 1 \quad ; \quad \mathrm{Tr}\varrho H \equiv \sum_n p_n \varepsilon_n = U \quad . \tag{10.2.7}$$

The proof has thus been reduced to a classical problem, which can be solved by the Lagrange multiplier method. The argument indeed can now follow step-by-step that given in our discussion (5.3.23–26). Specifically, upon equating to zero each of the partial derivative of $f(\varrho) \equiv S(\varrho) - k\beta \mathrm{Tr}\varrho H - \mu \mathrm{Tr}\varrho$ with respect to p_n, we obtain

$$-k(\ln p_n + 1) - k\beta\varepsilon_n - \mu = 0 \quad , \qquad (10.2.8)$$

i.e. upon taking into account the constraint $\sum_n p_n = 1$:

$$p_n = Z^{-1}\exp(-\beta\varepsilon_n) \quad \text{with} \quad Z = \sum_m \exp(-\beta\varepsilon_m) \quad . \qquad (10.2.9)$$

Here again, we obtain that the value of β is uniquely determined by the constraint $\mathrm{Tr}\varrho H = U$, since the LHS is a monotonically strictly decreasing function of β; indeed we check again that

$$\partial_\beta \langle H \rangle_\beta = -\langle (H - \langle H \rangle)^2 \rangle_\beta \quad \text{where} \quad \langle A \rangle_\beta \equiv \frac{\mathrm{Tr}\,e^{-\beta H}A}{\mathrm{Tr}\,e^{-\beta H}} \quad . \qquad (10.2.10)$$

q.e.d.

On the semantic level – with H interpreted as the microscopic energy, S as the entropy, $U = \langle H \rangle_\beta$ as the thermodynamical internal energy, and β as $1/kT$, with T the absolute temperature – note that (10.2.10) is to be interpreted as stating that the internal energy U is a strictly increasing function of the temperature, i.e. that the quantum specific heat is positive, as was its classical forerunner; see (10.1.23). Thm. 10.2.1, together with the above interpretation, provide the motivation for the following definition.

Definition 10.2.1. *The map*

$$\phi_\beta : A \in \mathcal{B}(\mathcal{H}) \mapsto \langle \phi_\beta; A \rangle \equiv \mathrm{Tr}\,\varrho_\beta A \in \mathbb{C} \quad ,$$

associated to the density matrix

$$\varrho_\beta \equiv Z_\beta^{-1}\exp(-\beta H) \quad \text{with} \quad Z_\beta = \mathrm{Tr}\exp(-\beta H) \quad \text{and} \quad \beta \in (0,\infty)$$

defined as in Thm. 10.2.1, is called the quantum canonical equilibrium state *w.r.t. the absolute temperature $T = 1/k\beta$ and the Hamiltonian H.*

We shall see in the next section that this interpretation works well in elementary applications, and was indeed the cornerstone that lay at the historical foundations of quantum mechanics.

10.2.2 The KMS Condition

After the initial successes garnered by the nascent quantum statistical mechanics, it became apparent nevertheless that some further extension of the canonical equilibrium formalism had become necessary. This extension takes off from Thm. 10.2.2 below, first recognized by [Haag, Hugenholtz, and Winnink, 1967] from among rather technical related materials that had belonged for some time to the toolbox of the quantum statistical mechanics practitioners [Kubo, 1957, Martin and Schwinger, 1959, Bonch-Bruevich and Tyablikov, 1962].

To motivate our next definition, we consider for any pair of observables A and B the following two time-correlation functions

$$f_{A,B}(t) = \mathrm{Tr}(\varrho_\beta \, A \, \alpha_t[B]) \quad \text{and} \quad g_{A,B}(t) = \mathrm{Tr}(\varrho_\beta \, \alpha_t[B] \, A) \quad , \qquad (10.2.11)$$

where we describe the time-evolution in the Heisenberg picture by

$$B \to \alpha_t[B] \equiv \mathrm{e}^{\mathrm{i}Ht} B \mathrm{e}^{-\mathrm{i}Ht} \quad . \qquad (10.2.12)$$

Notice that this is consistent with the Schrödinger picture

$$\Psi_t \to \Psi_t \equiv \mathrm{e}^{-\mathrm{i}Ht}\Psi$$

which is indeed the integral form of the Schrödinger equation

$$i \, \partial_t \Psi_t = H\Psi_t$$

where H is the Hamiltonian of the system considered. We have indeed

$$(B\Psi_t, \Psi_t) = (B_t\Psi, \Psi) \quad .$$

A "naïve" manipulation, using repeatedly the commutativity of the trace Tr, leads to:

$$\begin{aligned}
g_{A,B}(t) &= \mathrm{Tr}\mathrm{e}^{-\beta H}\mathrm{e}^{\mathrm{i}Ht} B \mathrm{e}^{-\mathrm{i}Ht} A \; / \; \mathrm{Tr}\mathrm{e}^{-\beta H} \\
&= \mathrm{Tr} A \mathrm{e}^{\mathrm{i}H(t+\mathrm{i}\beta)} B \mathrm{e}^{-\mathrm{i}H(t+\mathrm{i}\beta)} \mathrm{e}^{-\beta H} \; / \; \mathrm{Tr}\mathrm{e}^{-\beta H} \\
&= \mathrm{Tr}\mathrm{e}^{-\beta H} A \mathrm{e}^{\mathrm{i}H(t+\mathrm{i}\beta)} B \mathrm{e}^{-\mathrm{i}H(t+\mathrm{i}\beta)} \; / \; \mathrm{Tr}\mathrm{e}^{-\beta H} \\
&= \langle \phi_\beta; A \, \alpha_{(t+\mathrm{i}\beta)[B]} \rangle
\end{aligned}$$

namely – recall (10.2.11) – the following remarkable equality:

$$g_{A,B}(t) = f_{A,B}(t + \mathrm{i}\beta) \quad . \qquad (10.2.13)$$

The derivation of (10.2.13) may be "naïve", in the sense that it freely manipulates the operators $\exp(\beta H)$ and $\alpha_{t+\mathrm{i}\beta)}[B]$, but these technicalities can be fixed and (10.2.13) suggests the following definition.

Definition 10.2.2. *A density matrix ϱ is said to satisfy the KMS condition for the inverse temperature β and the Hamiltonian H, whenever, for every $A, B \in \mathcal{B}(\mathcal{H})$ there exists a complex-valued function $f_{A,B}$, bounded and continuous on the strip*

$$\Omega_\beta = \{z \mid 0 \le \mathrm{Im}\, z \le \beta\} \quad ,$$

analytic in the interior of that strip, and such that

$$f_{A,B}(t) = \text{Tr}(\varrho\, A\, \alpha_t[B]) \quad \text{and} \quad f_{A,B}(t+\mathrm{i}\beta) = \text{Tr}(\varrho\, \alpha_t[B]\, A)$$

where

$$\alpha_t[B] \equiv \mathrm{e}^{\mathrm{i}Ht} B\, \mathrm{e}^{-\mathrm{i}Ht} \quad .$$

The immediate interest of this definition is that it can serve as an alternate syntactic characterization of the canonical equilibrium state ϕ_β of the semantically meaningful Def. 10.2.1. Indeed, we have:

Theorem 10.2.2. *The following conditions on $\varrho \in \mathcal{R}(\mathcal{H})$ are equivalent*

1. *ϱ satisfies the KMS condition w.r.t. β and H.*
2. *ϱ is the canonical equilibrium density matrix*
   $\varrho_\beta = \exp(-\beta H) \,/\, \text{Tr}\exp(-\beta H)$.

Proof: The entailment (2) $\models$ (1) served as our motivation; see (10.2.11) and (10.2.13); we are thus interested now in proving the converse. We first note that if $A = I$ the corresponding function $f_{I,B}$ satisfies

$$\forall t \in \mathsf{R} \quad : \quad f_{I,B}(t) = f_{I,B}(t+\mathrm{i}\beta) \quad . \tag{10.2.14}$$

Since, by definition $f_{I,B}$ is continuous, bounded on the strip Ω_β, and analytic in this strip, (10.2.14) allows to extend it to a function, defined on the whole complex plane C, continuous, bounded, of period $\mathrm{i}\beta$, and analytic in each of the strips $\Omega_{n,\beta} = \{z \mid n\beta < \text{Im}\,z < (n+1)\beta\}$. It then follows from classical complex analysis, that this extended function is in fact entire, i.e. is analytic on the whole complex plane C; consequently, it satisfies the assumption of the Liouville theorem: it is a bounded entire function, and thus must be constant on C. Hence, in particular:

$$\forall t \in \mathsf{R} \text{ and } \forall B \in \mathcal{B}(\mathcal{H}) \quad : \quad \text{Tr}\varrho B = \text{Tr}\varrho\alpha_t[B]$$

i.e.

$$\forall t \in \mathsf{R} \text{ and } \forall B \in \mathcal{B}(\mathcal{H}) \quad : \quad \text{Tr}\varrho B = \text{Tr}\varrho\, \mathrm{e}^{\mathrm{i}Ht} B\, \mathrm{e}^{-\mathrm{i}Ht}$$

and thus:

$$\forall t \in \mathsf{R} \quad : \quad \mathrm{e}^{-\mathrm{i}Ht}\varrho\, \mathrm{e}^{\mathrm{i}Ht} = \varrho \quad . \tag{10.2.15}$$

Hence there exists an orthonormal basis $\{\Psi_n\}$ for $\mathcal{H}$ which diagonalizes simultaneously ϱ and H, i.e. for which

$$\varrho\Psi_n = \varrho_n\Psi_n \quad \text{and} \quad H\Psi_n = \varepsilon_n\Psi_n \quad . \tag{10.2.16}$$

With m and n such that $\varepsilon_m \neq \varepsilon_n$, let

$$A_{m,n} : \Psi \in \mathcal{H} \to (\Psi, \Psi_n)\Psi_m \quad \text{and} \quad B_{m,n} = A_{m,n}{}^* : \Psi \in \mathcal{H} \to (\Psi, \Psi_m)\Psi_n \quad ;$$

we compute the values of $f_{A_{m,n},B_{m,n}}$ independently on each of the two boundaries of the strip Ω_β :

$$
\left.
\begin{array}{lll}
\text{(a)} & f_{A_{m,n},B_{m,n}}(t) & = \varrho_m e^{i(\varepsilon_n - \varepsilon_m)t} \\
\text{(b)} & f_{A_{m,n},B_{m,n}}(t + i\beta) & = \varrho_n\, e^{i(\varepsilon_n - \varepsilon_m)t}
\end{array}
\right\} \quad .
\tag{10.2.17}
$$

Upon comparing (b) with the analytic continuation of (a), we obtain:

$$
\forall\, n, m \text{ with } \varepsilon_m \neq \varepsilon_n \quad : \quad \varrho_m e^{-(\varepsilon_n - \varepsilon_m)\beta} = \varrho_n
$$

i.e. there exists a constant λ, determined by the normalization $\mathrm{Tr}\varrho = 1$, such that

$$
\forall\, m \quad : \quad \varrho_m = \lambda e^{-\beta \varepsilon_m} \quad .
\tag{10.2.18}
$$

This is precisely saying that $\varrho = \varrho_\beta$, i.e. corresponds to the quantum canonical equilibrium state w.r.t. the absolute temperature $T = 1/k\beta$ and the Hamiltonian H.   **q.e.d.**

The interpretation of this theorem is a fascinating study in contrasts.

First of all, the theorem asserts the equivalence of two conditions, and thus it establishes an analytic criterion for canonical equilibrium, based solely on the analytic properties of the time-correlation functions. This fact however has a dark side: its very rigidity seems to preclude a description of coexisting thermodynamical phases in the traditional framework of quantum mechanics sketched at the beginning of this section. Indeed, thermodynamical phases should be equilibrium states, and the theorem asserts that there is only one such state. Nevertheless, there is a way out. In their paper on the infinite Bose gas, Araki and Woods [1963] laid the technical groundwork to control the thermodynamical limit in a way that would have made it ideally suited to providing a specific model where they could have shown explicitly that the KMS condition is robust enough to extend beyond the confines of the traditional formalism for quantum mechanics, and into the new structural landscape provided by the work of Haag and co-workers on the algebraic approach to the foundations of the many-body quantum physics; for systematic presentation of the latter approach, see [Emch, 1972a, Sewell, 1986, Haag, 1996]. By a sad irony, Araki and Woods did not notice the connection of their work with [Kubo, 1957, Martin and Schwinger, 1959]. This connection was not missed in [Haag, Hugenholtz, and Winnink, 1967], and it was an essential part of the initial exposure this form of the KMS condition received before a wide audience of physicists through [Winnink, 1967].

At this point, it becomes necessary to sketch the program involved in the catch-word *thermodynamical limit*. While we present this program here in the context of quantum theory, we ought to mention that its first version was systematically formulated in the classical context [Ruelle, 1969].

- The configuration space of the infinite system is taken to be $\Omega = \mathbb{R}^n$ or $\mathbb{Z}^n$, with $n = 1, 2,$ or 3. An increasing net $\mathcal{F}$ of bounded open regions $\Lambda \subset \Omega$ is chosen, with the partial ordering provided by the set inclusion.

It is required that every point $\omega \in \Omega$ belongs to some Λ; and that for any two Λ_1 and Λ_2 of $\mathcal{F}$, $\Lambda_1 \cup \Lambda_2 \in \mathcal{F}$.

- To every Λ is associated a C*−algebra – see Def. D.2.9 – $\mathcal{A}_\Lambda$ to be inter-preted as the algebra of observables relative to this region; in agreement with the traditional framework of quantum mechanics, one has typically $\mathcal{A}_\Lambda = \mathcal{B}(\mathcal{H}_\Lambda)$, the algebra of all bounded linear operators on the Hilbert space $\mathcal{H}_\Lambda = \mathcal{L}^2(\Lambda)$ of square-integrable functions on Λ. If $\Lambda_1 \subset \Lambda_2$, then $\mathcal{A}_{\Lambda_1} \subset \mathcal{A}_{\Lambda_2}$, and this injection is a *−algebraic homomorphism (note that it is then automatically norm-preserving also). The C* inductive limit theorem asserts the existence of a smallest C*−algebra $\mathcal{A}$ containing all $\mathcal{A}_\Lambda$ as sub-C*−algebras. $\mathcal{A}$ is called the *algebra of quasi-local observables,* thus distinguishing it from its dense sub-algebra $\mathcal{A}_o = \cup_{\Lambda \in \mathcal{F}} \mathcal{A}_\Lambda$, which is referred to as the *algebra of local observables.*

- It is assumed that for each Λ one has a Hamiltonian H_Λ that determines, on $\mathcal{A}_\Lambda$, an evolution $A \to \alpha_\Lambda(t)[A] = \exp(iH_\Lambda t)A\exp(-iH_\Lambda t)$, a canoni-cal density matrix $\varrho_{\beta,\Lambda} = [\mathrm{Tr}\exp(-\beta H_\Lambda)]^{-1}\exp(-\beta H_\Lambda)$, and a canonical equilibrium state $\phi_{\beta,\Lambda} : A \in \mathcal{A}_\Lambda \mapsto \mathrm{Tr}\varrho_{\beta,\Lambda}A \in \mathbb{C}$.

- The above assignments are assumed to be consistent [in fact proven to be so case by case] in the sense that for every local observable $A \in \mathcal{A}_o$: $\alpha(t)[A] = \lim_{\Lambda \to \Omega} \alpha_\Lambda(t)[A]$ and $\langle \phi_\beta; A \rangle = \lim_{\Lambda \to \Omega} \langle \phi_{\beta,\Lambda}; A \rangle$ exist.

- In fact, a little more is required on these limits. Firstly, it is required that ϕ_β extends from $\mathcal{A}_o$ to a state ϕ – i.e. a positive, linear functional, normalized to 1 – on the algebra $\mathcal{A}$ of quasi-local observables. A standard procedure, known as the GNS construction, provides then
  1. a Hilbert space $\mathcal{H}$,
  2. a representation $\pi : A \in \mathcal{A} \mapsto \pi(A) \in \mathcal{B}(\mathcal{H})$,
  3. a unit vector $\Phi \in \mathcal{H}$ such that
     a) for all $A \in \mathcal{A}$, $\langle \phi; A \rangle = (\pi(A)\Phi, \Phi)$,
     b) $\{\pi(A)\Phi \mid A \in \mathcal{A}\}$ is dense in $\mathcal{H}$.
  Note that nowhere did we intimate that $\pi(\mathcal{A}) \equiv \{\pi(A) \mid A \in \mathcal{A}\}$ is even dense in $\mathcal{B}(\mathcal{H})$; in particular, (3b) does not say that ϕ is a pure state, and it is emphatically not. Let $\mathcal{N}$ denote the von Neumann algebra – see Def. D.2.11 – generated as the weak-operator closure $\{\quad\}''$ of $\pi(\mathcal{A})$ in $\mathcal{B}(\mathcal{H})$: $\mathcal{N} = \pi(\mathcal{A})''$.

- The second of the additional requirements we want to impose on the limits considered above is that $\{\alpha(t) \mid t \in \mathbb{R}\}$ extends to a group of automor-phisms of $\mathcal{N}$.

- Finally, we require that the KMS condition survives these limits; specifi-cally that the following condition be satisfied, which generalizes Def. 10.2.2 from $\mathcal{B}(\mathcal{H})$ to an arbitrary von Neumann algebra $\mathcal{N}$.

The existence of the limits considered above can be rather delicate to verify although help obtains from specific physical requirements; see e.g. Def. 12.1.1.

Definition 10.2.3. *A state ϕ on a von Neumann algebra $\mathcal{N}$ on which an action $\alpha : t \in \mathsf{R} \mapsto \mathrm{Aut}(\mathcal{N})$ has been defined, is said to satisfy the KMS condition with respect to the time-evolution α and the inverse temperature $\beta \in (0, \infty)$ whenever, for every $A, B \in \mathcal{N}$ there exists a complex-valued function $f_{A,B}$, bounded and continuous on the strip*

$$\Omega_\beta = \{z \mid 0 \leq \mathrm{Im}\, z \leq \beta\} \quad ,$$

analytic in the interior of that strip, and such that

$$\boxed{f_{A,B}(t) = \langle \phi; A\, \alpha_t[B] \rangle \quad \text{and} \quad f_{A,B}(t + \mathrm{i}\beta) = \langle \phi; \alpha_t[B]\, A \rangle} \quad .$$

Independently of the above explorations, conducted in the territory of physics, original strides were made in the mathematical theory of W^*–algebras – see Def. D.2.10 – that admit a normal faithful state, i.e. a state that is countably additive, and satisfies the condition that $N \in \mathcal{N}$ and $\langle \phi; N^*N \rangle = 0$ imply $N = 0$; [Tomita, 1967, Takesaki, 1970, 1973]; for standard textbook presentations, see For our purpose, note that the canonical equilibrium state of traditional quantum mechanics is normal and faithful, and that these properties are robust enough to survive the thermodynamical limit. [Pedersen, 1979, Kadison and Ringrose, 1983, 1986].

The GNS construction, starting from a faithful normal state ϕ on a W^*–algebra $\mathcal{W}$, realizes $\mathcal{W}$ as a von Neumann algebra $\mathcal{N}$, acting on a Hilbert space $\mathcal{H}$. Moreover the latter has then a unit vector Φ that is cyclic and separating for $\mathcal{N}$; i.e. $\mathcal{N}\Phi$ is dense in $\mathcal{H}$; and $\{N \in \mathcal{N}, N\Phi = 0\} \models N = 0$.

The existence of a cyclic and separating vector Φ on a von Neumann algebra $\mathcal{N}$ is already sufficient to prove two remarkable results.

1. $\mathcal{N}$ and its commutant $\mathcal{N}' \equiv \{B \in \mathcal{B}(\mathcal{H}) \mid NB = BN \ \forall\, N \in \mathcal{N}\}$ are spatially anti-isomorphic, i.e. there exists an anti-unitary, involutive operator $J : \mathcal{H} \to \mathcal{H}$ such that $N \in \mathcal{N} \mapsto JNJ \in \mathcal{N}'$ is bijective.
2. There exists a *unique* action $\alpha : t \in \mathsf{R} \mapsto \alpha_t \in \mathrm{Aut}(\mathcal{N})$ such that the state $\phi : N \in \mathcal{N} \mapsto (N\Phi, \Phi)$ satisfies the KMS condition w.r.t. α; the mathematicians assume here without loss of generality $\beta = 1$. This α is called the *modular* action associated to Φ.

We should also note, in the general setting of Def. 10.2.3, every state ϕ satisfying the KMS condition is stationary, as is seen immediately upon using an argument identical to that used to derive (10.2.15). From a semantic point of view, this in itself is evidently not sufficient in order to justify placing the KMS condition at the foundations of a new edifice of quantum statistical mechanics.

It is therefore very important, for foundational purposes, that the KMS condition was shown to be intimately linked to various notions of stability that seem quite germane to the ideas one forms about a state ϕ describing thermodynamical equilibrium *in the thermodynamical limit* of very large systems. Among these stability conditions, we want to mention here the following ones, the essence of which can easily be conveyed in plain English; throughout, ϕ is assumed tacitly to be stationary with respect to a specified dynamics α.

- **LTS,** *or local thermodynamical stability:* modifications of the state ϕ confined to bounded regions of space cannot increase the free-energy.
- **GTS** is a global version of the above stability requirement, formulated – now in terms of the free-energy spatial density – for states ϕ that are further assumed to be translation invariant.
- **LDS,** *or local dynamical stability:* arbitrarily small local perturbations of the dynamics α admit stationary states that are arbitrarily close to ϕ.
- **P,** *or passivity:* energy cannot be extracted from the system initially in the state ϕ by applying – for any finite amount of time – any local perturbation of the dynamics α.
- **RS,** *or reservoir stability:* The system in the state ϕ can serve as a thermal reservoir when weakly coupled to a finite system.

The above conditions – and some more – are reviewed in [Sewell, 1980] and revisited in [Sewell, 1986]. Both of these references contain an extensive bibliography, and provide the precise technical definitions necessary to further semantic discussions of the role played by the KMS condition in the foundations of quantum statistical mechanics; see also [Sewell, forthcoming].

In addition, the following feature of the mathematical theory is of much physical import. While we already learned that to every faithful normal state is associated exactly one group of automorphisms for which it satisfies the KMS condition, *the converse is emphatically not true:* contrary to the traditional formalism embodied in Def. 10.2.2 and Thm. 10.2.2, the generalized Def. 10.2.3 allows for the simultaneous existence of several KMS states with respect to the *same* temperature and the *same* time-evolution. Moreover, the collection $\mathcal{S}_{\alpha,\beta}$ of all the states that satisfy the KMS condition with respect to the same evolution α and the same temperature β form a Choquet simplex, i.e. every KMS state can be decomposed in a unique way – asides from measure-theoretical trimmings we do not need to enter into here – into extremal KMS states.

One of the essential differences between classical mechanics and traditional quantum mechanics is that a quantum state does not in general decompose uniquely into extremal states. The simplex property of KMS states is thus quite remarkable. Now, as KMS states are interpreted as temperature equilibrium states, then extremal KMS states are candidates to formalize the notion of pure thermodynamical phases; see e.g. [Emch, Knops, and Verboven, 1970, Emch and Knops, 1970]: a canonical equilibrium state decomposes uniquely into its pure thermodynamical phases. We should also report the observation that this decomposition is often accompanied by spontaneous symmetry breaking: the KMS states ψ appearing in the decomposition of a KMS state ϕ do not necessarily inherit the full symmetry of ϕ.

Although the concept of spontaneous symmetry breaking came into the forefront of the theoretical physics discourse – in elementary particles as well as in statistical mechanics – during the second half of the 20th century, some versions of the phenomenon had been recognized already by the Bernoullis and Euler in the 18th century, and by Jacobi, G.H. Darwin, Lyapunov and Poincaré in the 19th century.

In the former case, a vertical elastic rod with cylindrical symmetry is subject to a load P; at first the rod is simply compressed; but there exists a critical value of the load above which the straight compressed solution becomes unstable and the rod "buckles", thus breaking the cylindrical symmetry. Euler computed this critical value $P_c = \pi^2 D L^{-2}$, (where D is a measure of the global elasticity of the rod, and L its length. Euler also gave a complete classification of the figures of equilibrium of this, the "planar elastica" [Truesdell, 1960, 1983].

The second classical model of spontaneous symmetry breaking obtained in discussions of the origin of the Earth-Moon system and the creation of double-stars from the cooling down of a self-gravitating homogenous nebula rotating around a fixed axis. At small angular velocity, the figure of equilibrium is an oblate ellipsoid, as known already to Newton and MacLaurin in their discussion of the shape of the Earth. As the temperature of our nebula decreases, due to radiation loss, the fluid contracts, and its angular velocity increases. There exists then a critical velocity above which this solution becomes unstable, and the new stable equilibrium solutions are rotating elongated ellipsoids with three unequal axes, the so-called Jacobi ellipsoids. A second bifurcation occurs at a higher, second critical velocity where the Jacobi ellipsoids in turn become unstable and the new equilibrium figures are pear-like shapes which would ultimately develop into two distinct bodies of roughly comparable size, the double-stars or the Earth-Moon system. As with the elastica of Euler, the original symmetry of the problem is lost only when one looks at the figures of equilibrium separately – here at a fixed time – but is still manifest in the manifold of the solutions [Poincaré, 1885, Chandrasekhar, 1969, Bertin and Radicati, 1976].

As for the above two classical models, spontaneous symmetry breakings occur in equilibrium statistical mechanics. At high temperatures the thermodynamical limit – with periodic boundary conditions – leads to a unique equilibrium state, and this state carries the full symmetry G of the Hamiltonian; immediately below a certain critical temperature T_c the equilibrium state is not unique anymore, but decomposes into pure phases, each of which may be invariant only under a subgroup $G_o \subset G$. The Reader will find specific models of this situation in Subsection 11.3.3 (the Weiss model); Chap. 12 (the Ising model, see in particular the discussion at the end of Sect. 12.2); and in Subsect. 14.1.2 (the BCS model).

For an application of these ideas to the existence of crystals in an Euclidean invariant theory and for general presentations of the theory, see [Emch, Knops, and Verboven, 1970, Emch, 1977, 1984].

We still want to comment on the sense in which the symmetry is *spontaneously* broken. In the two classical models just discussed, an external parameter determine the equilibrium state of the system (the load on the rod; the angular velocity for the rotating fluid). At a certain critical value of this parameter, a bifurcation occurs: a state that was stable (the straight rod; the MacLaurin ellipsoid) becomes unstable and, under the slightest asymmetric perturbation or fluctuation – however weak it may be – the system falls in another equilibrium state which is now stable (the bent rod; the precessing Jacobi ellipsoid). In the statistical models, the external control parameter is the temperature. The analog of the arbitrary small asymmetric perturbations in the case of the ferromagnetic models are the asymmetric boundary conditions at infinity. For the BCS model, the situation is less immediately visualizable as the symmetry that is broken is a gauge group; but there again, the original symmetry group acts transitively on the manifold of the pure phases.

Lest some Readers may suspect to have heard us say that the coexistence of several thermodynamical phases can only occur in the thermodynamical limit, we want to reassure them. We, too, have set water boiling in a finite kettle, and have watched ice cubes melt in our finite drink. Nevertheless, when we return to our study and theorize about what we have seen and/or done, we find that the idealization provided by the thermodynamical limit enables us to focus on those actual aspects of the phenomena that would otherwise be masked by finite-size effects. The only statement we are ready to make is that those aspects that are isolated in the process of taking the thermodynamical limit do capture the essence of a description of coexisting thermodynamical phases, although there is obviously no denying that many other phenomena do float around.

10.3 Early Successes

The harmonic oscillator played an important role in the development of quantum mechanics, and its simplicity is still reflected in the didactic expositions of the theory. Both of these statements hold particularly true in the case of statistical mechanics. We use this model first to explore and compare predictions based on either classical or quantum canonical equilibrium.

The one-dimensional harmonic oscillator is an Hamiltonian system with Hamiltonian

$$H = \frac{1}{2m}p^2 + \frac{1}{2}m\omega^2 q^2 \qquad (10.3.1)$$

where m is the mass, p the momentum, and q the position of a point particle constrained to move on the line R; the phase-space of the system is then

$\Gamma = \mathbb{R}^2$; as for ω , we will see shortly that the motion is periodic with period $\tau = 2\pi/\omega$.

To prepare for the solution of the quantum problem, we first look at the classical case and make a change of variables in the phase-space $\Gamma = \mathbb{R}^2 \approx \mathbb{C}$, namely

$$\left.\begin{array}{l} a \;=\; (m\omega/2)^{\frac{1}{2}}\, q + i\,(2m\omega)^{-\frac{1}{2}}\, p \\[2mm] a^* = (m\omega/2)^{\frac{1}{2}}\, q - i\,(2m\omega)^{-\frac{1}{2}}\, p \end{array}\right\} \qquad \text{and}$$

i.e. $\qquad\qquad\qquad\qquad\qquad\qquad\qquad\qquad\qquad\qquad$ (10.3.2)

$$\left.\begin{array}{l} p = i\,(m\omega/2)^{\frac{1}{2}}\,(a^* - a) \\[2mm] q =)(2m\omega)^{-\frac{1}{2}}\,(a^* + a) \end{array}\right\} \qquad \text{and} \quad .$$

With this change of variables the Hamiltonian (10.3.1) takes the form

$$H = \omega\, a^* a \quad . \tag{10.3.3}$$

With the Poisson bracket defined for any pair f, g of observables – i.e. real-valued functions on Γ – by

$$\{f, g\} = \partial_q f\, \partial_p g - \partial_q g\, \partial_p f \quad , \tag{10.3.4}$$

we have:

$$\{a^*, a\} = i\,1 \qquad \text{and} \qquad \{p, q\} = -1 \quad ; \tag{10.3.5}$$

and the Hamilton equations of motion in the new variables, namely:

$$\dot{a} = \{a, H\} = -i\omega a \qquad \text{or} \qquad \dot{a}^* = \{a^*, H\} = i\omega a^* \tag{10.3.6}$$

are immediately integrated to:

$$a(t) = e^{-i\omega t} a_o \qquad \text{or} \qquad a^*(t) = e^{i\omega t} a_o^* \quad . \tag{10.3.7}$$

Upon inserting this into (10.3.2), we find the familiar solutions

$$p(t) = p_o \cos\omega t - m\omega q_o \sin\omega t$$
$$q(t) = \tfrac{1}{2m} p_o \sin\omega t + q_o \cos\omega t \tag{10.3.8}$$

which satisfy indeed the usual Hamilton equation of motion

$$\dot{p} = \{p, H\} = -m\omega^2 q \qquad \dot{q} = \{q, H\} = \frac{1}{m} p \quad . \tag{10.3.9}$$

The dynamics of our harmonic oscillator is thus completely under control.

We now apply to this system the classical canonical equilibrium formalism. The canonical partition function of Def. 10.1.2, namely

$$Z_\beta^{(cl)} \equiv \int_\Gamma dp\, dq\, e^{-\beta H(p,q)} \quad ,$$

computed for the Hamiltonian H (10.3.1), is:

$$Z_\beta^{(cl)} = \frac{2\pi}{\omega} \frac{1}{\beta} \quad . \tag{10.3.10}$$

From this, we compute the free energy and its dispersion, namely

$$U_\beta^{(cl)} \equiv \frac{1}{Z_\beta^{(cl)}} \int_\Gamma dp dq\, e^{-\beta H(p,q)} \, H(p,q)$$

$$\langle (\Delta H)^2 \rangle_\beta^{(cl)} \equiv \frac{1}{Z_\beta^{(cl)}} \int_\Gamma dp dq\, e^{-\beta H(p,q)} \, [H(p,q) - U_\beta^{(cl)}]^2$$

$$U_\beta^{(cl)} = -\partial_\beta \ln Z_\beta^{(cl)} \quad \text{and} \quad \langle (\Delta H)^2 \rangle_\beta^{(cl)} = \partial^2{}_\beta \ln Z_\beta^{(cl)} \tag{10.3.11}$$

which give

$$U_\beta^{(cl)} = \frac{1}{\beta} \quad \text{and} \quad \langle (\Delta H)^2 \rangle_\beta^{(cl)} = \frac{1}{\beta^2} \tag{10.3.12}$$

and thus

$$\langle (\Delta H)^2 \rangle_\beta^{(cl)} = \left[U_\beta^{(cl)} \right]^2 \quad . \tag{10.3.13}$$

Note, that even in its classical version the harmonic oscillator involves several idealizations; for instance, the quintessential example of the Galilei pendulum ignores higher order terms which appear in the Hamiltonian when one wants to describe oscillations that are not so small as to allow to replace $q = L \sin \theta$ by $L\theta$, where L is the length of the pendulum, and θ the angle it makes with the vertical; it also ignores all friction terms. These idealizations are in part responsible for the ease with which the quantum version of the harmonic oscillator can be obtained.

In general the "quantization" of a system, i.e. the finding of its proper quantum description is a hard problem. At first sight, one could hope to construct canonically a solution of the *Dirac problem,* which consists in finding a bijective correspondence π between the classical observables, described as real-valued functions defined on a phase-space Γ, and the quantum observables, described as self-adjoint operators acting on a Hilbert space $\mathcal{H}$; it would seem reasonable to request such a correspondence to satisfy the following conditions:

1. to the trivial function 1 should correspond the trivial operator I;
2. the correspondence should be linear;
3. to the Poisson bracket between functions should correspond a commutator between operators; specifically

$$\pi(\{f,g\}) = -i\frac{1}{\hbar}[\pi(f), \pi(g)] \qquad \text{where} \qquad [A, B] = AB - BA \quad ;$$

4. the set $\{\pi(f) \mid f \in \mathcal{C}\}$ should be irreducible.

The meaning of the first two conditions is immediate. The third condition originates in the analogy between the classical relation (10.3.4) and the quantum Heisenberg canonical commutation relation

$$[p, q] = -i\hbar I \qquad (10.3.14)$$

where $\hbar = \frac{h}{2\pi}$ is a universal constant: $h = 6.6256 \cdot 10^{-34}\,\mathrm{J\,sec} = 6.6256 \cdot 10^{-27}\,\mathrm{erg\,sec}$ is known as the *Planck constant;* theoretical physicists often choose units in which $\hbar = 1$ and we followed them in this usage in other parts of this work; here, however, we keep $\hbar$ in our formula for two reasons: (i) we intend to comment on the so-called classical limit; and (ii) we want to cover Planck's radiation formula, where this constant was first introduced.

Note that the Poisson bracket and the commutator share three essential properties: they are linear in each of their factor, they are derivations, and they satisfy the Jacobi identity:

$$\{\alpha f + \beta g, h\} = \alpha\{f, h\} + \beta\{g, h\}$$

$$\{fg, h\} = \{f, h\}g + f\{g, h\}$$

$$\{\{f, g\}, h\} + \{\{g, h\}, f\} + \{\{h, f\}, g\} = 0$$

and similarly for the quantum bracket $-i\hbar\,[\,.\,,\,.\,]$.

Finally, the last of the four conditions tentatively imposed on the quantization map π is motivated by the remark that in the usual formalism of quantum mechanics, the constant multiples of the identity operator are the only observables that commute with all the other observables.

This program can be traced back to a series of papers by Dirac [1925]. Interestingly enough, this program was **proven impossible** to achieve [Gronewold, 1946]; the obstruction persists even when the fourth requirement is relaxed from irreducibility to finite multiplicity [Chernoff, 1981]; and it would require infinite multiplicity to allow for the simultaneous fulfillment of the program [Van Hove, 1951]; for didactic accounts, see [Abraham and Marsden, 1978, Emch, 1984]; and for the establishment of an unexpected connection between the KMS structures and the version of the van Hove proposal pursued in the geometric quantization program, see [Emch, 1981]. An inspection of the proof of the no-go theorem shows that difficulties with the Dirac program occur as soon as third-degree polynomials in the operators p and q need to be considered. The harmonic oscillator is special precisely in that it involves, in its formulation, only quadratic polynomials. We can therefore safely forge ahead.

For this purpose, we first describe a realization of the operators corresponding to the variables introduced in (10.3.2).

Let $\{\Psi_n \mid n \in \mathbf{Z}^*\}$ be an orthonormal basis in the separable Hilbert space $\mathcal{H}$, and define the Hermitian conjugate operators

$$a\,\Psi_n = \begin{cases} \sqrt{n\hbar}\,\Psi_{n-1} & \text{when} \quad n \geq 1 \\ 0 & \text{when} \quad n = 0 \end{cases}, \qquad (10.3.15)$$

and thus

$$a^* \Psi_n = \sqrt{(n+1)\hbar}\, \Psi_{n+1} \quad \forall\, n \quad . \tag{10.3.16}$$

These operators satisfy indeed

$$[a^*, a] = -\hbar I \quad , \tag{10.3.17}$$

which satisfies the correspondence rule (3): compare indeed with the classical relation (10.3.5). This representation is also irreducible, and thus satisfies the correspondence rule (4). Upon applying to the operators a and a^*, the linear transformation (10.3.2) we obtain the operators

$$\left.\begin{aligned} p &= i\,(m\omega/2)^{\frac{1}{2}}\,(a^* - a) \\[2mm] q &= (2m\omega)^{-\frac{1}{2}}\,(a^* + a) \end{aligned}\right\} \tag{10.3.18}$$

which satisfy the canonical commutation relations [$\equiv$ CCR] (10.3.5).

When we turn to the quantization of the Hamiltonian, we must make a choice between (10.3.1) or (10.3.3) that are equivalent in the classical theory: at the quantum level we have indeed

$$\frac{1}{2m}p^2 + \frac{1}{2}m\omega^2 q^2 = \frac{1}{2}\omega(a * a + aa^*) = \omega a^* a + \frac{1}{2}\hbar\omega I \quad . \tag{10.3.19}$$

Since adding a constant multiple – here the "zero–point energy" $\frac{1}{2}\hbar\omega$ of the identity operator to the Hamiltonian does not change any of the predictions of the theory concerning the evolution and the canonical equilibrium value of the observables, we are free here to choose the simpler form, namely

$$H = \omega\, a^* a \quad \text{with} \quad a \text{ and } a^* \text{ as in} \quad (10.3.15 - 16). \tag{10.3.20}$$

As far as the dynamics is concerned, all the results obtained in the classical case transpose without changes to quantum case.

The comparison is more interesting when we consider the canonical equilibrium state. The representation we just constructed for the operators a and a^* diagonalizes the Hamiltonian (10.3.20):

$$H\Psi_n = \varepsilon_n \Psi_n \quad \text{with} \quad \varepsilon_n = n\,\hbar\omega \quad . \tag{10.3.21}$$

Hence, the quantum canonical partition function of Def. 10.2.1, namely

$$Z_\beta^{(qu)} \equiv \mathrm{Tr}\exp(-\beta H) \quad ,$$

simply obtains as the sum of a geometric progression:

$$Z_\beta^{(qu)} = \left[1 - e^{-\beta\hbar\omega}\right]^{-1} \quad . \tag{10.3.22}$$

We now revise for the quantum situation the argument given above for the classical case. The free energy and its dispersion

$$U_\beta^{(qu)} \equiv \frac{1}{Z_\beta^{(qu)}} \operatorname{Tre}^{-\beta H} H$$

$$\langle (\Delta H)^2 \rangle_\beta^{(qu)} \equiv \frac{1}{Z_\beta^{(qu)}} \operatorname{Tre}^{-\beta H} [H - U_\beta^{(qu)}]^2$$

can again be computed directly from the partition function, here (10.3.22), since, as we saw in the course of the proof of Thm. 10.2.1, the classical relations (10.3.11) transpose without modifications to the quantum case, namely:

$$U_\beta^{(qu)} = -\partial_\beta \ln Z_\beta^{(qu)} \quad \text{and} \quad \langle (\Delta H)^2 \rangle_\beta^{(qu)} = \partial^2_\beta \ln Z_\beta^{(qu)} \tag{10.3.23}$$

from which we obtain

$$U_\beta^{(qu)} = \hbar\omega \frac{1}{e^{\beta\hbar\omega} - 1} \quad \text{and} \quad \langle (\Delta H)^2 \rangle_\beta^{(qu)} = \left[\hbar\omega \frac{1}{e^{\beta\hbar\omega} - 1}\right]^2 e^{\beta\hbar\omega} \tag{10.3.24}$$

i.e.

$$\langle (\Delta H)^2 \rangle_\beta^{(qu)} = \hbar\omega \left[U_\beta^{(qu)}\right] + \left[\left(U_\beta^{(qu)}\right)\right]^2 . \tag{10.3.25}$$

We now compare the predictions thus obtained from the classical and quantum canonical equilibrium states.

We first notice that the quantum relation (10.3.24) between the free-energy and its dispersion is qualitatively different from its classical predecessor (10.3.13). When $\beta\hbar\omega$ is very small, we have $[U_\beta^{(qu)}]^2 >> U_\beta^{(qu)}$ so that the second term dominates, making (10.3.25) resemble (10.3.13), its classical predecessor. On the contrary, when $\beta\hbar\omega$ becomes large, $[U_\beta^{(qu)}]^2 << U_\beta^{(qu)}$ so that, now, it is the first term that dominates, a purely quantum manifestation without classical analogue.

Pursuing this comparison further, we note that

$$\left.\begin{array}{c} \lim_{\beta\hbar\omega \to 0} U_\beta^{(qu)} = U_\beta^{(cl)} \\[2ex] \lim_{\beta\hbar\omega \to 0} \langle (\Delta H)^2 \rangle_\beta^{(qu)} = \langle (\Delta H)^2 \rangle_\beta^{(cl)} \end{array}\right\} . \tag{10.3.26}$$

For reasons that will become clear when we study the Planck radiation formula, this limit is called the *Rayleigh–Jeans limit*. In this limit, the quantum and classical predictions tend to agree. There are three ways to interpret this limit.

The first way is to view it as a formal procedure, the loosely called classical limit, in which $\hbar \to 0$. As $\hbar$ is a fixed universal constant, the latter limit is, *stricto sensu,* not an option available to the human laboratory assistants; since $\hbar$ has the dimension of an action, the limit $\hbar \to 0$ should be understood as a way to bring into focus the phenomena that occur at energies much higher than the quantum energies, measured for the harmonic oscillator in terms of $\hbar\omega$.

The second interpretation is to implement this limit by letting $\beta \to 0$, i.e. $T \to \infty$: the *high temperature limit* of the quantum theory coïncides with its classical limit; this seems to be a general feature. As both the classical and the quantum quantities considered now actually tend to infinity in this limit – see (10,3.12) – one should view this limit as an asymptotic limit i.e., in this particular case, rewrite (10.3.26) as

$$\lim_{\beta\hbar\omega\to 0} \frac{U_\beta^{(qu)}}{U_\beta^{(cl)}} = 1 \quad \text{and} \quad \lim_{\beta\hbar\omega\to 0} \frac{\langle(\Delta H)^2\rangle_\beta^{(qu)}}{\langle(\Delta H)^2\rangle_\beta^{(cl)}} = 1 \quad . \tag{10.3.27}$$

The third way is to look at this limit as $\omega \to 0$, i.e. as limit of very small frequencies, or very large wave-lengths; hence the name *infrared limit*.

The situation changes drastically in the *Wien limit*, where we have:

$$\lim_{\beta\hbar\omega\to\infty} \frac{U_\beta^{(qu)}}{U_\beta^{(cl)}} = 0 \quad \text{and} \quad \lim_{\beta\hbar\omega\to\infty} \frac{\langle(\Delta H)^2\rangle_\beta^{(qu)}}{\langle(\Delta H)^2\rangle_\beta^{(cl)}} = 0 \quad . \tag{10.3.28}$$

Hence, in the limit (10.3.28) the classical theory becomes an unreliable predictor for the quantum results; the harmonic oscillator model suggests – correctly as it turns out – that this limit becomes relevant when the frequencies are very large, i.e. the wave-lengths are very small; hence, in more general context, this failure of the classical theory is referred to as the *ultraviolet catastrophe*; the limit considered in (10.3.28), namely $\beta\hbar\omega \to \infty$, is also focusing attention on low temperature – $\beta \to \infty$ i.e. $T \to 0$ – discrepancies between predictions from classical theory and some laboratory evidences.

The above terms – *Rayleigh–Jeans limit, Wien limit, infrared and ultraviolet limits* – are rooted in Planck's study of the black-body radiation at the dawn of the 20th century; the term *ultraviolet catastrophe,* so suggestive of the disarray created by the attempts to account for the empirical data collected at the time, was apparently coined shortly thereafter by Ehrenfest; see [Klein, 1970][p. 249].

We now want to report briefly on the black-body radiation, the photoelectric effect and the specific heat of solids, which are three such situations where quantum theory succeeded in overcoming failures of the classical theory.

In the context of discovery, the black-body radiation is much more than a success of quantum mechanics: it is its birthplace.

A *black-body* is an idealized material that absorbs all the electromagnetic radiation – in particular light – that impinges on it. Physical approximations of this idealization include a screen that has been blackened with candle smoke, or the surface of the stars, and in particular the Sun. The black-body radiation problem is *to describe* the state of the electromagnetic radiation in equilibrium with such a body. For instance, it is generally accepted that the temperature of the surface of the Sun is about 6000 °K, and it is every-day experience that a significant part of its radiation is in the visible light spectrum, characterized by wave-lengths that range roughly between 4000Å (violet) and 8000Å (red) [Note: 1Å $= 10^{-10}$m $= 10^{-4}\mu$] .

It was soon agreed that the object of interest for this description is $\varrho(\nu, T)$, i.e. the spectral density – or distribution with respect to the frequency ν of the radiation – of the energy of the radiation per unit volume, when the radiation is in equilibrium at temperature T. This has the advantage of being accessible to experiment and specific enough to be the subject of theoretical analysis.

Planck [1900a] proposed a formula that would fit the data, and then proposed a theoretical justification [Planck, 1900b], the interpretation of which was at the center of much controversy, in particular with Einstein [1906], who in this matter received much support from the criticism of [Ehrenfest, 1911]. This epic has been recounted many times, and often very well; see e.g. [Jammer, 1966][Chap. 1] [Klein, 1970][Chap. 10]; therefore, here is not the place to do it again, and we propose, rather, to discuss in retrospect its context of justification. The Planck formula is:

$$\varrho(\nu, T) = \frac{8\pi}{c^3}\, h\nu^3\, \frac{1}{e^{h\nu/kT} - 1} \qquad . \qquad (10.3.29)$$

As Planck [1958] described it himself, he obtained this out of a *"lucky intuition"* followed by *"weeks of the most strenuous work in my life"*; the first assertion, at least, has to be taken *cum grano salis:* Planck's was an informed guess that followed several unsuccessful attempts by himself and others.

At the pragmatic level, precise observations were made concurrently in Berlin, and the formula fitted them well enough for Planck to ascertain from these observations reliable values for the two constants that appear in (10.3.29). From the theoretical background from which he drew the formula, Planck identified k with the Boltzmann constant which is linked to the gas constant R by the definition $k = R/N_{Av}$ where N_{Av} is the Avogadro number, the value of which he computed from the value of k obtained from black-body radiation experimental data and the routinely used value of R. For the constant h, he had obviously no compelling classical interpretation beyond the emphasis on the fact that it appeared to be universal in the sense that the same constant covered all the available data.

As we need these constants, as well as the speed of light, in the immediate sequel, we give here their modern values (in parenthesis, we note the values found by Planck)

$$
\left.\begin{array}{llll}
k & = 1.38054 \ (1.346) \ 10^{-16} & \text{erg deg}^{-1} \\
R & = 8.3143 & 10^{7} & \text{erg deg}^{-1}\text{mole}^{-1} \\
N_{Av} & = 6.0225 \ (6.175) \ 10^{23} & \text{mole}^{-1} \\
h & = 6.6256 \ (6.55) \ 10^{-27} & \text{erg sec} \\
c & = 2.9979 & 10^{8} & \text{m sec}^{-1}
\end{array}\right\} . \qquad (10.3.30)
$$

From (10.3.29), follows mathematically that at fixed temperature T the energy density ϱ has exactly one maximum, reached for

$$
\frac{h\nu_{max}}{kT} \approx 2.8215 \quad \text{(i.e.} \quad \lambda_{max} T \approx 5.0995 \ 10^{7} \ \text{Å} \ {}^{\circ}\text{K)} \qquad (10.3.31)
$$

which shows that this is well within experimental range; for instance, according to this formula, the surface temperature of the Sun $T \approx 6000^{\circ}K$ corresponds to $\lambda_{max} \approx 8500\text{Å}$ which is in the very near UV.

It is of theoretical interest that Planck's formula gives a finite result for the total energy of the radiation in an enclosure of volume V, namely

$$
E(T) \equiv V \int_{o}^{\infty} d\nu \varrho(\nu, T) \quad . \qquad (10.3.32)
$$

Note that since the Planck energy density satisfies the *Wien displacement law* [Wien, 1894]:

$$
\varrho(\nu, T) = \nu^{3} \, f(\frac{\nu}{T}) \qquad (10.3.33)
$$

where f is a function only of the combined variable ν/T; from (10.3.33), the *Stefan–Boltzmann law* [Stefan, 1879, Boltzmann, 1884] follows immediately upon a change of variable of integration in the definition (10.3.32), namely

$$
E(T) = \sigma \, T^{4} \qquad (10.3.34)
$$

where σ is a constant, now called the Stefan–Boltzmann constant; the value of this constant can now be computed from the Planck distribution. Hence, the Planck distribution postdicts – rather than predicts – and sharpens two of the then well-established laws of black-body radiation.

There was also a third law – an alternate formula for the spectral density – namely [Wien, 1896]:

$$
\varrho_{W}(\nu, T) = \alpha \, \nu^{3} \, \exp(-\gamma \frac{\nu}{T}) \qquad (10.3.35)
$$

with α and γ being two constants. By design this distribution had been proposed by Wien to satisfy his displacement law (10.3.33), and thus it leads to the Stephan–Boltzmann law (10.3.34). Planck's modification (10.3.29) of the Wien spectral density formula was purposefully wrought to preserve these properties.

While Planck, at first, had given a new theoretical derivation of the Wien spectral density, mounting experimental evidence collected for low frequencies

– i.e. long wave-lengths – motivated Planck to propose his formula (10.3.29) as an "improvement", which it is: for instance, Wien's law predicts exactly $h\nu_{max}/kT = 3$ instead of the result given in (10.3.31); such differences, as well as other features of the spectral density were experimentally detectable by then, and the observations were in favor of Planck's formula, although there were at first some discussions on whether the experimental circumstances really conformed to the black-body idealization.

After these necessary pragmatic preliminaries, we turn to some of the theoretical underpinnings of the Planck formula (10.3.29).

The formula is better understood when written in the form

$$\varrho(\nu, T) = n(\nu)\, u(\nu, T) \tag{10.3.36}$$

with

$$n(\nu) = \frac{8\pi\nu^2}{c^3} \quad \text{and} \quad \boxed{u(\nu, T) = \frac{h\nu}{e^{h\nu/kT} - 1}} \tag{10.3.37}$$

where c is the speed of light.

The first term is derived from the number $N(\nu) = V n(\nu)$ of electromagnetic modes at frequency ν in an enclosure of volume V. This number is a classical result, an adaptation of an argument in [Rayleigh, 1877–78] on standing sound waves; the only two changes are numerical: (i) substituting of the speed of light c to the speed of sound s; and (ii) taking into account the fact that electromagnetic waves are transversal waves, thus with *two* polarizations at the same frequency, while sound in air is a compression wave, with therefore only *one* mode for each frequency.

The second term appeared as a pure manifestation of Planck's genius. With the substitutions $\hbar = h/2\pi$, $\omega = 2\pi\nu$ and $\beta = 1/kT$, this term coïncides with the formula (10.3.24) we obtained for the average energy $U_\beta^{(qu}$ of a quantum oscillator ... except that there were no *quantum* oscillator formalism waiting for Planck at the time!

Alternates to this new form for $u(\nu, T)$ appeared, both before and after Planck's proposal; they now appear as limiting cases – or approximations within limited domain – of the Planck spectral density formula (10.3.29)

$$\left. \begin{aligned} (W) \ \text{for } \tfrac{h\nu}{kT} \text{ large} \ : \quad & \varrho(\nu, T) \approx \varrho_W(\nu, T) = \tfrac{8\pi}{c^3}\, h\nu^3\, e^{-h\nu/kT} \\[2mm] (RJ) \ \text{for } \tfrac{h\nu}{kT} \text{ small} \ : \quad & \varrho(\nu, T) \approx \varrho_{RJ}(\nu, T) = \tfrac{8\pi}{c^3}\, \nu^2\, kT \end{aligned} \right\} \tag{10.3.38}$$

(in these expressions read $LHS \approx RHS$ as LHS "is approximated by" RHS). The first of these expressions is the Wien spectral density (10.3.35) [with now the arbitrary constants α and γ determined from Planck's formula]. The second is the Rayleigh–Jeans law [Rayleigh, 1900, Jeans, 1905]; it was proposed on the premise that each mode should carry the same energy, namely kT, the familiar classical equipartition of the energy among the degrees of freedom of the system. Asides from the fact that this distribution only fits the experimental data for long wave lengths, even theoretically it must be doomed since its behavior at high frequency would lead to a divergence of the total energy, the Ehrenfest ultraviolet catastrophe we mentioned already. Indeed, upon comparison with (10.3.27) – including (10.3.12) – the Rayleigh–Jeans formula appears as the "classical" limit of the Planck's formula. It is thus a mild irony of history, due to the peculiar development of experimental techniques, that a correction – the Planck formula – was needed in order to adapt for long wave length (the classical realm) a prior formula – the Wien's law – that becomes exact in the extreme quantum domain when the frequency becomes very large, and thus the wave-length becomes very short (UV range).

Planck introduced the discrete energy "quantum" by hand, so to speak, and attempted to justify it from a model of the interaction of the black-body radiation with vibrating dipoles in the walls of the cavity containing the radiation, the discreteness of the energy distribution being traced to a property of the dipoles. However, the theoretical formula for the energy distribution – as well as in the experimental data it describes – does not contain any parameter pertaining to the walls, beyond their temperature. This fact allowed an alternate explanation: to see the quanta as a property of light itself, with the role of the walls limited to their serving as a thermal reservoir. This view was proposed by Einstein and he pursued it in a series of papers [Einstein, 1905b, 1906, 1909b,a], where a new concept was advanced: the *light quantum*, later called the *photon* [Lewis, 1926].

The first paper of the series, Einstein [1905b], offers an empirical test for this proposal: an explanation of the photo-electric effect. The effect itself had been discovered some twenty years before: light impinging on a metal may cause negatively charged particles to be ejected from the metal. Just before the turn of the 20th century, these particles had been shown to have the same charge/mass ratio as the particles found in cathode rays and thermo-ionic emission: they are *electrons*; the existence of an elementary "grain of electricity" had been postulated as far back as the 1830s by Faraday, but its existence became a physical reality only towards the dawn of the 20th century, [Lorentz, 1909]; and for an overview, see [Anderson, 1964].

In relation to the photo-electric effect, it is important to realize that, according to Maxwell electromagnetic waves theory, when the intensity of the impinging radiation is increased, so should the energy of the emitted photons, and that this classical theory fails to explain the experimental evidence obtained in particular by Ladenburg [1903]. First, the kinetic energy of the individual electrons is independent from the intensity of light; second, at fixed frequency of the incident electromagnetic radiation, an increase of the intensity of the radiation results only in an increase of the number of the emitted electrons; and third, the photo-electric effect disappears altogether when the frequency of the incident radiation falls below a specific threshold that depends on the metal used in the experiment.

Einstein proposed that if light occurs in indivisible packages of energy $\varepsilon = h\nu$, where h is the Planck constant, and ν the frequency of the light, then the kinetic energy $\frac{1}{2}mv^2$ of the photons emitted in the photo-electric should be given by

$$\boxed{\frac{1}{2}mv^2 = h\nu - e\Phi} \qquad (10.3.39)$$

where e is the charge of the electron, and Φ is the "extraction potential", i.e. $e\Phi$ is the minimal work necessary to extract an electron from the metal, and $\nu_o = e\Phi/h$ gives the frequency threshold below which no electrons are emitted; Φ depends on the metal considered. According to this prediction, the kinetic energy of the emitted electrons should be a linear function of the frequency, with slope equal to the Planck constant.

The experimental data [Hughes, 1912, Richardson and Compton, 1912, Millikan, 1915] – however delicate they were to collect at the time – came in to vindicate (10.3.39), even though Millikan, for one, appears to have set up his experiments to disprove a formula in which he could not believe.

The Nobel Prize in Physics recognized Einstein in 1921, specifically for his theory of the photo-electric effect; Millikan received the Prize in 1923 for his determination of the value of the Planck constant – obtained as a by-product of his experimental verification of Einstein's formula (10.3.39) – and, naturally, for his famous oil-drop experiment to measure the charge of the electron. Physicists today seem to agree routinely in viewing the photo-electric effect as direct and convincing evidence for the objective meaning of Einstein's concept of the light quantum. Yet, general acceptance was not so forthcoming at first, providing us with a historical illustration of the conflict between the interpretation of experimental data as *confirmation vs. falsification* of theoretical ideas. We already mentioned the initial reticence Millikan had to overcome; moreover, and in the same line, here is another case.

Four of the most eminent German physicists of the time – Planck, Nernst, Rubens and Warburg – wrote in a joint letter, dated June 12, 1913, supporting Einstein's election to the Royal Prussian Academy of Sciences, see [Kahan, 1962]. Their praise for the corpus of Einstein's contributions to Physics is captured in the following lines:

> *In summary, it can be said that, among the great problems, in which modern physics is so rich, there is hardly one to which Einstein has not made important contributions.*

And yet, they found it necessary to add immediately:

> *It should not be held too severely against him that he also happened to have occasionally overshot the mark in his speculations, as for example in his hypothesis about light quanta: even in the most exact sciences, no important new ideas can be advanced without ever taking any risk.*

Nevertheless, earlier in the letter, they had accepted that:

With the photoelectric and photochemical effect, he has brought the quan-
tum hypothesis to bear on devising new, interesting laws that are control-
lable by measurement ...

There seems therefore to be in their minds a distinction between empirically successful predictions – the law governing the photo-electric effect – and the theoretical hypotheses from which they are derived – the light quanta. We return to this letter later in this section.

The controversy around the photon continued well into the mid 1920s, with such leading figures as Bohr still expressing very strong doubts, see e.g. the vivid and detailed descriptions offered in [Dresden, 1987][Chaps. 3 and 13]. We do not pursue that line here, as it pertains more to the quantum theory of individual atoms in interaction with monochromatic electromagnetic radiation, rather than to the quantum statistical mechanics of many-body systems. Let us close, therefore, by mentioning only: on the theoretical level, the quantum mechanics – emerging from the work of Einstein, de Broglie, Schrödinger, and Heisenberg – brought about a quantum coexistence between the classical concepts of waves and particles; on the experimental level, conclusive evidence of this duality was provided by the diffraction patterns obtained from beams of electrons [Davisson and Germer, 1927].

Returning thus to statistical mechanics, we note that Einstein devised an analogy between his light quanta and what later became known as the *phonon,* an analogy he exploited to compute the temperature dependence of the specific heat of solids [Einstein, 1907, 1911, Debye, 1912].

The classical theory is summarized in the law of Dulong and Petit [1819] – a form of the law of equipartition of the energy – which implies that the total vibration energy of a solid with n sites should be $3nkT$, or in the language that predates the estimate of the Avogadro number N_{Av} and the Boltzmann constant k : the number $3NRT$ where N is the number of moles, so that the specific heat should be $3R$ per mole and per degree, i.e. – upon taking into account the mechanical equivalent of heat 10^7 ergs = 4.185 calories – the classical theory gives, for the specific heat per mole: $C_V \approx 6$ cal. mole^{-1} deg.$^{-1}$.

Until the advent of laboratory techniques allowing the measurement of specific heat at even moderately low temperatures, this prediction seemed to hold; but, when the specific heat of diamond was measured at $-50°$C, and was found [Weber, 1872] to be $C_V = .86$ cal. mole^{-1} deg.$^{-1}$, such a discrepancy could not be ignored, all the more so that mounting experimental evidence obtained for various solids pointed to the empirical fact that their specific heat was indeed temperature dependent.

The Einstein–Debye theory proposed to view the harmonic oscillations of a crystalline solid – one longitudinal and two transversal modes – as "quanta" which, when in thermal equilibrium, could be described by a formula analogous to the Planck radiation formula (10.3.29). The transfer of context required a few adaptations.

1. The specific heat of the solid obtains from its total energy

$$U(T) = \int_o^\infty d\nu \, f(\nu) \, u(\nu, T) \qquad (10.3.40)$$

where f is the distribution of modes of frequency ν in a solid of volume V; and $u(\nu, T)$ is the equilibrium value of the energy of the mode ν at temperature T.

2. Debye proposed to use for the density $n(\nu) = f(\nu)/V$ the same formula as that appearing in (10.3.37), with three adjustments, the first two of which are obvious: replace the speed of light c by the speed of sound s, and take into account the three modes of vibrations in the solid, hence replace the factor 8π by 12π.

3. Moreover, in order to have a finite result for the total number of modes, Debye proposed to introduce a cut-off at some value ν_o of the frequency, determined by the normalization

$$\int_o^{\nu_o} d\nu \, f(\nu) = 3n \quad .$$

Conditions (2) and (3) are satisfied by the distribution

$$f(\nu) = \begin{cases} 12\pi\nu^2 V/s^3 & 0 \le \nu \le \nu_o \\ 0 & \nu > \nu_o \end{cases} \quad \text{with } \nu_o = s \left(\frac{3}{4\pi} \frac{n}{V} \right)^{\frac{1}{3}} . \qquad (10.3.41)$$

Upon inserting this distribution in (10.3.40), and using for the spectral density of the phonons (the modes of vibrations) *exactly the same distribution* as for the photons (the modes of the electromagnetic radiation), namely the form appearing for $u(\nu, T)$ in (10.3.37), we receive

$$\left. \begin{aligned} U(T) &= 3nkT \, D(\tfrac{\Theta}{T}) \qquad \text{with} \\[2mm] k\Theta &= h\nu_o \quad \text{and} \quad D(x) = \tfrac{3}{x^3} \int_o^x dt \, \tfrac{t^3}{e^t - 1} \end{aligned} \right\} . \qquad (10.3.42)$$

The constant Θ is called the *Debye temperature* of the solid. The universal function D is called the *Debye function;* its values need to be computed numerically, and are tabulated; for our purpose, it suffices to know that D interpolates smoothly between the two extreme regimes described by:

$$D(x) \approx \begin{cases} 1 & \text{for } 0 < x \ll 1 \\[2mm] \frac{1}{5} \pi^4 x^{-3} & \text{for } x \gg 1 \end{cases} . \qquad (10.3.43)$$

Upon taking the derivative of (10.3.42) with respect to the temperature T, and adjusting the units, we obtain for the specific heat per mole and per degree:

$$C_V = 12R\,D(\Theta/T) - 9R\,\Theta/T \left[e^{\Theta/T} - 1\right]^{-1} \qquad (10.3.44)$$

i.e., upon using the asymptotic expressions (10.3.43):

$$C_V \approx \begin{cases} 3R & \text{for} \quad T \gg \Theta \\ \frac{12}{5}\pi^4 R\,[T/\Theta]^3 & \text{for} \quad T \ll \Theta \end{cases} . \qquad (10.3.45)$$

The first line in (3.45) states that the high-temperature limit of the theory gives the classical result of Dulong–Petit. The second line emphasizes that the specific heat goes to zero as the temperature goes to absolute zero; this result – akin to Nernst's "third law of thermodynamics" according to which the thermodynamical entropy would vanish at absolute zero – seems to have played an essential role in converting Nernst from his initial scepticism towards quantum theory to the view that it may be more than just a computational trick and could really have some intrinsic value. For a warning on the putative universality of the third law of thermodynamics, see however the problem of the residual entropy of ice discussed at the end of Sect. 12.3.

Nevertheless (10.3.44) fits remarkably well the measured temperature dependence of the specific heat over a wide range of temperatures; the fit is so good in fact that the observed deviations, in the form of bumps or peaks, are often referred to as "anomalies" and the temperatures at which these occur have their own interpretation as the unfreezing of degrees of liberty or the setting on of some cooperative phenomena. The agreement between (10.3.44) and experimental data is in fact somewhat surprising, would it be only in view of the gross approximation involved in the choice of the distribution (10.3.41). It seems nevertheless that the cut-off frequency is genuine, since the interparticle distances impose a lower limit λ_o to the wave-length of the vibrations. When one fits Debye's Θ with the observed temperature dependence of the specific heat, the number one finds can be used to determine the cut-off frequency ν_o [see (10.3.43)]; since the speed of sound s in solids is also a laboratory datum, the ratio $\lambda_o = s/\nu_o$ is available for comparison with the interparticle distance, and the match is good. The coarseness of the approximation resides in the ν^2 behavior in (10.3.41); this represents a lucky, but drastic smoothing out of the frequency spectra that have been measured, or computed for simple crystals. We may comment that the high temperature limit is rather insensitive to the frequency distribution; for instance Einstein had first assumed that all the oscillations had the same frequency, i.e. instead of (10.3.41), he took a Dirac distribution, from which he already derived the validity of the Dulong–Petit rule at high temperature; compare with (10.3.26). At the lower end of the temperature scale, he also obtained that the specific heat approaches zero, but the rate of the approach was much too fast – compare the T^3 behavior in (10.3.45) and the exponential decay one would obtain from (10.3.24).

This failure is also alluded to in the recommendation letter for Einstein written by Planck et al. quoted above from [Kahan, 1962]:

> ... he derived from this [quantum] hypothesis a formula for the specific heat of solids, which although not completely confirmed, nevertheless provided already the correct foundation for the later development of the new kinetic atomistics.

The Debye cut-off was the idea that salvaged the model, making it reliable for low temperatures also, and thus enhancing the plausibility of the transfer of the expression (10.3.29) for the energy $u(\nu, T)$ from the photons to the phonons.

To place this remarkable success in the intellectual perspective of its time, one can hardly do better than quote Nernst, as in [Jammer, 1966][p. 59]:

> At present, quantum theory is essentially only a rule for calculation, of a nature that is apparently very strange, one might even say grotesque; but it has proven so fruitful by the work of Planck, as far as radiation is concerned, and by the work of Einstein, as far as molecular mechanics is concerned ... that it is the duty of science to take it seriously and to subject it to careful investigations.

[Nernst, 1911]

Had the quantum theory been restricted to the description of black-body radiation, however well the Planck formula (10.3.29) fitted with experimental data, the physics community might have regarded it as "just a model". The transpositions of the model, first to a completely novel interpretation of the photo-electric effect, and then to the computation of the specific heat of crystalline solids, *both* led to predictions – (10.3.37) and (10.3.45) – that were indeed so dramatically successful that they had to call for a general theory of many-body systems: the nascent quantum statistical mechanics and field theory, in which the classical theory would ultimately appear as the result of a limiting procedure.

We discuss in Chaps. 14 and 15 some quantum models in equilibrium and non-equilibrium statistical mechanics.

11. Phase Transitions: van der Waals to Lenz

11.1 Introduction

The present section is meant to serve as a global orientation through Chaps. 11 to 14, all four of them being devoted to the modeling of phase transitions.

When water boils and turns into steam under sufficient heat, we know that the substance in question, whatever it is, has undergone a transition from one form of existence – i.e. liquid – to another – i.e. gas. This transition from a liquid phase of the substance to a gaseous one, and the opposite process of condensation from gas to liquid, point just one of the many types of coexistence collectively known as "*phase transitions*". Freezing/melting is another such pair of transitions, familiar enough and yet mysterious enough to have prompted the ancient narrator of the Book of Job. The aim of the modern description of phase transition is to obtain an explanation of such phenomena in terms of the microscopic structures within certain specific constraints. The phenomenological constraints are the various sets of data which experimental physicists and/or chemists obtain from their observations of phase transitions, while the theoretical constraints are the thermostatistic theories of matter in bulk.

Serious experimentation and theoretical accounts of phase transitions and critical phenomena did not begin until the 19th century, when the works of Andrews, Clausius, Clapeyron, van der Waals, and Pierre Curie helped define a thermodynamic description. This we review in Sect. 11.2; we see how some universal properties of the critical exponents started to emerge from early experimental data. This universality is pursued through much of subsequent materials, especially in Chap. 13.

In Sect. 11.3 of the present chapter, we start with van der Waals' microscopic theory of phase transitions in fluids; this leads to his ingenious construction of the equation of corresponding states allowing to derive – albeit with the wrong individual values – some correct relations between critical exponents, and gives thus the first theoretical glimpse at universality. Weiss [1907] did for magnets what van der Waals did for fluids, namely, he proposed a *mean field theory* for ferromagnets. He did it by inserting a simple hypothesis into Langevin's theory for paramagnets. This achievement is also discussed in this section.

Physics had to wait thirteen years after Weiss to see a microscopic model of ferromagnetism. Today the model is known as the *Ising model* [Ising, 1925], although the idea was due to Lenz, Ising's research director. Adding to the irony, Ising solved a one-dimensional model, found – correctly – that there is no ferromagnetic phase transition in the one-dimensional system he considered, but he conjectured – incorrectly – that the same holds for the corresponding two- or three-dimensional models. Later studies corrected the latter misleading guess, and showed the extraordinary fruitfulness of the Ising model, which made it possible to understand, among other things, how a long-range order could emerge from, for instance, a nearest-neighbor interaction among spins – i.e. the possibility of genuine *cooperative phenomena*. The basic properties of the Ising model are studied in Sects. 12.1 (1 dimension), 12.2 (2 dimensions) and 12.3 (related models); for the sad story of Ising's life, see [Stutz and Williams, 1999].

With precise models in hand, from which a rigorous micro-explanation of ferromagnetism and other related phase transitions seems in sight, a daunting problem arises which threatens to sabotage the whole enterprise. It becomes quite simple to see, because of some obvious symmetry properties of their partition functions, that no finite systems of the Ising type are able to exhibit a phase transition. To resolve the problem, the idea of *thermodynamic limit* [Kramers and Wannier, 1941] was born; see Sect. 12.1. We show how taking the thermodynamic limit brings about the singularities to which the thermodynamical description hints; Sect. 12.2. This process may also be accompanied by spontaneous symmetry-breaking; Sects. 10.2, 12.2 and 14.1.

We discuss how the Griffiths inequalities allow one to conclude from the presence of phase transitions in the two-dimensional Ising model that similar models in higher dimensions do also necessarily exhibit phase transitions; Sect. 12.3.

Establishing the presence of phase transitions rigorously in specific microscopic models may be difficult, yet determining how a microscopic system behaves near a critical point is even harder. Indeed, while the mean field models (Sect. 11.3) allowed the pioneers to account for the occurrence of phase transitions, their computation of the critical exponents failed to agree with the experimental findings. Even after obtained the correct exponents for the two-dimensional Ising models, it seemed all but hopeless to try and solve any other specific models rigorously to get the corresponding critical exponents. It was not until around 1966 when Kadanoff [1966], Kadanoff, Götze, Hamblen, Hecht, Lewis, Palciauskas, Rayl, and Swift [1967] perfected the method of scaling – the *scaling hypothesis* – [Onsager, 1944] that the hope rose again to take advantage of the divergence of the correlation length to focus attention to the vicinity of the critical point and to obtain rigorously the correct values of the critical exponents. Wilson's work on *renormalization (semi-)group* [Wilson, 1975, Wilson and Kogut, 1974] finally rounded out the subject: some exponents can then be calculated whose values agree with the data, although it is extremely hard to prove that the computations contain no uncontrolled approximations. Chapter 13 is devoted to this story.

In Chap. 14, we consider phenomena for which no classical model seems adequate, e.g. the superconductivity and superfluidity; two quantum models are presented: the BCS model and the Bose–Einstein condensation.

11.2 Thermodynamical Models

Models for phase transitions. Even though it is difficult to give a general and precise definition of a pure phase [Stanley, 1971, Fisher, 1990], be it a liquid or a phase of magnetization, the phenomena of phase transition – a physical process unique to a transition from one phase to another – can be readily observed and generalized into data models. In a physics laboratory, some phase transitions can actually be seen with naked eyes, for instance, the transition from the liquid to the gaseous phase of some fluid appears as the presence of a mixture of gas and liquid. To characterize such occurrences theoretically but still phenomenologically (or macroscopically), an empirical or data model of the system in question proceeds in general as follows.

1. Identifying the relevant and connected variables for the system, e.g. p (pressure), V (volume), and T (temperature), which fully describe the system's thermodynamical states.
2. Measuring such variables and plotting the values into graphs; e.g. the p-V, isothermal curves.
3. Looking for non-smoothness, or breaking points, on such curves.

The signature of phase transitions is the occurrence of non-smooth, or more precisely, the non-analytic points in the macroscopic state space. Non-analyticity is an apt mathematical description of a drastic change in the 'rate' of some semi-stationary evolution of a system's states; for instance, the liquid-gas phase transition of a fluid below the critical temperature is bounded by two points in the system's p-V isotherms across which $[\partial_V p]_T$ is not continuous. Phase transition of the first order is one in which the first order derivative of a state function becomes discontinuous, while the second or higher order phase transitions are ones where only second or higher order derivatives of such are discontinuous. In what follows, we give a brief discussion of the different kinds of phase transitions in their thermodynamic state spaces.

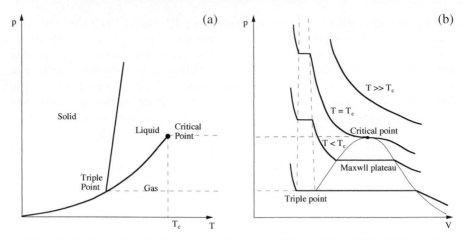

Fig. 11.1. Vapor-liquid phase transition: (**a**) $pVT \to pT$ and (**b**) $pVT \to pV$

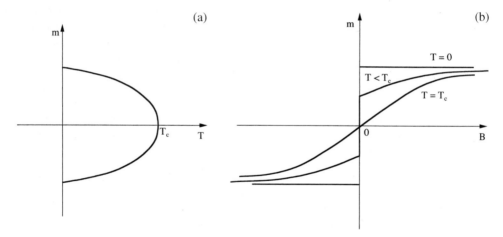

Fig. 11.2. Ferromagnetic phase transition: (**a**) $mBT \to mT$ and (**b**) $mBT \to mB$

Fig. 11.1(a) shows the boundaries of the two phases and the critical point, (p_c, T_c), and (b) shows the isotherms above, at, and below the critical temperature, T_c. One can hardly miss the phase transitions in Fig. 11.1(b), indicated by the coexistence plateau and the fact that, while the isotherms are continuous, their derivative are not. We mention moreover, later in this section, the phenomenon of critical opalescence.

The parallelism between fluids and magnets is unmistakable at this phenomenological level; note the analogy between the isotherms, especially below the critical temperature T_c, the presence of the Maxwell plateau of Fig. 11.1(b), and of the residual magnetization of Fig. 11.2(b) at $B = 0$. This cries out for an explanation. Coïncidence vs. having a common cause? This is a question that the microscopic models must (and did) answer (see Sect. 12.3).

These macroscopic models of phase transitions are clearly state-space models (cf. 2 in Sect. 1.3) in which continuous macroscopic magnitudes determine all states of a system. One can construct experiments to *directly test* (cf. 5 in Sect. 1.3) the models; for instance, [Andrews, 1869] observed a critical temperature for the fluid CO_2, and he was able to continuously vary the volume and observe the change of the pressure. He saw and drew, within the proper limit of error, what is shown in Fig. 11.1b. More on the functions of phenomenal models is said in our Chap. 16.

We note that the most basic assumptions under which such primitive models for phase transitions operate are of two types.

1. Assumptions concerning which or what kind of variables are chosen to describe the states of the systems: their justification is usually rooted in either a theoretical tradition or their obvious validity.
2. The assumption that in an immediate neighborhood of a "pure state", a quasi-static process ought to proceed continuously, i.e. for the set of state variables, the first (or the second) order derivatives of one variable against another – for at least some pairs of macroscopic variables – ought to be continuous. This might be one version of a definition of a *pure phase*.

With this, one part of the explanadum in any micro-explanation of phase transitions is determined – the other part being critical phenomena. The success of any microscopic models for phase transitions is measured in part by how well each aspect of the above-listed phenomena is anticipated.

Critical phenomena. *Critical opalescence* was first described by [Andrews, 1869], although, according to Andrews himself, the phenomena had been discovered earlier by a certain Dr. Miller. Critical opalescence can be described as a process unfolding in the following stages. Beginning with a temperature $T \gg T_c$, only a clear single phase is present, usually thought of as a gas; in particular, there are no local density fluctuations that would give rise to any noticeable light scattering; note that the latter statement does not require the postulation of any specific microscopic model. As the temperature is lowered to $T \approx T_c$, the gas takes on a glow of dramatic intensity; this sudden increase in light scattering is attributed to the condensation, in the gas, of very small liquid droplets, of the size of the wavelength of the light; this is accompanied by a great change in the compressibility, so that the density distribution becomes very sensitive to pressure gradients. Finally, as the temperature is further lowered to $T \ll T_c$, the glow disappears and a meniscus forms, separating the system in two regions: a gas phase and a liquid phase.

The phenomena observed by Andrews [1869] for CO_2 were found later to be *universal,* a discovery that was to be motivated in part by the form of the van der Waals equation (see Sect. 11.3). The only peculiarity of CO_2 in this regard is that its critical temperature is easily reachable experimentally:

Andrews measured it quite precisely to be $T_c = 30.92^\circ C$; for comparison, it is $374.14^\circ C$ for water; and $5.2^\circ K$ for Helium.

Critical opalescence is of special significance because it marks the transition between a homogeneous (i.e. single phase) fluid system and an heterogeneous (i.e. multi-phase). The same kind of transitions – to long-range order – also occurs in other systems, such as in ferromagnets; and the locale of such transitions are critical points; the local behavior of the fluid in the neighborhood of a critical point is known collectively under the name of *critical phenomena*; and these are numerically characterized by the *critical exponents* [Fisher, 1967b, 1965, Ma, 1976].

To study critical phenomena, we find out how a system's main thermodynamic features vary with $|T - T_c|$, where T_c is the critical temperature. Experimental observations of certain thermodynamical quantities reveal that, as the critical temperature is approached, these quantities vary according to an inverse power law: $A \propto |T - T_c|^{-x}$, defining x, the critical exponent relative to the observed quantity A.

From established practice and theory, we have the following critical exponents for those models we listed above, where, for simplicity, we use a dimensionless variable – the normalized temperature,

$$t = \frac{T - T_c}{T_c},$$

for the base of the exponents.

From Fig. 11.1(b) and 11.2(b), we know that the slopes of the isotherms are proportional to the inverse of isothermal compressibility, κ_T^{-1}, for the fluids and of isothermal susceptibility, χ_T^{-1}, for the magnets, both of which approach zero as $T \to T_c$. Experimental evidence suggests that there are constants γ and γ' with $\gamma \approx \gamma'$, such that:

for fluids $\qquad \kappa_T \equiv -\frac{1}{V}[\partial_p V]_T \propto \begin{cases} t^{-\gamma} & (t > 0) \\ (-t)^{-\gamma'} & (t < 0) \end{cases}$

for magnets $\qquad \chi_T \equiv [\partial_B m]_T \quad \propto \qquad t^{-\gamma} \qquad (t > 0)$

$$\left.\begin{array}{}\\ \\ \\ \end{array}\right\} \quad (11.2.1)$$

Similarly, below the critical temperature T_c, one finds for $\Delta\varrho \equiv \varrho_L - \varrho_G$ along the liquid-gas coexistence curve, and $m_{B=0}$ denoting the residual magnetization at $B = 0$, experimental evidence indicates that there is a constant β such that:

for fluids $\qquad \Delta\varrho \quad \propto (-t)^\beta \qquad (t < 0)$

for magnets $\qquad m_{B=0} \propto (-t)^\beta, \qquad (t < 0)$

$$\left.\begin{array}{}\\ \\ \end{array}\right\} \quad (11.2.2)$$

With ϱ_c and p_c denoting the density and pressure at the critical point, one finds experimentally that, along the critical isotherm $T = T_c$:

for fluids $\quad |\varrho - \varrho_c|^\delta \propto |p - p_c| \qquad (t = 0)$

for magnets $\quad m^\delta \qquad \propto B \qquad (t = 0)$

$$\left.\begin{array}{l}\\ \\ \end{array}\right\} . \qquad (11.2.3)$$

As for the specific heat, one finds:

for fluids $\qquad C_V \propto \begin{cases} t^{-\alpha} & (t > 0) \\ (-t)^{-\alpha'} & (t < 0) \end{cases}$

for magnets $\quad C_B \propto \begin{cases} t^{-\alpha} & (t > 0) \\ (-t)^{-\alpha'} & (t < 0) \end{cases}$

$$\left.\begin{array}{l}\\ \\ \\ \\ \end{array}\right\} . \qquad (11.2.4)$$

Finally, there is a pair of exponents, ν (ν') and η, associated with two functions that are not, strictly speaking, macroscopic, even though they are certainly thermodynamic functions, as we see later. ν and ν' are connected to the correlation length, ξ; and η to the correlation function, $G(\boldsymbol{r})$,

$$G(\boldsymbol{r}) \equiv \langle \varrho(\boldsymbol{0})\varrho(\boldsymbol{r}) \rangle - \langle \varrho(\boldsymbol{0}) \rangle \langle \varrho(\boldsymbol{r}) \rangle \quad \propto \quad exp(-\|\boldsymbol{r}\|/\xi) \qquad (11.2.5)$$

so that ξ is to be interpreted as the rate at which the correlation decays. We have then, near the critical point:

$$\xi \propto \begin{cases} |t|^{-\nu} & (t > 0) \\ |t|^{-\nu'} & (t < 0), \end{cases} \qquad (11.2.6)$$

and

$$G(\boldsymbol{r}) \propto 1/\|\boldsymbol{r}\|^{d-2+\eta}, \qquad (11.2.7)$$

where d is the dimension of the system. The sign ($\propto$) rather than equality in relations (2.1-7) indicates that the critical exponents are defined only as limiting power laws as $T \to T_c$, namely,

$$f(t) \propto t^x \qquad \text{means that} \qquad x = \lim_{t \to 0} \frac{\ln f(t)}{\ln t}. \qquad (11.2.8)$$

Extensive experimental work not only obtains values for the exponents – with greater and greater accuracy – but also shows that the exponents are not all independent of one another (via some equalities) and systems of radically different constitutions share the same values for their exponents (i.e. a hint of universality).

While the thermodynamic description of phase transitions in general points towards the direction in which micro-explanations of them can be had – such as the necessity of taking the thermodynamic limit; the description of critical phenomena provides a further and finer discrimination for the selection of the available models. For instance, as we see later, the Ising model must be preferred in this respect to the Weiss model.

11.3 Mean-Field Models

These models propose a first line of approach towards a microscopic description of the macroscopic occurrence of phase transitions and critical phenomena. The van der Waals equation model for the gas-liquid phase transition is first discussed from a purely thermodynamical standpoint; we then describe two of its derivations from microscopic assumptions. We finally review a similar treatment of the ferromagnetic phase transition. A quantum example of a mean free field theory – the BCS model – is described in Sect. 14.1.

11.3.1 The van der Waals Model of Fluids

In a dissertation [Van der Waals, 1873] submitted to the University of Leiden, – later expanded in book–form [Van der Waals, 1881], both under titles very reminiscent of the paper [Andrews, 1869] on the liquefaction of CO_2 – van der Waals offered his famous proposal for an equation of state that would account for the phenomena found experimentally by Andrews. As Maxwell recognized immediately, one of van der Waals' main conceptual achievement was to realize

> That the same substance at the same temperature and pressure can exist in two very different states, as a liquid and as a gas, is in fact of the highest scientific importance, for it is only by the careful study of the difference between these two states, the conditions of the substance passing from one to the other, and the phenomena which occur at the surface which separates a liquid from its vapor, that we can expect to obtain a dynamical theory of liquids.
>
> [Maxwell, 1874]

The equation of state proposed by van der Waals reads:

$$[p + a(\frac{N}{V})^2][V - Nb] = NkT \tag{11.3.1}$$

or equivalently:

$$[p + a v^{-2}][v - b] = kT \tag{11.3.2}$$

where p is the pressure in the fluid, V its volume, and T its temperature; k the Boltzmann constant; $v \equiv V/N$ with N the number of particles in the fluid; for each substance separately, a and b are supposed to be numerical constants; clearly, the van der Waals equation reduces to the Boyle/Gay-Lussac law (2.2.10) when a and b are neglected. Note that the second of the above equivalent forms of the van der Waals law is a truly (macroscopic) thermodynamical equation, involving only the intensive variables p, v, T. The two constants a and b are meant to account, respectively, for the effect of (a) the attractive interaction between the molecules, and (b) the hard-core that prevent the molecules from inter-penetrating each others.

Even before we ask how well the van der Waals equation describes any specific substance, we have to check whether it satisfies the basic laws of

thermodynamics. In particular, already on the basis of every-day experience
– and in fact, as we shall see, on the basis of the second law of thermodynamics
– we expect that if we compress a fluid, and keep its temperature constant,
the pressure in the fluid increases; i.e.

$$(\partial_v p)_T \leq 0 \qquad \forall \quad v > b \quad . \tag{11.3.3}$$

At fixed temperature T and over the whole range of v, this *stability condition*
is satisfied, for a van der Waals fluid, if but only if

$$kT \geq \max_{v>b} f(v) \quad \text{with} \quad f(v) = 2a\,\frac{(v-b)^2}{v^3} \quad . \tag{11.3.4}$$

This maximum happens at the critical value

$$v_c \equiv 3b \quad , \tag{11.3.5}$$

and the stability condition (11.3.3) is satisfied only when $T \geq T_c$, with

$$kT_c \equiv \frac{8a}{27b} \quad . \tag{11.3.6}$$

With v_c and T_c given by (11.3.5) and (11.3.6), the van der Waals equation of
state (11.3.2) is satisfied exactly by one value of p, namely

$$p_c \equiv \frac{a}{27b^2} \quad . \tag{11.3.7}$$

The *critical point* (p_c, v_c, T_c) presents two remarkable features which, for the
time-being, are only mathematical artifacts in search of a physical interpretation. Firstly, as the isotherm $T = T_c$ passes through this critical point, we
have both

$$(\partial_v p)_{T_c} = 0 \qquad \text{and} \qquad (\partial_v{}^2 p)_{T_c} = 0 \quad . \tag{11.3.8}$$

Secondly,

$$\boxed{\frac{p_c v_c}{kT_c} = \frac{3}{8}} \quad ; \tag{11.3.9}$$

this mathematical quantity is *universal* in the context of the van der Waals
equation in the sense that it is *independent* of a and b. This suggests the
introduction of the *reduced variables*

$$\tilde{p} = \frac{p}{p_c} \quad , \quad \tilde{v} = \frac{v}{v_c} \quad \text{and} \quad \tilde{T} = \frac{T}{T_c} \tag{11.3.10}$$

in terms of which the van der Waals equation takes the *universal form*

$$\boxed{[\,\tilde{p} + 3\,\tilde{v}^{-2}\,]\,[\,\tilde{v} - \frac{1}{3}\,] = \frac{8}{3}\,\tilde{T}} \quad . \tag{11.3.11}$$

No contingent constants, such as a and b, appear in this equation, although these are obviously hidden in (11.3.5–7) and (11.3.10); for this reason, the substitution (11.3.10) or the equation (11.3.11) are referred to as the *law of corresponding states*. This universality is perhaps the most consequential prediction of the van der Waals equation. It was announced by van der Waals in 1880 before the Dutch Academy of Sciences [Kipnis, Yavelov, and Rowlinson, 1996], and it was almost simultaneously prepared for inclusion in [Van der Waals, 1881].

From the experimental data collected in [Stanley, 1971], the first indication is that the classical theory prediction (11.3.9) of the "universal ratio" $3/8 = .375$ is fairly well verified in real substances as diverse as Water (.230), Carbon dioxide (.275), Hydrogen (.304) and Helium (.308). This should be appreciated in comparison to the fact that the substance-specific constant a which enters the original van der Waals equation varies 200-fold among the substances just listed, even though the other constant b is more tame: $1 < b < 1.5$.

The superposition of the empirical isotherms, expressed in the reduced variables $\{\tilde{p}, \tilde{v}, \tilde{T}\}$, also shows a remarkable agreement with the universal character of these reduced variables across a similar variety of substances.

Moreover, in van der Waals' times it was thought that some gases, dubbed "permanent", were not liquefiable; for instance Maxwell [1874], speaking of hydrogen, writes: *"This gas has never been liquefied* [which was true at the time], *and it is probable that it never will be liquefied, as the attractive forces are so weak."* In fact, the belief in the universality of van der Waals' law of corresponding states was to play an important role in motivating the quest for the liquefaction of Helium successfully accomplished in Kammerlingh Onnes laboratory in 1908 [Uhlenbeck, 1966, Kamerlingh Onnes, 1913].

We are still left here with the problem of what to do with the isotherms $T < T_c$, along which a *thermodynamical instability* violating (11.3.3) develops on a finite segment (E, C), where $v_E \equiv V_E/N$ and $v_C \equiv V_C/N$ are the two distinct solutions of $(\partial p/\partial v)_T = 0$ that exist for every value of $T < T_c$, since the function f defined in (11.3.4) has exactly one maximum in the interval (b, ∞), is positive there, and satisfies $0 = f(b) = \lim_{v \to \infty} f(v)$. Note that every isotherm for $T < T_c$ develops a "wiggle" as drawn in Fig. 11.3. A thermodynamical cure to this thermodynamical disease is the rule named after Maxwell [1875]. Geometrically, the construction consists in three steps: *(i)* draw a straight-line at $p_T < p_c$; this intersects the isotherm at the points F and B; *(ii)* between F and B replace the wiggly isotherm by the straight-line $p = p_T$; *(iii)* adjust p_T in such a manner as to ensure the equality of the areas between this line and the wiggly isotherm. Analytically, this means that

$$p_T(V_B - V_F) = \int_{V_F}^{V_B} p_w(V, T)dV \qquad (11.3.12)$$

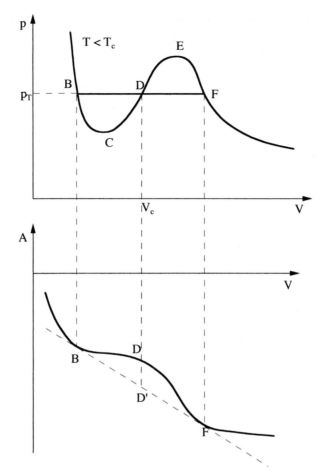

Fig. 11.3. van der Waals isotherm for $T < T_c$, in terms of p and A

where p_w indicates that the integral is taken along the original wiggly isotherm. The relation (11.3.12) is the key to the thermodynamical interpretation of the geometry of this construction, for which Maxwell [1875] offers the following justification:

> *Let us now suppose the medium to pass from B to F along the hypothetical curve BCDEF in a state always homogeneous, and to return along the straight-line BF in the form of a mixture of liquid and vapor. Since the temperature has been constant throughout, no heat can have been transformed into work. Now the heat transformed into work is represented by the excess of the area FDE over BCD. Hence the condition that determines the maximum pressure of the vapor at given temperature is that the line BF cuts off equal areas from the curve above and below.*
>
> [Maxwell, 1875]

Notice the flavor of the cycle-analysis, typical of traditional Thermostatics (Chap. 2). Maxwell's argument is, in fact, equivalent to a slightly more modern version of the second law, requiring that the Helmholtz free energy $A = U - TS$ be convex. Indeed, the physical meaning of A is that it gives the maximum work that can be extracted from the system along any isothermal transformation. In fact, since $dU = -p\,dV + T\,dS$:

$$(\partial_V A)_T = -p \quad . \tag{11.3.13}$$

Upon following the isotherm, we find for the Helmholtz energy A the graph represented on the lower part of Fig. 11.3; its convex envelope replaces the bump FDB by the tangent straightline $FD'B$. The tangency condition at points F and B reads thus

$$p_T = -(\partial_V A)_T = -\frac{A(V_B) - A(V_F)}{V_B - V_F}$$

i.e.

$$p_T(V_B - V_F) = -[A(V_B) - A(V_F)] \tag{11.3.14}$$

which is (11.3.12), by the fundamental theorem of calculus and (11.3.13). Note also that the Maxwell construction does not depend on the specific form of the van der Waals equation, but only on the fact that the continuous isotherms have an unstable segment; indeed Maxwell [1875] presents his construction immediately *before* he mentions van der Waals' work, with which he was familiar [Maxwell, 1874] [the latter however makes no reference to the above construction.]

The van der Waals equation (11.3.2), together with the Maxwell construction (11.3.12), is referred to as the *classical equation of state*. We are now in position to examine its relevance as a model of the liquid-gas phase transition.

We should first notice that, for all temperatures T, it describes qualitatively well the corresponding isotherms observed, and drawn so delicately, by Andrews [1869]. For $T > T_c$, the monotonic decrease of the pressure along the isotherms confirms the classical equation of state (11.3.2), and these isotherms are smooth so that there is no way to differentiate in the fluid a gas or a liquid. For $T < T_c$, Andrews saw rounded corners on the liquid side of the Maxwell plateau; he already correctly attributed these to a fluke due to the presence of some *"traces of air (about $\frac{1}{500}$ part) in the carbonic acid"*. Thereafter we refer to the locus of the extreme points of the Maxwell plateau – namely $(v_l = V_F/N)$ and $(v_g = V_B/N)$ – as the *coexistence curve*, and to v_l (resp, v_g) as the specific volume of the liquid (resp. gas) phase at coexistence.

Troubles however start with the coexistence curve traced by the extreme points of the Maxwell plateau. Here again, one has a wide variety of substances, all giving the same graph, when expressed in the reduced variables (11.3.10) with $\tilde{T}$ plotted as a function of $\tilde{\varrho}$. So far this is in satisfactory conformity with the law of corresponding states. *However,* while this experimental curve appears to be well-fitted by $\tilde{\varrho} - 1 \approx (\tilde{T} - 1)^\beta$ with $\beta = 1/3$, the classical theory predicts $\beta = 1/2$ around the critical point, as we are now going to see.

For this purpose, let us rewrite the reduced equation (11.3.11) in term of variables π, ν, ε which are better suited to a discussion focusing on a neighborhood of the critical point; these are defined by the substitutions

$$\tilde{p} = 1 + \pi \qquad \tilde{v} = 1 + \nu \qquad \tilde{T} = 1 + \varepsilon \tag{11.3.15}$$

so that the critical point is given by $\pi = 0, \nu = 0, \varepsilon = 0$. We obtain an exact, but ugly-looking, equation which we give here only for the record:

$$\pi[1 + a_1\nu + a_2\nu^2 + a_3\nu^3] = 8\varepsilon[1 + 2\nu + \nu^2] - \frac{3}{2}\nu^3 \quad . \tag{11.3.16}$$

We only need to know that the coefficients a_i are (small) real numbers, and contain no dependence on the variables π, ν, ε. As a consequence, for small values of ν and ε

$$\pi \approx b_o + b_1\nu + b_2\nu^2 + b_3\nu^3$$

where b_o, b_1 and b_2 are proportional to ε, while b_3 is a constant. This tells us two things: first, for T approaching T_c from below, the wiggly part of the original van der Waals isotherm is almost symmetric around the critical isochore $v = v_c$; and then the abscissa of the extremities of the Maxwell plateau satisfy $|\nu| \approx \varepsilon^\beta$ with $\beta = 1/2$, i.e. in terms of $\varrho_L = 1/v_L$ and $\varrho_G = 1/v_G$:

$$\varrho_L - 1 \approx (T - T_c)^\beta \quad \text{and} \quad 1 - \varrho_G \approx (T - T_c)^\beta \qquad \text{with} \quad \beta = \frac{1}{2} \quad . \tag{11.3.17}$$

Let us look further into the behavior of the van der Waals classical equation of state in the neighborhood of the critical point. From (11.3.16) again, we note immediately that, since $\varepsilon = 0$ along the critical isotherm, we have there:

$$1 - p \approx (1 - v)^\delta \approx (\varrho - 1)^\delta \qquad \text{with} \quad \delta = 3 \quad . \tag{11.3.18}$$

From (11.3.11), we see that

$$\left(\frac{\partial \tilde{p}}{\partial \tilde{v}}\right)_{\tilde{T}} = 6(\frac{1}{\tilde{v}})^3 [1 - 4(\frac{1}{3\tilde{v} - 1})^2 \tilde{T}]$$

so that along the critical isochore $\tilde{v} = 1$:

$$\left(\frac{\partial p}{\partial v}\right)_T \approx (1 - T)^\gamma \qquad \text{with} \quad \gamma = 1 \quad . \tag{11.3.19}$$

Finally, one verifies that the classical theory gives for the specific heat:

$$C_V = \frac{3}{2}k + \begin{cases} 0 & \text{for} \quad T > T_c \\ \frac{9}{2}k[1 - \frac{28}{25}\frac{T_c-T}{T_c} + \cdots] & \text{for} \quad T < T_c \end{cases} \quad ; \qquad (11.3.20)$$

hence the coefficient α, defined in (11.2.4), vanishes; $\alpha = 0$ expresses that the specific heat exhibits only a finite discontinuity along the critical isochore.

The values obtained in (11.3.20), (11.3.17), (11.3.19) and (11.3.18) for the coefficients α, β, γ, and δ satisfy the two equalities

$$\boxed{\alpha + 2\beta + \gamma = 2 \quad ; \quad \alpha + \beta(1 + \delta) = 2} \qquad (11.3.21)$$

While the experimental evidence obtained separately for the *individual* critical exponents $(\alpha, \beta, \gamma, \delta)$ appearing in (11.3.21), does *not* conform to the values (11.3.17–20) predicted by the van der Waals equation of state, the *relations* (11.3.21) *themselves* seem nevertheless to be confirmed in the laboratory. Such coïncidences beg for elucidation.

The original derivations of the van der Waals equation raised a few eyebrows; see for instance [Maxwell, 1874, 1875,/]; all three refer to the work of van der Waals with both praise and reservations; and both seem to come from Maxwell's requirement that the equation of state should provide reliable information on the intermolecular forces. Thus the praise is mostly for two reasons: first, the originality of the work, and second Maxwell's perception of its essential correctness in its account of the attractive part of the intermolecular forces by adding to the pressure a term proportional to the square of the density $\varrho \equiv v^{-1}$. The exceptions that Maxwell take are concerned also with two points: the first is that he feels the repulsive part of the intermolecular forces has not be given a careful treatment; the second addresses the question of the deeper meaning of the work. Maxwell professes not to be convinced by a mere agreement with experimental data. In spite of the theoretical arguments advanced by van der Waals in his thesis, Maxwell regarded the van der Waals equation still as an empirical formula; specifically:

> An empirical formula may be defined as one so framed as to give results consistent with experiments within the range of these experiments, but which we have no reason, founded on general principles of physics, for believing to be applicable beyond that range. The degree of accuracy of an empirical formula is a test of its merit, but not of whether it is empirical or not.

[Maxwell, 1875/6]

The next question therefore is to understand the theoretical limits in which the van der Waals equation is to be expected to capture the physics of the problems at hand. A partial answer is proposed in the next subsection.

11.3.2 The van der Waals Equation from Statistical Mechanics

In this subsection, we review the original Ornstein derivation of the van der Waals equation. We then mention a modern derivation, based on more satisfying premises, and known as the Kac–Baker model. Let

$$\mathcal{H} = K + V \quad \text{with} \quad K = \frac{1}{2m} \sum_{i=1}^{N} p_i^2 \quad \text{and} \quad V = \Phi(x_1, \ldots, x_N) \quad (11.3.22)$$

be the Hamiltonian of a system of N spherical molecules, confined in a region Λ in the configuration space R^d, and interacting via two-body forces, i.e.

$$\Phi(x_1, \ldots, x_N) = \frac{1}{2} \sum_{i \neq j} \phi(|x_i - x_j|) \quad . \quad (11.3.23)$$

The partition function for this system

$$Z_N(V, T) = \frac{1}{N!} \int_{\mathsf{R}^{Nd}} dp_1 \cdots dp_N \int_{\Lambda^N} dx_1 \cdots dx_N E^{-\frac{1}{kT}\mathcal{H}} \quad (11.3.24)$$

splits into two parts

$$\left. \begin{aligned} Z_N(V, T) &= Z_N^{kin}(V, T) + Z_N^{pot}(V, T) \\[2mm] \text{with} \quad Z_N^{kin}(V, T) &= \left[\int_{\mathsf{R}^d} dp\, e^{-p^2/2mkT} \right]^N \\[2mm] \text{and} \quad Z_N^{pot}(V, T) &= \tfrac{1}{N!} \int_{\Lambda^N} dx_1 \cdots dx_N\, e^{-\Phi(x_1, \cdots, x_N)/kT} \end{aligned} \right\} \quad (11.3.25)$$

The kinetic part is already known to us from the Maxwell–Boltzmann kinetic theory of gases (Chap. 3):

$$Z_N^{kin}(V, T) = \lambda^{-Nd} \quad \text{with} \quad \lambda = (2mkT)^{-\frac{1}{2}} \quad . \quad (11.3.26)$$

The contribution of the potential part is much more difficult to evaluate, as one needs more information on the pair potential ϕ appearing in (11.3.23). We assume that ϕ consists in a repulsive hard-core that is short range; and an attractive, weak, long-range tail:

$$\phi(|x_i - x_j|) = \begin{cases} \infty & |x_i - x_j| < \delta \\[2mm] \phi^{attr}(|x_i - x_j|) & \text{otherwise} \end{cases} \quad (11.3.27)$$

where $\delta/2$ is the radius of the molecules. Moreover, we *assume* that the attractive part of the potential is so long-range that one can *approximate* its contribution to the partition function by simply replacing it by its average:

$$a \equiv \frac{1}{V} \int_{\Lambda} dx\, \phi^{attr}(x) \quad \text{with} \quad V = |\Lambda| \quad . \quad (11.3.28)$$

In this approximation

$$Z_N^{pot}(V,T) \approx e^{-\frac{1}{kT} a \frac{N^2}{V}} \frac{1}{N!} \int_{\Lambda^N} \prod_{i \neq j} \Theta(|x_i - x_j|)$$

where

$$\Theta(|x|) = \begin{cases} 1 & |x| < \delta \\ \\ 0 & \text{otherwise} \end{cases} \quad .$$

Notice further that in one dimension ($d = 1$ and $|\Lambda| = L$):

$$\left. \begin{array}{l} \frac{1}{N!} \int_{\Lambda^N} \prod_{i \neq j} \Theta(|x_i - x_j|) = \\ \\ \int_0^{L-N\delta} dx_1 \int_{x_1+\delta}^{L-(N-1)\delta} dx_2 \cdots \int_{x_{N-1}+\delta}^{L-\delta} dx_N = \frac{1}{N!} (L - N\delta)^N \end{array} \right\} \quad ; \quad (11.3.29)$$

this suggests, for higher dimensions, the approximation:

$$\frac{1}{N!} \int_{\Lambda^N} \prod_{i \neq j} \Theta(|x_i - x_j|) = \frac{1}{N!} (V - Nb)^N \quad .$$

We therefore approximate $Z_N(V,T)$ by

$$\overline{Z}_N(V,T) = \frac{1}{N! \lambda^{Nd}} e^{-\frac{1}{kT} a \frac{N^2}{V}} (V - Nb)^N \quad . \qquad (11.3.30)$$

Upon introducing $A_N(V,T) \equiv -kT \ln \overline{Z}_N(V,T)$, we obtain

$$p \equiv -(\partial A/\partial V)_T = -a\, v^{-2} + kT \frac{1}{v - b}$$

which coïncides with the van der Waals equation (11.3.2).

The above derivation, due to Ornstein [1908], has been criticized on three accounts: *(i)* the hard-core potential is *not* properly treated in dimensions greater than 1; *(ii)* the replacement (11.3.28) of the attractive potential by its average is not a controlled approximation; and *(iii)* the Maxwell plateau thermodynamical construction has still to be added on, rather than coming out as a derived consequence of statistical mechanics.

The first objection was raised with the hope to derive, in cases where V becomes comparable with Nb, a description of the liquid-solid transition – even, perhaps, in the absence of an attractive potential (i.e. when $a = 0$); for early attempts see [Tonks, 1936, Kirkwood and Monroe, 1941, Kirkwood, Mann, and Adler, 1959]. While one does expect that bulk matter under extreme pressure and very low temperature would show some kind of crystalline order, to control theoretically the underlying mechanism has turned out to be a notoriously hard problem, see e.g. [Uhlenbeck, 1968] or [Anderson, 1984]; for a modern mathematical perspective, see [Radin, 1991, 1987] and references therein.

The second and third objections to the Ornstein derivation have been addressed in a series of papers covering the *one-dimensional* Kac–Baker model,

[Kac, 1959a, 1962, Baker, Jr., 1961, 1962, Kac, Uhlenbeck, and Hemmer, 1963, Hemmer, 1965, Helfand, 1964a, Kac, 1968]; this continuous model also admits a simpler lattice version developed in [Kac and Helfand, 1963, Helfand, 1964b]. At the end of this section, we briefly comment on an interesting resurgence of the continuous model, due to [Lebowitz, Mazel, and Presutti, 1998]. In the original model, the potential is as in (11.3.27), with now ϕ^{attr} defined, for any $\gamma > 0$ by:

$$\phi^{attr}(|x|) = -\alpha_o \, \gamma \, e^{-\gamma|x|} \quad . \tag{11.3.31}$$

This model can be solved by taking advantage of two features. First, its being one-dimensional with hard-core as in (11.3.27) allows to consider a fixed ordering $0 < x_1 < x_2 < \cdots < x_{N-1} < x_N$ as in (11.3.29). Second, the attractive part of the potential in (11.3.31) is chosen to espouse the form of the covariance of a stationary, Gaussian, Markov process (or O–U process): $E\{(X(t)X(t')\} = e^{-\gamma|t-t'|}$.

An equation of state can be derived from the partition function through the thermodynamical limit ($N \to \infty, L \to \infty$; with $L/N \equiv l$ fixed). The pressure p is a monotonically strictly decreasing smooth function of the specific "volume" l; this illustrates general theorems (See Sect. 12.1) to the effect that no phase transitions are to be expected in one-dimensional system of particles interacting via two-body potentials with a repulsive hard core, and a finite range residual interaction; indeed, we have here, for every finite $\gamma > 0$:

$$\left.\begin{array}{ll} \text{(a)} & \overline{\phi} \equiv \int_0^\infty dx\, \phi^{attr}(x) = -\alpha_o \quad ; \\[2mm] \text{(b)} & \overline{x} \equiv (\overline{\phi})^{-1} \int_0^\infty dx\, x\, \phi^{attr}(x) = \gamma^{-1} \, . \end{array}\right\} \tag{11.3.32}$$

It is at this point that an additional limiting procedure is introduced, namely to let $\gamma \to 0$. As this limit is approached, the attractive potential becomes ever weaker [see (11.3.31)] and of longer range [see (11.3.32b)], while its average remains finite [see (11.3.32a)]. *In this limit, the van der Waals equation is controllably recovered,* moreover *with a major improvement:* below the critical temperature, the Maxwell plateau now appears as a consequence of the statistical formalism, rather than having to be superimposed by a supplementary thermodynamical argument. At a technical level, and with a view to what happens also for other models exhibiting phase transitions – e.g. the 2-d Ising model of Sect. 12.2 – we mention that the free-energy, and thus the thermodynamical behavior, is governed by the maximal eigenvalue of a certain operator; and that the discontinuity of l as a function p, corresponding to the pressure remaining constant in the coexistence region in l, is linked to a degeneracy of this maximal eigenvalue, appearing here only as $\gamma \to 0$.

11.3.3 The Weiss Model for Ferromagnets

The Weiss model [Weiss, 1907] was intended to give a microscopic explanation for ferromagnetism, namely, to show the spontaneous magnetization below a critical temperature T_c which is called the *Curie temperature*. A Physicist's presentation can be found in [Wannier, 1966, 1987][Sect. 15-2]. The Weiss model consists of the following basic elements. **(i)** Dipoles, as tiny compasses, are fixed in place, and the assembly is embedded in a spacially uniform magnet field B; the effect of this field is to tend to align the directions of the dipoles, a tendency which runs against the thermal agitation. **(ii)** The assembly of dipoles is divided into relatively large blocks, called *domains;* each dipole in a domain is affected only by other dipoles in the same domain. **(iii)** On a given dipole, the collective effect of all the other dipoles in the same domain is mimicked as an additional *"internal"* magnetic field B_{int}, obtained as an average of the magnetic fields due to each dipole in the domain; hence the name of *"mean-field models"* attributed to models of this type.

Let us now construct a model that meets these specifications. Since the domains are not interacting with one another, we limit our attention to a single domain. While the method applies to any Hamiltonian of the general form

$$H = -B \cdot \sum_i \sigma_i - \sum_{i \neq j} J_{i,j} \, \sigma_i \cdot \sigma_j \qquad (11.3.33)$$

it is convenient to restrict the exposition by the following assumptions. The dipoles are restricted to pointing either "up" or "down" in the fixed direction of the magnetic field B; thus they are viewed as random variables, indexed by i and taking values ± 1. The dipoles are restricted to the sites of a regular lattice L. The dipole-dipole interaction is translation invariant and symmetric, i.e.

$$\forall \, i, j, k \in L \; : \; \text{(a) } J_{i+k,j+k} = J_{i,j} \; ; \; \text{(b) } J_{j,i} = J_{i,j} \; ; \; \text{(c) } J_{i,i} = 0 \, . \quad (11.3.34)$$

The interaction is summable, i.e.

$$\sum_i J_{i,j} < \infty \quad . \qquad (11.3.35)$$

The mean-field approximation now consists in substituting for the Hamiltonian H in (11.3.33) the "effective Hamiltonian" $\tilde{H}$ defined as follows :

$$\tilde{H} = -\tilde{B} \sum_i \sigma_i$$

where
$$\tilde{B} \equiv B + B_{int} \quad \text{with} \quad B_{int} \equiv \sum_j J_{i,j} \, \langle \sigma_j \rangle \qquad (11.3.36)$$

and
$$\langle \sigma_j \rangle \equiv Z_\beta^{-1} \, Tr \, e^{-\beta \tilde{H}} \sigma_j$$

As usual, $Z_\beta = Tr \exp(-\beta \tilde{H})$ is the partition function that normalizes the canonical equilibrium state $\varrho_\beta = Z_\beta^{-1} \, e^{-\beta \tilde{H}}$; and $\beta = (kT)^{-1}$ where T denotes the temperature.

Note that the translation invariance (11.3.34 a) of the interaction was assumed so that the internal magnetic field B_{int} defined in (11.3.36) and the magnetization $m \equiv \langle \sigma_i \rangle$ are the same at all sites i.

Upon solving the self-consistency equations (11.3.36) – i.e. upon inserting $\tilde{H}$ into the definition of $m \equiv \langle \sigma_i \rangle$ – we find:

$$m = \tanh \beta \left(B + \lambda m \right) \qquad \text{with} \quad \lambda = \sum_j J_{i,j} \quad . \qquad (11.3.37)$$

Note that the slope of the tangent to the graph of the function $f(x) = \tanh \beta \lambda x$ is smaller or equal to $\beta \lambda$, with equality holding exactly at $x = 0$; hence, at vanishing external field $B = 0$, and as long as $\beta \lambda \geq 1$, (11.3.37) admits only the trivial solution $m = 0$. However, as soon as $\beta \lambda < 1$, two new solutions appear $m = \pm m(0, T) \neq 0$.

In physical language, this tells us that there exists a critical temperature, namely $T_c = 1/(k\beta_c)$ with $\beta_c \lambda \equiv 1$, such that a non-zero spontaneous magnetization appears exactly when $T < T_c$. The isotherms at $T < T_c$ present a wiggle akin to the instability we discussed in the case of the van der Waals model for fluids; it can also be disposed of by a thermodynamical argument: for the Weiss model, the equivalent of the construction of Maxwell plateau gives a straightline segment at $B = 0$, along which one has a mixture of the two thermodynamical phases, with the magnetization "up" in one of them and "down" in the other, a residual manifestation of the flip-flop symmetry of the model, whereby for all i simultaneously: $\sigma_i \to -\sigma_i$.

Again in complete similarity with the discussion of the van der Waals equation, the critical exponents can be computed readily from (11.3.37). For instance, upon expanding $\tanh x$ to second order in x, we receive that the spontaneous magnetization $m_T \equiv m(B = 0, T)$ is governed asymptotically by

$$m_T = \frac{T_c}{T} \, m_T [1 + \frac{1}{2} (\frac{T_c}{T} \, m_T)^2 + \cdots]^{-1}$$

i.e.

$$m_T \propto (T_c - T)^{\frac{1}{2}} \quad \text{as} \quad T_c \nearrow T \qquad (11.3.38)$$

which is identical to the value found in (11.3.17) for the critical coefficient β, obtained from the van der Waals equation. The discussion of the other critical exponents leads to exactly the same conclusions as those we found in analyzing the consequences of the van der Waals equation, the translation being given by the dictionary (11.2.1–4).

The Weiss and van der Waals models are characteristic of mean-field methods. They are insensitive to dimensionality and they are unreliable for short-range interactions: we show in Sect. 12.1 that the $1-d$ Ising model with nearest-neighbor interactions, does *not* exhibit a ferromagnetic phase transition, and *yet* the application of the mean-field method to this model would predict a phase transition.

The mean-field models start with a microscopic description, but the explanation is only semi-microscopic: the thermodynamic models are not reduced – and hence, explained – by purely microscopic elements. Indeed, the "mean field" in (11.3.41) is not a micro-element in the model. One may think that it is nothing more than the average magnetic effect of the other dipoles in the domain. The problem is that replacing the interaction by its average is a very drastic substitution: it gives the system a rigidity that is not there when the interaction is reasonably short range, as is the case for electromagnetic dipole-dipole interaction. As a consequence, a computation of B_{int} involving a realistic $J_{i,j}$ would give a critical temperature that is several orders of magnitude off the mark. Therefore models of the Weiss-type are better used for some restricted qualitative purposes where T_c is left as a macro-parameter to be adjusted from the experimental results. In this sense, the micro-explanation is not thoroughly carried out.

While the Ornstein–Weiss mean-field theory recognizes correctly that phase transitions are cooperative phenomena, it does overshoot: the substitution of a mean field in place of more subtle interactions that have *specific structures in configuration space,* leads to an over-emphasis that replaces a hopefully non-circular derivation of the cooperative behavior by a brutal *de facto* assumption that is not controlled by microscopic arguments. Nevertheless, one should not throw the baby with the bath water: the Weiss model and the van der Waals equation do make some correct predictions, notably the universal consequence of the van der Waals equation known as the law of corresponding states (11.3.9–11). Moreover, in spite of their predicting the wrong values for the critical exponents, they do point out towards some relations – see e.g. (11.3.21)– between these exponents, relations that seem to be universality satisfied, both by experiments and by exact microscopic models. This is pursued in our Chap. 13 on scaling and renormalization methods.

Therefore the challenge is first to isolate features that allow to discriminate between the spectacularly correct and the dismally incorrect predictions of the mean-field theory. The next three chapters present models that dispense entirely with the mean-field approach. Finally, we should mention that in dimension $d \geq 2$, a variant of the Kac–Baker rescaling [see (11.3.31–32)]

$$\phi^{attr}(r) \quad \rightarrow \gamma^d \phi^{attr}(\gamma r) \qquad \text{with} \quad \gamma \rightarrow 0 \qquad (11.3.39)$$

has been proven reliable as a zeroth-order approximation in a perturbative approach that allows qualitative, yet rigorous, predictions on the behavior of the system for sufficiently small, but non-vanishing γ; the outline in [Lebowitz, Mazel, and Presutti, 1998] is very instructive.

12. Ising and Related Models

Why am I lost,
Down the endless road to infinity toss'd?
[Walton and Sitwell, 1951]

12.1 The 1-d Ising Model. No-Go Theorems

The model was invented with the purpose of deriving ferromagnetism solely from the tenets of statistical mechanics. The first attempt was to consider a one-dimensional version of the model; we review this in the present section to try and learn how and why it was unsuccessful. Successful, higher dimensional versions of the model are studied in the next two sections; some related models are also mentioned in Sect. 12.3. The Ising model for ferromagnetism satisfies the following four defining conditions:

(i)   It consists of $|\Lambda|$ fixed sites which form a regular lattice Λ; in dimension $d = 1$, Λ is a linear chain.

(ii) Each of the sites is occupied by a classical two-level system; the states of the individual system at the site $i \in \Lambda$ are denoted by the value ± 1 of a variable σ_i. The configuration space of the whole lattice is therefore the set $\{-1, +1\}^{|\Lambda|}$. To make contact with the physical world, Λ can be viewed as a crystal lattice, each vertex of which is occupied by a spin-half particle; this would correspond to what is known as the Heisenberg model; the Ising model makes one further idealization: each of the spin is constrained to point only "up" or "down".

(iii) An external homogeneous "magnetic field" $B > 0$ tends to align the "spins" in the "up" position.

(iv) An internal two-body interaction $J > 0$ is assumed, which tends to align the "spins" parallel to one another. This interaction is restricted to pairs of nearest-neighbor "spins" in the lattice Λ. The Hamiltonian of the model, therefore, is given by:

$$H_\Lambda = -B \sum_i \sigma_i - J \sum_{<i,j>} \sigma_i \sigma_j \quad , \qquad (12.1.1)$$

where $< i, j >$ means that the sum carries over (unordered) pairs of nearest neighbors.

For convenience, in the 1−d version of the model, we impose the periodic boundary condition $\sigma_{|\Lambda|+1} = \sigma_1$ and we thus rewrite now (12.1.1) in the form

$$H_\Lambda = -B \sum_{i=1}^{|\Lambda|} \sigma_i - J \sum_{i=1}^{|\Lambda|} \sigma_i \, \sigma_{i+1} \quad . \tag{12.1.2}$$

The statistical mechanics of the model can be solved explicitly. Since the total number of the spins in an Ising system is constant, we use the partition function of the canonical ensemble, i.e.

$$Z_\Lambda(J, B, \beta) = \sum_{\sigma_1 = \pm 1} \cdots \sum_{\sigma_{|\Lambda|} = \pm 1} \exp\left(-\beta \, H_\Lambda\right) \quad . \tag{12.1.3}$$

We compute this by the 1−d version of *transfer matrix formalism* introduced by [Kramers and Wannier, 1941] for the 2−d Ising model. We start by rewriting (12.1.3) as:

$$Z_\Lambda(J, B, \beta) = \sum_{\sigma_1 = \pm 1} \cdots \sum_{\sigma_{|\Lambda|} = \pm 1} V_{\sigma_1 \sigma_2} \, V_{\sigma_2 \sigma_3} \cdots V_{\sigma_{|\Lambda|-1} \sigma_{|\Lambda|}} \, V_{\beta J \sigma_{|\Lambda|} \sigma_1} \tag{12.1.4}$$

with

$$V_{\sigma_i \sigma_{i+1}} = \exp\left(\frac{1}{2}\beta B \sigma_i + \beta J \sigma_i \sigma_{i+1} + \frac{1}{2}\beta B \sigma_{i+1}\right) \quad . \tag{12.1.5}$$

Hence, the partition function for the one-dimensional Ising model takes the form:

$$Z_\Lambda(J, B, \beta) = \operatorname{Tr} V^{|\Lambda|} \tag{12.1.6}$$

with

$$V = \begin{pmatrix} e^{\beta(J+B)} & e^{-\beta J} \\ e^{-\beta J} & e^{\beta(J-B)} \end{pmatrix} \quad . \tag{12.1.7}$$

The matrix V is symmetric and can therefore be diagonalized; a straightforward computation of its eigenvalues delivers:

$$\lambda_\pm = e^{\beta J} \cosh \beta B \pm \left(e^{2\beta J} \sinh^2 \beta B + e^{-2\beta J}\right)^{\frac{1}{2}} \tag{12.1.8}$$

in terms of which we obtain the explicit form of the partition function, namely:

$$Z_\Lambda(J, B, \beta) = \lambda_+{}^{|\Lambda|} + \lambda_-{}^{|\Lambda|} \quad . \tag{12.1.9}$$

This is a smooth expression in terms of the external parameters β and B; it rules out the appearance of a discontinuous isothermal magnetization: so far, *no* phase transition.

Around the mid 1930s, the feeling began to dawn on the physics community that phase transitions could perhaps be better understood from models considered in the thermodynamical limit. H. A. Kramers, one of the very first proponents of the suggestion, was chairing one of the sessions at the famous 1937 van der Waals Congress held in Amsterdam *and in a whimsical mood he put the question to a vote. The vote was never recorded because it was not an official act of the congress.*

[Dresden, 1988].

With a recorded outcome or not, this unorthodox intrusion of the democratic process to decide a scientific truth is reported here to give solace to a finitist Reader. Two observations on motivation (if not justification, yet) are in order.

First, the physical motivations. (a) From a theoretical point of view, the actual systems in which ferromagnetism is to be observed – even at the level of the Curie domains – are so very large compared to the microscopic lattice spacings that the approximation provided by the *thermodynamical limit* may point to some new understanding that would otherwise remain blurred in the complexities of the finite model; for instance tracing analytically, without approximation, the isotherms resulting from (12.1.9) for $|\Lambda| < \infty$ would be more complicated than genuinely instructive. (b) From a pragmatic point of view, the bulk properties of the system – e.g. its equation of state – appear to be insensitive to the actual size of the system, provided the latter has macroscopic dimensions.

Second, the mathematical motivation: unless some stringent uniformity condition is proven to hold, the limit of continuous functions cannot be expected to be everywhere continuous; the hope is that such a condition is not satisfied.

While the 1−d Ising model is shown presently *not* to exhibit a phase transition even in the thermodynamical limit, the limiting procedure is illustrated on this model with a view to apply the technique later to the more difficult 2−d version of the model – which itself *does* exhibit a phase transition. The free-energy per site, $f_\Lambda(J, B, \beta)$, obtained from (12.1.9) is

$$-\frac{1}{|\Lambda|}\frac{1}{\beta}\ln Z_\Lambda(B, \beta) = -\ln\lambda_+ - \frac{1}{|\Lambda|}\ln\left[1 + (\frac{\lambda_-}{\lambda_+})^{|\Lambda|}\right] \qquad (12.1.10)$$

so that in the thermodynamical limit

$$f(J, B, \beta) \equiv \lim_{|\Lambda|\to\infty} f_\Lambda(J, B, \beta) = -\frac{1}{\beta}\ln\lambda_+ \quad . \qquad (12.1.11)$$

Upon inserting in this limit the expression obtained for λ_+ in (12.1.8), we find for the magnetization per site, $m(J, B, \beta) \equiv -\partial_B f(J, B, \beta)$:

$$\boxed{m(J, B, \beta) = \tanh\beta B \left[1 - (1 - e^{-4\beta J})\cosh^{-2}\beta B\right]^{-\frac{1}{2}}} \quad . \qquad (12.1.12)$$

This expression is *exact* in the thermodynamical limit. We note that this magnetization has the following properties:

$$(1) \quad m(J, -B, \beta) = -m(J, B, \beta)$$

$$(2) \quad \partial_B m(J, B, \beta) > 0 \; ; \; m(J, 0, \beta) = 0$$

$$(3) \quad \lim_{B \to \pm\infty} m(J, B, \beta) = \pm 1 \qquad \Bigg\} . \qquad (12.1.13)$$

$$(4) \quad m(0, B, \beta) = \tanh \beta B$$

$$(5) \quad \forall B \neq 0 : |m(J, B, \beta)| > |m(0, B, \beta)|$$

The first line reflects *flip-flop* symmetry: the Hamiltonian (12.1.1) is invariant when the signs of B and of all the σ_i's are changed simultaneously. Lines 2 and 3 tell us that the magnetization, as function of the external magnetic field B, increases monotonically from -1 to $+1$, passing through the value 0 when the magnetic field vanishes: no discontinuity, no wiggles, no phase transition, although the thermodynamical limit has been taken. Ernst Ising had a good reason to feel discouraged. We now know better; see the next section. Lines 4 and 5 confirm two natural expectations: (i) in the absence of interaction, i.e. for $J = 0$, the magnetization is indeed equal to the canonical equilibrium value of an isolated spin; (ii) any interaction with $J > 0$ increases the tendency of the spins to align with one another, so that at any fixed temperature β and magnetic field $B > 0$ (resp. $B < 0$), the magnetization is larger (resp. smaller) that it would have been if the spin-spin interaction were to vanish. In particular, we have for the slope of the tangent at $B = 0$ to the isotherm at inverse temperature β :

$$\chi(J, 0, \beta) \equiv [\partial_B m(J, B, \beta)]_{B=0} = \beta e^{2\beta J} \quad . \qquad (12.1.14)$$

These results are in the direction of what we are looking for, but the enhancing effect of the interaction is not yet strong enough to cause a discontinuity.

A related indication of the cooperative trend reflected by this enhancement is the behavior of the space-correlation function. Let indeed A and B be two one-site observables, i.e. functions of some σ; in particular $A_k = A(\sigma_k)$ is an observable at the site k and $B_{k+n} = B(\sigma_k) \equiv \alpha_n[B_k]$ is an observable at the translated site $k + n$. The method we used to compute the partition function from the transfer matrix V in (12.1.7) – which, incidentally owes this name to the third line of (12.1.15) below – can be applied to the following computations.

$$\langle A\alpha_n[B] \rangle \equiv \lim_{|\Lambda| \to \infty} Z_\Lambda^{-1} \sum_{\sigma\{\Lambda\}} \exp(-\beta H_\Lambda) A_k B_{n+k}$$

$$= \lim_{|\Lambda| \to \infty} \left[\mathrm{Tr} V^{|\Lambda|} \right]^{-1} \left[\mathrm{Tr}\, V^{|\Lambda|} A \gamma_n[B] \right] \qquad \Bigg\} . \qquad (12.1.15)$$

$$\gamma_n[B] \equiv V^n B V^{-n}$$

With (12.1.9) and $\phi_\pm$ defined by $V\phi\pm = \lambda_\pm \phi_\pm$, we have:

$$\text{Tr } V^{|\Lambda|}\, A\, V^n\, B\, V^{-n} =$$

$$\lambda_+{}^{|\Lambda|} \left[(A\phi_+, \phi_+)(B\phi_+, \phi_+) + \left(\tfrac{\lambda_-}{\lambda_+}\right)^n (A\phi_-, \phi_+)(B\phi_+, \phi_-) \right]$$

$$+ \lambda_-{}^{|\Lambda|} \left[(A\phi_-, \phi_-)(B\phi_-, \phi_-) + \left(\tfrac{\lambda_-}{\lambda_+}\right)^n (A\phi_+, \phi_-)(B\phi_-, \phi_+) \right] \quad .$$

Hence, in the limit $|\Lambda| \to \infty$, the space-correlation function is:

$$\langle A\alpha_n[B] \rangle - \langle A \rangle \langle B \rangle = \exp(-\xi^{-1} n)\, C_{A,B} \qquad (12.1.16)$$

where the equilibrium values of A and B, and the constants $C_{A,B}$ and ξ are:

$$\left. \begin{array}{l} \langle A \rangle = (A\phi_+, \phi_+) \quad ; \quad \langle B \rangle = (B\phi_+, \phi_+) \\[2mm] C_{A,B} = (A\phi_-, \phi_+)(B\phi_+, \phi_-) \quad ; \quad \xi^{-1} = \ln\lambda_+ - \ln\lambda_- \end{array} \right\} \quad . \quad (12.1.17)$$

The correlation length ξ remains finite for all finite values of the external magnetic field B, the coupling constant J, and the temperature β; (12.1.16–17) therefore describe a true exponential decay, yet another manifestation of the fact that, even in the thermodynamical limit, the one-dimensional Ising model does not exhibit any phase transition: the interaction is not strong enough to create long-range order. Yet, from (12.1.8):

$$\text{for} \quad B = 0 : \quad \frac{\lambda_+}{\lambda_-} = \frac{e^{\beta J} + e^{-\beta J}}{e^{\beta J} - e^{-\beta J}} \quad \text{hence as } J \nearrow \infty \;\; \xi \propto e^{2\beta J} \quad (12.1.18)$$

i.e. in the absence of magnetic field, the correlation length increases exponentially with the coupling constant when the latter becomes asymptotically large – so does the susceptibility (12.1.14).

In summary, we learned that while the largest eigenvalue of the transfer matrix V controls the equilibrium behavior of the thermodynamical functions such as the free energy – or the magnetization – per site, the ratio of its two eigenvalues controls the spin-spin correlation length; long-range order would set in, if only the largest eigenvalue were to become degenerate. This is precisely what occurs in the 2−d Ising model; see Sect. 12.2.

Before we close the present section, however, we must circumscribe the obstructions any model has to meet if it is to exhibit the truly cooperative behavior that could entail phase transitions. For this purpose, we consider a $n−$dimensional version of the Ising model where Λ is a finite subset of $\mathbf{Z}^d$ with a spin-spin interaction that is not necessarily limited to the nearest neighbors [compare with (12.1.1)]:

$$H_\Lambda = -B \sum_{i \in \Lambda} \sigma_i - \frac{1}{2} \sum_{i,j \in \Lambda} J_{i,j} \sigma_i \sigma_j \qquad (12.1.19)$$

where the coupling $J : i, j \in \mathbf{Z}^n \times \mathbf{Z}^n \mapsto J_{i,j} \in \mathbf{R}$ satisfies the following three conditions for all i and $j \in \mathbf{Z}^n$:

$$(a) \; J_{i,j} = J_{j,i} \quad ; \quad J_{i,i} = 0$$

$$(b) \; \forall \, k \in \mathsf{Z}^n \; : \quad J_{i+k,j+k} = J_{i,j} \tag{12.1.20}$$

$$(c) \; \sum_{j \in \mathsf{Z}^n} J_{i,j} < \infty$$

The first of these relations states that the interaction is between pairs of distinct spins; the next two are imposed with an eye to the thermodynamical limit: the second is to make the model translation invariant and the third ensures that the interaction energy of any spin with all the others remains finite when Λ tends to infinity.

The first no-go theorem rules out the occurrence of phase transition for finite models, independently of the dimensionality of the model.

Theorem 12.1.1. *For all finite $\Lambda \subset \mathsf{Z}^d$ and all inverse temperatures $\beta = (kT)^{-1}$, the canonical partition function ensures that the free-energy per site*

$$f_\Lambda(J, B, \beta) = -\frac{1}{\beta} \ln Z_\Lambda(J, B, \beta) \quad \text{with} \quad Z_\Lambda(J, B, \beta) \equiv \sum_{\sigma \in \{1, +1\}^{|\Lambda|}} e^{-\beta H_\Lambda(\sigma)}$$

is analytic and concave in all J_{ij} and B.

Recall that a function g defined on an interval (a, b) is said to be *concave* whenever

$$\forall \, x, y \in (a, b) \text{ and } \forall \, \lambda \in [0, 1] \; : \; g(\lambda x + (1 - \lambda)y) \geq \lambda g(x) + (1 - \lambda) g(y) \quad ,$$

i.e. the graph of the function for any interval $[x, y]$ remains above the straight-line joining the points $(x, g(x))$ to $(y, g(y))$. Another way to express this condition is:

$$\forall \, a < s < t < u < b \quad : \quad \frac{g(u) - g(t)}{u - t} \leq \frac{g(u) - g(s)}{u - s} \leq \frac{g(t) - g(s)}{t - s} \quad .$$

We may add that a function h is said to be *convex* whenever $(-h)$ is concave, i.e. whenever the inequalities in the above equivalent defining relations are reversed. Corollary 12.1.1 below is only one of the many arguments in statistical mechanics based on convexity properties [Israel, 1979]. It is easy to verify the following basic properties of concave functions.

Scholium 12.1.1.

1. *Every concave function is continuous.*
2. *For a differentiable function g, the following conditions are equivalent:*
   (i) g is concave ;
   (ii) the derivative g' of g is monotonically decreasing.

The *proof* of the concavity of the free-energy – asserted in Thm. 12.1.1 – is an immediate consequence of a mathematical fact, known as the Hölder inequality, namely:

$$\forall\, p,q \quad \text{with} \quad \frac{1}{p} + \frac{1}{q} = 1 \quad : \quad \sum_n |x_n y_n| \leq \left[\sum_n |x_n|^p \right]^{\frac{1}{p}} \left[\sum_n |y_n|^q \right]^{\frac{1}{q}} \quad .$$

Upon substituting

$$B^{(0)} = \lambda B^{(1)} + (1 - \lambda) B^{(2)} \quad \text{and} \quad J_{i,j}^{(0)} = \lambda J_{i,j}^{(1)} + (1 - \lambda) J_{i,j}^{(2)}$$

in

$$Z_\Lambda(J^{(0)}, B^{(0)}, \beta) \equiv \sum_{\sigma_1, \cdots \sigma_{|\Lambda|}} \exp(-\beta H_\Lambda(J^{(0)}, B^{(0)})(\sigma_1, \cdots, \sigma_{|\Lambda|}))$$

the RHS is brought to a form where we can apply the Hölder inequality:

$$\sum_{\sigma_1, \cdots \sigma_{|\Lambda|}} \left[\exp(-\beta H_\Lambda(J^{(1)}, B^{(1)})(\sigma_1, \cdots, \sigma_{|\Lambda|})) \right]^\lambda$$

$$+ \sum_{\sigma_1, \cdots \sigma_{|\Lambda|}} \left[\exp(-\beta H_\Lambda(J^{(2)}, B^{(2)})(\sigma_1, \cdots, \sigma_{|\Lambda|})) \right]^{1-\lambda}$$

$$\leq \left[\sum_{\sigma_1, \cdots \sigma_{|\Lambda|}} \exp(-\lambda \beta H_\Lambda(J^{(1)}, B^{(1)})(\sigma_1, \cdots, \sigma_{|\Lambda|})) \right]^\lambda$$

$$+ \left[\sum_{\sigma_1, \cdots \sigma_{|\Lambda|}} \exp(-\beta H_\Lambda(J^{(2)}, B^{(2)})(\sigma_1, \cdots, \sigma_{|\Lambda|})) \right]^{1-\lambda}$$

and thus

$$Z_\Lambda(J^{(0)}, B^{(0)}, \beta) \leq \left[Z_\Lambda(J^{(1)}, B^{(1)}, \beta) \right]^\lambda \left[Z_\Lambda(J^{(2)}, B^{(2)}, \beta) \right]^{1-\lambda} \quad .$$

Hence, with

$$f_\Lambda(J, B, \beta) \equiv -\frac{1}{\beta |\Lambda|} \ln Z_\Lambda(J, B, \beta)$$

we have

$$f_\Lambda(J^{(0)}, B^{(0)}, \beta) \geq \lambda f_\Lambda(J^{(1)}, B^{(1)}, \beta) + (1 - \lambda) f_\Lambda(J^{(2)}, B^{(2)}, \beta) \quad .$$
q.e.d.

The second assertion in Scholium 12.1.1 leads now to the following consequence of Thm. 12.1.1:

Corollary 12.1.1. *Under the conditions of Thm. 12.1.1, the magnetization per site*

$$m_\Lambda(J, B, \beta) \equiv -\partial_B f_\Lambda(J, B, \beta)$$

is analytic and satisfies the stability condition

$$\partial_B m_\Lambda(J, B, \beta) > 0 \quad .$$

Hence, no discontinuity of any kind and no wiggles; i.e. no phase transition in any dimension, for *finite* Ising-type models.

Next, to ensure the existence and uniqueness of *thermodynamical limit*, some restrictions need to be imposed on the way the region Λ increases to infinity in order for the model to capture the bulk properties of the material considered: the surface is to become small with respect to the volume; the vernacular counterexample would be a sponge; the precise conditions are spelled out below.

Firstly, a collection $\mathcal{F}$ of finite regions $\Lambda \subset \mathsf{Z}^d$ is a *directed set* under inclusion, whenever:

$$\Lambda_1, \Lambda_2 \in \mathcal{F} \quad \models \quad \exists\, \Lambda \in \mathcal{F} \text{ such that } \{\, \Lambda_1 \subseteq \Lambda \text{ and } \Lambda_2 \subseteq \Lambda \,\}. \quad (12.1.21)$$

We say that a function $f : \Lambda \in \mathcal{F} \mapsto f_\Lambda \in \mathsf{R}$ admits a limit $f \in \mathsf{R}$ over $\mathcal{F}$, a fact we denote

$$f = \lim_{\Lambda \in \mathcal{F}} f_\Lambda \quad,$$

whenever

$$\forall\, \varepsilon > 0 \quad \exists\, \Lambda_\varepsilon \in \mathcal{F} \quad \text{such that} \quad \{\, \Lambda \supset \Lambda_\varepsilon \quad \models \quad |f - f_\Lambda| < \varepsilon \,\} \quad.$$

Secondly, Λ is said to be asymptotically large, whenever

$$\lim_{\Lambda \in \mathcal{F}} |\Lambda|^{-1} = 0 \qquad\qquad (12.1.22)$$

where $|\Lambda|$ denotes as usual the number of sites in $|\Lambda|$.

Finally, let us denote by $d(x, \Lambda^c)$ the distance of $x \in \Lambda$ to Λ^c, the complement of Λ in Z^d. The boundary of Λ is said to be relatively asymptotically small whenever

$$\forall\, h \in \mathcal{Z}^+ : \quad \Lambda_h \equiv \{x \in \Lambda \,|\leq h\} \models \lim_{\Lambda \in \mathcal{F}} |\Lambda|^{-1}|\Lambda_h| \quad. \qquad (12.1.23)$$

Alternatively, for any $a \in (\mathsf{Z}^+)^d$ consider the partition $\mathcal{P}_a = \{P_\nu(a) \mid \nu \in \mathsf{Z}^d\}$ of Z^d in cells $P_\nu(a) = \{x \in \mathsf{Z}^d \mid x^k \in [\nu^k a^k, (\nu^k - 1)a^k]\, k = 1, \cdots, n\}$. Define then, for every finite region $\Lambda \subset \mathsf{Z}^d$:

$$N_a{}^- \equiv \text{card}\{\nu \in \mathsf{Z}^d \mid P_\nu(a) \subseteq \Lambda\} \text{ and } N_a{}^+ \equiv \text{card}\{\nu \in \mathsf{Z}^d \mid P_\nu(a) \cap \Lambda \neq \emptyset\} \,.$$

In terms of these, the condition that the boundary be relatively asymptotically small is:

$$\forall\, a \in \mathsf{Z}^{+d} : \quad \lim_{\Lambda \in \mathcal{F}} \frac{N_a{}^-}{N_a{}^+} = 1 \quad. \qquad (12.1.24)$$

Definition 12.1.1. *A collection $\mathcal{F} = \{\Lambda\}$ of finite regions in Z^d is said to be admissible in the sense of van Hove whenever it satisfies (12.1.21) and (12.1.22), together with (12.1.23) or (12.1.24).*

Theorem 12.1.2. *Let $\mathcal{F} = \{\Lambda\}$ be a collection of finite regions in $\mathbb{Z}^d$ admissible in the sense of van Hove; then the free energy per site $f_\Lambda(J, B, \beta)$ admits a limit*

$$f(J, B, \beta) = \lim_{\Lambda \in \mathcal{F}} f_\Lambda(J, B, \beta) \quad ;$$

this limit is continuous and is concave in the variables $J_{i,j}$ and B.

Although this theorem guaranties the existence of the thermodynamical limit of the free-energy per site, it does *not* ensure that it be differentiable, so that the magnetization per site could be even defined. For this, one needs that the interaction J decays at infinity faster than is imposed by condition (12.1.20c), which in one-dimension essentially requires only that, at infinity, $J_{i,j} \approx |i - j|^{-\alpha}$ with $\alpha > 1$.

The following is the no-go theorem that informs us that, in one dimension, and when the range of the interaction does not decay slowly enough at infinity, the thermodynamical limit does *not* allow to escape the situation met in the finite case described by Thm. 12.1.1 and its Corollary.

Theorem 12.1.3. *For any one-dimensional model of the Ising type, where the interaction decreases at infinity at least exponentially, the free-energy per site*

$$f(J, B, \beta) = \lim_{\Lambda \in \mathcal{F}} f_\Lambda(J, B, \beta)$$

exists, is analytic and is concave. In particular, the magnetization per site

$$m(J, B, \beta) \equiv \lim_{\Lambda \in \mathcal{F}} m_\Lambda(J, B, \beta)$$

exists; m satisfies the thermodynamical relation

$$m(J, B, \beta) = -\partial_B f(J, B, \beta) \quad ;$$

it is analytic, and it satisfies the stability condition

$$\partial_B m(J, B, \beta) > 0 \quad .$$

Remarks and bibliographical notes. The line of arguments reported above was initiated by [Van Hove, 1949, 1950], where the idea embodied in Def. 12.1.1 was introduced; van Hove discussed in particular one-dimensional continuous models with an interaction that is the sum of: (a) a repulsive hard-core – so that the particles cannot approach each other closer than a distance r_o – and (b) a potential that is of strictly finite-range – i.e. models for which there exists some finite number r_1 such that the interaction is *strictly* zero when the interparticle distance is larger than r_1. The Reader interested in tracing further back the idea of ruling out phase transitions in one-dimension will note that van Hove mentioned his being made aware, after the fact, of the existence of a precedent [Takahashi, 1942]. This condition of strictly finite range interactions is evidently satisfied by the original nearest neighbor Ising model which satisfies the condition under which Thm. 12.1.3 is stated. This theorem was established by [Araki, 1969] as a by-product of his study of one-dimensional *quantum* systems.

Earlier, [Ruelle, 1968] had considered a condition less stringent than exponential decay of the interaction at infinity, namely that the interaction have finite first moment, i.e. $\sum_i |i - j| J_{i,j} < \infty$, which amounts to impose that, at infinity, $J_{i,j} \approx |i - j|^{-\alpha}$ with $\alpha > 2$; under this condition, $f(J, B, \beta)$ exists and is concave – as requested by Thm. 12.1.2 – but its analyticity must be replaced by the weaker conclusion that it admits a continuous first derivative, i.e. that m, the magnetization per site, is continuous; note that the concavity of f implies that m is increasing in B : i.e. under Ruelle's condition on the interaction, first-order phase transitions are still ruled out. For a didactic exposition, with proofs, that extends beyond the 2-body interactions J to cover also general many-body interactions Φ; see [Ruelle, 1969]. The role of *convexity* in equilibrium statistical mechanics is expounded in [Israel, 1979], with the help of a specific family of models; this work also contains a 85-page historical introduction by A. S. Wightman, providing a broad conceptual motivation. These results are (almost) optimal for two-body interactions $J_{i,j}$. Indeed, it follows in particular from sharp estimates obtained by [Dyson, 1969] that, even in one-dimension, a phase transition *does* occur when $J_{i,j}$ decays at infinity as slowly as $|i - j|^{-\alpha}$ with $1 < \alpha < 2$. Dyson's results were obtained by comparison with a class of somewhat artificial toy-models – the so-called Dyson hierarchical models. The ideas behind the construction of these models become most useful to help analyze the critical behavior of higher dimensional models that are otherwise known to exhibit phase transitions; see Sect. 13.2.

It is true that the Kac–Baker model – sketched in Sect. 11.3 – shows a phase transition in one dimension, but it must be realized that this is the result of a peculiar limiting procedure $\gamma \to 0$ whereby the interaction becomes simultaneously asymptotically long-range and asymptotically weak, thus escaping the stricture of Thm. 12.1.3. This feature of the mean-field method persists in the quantum realm; for a quantum toy-model, where the Weiss approximation becomes exact in the thermodynamical limit, see [Emch and Knops, 1970].

In conclusion, we have learned two lessons in this section. The first is that our discussion of models of the Ising-type has turned out no physically reasonable one-dimensional model that would exhibit ferromagnetic phase transitions. A second lesson, nevertheless, derives from our having seen explicitly how the trend towards long range order and/or very large magnetization in very small applied magnetic field does develop – even in the regime where correlations still decay exponentially – when the interactions become strong enough to entail cooperative behavior. We argue in the next section how such cooperative behavior – resulting in long-range order – is actually possible in higher dimensions, even when only short range interactions are present.

12.2 The 2-d Ising Model

The 2-dimensional Ising model is the first model establishing that no *ad hoc* assumptions need to be added to the basic tenets of statistical mechanics in order to derive the existence of a ferromagnetic phase transition.

The history of this result span a generation of efforts, from the awe surrounding its almost forbidding presentation [Onsager, 1944] – improved, but only partially demystified, in [Kaufman, 1949] – to an alternate derivation in [Yang, 1952]; and finally, to the instructive elucidation by [Schultz, Mattis, and Lieb, 1964]. The latter rightfully became a classic, which we now follow, as do textbooks such as [Wannier, 1966, 1987] [advanced undergraduate physics] or [Gallavotti, 1999] [introductory graduate mathematical physics].

Even though the argument can now be regarded as straightforward, it is sufficiently intricate to warrant a close look. In a nutshell, the ultimate aim is to establish the existence of a critical temperature – given by (12.2.30) – under which a phase transition occurs characterized by long-range order and a non-zero spontaneous magnetization that persists in the absence of external magnetic field; hence the name *spontaneous*. This is the content of the famous Onsager formula – here (12.2.57).

As for the technicalities in the argument, beyond some perseverance, what is useful is an elementary knowledge of linear algebra, namely: symmetric matrices can be diagonalized, and there are ways to determine whether their maximal eigenvalue is degenerate or not, i.e. whether the corresponding eigenvector is unique or not. How this bears on the existence of phase transition is the story we want to tell step-by-step in this section. Some bibliographical notes are collected at the end. For supplementary references, see [Schultz, Mattis, and Lieb, 1964] and [Gallavotti, 1999].

Step 1. Setting up the problem: the Hamiltonian and the transfer matrix. The model is defined on a rectangular lattice

$$Z_N \times Z_M = \{ (n,m) \mid n = 1, \cdots, N \; ; \; m = 1, \cdots, M \} \tag{12.2.1}$$

of N rows, numbered by n; each row n has M sites, numbered by the indices (n,m). To each site is attached a variable $\sigma_{n,m}$ taking one of the two values ± 1; hence, to the $n-$th row is attached a variable $\sigma^{(n)} \equiv (\sigma_{n,1}, \cdots, \sigma_{n,M})$ taking one of 2^M possible values. More generically, we write $\underline{\sigma} \equiv (\sigma_1, \cdots, \sigma_M)$ for any vector-variable, each of the M components σ_m of which has possible values ± 1. In order to obtain a model that is separately invariant under translations along rows and columns, we choose to impose periodic boundary conditions, i.e. to wrap our rectangle over a torus, thus setting $\underline{\sigma}^{(N+1)} = \underline{\sigma}^1$ and, $\forall \, n$, $\sigma_{n,M+1} = \sigma_{n,1}$.

The Hamiltonian of the model is taken to be

$$H_{N,M}(\underline{\sigma}^{(1)}, \cdots, \underline{\sigma}^{(N)}) \equiv \sum_{n=1}^{N} \left[H^{(v)}(\underline{\sigma}^{(n)}, \underline{\sigma}^{(n+1)}) \; + \; H^{(h)}(\underline{\sigma}^{(n)}) \right] \tag{12.2.2}$$

where $H_n^{(v)}$ describes the interaction between the consecutive rows n and $n + 1$; and $H_n^{(h)}$ describes the interaction between consecutive sites in row n, with coupling constants J_1 and J_2 strictly positive. Specifically:

$$\left. \begin{aligned} H^{(v)}(\underline{\sigma}, \underline{\sigma}') &\equiv -J_1 \sum_{m=1}^{M} \sigma_m \sigma_m' \\ H^{(h)}(\underline{\sigma}) &\equiv -J_2 \sum_{m=1}^{M} \sigma_m \sigma_{m+1} \end{aligned} \right\}. \tag{12.2.3}$$

(12.2.2) thus becomes:

$$\left. \begin{aligned} H_{N,M}(\underline{\sigma}^{(1)}, \cdots, \underline{\sigma}^{(N)}) = \\ -J_1 \sum_{n=1}^{N} \sum_{m=1}^{M} \sigma_{n,m} \sigma_{n+1,m} - J_2 \sum_{n=1}^{N} \sum_{m=1}^{M} \sigma_{n,m} \sigma_{n,m+1} \end{aligned} \right\}. \tag{12.2.4}$$

The canonical partition function of the model

$$Z_{NM} \equiv \sum_{\underline{\sigma}^{(1)}, \cdots, \underline{\sigma}^{(N)}} \exp\left[-\beta H_{N,M}(\underline{\sigma}^{(1)}, \cdots, \underline{\sigma}^{(N)})\right] \tag{12.2.5}$$

can therefore be rewritten in the form

$$Z_{NM} = \sum_{\underline{\sigma}^{(1)}, \cdots, \underline{\sigma}^{(N)}} \prod_{n=1}^{N} V(\underline{\sigma}^{(1)}, \cdots, \underline{\sigma}^{(n)}) = \mathrm{Tr}\, V^N \tag{12.2.6}$$

where the *transfer matrix* V is a symmetric matrix with entries

$$\boxed{ \begin{aligned} \text{where} \quad V(\underline{\sigma}, \underline{\sigma}') &= V_2^{\frac{1}{2}}(\underline{\sigma})\, V_1(\underline{\sigma}, \underline{\sigma}')\, V_2^{\frac{1}{2}}(\underline{\sigma}') \\ V_1(\underline{\sigma}, \underline{\sigma}') &= \exp\left[\beta J_1 \sum_{m=1} \sigma_m \sigma_m'\right] \\ V_2(\underline{\sigma}) &= \exp\left[\beta J_2 \sum_{m=1} \sigma_m \sigma_{m+1}\right] \end{aligned} } \tag{12.2.7}$$

At this point, let us pause and examine the similarities and differences between this and the corresponding result obtained for the 1-dimensional model (Sect. 12.1): the sites are replaced by rows; and the transfer matrix (12.1.7) – which was a 2 by 2 matrix – has now become a 2^M by 2^M matrix, the size of which depends thus on the length M of the rows. As with (12.1.7) this matrix is symmetric, and all its entries are positive; *hence,* all its eigenvalues are real, and its maximal eigenvalue is positive and *non-degenerate,* a particular consequence of a theorem on "positive matrices" first proven by [Perron, 1907] and [Frobenius, 1908]; for a standard textbook reference, see [Gantmacher, 1959][Thms.XIII.1&2].

At M fixed, and as $N \to \infty$, the thermodynamical behavior is controlled by this eigenvalue. An argument akin to that behind Thm. 12.1.2 shows that no phase transition can occur, yet. The situation can only be saved by taking the thermodynamical limit in the sense of van Hove, with *both N and M* becoming very large. Indeed, the hope is that, as $M \to \infty$, the maximal eigenvalue of V becomes degenerate – below a certain critical temperature to be determined. This double limit was evidently not an option available in the one dimensional model. This is the program; its implementation nevertheless requires to diagonalize the transfer matrix, and to control this diagonalization as $M \to \infty$.

Step 2. From Pauli matrices to fermions: the Jordan–Wigner transform. Although the problem is classical, we make use in the sequel of the Pauli matrices

$$\sigma_x \equiv \begin{pmatrix} 0 & 1 \\ 1 & 0 \end{pmatrix} \quad ; \quad \sigma_y \equiv \begin{pmatrix} 0 & -i \\ i & 0 \end{pmatrix} \quad ; \quad \sigma_z \equiv \begin{pmatrix} 1 & 0 \\ 0 & -1 \end{pmatrix} \quad . \tag{12.2.8}$$

We denote by σ_m^k $(k = x, y$ or $z)$ the 2^M by 2^M matrices

$$\sigma_m^k \equiv I \otimes I \otimes \cdots \otimes I \otimes \sigma^k \otimes I \otimes \cdots \otimes I \tag{12.2.9}$$

where σ^k enters as the $m-$th factor in this $M-$fold tensor product of 2 by 2 matrices; and $I \equiv (\delta_{ij})$ denotes the identity matrix.

First, we take our clue from our study of the 1-dimensional version of the model, to rewrite the transfer matrix in a more manageable form; namely, we note that V_1 can be rewritten as:

$$V_1(\underline{\sigma}, \underline{\sigma}') = \prod_{m=1}^{M} v(\sigma_m, \sigma_m')$$

with

$$v(\sigma_m, \sigma_m') = \exp(\beta J_1) \, \sigma_m \sigma_m' = \begin{pmatrix} \exp(\beta J_1) & \exp(-\beta J_1) \\ \exp(-\beta J_1) & \exp(\beta J_1) \end{pmatrix} =$$

$$\exp(\beta J_1) \, I + \exp(\beta J_1) \, \sigma^x \equiv A \left[\cosh(\beta J_1^*) \, I + \sinh(\beta J_1^*) \, \sigma^x \right] = A \exp\left[\beta J_1^* \sigma_x \right]$$

where we introduced the positive constants J_1^* and A defined by:

$$\sinh(2\beta J_1) \sinh(2\beta J_1^*) = 1 \quad \text{and} \quad A = (2 \sinh(2\beta J_1^*))^{\frac{1}{2}} \tag{12.2.10}$$

This allows to rewrite the transfer matrix (12.2.7) as:

$$V = V_2^{\frac{1}{2}} V_1 V_2^{\frac{1}{2}}$$

where

$$\left. \begin{aligned} V_1 &= A^M \exp\left[\beta J_1^* \sum_{m=1}^{} \sigma_m^x \right] \\ V_2 &= \exp\left[\beta J_2 \sum_{m=1}^{} \sigma_m^z \sigma_{m+1}^z \right] \end{aligned} \right\} \tag{12.2.11}$$

which the following substitution brings to a more suggestive form. We let

$$\sigma^{\pm} \equiv \frac{1}{2}(\sigma^z \pm i\sigma^y) \qquad (12.2.12)$$

which imply

$$\sigma^x = -(\sigma^+\sigma^- - \sigma^-\sigma^+) \quad \text{and} \quad \sigma^z = \sigma^+ + \sigma^- \quad . \qquad (12.2.13)$$

Upon inserting (12.2.13) in (12.2.11), we receive:

$$\left.\begin{array}{rl} V &= V_2^{\frac{1}{2}} V_1 V_2^{\frac{1}{2}} \\[2mm] \text{where} \quad V_1 &= A^M \exp\left[-\beta J_1^* \sum_{m=1}(\sigma_m^+\sigma_m^- - \sigma_m^-\sigma_m^+)\right] \\[2mm] V_2 &= \exp\left[\beta J_2 \sum_{m=1}(\sigma_m^+ + \sigma_m^-)(\sigma_{m+1}^+ + \sigma_{m+1}^-)\right] \end{array}\right\} \qquad (12.2.14)$$

in which the exponents are *bilinear* in the matrices σ_k^+ and σ_l^-. However, this is hard to exploit directly, due to the fact that these matrices have awkwardly "mixed" commutation/anticommutation relations, e.g.

$$\left.\begin{array}{l} \text{for } k \neq l \ : \ \sigma_k^+\sigma_l^- - \sigma_l^-\sigma_k^+ = 0 \\[3mm] \text{for } k = l \ : \ \sigma_k^+\sigma_k^- + \sigma_k^-\sigma_k^+ = I \\[3mm] \forall\, k, l \ : \ \sigma_k^+\sigma_l^+ - \sigma_l^+\sigma_k^+ = 0 \quad \text{and} \quad \sigma_k^-\sigma_l^- - \sigma_l^-\sigma_k^- = 0 \end{array}\right\} \quad . \qquad (12.2.15)$$

This is where [Schultz, Mattis, and Lieb, 1964] recognized that an old trick, due to [Jordan and Wigner, 1928], could be called to the rescue.

For each $m = 1, \cdots, M$, let

$$\left.\begin{array}{c} a_m \equiv \sigma_m^- U_m \quad \text{and} \quad a_m^* \equiv \sigma_m^+ U_m \quad \text{where} \\[3mm] U_m \equiv \begin{cases} 1 & m = 1 \\[2mm] \prod_{\mu=1}^{m-1} \exp(i\pi\sigma_\mu^+\sigma_\mu^-) = \prod_{\mu=1}^{m-1} \sigma_\mu^x & m > 1 \end{cases} \quad \text{for} \end{array}\right\} \quad . \qquad (12.2.16)$$

Since the Pauli matrices are hermitian, σ_m^- is the Hermitian conjugate of σ_m^+ matrices, and thus a_m^* is the conjugate of a_m. For all m and n, these matrices satisfy the so-called *fermion anticommutation relations*:

$$\left.\begin{array}{c} a_m^* a_n + a_n a_m^* = \delta_{m,n} I \\[3mm] a_m a_n + a_n a_m = 0 \quad \text{and thus} \quad a_m^* a_n^* + a_n^* a_m^* = 0 \end{array}\right\} \quad . \qquad (12.2.17)$$

Note the consistency of the $+$ signs, as compared to (12.2.15). The substitution (12.2.16) in the transfer matrix (12.2.14) is based on the relations (12.2.18–21) below, which are direct consequences of (12.2.16). For $m = 1, \cdots, M$:

$$a_m^* a_m = \sigma_m^+ \sigma_m^- \quad \text{and} \quad a_m^* a_m^* = \sigma_m^+ \sigma_m^+ = 0 \quad . \tag{12.2.18}$$

For $m = 1, \cdots, M - 1$:

$$a_m^* a_{m+1} = \sigma_m^+ \sigma_{m+1}^- \quad \text{and} \quad a_{m+1}^* a_m = \sigma_{m+1}^+ \sigma_m^- \tag{12.2.19}$$

(upon taking the Hermitian conjugate of these relations, recall that at different sites the fermionic matrices anticommute while the corresponding Pauli matrices commute). However, the boundary term presents a peculiarity, namely:

$$\sigma_M^+ = \exp\left[i\pi N_{(M)} + 1\right] a_M^* \quad \text{and} \quad \sigma_M^- = \exp\left[i\pi N_{(M)}\right] a_M \tag{12.2.20}$$

where

$$N_{(M)} \equiv \sum_{\mu=1}^{M} a_\mu^* a_\mu \quad . \tag{12.2.21}$$

This can be seen upon rewriting σ_M^+ as

$$\sigma_M^+ = \left[\exp i\pi \left(\sum_{\mu=1}^{M} a_\mu^* a_\mu - a_M^* a_M\right)\right] a_M^*$$

where we used that for any pair of indices $\{\mu, \nu\}$: $a_\mu^* a_\mu$ commutes with $a_\nu^* a_\nu$. Furthermore, $(a_\mu^* a_\mu)^2 = a_\mu^* a_\mu$ implies

$$\exp(-i\pi a_M^* a_M) \quad = I + [\exp(i\pi) - 1] a_M^* a_M \; ;$$
$$\exp(-i\pi a_M^* a_M) a_M = a_M \quad \text{and} \quad \exp(-i\pi a_M^* a_M) a_M^* = -a_M^* .$$

Since each of the mutually commuting Hermitian matrices $a_\mu^* a_\mu$ satisfies $(a_\mu^* a_\mu)^2 = a_\mu^* a_\mu$, each has eigenvalues 0 and 1; and thus the eigenvalues of the matrices $N_{(M)}$ are positive integers, so that the matrix $\exp[i\pi N_{(M)}]$ has eigenvalues ± 1; we denote the corresponding eigenspaces by $\mathcal{H}^\pm$.

Upon inserting (12.2.18–20) into the factors V_1 and V_2 in (12.2.14), we notice that their exponents are bilinear in the matrices a_k^* and a_l; hence V_1 and V_2 both commute with the matrix $\exp\left[i\pi N_{(M)}\right]$, so that the subspaces $\mathcal{H}^\pm$ are stable under these matrices. We denote by $V_1^\pm$ (resp. $V_2^\pm$ and $V^\pm$) the restriction of V_1 (resp. $V_2^\pm$ and $V^\pm$) to $\mathcal{H}^\pm$.

Taking now (12.2.20) into account, we define:

$$a_{M+1} \equiv \mp a_1 \quad \text{i.e.} \quad a_{M+1}^* \equiv \mp a_1^* \quad \text{in} \quad \mathcal{H}^\pm \tag{12.2.22}$$

which allow to rewrite the transfer matrix (12.2.14) in the concise form

where
$$V = V^+ \oplus V^- \quad \text{with} \quad V^\pm = (V_2^\pm)^{\frac{1}{2}} V_1^\pm (V_2^\pm)^{\frac{1}{2}}$$

$$V_1^\pm = A^M \exp\left[-\beta J_1^* \sum_{m=1}(a_m^* a_m - a_m a_m^*)\right]$$

$$V_2^\pm = \exp\left[\beta J_2 \sum_{m=1}(a_m^* - a_m)(a_{m+1}^* + a_{m+1})\right]$$
$$\tag{12.2.23}$$

We can then diagonalize separately V^+ and V^-. The main point is that we now have, for the exponents of each of the four matrices $V_1^{\pm}$ and $V_2^{\pm}$, expressions that are *bilinear* in the fermionic matrices a_k^* and a_l. As we shall presently see, these expressions can always be diagonalized by suitably chosen linear transformations.

Before doing so, we ought to emphasize that two special assumptions allowed us to keep bilinearity in the substitution from (12.2.14) to (12.2.23): the first was already necessary for (12.2.14), namely the absence of external magnetic field; this causes some additional problems when we come to define the spontaneous magnetization. The second restrictive assumption that is now essential for the Jordan–Wigner transform (12.2.16) not to drag in non-quadratic factors is that the interactions are restricted to nearest neighbors, along rows as well as along columns – see again the Hamiltonian (12.2.3).

Step 3. Diagonalization of the transfer matrix: the Bogoliubov transform. To guide the choice of the linear transformations diagonalizing (12.2.23), we first appeal to the translation invariance of the theory. Since we ultimately want to take the thermodynamical limit, we do not loose any generality by restricting our attention to models in which each row contains an even number M of sites. We consider first the cyclic case, $a_{M+1} = a_1$, i.e. the restriction V^- of V to $\mathcal{H}^-$ – see (12.2.23). In this space, we introduce the matrices η_q and their Hermitian conjugates η_q^*, defined by

$$\eta_q \equiv M^{-\frac{1}{2}} \exp(i\frac{1}{4}\pi) \sum_{m=1}^{M} e^{-iqm} a_m \qquad (12.2.24)$$

with

$$q \in Q^- \equiv \{ q = 0, \pm\frac{2}{M}\pi, \pm\frac{4}{M}\pi, \cdots, \pm\frac{M-2}{M}\pi, \pi \} \quad . \qquad (12.2.25)$$

These satisfy again the fermion anticommutation relations (12.2.17), namely:

$$\left. \begin{array}{l} \eta_q^* \eta_{q'} + \eta_{q'} \eta_q^* = \delta_{q,q'} I \\[2mm] \eta_q \eta_{q'} + \eta_{q'} \eta_q = 0 \quad \text{and thus} \quad \eta_q^* \eta_{q'}^* + \eta_{q'}^* \eta_q^* = 0 \end{array} \right\} \quad . \qquad (12.2.26)$$

Notice also that the choice of Q^- ensures that the translation $\tau[a_m] = a_{m+1}$ implies by linearity that $\tau[\eta_q] = e^{iq}\eta_q$, and thus $\tau[\eta_q^*\eta] = \eta_q^*\eta$. The pragmatic reason for introducing these matrices is that they allow to rewrite V^- in (12.2.23) as

$$V^- = A^M \prod_{q\in Q_o^-} V(q) \quad \text{with} \quad Q_o^- \equiv \{q \in Q^-; q \neq 0, \pi\} \qquad (12.2.27)$$

where the $V^-(q)$ are *mutually commuting* 4 by 4 matrices that can therefore be diagonalized separately; this is obviously be *a tremendous simplification*. Specifically these matrices are given by

$$V(0) = \exp\left[-\varepsilon_o\left(\eta_o^*\eta_o - \tfrac{1}{2}\right)\right] \quad \text{with} \quad \varepsilon_o = 2\beta(J_1^* - J_2)$$

$$V(\pi) = \exp\left[-\varepsilon_\pi\left(\eta_\pi^*\eta_\pi - \tfrac{1}{2}\right)\right] \quad \text{with} \quad \varepsilon_\pi = 2\beta(J_1^* + J_2) \qquad (12.2.28)$$

and for $q \neq 0, \pi$:

$$V(q) = V_2(q)^{\frac{1}{2}} V_1(q) \left(V_2(q)\right)^{\frac{1}{2}}$$

$$V_1(q) = \exp\left[-2\beta J_1^*\left(\eta_q^*\eta_q + \eta_{-q}^*\eta_{-q} - 1\right)\right] \qquad (12.2.29)$$

$$V_2(q) = \exp\left[2\beta J_2\left\{(\cos q)(\eta_q^*\eta_q + \eta_{-q}^*\eta_{-q}) + (\sin q)(\eta_q\eta_{-q} + \eta_{-q}^*\eta_q^*)\right\}\right]$$

Let us briefly interrupt the flow of the argument here to point out, from (12.2.10), that J_1^* is a function of the natural temperature β; that βJ_1 is monotonically decreasing in β, and that its range – like that of β – covers the open interval $(0, \infty)$. Hence, while ε_π is strictly positive, the equation $\varepsilon_o = 0$ is equivalent, via the definition of J_1^* in (12.1.10), to:

$$\boxed{\sinh(2\beta_c J_1) \sinh(2\beta_c J_2) = 1} \qquad (12.2.30)$$

which has a unique solution $\beta_c \in (0, \infty)$; we have then:

$$\left.\begin{array}{ccc} \beta < \beta_c & \models & \varepsilon(0) > 0 \\ \beta > \beta_c & \models & \varepsilon(0) < 0 \end{array}\right\} \qquad (12.2.31)$$

First isolated in the isotropic case $J_1 = J_2$, (12.2.30) and its pivotal importance were recognized by [Kramers and Wannier, 1941] : $T_c = (k\beta_c)^{-1}$ *is the critical temperature at which a ferromagnetic phase transition occurs.* This connection is made more explicit in the sequel; see e.g. (12.2.46).

We now return to our diagonalization problem, which has been reduced to the independent diagonalization of 4 by 4 matrices $V(q)$ in (12.2.29). This task is aided by the remark that the relations $\eta_q \Phi = 0 = \eta_{-q}\Phi$ determine a one-dimensional subspace of $\mathcal{H}^-(q)$. Choose a vector Φ_o in this subspace, and note that the four vectors

$$\{\Phi_o,\ \Phi_q \equiv \eta_q^*\Phi_o,\ \Phi_{-q} \equiv \eta_{-q}^*\Phi_o,\ \Phi_{-q,q} \equiv \eta_{-q}^*\eta_q^*\Phi_o\}$$

define a basis in $\mathcal{H}^-$, and that Φ_q and Φ_{-q} are eigenvectors of both $V_1(q)$ and $V_2(q)$, and thus of $V(q)$, with

$$V(q)\Phi_{\pm q} = \exp[2\beta J_2 \cos q]\,\Phi_{\pm q} \qquad (12.2.32)$$

Hence the problem is effectively reduced to diagonalizing the 2 by 2 matrix $V^\perp(q)$ obtained by restricting $V(q)$ to the stable subspace $\mathcal{H}^\perp$ spanned by

Φ_o and $\Phi_{-q,q}$. It is easy to compute the determinant of the corresponding restrictions of $V_1(q)$ and $V_2(q)$, and to verify that $\det[V^\perp(q)] = \exp[4\beta J_2 \cos q]$ from which one concludes immediately that there exist: a number $\varepsilon_q > 0$, and two vectors Ψ_o and $\Psi_{-q,q}$, linear combinations of Φ_o and $\Phi_{-q,q}$ such that

$$\left. \begin{aligned} V(q)\,\Psi_o &= \exp[2\beta J_2 \cos q - \varepsilon_q]\,\Psi_o \\[2mm] V(q)\,\Psi_{-q,q} &= \exp[2\beta J_2 \cos q + \varepsilon_q]\,\Psi_{-q,q} \end{aligned} \right\} . \tag{12.2.33}$$

Upon putting (12.2.32) and (12.2.33) together, we get:

$$V(q) = \exp[2\beta J_2 \cos q] \begin{pmatrix} e^{\varepsilon_q} & & & \\ & 1 & & \\ & & 1 & \\ & & & e^{-\varepsilon_q} \end{pmatrix} . \tag{12.2.34}$$

It is somewhat less immediate – but still straightforward, as it involves only computations with 2 by 2 matrices – to find the values of the angle θ_q such that

$$\left. \begin{aligned} \Psi_o &= \cos\theta_q\,\Phi_o + \sin\theta_q\,\Phi_{-q,q} \\[2mm] \Psi_{-q,q} &= -\sin\theta_q\,\Phi_o + \cos\theta_q\,\Phi_o \end{aligned} \right\} \tag{12.2.35}$$

and to see, by the same token, that ε_q is the unique positive root of the equation:

$$\cosh(\varepsilon_q) = \cosh(2\beta J_2)\,\cosh(2\beta J_1^*) - \sinh(2\beta J_2)\,\sinh(2\beta J_1^*)\,\cos q . \tag{12.2.36}$$

While (12.2.36) is derived only for $q \in Q_o^-$, we can still inquire what it would give for $q = 0$ and $q = \pi$, keeping in mind the positivity condition; we get then, for $q = 0$ a value, which we call ε_{min} and which is given by $\beta|J_2 - J_1^*|$; and for $q = \pi$ the solution of (12.2.36) coïncides with ε_π already obtained in (12.2.28). Note further that, at fixed β :

$$0 < q < q' < \pi \quad \models \quad 0 \le |\varepsilon_o| = \varepsilon_{min} < \varepsilon_q < \varepsilon_{q'} < \varepsilon_\pi \tag{12.2.37}$$

where ε_o is the value obtained in (12.2.28); recall from (12.2.31) that its sign is the sign of $\beta_c - \beta$, i.e. the sign of $T - T_c$.

The diagonalization of $V(q)$ for every $q \in Q^-$, and thus of V^-, has been completed.

While not essential to our immediate purpose, namely to compute the maximal eigenvalue of the transfer matrix, it is interesting to point out that (12.2.32–34) can be summarized by writing, for $q \ne 0, \pi$:

$$V(q) = \exp\left(2\beta J_2 \cos q\right) \exp\left[-\varepsilon_q\left(\xi_q^* \xi_q + \xi_{-q}^* \xi_{-q} - 1\right)\right] \tag{12.2.38}$$

where the matrices ξ_q, ξ_{-q}^* and their Hermitian conjugates ξ_q^*, ξ_{-q} are defined by

$$
\left.\begin{aligned}
\xi_q &= \cos\theta_q\,\eta_q + \sin\theta_q\,\eta^*_{-q} \\
\xi^*_{-q} &= -\sin\theta_q\,\eta_q + \cos\theta_q\,\eta^*_{-q}
\end{aligned}\right\}
\tag{12.2.39}
$$

with θ_q as in (12.2.33). These satisfy – again – the fermion anticommutation relations, and moreover:

$$
\left.\begin{aligned}
\xi_q\Psi_o = \xi_{-q}\Psi_o = 0 \quad &; \quad \xi^*_{-q}\xi^*_q\Psi_o = \Psi_{-q,q} \\
\xi^*_q\Psi_o = \eta^*_q\Phi_o \quad &; \quad \xi^*_{-q}\Psi_o = \eta^*_{=q}\Phi_o
\end{aligned}\right\}
\tag{12.2.40}
$$

This passage from the fermion matrices η_q's to the new fermion matrices ξ_q's, with its pairing of stared and non-stared matrices (referred to as fermion creation and annihilation operators) was first used, in the context of the BCS model for superconductivity, by [Bogoliubov, 1958] and [Valatin, 1958].

Upon defining $\xi_o \equiv \eta_o$ and $\xi_\pi \equiv \eta_\pi$, recalling that $\cos(q = 0) = 1$, $\cos q = \pi = -1$ and that $\sum_{q\in Q^-}\cos q = 0$, we finally obtain:

$$
V^- = A^M \sum_{q\in Q^-} \exp\left[-\varepsilon_q\left(\xi^*_q\xi_q - \frac{1}{2}\right)\right]
\tag{12.2.41}
$$

A similar argument can be carried over to V^+, the only modification required being to replace, in (12.2.25-41), Q^- by

$$
Q^+ \equiv \{\pm\frac{1}{M}\pi, \pm\frac{3}{M}\pi, \cdots, \pm\frac{M-1}{M}\pi\} \quad ;
\tag{12.2.42}
$$

in contrast to Q^-, Q^+ does not include $q = 0$ and $q = \pi$ do not belong to Q^+. We have then:

$$
V^+ = A^M \sum_{q\in Q^+} \exp\left[-\varepsilon_q\left(\xi^*_q\xi_q - \frac{1}{2}\right)\right]
\tag{12.2.43}
$$

Step 4. Occurrence of an asymptotic degeneracy below T_c.
From (12.2.43), we obtain immediately that the largest eigenvalue

$$
\lambda^+_{max} = A^M \exp\frac{1}{2}\left[\sum_{q\in Q^+}\varepsilon_q\right]
\tag{12.2.44}
$$

of V^+ corresponds to the eigenvector $\Psi^+_o \in \mathcal{H}^+$ defined (uniquely, up to an irrelevant global phase) by: $\xi_q\Psi^+_o = 0 \; \forall \, q \in Q^+$.

For V^- the matter is a little more subtle, due to the fact that, from its very definition, $\mathcal{H}^-$ contains exactly the vectors Ψ^- for which $N_{(M)}\Psi^- = (2n+1)\Psi^-$ with n an integer. Consequently, the maximum eigenvalue of V^- corresponds to the eigenvector $\Psi^-_{q=0} \equiv \xi_{q=0}\Psi^-_o \in \mathcal{H}^-$, and it is given by

$$\lambda_{max}^{-} = A^{M} \exp \frac{1}{2} \left[-\varepsilon_{o} + \sum_{q \in Q^{-}}{}' \varepsilon_{q} \right]$$

where $\sum'$ means that $q = 0$ has been omitted. Upon taking into account (12.2.31), we can rewrite this as

$$\lambda_{max}^{-} = A^{M} \exp \frac{1}{2} \left[\sum_{q \in Q^{-}} \varepsilon_{q} \right] \begin{cases} e^{-|\varepsilon_{o}|} & \text{when} \quad \beta < \beta_{c} \\ 1 & \text{when} \quad \beta > \beta_{c} \end{cases} . \qquad (12.2.45)$$

Hence the quotient of (12.2.45) by (12.2.44) gives:

$$\left. \begin{aligned} (\lambda_{max}^{-}/\lambda_{max}^{+}) &= R_{M} \begin{cases} e^{-|\varepsilon_{o}|} & \text{when} \quad \beta < \beta_{c} \\ 1 & \text{when} \quad \beta > \beta_{c} \end{cases} \\ \text{where} \quad & \\ R_{M} &= \exp \tfrac{1}{2} \left[\sum_{q \in Q^{-}} \varepsilon_{q} - \sum_{q \in Q^{+}} \varepsilon_{q} \right] \\ \text{with} \quad & \\ \lim_{M \to \infty} R_{M} &= 1 \end{aligned} \right\} . \qquad (12.2.46)$$

This is the result we were looking for: the critical temperature $T_{c} = 1/k\beta_{c}$, obtained from (12.2.30), is such that for $T < T_{c}$ the maximal eigenvalue of the transfer matrix becomes asymptotically degenerate as $M \to \infty$, while the ratio of the two largest eigenvalues remains different from 1 for $T > T_{c}$. A more detailed analysis would have shown that the limit in (12.2.46) is approached exponentially fast as $M \to \infty$.

Step 5. Long-range order and spontaneous magnetization.
The most dramatic consequence of the asymptotic degeneracy deployed in Step 4 concerns the long-range correlation between "spins" at mutually far away lattice sites:

$$\left. \begin{aligned} g_{N,M}(\nu, \mu) &\equiv \\ \langle \sigma_{n+\nu N, m+\mu M} \, \sigma_{n,m} \rangle_{N,M} &- \langle \sigma_{n+\nu N, m+\mu M} \rangle_{N,M} \, \langle \sigma_{n,m} \rangle_{N,M} \end{aligned} \right\} \qquad (12.2.47)$$

where $0 < \mu, \nu < 1$ and

$$\langle A \rangle_{N,M} \equiv [Z_{N,M}]^{-1} \sum [A \exp(-\beta H_{N,M})] \qquad (12.2.48)$$

denotes the canonical equilibrium value of the observable A; and $Z_{N,M}$ is the canonical partition function defined in (12.2.5); the above summation carries over all 2^{NM} configurations of the lattice $\mathsf{Z}_{N} \times \mathsf{Z}_{M}$.

Because of the invariance of the Hamiltonian (12.2.4) under a simultaneous flip $\sigma_{n,m} \mapsto -\sigma_{n,m}$ for all (n, m) – the so-called *flip-flop symmetry* –

$$\forall\,(n,m)\in Z_N\times Z_M\ :\qquad \langle\sigma_{n,m}\rangle_{N,M}=0\ . \tag{12.2.49}$$

Hence $g_{N,M}(\nu,\mu)$ in (12.2.47) reduces to its first term only. To compute this term, we use the transfer matrix V

$$\left.\begin{array}{c}\langle\sigma_{n+\nu N,m+\mu M}\,\sigma_{n,m}\rangle_{N,M}=\\[2mm] [\mathrm{Tr}V^N]^{-1}\mathrm{Tr}[\sigma^z_{m+\mu}\,V^{\nu N}\,\sigma^z_m\,V^{(1-\nu)N}]\end{array}\right\}. \tag{12.2.50}$$

Upon writing the transfer matrix in terms of its spectral decomposition, denoting by λ_α its eigenvalues, and by Ψ_α the corresponding eigenvectors, we have

$$\mathrm{Tr}V^N=\sum_\alpha\lambda_\alpha^N=\lambda_{max}^N\sum_\alpha\left(\frac{\lambda_\alpha}{\lambda_{max}}\right)^N \tag{12.2.51}$$

and similarly:

$$\left.\begin{array}{c}\mathrm{Tr}[\sigma^z_{m+\mu N}V^{\nu N}\,\sigma^z_m\,V^{(1-\nu)N}]=\\[2mm] \lambda_{max}^N\sum_{\alpha,\beta}(\sigma^z_{m+\mu}\Psi_\alpha,\Psi_\beta)\left(\frac{\lambda_\beta}{\lambda_{max}}\right)^{\nu N}\times\\[2mm] \times\,(\sigma^z_m\Psi_\beta,\Psi_\alpha)\left(\frac{\lambda_\alpha}{\lambda_{max}}\right)^{(1-\nu)N}\end{array}\right\}. \tag{12.2.52}$$

The following five remarks are helpful to the computation of the thermodynamical limit of the correlation function (12.2.47) which, because of (12.2.49), reduces to (12.2.50).

1. For large values of N, the factors

$$\left(\frac{\lambda_\beta}{\lambda_{max}}\right)^{\nu N}\quad\text{and}\quad\left(\frac{\lambda_\alpha}{\lambda_{max}}\right)^{(1-\nu)N}$$

   become negligible unless α and β are associated to the maximal eigenvalue λ_{max}.
2. Below the critical temperature T_c, this maximal eigenvalue is asymptotically degenerate as M becomes very large, with eigenvectors Ψ_o^+ and $\Psi_{q=0}^-$.
3. The $\sigma_m^z=\sigma_m^++\sigma_m^-$ – recall our unconventional definition (12.2.12) – only have matrix elements between vectors $\Psi^+\in\mathcal{H}^+$ and $\Psi^-\in\mathcal{H}^-$.
4. Because of translation invariance, $(\sigma_{m+\mu}\Psi_o^+,\Psi_{q=0}^-)=(\sigma_m\Psi_o^+,\Psi_{q=0}^-)$.
5. The relation (12.2.37), with $\varepsilon_{min}>0$ for $T\neq T_c$ ensures that sum in α,β not corresponding to the maximal eigenvalue add up to terms that become exponentially small in the thermodynamical limit.

One then obtains for $T<T_c$:

$$g(\nu,\mu)=\lim_{N,M\to\infty}g_{N,M}(\nu,\mu)=\lim_{N,M\to\infty}\left|(\sigma_m^z\Psi_o^+,\Psi_{q=0}^-)\right|^2\ . \tag{12.2.53}$$

As conducted so far, this holds for all $0 < \mu, \nu < 1$ but a review of the argument shows that it extends without modification to the case $\mu = 0$; upon exchanging throughout – including in the definition of the transfer matrix – the role of the rows and columns, we conclude that (12.2.53) is also valid for $\nu = 0$ when $\mu > 0$; hence, $g(\nu, \mu)$ is constant over the whole quadrant $0 \leq \mu, \nu \leq 1$, except evidently the origin $\mu = \nu = 0$; hence the notation g we use forthwith instead of $g(\nu, \mu)$. The same analysis shows that

$$T > T_c \quad \models \quad g = 0 \quad . \tag{12.2.54}$$

Indeed, when $T > T_c$ the maximal eigenvalue is non-degenerate; thus, in view of property (3) above, no non-diagonal matrix element from (12.2.52) survives in the thermodynamical limit.

There are several ways to compute the RHS in (12.2.53). All boil down to explicitly un-raveling the transformations

$$(\sigma_m^+, \sigma_{m'} - \to (a_m^*, a_{m'}) \to (\eta_q *, \eta_{q'}) \to (\xi_q *, \xi_{q'})$$

described in step 3 above; this requires in particular the knowledge of the angles θ_q appearing in the Bogoliubov transform (12.2.35); see [Schultz, Mattis, and Lieb, 1964] who completed, along these lines, the computation of the long-range order :

$$g = \begin{cases} \left[1 - (\sinh 2\beta J_1 \cdot \sinh 2\beta J_2)^{-2} \right]^{\frac{1}{4}} & \beta > \beta_c \\ 0 & \beta < \beta_c \end{cases} \quad . \tag{12.2.55}$$

Note, incidentally, that this expression – like (12.2.30) – is symmetric in J_1 and J_2, and thus under an interchange of the rows and columns.

In order to model, as close as possible, the situation encountered in the laboratory, the spontaneous magnetization should be defined as:

$$m_L = \lim_{B \to 0} \lim_{N, M \to \infty} \frac{1}{NM} \langle \sigma_{n,m} \rangle_{(J_1, J_2, B, N, M)} \tag{12.2.56}$$

where $\langle \cdots \rangle_{(J_1, J_2, B, N, M)}$ denotes the canonical equilibrium state at natural temperature $\beta = 1/kT$, for the finite model with N rows and M columns, spin-spin coupling constants J_1, J_2, and magnetic field $B \neq 0$. Recall indeed that the Hamiltonian with $B = 0$ (12.2.4) is invariant under the flip-flop operation, i.e. the simultaneous reversal of all spins $\sigma_{n,m} \to -\sigma_{n,m}$, $\forall \, n, m$; thus $\langle \sigma_{n,m} \rangle_{J_1, J_2, B=0, N, M} = 0$. Hence the definition (12.2.56).

Unfortunately, this m_L cannot be computed directly from its definition (12.2.56) since one does not know how to compute the partition function for the Ising model in the presence of a non-zero magnetic field. Nevertheless, there are several ways to argue that the spontaneous magnetization should be equal to the square-root of the long-range order g thus giving, together with (12.2.55), the *Onsager formula* for the spontaneous magnetization below the critical temperature:

$$
m_O = \left[1 - (\sinh 2J_1/kT \cdot \sinh 2J_2/kT)^{-2}\right]^{\frac{1}{8}} \quad \text{for} \quad T < T_c \qquad . \text{(12.2.57)}
$$

The spontaneous magnetisation vanishes for all temperatures $T > T_c$. One intuitive way to argue in favor of $m^2 = g$ proceeds in two steps.

Firstly, one uses the translation invariance of the canonical equilibrium state – with $B = 0$, $\beta > \beta_c$, $J_1, J_2 > 0$ – to write:

$$
\langle(\tfrac{1}{NM}\textstyle\sum_{n,m}\sigma_{n,m})^2\rangle_{N,M} \equiv (\tfrac{1}{NM})^2 \langle\textstyle\sum_{n,m;n',m'}\sigma_{n,m}\sigma_{n',m'}\rangle_{N,M}
$$
$$
= \tfrac{1}{NM}\textstyle\sum_{n',m'}\langle\sigma_{n,m}\sigma_{n',m'}\rangle_{N,M}
$$
(12.2.58)

from which one can conclude:

$$
g = \lim_{N,M\to\infty} \langle(\frac{1}{NM}\sum_{n,m}\sigma_{n,m})^2\rangle_{N,M} \qquad . \text{(12.2.59)}
$$

Secondly, one conjectures that in the absence of magnetic field and in the thermodynamical limit, the canonical equilibrium state $\langle\cdots\rangle$ of the model is a mixture $\langle\cdots\rangle = \frac{1}{2}\langle\cdots\rangle_+ + \frac{1}{2}\langle\cdots\rangle_-$ of two states $\langle\cdots\rangle_\pm$ that inherit the translation invariance of $\langle\cdots\rangle$ and cannot be decomposed further into mixtures of translational invariant states. These conditions imply that the states $\langle\cdots\rangle_\pm$ are dispersion-free on the macroscopic observables obtained as average over the translations. Moreover, to preserve the flip-flop invariance of the canonical equilibrium state, we have one of the following two possibilities: either these states are also invariant under the action of the flip-flop symmetry, or this action exchanges the two states $\langle\cdots\rangle_\pm \to \langle\cdots\rangle_\mp$; we assume the latter. We have then with

$$
\sigma_{M,N} \equiv \frac{1}{NM}\sum_{n,m}\sigma_{n,m} :
$$

$$
\left.\begin{array}{l}
\lim_{N,M\to\infty}\langle\sigma_{M,N}{}^2\rangle = \frac{1}{2}\lim_{N,M\to\infty}\left[\langle\sigma_{M,N}{}^2\rangle_+ + \langle\sigma_{M,N}{}^2\rangle_-\right] \\[2mm]
\qquad\qquad = \frac{1}{2}\lim_{N,M\to\infty}\left[\langle\sigma_{M,N}\rangle_+{}^2 + \langle\sigma_{M,N}\rangle_-{}^2\right]
\end{array}\right\} . \text{(12.2.60)}
$$

Hence, upon taking into account the flip-flop symmetry

$$
m_+ \equiv \lim_{N,M\to\infty}\langle\sigma_{N,M}\rangle_+ = -\lim_{N,M\to\infty}\langle\sigma_{N,M}\rangle_- \equiv m_- \quad , \qquad \text{(12.2.61)}
$$

one obtains that the magnetization $m_\pm$ in the two phases $\langle\cdots\rangle_\pm$ satisfy:

$$
m_+{}^2 = \lim_{N,M\to\infty}\langle(\frac{1}{NM}\sum_{n,m}\sigma_{n,m})^2\rangle
$$

and thus, upon taking into account (12.2.59) and (12.2.55)

$$m_{\pm} = \pm \left[1 - (\sinh 2J_1/kT \cdot \sinh 2J_2/kT)^{-2} \right]^{\frac{1}{8}} \quad \text{for} \quad T < T_c \quad (12.2.62)$$

which gives the desired interpretation of (12.2.57). When $T < T_c$ approaches the critical temperature T_c, m tends to zero; specifically (12.2.57) gives:

$$m_O \propto [T_c - T]^{\frac{1}{8}} \quad \text{as} \quad T \nearrow T_c \quad (12.2.63)$$

which thus fixes to $\frac{1}{8}$ the value of the critical exponent called β; while the 3-dimensional Ising model cannot be solved exactly, some reliable approximation lead to $\beta \approx \frac{5}{16}$. Recall from (11.3.17), that the mean-field theory – the Weiss model – gave $\beta = \frac{1}{2}$, a value that is dimension independent. Actual magnets seem to show $\beta \approx \frac{1}{3}$.

Bibliographical notes. The first convincing demonstration that the 2-dimensional Ising model has a phase transition is due to [Peierls, 1936] who pointed out that, for sufficiently low temperatures, the model develops a sensitivity to the boundary conditions; in essence, if Λ is a finite subregion of the lattice, with N rows of N sites each; and if one imposes that all the spins on the boundary are "up" (resp. "down"), then for sufficiently low temperatures, the average magnetization per spin $\langle \sigma \rangle_+$ is necessary larger (resp. smaller) than a number $m > 0$ (resp. $m < 0$), independent of N. Peierls' argument was essentially correct, and it was made into a rigorous proof in [Griffiths, 1964, Dobrushin, 1965]. The dependence of the initial Peierls argument on the symmetry of the Ising model interaction was later relaxed, [Pirogov and Sinai, 1975, 1976, Sinai, 1982], providing a key to the solution of a variety of models [Lebowitz and Mazel, 1998, Lebowitz, Mazel, and Presutti, 1999, Pirogov, 2000, Zahradnik, 2000]. The beauty of the Peierls argument – in its various avatars – is that it does not require the computing of the magnetization, only the existence of a bound m; since this is only a bound, it does not give the value of the critical temperature at which this instability begins to appear.

The first sharp determination of value of the critical temperature for the occurrence of a phase transition, namely (12.2.30), appeared in [Kramers and Wannier, 1941].

The value of the spontaneous magnetization (12.2.57) is an amusing tale that is told at the beginning of [Montroll, Potts, and Ward, 1963]; the result itself was published by [Onsager, 1949a], without proof, and as the record of a seemingly incidental conference comment. Recall, however, that since no exact computation of the partition function of the 2−dimensional Ising model *in the presence of a fixed magnetic field* $B \neq 0$ is known, the spontaneous magnetization *cannot* be computed directly from its genuine pragmatic definition, namely (12.2.56), i.e. as the residual magnetization when the external magnetic field is slowly turned off from a non-zero value. [Yang, 1952] recovered the Onsager value (12.2.57) from an alternate definition of the spontaneous magnetization, and through a series of steps the author himself characterizes as "*very complicated*", although "*attempts to find a simpler way ... have failed.*" The idea was to realize that, while the magnetization in finite magnetic field $B > 0$ cannot be computed directly, sufficient control can be gained on the finite lattice free-energy at natural temperature $\beta > \beta_c$ – and thus the magnetization $m_{NM}(\beta, B)$ – when the external magnetic field B is of the order of $1/M$ so that the following limit can be computed:

$$m_Y = \lim_{b \searrow 0} \lim_{N,M \to \infty} m_{NM}(\beta, b/M) \quad . \tag{12.2.64}$$

[Yang, 1952]'s derivation was made more transparent as a by-product of the introduction of the fermion matrices in [Schultz, Mattis, and Lieb, 1964].

[Montroll, Potts, and Ward, 1963] identified the Onsager value of the spontaneous magnetization as the square-root of the long-range order along rows, and [Schultz, Mattis, and Lieb, 1964] proved that the latter did not depend on the direction – across the lattice – along which the spins are taken apart from one another, and is given by (12.2.55) via (12.2.53) through the definition (12.2.47). These results were obtained with periodic boundary conditions, as we did consistently in the main body of the present section.

Up to that point, it was somewhat of a mystery that these alternate definitions of the spontaneous magnetization should lead precisely to the same numerical value. The resolution came from a consolidation of the Peierls argument, namely the instability of the model under changes of the boundary conditions. The argument – proposed in [Emch, Knops, and Verboven, 1968] – which we used to pass from (12.2.60) to (12.2.62), leaves open the pragmatic interpretation of the states $\langle \cdots \rangle_\pm$; these are the canonical equilibrium states obtained from the two extreme boundary conditions we described in the above presentation of the Peierls argument, namely the spins on the boundary being clamped to be all "up" or all "down"; this was settled in [Benettin, Gallavotti, Jona-Lasinio, and Stella, 1973, Abraham and Martin-Löf, 1973]; see also the related papers [Lebowitz and Martin-Löf, 1972, Lebowitz, 1972].

These results allow to ask cogently another question, namely whether the mathematical assertion that the Gibbs state of the system is a mixture, i.e. a convex combination $\langle \cdots \rangle = \lambda \langle \cdots \rangle_+ + (1 - \lambda) \langle \cdots \rangle_-$ (with $0 \le \lambda \le 1$) of the two extreme equilibrium states $\langle \cdots \rangle_+$ and $\langle \cdots \rangle_-$ is anything more than a statistical statement; specifically, can the 2-d Ising model exhibit side-by-side coexistence of pure thermodynamical phases that are *spatially segregated* by macroscopic boundary lines, as one may expect from the conjunction of (a) a ferromagnetic interaction which tends to align each spin with its neighbors, and (b) mixed boundary conditions, say with m_+ on the right, and m_- on the left; see [Gallavotti and Martin-Löf, 1972, Gallavotti, 1972, Abraham and Reed, 1976, Messager and Miracle-Sole, 1977, Russo, 1979]. However, [Aizenmann, 1980] argued that the two-dimensional nature of the model allows fluctuations to frustrate this conjecture, whereas [Dobrushin, 1972] had produced an argument according to which, in the three-dimensional model, mixed boundary conditions do induce spacial separation with an actual interface; see also [Van Beijeren, 1975]. This is yet another manifestation of the importance of dimensionality for the occurrence, and degree, of cooperative behavior in a system with short-range interactions.

We mentioned towards the end of Sect. 12.1 that one-dimensional chains, with a two-body interaction $J_{i,j}$ decaying as slowly as $|i-j|^\alpha$ with $1 < \alpha < 2$, overcome the one-dimensionality of these systems and exhibit a phase transition. [Johansson, 1995] established that a phase segregation does show up as well, in the sense that for sufficiently low temperatures, with high probability the thickness of the interface – between the spin-up phase and the spin-down phase – is small on macroscopic length scales; note moreover that the arguments presented there extend from lattice to continuous systems.

The problem of phase segregation had been posed in general terms for Ising models in any dimension (≥ 2) by [Minlos and Sinai, 1967] where a contour separating phases was proven to exist provided the temperature is very low and the volume is very large; their result, in fact, is asymptotic and involves a double limit where both the inverse temperature β and the number of particles N tend *simulatneously* to infinity.

This deep result, however, left open the problem of replacing the zero temperature limit by a condition that the temperature be sufficiently small. The breakthrough was provided by [Dobrushin, Kotecki, and Shlosman, 1992] who realized the similarity of the problem with the problem of describing the growth of crystals; the latter had been formulated at the phenomenological macroscopic level by [Wulff, 1901]; see also [Dinghas, 1944]. In a remakable tour de force, Dobrushin, Kotecki, and Shlosman [1993] derived the Wulff variational formulation from a microscopic foundation: the 2-d Ising model. The essential tool, but by no means the only one, was to go beyond the existence of the surface tension in the thermodynamical limit – which here is the function entering in the Wulff integral – and to obtain detailled information on the rate at which this limit is approached. For a didactic presentation of these ideas and the essential role played in the analysis by the theory of large deviations, applied here to the magnetization $|\Lambda|^{-1} \sum_{k \in \Lambda} \sigma_k$, see [Pfister, 1991].

This still left open a few problems: to extend the result on a range of temperatures that approaches the critical temperature [Cesi, Guadagni, Martinelli, and Schonmann, 1996, Lebowitz, Mazel, and Suhov, 1996]; to illustrate with simpler model the emergence of the macroscopic behavior from its microscopic substratum [Miracle-Solé and Ruiz, 1994]; to estimate the fluctuations around the coexistence boundary [Dobrushin and Hryniv, 1997, Pfister and Velenik, 1999]; and to extend the formalism to three dimensions [Bodineau, 1999, Cerf, 2000]. For a return to the original intent of the Wulff construction, namely the application to crytals, see [Messager, Miracle-Solé, and Ruiz,

1992, Miracle-Solé, 1995, Bodineau, 2000]. For a discussion of interfaces in *quantum* lattice systems, see [Nachtergaele, 2000].

12.3 Variations on the Theme of the Ising Model

Many variations have been drawn from the theme of the Ising model; see *inter alia* [Stanley, 1971, Thompson, 1972, Baxter, 1982]. We present below only three classes of variations: on the interpretation of the variables, on the range of these variables, and on the range of their interactions.

12.3.1 Variations on the Interpretation of the Variables

The lattice gas is a caricature of a fluid where the particles move in a container $\Lambda \subset \mathbb{R}^d$ and interact via a two-body potential

$$\phi(r) = B r^{-\beta} - A\, r^{-\alpha} \tag{12.3.1}$$

where $A, B > 0$ and $\beta >> \alpha > 0$. The caricature emphasizes the steepness (β) of the repulsive part, and the shortness (α) of the range of the attractive part of the potential, noting that for a realistic potential such as Lennard–Jones, one has $\beta = 12$ and $\alpha = 6$. Firstly, one thus replaces (12.3.1) by:

$$\phi(r) = \begin{cases} \infty & 0 \leq r < a \\ -\varepsilon & a \leq r \leq 2a \\ 0 & 2a < r \end{cases} \tag{12.3.2}$$

with $a, \varepsilon > 0$ (for simplicity, we choose length units such that $a = 1$). Secondly, once this substitution is accepted, one views the fluid as moving on a discrete sublattice $\Lambda \subset \mathbb{Z}^d$, a step that is compatible with ignoring the kinetic energy of the particles. The effect of the hard core $0 \leq r < a$ is to impose that any one site of the lattice either is occupied by exactly one particle, or is empty. The effect of cutting-off the attractive part for $r > 2a$ is to limit the interaction to particles that are on neighboring sites. The system so defined is called the *lattice gas*. A configuration of the lattice gas is thus a random variable

$$\nu : i \in \Lambda \mapsto \nu_i \in \{0, 1\} \quad . \tag{12.3.3}$$

The total number N of particles and the energy H_N of the configuration ν are thus:

$$N(\nu) = \sum_i \nu_i \quad ; \quad H_N(\nu) = -\varepsilon \sum_{<i,j>} \nu_i \nu_j \tag{12.3.4}$$

where $< i, j >$ indicates that the sum carries over pairs of nearest neighbors only. The grand canonical partition function is given then by the discrete analogue of definition 10.1.3, namely

$$\mathcal{Z} \equiv \sum_{\nu \in \{0,1\}^{|\Lambda|}} \exp\left(-\beta\left[-\mu N(\nu) + H_N(\nu)\right]\right)$$

i.e. upon inserting (12.3.4):

$$\mathcal{Z} = \sum_{\nu \in \{0,1\}^{|\Lambda|}} \exp\left(\beta\left[\mu \sum_i \nu_i + \varepsilon \sum_{<i,j>} \nu_i \nu_j\right]\right) \tag{12.3.5}$$

where $T = 1/k\beta$ is the temperature, and μ is the chemical potential. With the change of variables:

$$\sigma : i \in \Lambda \mapsto 2\nu_i - 1 \in \{-1, +1\} \tag{12.3.6}$$

the grand-canonical partition function $\mathcal{Z}$ of the lattice gas can be rewritten in terms of the canonical partition function Z_I of an Ising model, namely

$$\mathcal{Z} = e^{\beta K} Z_I \tag{12.3.7}$$

with

$$\left. \begin{array}{l} Z_I = \sum_{\sigma \in \{-1, +1\}^{|\Lambda|}} \exp\left(\beta\left[B \sum_i \sigma_i + J \sum_{<i,j>} \sigma_i \sigma_j\right]\right) \\[2mm] B = \frac{1}{4}(c\varepsilon + 2\mu) \quad ; \quad J = \frac{1}{4}\varepsilon \quad ; \quad K = \frac{1}{4}|\Lambda|(\frac{1}{2}c\varepsilon + 2\mu) \end{array} \right\} \tag{12.3.8}$$

and c is the number of nearest neighbors to a site: $c = 2^d$ for the square $d-$dimensional lattice Z^d.

The relations (12.3.6–8) give readily a complete dictionary between the thermodynamical observables of the lattice gas and of the corresponding Ising model. In particular, let P and ϱ denote the pressure and density for the lattice gas defined by:

$$P \equiv (\beta|\Lambda|)^{-1} \ln \mathcal{Z} \quad ; \quad \varrho \equiv (|\Lambda|)^{-1}\langle N \rangle = \partial_\mu P \quad ; \tag{12.3.9}$$

and let f and m denote the free energy per site and the magnetization per site of the corresponding Ising model, defined by

$$f \equiv -(\beta|\Lambda|)^{-1} \ln Z_I \quad ; \quad m \equiv |\Lambda|)^{-1}\langle \sum_i \sigma_i \rangle = -\partial_B f \quad . \tag{12.3.10}$$

We have then

$$P = -f + B - \frac{1}{2}cJ \quad ; \quad \varrho = \frac{1}{2}(m+1) \quad . \tag{12.3.11}$$

Consequently, the two models are equivalent, and the information obtained for one of them translates faithfully to information on the other.

Remarks: While the name "lattice gas" seems to have appeared first in the milestone paper [Yang and Lee, 1952, Lee and Yang, 1952], the origin of the interpretation of the Ising model as a model for a fluid has more ancient roots; [Brush, 1983][p. 176] mentions [Cernushi and Eyring, 1939].

In addition to being a model for *both* ferromagnetism and gas condensation, the Ising model can also be interpreted as a model for binary alloys such as β-brass. The latter is a mixture of a fixed number of Copper atoms and of Zinc atoms. At low temperatures, x-ray diffraction shows that the atoms are orderly arranged on a cubic-entered lattice with, say, one Cu atom sitting at each of the vertices of the elementary cell while the center of the cell is occupied by a Zn atom. Above a critical temperature $T_c = 742$ K $(469°C)$, the atoms are and remain thoroughly mixed with probability $\frac{1}{2}$ to be in the "right place".

As T_c is approached – from either below or above – measurements of the specific heat indicate that it diverges. The "lattice gas" model of a binary alloy consists again of a lattice of N sites among which are distributed fixed numbers N_A of atoms of species A and N_B of atoms of species B, with exactly one atom per site, so that $N = N_A + N_B$. A pair of atoms contributes to the total energy if and only if the atoms are on neighboring sites, and the contribution of such a pair is ε_{AA} if both atoms are of species A, ε_{BB} if both atoms are of species B, ε_{AB} if the two atoms are of different species. Hence, with ν as in (12.3.3) with $\nu_i = 1$ (resp. 0) if site i is occupied with a particle of species A (resp. B), the interaction energy $H_N(\nu)$ for this configuration is

$$\sum_{<i,j>} \{\varepsilon_{AA}\nu_i\nu_j + \varepsilon_{BB}(1-\nu_i)(1-\nu_j) + \varepsilon_{AB}[\nu_i(1-\nu_j) + (1-\nu_i)\nu_j]\}$$

which can be rewritten as

$$(\varepsilon_{AA} + \varepsilon_{BB} - 2\varepsilon AB)N_{AA}(\nu) + c[\varepsilon_{AB}N_A - \frac{1}{2}\varepsilon_{BB}(N_A - N_B)]. \quad (12.3.12)$$

Note that the only stochastic variable appearing in this expression is $N_{AA}(\nu)$, the number of pairs of nearest neighbors of the species A present in the configuration ν. This observable plays, for the binary alloy, the role played by the observable $N(\nu)$ – see (12.3.4) – for the lattice gas. Here again the two descriptions differ only by a constant – the second term in (12.3.12); as a consequence, one obtains again a dictionary of the type of (12.3.11). Hence, the Ising model can be viewed also as a model for a binary alloy.

12.3.2 Variations Involving the Range of the Variables

In Ising's own interpretation, his model had failed to account for ferromagnetism. Soon thereafter, [Heisenberg, 1928] proposed the quantum model that bears his name, namely

$$H = -\boldsymbol{B} \cdot \sum_i \boldsymbol{\sigma}_i - \sum_{i,j} J_{i,j} \; \boldsymbol{\sigma}_i \cdot \boldsymbol{\sigma}_j \qquad (12.3.13)$$

with the classical random variable $\sigma : i \in \Lambda \mapsto \sigma_i \in \{-1, +1\}$ of the Ising model now replaced by quantum variables

$$\left.\begin{array}{c} \boldsymbol{\sigma} : i \in \Lambda \mapsto \boldsymbol{\sigma}_i \in S \\[8pt] S \equiv \{A \in M^h(2, \mathsf{C}) \mid \mathrm{tr} A = 0 \, ; \, \det A = -1\} \end{array}\right\} \qquad (12.3.14)$$

where $M^h(2, \mathsf{C})$ is the four-dimensional real vector space that consists of the 2 by 2 Hermitian matrices with complex entries. Imposing $\mathrm{tr} A = 0$ reduces this to a three-dimensional vector space, $M_o^h(2, \mathsf{C})$ on which $\|A\| = -\det A$ defines a norm. In fact $M_o^h(2, \mathsf{C}) = \{A = a\sigma^x + b\sigma^y + c\sigma^z \mid a, b, c \in \mathsf{R}; \|A\| = a^2 + b^2 + c^2\}$ where the Pauli matrices $\sigma^x, \sigma^y, \sigma^z$ are defined as in (12.2.8). Hence S can be viewed as a unit sphere in R^3. The quantum nature of the objects just introduced comes from the fact that $M^h(2, \mathsf{C})$ is equipped with the structure of a Lie–Jordan algebra by the operations which associate to every pair A, B of elements in $M^h(2, \mathsf{C})$ two others of its elements, namely $-i[A, B] = -i(AB - BA)$ and $A \circ B = \frac{1}{2}\{(A + B)^2 - A^2 - B^2\}$. σ_i in (12.3.14) is thus interpreted as a *"quantum spin"* sitting at the site i in the lattice Λ.

Since Nature is supposed, at the microscopic level, to obey the laws of quantum mechanics, this model seems *a priori* a more realistic candidate from which to seek a microscopic explanation of ferromagnetism. However, the computation of the canonical partition function – Def. 10.2.1 – from the Hamiltonian (12.3.13) turned out to be *very hard*. [Araki, 1969] (see also [Araki, 1970]) managed to generalize for the 1–dimensional Heisenberg model the transfer matrix method described for the Ising model in Sects. 12.1–2 above. Although one of us has attempted to give another didactic account of this work in [Emch, 1972b], even that presentation would be beyond the bound of the present book. For our purpose here, the result is that for all $T > 0$ the state of the system in the thermodynamical limit remains an extremal KMS state, and the 1–dimensional Heisenberg model with short range interactions does not exhibit a phase transition, any more than did the Ising model.

In contrast to the Ising model, moreover, no spontaneous magnetization obtains for the 2–dimensional Heisenberg model. Specifically, [Mermin and Wagner, 1966] considered the case where

$$J_{i,j} = J(|i - j|) \qquad \sum_{n \in \mathsf{Z}^d} |n|^2 J(n) < \infty \qquad (12.3.15)$$

with $|\ldots|$ denoting the Euclidean norm: $|n|^2 \equiv n_1^2 + n_2^2 + \cdots n_d^2$ in Z^d. They showed that, in dimension $d \leq 2$, the magnetization per spin, in the direction – say z – of the magnetic field, satisfies:

$$\lim_{B \to 0} \lim_{\Lambda \nearrow \mathsf{Z}^d} |\Lambda|^{-1} \sum_{i \in \Lambda} \sigma_i^z = 0 \quad . \tag{12.3.16}$$

The proof depends on the clever use of a general result in linear algebra:

Scholium 12.3.1 (Bogoliubov inequality). *Let* A, C, X, Y, H *be* $n \times n$ *matrices, with* H *Hermitian; for* $\beta > 0$, *let* $\langle \cdot \rangle$ *denote the canonical equilibrium state for the temperature* β *and the Hamiltonian* H; *let* $[X, Y]$ *denote the commutator* $= XY - YX$; *and* X^* *denote the Hermitian conjugate of* X. *Then*

$$\frac{1}{2} \beta \langle A^* A + A\, A^* \rangle \, \langle [\, [C, H], C^*] \rangle \; \geq \; |\langle [A, C] \rangle|^2 \quad .$$

While this inequality was first proven as a consequence of the Schwarz inequality for a scalar product that looked somewhat *ad hoc*, it was placed later [Fannes and Verbeure, 1977] in the natural context of the KMS condition which was shown to imply several inequalities with similar flavor. The insight of [Mermin and Wagner, 1966] was to intuit the matrices A and C one should insert in the Bogoliubov inequality in order to obtain the desired result on the magnetization. It is interesting to note in anticipation that the dimension of the system enters through the behavior of the integral $\int d^d k \, k^{-2}$ over the first Brillouin zone: this integral diverges for $d \leq 2$. For details see [Ruelle, 1969][pp. 129–134] or [Griffiths, 1972][pp. 84–89].

Note that the argument is inconclusive for $d = 3$, thus leaving the door open for a phase transition to occur for a 3–dimensional Heisenberg model.

It could also be noted that [Mermin, 1967] succeeded in transposing the argument from the quantum to the classical case, with the Lie structure of quantum mechanics – namely the commutator – surviving in the classical limit as the Poisson bracket; for background, see Sect. 10.3, and [Landsman, 1998].

The paucity of 3–dimensional models that could be solved explicitly prompted [Berlin and Kac, 1952] to propose yet another variation on the nature of the stochastic variables appearing in the Ising model: the *"spherical model."* Here again, the system is defined first on a finite subset Λ of a lattice Z^d, but now the random variables that describe each of the configurations of the system are

$$s : i \in \Lambda \to s_i \in \mathsf{R} \qquad \text{with} \quad |\Lambda|^{-1} \sum_{i \in \Lambda} s_i^{\,2} = 1 \quad . \tag{12.3.17}$$

Hence, while the "spins" attached to the sites of the lattice take now their values in R, the constraint imposes that, in every configuration, the spatial average of their square is still 1; recall that in the Ising model a stronger condition prevails: $\sigma_i^2 = 1 \; \forall \; i \in \Lambda$. The energy of a configuration is chosen to be of the familiar form

$$H(s) = - B \sum_i s_i - J \sum_{<i,j>} s_i s_j \quad . \tag{12.3.18}$$

The canonical partition function is then

$$Z_\Lambda = \int_{R^{|\Lambda|}} \mathrm{d}^{|\Lambda|}s \, \mathrm{e}^{\beta[B\sum_i s_i + J\sum_{<i,j>} s_i s_j]} \delta(|\Lambda| - \sum_{i \in \Lambda} s_i^2) \quad . \qquad (12.3.19)$$

To evaluate this integral in the thermodynamical limit $|\Lambda| \to \infty$ involves a sequence of technical moves – including integration by steepest descent – for the details of which it is best to refer the Reader to [Baxter, 1982]. The central result is an implicit equation of state

$$\left. \begin{array}{c} 2\beta(1-m^2)J = g'(B/2mJ) \\[2mm] g'(z) = \int_0^\infty \mathrm{d}t \, \mathrm{e}^{-t(z+d)} \, [J_o(it)]^d \end{array} \right\} \qquad (12.3.20)$$

with

where J_o is the zeroth Bessel function. The relation (12.3.20) indeed expresses the magnetization per site m, in terms of the natural temperature $\beta = 1/kT$ and the external magnetic field B; J is the coupling constant defining the model; see (12.3.18). The existence of a non-zero spontaneous magnetization

$$m_o(\beta) \equiv \lim_{B \to 0} m(\beta, B) \neq 0$$

depends on whether the $g'_o \equiv \lim_{x \to 0} g'(x)$ exists and is finite; the latter condition can be proven to be satisfied if and only if $d > 2$. One concludes immediately from this argument that there exists then a finite critical temperature $T_c = 1/k\beta_c$ given by

$$2\beta_c J = g'_o \quad . \qquad (12.3.21)$$

Moreover, by the same token, one sees that

$$m_o \approx \left(\frac{T_c - T}{T_c}\right)^{\frac{1}{2}} \qquad \text{as} \quad T \nearrow T_c \quad . \qquad (12.3.22)$$

The other critical exponents can also be computed – see e.g. [Baxter, 1982]:

$$\left. \begin{array}{ccccc} & \alpha & \beta & \gamma & \delta \\[2mm] \text{for} \quad 2 < d < 4 : & -\frac{4-d}{d-2} & \frac{1}{2} & \frac{2}{d-2} & \frac{d+2}{d-2} \\[3mm] \text{for} \quad d > 4 : & 0 & \frac{1}{2} & 1 & 3 \end{array} \right\} \quad . \qquad (12.3.23)$$

We reproduced the values of the critical coefficients for two main reasons:

1. in contrast to the coefficients obtained in Sect. 11.3 for the mean-field theory, α, γ, δ depend explicitly on the dimension d of the lattice;
2. the coefficients (12.3.23) satisfy again the "universal" relations (11.3.21): this coïncidence begs for an explanation; see Chap. 13.

The literature on the spherical model and its variations was reviewed in [Joyce, 1972]. The constraint in (12.3.18) is a global condition, and therefore is of dubious direct physical significance; this problem has been addressed by [Kac and Thompson, 1971, Pearce and Thompson, 1977], who established that the spherical model is equivalent to a limit of a more acceptable model involving nearest neighbor interactions only. [Stanley, 1971][pp. 128–131] traces the genesis of this idea, to the early stage of which he had contributed.

12.3.3 Variations Modifying the Domain or Range of the Interactions

Let again

$$\sigma : i \in \Lambda \mapsto \sigma_i \in \{-1, +1\} \qquad (12.3.24)$$

denote the configuration of a domain $\Lambda \subset \mathsf{Z}^d$ of the $d-$dimensional lattice, and let the energy of any configuration be defined to be:

$$H_\Lambda(\sigma) = -B \sum_i \sigma_i - \sum_{(i,j)} J_{i,j} \sigma_i \sigma_j \qquad \text{with} \quad J_{j,i} = J_{i,j} \geq 0 \qquad (12.3.25)$$

where the summation $\sum_{(i,j)}$ carries over all pairs of sites in Λ; in the usual Ising model, this sum is restricted to nearest neighbor pairs – we wrote $\sum_{<i,j>}$ – a restriction that can alternatively be expressed by a supplementary condition in (12.3.25), namely $J_{i,j} = 0$ for all, but nearest neighbor pairs (i,j). Hence the variations considered in this subsection involve both the range of the interactions, and the dimension d of the domain $\Lambda \subset \mathsf{Z}^d$; the sites are still occupied by classical spins with values ± 1.

The purpose now is to ground in mathematical theorems the intuitive feeling that the positivity condition $J_{i,j} \geq 0$ on the interactions fosters a tendency of the spins to align with one another, and that this tendency is increased when the temperature T is decreased, or when the interactions $J_{i,j}$ are increased; note, in this connection, that T and J occur together in the partition function in the combination $\beta J = J/kT$. One would expect things like:

$$\langle \sigma_i \sigma_j \rangle \geq 0 \; ; \; \partial_{J_{i,j}} \langle \sigma_k \sigma_l \rangle = \beta \left[\langle \sigma_i \sigma_j \sigma_k \sigma_l \rangle - \langle \sigma_i \sigma_j \rangle \langle \sigma_k \sigma_l \rangle \right] \geq 0 \; . \quad (12.3.26)$$

This is precisely what was proven by [Griffiths, 1967]. It was generalized by [Kelly and Shermann, 1968], and the latter's proof was simplified by [Ginibre, 1969]; see also [Griffiths and Lebowitz, 1968, Ginibre, 1970], and for didactic proofs [Ruelle, 1969, Griffiths, 1972].

Theorem 12.3.1 (Griffiths–Kelly–Sherman inequalities). *Let Λ be a finite domain in Z^d, and denote by $\mathcal{P}(\Lambda)$ the collection of all subsets of Λ; for every configuration $\sigma : i \in \Lambda \mapsto \sigma_i \in \{-1, +1\}$ and every $\Xi \in \mathcal{P}(\Lambda)$, define $\sigma^\Xi \equiv \prod_{i \in \Xi} \sigma_i$; and for every coupling function $K : \Xi \in \mathcal{P}(\Lambda) \mapsto K_\Xi \in [0, \infty)$ with $K_\emptyset = 0$, and every configuration σ on Λ, let*

$$H_{\Lambda,K}(\sigma) - \sum_{\Xi \subset \Lambda} K_{\Xi}\,\sigma^{\Xi}$$

be the energy of the configuration. Let finally

$$\langle \sigma^{\Xi} \rangle \equiv Z^{-1} \sum_{\sigma} \sigma^{\Xi}\,\mathrm{e}^{-H_{\Lambda,K}(\sigma)} \qquad \text{where} \qquad Z \equiv \sum_{\sigma} \mathrm{e}^{-H_{\Lambda,K}(\sigma)}\,.$$

Then, for every $\Xi, \Xi' \in \mathcal{P}(\Lambda)$:

1. $\partial_{K_{\Xi}} \langle \sigma^{\Xi'} \rangle = \langle \sigma^{\Xi}\sigma^{\Xi'} \rangle - \langle \sigma^{\Xi} \rangle \langle \sigma^{\Xi'} \rangle$;
2. $\langle \sigma^{\Xi} \rangle \geq 0$;
3. $\langle \sigma^{\Xi}\sigma^{\Xi'} \rangle - \langle \sigma^{\Xi} \rangle \langle \sigma^{\Xi'} \rangle \geq 0$.

The $H_{\Lambda,K}(\sigma)$ of the theorem covers (12.3.25) for $K_{\{i\}} = \beta B, K_{\{i,j\}} = J_{i,j}$, and $K_{\Xi} = 0$ for all $\Xi \subset \Lambda$ with more than three sites. In addition, it allows for inhomogenous external magnetic fields, and for more than 2−body interactions. Clearly, the conclusions of the theorem cover the inequalities (12.3.26) as particular cases. Hence, the intuitive motivation presented for the theorem was correct. A model theorist will see in the GKS inequalities a recipe for theory or model reduction in the sense of Sect. 1.7.

Within the class of models with a Hamiltonian of the form described in the theorem, the proof of the existence of long-range order – or of a ferromagnetic phase with non-zero spontaneous magnetization – in one model is reduced to proving that the corresponding property is satisfied by another model with a coupling function K that is no greater than that of the model one is interested in; the latter model can thus be viewed as a generalization of the former. In respect to this generalization relation, GKS inequalities insure that, if the alignment property among pairs of spins obtains in one model, then it obtains in all other models which are such generalizations of it. The methodological upshot is quite obvious. Consider, for instance, an Ising model defined on $\mathbf{Z}^3$ with interactions satisfying the assumptions of the theorem. Use first the GKS to reduce this model to one with nearest-neighbor interactions. Second, as $\mathbf{Z}^3 = \mathbf{Z}^2 \times \mathbf{Z}$, one can view the simplest 3−dimensional Ising model with nearest neighbor interaction as an assemblage of mutually interacting 2−dimensional Ising model; the coupling function for this model is larger than that of a model where the cross interaction between the neighboring slices has been switched off; since the latter model exhibits a ferromagnetic phase transition, so do the intermediate model and thus again the general Ising model in 3−dimensions. This conclusion by-passed any reference to a putative computation of the partition function of 3−dimensional Ising model. Hence, the two dimensional Ising model solved in Sect. 12.2 can really be said to have provided an explanation of ferromagnetic phenomenon without being even remotely resembling any real ferromagnetic materials.

Conclusions and further readings. In Sect. 12.1, we argued that the thermodynamical limit was an essential ingredient to focus on the occurence of phase transitions.

We discussed in Sect. 12.2 the idea that phase transitions are cooperative phenomena linked to the establishment of long-range order. This idea can be traced back at least to [Zernike, 1940], and was pursued by several authors, notably [Ashkin and Lamb, 1943] who emphasize the further connection with the degeneracy of the maximal eigenvalue of a certain operator; [Kaufman, 1949] also draws attention to [Lassettre and Howe, 1941].

We attributed the method of the transfer matrix to [Kramers and Wannier, 1941]. Note that [Kaufman, 1949] also credits [Montroll, 1941]; the latter quite fairly mentions that his work was motivated by what he heard from Wannier, as "the present state of affairs of the world delayed publication" of Wannier's contribution with Kramers, a fact that is confirmed obliquely by Wannier in a footnote at the beginning of their joint paper. Montroll's paper is more didactic, whereas the paper by Kramers & Wannier is very thorough and detailed. Asides from their brilliant use of the Jordan–Wigner transformation to reduce earlier works on the Ising model – in particular [Onsager, 1944, 1949a, Kaufman, 1949] and [Yang, 1952, Montroll, Potts, and Ward, 1963] – to a many-Fermion problem that is definitely more tractable by methods of common knowledge among physicists, [Schultz, Mattis, and Lieb, 1964] also offers some bibliographical references beyond those given here.

Section 12.3 presented a few variations on the theme of the Ising model; while these particular illustrations were chosen with a view to provide the beginning of an explanation for phase transitions, they do not exhaust the panoply of solvable models. In particular, the use of the transfer matrix method – first introduced in Sect. 12.1, and then used extensively in Sect. 12.2 – was extended to the rigorous analysis of models for antiferromagnetism [Lieb and Wu, 1972]. This paper also reviewed how the famous problem of the *residual entropy of ice* fits in this class of models. Consider a rectangular lattice, each of the MN vertices of which is occupied by a water molecule; the H–O–H angle is assuned to be $90°$, so that the Hydrogen bounds can be aligned with the edges of the lattice. The so-called "ice condition" is that at each vertex there are exactly two ingoing and two outgoing bonds; the six possible site-configurations are

$$\rightarrow \updownarrow \rightarrow \qquad \rightarrow \updownarrow \rightarrow \qquad \leftarrow \updownarrow \leftarrow \qquad \leftarrow \updownarrow \leftarrow \qquad \rightarrow \updownarrow \leftarrow \qquad \leftarrow \updownarrow \rightarrow \ . \qquad (12.3.27)$$

A configuration of the whole system is thus a consistent assigment of a direction to each edge of the lattice. The question is to find an asymptotic formula for the number Z_{MN} of configurations when both M and N are large. The residual entropy of ice, i.e. the entropy per site, at zero temperature, is then:

$$S = k \ln W \qquad \text{with} \qquad W = \lim_{M,N \to \infty} [Z_{MN}]^{\frac{1}{MN}} \ . \qquad (12.3.28)$$

Of course, real ice is a 3-dimensional crystal, but its coordination number is four, as it is in the square lattice considered here. Under these circumstances, note with [Pauling, 1935] that *if* the lattice edges and the vertex configurations were independent, one would have

$$Z_{MN} = 2^{2MN}(\frac{6}{16})^{MN} \quad \text{i.e.} \quad W_{Pauling} = 1.5 \qquad (12.3.29)$$

where $\frac{6}{16}$ is the fraction of site configurations (6) allowed by the "ice rule" over the number possible configurations ($2^4 = 16$) one would have in the absence of the rule. What was meant to be only a rough approximation was confirmed, first by the numerical estimates of [Nagle, 1966] namely

$$W_{Nagle} = 1.540 \pm .001 \qquad (12.3.30)$$

and then by the exact solution obtained by [Lieb, 1967] namely

$$W_{exact} = (\frac{4}{3})^{\frac{3}{2}} \quad \text{note} \quad (\frac{4}{3})^{\frac{3}{2}} \approx 1.5396007\dots \qquad (12.3.31)$$

There are several reasons for us to mention this result here. First, it gives one more illustration of the wide applicability of the transfer matrix formalism to asymptotic problems: here Z_{MN} can again be put in the form $\mathrm{Tr}V^M$ where the matrix elements of V – taken between configurations of successive rows – are non-negative, so that the Perron–Frobenius theorem can be used again.

Second, the methods developed for this and related models had interesting extensions to other fields of physics and mathematics, such as in the study of percolation problem, or in graph- and knot-theories; see e.g. [Temperley and Lieb, 1971].

Third, the model disproves the universal validity of a *third law of thermodynamics* which would claim that the entropy of a chemically simple, perfectly crystaline body should vanish at absolute zero temperature. Recall that the belief in the validity of Nernst's law contributed to the recognition of the fundamental role of quantum theory in statistical mechanics; see e.g. our discussion of the Einstein–Debye theory of the specific heat of solids; see our Sect. 10.3 and, in particular, formula (10.3.45). For a physical discussion of various tentative versions of the third law, see [Wannier, 1966, 1987, pp. 111–114].

And fourth, the exact solvability of the model can be used as a historical marker; indeed, it came in a changing climate when statistical mechanics started to appear in a new light: deep physical conjectures and/or insights, which often had been expressed in uncontolled approximations, were beginning to be settled and often confirmed by the mathematically rigorous analysis of models. Note that, in line with our general delineation of the role of models, we had to say *confirmed* and not *replaced,* thus addressing an euphoric temptation to overstep the boundaries and reverse the proper order of priorities. The enthusiasm and some of the tensions of the period are captured well through the brief reminiscences preserved in [Lieb, 1997]. Other papers in the same issue of that Journal – the procceedings of a conference on exactly soluble models – will help the Reader measure the distance that has been traveled since Onsager's mystical offering for the solution of the two-dimensional Ising model in the absence of magnetic field.

For more information on the material of this chapter, we already cited [Lieb and Mattis, 1966, Stanley, 1971, Thompson, 1972, Joyce, 1972, Griffiths, 1972, Baxter, 1982, Sewell, 1986, Gallavotti, 1999]; the Reader will find further insights and vistas in [Domb and Green, 1972, Dubin, 1974, Israel, 1979, Sinai, 1982, Dorlas, 1999].

13. Scaling and Renormalization

13.1 Scaling Hypotheses and Scaling Laws

We begin with the *static scaling laws* which have the objective of reducing the number of independent critical exponents to a couple, and the proposal is to achieve this by focussing on ever smaller neighborhoods of the critical point. The basic ideas of scaling are to be traced back to [Essam and Fisher, 1963, Widom, 1965, Kadanoff, 1966, Fisher, 1967a]; their justification was found later in the renormalization program we discuss in the next section. In a nutshell, the idea was that in the neighborhood of the critical point, the thermodynamic functions can be *assumed* to be generalized homogenous functions; see e.g. [Stanley, 1971, Fisher, 1983, Cardy, 1996].

For didactic reasons, we consider the simplest case of a spin system on a regular lattice – much like the Ising model discussed earlier, but now with arbitrary interactions between the spins, subject only to the condition that the thermodynamical limit exists. We suppose this system to be in canonical thermal equilibrium near a critical point $(T = T_c, B = 0)$. We write the free-energy per spin as a function of the two independent variables $t = (T - T_c)/T_c$ – the reduced temperature – and $b = B/kT_c$ – the reduced external magnetic field.

First scaling hypothesis: *There exist two scaling parameters p and q such that the free energy per spin can be approximated in the neighborhood of the critical point by a function f of t and b, such that for every $\lambda \in \mathsf{R}$:*

$$f(\lambda^p t, \lambda^q b) = \lambda f(t, b) \quad . \tag{13.1.1}$$

Note first that

$$(\partial_t^2 f)(t, b) = \lambda^{(2p-1)}(\partial_{\lambda^p t}^2 f)(\lambda^p t, \lambda^q b) .$$

Since this holds for every λ, we can freely impose $\lambda^p |t| = 1$ to obtain for $b = 0$:

$$(\partial_t^2 f)(t,0) = |t|^{-\frac{1}{p}(2p-1)}(\partial_{\lambda^p t}^2 f)(t/|t|,0)$$

i.e.

$$C_{B=0} \approx |t|^{-\alpha} \qquad \text{with} \qquad \alpha = 2 - \frac{1}{p} \quad . \tag{13.1.2}$$

Upon repeating the argument that led to (13.1.2), with now the first derivative of f with respect to b, we obtain:

$$(\partial_b f)(t,b) = \lambda^{(q-1)}(\partial_{\lambda^q b} f)(\lambda^p t, \lambda^q b) \, .$$

Since this holds for every λ, we can freely impose $\lambda^p t = -1$ [resp. $\lambda^q b = 1$] to obtain for $b = 0$ [resp. $t = 0$]:

$$(\partial_b f)(t,0) = (-t)^{-\frac{1}{p}(q-1)}(\partial_{\lambda^q b} f)(t/|t|,0)$$

and

$$(\partial_b f)(0,b) = b^{-\frac{1}{q}(q-1)}(\partial_{\lambda^q b} f)(0,1)$$

i.e.

$$m(t,0) \approx (-t)^{\beta} \qquad \text{with} \qquad \beta = \frac{1-q}{p} \tag{13.1.3}$$

and

$$m(0,b) \approx b^{\frac{1}{\delta}} \qquad \text{with} \qquad \delta = \frac{q}{1-q} \quad . \tag{13.1.4}$$

The second derivative of (13.1.1) with respect to b gives similarly

$$\chi_T(t,0) \approx |t|^{\gamma} \qquad \text{with} \qquad \gamma = \frac{2q-1}{p} \quad . \tag{13.1.5}$$

Remarks A:

1. The relations (13.1.2–1.5) express the *four* critical exponents α, β, γ and δ – compare with (11.2.1–4) – in terms of the *two* scaling parameters p and q.
2. The scaling hypothesis only assumes the *existence* of some scaling parameters, but gives no indication about their values.
3. In view of the above two remarks, we eliminate the scaling parameters p and q by writing two independent relations between the four critical exponents. This is achieved by noticing that (13.1.2–1.5) entail the two universal equalities:

$$\alpha + 2\beta + \gamma = 2 \quad \text{and} \quad \alpha + \beta(1+\delta) = 2 \quad . \tag{13.1.6}$$

4. This trivial mathematical trick is physically relevant since the relations (13.1.6) are *universal* and exact consequences of the scaling hypothesis alone: they do not depend on any particular value of the scaling parameters.
5. The relations (13.1.6) are satisfied in diverse models.

a) The van der Waals model of fluids – (11.3.21) – and thus the Weiss model for ferromagnetism since the two models have exactly the same critical exponents – see (11.3.38) and the discussion following it.
b) The exactly solvable spherical model – see Sect. 12.3.
c) For the Ising models in dimension 2 and 3, one knows from Onsager's rigorous solution, the exact value of one of the critical exponents, namely $\beta = \frac{1}{8}$ – see (12.2.63)– for the two-dimensional lattice, with $\alpha = 0$ and $\gamma = 7/4$. In three dimensions also, approximate results seem to agree with (13.1.6).
d) Experimental data obtained very early in the game gave strong indications that many other systems show critical exponents that satisfy (13.1.6) as well.

6. The equalities (13.1.6) first appeared in the thermodynamical literature in a weaker form, namely as the *Rushbrooke inequality* $[\alpha + 2\beta + \gamma \geq 2]$ and the *Griffiths inequality* $[\alpha + \beta(1 + \delta) \geq 2]$; cf. [Rushbrooke, 1963, Griffiths, 1965], and for context: [Stanley, 1971][Chap. 4].

7. For any $d > 0$, the change of variables $(\lambda, p, q) \longrightarrow (\lambda^d, p/d, q/d)$ brings (13.1.1) to the equivalent form:

$$f(\lambda^p t, \lambda^q b) = \lambda^d f(t, b) \quad ; \tag{13.1.7}$$

with respect to these new parameters (13.1.2–1.5) become:

$$\alpha = 2 - \frac{d}{p} \quad , \quad \beta = \frac{d - q}{p} \quad , \quad \gamma = \frac{2q - d}{p} \quad , \quad \delta = \frac{q}{d - q} \quad . \tag{13.1.8}$$

The critical exponents η and ν describe the decay of the correlation function

$$G(r, T) = <\sigma_x \sigma_{x+r}> - <\sigma_x><\sigma_{x+r}>$$

where $< ... >$ indicates the canonical equilibrium state at temperature T and in the absence of external magnetic field, $B = 0$; the translation invariance of the Hamiltonian implies that G is independent of the site $x \in \Lambda$ chosen to compute it. Since G is not a macroscopic observable as were the specific heat C_B, the magnetization m, and the susceptibility χ_T, we need to adapt – [Goldenfeld, 1992, Cardy, 1996] – the first scaling hypothesis (13.1.1) to this microscopic situation:

Second scaling hypothesis: *With p and q as before, the correlation function can be approximated in the neighborhood of the critical point by a function g of r and t, such that for every $\lambda \in \mathrm{R}$:*

$$g(r, t) = \lambda^{2(q-1)d} g(\lambda^{-1} r, \lambda^{pd} t) \tag{13.1.9}$$

where d is the dimensionality of the model.

Again, upon taking advantage of the fact that λ is arbitrary, we choose to insert in (13.1.9) a particular value, namely $\lambda = r$ to obtain:

$$g(r,t) = r^{2(q-1)d}g(1, r^{pd}t) \qquad (13.1.10)$$

and thus, in particular, at the critical temperature: $g(r,0) = r^{2(q-1)d}g(1,0)$,
i.e. with the definition of η in (11.2.7):

$$g(r,0) \approx r^{-(d-2+\eta)} \qquad \text{with} \qquad \eta = d + 2 - 2qd \quad . \qquad (13.1.11)$$

In the vincinity of the critical temperature, but not exactly at $t = 0$ an exponential decay of the correlation function obtains only if

$$g(r,t) \approx e^{-r/\xi} \quad \text{with} \quad \xi \approx t^{-\nu} \quad \text{and} \quad \nu = \frac{1}{pd} \quad . \qquad (13.1.12)$$

Remarks B:

1. Upon combining the expressions obtained for ν and η with (13.1.2) and (13.1.5), we obtain the two new *universality* relations that we want to retain together with the two relations (13.1.6), namely:

$$2 - \alpha = \nu d \quad \text{and} \quad \gamma = \nu(2 - \eta) \quad . \qquad (13.1.13)$$

2. The first of the equalities (13.1.13)is known as the *hyperscaling law*, because it is the only such law that involves the dimensionality, d.

3. Here again – as for (13.1.6) – the equalities (13.1.13) first appeared in the thermodynamical literature in a weaker form, namely as the *Josephson inequality* $[2 - \alpha \leq \nu d]$ and the *Fisher inequality* $[\gamma \leq \nu(2 - \eta)]$; cf. [Josephson, 1967, Fisher, 1969] and [Stanley, 1971][Chap. 4].

To summarize, here is a preliminary balance sheet for the exercise.

1. On the positive side:
   a) The scaling hypotheses are compatible with the tenets of thermodynamics in the sense that their central predictions – the universality equalities (13.1.6) and (13.1.13) are contained in the general thermodynamical inequalities of Rushbrooke, Griffiths, Josephson and Fisher; see Rems. A.6 and B.3 above.
   b) The theoretical evidence available – sharp for the few existing solvable models, and approximate for the others – as well as the laboratory data seem to confirm that the universality equalities are indeed realized.
2. On the negative side:
   a) The hypotheses are not anchored on microscopic first principles; rather they propose an *ad hoc* formalization of phenomenological observations indicating that near the critical point things seem to rescale. Compare this to a similar situation encountered in Sect. 3.1: the formalization provided by the diffusion equation for the phenomenological observations that sequences of pictures taken in the course of Brownian motion look the same if one rescales consistently space-unit (to the power 2) and time-unit (to the power 1).

b) The scaling hypotheses in themselves carry no information on the rescaling parameters p and q and therefore none of the results (13.1.2–5) and (13.1.12–13) can be used to compute the individual values of the critical exponents.

The renormalization program proposes to give a more fundamental explanation for the scaling hypotheses, one that would alleviate these criticisms.

13.2 The Renormalization Program

To present the basic idea in a simple form, we chose two didactic models where the program can be exactly implemented. The first is again the one-dimensional Ising model, for which the result is evidently trivial, but nevertheless instructive; see in particular the renormalization basic step (13.2.8), and its consequence (13.2.9). The second illustration is provided by the Dyson hierarchical model; here, the program requires serious analytic work, the main stages centering around: the basic iteration formula, (13.2.19); the discussion summarized in Fig. 13.3; and the values of the critical indices (13.2.40). Accordingly, the treatment of the model is broken up into three subsections: definitions (see Fig. 13.2), critical fixed points, and critical exponents.

Consider now the class of systems of the Ising-type, defined on a finite, d–dimensional lattice, with periodic boundary conditions, $\Lambda = (\mathsf{Z}_N)^d$; for convenience, we assume that $N = M^{m\lambda}$ with $\lambda = 2\kappa + 1$, M, m and κ positive integers, and m large. On each site k of this lattice sits a classical "spin" σ_k with possible values ± 1. Hence a configuration of this system is a function $\sigma : k \in \Lambda \to \sigma_k \in \{-1, +1\}$ so that the state-space of the system, i.e. the collection of all such configurations, is $\{-1, +1\}^\Lambda$. The Hamiltonian of any of these systems is taken to be of the form:

$$H : \sigma \in \{-1, +1\}^\Lambda \to H(\sigma) = \sum_A K_A \sigma_A \qquad (13.2.1)$$

with A running over $\mathcal{P}(\Lambda)$, the collection of all subsets of Λ; and $\sigma_A = \prod_{k \in A} \sigma_k$; K_A is the strength of the interaction between the spins in A.

For convenience, we assume that the natural temperature $\beta = 1/k_{Boltz}T$ as been incorporated in the definition of K. For instance, for the ordinary Ising Hamiltonian – discussed in Sects. 12.1–2 for $d = 1, 2$, – the only $K_A \neq 0$ are $K_{\{k\}} = -\beta B$, the external magnetic field, and $K_{\{j,k\}} = -\beta J$ when j and k are nearest neighbors. Hence to specifiy a Hamiltonian in the class (13.2.1) is to specify the function $K : A \in \mathcal{P}(\Lambda) \to K_A \in \mathsf{R}$.

Let us now divide the original lattice in *cells* over which we take average. We thus introduce the *sublattice* $\Lambda' = (\lambda \mathsf{Z}_{N'})^d$ – with $N' = \lambda^{-1}N = M^{(m-1)\lambda}$ – and its complement $\Lambda'' = \{k \in \Lambda \mid k \notin \Lambda'\}$. We then denote by σ' [resp. σ'']

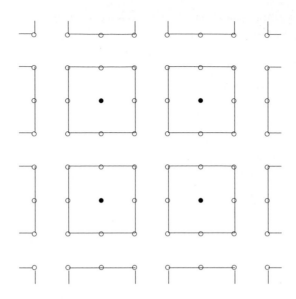

Fig. 13.1. Decimation and majority rules: Sites marked • belong to Λ' ; sites marked ○ belong to Λ''. Decimation retains only •. The majority rule attributes to • a value determined by the surrounding ○. Lines mark the boundaries of blocks

the restriction of the configuration σ to Λ' [resp. Λ''] , i.e. $\sigma' : k' \in \Lambda' \to \sigma'_{k'} = \sigma_{k'} \in \{-1, +1\}$; and similarly for σ''.

The first step of the method known as *renormalization by decimation* (see Fig. 13.1) is to rewrite the partition function in the form:

$$\left.\begin{array}{c} Z = \sum_\sigma e^{-H(\sigma)} = \sum_{\sigma'} e^{-H(\sigma')} \\ \text{with} \\ e^{-H(\sigma')} = \sum_{\sigma''} e^{-H(\sigma' \cup \sigma'')} \end{array}\right\} \qquad (13.2.2)$$

which defines the *coarse-grained* Hamiltonian

$$H(\sigma') = \sum_{A'} K'_{A'} \, \sigma'_{A'} \qquad (13.2.3)$$

where A' runs over the subsets of Λ' ; and $\sigma'_{A'} = \prod_{k' \in A'} \sigma'_{k'}$.

On the space of the coupling constants $\mathcal{K}$, this defines a map $\mathcal{R}[K] = K'$. One iterates this procedure to produce successive Hamiltonians $H^{(n)}$ with coupling constants $K^{(n)}$, and the corresponding sequence of maps $\{\mathcal{R}^n \,|\, n \in \mathsf{Z}^+\}$ is called a *renormalization semigroup*, on account of the fact that the composition $\mathcal{R}^{n_1} \circ \mathcal{R}^{n_2} = \mathcal{R}^{n_1+n_2}$, defined for any pair of non-negative integers (n_1, n_2), is an associative binary relation, with unit $\mathcal{R}^0 = Id$.

Lest the above looks too abstract, consider the very particular case where the initial Hamiltonian H is the one-dimensional Ising Hamiltonian with nearest-neighbor interactions – $K_{\{j,k\}} = -\beta J$ when j and k are nearest neighbors, with all the other coupling constants being equal to zero, including the external magnetic field $K_{\{k\}} = -\beta B = 0$:

$$H(\sigma) = -K \sum_{k \in \Lambda} \sigma_k \sigma_{k+1} \tag{13.2.4}$$

and thus

$$e^{-H(\sigma)} = \prod_{k \in \Lambda} e^{K \sigma_k \sigma_{k+1}} = \prod_{k \in \Lambda} V(\sigma_k, \sigma_{k+1}) \tag{13.2.5}$$

where the transfer matrix V is rewritten in the form – compare with (12.1.7) – with the notation $\tilde{K} = \tanh[\beta J]$:

$$V = \begin{pmatrix} e^K & e^{-K} \\ e^{-K} & e^K \end{pmatrix} = \cosh K \begin{pmatrix} 1 + \tilde{K} & 1 - \tilde{K} \\ 1 - \tilde{K} & 1 + \tilde{K} \end{pmatrix} \quad . \tag{13.2.6}$$

Hence upon summing over the configurations attached to the sites that belong to Λ'' we obtain:

$$\left.\begin{array}{l} \sum_{\sigma''} e^{-H(\sigma)} = \prod_{k' \in \Lambda'} V^\lambda(\sigma_{k'}, \sigma_{k'+1}) \\[2mm] \text{with} \\[2mm] V^\lambda = 2^{\lambda-1}(\cosh K)^\lambda \begin{pmatrix} 1 + (\tilde{K})^\lambda & 1 - (\tilde{K})^\lambda \\ 1 - (\tilde{K})^\lambda & 1 + (\tilde{K})^\lambda \end{pmatrix} \end{array}\right\} \tag{13.2.7}$$

which we can rewrite

$$\left.\begin{array}{l} e^{-H(\sigma')} = \sum_{\sigma''} e^{-H(\sigma' \cup \sigma'')} \\[2mm] \text{with} \\[2mm] H(\sigma') = -K' \sum_{k' \in \Lambda'} \sigma'_{k'} \sigma'_{k'+1} + C \\[2mm] \text{and} \\[2mm] K' = \beta J' = \tanh^{-1}\{[\tanh(\beta J)]^\lambda\} \quad \text{i.e.} \quad \tilde{K}' = \tilde{K}^\lambda \end{array}\right\} \tag{13.2.8}$$

where the constant C does not affect the computation of expectation values; it only betrays the presence of the multiplicative constant $2^{\lambda-1}(\cosh K)^\lambda$ in (13.2.7) rather than $\cosh K'$, and we can ignore it here.

We know – see Sect. 12.1 – that, even in the thermodynamical limit, the 1–d Ising model with nearest-neighbor interaction does not exhibit a phase transition at any finite temperature. The renormalization semigroup confirms this. Indeed, for this particular model, the map $\mathcal{R} : \tilde{K} \in [0,1] \to \mathcal{R}[\tilde{K}] = \tilde{K}' \in [0,1]$ has exactly two fixed points: 0 and 1, with $0 \leq \tilde{K} < 1 \models \lim_{n \to \infty} R^n[\tilde{K}] = 0$.

Upon recalling the notation $\tilde{K} = \tanh \beta J$, we see that the fixed point $\{0\}$ corresponds to the high temperature limit $T = \infty$, where the system behaves as if there were no interaction; and the fixed point $\{1\}$ corresponds to the low temperature limit $T = 0$, where everything is frozen: either all the spins are up, or all are down. Let us briefly examine what the renormalization analysis has to add.

We note: (i) the lucky fact that the coarse-grained Hamiltonian (13.2.8) and the original Hamiltonian (13.2.4) both are Ising model with nearest-neighbor interactions only; (ii) the coupling constants of these two systems are related by the third relation in (13.2.8); and (iii) λ is the ratio of the lattice spacing in the coarse-grained system over the lattice spacing of the original system. We then conclude that the correlation length ξ satisfies:

$$\xi(\tilde{K}') = \lambda^{-1}\xi(\tilde{K}) \quad \text{i.e.} \quad \xi \propto [\ln\tanh K]^{-1} \tag{13.2.9}$$

in agreement with (12.1.18) as $T \to 0$; for finite temperatures, the correlation decays exponentially fast, but the rate of decay ξ^{-1} slows down to 0 as $e^{-\beta J}$ when the temperature approaches the absolute zero.

While not bringing new results yet, the approach is certainly more direct than that followed in Sect. 12.1.

Note that one can view renormalization by decimation as a selection process. Around each site $k' = k\lambda$ in the sublattice Λ', let us consider the box $B_{k'}$ of length $(\lambda - 1)$ centered on this site – recall that we chose λ to be odd, say, $\lambda = 2\kappa + 1$, so that $B_{k'} \equiv \{k \in \Lambda \,|\, k' - \kappa \le k \le k' + \kappa\}$. Note that we wrote this in one dimension only; if the lattice has more dimensions, these inequalities are required to hold for each components of the position vectors.

The idea of a representative selection process is now this: when the system is close to its critical point, the correlation length ξ is much larger than lattice spacing (chosen here to be $a = 1$); to focus on the behavior of the system in intermediate scales, one chooses λ such that $1 \ll \lambda \ll \xi$. The spins inside the same box $B_{k'}$ are therefore expected to be strongly correlated, and $\sigma_{k'}$ can therefore be considered as a "representative" of the box $B_{k'}$ in the middle of which it sits. This description suggests immediately other possible selection processes. For instance, a more "democratic" selection would be the *majority rule* by which one defines the spin variable $\sigma_{k'}$ the values ± 1 of which are determined by the condition $\sigma_{k'} \cdot \sum_{k \in B_{k'}} \sigma_k > 0 : \sigma_{k'}$ is $+1$ [resp. -1] when the majority of the spins in the box point up [resp. down]; this assignment is unambiguous, since we have assumed in effect that each box contains an odd number of sites, irrespectively of the dimensionality of the system. Still other assignments are possible, mostly chosen for computational convenience when searching for fixed points of the corresponding renormalization semigroup.

Yet another way to implement the idea of the renormalization method is to make it appear as a deviation from the Gaussian distribution of the central limit theorem – see Sect. 5.2 – when the random variables considered are *not* independent. Following [Collet and Eckmann, 1978], we present this approach in a case where it is implementable without approximation.

Definition of the hierarchical model. The original model was proposed by [Dyson, 1969]. For every positive integer N, let $\Lambda_N = \{1, 2, 3, \cdots, 2^N\}$ and $\{-1, +1\}^{\Lambda_N} = \{\sigma : k \in \Lambda_N \to \sigma_k \in \{-1, +1\}\}$. For each integer p with $0 \le p \le N$, consider the partition of Λ_N into 2^{N-p} blocks $B_{p,r}$ consisting of 2^p consecutive sites:

$$B_{p,r} = \{k \in \Lambda_N \mid (r-1)\, 2^p + 1 \le k \le r\, 2^p\} \; ; \; r = 1, 2, \cdots, 2^{N-p}. \quad (13.2.10)$$

The Hamiltonian then is defined as:

$$H_{\Lambda_N} = -\sum_{p=1}^{N} \frac{1}{2^{2p}+1} b_p \sum_{r=1}^{2^{N-p}} \Big(\sum_{k \in B_{p,r}} \sigma_k\Big)^2 \qquad (13.2.11)$$

where the coupling constants b_p are positive numbers, defined for all non-negative integers p, and satisfying the condition that the following sum converges:

$$\lim_{N \to \infty} E_N = \sum_{p=1}^{N} 2^{p-1} E_N(p) \quad \text{with} \quad E_N(p) = \sum_{q=p}^{N} 2^{-2q} b_q \quad . \qquad (13.2.12)$$

To see the reason for the appellation "hierarchical" and the convergence condition, note that, for all integers p with $1 \le p \le N$: the block $B_{p,r}$ of length 2^p is the disjoint union of two consecutive blocks of length 2^{p-1}, namely $B_{p-1,2r-1}$ and $B_{p-1,2r}$. Consequently

$$\left.\begin{aligned} \big(\textstyle\sum_{k \in B_{p,r}} \sigma_k\big)^2 = \big(\textstyle\sum_{k \in B_{p-1,2r-1}} \sigma_k\big)^2 + \big(\textstyle\sum_{l \in B_{p-1,2r}} \sigma_l\big)^2 \\ + \textstyle\sum_{(k,l) \in B_{p-1,2r-1} \times B_{p-1,2r}} \sigma_k \sigma_l \end{aligned}\right\} . \qquad (13.2.13)$$

Since the first and the second terms of the RHS are of the same form as the LHS, except for the fact that they involve blocks of length 2^{p-1} instead of a block of length 2^p , the Hamiltonian (13.2.11) can be rewritten as

$$H_{\Lambda_N} = \sum_{p=1}^{N} v_p \sum_{(k,l) \in V_p} \sigma_k \sigma_l \qquad (13.2.14)$$

where V_p is the set of all pairs of sites (k,l) for which k and l belong to the same block of length 2^p but to different blocks of length 2^{p-1} . Hence the Hamiltonian can be viewed as the sum of interactions between consecutive blocks, at all possible levels p. This is illustrated in Fig. 13.2. Note then that for any pair of sites $(k,l) \in \Lambda \times \Lambda$ there is a smallest integer $p(k,l)$ such that k and l belong to the same block. With this picture in mind, one sees with [Dyson, 1969], that E_N – in (13.2.12) – is the sum of the interactions coupling an arbitrary, but fixed, spin to all the others. Hence condition (13.2.12) expresses that this bound remains finite so that one can define an infinite hierarchical model, as $N \to \infty$. Note also that the condition is satisfied for the particular choices $b_p = c^p$ with $1 < c < 2$.

The primary concern in [Dyson, 1969] was to use the model to explore the effect of interactions, decaying slowly with distance, on the occurrence of long-range order in one-dimension – see the closing remarks in Sect. 12.1.

(a)

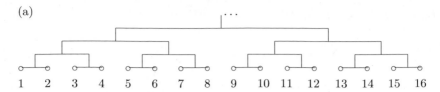

(b)

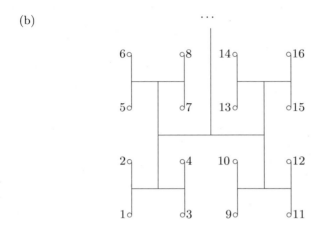

Fig. 13.2a,b. The one-dimensional hierarchical model. The successive blocks containing the site $\{1\}$ are shown: $B_{0,1} = \{1\}, B_{1,1} = \{1,2\}, B_{2,1} = \{1,2,3,4\},$ $B_{3,1} = \{1,2,\cdots,7\}, B_{4,1} = \{1,2,\cdots,16\}$ with the interactions binding each block indicated by a thin line.

The hierarchical model was re-invented by [Baker, Jr., 1972] for the explicit purpose of exhibiting an exactly solvable model to which the [Wilson, 1971] renormalization program did apply. The analysis of a class of models of this type was reworked with great mathematical care by [Bleher and Sinai, 1975, Gallavotti and Knops, 1975, Collet and Eckmann, 1978].

One has again a one-dimensional array Λ_N of 2^N sites, but the "spins" are assumed to take continuous rather than discrete values. The Hamiltonian is now:

$$\left.\begin{array}{l} H_{\{f,\Lambda_N\}}(\sigma) = H_{\{0,\Lambda_N\}}(\sigma) + \sum_{k\in\Lambda_N} f(\sigma_k) \\ \text{with} \\ H_{\{0,\Lambda_N\}}(\sigma) = -\sum_{p=0}^{N} \frac{c^p}{2^{2p+1}} \sum_{r=1}^{2^{N-p}} \left(\sum_{k\in B_{p,r}} \sigma_k\right)^2 \end{array}\right\} \quad (13.2.15)$$

where the value of the constant c and the form of the function f are still to be adjusted to ensure thermodynamical stability, in particular the existence of the thermodynamical limit. The hierarchical structure of the Hamiltonian is captured by performing in the Hamiltonian $H_{\{0,\Lambda_{N+1}\}}$ the change of variables

$$\left.\begin{array}{l} \tau_r = \lambda_c^{-1}\,(\sigma_{2r-1}+\sigma_{2r}) \\ \nu_r = \lambda_c^{-1}\,(\sigma_{2r-1}-\sigma_{2r}) \end{array}\right\} \quad \text{with} \quad r\in\Lambda_N \quad \text{and} \quad \lambda_c = 2\,c^{-\frac{1}{2}} \quad (13.2.16)$$

where the variable τ_r is proportional to the sum of the spins σ_k in the block $B_{1,r}$ of length 2 in the array Λ_{N+1}; the value of the scaling constant λ_c is chosen to ensure that (13.2.17) and (13.2.19) below are satisfied. We have:

$$H_{\{0,\Lambda_{N+1}\}}(\sigma) = H_{\{0,\Lambda_N\}}(\tau) - \frac{1}{2}\sum_{r\in\Lambda_N}(\tau_r)^2 \qquad (13.2.17)$$

where the variables ν_k do not appear. The scalar counter-term in the full Hamiltonian $H_{\{f,\Lambda_{N+1}\}}(\sigma)$ – see (13.2.15) – contains sums of functions of the one-spin variables σ_k and requires more focussed attention. For this purpose, consider only the equilibrium expectation

$$\langle A\rangle_\beta = \int \mathrm{d}S\, P_{\{\beta,f,\Lambda_N\}}(S)\, A(S)$$

of the macroscopic variables of the form $A(\sum_{k\in\Lambda_N}\sigma_k)$, where A is any mea-surable function, β denotes the natural temperature and P is the measure corresponding to the Hamiltonian $H_{\{f,\Lambda_N\}}$; i.e.

$$\left.\begin{array}{l} P_{\{\beta,f,\Lambda_N\}}(S) = \\ Z_{\{\beta,f,\Lambda_N\}}^{-1}\,\int\cdots\int \mathrm{d}\sigma_1\cdots\mathrm{d}\sigma_{2^N}\; \mathrm{e}^{-\beta H_{N,f}(\sigma)}\,\delta(S-\sum_{k=1}^{2^N}\sigma_k) \\[2mm] \text{with}\quad Z_{\{\beta,f,\Lambda_N\}} = \int\cdots\int \mathrm{d}\sigma_1\cdots\mathrm{d}\sigma_{2^N}\; \mathrm{e}^{-\beta H_{N,f}(\tau)} \end{array}\right\}. \quad (13.2.18)$$

Upon using the definition (13.2.15) and the recursion relation (13.2.17), one obtains that the probability distribution (13.2.18) satisfies the *renormalized scaling property*:

$$\left.\begin{array}{l} \text{(a)}\;\; P_{\{\beta,f,\Lambda_{N+1}\}}(S) = \lambda_c^{-1}\,P_{\{\beta,\mathcal{N}_{\beta,c}[f],\,\Lambda_N\}}(\lambda_c^{-1}S) \\[2mm] \text{(b)}\;\; \mathcal{N}_{\beta,c}[f](\tau) = \\ \quad -\tfrac{1}{\beta}\ln\{\lambda_c\,\mathrm{e}^{\frac{1}{2}\beta\tau^2}\int \mathrm{d}\nu\,\mathrm{e}^{-\beta[f(\frac{\lambda_c}{2}\tau+\nu)\,+\,f(\frac{\lambda_c}{2}\tau-\nu)]}\} \end{array}\right\}. \quad (13.2.19)$$

where

The most remarkable feature of the exact relation (13.2.19) implies that the Hamiltonian remains in the class defined by (13.2.15) and only involves tuning the one-spin function – f being replaced by $\mathcal{N}_{\beta,c}[f]$ – appearing in the scalar counter-term.

Remark: A motivating analogy with the central limit theorem. To bring up the sense in which the renormalization program can be viewed as a variation on the theme of the central limit theorem – Sect. 5.4 – suppose for an instant that the spin variables of our array of 2^N sites are *independent* and identically distributed, with one-spin density ϱ, with mean $m\equiv\,<\sigma>_\varrho$ and variance $v\equiv\,<(\sigma-m)^2>_\varrho^{\frac{1}{2}}$. || Then (13.2.18) is replaced by

$$P_{\{0,\Lambda_N\}}(S) =$$

$$\left. \int d\sigma_1 \, \varrho(\sigma_1) \int d\sigma_2 \, \varrho(\sigma_2) \cdots \int d\sigma_{2^N} \, \varrho(\sigma_{2^N}) \, \delta(S - \textstyle\sum_{k=1}^{2^N} \sigma_k) \right\} \qquad (13.2.20)$$

and the central limit theorem asserts that the following limit exists (in the weak sense of convergence of expectation values):

$$\Phi_0(S) = \lim_{N \to \infty} P_{\{0,\Lambda_N\}}\left(2^{\frac{1}{2}N}(S + 2^N m)\right) \qquad (13.2.21)$$

and is equal to the Gaussian probability distribution

$$\Phi_0(S) = \frac{1}{\sqrt{2\pi v^2}} \, e^{-\frac{1}{2}(S/v)^2} \quad . \qquad (13.2.22)$$

Note that (13.2.20) can be rewritten as the 2^N–fold convolution product

$$P_{\{0,2^N\}} = \varrho * \varrho * \cdots * \varrho \qquad (13.2.23)$$

where for any two $\mathcal{L}_1$–functions f and g the $\mathcal{L}_1$–function $f * g$ is defined by:

$$f * g(s) \equiv \int_{-\infty}^{\infty} u \, f(s-u) \, g(u) = \int_{-\infty}^{\infty} dv \, f(\tfrac{s}{2}+v) \, g(\tfrac{s}{2}-v) \quad .$$

In particular, when the one-spin distribution is Gaussian – as in (13.2,22) - the probability distribution for the array Λ_N is again a Gaussian, and does satisfy the central scaling property:

$$\left. \begin{array}{l} P_{\{0,\Lambda_N\}}(S) = \frac{1}{\sqrt{2\pi v_N{}^2}} \, e^{-\frac{1}{2}(S/v_N)^2} \qquad \text{with} \quad v_N = |\Lambda_N|^{\frac{1}{2}} v \, , \\[2ex] P_{\{0,\Lambda_{N+1}\}}(S) = \lambda^{-1} P_{\{0,\Lambda_N\}}(\lambda^{-1}S) \quad \text{with} \quad \lambda = 2^{\frac{1}{2}} \quad . \end{array} \right\} \qquad (13.2.24)$$

Compare this to (13.2.19); the comparison is pursued later on. In general, since the convolution product is associative, (13.2.23) entails

$$P_{\{0,\Lambda_{N+1}\}}(S) = [P_{\{0,\Lambda_N\}} * P_{\{0,\Lambda_N\}}](S) \qquad (13.2.25)$$

or equivalently, for any $\lambda > 0$:

$$\mathcal{R}_{\{0,\Lambda_N\}}(S) \equiv P_{\{0,\Lambda_N\}}(\lambda^N S) \quad \models$$

$$\mathcal{R}_{\{0,\Lambda_{N+1}\}}(S) = \lambda^N \int du \, \mathcal{R}_{\{0,\Lambda_N\}}(\tfrac{1}{2}\lambda S + u) \, \mathcal{R}_{\{0,\Lambda_N\}}(\tfrac{1}{2}\lambda S - u). \qquad (13.2.26)$$

The choice $\lambda = 2^{\frac{1}{2}}$ corresponds to the scaling that appears in the central limit theorem; hence (13.2.24) and (13.2.26) – with $\lambda = 2^{\frac{1}{2}}$ – can be viewed as recursion relations for which the limiting Gaussian distribution $P_{\{0,\Lambda_N\}}(2^{\frac{1}{2}N}S)$ is a fixed point.

For any model in which the spins *do* interact, and are thus *not independent* random variables, the central limit theorem cannot be expected to hold in the form just given. It is therefore remarkable that, for the hierarchical model (13.2.15), the following simple generalization of (13.2.26) holds:

$$
\left.
\begin{aligned}
& \mathcal{R}_{\{\beta,f,\Lambda_N\}}(S) \equiv P_{\{\beta,f,\Lambda_N\}}(\lambda_c^N S) \quad \vDash \\[8pt]
& \mathcal{R}_{\{\beta,f,\Lambda_{N+1}\}}(S) = \alpha_N\, e^{\frac{1}{2}\beta S^2} \quad \times \\[8pt]
& \lambda_c^N \int du\, \mathcal{R}_{\{\beta,f,\Lambda_N\}}(\tfrac{1}{2}\lambda_c S + u)\, \mathcal{R}_{\{\beta,f,\Lambda_N\}}(\tfrac{1}{2}\lambda_c S - u)
\end{aligned}
\right\} \quad (13.2.27)
$$

where λ_c is the scaling defined in (13.2.16). Equivalently, (13.2.27) can be rewritten as:

$$
P_{\{\beta,f,\Lambda_{N+1}\}}(S) = \alpha_N\, e^{\frac{1}{2}\beta[\lambda_c^{-(N+1)}S]^2}\, [P_{\{\beta,f,\Lambda_N\}} * P_{\{\beta,f,\Lambda_N\}}](S) \quad . \quad (13.2.28)
$$

Hence the only difference between the recursion relation (13.2.27) – valid for the hierarchical model – and the relation (13.2.26) – valid for the case where the spin variables are independent – is the factor $D(S) \equiv e^{\frac{1}{2}\beta S^2}$; α_N only ensures that the LHS of (13.2.28) is still a probability distribution – specifically that its integral is still equal to 1. In particular, when the coupling constant $c = 0$, $\lambda_c^{-1} = 0$, and (13.2.28) reduces to (13.2.25) as it should. Moreover, in the high temperature limit $\beta \to 0$ the factor in $D(S)$ disappears confirming that the hierarchical model behaves properly: in this limit the thermal agitation prevails on the mechanical interaction, and the spins become independent of one another. Furthermore, as already pointed out, when f is a fixed point of the map $f \to \mathcal{N}_{\beta,c}[f]$ in (13.2.19b), the scaling relation (13.2.19a) is analogous to (13.2.24).

The analogy with the central limit theorem therefore suggests that in the study of the asymptotic behavior of the hierarchical model one should expect (13.2.21) to be replaced by

$$
\Phi_\beta(S) = \lim_{|\Lambda_N|\to\infty} P_{\{\beta,f,\Lambda_N\}} \left(|\Lambda_N|^{\frac{1}{2}\tau}(S + |\Lambda|m) \right) \; ; \; \tau = 2 - \log_2 c \quad (13.2.29)
$$

where we used the substitutions $|\Lambda| = 2^N$, and $\lambda_c = 2\,c^{-\frac{1}{2}}$ as defined in (13.2.16). When the limit (13.2.29) exists and is *not* Gaussian, it is called *critical*.

Critical fixed points for the hierarchical model. To complete the illustration of the renormalization method for the case of the hierarchical model, the existence and properties of fixed points have to be established. For this purpose, it is useful to replace the renormalization operator $\mathcal{N}_{\beta,c}$ – see (13.2.19b) – by an equivalent, but more manageable operator. Consider indeed the following two bijective transformations:

$$\mathcal{E}_\beta : f \to f_\beta \quad \text{and} \quad \mathcal{S}_\beta^{-1} : f_\beta \to \varphi \quad \text{defined by}$$

$$f_\beta(s) = \mathrm{e}^{-\beta f(s)} \quad \text{and}$$

$$\varphi(s) = [4\pi(2-c)/c\beta]^{\frac{1}{2}} \, \mathrm{e}^{\frac{1}{2}s^2} \, f_\beta([(2-c)/c\beta]^{\frac{1}{2}} s) \Bigg\} . \qquad (13.2.30)$$

One verifies that f is a fixed point of $\mathcal{N}_{\beta,c}$ if and only if f_β satisfies

$$f_\beta(s) = \lambda_c \, \mathrm{e}^{\frac{1}{2}\beta s^2} \, [f_\beta * f_\beta](\lambda_c \, s) \qquad (13.2.31)$$

and this happens if and only if φ is a fixed point of the new operator $\mathcal{N}_c$ defined by:

$$\mathcal{N}_c[\varphi](s) = \frac{1}{\pi^{\frac{1}{2}}} \int \mathrm{d}u \, \mathrm{e}^{-u^2} \, \varphi(\tfrac{1}{2}\lambda_c s + u) \, \varphi(\tfrac{1}{2}\lambda_c s - u) \quad . \qquad (13.2.32)$$

Two features of this operator govern the following discussion. First, the operator $\mathcal{N}_c$ depends only on the strength c of the interactions, and not on the temperature β. The latter is reintroduced into the picture by the inverse of the transformations (13.2.30), but only as a deviation from a critical temperature β_c determined by c. Second, however, $\mathcal{N}_c$ still inherits – through the convolution (13.2.31) – the non-linear character of the original renormalization operator $\mathcal{N}_{\beta,c}$. Clearly the constant function $\varphi(s) = 1$ is a fixed point of $\mathcal{N}_c$ for every c in the range of interest to us, namely $1 < c < 2$, which are conditions that come in handy when one established the existence of the thermodynamical limit $[c < 2\,]$ and of a phase transition $[c > 1]$.

For orientation purposes, let us tentatively limit our attention to the linear approximation $\mathcal{D}\mathcal{N}_c(\varphi)$ of $\mathcal{N}_c$ around a solution φ of $\mathcal{N}_c[\varphi] = \varphi$ i.e. $\mathcal{N}_c[\varphi + \delta\varphi] \approx \varphi + \mathcal{D}\mathcal{N}_c(\varphi)[\delta\varphi]$. Mathematically, $\mathcal{D}\mathcal{N}_c(\varphi)$ is the tangent map to $\mathcal{N}_c$ at φ. Analytically, one verifies that $\mathcal{D}\mathcal{N}_c(\varphi)[\delta\varphi](s) = 2\pi^{-\frac{1}{2}} \int \mathrm{d}u \, \mathrm{e}^{-u^2} \varphi(\tfrac{1}{2}\lambda_c s + u)\, (\delta\varphi)(\tfrac{1}{2}\lambda_c s - u)$; thus, in particular, for the solution $\varphi = 1$ of $\mathcal{N}_c[\varphi] = \varphi$:

$$\mathcal{D}\mathcal{N}_c(1)[\delta\varphi](s) = 2\pi^{-\frac{1}{2}} \int \mathrm{d}u \, \mathrm{e}^{-(\frac{1}{2}\lambda_c s - u)^2} (\delta\varphi)(u) \quad . \qquad (13.2.33)$$

It is then straightforward to verify that for every $n \in \mathbf{Z}^+$:

$$\mathcal{D}\mathcal{N}_c(1)[\psi_c^{(n)}](s) = \lambda_c^{(n)} \, \psi_c^{(n)}(s) \qquad (13.2.34)$$

where the eigenfunctions $\psi_c^{(n)}$ are the normalized Hermite polynomials

$$\psi_c^{(n)}(s) = h_n((1 - c^{-1})^{\frac{1}{2}} s) \quad \text{with} \quad \begin{cases} h_n(s) = (2^n \, n!)^{\frac{1}{2}} H_n \\[2mm] H_n(s) = (-1)^n \mathrm{e}^{s^2} \frac{\mathrm{d}^n}{\mathrm{d}^n s} \mathrm{e}^{-s^2} \end{cases} ;$$

and where the eigenvalues are $\lambda_c^{(n)} = 2\,c^{-\frac{1}{2}n}$. In particular, for all values of c : $\lambda_c^{(o)} = 2$ and $\lambda_c^{(1)} = 2c^{-\frac{1}{2}} = \lambda_c$, where λ_c is precisely the scaling factor introduced in (13.2.16). Since $\{\psi_c^{(n)} \mid n \in \mathbb{Z}^+\}$ is an orthonormal basis in the Hilbert space $\mathcal{H}_c \equiv \mathcal{L}^2(\mathbb{R}, d\mu_c)$ with $d\mu_c(s) = e^{(1-c^{-1})s^2} ds$, $\mathcal{DN}_c(1)$ can be viewed as a self-adjoint operator, mapping $\mathcal{H}_c$ into itself, and with simple discrete spectrum

$$\mathrm{Sp}\,[\mathcal{DN}_c(1)] = \{\lambda_c^{(n)} = 2\,c^{-\frac{1}{2}n} \mid n \in \mathbb{Z}^+ \cdots\} \subset (0,2] \quad . \qquad (13.2.35)$$

For any $c \in (1,2)$, the spectrum (13.2.35) contains finitely many eigenvalues larger than 1 and infinitely many eigenvalues smaller than 1, $\varphi = 1$ is hyperbolic, with finitely many unstable directions and infinitely many stable directions.

In addition, there may exist one zero eigenvalue: this happens exactly when the coupling constant c hits one of the special values $\{2^{\frac{2}{n}} \mid n = 3, 4, \cdots\}$. For instance, if $c = 2^{\frac{2}{n}}$, there are n unstable directions, corresponding to $2 = \lambda_c^{(o)} > \lambda_c^{(1)} > \cdots > \lambda_c^{(n-1)} > 1$, while $\lambda_c^{(n)} = 1$, and $\forall k > n : \lambda_c^{(k)} < 1$.

The analysis of $\mathcal{DN}_c(1)$ therefore sugests that, at a particular value of the coupling constant c for which one of the $\lambda_c^{(n)}$ vanishes, a new solution of $\mathcal{N}_c[\varphi] = \varphi$ may bifurcate away from the trivial fixed point $\varphi = 1$. The proof that this really happens – namely that such a solution exists and has smooth analytic properties – requires some hard analysis that goes beyond the elementary perturbation theory of bifurcations; this analysis is provided in [Collet and Eckmann, 1978]. We summarize their results.

First of all, a disclaimer: the above conjecture is borne out only when n is even, i.e. for $n = 2k$ and thus $c \in \{2^{\frac{1}{k}} \mid k = 2, 3, \cdots\}$. The first putative point where a new solution may branch off is therefore $n = 4$ i.e. $c = 2^{\frac{1}{2}}$; since it turns out to be typical, we can lighten the notation, and restrict our attention to this branching point.

Second, the existence and uniqueness of a non-trivial fixed point can be proven only in a finite, but small, neighborhood of our branching point, and the notation $c(\varepsilon) = 2^{\frac{1}{2}(1-\varepsilon)}$ reminds us of this. We denote by φ_o the trivial fixed point $\varphi_0 = 1$.

The main result is that for all ε positive and sufficiently small, there exists a function $\varphi_\varepsilon : s \in \mathbb{R} \to \varphi_\varepsilon(s) \in \mathbb{R}$ smooth in ε and s, such that $\mathcal{N}_{c(\varepsilon)}[\varphi_\varepsilon] = \varphi_\varepsilon$ and $\varphi_\varepsilon \neq \varphi_o$, with φ_ε branching off smoothly from φ_o at $c(0) = 2^{\frac{1}{2}}$. φ_ε decreases like $e^{-\frac{1}{2}\varepsilon\theta s^4}$ as $|s| \to \infty$, for some constant $\theta \in (0,1)$ the precise value of which is known, but not essential beyond the central claim, namely that φ_ε is critical. Upon recalling the defining relations (13.2.30), the physicist as well as the probabilist will be interested in noting with [Collet and Eckmann, 1978] that the distribution density $e^{-\frac{1}{2}s^2}\varphi_\varepsilon$ is *not* infinitely divisible; thus it *cannot* be written as a limit of sums of *independent* random variables – for the definition and properties of infinitely divisible distributions in the context of the techniques associated to the central limit theorem, the Reader may consult [Feller, 1968, 1971].

Pursuing the analysis one step further, one can show that the spectrum of $\mathcal{DN}_{c(\varepsilon)}(1)$ changes controllably little when one passes from the trivial solution $\varphi = 1$ with $c = 2^{\frac{1}{2}}$ to the bifurcated solution φ_ε corresponding to $c(\varepsilon) = 2^{\frac{1}{2}(1-\varepsilon)}$ with $\varepsilon > 0$; the spectrum remains discrete, and there are still four eigenvalues larger than 1, corresponding thus again to four stable directions; perturbation computations provide these eigenvalues: $\{\lambda_{c(\varepsilon)}^{(n)} \approx 2c(\varepsilon)^{-\frac{1}{2}n} \mid n = 0, 1, 2, 3\}$. In the sequel, we will need to know that, for the first two eigenvalues, this approximation turns out to be exact: $\lambda_{c(\varepsilon)}^{(0)} = 2$, and $\lambda_{c(\varepsilon)}^{(1)} = \lambda_{c(\varepsilon)}$. For the eigenvalue $\lambda_{c(\varepsilon)}^{(2)}$ the approximation is good only for small ε; indeed while this eigenvalue is smooth in ε, starts exactly at $\lambda_{c(0)}^{(2)} = 2c(0)^{-1} = 2^{\frac{1}{2}}$, and at first increases for a little while, it soon reaches a maximum from which it decreases monotonically the limiting value $1 \neq 2 = 2c(1)^{-1}$ as ε approaches 1. Moreover, the fifth eigenvalue is now strictly smaller than 1, and thus all – but the first four – eigenvalues correspond to stable directions.

Additional – precise and technically important – information on the global properties of the flow of the renormalization map $\mathcal{N}_c$ is also presented in [Collet and Eckmann, 1978].

The critical exponents of the hierarchical model. The temperature has to be brought back into the picture. This is done in a somewhat round-about way by tuning the parameters c and f that define the Hamiltonian of the hierachical model (13.2.15) in such a manner that it has a preassigned critical temperature. To this effect, choose $\varepsilon > 0$ sufficiently small, and let $c(\varepsilon) = 2^{\frac{1}{2}(1-\varepsilon)}$; this fixes the first parameter in (13.2.15). Next choose a positive number $\beta_o > 0$ which will play the role of a critical temperature for the Hamiltonian still to be specified by a properly tuned f. One single restriction is imposed on the choice of β_o, namely that it be different from $4\pi e\,(2-c(\varepsilon))\,c(\varepsilon)^{-2}$. To complete the specification of f, consider a sufficiently small neighborhood $\mathcal{W}$ of the fixed point φ_ε. This manifold – the elements of which are functions φ on which the renormalization map acts – splits into a finite dimensional unstable manifold $\mathcal{U}$ and an infinite dimensional stable manifold $\mathcal{S}$; in particular, any function $\varphi_S \in \mathcal{S}$ gets closer and closer to φ_ε under the successive iterations of the renormalization map. One last technicality: it is possible to choose $\varphi_S \in \mathcal{S}$ such that it satisfies the following five conditions: (i) $\varphi_S > 0$, (ii) φ_S is continuously differentiable, (iii) and (iv) the functions $x \to x\frac{d}{dx}[\varphi_S](x)[\varphi_S(x)]^{-\frac{1}{2}}$ and $x \to \log[\varphi_S](x)x\frac{d}{dx}[\varphi_S](x)[\varphi_S(x)]^{-\frac{1}{2}}$ are measurable and essentially bounded, (v) the ess-sup norm (see Example (D.2.3) $\| \frac{d}{dx}[\varphi_S - \varphi_\varepsilon\|_\infty$ is small. Provided these conditions are satisfied, the ultimate results do not depend on the choice of φ_S, one of the nice features of the situation, known as *universality*. The function f_S is then defined by inverting the bijective transformations (13.2.30):

$$f_S = \mathcal{E}_{\beta_o}^{-1} \circ \mathcal{S}_{\beta_o}[\varphi_S] \ . \tag{13.2.36}$$

The Hamiltonian is specified by inserting in (13.2.15) $c(\varepsilon)$ for c and f_S for f.

The introduction of the temperature β with $\beta \in V_o$ where V_o is a small neighborhood of β_o changes $e^{-\beta_o fs}$ to $e^{-\beta fs}$ i.e. $\mathcal{E}_{\beta_o}[fs]$ to $\mathcal{E}_\beta[fs]$; this transformation is traced in the φ—function space by the transformation

$$\left.\begin{aligned}\varphi s \to \varphi_\beta &\equiv \mathcal{S}_\beta^{-1} \circ \mathcal{E}_\beta[fs] = \mathcal{S}_\beta^{-1} \circ \mathcal{E}_\beta \circ \mathcal{E}_{\beta_o}^{-1} \circ \mathcal{S}_{\beta_o}[\varphi s] \\ &= \mathcal{S}_\beta^{-1}\left(\mathcal{S}_{\beta_o}[\varphi s]\right)^{\frac{\beta}{\beta_o}}\end{aligned}\right\} . \qquad (13.2.37)$$

The curve traced in the φ—function space by $\beta \to \varphi_\beta$ can be shown to be differentiable and to intersect the stable manifold *transversally* at φ_S. Hence β can be used as a local coordinate in the φ—function space.

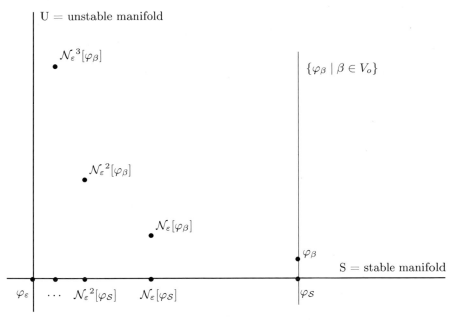

Fig. 13.3. The renormalization flow: straightened picture of the flow under $\mathcal{N}_\varepsilon$ around φ_ε; only one quadrant shown: similar patterns occur in all four quadrants depending on the choice of φ_o and the sign of $\beta - \beta_o$.

The argument leading to (13.2.37) allows to rewrite the renormalization map in the physical space of the f—functions as:

$$f \to \mathcal{N}^\beta[f]$$

with

$$\left.\mathcal{N}^\beta[f] = \mathcal{S}_\beta^{-1} \circ \mathcal{E}_\beta \circ \mathcal{N}_{c(\varepsilon)} \circ \mathcal{E}_\beta^{-1} \circ \mathcal{S}_\beta[f]\right\} \qquad (13.2.38)$$

which the following diagram describes:

$$\varphi_\beta \xrightarrow{\;\mathcal{N}_\varepsilon\;} \mathcal{N}_\varepsilon[\varphi_\beta]$$

$$\mathcal{S}_\beta{}^{-1}\mathcal{E}_\beta \uparrow \qquad\qquad \downarrow \mathcal{E}_\beta{}^{-1}\mathcal{S}_\beta$$

$$f_S \xrightarrow{\;\mathcal{N}^\beta\;} \mathcal{N}^\beta[f_S]$$

A pattern similar to that depicted in Fig. 13.3 holds then in the space of the structure functions f defining the Hamiltonian. Note that this plays here the role played elsewhere (see e.g. Sect. 13.1) by the space referred to as the space of coupling constants. As the renormalization map is repeatedly applied, the system grows larger and thermodynamical functions computed at a temperature β near β_o approach corresponding values moving away on the unstable manifold. To capture them back, one let simultaneously $\beta \to \beta_o$. Clearly different thermodynamical functions involve different directions in the unstable manifold, and thus different rates as measured by the different eigenvalues $\lambda^{(n)} > 1$ in the spectrum of $\mathcal{DN}_{c(\varepsilon)}(1)$. With the critical exponents defined generically for an observable A by

$$\lim_{\beta\to\beta_o} [\log \lim_{N\to\infty} \langle A \rangle_{\beta,f,\Lambda_N}][\log|\beta - \beta_o|]^{-1} \qquad (13.2.39)$$

where the limit as $N \to \infty$ is controlled by the renormalization map; [Collet and Eckmann, 1978] compute the following values for some of the traditional critical exponents

$$\boxed{\begin{aligned} \beta &= [\log c(\varepsilon)]/[2 \log \lambda^{(2)}(\varepsilon)] \\[2mm] \gamma &= [\log 2 - \log c(\varepsilon)]/[\log \lambda^{(2)}(\varepsilon)] \\[2mm] \delta &= [2 \log \lambda^{(1)}(\varepsilon)]/[\log c(\varepsilon)] \\[2mm] \eta &= 1 + [\log c(\varepsilon)]/[\log 2] \end{aligned}} \qquad (13.2.40)$$

Since $\lambda^{(1)}$ and $\lambda^{(2)}$ are computable functions of the coupling constant $c(\varepsilon)$, (13.2.40) gives formula for each of the critical exponents *separately*.

To make the connection with the scaling laws obtained in Sect. 13.1, we can eliminate $\lambda^{(1)} = 2c(\varepsilon)^{-\frac{1}{2}}$ and $\lambda^{(2)}$ from (13.2.40) to express two of these coefficients in terms of the other two, e.g.:

$$\boxed{\eta = \frac{4\beta + \gamma}{2\beta + \gamma} \quad \text{and} \quad \delta = \frac{\beta + \gamma}{\beta}} \qquad (13.2.41)$$

The relations (13.2.41) are not new, but they are precisely on target: indeed, they also result from the elimination of α and ν from the general scaling relations (13.1.6) and (13.1.13) with $d = 1$. Moreover, in the limit $\varepsilon \to 0$ and thus $c \to 2^{\frac{1}{2}}$ – but in this limit only – the critical exponents listed above reduce to the classical values: $\beta = \frac{1}{2}, \gamma = 1, \delta = 3$ which are the classical values of these exponents – see (11.3.17–19) – and $\eta = \frac{3}{2}$ which is not, a reflection of the price one must expect to pay ultimately for limiting one's attention to the $1-$d hierarchical model.

With these results we close our overview of the renormalization idea as they apply without uncontrolled approximation to the one-dimensional hierarchical model.

Nevertheless, we should mention that much more is done in [Collet and Eckmann, 1978], in particular the analysis in the large – rather than the local analysis reviewed here – of the global flow under the renormalization map; this bears among other things on the existence and properties of the thermodynamical limit and distinct thermodynamical phases. For an updated discussion, see [Bleher and Major, 1987].

The Reader will also find in [Baker, Jr., 1990] and references therein, results pertaining to the extension of the method to *higher-dimensional* versions of the hierarchical model. One may certainly argue that the results obtained for the critical exponents of the one-dimensional hierarchical model do not match the empirical values obtained in real materials. This however is not the point of our argument. Indeed we wanted to *illustrate the method with a particular model* which can be analyzed in all mathematical rigor; which does not involve any prescription beyond the specification (13.2.15) of the Hamiltonian; and which provides values for individual exponents rather than just putative relations between them, thus providing a check that these relations pertain genuinely to the realm of statistical mechanics.

The usefulness of hierachical couplings is not limited to the analysis of classical models; for instance, a generalization to quantum oscillators is presented in [Albeverio, Kondratiev, and Kozitsky, 1997].

The value of the renormalization program itself extends beyond this particular exactly solvable model. Other, more realistic models have been amenable through various approximations to a treatment by renormalization methods to provide empirically reliable information [Wilson, 1971, Wilson and Fisher, 1972, Wilson and Kogut, 1974, Wilson, 1975, Fisher, 1974]. A personal perspective on that line of research is offered in [Fisher, 1998]. The probabilistic approach to the renormalization program – which we chose to follow here – was initiated by [Di Castro and Jona Lasinio, 1969]; a synthesis of its developments is presented in [Benfatto and Gallavotti, 1995], the broad sweep of which informs a wide array of related ideas and techniques.

The more traditional physics literature offers a somewhat bewildering range of more or less pragmatic texts, such as [Pfeuty and Toulouse, 1977, Patashinskii and Pokrovskii, 1979, Amit, 1984, Parisi, 1988, Goldenfeld, 1992, Cardy, 1996]. These either build – or are built on – analogies between the process of iterating the renormalization map, which we reviewed in this section to account for the divergences that appear near the critical point, and a program originally invented for the purpose of dealing with the divergences of quantum field theory in four dimensions. Although the divergences themselves have different physical origins, the importance of formal relations between these two domains – as well as others, such as KAM theory – for a primary understanding of either of them is still debated; contrast for instance [Fisher, 1998], [Benfatto and Gallavotti, 1995] and the rich crop of supportive references cited there.

In order to place succinctly these in a historical perspective, one traces back the idea of a renormalization "group" in QFT to [Stuekelberg and Petermann, 1953, Gell-Mann and Low, 1954] via [Bogoliubov and Shirkov, 1959][Chap. VIII]; for a vernacular presentation of the original formulations of QFT renormalization, see [Schweber, 1994]; and for an overview of the related mathematical issues there – just before the cross-over to the study of critical exponents in classical statistical mechanics – see [Hepp, 1969].

The scope of renormalization methods has become so encompassing during the last thirty years, that it seems bound to stay as an integral part of our understanding of the coexistence of the different scales on which the world operates ... and thus must be apprehended and understood. In the lattice spin systems we considered, there are at least three such scales: the microscopic scale given by the lattice spacing, the mesoscopic scale characterized by the correlation length, and the macroscopic scale on which the naked eye usually operates. For the hierarchical model of phase transitions and critical phenomena, this coexistence is manifestly built in the Hamiltonian itself; however, for most physical systems, the focusing on a particular scale requires insights and the mediation of approximations that are more difficult to control. The renormalization program can be viewed as providing the "coarse graining" necessary to bridge from the micro- to the meso-scopic scales.

14. Quantum Models for Phase Transitions

In mathematics ambiguity is a sin, and contradiction a disaster. In physics a certain amount of ambiguity is a fact of life, and a contradiction may well be the beginning of an important new development.

[Kac, 1979]

14.1 Superconductivity and the BCS Model

14.1.1 Preliminaries to the BCS Model

The BCS model for superconductivity was proposed in [Bardeen, Cooper, and Schrieffer, 1957]. Two authoritative reviews on superconductivity were published just one year before the model appeared: [Serin, 1956] surveys the experimental evidence; and [Bardeen, 1956] presents the latest understanding of the phenomena. Both of these aspects as well as the immediate impact of the BCS model are discussed in the Nobel lectures [Bardeen, 1972, Cooper, 1972, Schrieffer, 1972] and the textbook [Schrieffer, 1964].

By the time of the experimental discovery of superconductivity [Kamerlingh Onnes, 1911], it had been recognized that electrons are the charge carriers responsible for the electric conductivity in metals, but not much more was known. [Lorentz, 1905] modeled their behavior as that of billiard balls bouncing against fixed atoms – or molecules. [Brush, 1983] mentions that the ideas of kinetic theory were brought to bear on this line of thought by [Bohr, 1911] in a doctoral dissertation, which however long remained unpublished; its importance is therefore mostly as a historical marker: if the electrons are hampered in their motion by thermal agitations in the metal, this could be the beginning of the explanation of a commonly observed decrease of resistivity in metals as the temperature is lowered. Both of these works resided within the confines of classical theory. That something was still missing is reflected in the presentation address made at the occasion of Kamerlingh Onnes' Nobel prize:

It has become more and more clear that a change in the whole theory of electrons is necessary. Theoretical work has already begun by a number of research workers, particularly Planck and Einstein.

[Nordström, 1913]

In particular, Lorentz and Bohr could hardly be expected to avail themselves of the interpretation of the specific heat of solids based on the quantum treatment of the vibrations of atoms arranged in a crystal lattice; see our account in Sect. 10.3 of the contributions of [Einstein, 1907, 1911, Debye, 1912]. Even further away stood the quantum theory of electrons in metals [Bloch, 1928], according to which a perfectly periodic crystal offers no resistance to an electric current, so that resistance cannot be attributed to their interaction directly with the atoms or molecules of the crystal but, rather, it should be sought in their interactions with the vibrations of the crystal at $T > 0$ K and with the foreign impurities that may be present in the crystal. In his Nobel lecture, [Kamerlingh Onnes, 1913] explains that, while collecting low-temperature data on the gradual decrease of the electrical resistivity of Mercury, he noticed that the latter vanishes below a certain temperature:

> the experiment left no doubt that, as far as accuracy of measurement went, the resistance disappeared. At the same time, however, something unexpected occurred. The disappearance did not take place gradually but ... abruptly. From 1/500 the resistance at 4.2 K drops to a millionth part. At the lowest temperature, 1.5 K , it could be established that the resistance had become less than a thousand-millionth part of that at normal temperature.

And he proceeds to suggest the name that has stuck since then:

> Thus the mercury at 4.2 K has entered a new state, which, owing to its particular electrical properties, can be called the state of superconductivity.

Kamerlingh Onnes already knew that this phenomenon was not limited to Mercury: by the time of his Nobel lecture, he already had ascertained that Tin and Lead also become superconductive at a transition temperatures he gave as 3.8 K $for Tin, and "probably''$ 6 K for Lead. He also verified, shortly afterwards and reported as an Addendum to his Nobel lecture, that the vanishing of the resistance implies that in the absence of any external source, currents will persist in a closed wire of superconductive material:

> The current, once produced, lasted for hours practically unchanged in the superconductive coil. Heating the latter above the transition point immediately destroys the electromagnetic field borne by the circuit.

This persistence certainly captured attention when a supercooled loop carrying such a current was transported soon thereafter from Leiden to Cambridge!

> [This] was later verified with an enormous accuracy. A lead ring carrying a current of several hundred amperes was kept cooled for a period of 2 1/2 years with no measurable change in the current.
>
> [Lundqvist, 1972]

Another discovery is also reported as a footnote in the printed text of Kamerlingh Onnes' Nobel lecture, namely that an external magnetic field, if strong enough, could destroy superconductivity:

> experiments then planned were carried out after this lecture was given and produced surprising results ... a threshold value (for lead at the boiling point

of helium 600 Gauss) [was observed] ... *In fields above this threshold value a relatively large resistance arises at once, and grows considerably with the field.*

The dependence of this critical field $H_c(T)$ on the temperature T was later found to follow closely the law:

$$\left[1 - \left(\frac{T}{T_c}\right)^2\right] - \frac{H_c(T)}{H_c(0)} = 0 \qquad \text{for} \quad T < T_c \quad . \tag{14.1.1}$$

While this law served as a stimulus for phenomenological, thermodynamical models of superconductivity, such as the *two-fluid* model of [Gorter and Casimir, 1934], which derives (14.1.1) and links it to the behavior of the specific heat in the superconductive phase, namely:

$$C_s \propto \left(\frac{T}{T_c}\right)^3 \quad . \tag{14.1.2}$$

[Bender and Gorter, 1952, Marcus and Maxwell, 1953](and ref. therein) showed that some adaptations of the two-fluid model were possible to accommodate some of the new experimental data [Keesom and Van Laer, 1938, Brown, Zemansky, and Boorse, 1952, Wexler and Corak, 1952, Corak and Satterthwaite, 1956]; for instance, small deviations – with an observed maximum ≈ 0.03 – were observed in (14.1.1) for which a theoretical upper bound was to be obtained later in [Bardeen, Cooper, and Schrieffer, 1957].

Let us back up for a little while and mention now that the persistent currents discovered by Kamerlingh Onnes did not come without their own set of problems.

Gedanken experiment 1. Suppose that we lower a small magnetized bar of ordinary ferromagnetic material down towards a bowl of superconducting material below its critical temperature. This would induce in the superconductor a large and persisting electric current which creates a magnetic field in the direction opposite to that of the ferromagnetic magnet, resulting in a repulsion that could keep the magnet levitating at a distance above the surface of the superconductor.

Gedanken experiment 2. Take a long, solid cylinder of superconducting metal – other than a ferromagnetic material – and maintain it *above* its critical transition temperature. Then apply a spatially homogeneous magnetic field to this normal conductor, perpendicular to the axis of the cylinder. By Faraday's law of induction, electric currents are induced and their associated magnetic field opposes the change; these currents however are very quickly dissipated due to the finite resistance of the metal. Thus, a magnetic flux would be established through the bulk of the material. Now, lower the temperature *below* its critical value. Finally, switch the applied magnetic field off. Surface currents would again be induced. However, the material being now a perfect conductor, these surface currents will persist forever and will trap the magnetic flux inside the material. Hence, the state of a superconducting metal below its critical temperature would depend on its history, and not just on the permanent value of external constraints – applied magnetic field and temperature – and the usual tenets of equilibrium thermodynamics would *not* suffice to describe its thermodynamical behavior.

Some twenty years after the discoveries of Kamerlingh Onnes, [Meissner and Ochsenfeld, 1933] showed how the second of the above Gedanken experiments does *not* model the empirical situation in the experimental physicist's laboratory. Specifically: (a) they verified that above the critical temperature, the magnetic lines of force pass through the material unhindered and present the expected pattern of unhindered parallel lines; (b) however, once the temperature has been lowered below the critical transition value, and before switching the applied magnetic field off, they observed that the magnetic lines of force in the vicinity of the superconductor bent around it – piling closer together at the equator – very much in the same way that do the laminar flow lines of a fluid around an obstacle; (c) they repeated the experiment with a hollow cylinder and found that inside the hole the parallel magnetic field persisted, while the field at the exterior of the cylinder presented the avoidance pattern just described. They concluded that while the experiment results obtained with a solid cylinder could have been explained by assuming that the permeability of the material had vanished, this assumption would imply, contrary to the observations, that no field line could end non-tangentially on the inner surface of the hole in the superconductor.

The phenomenon discovered by [Meissner and Ochsenfeld, 1933] is known as the *Meissner effect* and it is described by saying that, below their transition temperature, superconductors show *perfect diamagnetism*, by which one means that – except for a microscopic surface layer – no magnetic field can exist within the bulk of a superconductor.

Having provided an empirical refutation to the Gedanken experiment 2, the Meissner effect removes the principal objection to using the known thermodynamical formalism to describe superconductivity. The Meissner effect, however, sustains the Gedanken experiment 1 , which can be credibly implemented in the laboratory and has now become the favored demonstration of the Meissner effect: the Reader will find a photograph of such a levitating magnet in the article "Superconductivity" of the popular Microsoft Encarta 98 Encyclopedia.

As for the thickness of the superficial layer – known as the *penetration depth* – it is defined by:

$$\lambda \equiv [\int_{o}^{L/2} dx \ H(x)]^{-1} \int_{o}^{L/2} dx \ x \ H(x)$$

where x is the distance from the boundary and L is the linear dimension of the material; when the decay is exponential, i.e. when $H(x) = H(0) \exp(-x/\lambda)$

with $\lambda << L$, λ can be computed from $\lambda = [H(0)]^{-1} \int_o^\infty dx\, H(x)$ which is the expression used by theoretical physicists. The temperature dependence of the penetration depth, predicted from the two-fluid model, is given by:

$$\lambda(T) = \lambda(0) \left[1 - \left(\frac{T}{T_c} \right)^4 \right]^{-\frac{1}{2}} .\tag{14.1.3}$$

This behavior was confirmed experimentally – see e.g. the last paper in the series [Biondi, Garfunkel, and Coubrey, 1956] for Aluminium, for which it was found that $\lambda(0) \approx 5 \times 10^{-6} cm = 500$ Å . This is indeed very small compared to the bulk dimensions of the usual samples. Note also that (14.1.3) gives that $\lambda(T)$ becomes infinite as $T \nearrow T_c$: no magnetic flux is excluded from the bulk of the material at $T = T_c$.

Motivated explicitly by the Meissner effect, a phenomenological theory was proposed in [London and London, 1935], but in spite of its many virtues, it predicted a penetration depth that was too small by one or two orders of magnitude, Yet, the discovery of the Meissner effect produced valuable theoretical insights. In particular, by that time, much progress had been achieved in the understanding of metals from the point of view of quantum theory, e.g. [Bloch, 1928].

Perceptive speculations about the microscopic nature of phenomena had been advanced as early as [London, 1935] – which were taken up again in [London, 1950] – such as a vision that superconductivity involves *"a kind of solidification or condensation of the average momentum distribution"* or the almost prophetic expectation of an energy gap – the centerpiece of the theory that was twenty years away – namely that *the electrons be coupled by some form of interaction in such a way that the lowest state may be separated by a finite interval from the excited ones.* In addition, a prediction was made, according to which the flux passing through a hollow superconducting cylinder would be quantized in integral multiples of the universal constant $(hc/2e)$ where h is the Planck constant, c the velocity of light, and e the charge of the electron; in fact the factor 2 is a later refinement of London's prediction, due to [Onsager, 1961]; flux quantization was observed, including the factor 2, by [Deaver and Fairbank, 1961]; compare with [Doll and Näbauer, 1961].

Several developments then occurred precipitously. [Fröhlich, 1950] recognized the key to the "condensation" in the interaction between the charge carriers – the electrons – and the vibrations of the lattice – the phonons – and, specifically, that the

> *virtual emission or absorption of the quantum ... may give rise to an interaction between electrons.*

On that basis he predicted the *isotope effect,* that was observed independently but essentially simultaneously: [Maxwell, 1950, Reynolds, Serin, Wright, and Nesbitt, 1950, Serin, Reynolds, and Nesbitt, 1950]. The critical temperature T_c of four isotopes of Mercury was found to vary with the atomic mass M, and to satisfy

$$T_c\sqrt{M} = C \tag{14.1.4}$$

where C is nearly a constant. The atomic mass M is a quantity that belongs to the atoms of the metal, rather than to the charge carriers – the electrons – that are freely moving through the metal: it enters into the frequency of the vibrations of the lattice as $M^{-\frac{1}{2}}$.

For the model builders of the 1950s the isotope effect served as a guide towards the understanding of the nature of the excitations to which the energy gap was to be associated.

In spite of its success with the prediction of the isotope effect, the Fröhlich scheme presented some serious analyticity difficulties; it follows from [Schafroth, 1951] that the energy gap, in terms of the electron-phonon coupling constant g, behaves like $\exp(-1/g^2)$, a non-analytic function, all derivatives of which vanish at $g = 0$, rendering futile any attempt at a Taylor series expansion and thus ruling out any sensible approach to the Meissner effect by perturbation techniques to any order. Bardeen took up – and to some extent anticipated – the challenge. He developed a theoretical frame for the [Fröhlich, 1950] idea in a series of papers [Bardeen, 1950] that culminated into his treatment of the Meissner effect [Bardeen, 1955], where he argued that a correct account of the thermodynamical properties of superconductors "will most likely" include the understanding of the Meissner effect; the model presented there is a modified degenerated electron gas with a temperature-dependent energy gap, in a function of which the penetration depth could be evaluated.

Schafroth [1954] pointed out that electron pairs, held together by an attractive interaction, would obey Bose statistics and thus be capable of Einstein condensation (Sect. 14.2). A likely source for the requisite attractive interaction was the mechanism proposed by [Fröhlich, 1950]. A more detailed mechanism was proposed by [Cooper, 1956] for the attraction between electrons, due to their interactions with the phonons, resulting in the formation of independent pairs of bound electrons, the pairs that now bear his name; his theory had an energy gap of the order of kT_c.

While the theoretical activity just reported was going on, the early 1950s saw an avalanche of new experimental results; these demanded quite urgently a comprehensive model through which the nature of this energy gap could be formulated and analyzed quantitatively. We review briefly some of empirical evidence.

The measurement of the amplitude and temperature dependence of the energy gap proceeded through a variety of techniques.

The electronic specific heat in the superconductive phase was found [Corak, Goodman, Satterthwaite, and Wexler, 1956, Corak and Satterthwaite, 1956] to behave like

$$C_{es} \approx \exp(-b(T_c/T)) \quad \text{near} \quad T = 0\,\text{K} \tag{14.1.5}$$

rather than like the behavior (14.1.2) predicted by the two-fluid model. The exponential relation (14.1.5) was construed to favor instead a single-electron model of a superconductor involving a gap ε of the order of kT_c up to perhaps $3kT_c$ per electron in the spectrum of available energy levels; compare to [Bardeen, 1955].

High frequency electromagnetic radiation is absorbed by superconductors [Blevins, Gordy, and Fairbank, 1955, Biondi, Garfunkel, and Coubrey, 1956, Glover III and Tinkham, 1956, 1957], and it was immediately realized that a theoretical model for these experiments requires an energy gap [Cunningham, 1956, Tinkham, 1956]; for a review of the experimental evidence, see [Biondi, Forrester, Garfunkel, and Satterthwaite, 1958].

Further confirmation of the existence and properties of the energy gap were obtained by measurements of nuclear spin relaxation [Hebel and Slichter, 1957, Redfield, 1959]; here the experimental set-up had to be astute enough to finesse the Meissner effect. The experimental data so collected were seen to be in direct conflict with the two-fluid model. While the first data could be qualitatively explained by a one-electron energy-gap model, predictions of this model were found unable to account simultaneously for the experimental evidences from both nuclear spin relaxation and ultrasonic absorption; that problem went away with the appearance of the electron-pair BCS model, the predictions of which were found to be in essential agreement with both sets of laboratory data; see also [Morse, 1959].

Finally, electron tunneling between two films of superconductive material separated by a thin oxide layer allowed a precise measure of the temperature dependence of the energy gap [Giaever, 1960] which confirmed [Bardeen, Cooper, and Schrieffer, 1957]. This clever experimental technique has been recognized by yet another Nobel prize for work related to superconductivity [Giaever, 1973]; see also [Josephson, 1962].

At the time of the [Bardeen, Cooper, and Schrieffer, 1957] proposal, the experimental evidence was sharp enough to select among a few competing models.

14.1.2 The Essentials of the BCS Model

This subsection describes: (1) the ingredients of the model; (2) its mean-field approximation; (3) why this approximation becomes exact in the thermodynamical limit; (4) the gauge-invariance and spontaneous symmetry breaking.

We start with the finite-volume Hamiltonian of the BCS model where the system is enclosed in a finite cubic box Λ of edge with length $2L = V^{\frac{1}{3}}$:

$$H_\Lambda = \sum_{p,s} \varepsilon(p) a_s(p)^* a_s(p) + \sum_{p,q} b(p)^* \, \tilde{v}(p,q) \, b(q) \quad . \tag{14.1.6}$$

The $a_s(p)^*$ implement the creation of electrons of momentum p and spin s up ($\uparrow$) or down ($\downarrow$); and the $a_s(p)$ their annihilation. Thus $n_s(p) = a_s(p)^* a_s(p)$

gives the number of electrons of momentum p and spin s, and $\sum_{p,s} \varepsilon(p) n_s(p)$ gives the energy of the electrons when they are free. Similarly, $b(p)^* \equiv a_\uparrow(p)^* a_\downarrow(-p)^*$ creates a pair of two electrons of opposite momentum and spin, a so-called *Cooper pair,* and $b(p)^* \, \tilde{v}(p,q) \, b(q)$ describes the energy transfer during a collision between an impinging Cooper pair – annihilated by $b(q)$ – and an emerging Cooper pair – created by $b(p)^*$.

The electron-phonon interaction enters through the form of $\tilde{v}$. We assume only:

$$\left. \begin{array}{c} \tilde{v}(p,q) = V^{-1} \int \mathrm{d}\xi \mathrm{d}\eta \, e^{ip\xi} \, v(\xi,\eta) \, e^{-iq\eta} \quad \text{with} \\[2mm] \sum_q |\tilde{v}(p,q)| < \infty \; ; \; v \equiv \int \mathrm{d}\xi \mathrm{d}\eta |v(\xi,\eta)| < \infty \; ; \; v(\xi,\eta)^* = v(\eta,\xi) \\[2mm] \text{and thus}: \quad |\tilde{v}(p,q)| < (v/V) \to 0 \quad \text{as} \quad V \to \infty \;\; . \end{array} \right\} \quad (14.1.7)$$

Mathematically, the $a_s(p)^*, a_s(p)$ are elements of an involutive algebra with unit I. This algebra is generated by elements of the form $a_s(f)$ and their conjugate $a_s(f)^*$, with f running over $L^2(\mathbb{R}^3, \mathrm{d}x)$; in particular $a_s(p) \equiv a_s(f_p)$ with

$$f_p(x) = \begin{cases} V^{-\frac{1}{2}} e^{ixp} & x \in \Lambda \\ 0 & \text{otherwise} \end{cases} \quad \text{with} \quad p = n \frac{2\pi}{V^{\frac{1}{3}}} \; ; \; n \in \mathbb{Z}^3 \quad (14.1.8)$$

a is assumed to be linear in f, and the Fermi structure is specified by:

$$\{a_s(f), a_{s'}(g)\} = 0 \quad \text{and} \quad \{a_s(f), a_{s'}(g)^*\} = (f,g) \, \delta_{s,s'} \, I \quad (14.1.9)$$

where $\{.,.\}$ denotes the anti-commutator (or Jordan product) $\{A, B\} = AB + BA$; and (f, g) is the scalar product of the functions f and g in $L^2(\mathbb{R}^3, \mathrm{d}x)$. The algebra is assumed to be closed in the weak-operator topology induced by a representation π_ϱ which we leave unspecified at this point; the resulting object is the *Fermi algebra* $\mathcal{A}_\varrho$ associated with the representation.

Cooper pairs have a mixed character liking them to the putative bosons that were expected to condense in some of the earlier models mentioned in Subsection 14.1.1 above:

$$\left. \begin{array}{cccc} & [b(p), b(q)] = 0 & \text{and} & [b(p), b(q)^*] = 0 & \text{for} & p \neq q \\ \text{but} & [b(p), b(q)^*] = 1 - (n_\uparrow(p) + n_\downarrow(p)) & & & \text{for} & p = q \end{array} \right\} \quad (14.1.10)$$

where $[.,.]$ denotes the commutator $[A, B] = AB - BA$.

The above prescriptions thus specify the model and its interpretation. The mean-field approximation of the model is defined by replacing (14.1.6) by

$$H_{\Lambda, \varrho} = \sum_{p,s} \varepsilon(p) a_s(p)^* a_s(p) + \sum_p \left(b(p)^* \Delta_\varrho(p) + \text{h.c.} \right) \quad (14.1.11)$$

where $\Delta_\varrho(p)$ is some element – to be specified later – of the center of the algebra $\mathcal{A}_\varrho$; i.e. $\Delta_\varrho(p)$ belong to $\mathcal{A}_\varrho$, and commute with all the elements of $\mathcal{A}_\varrho$.

The Hamiltonian (14.1.11) can be diagonalized, i.e. be shown to generate the same evolution as the quasi-free Hamiltonian

$$H_{\Lambda,\varrho} = \sum_{p,s} E_\varrho(p)\, \gamma_s(p)^* \gamma_s(p) \quad . \tag{14.1.12}$$

[Note that the $\gamma_s(p)'s$ and their adjoints should also carry an index ϱ omitted here.]

The diagonalization is achieved by the the Bogoliubov–Valatin transformation:

$$\gamma_\uparrow(p)^* = u_\varrho(p)^* a_\uparrow(p)^* - v_\varrho(p)^* a_\downarrow(-p)$$

$$\gamma_\downarrow(-p) = v_\varrho(p)\, a_\uparrow(p)^* + u_\varrho(p)\, a_\downarrow(-p) \tag{14.1.13}$$

where the coefficients $u_\varrho(p)$ and $v_\varrho(p)$ belong to the center of $\mathcal{A}_\varrho$ and satisfy:

$$u_\varrho(p)^* u_\varrho(p) + v_\varrho(p)^* v_\varrho(p) = I \tag{14.1.14}$$

and

$$\left. \begin{aligned} u_\varrho(p)^* u_\varrho(p) &= \tfrac{1}{2}\{1 + [\varepsilon(p)/E_\varrho(p)]\}\, I \\[4pt] v_\varrho(p)^* v_\varrho(p) &= \tfrac{1}{2}\{1 - [\varepsilon(p)/E_\varrho(p)]\}\, I \\[4pt] u_\varrho(p)^* v_\varrho(p) &= -\tfrac{1}{2}[\Delta_\varrho(p)/E_\varrho(p)]\, I \end{aligned} \right\} \quad . \tag{14.1.15}$$

The *quasi-particles* created and annihilated by the $\gamma_s(p)^*$ and $\gamma_s(p)$ are said to be the *proper excitations* of the system. (14.1.14) ensures that they satisfy the same Fermion anti-commutation relations as did the electrons; see (14.1.9). (14.1.15) have been adjusted so that the Hamiltonian (14.1.11) reduces to (14.1.12); as a consequence, we obtain:

$$\boxed{E_\varrho(p) = \left[\, \varepsilon(p)^2 + \Delta_\varrho(p)^* \Delta_\varrho(p) \,\right]^{\frac{1}{2}}} \quad . \tag{14.1.16}$$

We now choose the primary representation π_ϱ associated with the canonical equilibrium state ϱ at the natural temperature β for the Hamiltonian (14.1.12). In line with the tenets of the mean-field method – see e.g. (11.3.36) – we then impose

$$\Delta_\varrho(p) = \sum_q \tilde{v}(p,q)\langle b(q)\rangle_\varrho \quad . \tag{14.1.17}$$

From (14.1.12), we then obtain:

$$\langle b(q)\rangle_\varrho = u_\varrho(q)^* v_\varrho(q) \tanh\left(\frac{1}{2}\beta E_\varrho(q)\right) \tag{14.1.18}$$

and thus, from (14.1.15), (14.1.17) and (14.1.18), the self-consistency equation:

$$\boxed{\Delta_\varrho(p) = -\sum_q \tilde{v}(p,q)\frac{\Delta_\varrho(q)}{2E_\varrho(q)}\tanh\left(\tfrac{1}{2}\beta E_\varrho(q)\right)} \qquad . \tag{14.1.19}$$

To understand the elementary reason why the form (14.1.12) of the mean-field Hamiltonian (14.1.11) becomes an exact diagonalization of the BCS Hamiltonian (14.1.6) in the thermodynamical limit, we note that the element $\Delta(p) \equiv \sum_q \tilde{v}(p,q)b(q) \in \mathcal{A}_\varrho$ satisfies:

$$\left.\begin{array}{l} [\Delta(p), a_s(p')] = 0 \\[2mm] [\Delta(p), a_s(p')^*] = \tilde{v}(p,p') \end{array}\right\} \quad \forall\, p' \qquad . \tag{14.1.20}$$

We now use (14.1.7): $\tilde{v}(p,p') \to 0$ as $V \to \infty$. Hence, the limit of $\Delta(p)$ belongs to the center of the algebra $\mathcal{A}_\varrho$, thus suggesting the introduction of the central element Δ_ϱ satisfying (14.1.17).

From the point of view of the symmetry of the model, this solution however is not yet satisfactory. Indeed the original Hamiltonian (14.1.6) is gauge-invariant, i.e. is invariant under the group of automorphism defined by

$$\alpha(\theta)[a_s(p)] = e^{i\theta}a_s(p) \tag{14.1.21}$$

while the mean-field Hamiltonian (14.1.12) is not, since

$$\alpha(\theta)[b(p)^*] = e^{-i2\theta}b(p)^* \,. \tag{14.1.22}$$

To restore the symmetry of the model, we have to consider a non-primary representation, in which the symmetry group is allowed to act non-trivially on the center. Taking our clue from (14.1.17) and (14.1.22), we impose:

$$\alpha(\theta)[\Delta_\varrho(p)] = e^{i2\theta}\Delta_\varrho(p) \tag{14.1.23}$$

and, accordingly:

$$\alpha(\theta)[u(p)^*] = e^{2i\theta}u(p)^* \quad ; \quad \alpha(\theta)[v(p)] = v(p)$$

$$\alpha(\theta)[\gamma_s(p)] = e^{-i\theta}\gamma_s(p) \tag{14.1.24}$$

Hence the representation of the model, together with its symmetry group, is a direct integral – over the gauge-group S^1 – of the primary representations, each of which corresponding to a pure thermodynamical phase. Considering only one phase at a time leads to a *spontaneous symmetry breaking*. The situation is therefore quite similar to that encountered in the Weiss–Ising model, except for the fact that here the gauge-group is a continuous group, while in the flip-flop symmetry the group is the discrete group Z_2 of two elements.

In our presentation of the mathematical structure of the BCS model, we used [Bogoliubov, 1958, Valatin, 1958, Yoshida, 1958]. For more details, see [Haag, 1962, Ezawa, 1964, Emch and Guenin, 1966, Thirring and Wehrl, 1967, Jelinek, 1968, Thirring, 1969, Dubin, 1974]; and also [Thouless, 1970].

The self-consistency equation (14.1.19) is known in the physics literature as the *gap equation;* its interpretation is given by the form (14.1.16) of the energy of the elementary excitations of the model. A non-zero solution occurs only if the interaction v is attractive, and even then it only occurs below a critical temperature T_c , the value of which can be determined from (14.1.19) in terms of the value of the energy gap at zero temperature: $\Delta(0) \approx 1.75kT_c$. More generally, (14.1.19) provides the basis from which [Bardeen, Cooper, and Schrieffer, 1957] accounted for the temperature dependence of the observed energy gap, and – together with the partition function readily obtained from (14.1.12) – for the thermodynamical properties of the model. Finally, knowing the value of the energy gap allows to capitalize on earlier work of Bardeen, in particular about the Meissner effect [Bardeen, 1955], thus justifying theoretically the empirical formula (14.1.3) for the penetration depth.

14.1.3 Superconductivity after BCS

In the five years that followed the quantitative as well as qualitative success of [Bardeen, Cooper, and Schrieffer, 1957], some limitations were starting to appear concerning the universality of the model.

Theoretical objections were raised as to the simplifying assumptions one makes *before* one settles down to write the simple Hamiltonian (14.1.6); and the possible ambiguities relative to its extension to include a consistent treatment of the interaction with an external electro-magnetic field; see [Schafroth, 1960], but also [Bardeen and Schrieffer, 1961] and the Nobel lectures [Bardeen, 1972, Cooper, 1972, Schrieffer, 1972].

The thrust of [Matthias, Geballe, and Compton, 1963] is more pragmatic: it warns that several kinds of superconductors ought to be distinguished. For instance, while the isotope effect – which played a central motivating role in the model building of the 1950s – had been well-documented for elements such as Hg, Zn, Cd, Sn, Tl, Pb; the effect had not been seen with any of the transition elements such as Ru, Os; nor with compounds such as Nb_2Sn, Mo_3Ir .

Up to 1986, even playing creatively with intermetallic compounds did not break a barrier of a T_c that is slightly over 20 K. A mostly empirical search to produce superconductive materials with a transition temperature that would be high enough to allow competitively priced practical applications – such as Meissner trains that would levitate without friction above superconducting rails – seemed to go nowhere. Therefore considerable fanfare greeted the discovery of superconductivity in new ceramic materials. The boundary temperature was 50% higher than ever before; these materials also did exhibit the Meissner effect as to be expected from a material with a vanishing electric resistance. By the time the inventors, J. G. Bednorz and K. A. Müller, received the Nobel prize, the record T_c had risen so much that they were able to illustrate their acceptance presentation [Bednorz and Müller, 1987] with a material – $YBa_2Cu_3O_{9-\delta}$ (whatever that is called) – having a critical temperature $T_c = 92$ K . Although this material is not properly described by the BCS model, this model was mentioned in the presentation for the reasoning that presided over the choices that had to be made to arrive at such an exotic material.

14.2 Superfluidity and Bose–Einstein Condensation

14.2.1 The Early Days

Research in Bose–Einstein condensation [BEC] and superfluidity has been framed by the confluence of groundbreaking low temperature technologies and two theoretical models, the Bose–Einstein description of an ideal quantum gas and the Gross–Pitaevsky nonlinear Schrödinger equation for the interacting gas. Both are discussed in the present section. The search for an adequate microscopic explanation of superfluidity gives us yet another example of the complexity of model building in thermostatistical physics. The initial search for a plausible analogy between superconductors and superfluids began a twisted path that turned up some models from quite unlikely corners.

In Sect. 14.1, we made several allusions to "boson condensation" and "two-fluid model for superfluidity". The most immediate analogy is that, in superfluids as in superconductors, there exists besides the "normal phase" another phase which appears to allow the frictionless motion of a fluid: electrons in superconductors or Helium atoms in the superfluid. While the developments of the two topics present analogies, they are not coextensive; the conceptual difficulties involved in such a sweeping analogy are instructively illustrated by one of the final conclusions in [Gorter, 1955]:

> The connection often sought between the superfluid properties in helium II and the Bose–Einstein condensation has incidentally led to the suspicion that in superconductors one might also be concerned with even units viz. with electron pairs. This suggestion has ... had little success so far.

When, in 1908, Kamerlingh-Onnes liquefied Helium at 4.2 K, he went further down in temperature, and noticed an additional feature beyond the classical gas-liquid transition:

> It is very noticeable that the experiments indicate that the density of the helium, which at first quickly drops with the temperature, reaches a maximum at 2.2 K approximately, and if one goes down further even drops again. Such an extreme could possibly be connected with the quantum theory.
>
> [Kamerlingh Onnes, 1913]

The full measure of this Delphian oracle did not dawn for another twenty-five years, while more empirical data were gathered. In some of the last experiments he conducted just before his death, Kamerlingh-Onnes refined his account to note that the maximum seemed to happen at a discontinuity of the derivative of the density in terms of the temperature. Kamerlingh Onnes' successor in Leiden pursued the quest and examined other thermodynamical features of the transition – such as the heat of vaporization and the surface tension – to confirm that something so special was indeed going on that he proposed to distinguish two states of liquid Helium: the normal phase – or Helium I – which alone exists between 4.2 K and 2.2 K; and the new phase that appears below 2.2 K – Helium II [Keesom and Wolfke, 1928]. Moreover, it was observed [Keesom and Keesom, 1932] that the behavior of the specific heat as a function of the temperature in the neighborhood 2.2 K has a dramatic cusp, presenting such a compelling *"resemblance ... with the Greek letter λ "* that the name $\lambda-point$ stuck for the transition point between Helium I and II. As evidenced by the title of later papers, the phenomenon still had to be considered as an anomaly.

The next experimental advance was achieved independently by [Kapitza, 1938, Allen and Misener, 1938] who established that He II could easily flow between tightly packed plates and through very thin capillaries which would completely obstruct the passage of ordinary fluids, e.g. He I. An experimental upper bound was placed on the viscosity of He II, at 1/1500th of the viscosity of He I. Kapitza coined the word *"superfluid"* for He II.

The same year, a spark was ignited, [London, 1938a], setting afire a new approach [Tisza, 1938a,b, London, 1938b, 1939]. The spark was London's short note suggesting that the $\lambda-point$ in Helium was reminiscent of the behavior of the specific heat predicted in earlier – but largely ignored and/or dismissed – proposals by [Bose, 1924, Einstein, 1924] describing what is known today as the Bose–Einstein condensation [BEC], which we briefly review now in order to discuss next its suitability as a model for superfluidity. The early history of BEC is traced in [Pais, 1982]; in particular, see the description of how Einstein saw a model for a material ideal gas – Einstein [1924] mentioned indeed Hydrogen, Helium, [and the electron gas *(sic)*] – while Bose's concern was focused on a new derivation of the Planck radiation law. Before we review the model of ideal Bose gas, let us note that what London and Tisza did in 1938 was quite unusual in terms of building theoretical models for explanation. To discern an analogy behind the disparity between a superfluid and a model for radiation (i.e. photon gas) is an intrepid feat of imagination. The success of BEC in explaining superfluidity is truly remarkable if not miraculous, begging the philosophical question of how a model for radiation can be the right *explanatory* model for the phenomenon of a fluid, even if one can derive from the former the salient observable features of the latter. One may use a model of cranks and pulleys to make rough predictions of an economy, but such a model presumably could not be seriously taken as providing any explanations for the economy.

We define what is now called the *ideal Bose gas* of particles of mass m enclosed in a finite region $\Lambda \subset \mathsf{R}^3$; we start with the quantum version of the grand canonical partition function – see Sects. 10.1–3:

$$\left. \begin{aligned} \mathcal{Z}(\Lambda, T, \mu) &= \operatorname{Tr} e^{-\beta(H_\Lambda - \mu N_\Lambda)} \\[2mm] \text{with} \quad H_\Lambda - \mu N_\Lambda &= \textstyle\sum_p [\varepsilon(p) - \mu]\, a(p)^* a(p) \end{aligned} \right\} \tag{14.2.1}$$

$\beta = 1/k_B T$ where T is the temperature, and k_B is the Boltzmann constant; $\varepsilon(p) = \frac{p^2}{2m}$ is the energy of a particle of momentum p; and $a(p)^*$, $a(p)$ are boson creation and annihilation operators, i.e. satisfy the canonical commutation relations:

$$[a(p), a(p')] = 0 \quad [a(p)^*, a(p')] = \delta_{p,p'}\, I \quad . \tag{14.2.2}$$

These (unbounded) operators generate – by limits of linear combinations of finite products, due care being taken with domains of definition – the so-called *Bose algebra*. For simplicity, we assume that Λ is a cubic box of side L so that, with k running over Z^3, the set of all

$$f_k(x) \equiv L^{-\frac{3}{2}} \begin{cases} \exp(i\frac{2\pi}{L}k \cdot x) & x \in \Lambda \\ 0 & x \notin \Lambda \end{cases} \tag{14.2.3}$$

defines an orthonormal basis in $L^2(\Lambda, dx)$, with the property that, for $x = (x_1, x_2, x_3)$, the momentum operator $p_i \equiv -i\hbar \partial_{x_i}$ satisfies $p_i f_k = \hbar k_i f_k$. This basis, therefore, diagonalizes $H_\Lambda - \mu N_\Lambda$, and thus makes immediate the computation of the trace in (14.2.1); we have then:

$$\mathcal{Z}(\Lambda, T, \mu) = \prod_{k \in \mathsf{Z}^3} \{1 - e^{-\beta[\varepsilon_k - \mu]}\}^{-1} \quad \text{with} \quad \varepsilon_k = \frac{h^2 |k|^2}{2m} \quad . \tag{14.2.4}$$

Hence the average occupation number $\langle n_k \rangle$ and two thermodynamical quantities, the *pressure* P and the occupation *density* $v^{-1} = \frac{1}{|\Lambda|}\sum_k \langle n_k \rangle$, are:

$$\langle n_k \rangle \equiv -\frac{1}{\beta} \partial_{\varepsilon_k} \ln \mathcal{Z}(\Lambda, T, \mu) = \{e^{\beta[\varepsilon_k - \mu]} - 1\}^{-1} \tag{14.2.5}$$

and

$$\left. \begin{aligned} P &\equiv \tfrac{1}{\beta}\tfrac{1}{|\Lambda|} \ln \mathcal{Z}(\Lambda, T, \mu) = -\tfrac{1}{\beta}\tfrac{1}{|\Lambda|} \textstyle\sum_k \ln\{1 - e^{-\beta[\varepsilon_k - \mu]}\} \\[2mm] v^{-1} &\equiv z\partial_z \tfrac{1}{|\Lambda|} \ln \mathcal{Z}(\Lambda, T, \mu) = \tfrac{1}{|\Lambda|} \textstyle\sum_k \{e^{\beta[\varepsilon_k - \mu]} - 1\}^{-1} \end{aligned} \right\}, \tag{14.2.6}$$

where, $z = \exp \beta\mu$ is the activity. For $0 \le z < 1$, the gaseous phase obtains by taking the thermodynamical limit $|\Lambda| \to \infty$, replacing – see (14.2.6) – the discrete $\sum_k \cdots$ by the continuous $4\pi \int_o^\infty dp\, p^2 \cdots$:

$$\beta P = \lambda_{dB}^{-3}\, g_{\frac{5}{2}}(z) \quad ; \quad v^{-1} = \lambda_{dB}^{-3}\, g_{\frac{3}{2}}(z) \tag{14.2.7}$$

where the functions g_s with $s = \frac{1}{2}, \frac{3}{2}$ are given by:

$$g_s(z) = z\,\zeta_s(z) \quad \text{with} \quad \zeta_s(z) = \frac{1}{\Gamma(s)}\int_o^\infty dt\, \frac{t^{s-1}}{e^t - z} = \sum_{n=0}^\infty \frac{z^n}{(n+1)^s} \tag{14.2.8}$$

and $\Gamma(s)$ are the standard interpolating functions for the factorials $n!$ with, in particular, $\Gamma(\frac{1}{2}) = \sqrt{\pi}$ and $\Gamma(\frac{3}{2}) = \frac{1}{2}\Gamma(\frac{1}{2})$, so that ζ_s are the classical *Lerch zeta functions*. Finally,

$$\lambda_{dB} = [2\pi\hbar^2/mk_BT]^{\frac{1}{2}} \tag{14.2.9}$$

is the de Broglie *thermal wavelength* and it measures the position uncertainty associated with the thermal momentum distribution.

When the temperature T is sufficiently high with respect to the density v^{-1} to make $\lambda_{dB}^3/v^{-1} << 1$, the quantum equation of state (14.2.7) reduces to the *classical ideal gas* Boyle/Gay-Lussac law: $Pv = kT$. Indeed, as $\beta \to 0$ the *Bose–Einstein distribution* (14.2.5) is approximated by the classical Boltzmann distribution $\langle n_k \rangle = z\,\exp(-\beta\varepsilon_k)$; recall our comparison in Sect. 10.3 of the classical and quantum harmonic oscillators in canonical equilibrium, e.g. (10.3.26).

For a review of some putative precursors of the Bose–Einstein distribution in earlier writings on ideal systems of undistinguishable particles – starting with Boltzmann's writings – see [Bach, 1990].

The full quantum structure however becomes manifest when the activity $z \to 1$. Indeed since $g_{\frac{3}{2}}$ is a smooth, strictly increasing function on the closed interval $[0, 1]$, (14.2.6) imposes the condition that $\lambda_{dB}{}^3 v^{-1} \le g_{\frac{3}{2}}(1) = 2.612\ldots$ and thus defines a critical temperature $T_c = T_c(v)$:

$$[2\pi\hbar^2/mk_BT_c]^{\frac{3}{2}}\, v^{-1} = g_{\frac{3}{2}}(1) \quad . \tag{14.2.10}$$

To get to temperature lower than T_c, one can let $z \to 1$ and compensate for the divergence of $\langle n_o \rangle = \left(\frac{z}{1-z}\right)$ by letting simultaneously $|\Lambda| \to \infty$ in such a manner that $|\Lambda|^{-1}\left(\frac{z}{1-z}\right) \to v_o$, thus giving a *macroscopic occupation of the ground state* $p = 0$, or *Bose–Einstein condensation [BEC]*. Under these circumstances, the thermodynamical limit of (14.2.6) replaces (14.2.7) by

$$\beta P = \lambda_{dB}^{-3}\, g_{\frac{5}{2}}(1) \quad ; \quad v^{-1} = \lambda_{dB}^{-3}\, g_{\frac{3}{2}}(1) + v_o^{-1} \tag{14.2.11}$$

with $z = 1$. These new equations of state define the *condensed phase* of the ideal Bose gas. At fixed temperature T, the pressure P is constant for $0 \le v \le v_c$ where v_c is obtained by solving (14.2.10) for $T_c(v) = T$. For $v \ge v_c$, the isotherm continues smoothly into the gaseous phase. At the fixed density v^{-1}, and thus at the fixed T_c, the temperature dependence of ground state occupation density is given by:

$$v_o^{-1} = v^{-1}\left[1 - \left(\frac{T}{T_c}\right)^{\frac{3}{2}}\right] \quad . \tag{14.2.12}$$

Finally, the specific heat is also easily computed to give, in the condensed phase:

$$C_v = \frac{15k_B}{4}g_{\frac{5}{2}}(1)\,\lambda_{dB}^{-3}\,v \quad ; \tag{14.2.13}$$

thus, it increases monotonically with the temperature like $T^{\frac{3}{2}}$ to reach a *finite* maximum $C_v(T_c) \approx 1.28\left(\frac{3}{2}kT\right)$. Beyond T_c, we are back to the gaseous phase and C_v, without any jump from its value at $T = T_c$, decreases monotonically with increasing temperature to approach the classical value $\frac{3}{2}k$ at $T \to \infty$.

This concludes our analysis of the ideal Bose gas model. While we restricted our attention to the partition function and its immediate consequences, the full algebraic structure of the model is presented in [Araki and Woods, 1963].

We should also mention that the existence of the condensation in this quantum ideal gas depends essentially on the boson commutation relations (14.2.2) of the particles; no such phenomenon would occur if they were to obey the Fermi anticommutation relations (14.1.9). The dimensionality $d = 3$ of the model is essential: Hohenberg [1967] showed that the homogeneous ideal Bose gas with $d = 1, 2$ does not exhibit long-range order for any $T > 0$. This result is not isolated, and related techniques have been used to prove the absence of ferromagnetism or antiferromagnetism in one- or two-dimensional isotropic Heisenberg model [Mermin and Wagner, 1966], as well as the absence of order – e.g. crystalline order in 2 dim. – in classical models [Mermin, 1967].

As for the adequacy of the ideal Bose gas as a model for superfluidity, it is first to be noticed that the constancy of the pressure along the isotherm in the region $0 \le v \le v_c$ – see (14.2.11) – is in line with the behavior of the superfluid phase of ^{4}He. Moreover, upon inserting in (14.2.10) the value of v corresponding to the mass density ϱ observed in liquid Helium, one obtains $T_c = 3.1$ K, which is not widely off the mark, 2.2 K. Whence comes Tisza's daring identification of the phase $p = 0$ with the superfluid part of a two-fluid model made up of non-interacting parts, the "normal" part of which corresponding to the summation over the states with $p \ne 0$.

However, while the specific heat of liquid Helium diverges at the critical temperature – the famous λ–point – we found only a kink in that of the ideal Bose gas, i.e. a finite maximum with only a discontinuity of the derivative. Hence, while the model is suggestive, it does not present the full story: atoms or molecules in a liquid *do* interact, so that deviations from the ideal gas behavior are to be expected. This was realized as early as in [London, 1939]. But a realistic estimate of the nature of the deviations is a hard theoretical

problem which was all the more harder as long as only one single substance
– Helium II – was available for empirical guidance.

Nevertheless, several theoretical schemes were proposed in markedly different directions throughout the 1940s, 50s and 60s; for a chronicle, see [Brush, 1983]. We only wish to highlight the following contributions:

- the quantum hydrodynamics approach of [Landau, 1941], a radical alternative to starting from an ideal Bose gas and adding interactions;
- the effort of synthesis represented by the concept of collective excitations discussed by [Bogoliubov, 1947];
- the unified description of the excitations energy spectrum in [Feynman, 1953];
- the recognition that superfluids – unlike ordinary fluids – sustain only quantized vortices, namely vortices with circulation limited by the condition:

$$\Gamma = \oint ds \cdot V = n \frac{h}{m} \quad \text{with} \quad n = 1, 2, \cdots \tag{14.2.14}$$

where h is the Planck constant and m the mass of the molecule; the existence of these quantized vortices was suggested in a footnote of [Onsager, 1949b], and was proposed independently by [Feynman, 1955]; quantized vortices were observed experimentally by [Vinen, 1958];

- the formulation of off-diagonal long range order [ODLRO], i.e.

$$\lim_{|x-y| \to \infty} \langle a^*(x)\, a(y) \rangle \neq 0 \tag{14.2.15}$$

where a^* and a are the boson creation and annihilation operators, and $\langle \ldots \rangle$ denotes the equilibrium average; [Penrose and Onsager, 1956, Yang, 1962], see also [Martin, 1964, Hohenberg and Martin, 1965, Anderson, 1966];

- the non-linear Schrödinger equation of [Gross, 1961, Pitaevsky, 1961] which we discuss below.

14.2.2 Modern Developments

At this point we must keep in mind that the original BEC model was an *ideal* gas, while the only empirical evidence of a BEC was gathered from experiments on ^{4}He at temperatures at which it is a liquid, and thus where one should expect the inter-particle interactions to play some role. How important a role ? This is a question the answer to which is very difficult to ascertain in the absence of empirical evidence from several substances that could help discriminate between competing theoretical models.

Here comes a new twist in our story. Instead of treating BEC almost as an analogical model for superfluidity in Helium, exploration was begun to find new systems in which the model of BEC can be applied transparently. This marked the beginning of a distinction between the manifestations of BEC and of superfluidity, an empirical discrimination that, to this day, is still in the process of being conceptually clarified. The paper [Hecht, 1959] seems to be the first published proposal to explore BEC in atomic Hydrogen, i.e. Hydrogen whose atoms are prevented from recombining into molecules, e.g., by a strong magnetic field. It was advocated that atomic Hydrogen would remain a gas all the way down to absolute $T = 0$, thus affording a new adjustable parameter, the density, which would allow a more detailed comparison with theoretical models, especially in view of the fact that at the densities to be expected, the Hydrogen atom gas would be very weakly interacting.

To situate the experimental difficulties, consider the following off-the-cuff estimates. Recall from (14.2.10) that the critical temperature for BEC is given by

$$T_c = \frac{h^2}{2\pi k_B m} \left(\frac{n}{2.612}\right)^{\frac{2}{3}} . \tag{14.2.16}$$

Here we wrote n for the density v^{-1}, and replace the theoretical constant $g_{\frac{3}{2}}(1)$ by its approximate value 2.612. The quantities involved in this expression can be visualized by noticing that in term of (14.2.9) – the de Broglie thermal wave-length – the relation (14.2.16) is equivalent to $n\lambda_{dB}^3 = 2.612$ and thus to saying, roughly speaking, that the inter-particle distance $D = n^{-\frac{1}{3}}$ and the de Broglie wavelength λ_{dB} are of the same order of magnitude; recall that λ_{dB} is a measure of the position uncertainty associated with the thermal momentum distribution of the particles; under these circumstances, one should expect indeed indistinguishability to play a dominant role and quantum effects, such as BEC, to come to the fore.

With $h = 6.626 \times 10^{-27}$ erg sec for the Planck constant, $k_B = 1.381 \times 10^{-16}$ erg/deg for the Boltzmann constant, $m = 1.673 \times 10^{-24}$ g for the mass of the Hydrogen atom, this gives

$$T_c \approx \left(\frac{n}{4.965 \times 10^{20}}\right)^{\frac{2}{3}} \tag{14.2.17}$$

The critical temperatures that now prevail are in a range between $30\,\text{mK}$ and $30\,\mu\text{K}$ would correspond respectively to the densities $n \approx 2.5 \times 10^{18}$ cm^{-3} and $n \approx 8 \times 10^{13}$ cm^{-3} ; alternatively, these densities can be expressed in terms of the average inter-particle distance $D \equiv n^{-\frac{1}{3}}$, namely $D \approx 8 \times 10^{-6}$ cm and $D \approx 2 \times 10^{-5}$ cm; these should be compared to the the mean free path λ which is inversely proportional to the density and is estimated for the two densities considered above to be $\lambda \approx 4 \times 10^{-4}$ cm and $\lambda \approx 12$ cm.

Thus, such a gas is very dilute indeed, and hence the inter-particle interactions are so very weak that one might expect this system to admit the ideal Bose gas as a reasonable model; deviations from the ideal model can now be monitored by the density of the gas, and be empirically compared with the adjustable strength of the interaction.

Experimentally, one first has to produce an atomic Hydrogen gas; the creation of spin-polarized Hydrogen was first reported in [Silvera and Walravan, 1980]; see also [Silvera and Walravan, 1986].

The next problem is to keep the gas long enough for the purpose of measurements, i.e. to prevent for as long as possible the Hydrogen atoms to recombine to form molecules. As the presence of the walls of any material container favorizes dramatically the recombination, material walls were replaced by magnetic traps [Hess, 1986, Doyle, Sandberg, Yu, Cesar, Kleppner, and Greytak, 1991], later supplemented by microwave traps [Agosta, Silvera, Stoof, and Verhaar, 1989, Silvera and Reynolds, 1991]. Still, within the gas, the recombination rate is proportional to n^3; keeping it in check tends to limit the density. Moreover, the energy freed during the recombination dissipates through the gas, overheats it, and stands in the way of lowering the temperature to reach T_c.

Meanwhile, a new direction was proposed in the search of flexible systems showing BEC, namely to investigate exciton gases in semiconductors, such as Cu_2O; for a didactic progress report, see [Wolfe, Lin, and Snoke, 1995]. A basic reference on excitons is [Knox, 1980].

For our purposes here, it is sufficient to recall that an exciton is a pair consisting of an electron in the conduction band and a hole in the valance band, and thus it is a composite boson made up of two fermions of spin $\frac{1}{2}$; consequently the excitons exist in two states: paraexciton with spin $S = 0$ and orthoexciton with $S = 1$. This quasi-particle is not stable, with the orthoexciton having a much shorter lifetime – several hundred nanoseconds – than the paraexciton – several microseconds – in the kind of nearly perfect crystals that are grown naturally, e.g. in Copper mines. The photon emitted in its decay is what makes the exciton visible to the experimentalist through the photoluminescence spectrum; the momentum distribution of the exciton is obtained from neutron scattering. The brief lifetimes nevertheless are sufficiently long to determine experimentally that the excitons form a weakly interacting Bose gas that, even at high densities, move without friction through the crystal over macroscopic distances. When opposed to a conservative interpretation of this fact on the ground of a putative analogy with the free motion of electrons in a perfect crystal, the BEC interpretation seems appealing. The interest of the excitons for BEC studies is easily understood from the relation (14.2.16): the mass of the excitons is of the order of twice the mass m_o of the electron; various corrections bring it in fact to 2.7 m_o. Since m_o is about 1800 times smaller than the mass of the proton, the effect of this disparity is that the critical temperature of the exciton gas is much larger than the critical temperature of an atomic Hydrogen gas at the same density: about 680 times for paraexcitons, and due to its three-fold degeneracy, about 320 for orthoexcitons. By the mid-nineties, exciton gas densities of the order of 2.5×10^{18} cm^{-3} (or higher) were experimentally realizable, and it was possible to confirm the phase diagram (14.2.16): around this density, the exciton gas in Cu_2O requires only to operate in the range of a few tens of K, typically between 1 and 100 K; and [Wolfe, Lin, and Snoke, 1995] could conclude that "paraexcitons do indeed undergo Bose–Einstein condensation."

By that time, BEC was reported to have been observed in dilute atomic gases, such as [87]Rb [Anderson, Ensher, Matthews, Wieman, and Cornell, 1995], [7]Li [Bradley, Suckett, Tollet, and Hulet, 1995], and [23]Na [Davis, Mewes, Andrews, Druten, Durfee, Kurn, and Ketterle, 1995]..

The atoms were trapped in inhomogeneous magnetic fields – acting on the unpaired electron of each atom – and cooled by a combination of laser and evaporative techniques. Laser cooling and trapping techniques were proposed by [Hänsch and Schawlow, 1975, Wineland and Dehmelt, 1975]; drastic refinements and decisive developments were recognized by the 1997 Physics Nobel prize, awarded to S. Chu, C. Cohen-Tannoudji and W. C. Phillips; see [Levi, 1997].

Evaporative cooling techniques to further lower the temperature were developed in part by Greytak, Kleppner and coworkers until they finally succeeded to get BEC for Hydrogen: the BEC transition was found at about $T_c = 50\mu K$ and $n = 1.8 \times 10^{14}$ cm^{-3} [Fried, Killian, Willmann, Landhuis, Moss, Kleppner, and Greytak, 1998]; for a brief account, see [Levi, 1998a]. For a more technical, but still broadly aimed, review of the experimental situation, see [Ketterle, 1999]. The following paper in the same journal sketches the theoretical landscape [Burnett, Edwards, and Clark, 1999]; see also [Wieman, Pritchard, and Wineland, 1999] and, for more detailed discussions, [Dalfovo, Giorgini, Pitaevski, and Stringari, 1999].

Before we return to modeling, let us ask one more phenomenological question, namely whether the new BEC condensates are superfluid. Direct evidence of the kind we saw Kapitza [1938] had obtained for [4]He seems beyond the frontiers of the current empirical realm. However, recall that – in contrast to ordinary fluids – superfluids like [4]He support quantized vortices. Recently, similar vortices have been excited in an atomic Rubidium BEC condensate, thus providing indirect evidence that the latter is indeed a superfluid [Madison, Chevy, Wohlleben, and Dalibard, 2000, Chevy, Madison, and Dalibard, 2000]; for a non technical report, see [Fitzgerald, 2000].

Nevertheless, one essential difference between the physics of BEC in the new atomic gases and the old Helium experiments is that the inter-particle interactions are now significantly much weaker; a manifestation of this fact is that in Helium the condensate to be depleted to less than 10% while more than 99% of the alkali atoms are in the condensate phase. An other essential difference is that the strength of the interaction can be varied in the laboratory from the repulsive to the attractive regimes – [Cornish, Claussen, Roberts, Cornell, and Wieman, 2000] thus providing the model builders with guidance that was unavailable when empirical data were limited to the sole Helium.

A heuristic modeling of the experimental situation should take into account the following empirical facts. The magnetic field that traps the gas must be highly non-homogeneous, so that the translational symmetry of the ideal gas is lost; the trap is definitely small so that the total number of particles is finite; the gas is very dilute and very cold; hence the atoms are expected to interact predominently through head-on collisions with an interaction potential very peaked at the origin and a strength proportional to the s-wave scattering length; this quantity is denoted by a and its precise definition is given by (14.2.22) below. These circumstances suggest to consider an approximation in which the actual many-body evolution is mimicked by the one-particle non-linear Schrödinger equation [Gross, 1961, Pitaevsky, 1961]:

$$i\partial_t \Psi(x,t) = \left[-\nabla^2 + V_{\text{trap}} + 4\pi a \left| \Psi(x,t) \right|^2 \right] \Psi(x,t) \qquad (14.2.18)$$

where units are chosen so that $\hbar = 1 = 2m$. Some practitioners in the field like to view this equation as another application of the mean-field method. This is correct in the sense that indeed it models a many-body interaction by a one-particle Hamiltonian. Yet, this appelation should come with a warning: in the familiar applications – see e.g. Sect. 11.3 – one approximates the interaction by a very weak two-body potential of very long-range; here, on the contrary, the range of the approximating two-body potential tends to zero while the coupling constant approaches infinity – see (14.2.24–25) below.

As for the equation itself, the traditional wisdom [Burnett, Edwards, and Clark, 1999] is:

(a) for N large the GP equation mimics adequately the many-body problem at hand; note the normalization (14.2.31), and the scaling relation (14.2.34) below;

(b) the GP equation admits a stationary fundamental solution $\Psi(x,t) = e^{-i\mu t}\Phi(x)$, so that

$$\left[-\nabla^2 + V_{\text{trap}}(x) + 4\pi a \left| \Phi(x) \right|^2 \right] \Phi(x) = \mu \, \Phi(x) \quad ; \qquad (14.2.19)$$

(c) in the limit when the nonlinear interaction term is much larger than the kinetic energy term associated to the gradient term $-\nabla^2$, one obtains the approximation

$$\left| \Phi(x) \right|^2 \approx \frac{\mu - V_{\text{trap}}(x)}{4\pi a} \qquad . \qquad (14.2.20)$$

To justify the limiting situation (c), a summary appeal is made sometimes to the Thomas–Fermi approximation; it seems in fact that the situation is more subtle than just that.

The empirical data to back up the ansatz (14.2.18) – together with (14.2.20) – come from several directions, the most spectacular of which may be the following experiment. Once the condensate has been formed and is in the state Φ, one turns off the trapping potential; the remaining interaction term in (14.2.18) is then repulsive – $a > 0$ – so very much so indeed that the condensate expands very fast – $\approx$ 60ms – to a size – $\approx$ 40μm – where its space-density profile can be mapped; one image produced in this way is reproduced in [Burnett, Edwards, and Clark, 1999] together with a comparison with a theoretical computation that involves no adjustable parameter: shape and order of magnitude are in impressive agreeement – 5% .

The model described by the GP equation leads furthermore to similarly convincing agreements with other empirical tests, such as: the measurements of specific heat; and of the frequency and damping of collective excitations created by modulating the magnetic confining potential (although there may be some problems still with the temperature dependence of the damping). Even interference patterns have been observed when two quantum condensates were made to pass through each other [Andrews, Townsend, Miesner, Durfee, Kurn, and Ketterle, 1997]!

BEC is now a major and very vibrant field of enquiry, as witnessed for instance by the the sustained vitality demonstrated on the dedicated websites maintained by Georgia Southern University, MIT or JILA:

`http:/amo.phys.gasou.edu/bec.html`;

`http://web.mit.edu/physics/greytak-kleppner`; or

`http://jilawww.colorado.edu/bec`.

Yet, the agreements observed between the predictions of theoreticians and the findings of experimentalists still leave open to mathematical physicists the foundational question of the sense in which the GP model can be justified from first principles. An answer has recently been obtained [Lieb, Seiringer, and Yngvason, 2000a], and we summarize their results below, distinguishing three major steps.

The first step is the microscopic theory, starting with the quantum $N-$particle Hamiltonian

$$H_N^{\mathrm{QM}} = \sum_{i=1}^{N} \left[-\nabla_i{}^2 + V_{\mathrm{trap}}(x_i) \right] + \sum_{1 \leq i < j \leq N} v(x_i - x_j) \qquad (14.2.21)$$

acting on totally symmetric, square-integrable wave functions $\Psi(x_1, \cdots, x_N)$ with $x_i \in \mathsf{R}^3$.

Since [Lieb, Seiringer, and Yngvason, 2000a] present a mathematical proof, precise assumptions on the trapping potential V_{trap} and the two-body interaction potential v must be made that enable the techniques of the proof; and these assumptions must be realistic enough to cover realistic situations. Specifically here, V_{trap} is assumed to be measurable, locally bounded, and to tend to infinity as $|x| \to \infty$ in the sense that $\inf_{|x| \geq X} V(x) \to \infty$ as $X \to \infty$. V is bounded below, and this minimum is assumed to be zero (a condition that can always be met by the proper choice of the zero of the energy scale). Typically, $V_{\mathrm{trap}}(x) \sim |x|^2$ accomodates both of the mathematical and the empirical concerns just expressed. The ground state of $-\nabla^2 + V$ provides a natural energy unit, $\hbar\omega$, and the corresponding length unit $\sqrt{\frac{\hbar}{m\omega}}$ is used throughout.

The particle interaction is assumed to be nonnegative, spherically symmetric, and short range in the sense that there exists a constant c such that $v(r) \leq cr^{-3}$ as $r \to \infty$. Let now u_o be the solution of the zero-energy scattering equation for such a potential:

$$-u''(r) + \frac{1}{2}v(r)u(r) = 0 \quad ; \quad u(0) = 0 \qquad (14.2.22)$$

where $\frac{1}{2}$ reflects the reduced mass of the 2-body problem. Then the scattering length a that entered the GP equation (14.2.18) is defined by:

$$a = \lim_{r \to \infty} \left(r - \frac{u_o(r)}{u_o'(r)} \right) \quad . \qquad (14.2.23)$$

For the limiting procedure involved in justifying the GP equation, [Lieb, Seiringer, and Yngvason, 2000a] consider a *fixed* potential $v_1(r)$ with scattering length a_1; they then define the potential

$$v(r) = (\frac{a_1}{a})^2 \, v_1(\frac{a_1}{a}r) \qquad (14.2.24)$$

with scattering length a, which is now a free parameter. The limit procedure involves considering

$$a = \frac{1}{N}\,a_1 \quad \text{with} \quad N \to \infty \quad \text{and} \quad a_1 \quad \text{fixed} \quad . \qquad (14.2.25)$$

Let us now return to (14.2.21) with V_{trap} fixed and v as in (14.2.24), and denote the corresponding $N-$particle Hamiltonian by $H_{N,a}^{\text{QM}}$. Let $E_{N,a}^{\text{QM}}$ denote its ground state energy, and $\Psi_{N,a}^{\text{QM}}$ the corresponding eigenfunction

$$H_{N,a}^{\text{QM}} = E_{N,a}^{\text{QM}}\,\Psi_{N,a}^{\text{QM}} \quad . \qquad (14.2.26)$$

If we were dealing with the ideal gas, where $v \equiv 0$, the ground state $\Psi_o^{\text{QM}}(x_1, \cdots, x_N)$ would be a product state $\prod_{i=1}^{N} \Phi_o(x_i)$, a property that persists in general when a non-identically zero inter-particle potential is present. The problem is thus, in part, to understand why the one-particle wave-function produced by the GP ansatz is in such a good agreement with empirical data on BEC in atomic gases. The key is found in the comparison of the GP data with true quantum-mechanical particle density:

$$\varrho_{N,a}^{\text{QM}}(x) = N \int_{R^{3(N-1)}} dx_2 \cdots dx_N \, |\Psi_{N,a}^{\text{QM}}(x, x_2, \cdots, x_N)|^2 \quad . \qquad (14.2.27)$$

The second step is to reformulate the Gross–Pitaevsky theory leading to the solution of the effective one-particle non-linear Schrödinger equation

(14.2.18). For this purpose, [Lieb, Seiringer, and Yngvason, 2000a] consider the problem of minimizing the energy functional

$$\mathcal{E}^{\mathrm{GP}}[\Phi] = \int_{\mathbb{R}^3} \mathrm{d}x^3 \left[|\nabla\Phi(x)|^2 + V_{\mathrm{trap}}(x)|\Phi(x)|^2 + 4\pi a|\Phi(x)|^4 \right] \qquad (14.2.28)$$

where Φ is a function defined on $\mathbb{R}^2$. In order to have a proper mathematical formulation of the problem, conditions are to be imposed on Φ. [Lieb, Seiringer, and Yngvason, 2000a] choose to impose that the domain of variation $\mathcal{D}_N$ be the set of all Φ satisfying the following four conditions: (a) $\nabla\Phi \in \mathcal{L}^2(\mathbb{R}^3)$; (b) $V|\Phi|^2 \in \mathcal{L}^1(\mathbb{R}^3)$; (c) $\Phi \in \mathcal{L}^4(\mathbb{R}^3) \cap \mathcal{L}^2(\mathbb{R}^3)$; and (d) $\int_{\mathbb{R}^3} \mathrm{d}x^3 \, |\Phi(x)|^2 = N$. Under these conditions, they first establish that:

$$E_{N,a}^{\mathrm{GP}} \equiv \inf_{\Phi \in \mathcal{D}_N} \mathcal{E}^{\mathrm{GP}}[\Phi] \qquad (14.2.29)$$

is a minimum, i.e.

$$\text{there exists} \quad \Phi_{N,a}^{\mathrm{GP}} \in \mathcal{D}_N \quad \text{such that} \quad E_{N,a}^{\mathrm{GP}} = \mathcal{E}^{\mathrm{GP}}[\Phi_{N,a}^{\mathrm{GP}}] \quad ; \qquad (14.2.30)$$

the minimizer $\Phi_{N,a}^{\mathrm{GP}}$ is unique up to a phase which can be chosen so that this function is strictly positive; and it solves the Gross–Pitaevsky equation:

$$\left[-\nabla^2 + V_{\mathrm{trap}}(x) + 4\pi a \, |\Phi^{\mathrm{GP}}(x)|^2 \right] \Phi^{\mathrm{GP}}(x) = \mu \, \Phi^{\mathrm{GP}}(x) \qquad (14.2.31)$$

where

$$\mu = \frac{\mathrm{d}}{\mathrm{d}N} E_{N,a}^{\mathrm{GP}} = \frac{1}{N} E_{N,a}^{\mathrm{GP}} + 4\pi a \, n^{\mathrm{GP}} \qquad (14.2.32)$$

with the mean density

$$n^{\mathrm{GP}} = \overline{\varrho}^{\mathrm{GP}} = \frac{1}{N} \int_{\mathbb{R}^3} \mathrm{d}x \, \varrho_{N,a}^{\mathrm{GP}}(x) \quad \text{and} \quad \varrho_{N,a}^{\mathrm{GP}}(x) = \left(\Phi_{N,a}^{\mathrm{GP}}(x) \right)^2 . \qquad (14.2.33)$$

As for the smoothness of the ground state wave function $\Phi_{N,a}^{\mathrm{GP}}$, it is at least once continuously differentiable in x; in fact when the trapping potential V_{trap} is infinitely differentiable, so is Φ. The energy $E_{N,a}^{\mathrm{GP}}$ is continuously differentiable in a and in N. And the following scaling properties hold:

$$E_{N,a}^{\mathrm{GP}} = N \, E_{1,Na}^{\mathrm{GP}} \quad \text{and} \quad \varrho_{N,a}^{\mathrm{GP}}(x) = N \, \varrho_{1,Na}^{\mathrm{GP}}(x) \quad . \qquad (14.2.34)$$

The third and crowning step in [Lieb, Seiringer, and Yngvason, 2000a] is to link the above variational form of the Gross–Pitaevsky ansatz with the fundamental, microscopic theory, namely to succeed in showing that:

$$E_{1,a_1}^{\mathrm{GP}} = \lim_{N \to \infty} \frac{1}{N} E_{N,\frac{a_1}{N}}^{\mathrm{QM}} \quad \text{and} \quad \varrho_{1,a_1}^{\mathrm{GP}}(x) = \lim_{N \to \infty} \frac{1}{N} \varrho_{N,\frac{a_1}{N}}^{\mathrm{QM}}(x) \qquad (14.2.35)$$

where the convergence is uniform on bounded intervals of a_1 for the first limit; and is in the weak $\mathcal{L}^1$ sense for the second limit.

We should note in passing that in a subsequent paper [Lieb, Seiringer, and Yngvason, 2000b] show that the analysis can be transferred, from the three-dimensional case reported above, to the two-dimensional situation, provided one replaces a in the Gross–Pitaevsky energy functional (14.2.28) by $g = |\ln(\bar{\varrho}^{GP}a^2)|^{-1}$. This result cuts across several competing recent speculations.

The above rigorously controlled limits, from the microscopic N–body theory to an effective one-particle theory, show precisely the sense in which the quantum theory of a weakly interacting Bose gas can account for some important empirical BEC data obtained with dilute atomic gases at very cold temperatures, namely those data that are modeled so well by the Gross–Pitaevsky ansatz.

Nevertheless, two limitations of the above derivation should be noted. First, the one-particle configuration-space density $\varrho^{QM}_{N,Na}$ defined in (14.2.27) is only the diagonal part of the full one-particle density matrix, namely the operator given by the kernel

$$
\left.
\begin{aligned}
\gamma^{QM}_{N,Na}(x,y) = N \int_{\mathbf{R}^{3(N-1)}} dx_2 \cdots dx_n \\
\Psi^{QM}_{N,Na}(x, x_2, \cdots, x_N)\, \Psi^{QM}_{N,Na}(y, x_2, \cdots, x_N)
\end{aligned}
\right\}
\tag{14.2.36}
$$

for which we have:

$$
\mathrm{Tr}\, \gamma^{QM}_{N,Na} = \int dx\, \gamma^{QM}_{N,Na}(x,x) = \int dx\, \varrho^{QM}_{N,Na}(x) = N \quad .
\tag{14.2.37}
$$

While the second equality in (14.2.35) shows the sense in which the Gross–Pitaevsky functional approximates the one-particle density in configuration space, namely $\varrho^{QM}_{N,Na}(x)$, the one-particle density in momentum space

$$
\tilde{\varrho}^{QM}_{N,Na}(p) = \iint dx\, dy\; e^{ip(x-y)}\, \gamma^{QM}_{N,Na}(x,y)
\tag{14.2.38}
$$

requires a knowledge of $\gamma^{QM}_{N,Na}(x,y)$, which is still missing. It is not even known whether the particular value of (14.2.38) for $p = 0$ satisfies

$$
\iint dx\, dy\; \gamma^{QM}_{N,Na}(x,y) = 0(N) \quad ,
\tag{14.2.39}
$$

although a proof may well not be far beyond the horizon of the powerful analytic tools developed by [Lieb, Seiringer, and Yngvason, 2000a]. By the same token, the above formalism does not yet establish that ODLRO holds.

A further limitation of their formalism is that it focuses exclusively on the ground state, i.e. the situation where the absolute temperature strictly vanishes $T = 0$ K rather than temperature states with $T > 0$ K. For instance, it remains difficult to evaluate the effect of the inter-particle interactions on the transition temperature.

More generally, it seems fair to say that an orderly account of the bountiful harvest of empirical data has not been completed yet. In particular, in spite of earlier arguments to the contrary, it is not clear theoretically how – and even whether – BEC and superfluidity are co-extensive; see e.g. [Huang, 1995, Leggett, 1999]. More empirical evidence on the situations where these two phenomena are observed – or not observed – concurrently seems also required for a comprehensive resolution of this and the larger issues concerning the extent to which transport phenomena can be understood from the concepts of equilibrium statistical mechanics only, i.e. without some discussion of the non-equilibrium aspects of the physically extreme situations now made available by the technological advances pioneered in the 1990s.

For several of the other issues that guided BEC research through the 1990s, the Reader will find a comprehensive overview in [Griffin, Snoke, and Stringari, 1995] from which we already cited several contributions; we want to mention also the following papers for their suitablity to further BEC related case-studies on the role of models:

- genuine BEC not being an ideal gas effect [Nozières, 1995]
- symmetry breaking in BEC [Leggett, 1995];
- kinetics of BEC [Kagan, 1995, Stoof, 1995];
- BEC in nuclear physics [Iachello, 1995, Hellmich, Röpke, Schnell, and Stein, 1995];
- 2D-BEC [Matsubara, Atai, Hotta, Kothonen, and Mizusaki, 1995];
- relation between BCS and BEC [Randeira, 1995].

The last two entries require brief comments. We would be remiss indeed if we did not allude to the two-dimensional models proposed by [Kosterliz and Thouless, 1973] for magnetism, solid-liquid transition and neutral superfluids; see in particular its $XY-$model version [Nelson and Kosterlitz, 1977].

A transition exists there to a low temperature phase exhibiting super-fluidity though no BEC is present (since $d = 2$; see above); this type of "topological" transition had been confirmed in laboratory investigations of thin Helium films [Rudnick, 1978, Bishop and Reppy, 1978]; [Fröhlich and Spencer, 1981] rigorously established the Kosterlitz–Thouless transition for a class of 2–dimensional models; and [Kennedy, Lieb, and Shastry, 1984] used the XY model to establish rigorously the existence of a system of interacting particles exhibiting a condensate identified with ODLRO.

Finally, it should be mentioned that ^{3}He was observed in a superfluid condensed phase [Osheroff, Richardson, and Lee, 1972] suggesting some pseudo-BEC akin to the BCS mechanism between Cooper pairs, here *fermionic* Helium atoms [Leggett, 1980]. The circumstances of this experimental discovery are summarized in [Lubkin, 1996] as they were recognized by yet another Nobel prize in the rich field that occupied us in this chapter.

15. Approach to Equilibrium in Quantum Mechanics

Wait till free from pain and sorrow
He has gained his final rest.

[Sophocles, ca. 425 BC]

15.1 Introduction

A motivating purpose for the study of non-equilibrium statistical mechanics is to obtain equations governing transport processes and to relate the value of the transport coefficients to the microscopic structure of the materials considered – matter or radiation. In this sense, non-equilibrium statistical mechanics is a *continuation* of the Boltzmann kinetic theory of gases; see Chap. 3. In another sense, it is also an *extension* of the theory, which takes into account the empirical fact that at the microscopic level the laws of quantum mechanics prevail over their classical precursors. Recall in this context the equilibrium counterpart provided by the Einstein–Debye quantum theory of the specific heat of solids, which we reviewed at the end of Chap. 10.

The present chapter is divided into three parts. After this introduction, we discuss in Sect. 15.2 the position occupied by the Markovian Pauli master equation in the framework of the generalized master equations; the necessity of a rescaling in time is presented. Then, in Sect. 15.3, these ideas are illustrated with the help of five specific models.

In regards to the scope of the program just sketched, we ask the Reader to suspend their judgement on at least two counts, as we realize that we could be reproached to bring too much into the problem ... or too little. In particular, it is certainly a legitimate question to ask whether it is really necessary to bring *quantum* considerations into the picture. In truth, we do not think quantum considerations are indispensible, although a fundamental obstruction against including these considerations would be catastrophic. Hence our focus on models in which the tenets of *quantum* statistical mechanics can be implemented. As for our taking too small a scoop, we recognize that there are other – some would say more fundamental – motivations for the study of non-equilibrium statistical mechanics, such as the search for the inexorable march of Fate, or for an all-encompassing microscopic explanation of the macroscopic arrow of time, or still for a general understanding that *either* irreversible thermodynamics is compatible with reversible microscopic mechanics, *or* it is necessary that at some cosmic level the world be not symmetric under time reversal. Such speculations are discussed in [Sklar, 1993, Savitt, 1995, Prigogine, 1997].

In keeping with the perspective of the present work, we examine with the help of specific models what can be learned from partial – but rigorously controlled – descriptions of quantum microscopic dynamical systems.

Here is a list of the ingredients we bring to bear.

• A microscopic autonomous quantum description of the dynamics. The observables form a non-commutative algebra $\mathcal{A}_\Lambda$, originally viewed as an algebra of operators acting on a Hilbert space. Subsequently, we discuss: (a) what happens to this algebra when the thermodynamical limit is considered; and (b) the notions of closed and open systems.

• $\mathcal{S}_\Lambda$, the collection of all states pertaining to this microscopic description.

• A conservative microscopic evolution described by a continuous one-parameter group of unitary operators acting on the Hilbert space on which the observables are defined, with a generator that can be interpreted as the Hamiltonian of the system. In the so-called Heisenberg picture, the evolution is traced by the observables: $A_t = \alpha(t)[A] = U(t)A\,U(-t)$. The evolution, when traced by the states – i.e. in the Schrödinger picture – is denoted by $\alpha^*(t)$; for all $\phi \in \mathcal{S}$ and all $A \in \mathcal{A}_\Lambda$, the evolution of the expected value of the observable A in the state ϕ is given by $\langle \phi; \alpha(t)[A] \rangle = \langle \alpha^*(t)[\phi]; A \rangle$. When the thermodynamical limit is considered, the primary object is a one-parameter group of automorphisms of a non-commutative algebra.

• A macroscopic level of description determined by a particular collection $\mathcal{A}_\Omega$ of observables; usually we have $\mathcal{A}_\Omega \subset \mathcal{A}_\Lambda$. To this corresponds an operation of *coarse-graining,* or *conditional expectation,* which we denote by $\mathcal{E}$ and define precisely. The injection of $\mathcal{A}_\Omega$ into $\mathcal{A}_\Lambda$ is denoted by $\mathcal{J}$.

• A dissipative semi-group $\{\gamma(\tau) \mid \tau \in \mathsf{R}^+\}$ describing the evolution as viewed from the macroscopic observables. We argue that $\gamma(\tau)$ should be completely positive maps; and we present some empirical consequences of this conclusion.

• A class $\mathcal{S}_o$ of admissible initial states or preparations of the microscopic system.

• A rescaling of time $t \to \tau$, allowing to focus on the proper description of genuinely macroscopic phenomena, which we expect to unfold much more slowly than do the microscopic fluctuations.

The relations between the objects defined above is schematically summarized in the following diagram, where the top level symbolizes the microscopic dynamics, and the bottom level the macroscopic dynamics; while the diagram refers explicitly to the Heisenberg picture, a similar diagram holds in the Schrödinger picture. [Compare also with Fig. 8.4 in Rem. F of Sect. 8.2].

$$\mathcal{A}_\Lambda \xrightarrow{\ \alpha(t)\ } \mathcal{A}_\Lambda$$

$$\mathcal{J}\uparrow \qquad\qquad \downarrow\varepsilon$$

$$\mathcal{A}_\Omega \xrightarrow{\ \gamma(\tau)\ } \mathcal{A}_\Omega$$

15.2 Master Equations

The generic name, *master equation,* covers a class of differential equations that have in quantum statistical physics a role similar to the one we have seen played – in Sect. 3.3 – in classical theory by the Boltzmann equation. The archetype was derived in [Pauli, 1928] under seemingly *ad hoc* assumptions and/or approximations that left considerable room for further justifications. The Pauli master equation is akin to the Chapman–Kolmogorov equation encountered in the study of Markovian stochastic processes.

In the present context, this equation takes the form of equations (15.2.22) or (15.2.40). As with the classical Boltzmann's Stosszahlansatz, the main point of contention in the derivation of the quantum master equation was the unwarranted use of various form of the so-called *repeated random phase approximation*.

A renewed attack was developed in a series of papers [Van Hove, 1955, 1957, 1959]; the claim was that only an initial random phase assumption was needed, provided one restricts one's attention to systems where (a) some specific properties of the Hamiltonian are satisfied, which van Hove listed; and (b) the non-equilibrium regime can be characterized by a time-rescalling that is known today as the van Hove $\lambda^2 t$–limit. The approach however relied heavily on cumbersome perturbation techniques. Nevertheless, van Hove was able to illustrate how his formalism should/could be applied to various transport phenomena, such as: heat conduction in metals, ferromagnetic relaxation, electric resistivity in metals due to the interactions of the electrons with the phonons or with impurities. An independent, but often similar approach, was pioneered by [Prigogine and Résibois, 1961, Prigogine, 1962] and their collaborators.

Non-perturbative approaches recovering a structure akin to van Hove's generalized master equation were then proposed, in particular by [Nakajima, 1958, Zwanzig, 1960, Swenson, 1962, Zwanzig, 1964, Emch, 1964]. A general agreement on the status of various master equations seemed elusive for some time, as witnessed by the following articles: [Van Hove, 1955, 1957, 1959, 1961, 1962, Van Kampen, 1962, Montroll, 1962, Emch, 1966b, Lanz, Lanz, and Scotti, 1967, Kalashnikov and Zubarev, 1971].

Master equations are attempts to reduce the dissipative evolution of some systems of interest to a purposefully incomplete account of the conservative evolution of some underlying microscopic systems.

The case where both systems are classical belongs to the study of classical stochastic processes. In this section, we want to limit our attention to the cases where the microscopic systems are quantum systems; this study splits in two classes, depending on whether the dissipative level of description is classical or quantum. We start with cases where the *classical* systems of interest are viewed as part of a *quantum* microscopic system; and we consider the embeddings where both levels are quantum in the last part of this section. The corresponding generalized master equations are given by Thm. 15.2.1 and 2.2 below. Note that, in contrast with the differential equation of Pauli, these equations are integro-differential equations: the instantaneous rate of change of the state of the system of interest is expressed in terms of the whole past history of the system rather than just the instantaneous gain-loss mechanism of equations (15.2.22) and (15.2.40). Time-rescalling is devised in part to collapse this drawback.

Since the study of spin-lattice systems served us well in the previous chapters, it is tempting to turn to them again as putative models for the conservative microscopic description of quantum systems. Hence, we present the theory first in the case of a finite lattice $\Lambda = \mathsf{Z}^{N^d} = \{\, n = (n_1, n_2, \cdots, n_d) \mid n_\nu = 0, 1, \cdots, N - 1 \text{ and } \nu = 1, 2, \cdots, d \,\}$; as usual, we denote by $|\Lambda| = N^d$ the number of sites in the lattice Λ. At each site n of the lattice is sitting a quantum spin, $\sigma_n = (\sigma_n^{(x)}, \sigma_n^{(y)}, \sigma_n^{(z)})$, where the $\sigma_n^{(k)}$ ($k = x, y, z$) are copies of the Pauli matrices

$$\sigma^{(x)} = \begin{pmatrix} 0 & 1 \\ 1 & 0 \end{pmatrix} \quad ; \quad \sigma^{(y)} = \begin{pmatrix} 0 & -i \\ i & 0 \end{pmatrix} \quad ; \quad \sigma^{(x)} = \begin{pmatrix} 1 & 0 \\ 0 & -1 \end{pmatrix} \quad .$$

The three components of the spin at the site n generate the algebra $\mathcal{A}_n = M(2, \mathsf{C})$ of two-by-two matrices with complex entries. We next identify this algebra with the subalgebra of the algebra of the whole system, $\mathcal{A}_\Lambda = M(2^{|\Lambda|}, \mathsf{C})$, via the injection:

$$A_n \in \mathcal{A}_n \rightarrow j(A_n) = I \otimes I \otimes \cdots \otimes I \otimes A_n \otimes I \otimes \cdots \otimes I \in \mathcal{A}_\Lambda \qquad (15.2.1)$$

where A_n enters as the $n-$th factor in this $|\Lambda|-$fold tensor product; I denotes the identity matrix. Explicit mention of the injection j is usually dropped.

For any state ϕ and any observable $A \in \mathcal{A}_\Lambda$, let $f(t) = \langle \phi; A(t) \rangle$, with $A(t) = \alpha(t)[A]$, where $\{\alpha(t) \mid t \in \mathsf{R}\}$ is a group of automorphisms of $\mathcal{A}_\Lambda$. The fact that the Hilbert space $\mathcal{H}_\Lambda = \mathcal{C}^{2^{|\Lambda|}}$ is finite-dimensional brings in several simplifications.

First, to every state $\phi : A \in \mathcal{A}_\Lambda \mapsto \langle \phi; A \rangle \in \mathsf{R}$ corresponds a density matrix ϱ such that

$$\forall\, A \in \mathcal{A}_\Lambda \; : \; \langle \phi; A \rangle = \mathrm{Tr} \varrho A \qquad (15.2.2)$$

where, as usual, Tr denote the trace; we do not want at this point to specify whether ϕ is a pure state or a mixture.

Second, $\{\alpha(t) \mid t \in \mathsf{R}\}$ is unitarily implemented, i.e.

$$\forall A \in \mathcal{A}_\Lambda : \; \alpha(t)[A] = e^{iHt} A e^{-iHt} . \tag{15.2.3}$$

Third, the Hamiltonian H is bounded, and thus the following limit exists

$$\forall A \in \mathcal{A}_\Lambda : \; \mathcal{L}_H[A] \equiv -i\frac{d}{dt}\alpha(t)[A]|_{t=0} = HA - AH . \tag{15.2.4}$$

Fourth, the Hamiltonian H and the Liouville operator $\mathcal{L}_H$ have discrete spectrum

$$\left.\begin{array}{l} \mathrm{Spec}(H) = \{\nu_k \mid k = 1, 2, \cdots, K \le 2^{|\Lambda|} \} \\[2mm] \mathrm{Spec}(\mathcal{L}_H) = \{\lambda_{k,k'} = \nu_k - \nu_{k'} \mid k, k' = 1, 2, \cdots, K\} \end{array}\right\} . \tag{15.2.5}$$

Note that the spectrum of the Liouville operator is symmetric: together with $\lambda_{k,k'}$ it contains $\lambda_{k',k} = -\lambda_{k,k'}$.

Let us go back to $f(t) = \langle \phi ; \alpha(t)[A] \rangle$ and note that every such function is a superposition of periodic functions with frequencies proportional to the differences $(\lambda_{k,k'} = \nu_k - \nu_{k'})$; hence, the function $f(t)$ necessarily is *almost periodic*, namely, given any time t_o and any tolerance $\varepsilon > 0$, there exists a finite time T such that – after leaving the neighborhood $(f(t_o) - \varepsilon, f(t_o) + \varepsilon)$ – $f(t)$ returns into this neighborhood before the time T has elapsed: *recurrences are build into these models*. Nevertheless, from what we learned from the classical dog-flea model – Sect. 3.4 – we know that this should not necessarily preclude us from discerning some tendency towards an approach to equilibrium, as recurrence times are expected to become extremely long when the size of the system becomes larger. Hence, we should be prepared to have to bring into play the thermodynamical limit $|\Lambda| \to \infty$. So much for the "microscopic" level.

Suppose now that we are interested in a *classical* family $\mathcal{A}_\Omega$ of observables defined on a finite configuration space Ω :

$$\mathcal{A}_\Omega \equiv \{A : \omega \in \Omega \mapsto A(\omega) \in \mathsf{R}\} . \tag{15.2.6}$$

Our problem is now to see what kind of evolution is defined on this system if one considers it as part of the spin-lattice microscopic system described above.

To achieve this embedding, we select $\{E_\omega\}$, a partition of the identity I into mutually orthogonal projectors, i.e. $E_\omega = E_\omega{}^2 = E_\omega{}^*$; $E_\omega E_{\omega'} = \delta_{\omega,\omega'} E_\omega$; and $\sum_\omega E_\omega = I$. We denote by the same symbol E_ω the projector and its range, the subspace $\{\Psi \in \mathcal{H}_\Lambda \mid E_\omega \Psi = \Psi\}$; and by $G_\omega = TrE_\omega$ the dimension of this subspace. The subspaces E_ω are referred to as *cells*. We define then the injective map

$$\mathcal{J} : A \in \mathcal{A}_\Omega \mapsto \mathcal{J}[A] \in \mathcal{A}_\Lambda \quad \text{with} \quad \mathcal{J}[A] = \sum_{\omega \in \Omega} A(\omega) E_\omega . \tag{15.2.7}$$

Hence, once the choice of the partition $\{E_\omega\}$ has been made, a unique microscopic observable $\mathcal{J}[A] \in \mathcal{A}_\Lambda$ is associated to each A in the algebra $\mathcal{A}_\Omega$ of observables of interest. It is in this sense that the partial, classical description of the system is embedded in a complete, microscopic description, i.e. that the classical system of interest is viewed as part of a quantum microscopic system. At this point, we are not ready to commit on the specific choice of a collection $\mathcal{J}[\mathcal{A}_\Omega]$ of "macroscopic" observables that could lend themselves to a self-contained dynamics. A tentative example could obtain from observables that are space-averages of microscopic observables over a region sufficiently large to smooth out the rapid microscopic fluctuations, and yet small enough to allow for a description of realistic transport phenomena. For the moment, we wish to assume ideally that some $\mathcal{A}_\Omega$ and some $\mathcal{J}$ have been chosen.

The next step is to introduce a coarse-graining operation with respect to $\mathcal{A}_\Omega$. Formally, this is achieved by the conditional expectation

$$\mathcal{E} : A \in \mathcal{A}_\Lambda \mapsto \mathcal{E}[A] \in \mathcal{A}_\Omega \ \text{ with } \ \mathcal{E}[A] : \omega \in \Omega \mapsto \frac{1}{G_\omega} Tr(A \ E_\omega) \in \mathsf{R} .$$

(15.2.8)

This conditional expectation satisfies $\mathcal{E} \circ \mathcal{J}[A] = A$ for all $A \in \mathcal{A}_\Omega$. Let us also define

$$\mathcal{D} : A \in \mathcal{A}_\Lambda \mapsto \mathcal{J} \circ \mathcal{E}[A] = \sum_{\omega \in \Omega} Tr(A \varrho_\omega) \, E_\omega \in \mathcal{A}_\Lambda \quad \text{with} \quad \varrho_\omega \equiv \frac{E_\omega}{G_\omega} \ \ (15.2.9)$$

which describes the smoothing out of the microscopic observables over each cell E_ω; for this reason $\mathcal{D}[A]$ is referred to as the *coarse-graining of the microscopic observable A*.

As usual, we should be able to view things from the point of view of the states as well as that of the observables. The state-spaces on our two systems are:

$$\mathcal{S}_\Omega \equiv \{ \, p : A \in \mathcal{A}_\Omega \mapsto \langle p; A \rangle = \textstyle\sum_{\omega \in \Omega} p(\omega) A(\omega) \in \mathsf{R} \, \}$$

$$\mathcal{S}_\Lambda \equiv \{ \, \phi : A \in \mathcal{A}_\Lambda \mapsto \langle \phi; A \rangle = Tr(\varrho A) \in \mathsf{R} \, \}$$

(15.2.10)

where p is a probability distribution, and ϱ is a density matrix. We then define by duality from (15.2.7) and (15.2.8):

$$\mathcal{J}^* : \phi \in \mathcal{S}_\Lambda \mapsto \mathcal{J}^*[\phi] \in \mathcal{S}_\Omega \ \text{ with } \ \mathcal{J}^*[\phi](\omega) = \langle \phi; E_\omega \rangle$$

$$\mathcal{E}^* : \ p \in \mathcal{S}_\Omega \mapsto \mathcal{E}^*[p] \in \mathcal{S}_\Lambda \ \text{ with } \ \mathcal{E}^*[p] = \textstyle\sum_{\omega \in \Omega} p(\omega) \frac{1}{G_\omega} E_\omega$$

(15.2.11)

The meaning of the first of these operations is that $\mathcal{J}^*[\phi]$ is simply the restriction of the microscopic state to the observables of interest; and we have $\mathcal{J}^* \circ \mathcal{E}^*[p] = p$ for all $p \in \mathcal{S}_\Omega$. In the second operation in (15.2.11), we lightened the notation by identifying the microscopic states ϕ and their associated density matrix ϱ. The operation $\mathcal{E}^*$ is more subtle. Consider indeed the following equivalence relation between microscopic states

$$\phi_1 \equiv \phi_2 \bmod \mathcal{A}_\Omega \ \text{ whenever } \ \forall A \in \mathcal{A}_\Omega : \langle \phi_1; \mathcal{J}[A] \rangle = \langle \phi_2; \mathcal{J}[A] \rangle , \quad (15.2.12)$$

i.e. two microscopic states belong to the same equivalence class exactly when they cannot be distinguished from one another by any observable $A \in \mathcal{A}_\Omega$. For any microscopic state ϕ, the microscopic state

$$\mathcal{D}^*[\phi] \equiv \mathcal{E}^* \circ \mathcal{J}^*[\phi] = \sum_{\omega \in \Omega} \langle \phi; E_\omega \rangle \varrho_\omega \quad \text{with} \quad \varrho_\omega = \frac{1}{G_\omega} E_\omega \qquad (15.2.13)$$

belongs to the equivalence class of ϕ, and contains exactly the relevant information pertaining to ϕ as viewed from the observables of interest. For this reason, the representative $\mathcal{D}^*[\phi]$ of the equivalence class of ϕ is referred to as the *coarse-graining of microscopic state* ϕ. In particular, in the extreme case where all the projectors E_ω are one-dimensional, and thus define an orthonormal basis $\{\Psi_k\}$ in the Hilbert space $\mathcal{H}_\Lambda$, $\mathcal{E}^* \circ \mathcal{J}^*[\varrho]$ is the diagonal part of ϱ in that basis: $\mathcal{D}^*[\varrho] = \sum_k (\varrho \Psi_k, \Psi_k) P_{\Psi_k}$. In this case, one speaks of *fine-graining*.

With these notions and notations in place, we can now formulate the strategy.

1. View the system of interest as part of the microscopic system by associating to every observable $A \in \mathcal{A}_\Omega$ the microscopic observable $\mathcal{J}[A] \in \mathcal{A}_\Lambda$.
2. Let this microscopic observable evolve according to the microscopic evolution $\{\alpha(t) \mid t \in \mathsf{R}\}$, i.e. $\mathcal{J}[A] \xrightarrow{t} \alpha(t) \circ \mathcal{J}[A]$.
3. Associate the coarse-grained microscopic state $\mathcal{E}^*[p_{t_o}] \in \mathcal{S}_\Lambda$ to the initial state $p_{t_o} \in \mathcal{A}_\Omega$ of the system of interest. Hence $\mathcal{S}_o \equiv \mathcal{E}^*[\mathcal{S}_\Omega]$ is the class of microscopic states that we can accept as initial states on the basis of the information we have from the observables of interest; another way to phrase this is to say that we consider these states as resulting from initial preparations that average over all microscopic states that are equivalent w.r.t. the observables of interest.
4. Compute, for all $t > t_o$,

$$\langle p_t; A \rangle \equiv \langle \mathcal{E}^*[p_{t_o}]; \alpha(t - t_o) \circ \mathcal{J}[A] \rangle \quad . \qquad (15.2.14)$$

We obtain:

$$\langle p_t; A \rangle = \sum_{\omega'} p_t(\omega') A(\omega') \quad \text{with} \quad p_t(\omega') = \sum_\omega P_{t,t_o}(\omega', \omega) p_{t_o}(\omega)$$
$$\text{where} \quad P_{t,t_o}(\omega', \omega) = \text{Tr}(E_{\omega'} \alpha(t - t_o)[E_\omega]) \frac{1}{G_\omega} \qquad (15.2.15)$$

Equation (15.2.15) describes a homogeneous stochastic process, i.e. P_{t,t_o} depends only on the difference $t - t_o$; thus, we simply write P_{t-t_o}, and make the convention $t_o = 0$.

The first question is to determine whether this process can be Markovian, i.e. whether it is possible to have

$$\forall t_2 \geq t_1 \geq 0 \quad : \quad P_{t_2}(\omega'', \omega) = \sum_{\omega' \in \Omega} P_{t_2 - t_1}(\omega'', \omega') P_{t_1}(\omega', \omega) \quad . \qquad (15.2.16)$$

This condition is equivalent to the condition that

$$\gamma(t) = \mathcal{E} \circ \alpha(t) \circ \mathcal{J} \tag{15.2.17}$$

is a semi-group, i.e.

$$\forall\, t_2 \geq t_1 \geq 0 \quad : \quad \gamma(t_2) = \gamma(t_2 - t_1) \circ \gamma(t_1) \quad . \tag{15.2.18}$$

Scholium 15.2.1 (No-go theorem). *The stochastic process (15.2.15) cannot be Markovian unless it is trivial, i.e. unless any – and thus all – of the following three equivalent conditions are satisfied:*

1. *the $p_t(\omega)$ are constant in time;*
2. *$\forall\, t \geq 0$: $P_t(\omega', \omega) = \delta_{\omega',\omega}$;*
3. *$\forall\, t \geq 0$: $\gamma(t)$ is the identity map from $\mathcal{A}_\Omega$ onto itself.*

Proof: Condition (15.2.18) entails that for all $\tau > 0$,

$$\gamma(t+\tau) - \gamma(t) = (\gamma(\tau) - I) \circ \gamma(t),$$

and thus:

$$\frac{\mathrm{d}}{\mathrm{d}t}\gamma(t) = i\,\mathcal{L}_o \circ \gamma(t) \quad \text{with} \quad \begin{cases} \mathcal{L}_o = \mathcal{E} \circ \mathcal{L}_H \circ \mathcal{J} \quad \text{and} \\[2mm] \mathcal{L}_H[A] = HA - AH \end{cases} \quad . \tag{15.2.19}$$

However, a straightforward computation shows that for all $A \in \mathcal{A}_\Omega$:

$$\mathcal{E} \circ \mathcal{L} \circ \mathcal{J}\,[A](\omega) = \sum_{\omega',\omega} \frac{1}{G_\omega}\mathrm{Tr}(A\,E_\omega)\,\mathrm{Tr}(\mathcal{L}_H[E_\omega]\,E_{\omega'})$$

where for any $A, B \in \mathcal{A}$: $\mathcal{L}_A[B]$ denotes the commutator $AB - BA$. Upon noticing that for all $A, B, H \in \mathcal{A}$: $\mathrm{Tr}\mathcal{L}_H[A]\,B = \mathrm{Tr}H\mathcal{L}_A[B]$ and that – by definition – the E_ω commute among themselves, we see that the middle term in the RHS of the above equation vanishes; hence:

$$\mathcal{E} \circ \mathcal{L}_H \circ \mathcal{J} = 0 \tag{15.2.20}$$

Upon feeding (15.2.20) in (15.2.19) we obtain that $\gamma(t)$ is constant in time; the proof is thus completed by recalling that $\gamma(0) = I$. **q.e.d.**

The situation at this point can be summarized as follows:

> *there cannot be a rigorous mathematical derivation of the macroscopic equations from the microscopic ones. Some additional information or assumption is indispensable. One cannot escape from this fact by any amount of mathematical funambulism.*
>
> [Van Kampen, 1962]

Although it is true that there are physical systems that approach equilibrium in a non-Markovian manner – see Model 1 in Sect. 15.3 – it nevertheless appears that in the laboratory there are large ranges of times for which exponential decay is the rule rather than the exception. Hence, we cannot accept the above Scholium without further ado. The missing ingredient in the above Scholium is not entirely with the mathematics but rather has to be looked for in a better appraisal of the physics responsible for the approach to equilibrium.

We recall that there are no objection of principle to having a dissipative semi-group imbedded as part of a conservative dynamics, and we already saw an archetypical exemple in Model 4′ [Sect. 8.2]. Indeed, since the space $\mathcal{H}_\Lambda$ is finite-dimensional, $\mathcal{A}_\Lambda$ comes naturally equipped with a scalar product, namely: $\forall A, B \in \mathcal{A}_\Lambda \, (A, B) \equiv \mathrm{Tr} AB^*$. Let further $\mathcal{H}_\Omega$ denote the subspace spanned by $\{E_\omega \mid \omega \in \Omega\}$; compare $\{\mathcal{H}_\Omega, \mathcal{A}_\Lambda, \gamma, \alpha, \mathcal{E}, \mathcal{J}\}$ with $\{\mathcal{H}_S, \mathcal{H}, S, U, E, i\}$ in Model 8.2.4′.

There are several lessons to be learned from this model. First, $\mathrm{Span}\{U(t) [i[\mathcal{H}_o] \mid t \in \mathsf{R}\}$ is dense in $\mathcal{H}$, and thus the *dilation* of the semigroup S to the group U is minimal; it is unique up to unitary isomorphism; see [Riesz and Sz.-Nagy, 1955][Appendice], who show that such minimal dilation of a contraction semigroup acting on a Hilbert space is always possible. Second, the Hilbert space $\mathcal{H}$ is infinitely dimensional. The role of $\mathcal{L}$ is played here by the generator H of the unitary group $U(t)$; the action of this operator is given by: $H\Psi(x) - x\Psi(x)$; thus H is unbounded and Ψ_o is not in its domain; hence this example is not subject to the obstruction (15.2.20). The third, and related, lesson is that the spectrum of H is continuous. Hence, this typical model not only shows us what went wrong in Scholium 15.2.1, but it also indicates again that something like the thermodynamical limit needs to be taken, in order to allow the spectrum of our $\mathcal{L}_H$ to become denser and denser as to cover ultimately the whole of R.

Model 8.2.4′ thus certainly indicates that there is a hope to beating the Scholium. Nevertheless, as the thermodynamical limit is approached, very large eigenvalues of the Hamiltonian H and thus of the Liouville operator $\mathcal{L}$ appear, so that one has to contend with microscopic fluctuations of very high frequency. Let us examine thus how to implement in the present scheme the fact that measurements have a finite duration $\bar{t}$ over which it seems reasonable to consider a *time-smoothing* operation consisting, for instance, in taking some average over time, and thus to replace p_t in (15.2.15) by

$$\langle \hat{p}_t; A \rangle = \frac{1}{\bar{t}} \int_t^{t+\bar{t}} \mathrm{d}s \langle p_s; A \rangle \quad . \tag{15.2.21}$$

In contrast with (15.2.20), $\mathcal{E}\mathcal{L}_H \mathcal{M}_{\bar{t}} \mathcal{J}$ with $\mathcal{M}_{\bar{t}}[.] = \frac{1}{\bar{t}} \int_0^{\bar{t}} \mathrm{d}s \, \alpha(s)[.]$ does not necessarily vanish, thus lifting the obstruction which forbids the stochastic process $\{\hat{p}_t \mid t \in \mathsf{R}^+\}$ to be a non-trivial Markov process. Such a process would then satisfy the gain-loss Chapman–Kolmogorov equation

$$\frac{d}{dt}\hat{p}_t(\omega) = \sum_{w \in \Omega} P(\omega',\omega) \left(\frac{1}{G_\omega}\hat{p}_t(\omega) - \frac{1}{G_{\omega'}}\hat{p}_t(\omega') \right) \tag{15.2.22}$$

with

$$P(\omega',\omega) = -i \, \text{Tr}\left(E_{\omega'} \, \mathcal{L}_H \circ \mathcal{M}_{\bar{t}}[E_\omega]\right) \quad . \tag{15.2.23}$$

In order to explore the import of this result, let $\bar{t}$ be strictly positive but sufficiently small for the following approximation of (15.2.23) to be plausible:

$$P(\omega',\omega) \approx -\frac{1}{2}\text{Tr}(E_{\omega'} \mathcal{L}_H{}^2[E_\omega])\bar{t} \tag{15.2.24}$$

so that the kernel $P \leq 0$, thus confirming the tendency to an approach to equilibrium. Moreover, for any splitting $H = H_o + \lambda V$ of the Hamiltonian, where H_o commutes with all E_ω, we can replace $\mathcal{L}_H$ in (15.2.24) by $\mathcal{L}_{\lambda V} = \lambda \mathcal{L}_V$. Upon noticing further that the diagonal part of the P does not contribute in (15.2.22), we can then replace (15.2.24) by

$$\tilde{P}(\omega',\omega) = K(\omega',\omega)\lambda^2\bar{t} \quad \text{with} \quad K(\omega',\omega) = \text{Tr}\left(E_{\omega'} V E_\omega V\right) \quad . \tag{15.2.25}$$

Thus, in the particular case where all the E_ω are one-dimensional projectors, we can choose an orthonormal basis $\{\Psi_\omega \mid \omega \in \Omega\}$ in $\mathcal{H}_\lambda$ such that for all $\omega \in \Omega : E_\omega \Psi_\omega$; we obtain then:

$$\tilde{P}(\omega',\omega) = |(V\Psi_{\omega'},\Psi_\omega)|^2 \lambda^2\bar{t} \tag{15.2.26}$$

which is precisely the term that appears in the original master equation [Pauli, 1928]. To derive his equation, Pauli makes the pivotal *repeated random phase approximation*.

As for the *Stosszahlansatz* invoked in the classical derivation of the Boltzmann equation – see Sect. 3.3 – this superficially reasonable assumption begs for a rigorous justification; here lies the bone of contention in most of the subsequent arguments about the Pauli master equation; see [Van Kampen, 1962] for a thorough discussion of the nature of the problem.

Aside from the fact that the introduction of time-smoothing allowed us to avoid the pitfall pointed out in the Scholium 15.2.1, our main assumption – namely to impose that the processus be Markovian – is of the same vein as Pauli's *repeated* random phase assumption, and time smoothing is in line with van Kampen's argument that one must take advantage of a putative macroscopic time-scale to wipe out rapid microscopic fluctuations. Nevertheless ours is not a full justification, but rather should be taken mainly as an indication of the direction in which one ought to look for an explanation.

A more systematic approach is to shelve temporarily the project to derive immediately from quantum mechanics a Markovian differential equation like (15.2.22–26), but to settle first for an admittedly less transparent equation of which one controls all the ingredients, i.e. such that one justifies all the steps involved in its derivation. Once such an equation is obtained, the problem is still to isolate the limiting regime(s) that allow to extract the ultimate Markovian behavior characteristic of so many transport phenomena.

Theorem 15.2.1. *The stochastic process (15.2.15) – with the convention $t_o = 0$ – satisfies the integro-differential equation*

$$\frac{d}{dt} p_t(\omega') = \int_o^t ds \sum_{\omega \in \Omega} P_{t-s}(\omega', \omega) \left(\frac{1}{G_\omega} p_s(\omega) - \frac{1}{G_{\omega'}} p_s(\omega') \right)$$

where

$$P_t(\omega', \omega) = -\operatorname{Tr} E_{\omega'} \left(\mathcal{L}_H \circ e^{-i\mathcal{L}_{I-\mathcal{D}} t} \circ \mathcal{L}_H \right) [E_\omega]$$

and for all $A \in \mathcal{A}_\Lambda$

$$\mathcal{L}_{I-\mathcal{D}}[A] \equiv (I - \mathcal{D}) \circ \mathcal{L}_H \circ (I - \mathcal{D}) [A] \quad .$$

To invite comparison with (15.2.22–26), the evolution equation claimed in the theorem is called a *generalized master equation*. The essential difference is that instead of depending only of the instantaneous state of the system at time t – as in (15.2.22) – the rate of change $\frac{d}{dt} p_t$ depends now – through the integration in s – on the whole history of the system between the initial time $t_o = 0$ and the final time t.

Proof of Thm. 15.2.1. The Reader should perhaps be warned that the proof is a little more technical than most of the materials in this work; therefore, we emphasize the main articulations, thus leaving auxilliary computations aside. To save on notation, we write $\mathcal{L}$ for $\mathcal{L}_H$, $\mathcal{U}(t) = \alpha^*(t) = e^{-i\mathcal{L}} t$ and $\mathcal{U}_{I-\mathcal{D}}(t) = e^{-i\mathcal{L}_{I-\mathcal{D}} t}$. We also make two preliminary remarks.

First, the kernel P_t satisfies the *detailed balancing relation:*

$$\sum_{\omega \in \Omega} P_t(\omega', \omega) = 0$$

so that

$$\left. \begin{array}{c} \sum_{\omega \in \Omega} P_{t-s}(\omega', \omega) \left(\frac{1}{G_\omega} p_s(\omega) - \frac{1}{G_{\omega'}} p_s(\omega') \right) = \\[2mm] \sum_{\omega \in \Omega} P_{t-s}(\omega', \omega) \frac{1}{G_\omega} p_s(\omega) \end{array} \right\} \quad . \qquad (15.2.27)$$

Second, the structure of the kernel reflects the fact – see (15.2.14-15) – that the coarse-grained evolution of the states is driven by

$$\mathcal{D}\mathcal{U}(t)\mathcal{D} \quad . \qquad (15.2.28)$$

It is then sufficient to show that

$$\mathcal{D}\mathcal{U}(t)\mathcal{D} = \mathcal{D} - \int_o^t ds \int_o^s du \, \mathcal{D} \mathcal{L} \mathcal{U}_{I-\mathcal{D}}(s - u) \mathcal{L} \mathcal{D} \, \mathcal{U}(s) \mathcal{D} \qquad (15.2.29)$$

since, upon feeding this into (15.2.15), we obtain the RHS of (15.2.27) with P_t as stated in the theorem.

The convolution appearing in (15.2.29) suggests then the use of the Laplace transforms

$$\mathcal{R}(z) = \int_o^\infty dt e^{-izt} \mathcal{U}(t) \quad \text{and} \quad \mathcal{R}_{I-\mathcal{D}}(z) = \int_o^\infty dt e^{-izt} \mathcal{U}_{I-\mathcal{D}}(t) \quad (15.2.30)$$

which are well-defined for z outside of the spectra of $\mathcal{L}$ and $\mathcal{L}_{I-\mathcal{D}}$.

Since the *resolvant* $\mathcal{R}(z) = (z + i\mathcal{L})^{-1}$ and similarly for the resolvant $R_{I-\mathcal{D}}(z)$, we have

$$R(z) - R_{I-\mathcal{D}}(z) = -i R_{I-\mathcal{D}}(z) (\mathcal{L}_{I-\mathcal{D}} - \mathcal{L}) R_{I-\mathcal{D}}(z) \qquad (15.2.31)$$

and since $\mathcal{DLD} = 0$, which is (15.2.20), we obtain, upon taking the inverse Laplace transform of (15.2.31):

$$\mathcal{U}(t) = \mathcal{U}_{I-\mathcal{D}} - i \int_o^t ds \, \mathcal{U}_{I-\mathcal{D}}(t - s) \, (\mathcal{DL} + \mathcal{LD}) \, \mathcal{U}(s) \quad . \qquad (15.2.32)$$

We use this relation twice, to compute first

$$\mathcal{DU}(t)\mathcal{D} = \mathcal{D} - i \int_o^t ds \, \mathcal{DU}_{I-\mathcal{D}}(t - s) \, (\mathcal{DL} + \mathcal{LD}) \, I \, \mathcal{U}(s) \, \mathcal{D} \qquad (15.2.33)$$

(where the operator I serves only as a marking to be used shortly); similarly, we obtain from (15.2.33) – with the change of variables $(t, s) \to (s, u)$:

$$(I - \mathcal{D})\mathcal{U}(s)\mathcal{D} = -i \int_o^s du (I - \mathcal{D})\mathcal{U}_{I-\mathcal{D}}(s - u) \, (\mathcal{DL} + \mathcal{LD})\mathcal{U}(u)\mathcal{D}. \quad (15.2.34)$$

Upon making the substitution $I = \mathcal{D} + (I - \mathcal{D})$ in the placed marked I in (15.2.33), and using repeatedly $\mathcal{DLD} = 0$, we obtain:

$$\mathcal{DU}(t)\mathcal{D} = \mathcal{D} - i\lambda \int_o^t ds \, \mathcal{D} \, \mathcal{L}(I - \mathcal{D})\mathcal{U}(s) \, \mathcal{D} \quad . \qquad (15.2.35)$$

Finally, upon replacing the last three factors in (15.2.35) by their expression in (15.2.34), we obtain the desired relation (15.2.29).   **q.e.d.**

The generalized master equation obtained in the theorem has been derived without any approximation, or any assumption beyond those we needed to specify the microscopic dynamics and the initial state. No recourse to any perturbation expansion was required. It is a mathematical identity which raises above the status of a mere tautology only if it suggests a limiting regime where the description by the observables of interest recovers some Markovian character. It is this limiting procedure to which van Hove arrived using his perturbative approach.

To make van Hove's reasoning apply in the non-perturbative context of the above theorem, note that it is always possible to write the Hamiltonian in the form

$$H = H_o + \lambda V \quad \text{with} \quad \forall \, \omega \in \Omega \; : \; H_o E_\omega - E_\omega H_o = 0 \quad . \tag{15.2.36}$$

For instance, this condition is satisfied by $H_o \equiv \mathcal{D}[H_o]$, and $\lambda V \equiv H - H_o$ where λ is any strictly positive number. These objects model here the quantities physicists refer to as the *free Hamiltonian* H_o, the *interaction* V, and the *coupling constant* λ. van Hove's idea was that the time-scale of the putative macroscopic evolution $\gamma(\tau)$ is determined by the strength λ of the interaction; his perturbation expansion suggested that one takes $\tau = \lambda^2 t$ where t is the time that enters in the microscopic evolution $\alpha_\lambda(t)$, corresponding to the Hamiltonian $H = H_o + \lambda V$. Therefore, in order to focus on this regime he suggested to emphasize the presence of the coupling constant λ in the stochastic process, by writing p_t^λ instead of p_t, and to consider:

$$p_\tau \equiv \lim_{\lambda \to 0} p_t^\lambda \quad \text{with} \quad \tau = \lambda^2 t \quad . \tag{15.2.37}$$

To see how this shows up in the non-perturbative approach followed up here, we introduce the notations $\mathcal{L}_o[A] = H_o A - A\,H_o$, $\mathcal{L}_V[A] = V A - A\,V$, and $\mathcal{U}_o(t) = e^{-i\mathcal{L}_o t}$; also, we continue – as in the proof of the Thm. 15.2.1 – to omit the composition signs $\circ$.

Condition (15.2.36) entails the identities

$$\left. \begin{array}{l} \mathcal{D}\mathcal{L}_o = 0 = \mathcal{L}_o \mathcal{D} \;\; ; \;\; \mathcal{D}\mathcal{L}_H = \lambda \mathcal{D}\mathcal{L}_V \;\; ; \;\; \mathcal{L}_H \mathcal{D} = \lambda \mathcal{L}_V \mathcal{D} \\[2mm] (I - \mathcal{D})\mathcal{L}_H(I - \mathcal{D}) = \mathcal{L}_o + \lambda(I - \mathcal{D})\mathcal{L}_V(I - \mathcal{D}) \end{array} \right\} \quad . \tag{15.2.38}$$

Upon inserting these in the expression of the kernel given by Thm. 15.2.1, we obtain:

$$\left. \begin{array}{l} P_t^\lambda(\omega', \omega) = -\lambda^2 \, K_t^\lambda(\omega', \omega) \\[2mm] K_t^\lambda(\omega', \omega) = \mathrm{Tr} E_{\omega'} \, \mathcal{K}_t^\lambda[E_\omega] \;\; ; \;\; \mathcal{K}_t^\lambda = \mathcal{D}\,\mathcal{L}_V \, \mathcal{U}^\lambda{}_{I-\mathcal{D}}(t)\,\mathcal{L}_V\,\mathcal{D} \\[2mm] \mathcal{U}^\lambda{}_{I-\mathcal{D}}(t) = e^{-i\mathcal{L}_{I-\mathcal{D}}t} \;\; ; \;\; \mathcal{L}_{I-\mathcal{D}} = \mathcal{L}_o + \lambda(I - \mathcal{D})\mathcal{L}_V(I - \mathcal{D}) \end{array} \right\} \tag{15.2.39}$$

where, in order to emphasize the presence of the coupling constant λ, we wrote P_t^λ instead of P_t. While (15.2.39) may seem somewhat cumbersome, it is worth reading carefully, as it emphasizes the separate roles played by the free evolution generated from H_o and the interaction V. In particular, if we could carry out the van Hove limit, we would expect that the stochastic process obtained in the limit (15.2.37) would satisfy the Markovian master equation

$$\left. \begin{array}{l} \frac{\mathrm{d}}{\mathrm{d}\tau} p_\tau(\omega') = \sum_{\omega \in \Omega} P(\omega', \omega) \left(\frac{1}{G_\omega}\, p_s(\omega) - \frac{1}{G_{\omega'}}\, p_s(\omega') \right) \\[2mm] \text{with} \\[2mm] P(\omega', \omega) = \int_o^\infty \mathrm{d}s \, K_s^o(\omega', \omega) \end{array} \right\} \quad . \tag{15.2.40}$$

Notice in particular that the kernel P just obtained would now only involve an integration over the free evolution $\mathcal{U}_o(s)$: in the time-scale considered, the rapid microscopic fluctuations would have been washed out, and the Markovian equation (15.2.40) would only describe their long-time cumulative effect.

The Reader will have noticed that up to and including the derivation of (15.2.39) we made claims we could mathematically sustain, whereas after this, we use a more guarded conditional mode. The reason is that, since τ is kept finite, and $\lambda \to 0$, the van Hove limit subsumes a limit $t \to \infty$.

van Hove was well-aware that the existence of this limit is precluded by the recurrences inherent to finite systems, and his presentation involves a concurrent passage to the thermodynamical limit in which the size of the system tends to infinity. This pre-emptive limit is in fact part and parcel of the way he conducts his perturbation expansion, selecting the "most divergent" diagrams in a way that was bound to raise many eyebrows and may have been part of the reputation of "mathematical funambulism" we quoted earlier. van Hove also realized that such questions could not be resolved in general with the mathematical technology available at the time; consequently, he saw that the problems should be explored with the help of specific models. A close reading of van Hove's papers seems to indicate that he was enclined to put these difficulties on the account of insufficient mathematics rather than on any misinterpretation of the physics; this impression was confirmed on several occasions during conversations one of us had the priviledge to have with van Hove from the early sixties to the mid-seventies. We try and follow van Hove's admonitions in our presentation of some of the models in Sect. 15.3.

Before we pass to the study of models, we want to indicate how the master equation approach can be extended to situations where the observables of interest describe a *quantum* system interacting with another quantum system which we refer to as the *reservoir*. For more details, see [Davies, 1974, 1976, Martin, 1979, Spohn and Dümcke, 1979, Spohn, 1980].

Here both of these systems are modeled by lattices at the sites of which sit quantum spins; we have then, for the system of interest: $\mathcal{A}_\Omega = M(2^{\Lambda_1}, \mathbb{C})$, for the reservoir $\mathcal{A}_R = M(2^{\Lambda_2}, \mathbb{C})$, and for the composite system $\mathcal{A}_\Lambda = M(2^\Lambda, \mathbb{C})$ with $\Lambda = \Lambda_1 \cup \Lambda_2$.

Before the two systems are brought into contact they evolve separately, with Hamiltonian evolutions $A_\Omega \in \mathcal{A}_\Omega \mapsto \alpha_\Omega(t)[A_\Omega] = e^{iH_\Omega t} A_\Omega e^{-iH_\Omega t} \in \mathcal{A}_\Omega$ and $A_R \in \mathcal{A}_R \mapsto \alpha_R(t)[A_R] = e^{iH_R t} A_R e^{-iH_R t} \in \mathcal{A}_R$. The initial state ϕ_Ω of the system of interest is arbitrary, while the initial state ϕ_R is assumed to be stationary : $\alpha_R^*[\phi_R] = \phi_R$. We now embed the system of interest into the composite system, and define:

$$\left. \begin{array}{l} \mathcal{J} : A_\Omega \in \mathcal{A}_\Omega \mapsto \mathcal{J}[A_\Omega] = A_\Omega \otimes I \in \mathcal{A}_\Lambda \\[2mm] \mathcal{E}^* : \phi_\Omega \in \mathcal{S}_\Omega \mapsto \mathcal{E}^*[\phi_\Omega] = \phi_\Omega \otimes \phi_R \in \mathcal{S}_\Lambda \end{array} \right\} \quad ; \qquad (15.2.41)$$

recall that, in the above, ϕ_R is a fixed state, part of the specification of the reservoir.

By duality, we obtain:

$$\left.\begin{array}{l} \mathcal{J}^* : \ \phi_\Lambda \in \mathcal{S}_\Lambda \mapsto \mathcal{J}[\phi_\Lambda] = \mathrm{Tr}_R[\varrho_\Lambda] \in \mathcal{S}_\Omega \\[2mm] \mathcal{E} : \ A_\Lambda \in \mathcal{A}_\Lambda \mapsto \mathcal{E}[A_\Lambda] = \mathrm{Tr}_R[A_\Lambda \cdot I \otimes \varrho_R] \in \mathcal{A}_\Omega \end{array}\right\} \ . \tag{15.2.42}$$

The notation in the above two relations may require some explanation. Firstly, we freely identify a state ϕ (i.e. a normalized positive linear functional on an algebra) with the corresponding density matrix ϱ. Second, we have four different traces in the present situation; the first three are the ordinary traces: Tr_Ω over matrices in $\mathcal{A}_\Omega$, Tr_R over matrices in $\mathcal{A}_R$, and Tr_Λ over matrices in $\mathcal{A}_\Lambda$; when no confusion is likely to occur, we denote all three traces by the symbol Tr. And there is a new object, denoted in (15.2.42) $\mathrm{Tr}_R[\cdot]$, the *partial trace:*

$$\mathrm{Tr}_R[\ ] : X \in \mathcal{A}_\Lambda \mapsto \mathrm{Tr}_R[X] \in \mathcal{A}_\Omega$$

which can be visualized as follows; recall that since all algebras are finite dimensional, every element $X \in \mathcal{A}_\Lambda = \mathcal{A}_\Omega \otimes \mathcal{A}_R$ can be written (not uniquely !) as $X = \sum_k Y_k \otimes Z_k$ with $Y_k \in \mathcal{A}_\Omega$ and $Z_k \in \mathcal{A}_R$; with this decomposition, one has $\mathrm{Tr}_R[X] \equiv \sum_k (\mathrm{Tr} Z_k) Y_k$ where the last trace is the ordinary trace on $\mathcal{A}_R$; one then shows that the object thus defined does not depend on the decomposition chosen to define it. Equivalently, with respect to the orthonormal bases $\{\Phi_m\}$ in $\mathcal{H}_\Omega$, $\{\chi_n\}$ in $\mathcal{H}_R$ and $\{\Psi_{mn} = \Phi_m \otimes \chi_n\}$ in $\mathcal{H}_\Lambda$, we have, in terms of matrix elements relative to these bases: $\mathrm{Tr}_R[X]_{m,m'} \equiv \sum_n X_{mn,m'n}$.

As before, we introduce the coarse-graining operations $\mathcal{D} \equiv \mathcal{J} \circ \mathcal{E}$ on the observables, and $\mathcal{D}^* \equiv \mathcal{E}^* \circ \mathcal{J}^*$ on the states; these now take the form:

$$\left.\begin{array}{l} \mathcal{D} : A \in \mathcal{A}_\Lambda \mapsto \mathcal{D}[A] = \mathrm{Tr}_R[A \cdot I \otimes \varrho_R] \otimes I \in \mathcal{A}_\Lambda \\[2mm] \mathcal{D}^* : \varrho \in \mathcal{S}_\Lambda \mapsto \mathcal{D}^*[\varrho] = \mathrm{Tr}_R[\varrho] \otimes \varrho_R \in \mathcal{S}_\Lambda \end{array}\right\} \ . \tag{15.2.43}$$

In particular:

$$\forall \ \varrho_\Omega \in \mathcal{S}_\Lambda : \ \mathcal{D}^*[\varrho_\Omega \otimes \varrho_R] = \varrho_\Omega \otimes \varrho_R \ . \tag{15.2.44}$$

Since $\mathcal{A}_\Lambda$ is finite-dimensional, we can consider $\mathcal{S}_\Lambda$ as a full positive cone in $\mathcal{A}_\Lambda$, and thus we can extend uniquely by linearity $\mathcal{D}^*$ to a linear operator

$$\mathcal{P} : X \in \mathcal{A}_\Lambda \mapsto \mathcal{P}[X] = \mathrm{Tr}_R[X] \otimes \varrho_R \in \mathcal{A}_\Lambda \ . \tag{15.2.45}$$

It is important for the sequel to note that $\mathcal{P}^2 = \mathcal{P}$.

At time $t_o = 0$ the two systems are brought into contact and subsequently evolve jointly according to $A \in \mathcal{A}_\Lambda \mapsto \alpha^\lambda(t)[A] = e^{iH^\lambda t} A e^{-iH_\lambda t} \in \mathcal{A}$ where the total Hamiltonian is

$$\left.\begin{array}{l} H_\lambda \equiv H_o + \lambda V \quad \text{with} \\[2mm] H_o = H_\Omega \otimes I + I \otimes H_R \quad \text{and} \quad \forall \ \phi_\Omega \in \mathcal{S}_\Omega : \ \mathcal{E}[V] = 0 \end{array}\right\} \ . \tag{15.2.46}$$

The situation we have just described is sometimes referred to by saying that the quantum system described by the observables in $\mathcal{A}_\Omega$ is *an open system* in *interaction* with a *reservoir*, the latter being described by the observables in $\mathcal{A}_R$ and the *specified* time-invariant state ϱ_R. For the term "reservoir" to make sense, we assume at a later stage of the study that the reservoir R is much bigger than the system of interest Ω so that the latter is driven to equilibrium by the former whereas the effect of the interaction on the reservoir is negligible, i.e. that the reservoir remains "practically unaffected". How to implement – and formulate precisely – these further requirements is discussed on specific models in the next section. The following theorem does not depend on these as yet unformulated assumptions; its applications however will.

Theorem 15.2.2. *For every initial state $\phi_\Omega \in \mathcal{S}_\Omega$ of the system of interest, let ϱ_t^λ be the coarse-grained density matrix associated with the state*

$$\forall\ A_\Omega \in \mathcal{A}_\Omega\ :\ \langle \phi_\Omega^\lambda(t)\,;\, A_\Omega \rangle\ =\ \langle \mathcal{E}^*[\phi_\Omega]\,;\, \alpha(t) \circ \mathcal{J}[A_\Omega] \rangle\quad.$$

Then ϱ_t^λ satisfies the generalized master equation:

$$\frac{\mathrm{d}}{\mathrm{d}t}\varrho_t^\lambda = -\mathrm{i}\,\mathcal{L}_{H_o}[\varrho_t^\lambda] - \lambda^2 \int_o^t \mathrm{d}s\, \mathcal{K}_{t-s}^\lambda\, \varrho_s^\lambda$$

where

$$\mathcal{K}_t^\lambda = \mathcal{P}\,\mathcal{L}_V\,\mathcal{U}^\lambda{}_{I-\mathcal{P}}(t)\,\mathcal{L}_V\mathcal{P}\quad \text{with}\quad \mathcal{U}^\lambda{}_{I-\mathcal{P}}(t) = \mathrm{e}^{-\mathrm{i}\,\mathcal{L}^\lambda{}_{I-\mathcal{P}}\,t}$$

and for all $A \in \mathcal{A}_\Lambda$

$$\mathcal{L}_{H_o}[A] \equiv [H_o, A] = H_o A - A\,H_o\ ;\quad \mathcal{L}_V[A] \equiv [V, A] = V A - A\,V$$

$$\mathcal{L}^\lambda{}_{I-\mathcal{P}}[A]\ \equiv\ \mathcal{L}_{H_o}[A] + \lambda(I - \mathcal{P})\mathcal{L}_V(I - \mathcal{P})[A]$$

Proof of Thm. 15.2.2. We proceed as in the proof of Thm. 15.2.1. We first note that the coarse-grained evolution is governed by $\mathcal{P}\mathcal{U}^\lambda(t)\mathcal{P}$ with $\mathcal{U}^\lambda(t) = \mathrm{e}^{-\mathrm{i}\,\mathcal{L}_{H^\lambda}\,t}$. The main technical difference is that we only have now $\mathcal{P}\mathcal{L}_{H_o} = \mathcal{L}_{H_o}\mathcal{P}$ which does not vanish in general; nor does $\mathcal{P}\mathcal{L}_{H_\lambda}\mathcal{P}$. Consequently, the last relation in (15.2.38) does not hold here. Nevertheless, the condition imposed on V in (15.2.46) entails that we still have $\mathcal{P}\mathcal{L}_V\mathcal{P} = 0$; hence the definition of $\mathcal{L}^\lambda{}_{I-\mathcal{P}}$ chosen in the statement of Thm. 15.2.2. We can then use the same resolvent technique as in Thm. 15.2.1 to show that (15.2.32) is replaced now by:

$$\mathcal{U}^\lambda(t) = \mathcal{U}^\lambda{}_{I-\mathcal{P}} - \mathrm{i}\lambda \int_o^t \mathrm{d}s\,\mathcal{U}^\lambda{}_{I-\mathcal{P}}(t - s)\,(\mathcal{P}\mathcal{L}_V + \mathcal{L}_V\mathcal{P})\mathcal{U}^\lambda(s)\quad.\quad(15.2.47)$$

The difference just mentioned shows up only in the fact that in the RHS of (15.2.33) and (15.2.35) now appears the free evolution; we have indeed:

$$\mathcal{P}\mathcal{U}^\lambda(t)\mathcal{P} =$$

$$\left.\begin{array}{l} \mathcal{P}\mathcal{U}^o(t) - \mathrm{i}\lambda \int_o^t \mathrm{d}s\, \mathcal{P}\mathcal{U}^\lambda{}_{I-\mathcal{P}}(t-s)(\mathcal{P}\mathcal{L}_V + \mathcal{L}_V\mathcal{P})I\mathcal{U}^\lambda(s)\,\mathcal{P} = \\[3mm] \mathcal{P}\mathcal{U}^o(t) - \mathrm{i}\lambda \int_o^t \mathrm{d}s\, \mathcal{U}^o(t-s)\,\mathcal{P}\,\mathcal{L}_V(I-\mathcal{P})\mathcal{U}^\lambda(s)\,\mathcal{P} \end{array}\right\} . \quad (15.2.48)$$

(15.2.34) becomes similarly

$$(I - \mathcal{P})\mathcal{U}^\lambda(s)\mathcal{P} = -\mathrm{i}\,\lambda \int_o^t \mathrm{d}u\, \mathcal{U}^\lambda{}_{I-\mathcal{P}}(s-u)\,(I-\mathcal{P})\,\mathcal{L}_V\,\mathcal{U}^\lambda(u)\,\mathcal{P} \quad . \quad (15.2.49)$$

Upon inserting (15.2.49) in (15.2.48), and letting this act on our coarse-grained initial state $\varrho_o = \varrho_\Omega \otimes \varrho_R = \mathcal{P}\varrho_o$, we receive for $\varrho_t^\lambda = \mathcal{P}\mathcal{U}^\lambda(t)\varrho_o$

$$\varrho_t^\lambda = \mathcal{U}^o(t)\varrho_o - \lambda^2 \int_o^t \mathrm{d}s \int_o^s \mathrm{d}u\, \mathcal{U}^o(t-s)\,\mathcal{P}\,\mathcal{L}_V\,\mathcal{U}^\lambda{}_{I-\mathcal{P}}(s-u)\,\mathcal{L}_V\,\mathcal{P}\,\varrho_u^\lambda \quad . \quad (15.2.50)$$

Theorem 15.2.2 follows upon taking the derivative of this expression with respect to t.   **q.e.d**

The form of the Kernel $\mathcal{K}_t^\lambda$ in the present Thm. 15.2.2 is quite similar to the form of the operator denoted by the same symbol in (15.2.39). This analogy is not accidental: as for the case where the variables of interest were classical one may reasonably hope, in the quantum case also, that the weak-coupling limit would lead to a Markovian behavior, i.e. the coarse-grained state ϱ_t^λ could be used to define

$$\forall\, t \in \mathsf{R}^+ \; : \; \varrho_\Omega(\tau) = \lim_{\lambda \to 0} \varrho_t^\lambda \quad \text{with} \quad \tau = \lambda^2 t \qquad (15.2.51)$$

which would satisfy

$$\frac{\mathrm{d}}{\mathrm{d}\tau}\varrho_\Omega(\tau) = \tilde{K}[\varrho_\Omega(\tau)] \quad \text{where} \quad \tilde{K} = \int_o^\infty \mathrm{d}s\, \mathcal{K}_s^o \quad . \qquad (15.2.52)$$

This conjecture is equivalent to asserting that the evolution of the composite system, when viewed from the quantum observables of interest only, is described in this regime by a semi-group with generator $\tilde{K}$. [Spohn and Dümcke, 1979] argued that the correct form of the generator of the semigroup for an admissible transport equation should involve an additional ergodic average over $\tilde{K}(s) = \exp(\mathrm{i}\mathcal{L}_\Omega s)\,\tilde{K}\,\exp(-\mathrm{i}\mathcal{L}_\Omega s)$. This correction – interpreted in the interaction picture – has been confirmed from a different perspective by [Jaksic and Pillet, 1997].

Although it is of quantum origin, the Pauli master equation (1928) describes in effect a classical continuous-time Markov process on a *discrete* state space. Within the confines of classical theory the study of Markov processes in *continuous* state space had been in gestation since the conception of the Boltzmann equation (1872), the largely ignored model of the stockmarket by Bachelier (1900), and the discussion of Brownian motion by Einstein–Smolukowski (1905); see Chap. 3. As detailed in [Von Plato, 1994] a burst of mathematical activity took shape in the (late) 1920s; it gained a firm foundation in [Kolmogorov, 1931, 1933], and blossomed soon into a full-fledged mathematical theory of stochastic processes, including in a natural setting the theory Markov processes on a continuous state space Ω and in continuous time $t \in \mathsf{R}$; see [Feller, 1968, 1971]. There the corresponding Markovian master equation is typically of the form

$$\partial_t\, p(t,\omega) = \int_\Omega \mathrm{d}\omega'\, [W(\omega',\omega)p(t,\omega') - W(\omega,\omega')p(t,\omega)] \qquad (15.2.53)$$

where $W(\omega,\omega')$ is the *transition probability* from the state $\omega \in \Omega$ to the state $\omega' \in \Omega$. Upon letting Ω be an additive group with invariant measure, say R^n, assuming that every thing in sight is smooth, expanding the first integral into a power series in $\nu = \omega - \omega'$, and making the uncontrolled approximation that consists in cutting off the series after the quadratic term, one receives:

$$\left.\begin{array}{l} \partial_t\, p(t,\omega) = -\partial_\omega\, [a_1(\omega)p(t,\omega)] + \tfrac{1}{2}\partial_\omega{}^2\, [a_2(\omega)p(t,\omega)] \\[2mm] \text{with} \qquad\qquad a_n(\omega) = \tfrac{1}{n!} \int_\Omega \mathrm{d}\nu\, W(\omega,\omega+\nu)\, \nu^n \end{array}\right\} \quad . \qquad (15.2.54)$$

This equation was first proposed by [Fokker, 1914, Planck, 1917] and bears their names. For the place of this equation in the mathematical theory, see [Feller, 1968, 1971]. The Fokker–Planck equation is usually presented in the physics literature in connection with the Boltzmann equation; see e.g. [De Groot and Mazur, 1962, Mazo, 1967]; or for a presentation that goes more in depth, see e.g. [Carmichael, 1999]; in particular, the equation admits a natural generalization to the quantum realm by using the Wigner function associated to the kernel of the density operator describing the state of a quantum system.

Whether classical or quantum, the Fokker–Planck equation begs the same question as the Pauli master equation, namely whether it can be derived rigorously as a description of certain priviledged observables pertaining to a more complex Hamiltonian system that has a recognizable microscopic interpretation. [Castella, Erdos, Frommlet, and Markowich, 2000] indicates that this is indeed possible.

The two theorems presented so far in this section suggest already a common program for both the quantum and the classical cases, namely to focus on a time scale that is determined by the strength of the interaction responsible for the putative approach to equilibrium. This program is now to be tested on elementary specific quantum models.

15.3 Approach to Equilibrium: Quantum Models

The itinerary for this section is provided by a collection of simple models that mark stations in a personal pilgrimage seeking to extract a Markovian approach to equilibrium from reversible microscopic Hamiltonian dynamics. Some references for further travels along other routes are also given.

Model 1: Non-Markovian approach to equilibrium. This model was proposed independently by [Heims, 1965, Emch, 1966a] and revisited in [Thirring, 1983][Sect. 1.1.1]. It was motivated by an actual laboratory experiment, called free-induction relaxation, and interpreted as the result of dipolar interactions between nuclear spins attached on the sites of a rigid lattice [Lowe and Nordberg, 1957].

The experiment proceeds as follows: (a) A CaF_2 crystal lattice is placed in a magnetic field B pointing in the direction z; (b) the system is given time to reach equilibrium at temperature T; (c) a radio-frequency pulse is applied to turn the magnetic moment of the spins of the F−nuclei in the direction x; (d) the time evolution of the nuclear magnetization is observed: from an initial non-zero value, it shows an oscillatory decay.

The model is a one-dimensional finite lattice Λ_N wrapped on a ring; i.e. we assume periodic boundary conditions, identifying k and $k + N$ in Λ. For convenience, we also assume $|\Lambda| = N$ even. The Hamiltonian is taken to be:

$$H_N = -B \sum_{k=1}^{N} \sigma_k^z + \sum_{k=1}^{N} \sum_{n=1}^{N/2} J_n \, \sigma_k^z \sigma_{k+n}^z \qquad (15.3.1)$$

where B is a constant number – mimicking the magnetic field – σ_k^z are copies of the usual Pauli matrix σ^z – mimicking the z−component of the nuclear spin at the site $k \in \Lambda$ – and the coupling J is a real-valued function which is non-increasing with the distance n, with $\lim_{n \to \infty} J(n) = 0$; later on, we will find it convenient to impose more specific conditions of J; see (15.3.6).

The initial state of the system is given by the canonical equilibrium density matrix

$$\left. \begin{aligned} \varrho_N(0) &= Z_\zeta^{-1} \, e^{-\zeta \sum_{k=1}^{N} \sigma_k^x} \\[2mm] \text{with} \quad Z_\zeta &= \mathrm{Tr} \, e^{-\zeta \sum_{k=1}^{N} \sigma_k^x} \quad \text{and} \quad \zeta = -B/kT \end{aligned} \right\} . \qquad (15.3.2)$$

The observables of interest are the three components of the magnetization

$$S_N{}^x = \frac{1}{N} \sum_{k=1}^{N} \sigma_k^x \qquad (15.3.3)$$

and similarly for the other two components $S_N{}^y$ and $S_N{}^z$. The evolution of these intensive observables of interest under the Hamiltonian (15.3.1), viewed from initial state (15.3.2) is described by

$$\langle S_N{}^x \rangle(t) = \mathrm{Tr}\left(\varrho_N(0)\, e^{iH_N t}\, S_N{}^x\, e^{-iH_N t}\right) \tag{15.3.4}$$

and similarly for the other two components. Since all the terms in the Hamiltonian (15.3.1) commute with one another, the computation of these expectation values is straightforward; without involving *any* further assumption or approximation, it gives:

Scholium 15.3.1.

$$\left.\begin{array}{rl}
\langle S_N{}^x \rangle(t) =& \langle S_N{}^x \rangle(0)\cos(2Bt)f_N(t) \\
\langle S_N{}^y \rangle(t) =& -\; \langle S_N{}^y \rangle(0)\sin(2Bt)f_N(t) \\
\langle S_N{}^z \rangle(t) =& \langle S_N{}^z \rangle(0)
\end{array}\right\} . \tag{15.3.5}$$

$$\text{where} \quad f_N(t) = \prod_{k-1}^{N/2} \cos^2(2J_n t) .$$

The periodic Larmor precession of the magnetization around the direction z of the magnetic field B is not of interest to us, nor is the third equality which just confirms that $S_N{}^z$ is a constant of the motion with respect to the Hamiltonian (15.3.1). The interest of the model is the term $\left[\prod \cdots\right]$. To draw out its meaning, we consider a particular case, where

$$J_n = 2^{-n} J_o . \tag{15.3.6}$$

Indeed, a product formula due to Euler gives in that particular case:

$$f(t) \equiv \lim_{N\to\infty} \prod_{k-1}^{N/2} \cos^2(2J_n t) = \left[\frac{\sin(J_o t)}{J_o t}\right]^2 \tag{15.3.7}$$

and thus

$$f_N(t) = f(t)\cdot W_N(t)^{-1} \quad \text{with} \quad W_N(t) = \left[\frac{\sin(J_o 2^{-N/2} t)}{J_o 2^{-N/2} t}\right]^2 . \tag{15.3.8}$$

Hence, one can control immediately the limit when the size of the system is let go to infinity; for $k = x, y, z$, the limits $\langle S^k \rangle$ for $\lim_{N\to\infty}\langle S_N{}^k \rangle$, exist and are given by

$$\left.\begin{array}{rl}
\langle S^x \rangle(t) =& \langle S^x \rangle(0)\cos(2Bt)f(t) \\
\langle S^y \rangle(t) =& -\langle S^y \rangle(0)\sin(2Bt)f(t) \\
\langle S^z \rangle(t) =& \langle S^z \rangle(0)
\end{array}\right\} \quad \text{with} \quad f(t) = \left[\frac{\sin(J_o t)}{J_o t}\right]^2 ; \tag{15.3.9}$$

the long-time limit of these expectation values exists, since

$$\lim_{t\to\infty} f(t) = 0 . \tag{15.3.10}$$

Moreover, since $\langle S^z \rangle(0) = 0$, the system – as viewed from the three intensive observables $\langle S^i \rangle$ – approaches the microcanonical ensemble. Nevertheless,

(15.3.7) entails that this limit is approached in an oscillatory manner: while $f(t) \geq 0$, it actually vanishes exactly when $t \in \{t_n = nJ_o^{-1}\pi \mid n \in \mathsf{Z}\}$; thus, even when the infinite size limit has been taken, *this time-evolution describes an approach to equilibrium that is not Markovian*.

The model is so simple that we can even describe in detail the behavior of the *finite* system. First f_N as well as f is symmetric in time:

$$\forall\, t \in \mathsf{R} \quad : \quad f_N(-t) = f_N(t) \tag{15.3.11}$$

and second:

$$\text{with} \quad \Theta_N = 2^{N/2} J_o^{-1}\pi \quad \text{and} \quad n \in \mathsf{Z} \quad : \quad \lim_{t \to n\Theta_N} f_N(t) = 1 \quad . \tag{15.3.12}$$

These relations are quantum equivalents of the Loschmidt and Zermelo objections to the classical Boltzmann kinetic theory; see Sect. 3.3. While some may want to ignore (15.3.11), since one does not perform laboratory experiment backwards in time, we prefer to view it as one more manifestation that the task of reconstructing the past from the present is as hard as that of anticipating the future. (15.3.12) indicates the presence of recurrences; the Reader will notice that in this model as in the dog-flea model discussed in Sect. 3.4, the recurrence time Θ_N grows exponentially with N, the size of the system.

Several variations on the model have been devised.

The first is to note that what matters for the computations below is that $\varrho(0)$ be diagonal in a basis that diagonalizes all σ_k^x; hence, we can replace the inital state (15.3.2) by the canonical equilibrium state corresponding to the rotated Hamiltonian (15.3.1) where all σ_k^z have been turned to σ_k^x. This does not affect (15.3.5) and thus the consequences it entails.

A second variation is to replace the intensive variables corresponding to the three components of the magnetization S by a single macroscopic spin σ_κ; the relations (15.3.5) are formally invariant under this change.

In a third variation, we can decide to consider only the component σ_κ^z of the spin σ_κ. The Hilbert space of the finite system Λ decomposes then into the orthogonal sum of the two eigensubspaces $\{E_\omega \mid \omega \in \Omega \equiv \{-1, 1\}\}$. And we obtain

$$\langle \sigma_\kappa^x \rangle = \sum_{\omega \in \Omega} \omega\, p_t(\omega) \quad \text{with} \quad p_t(\pm) = \frac{1}{2}[1 \pm a\, f_N(t)] \tag{15.3.13}$$

where a is read from the inital conditions. This probability distribution satisfies a generalized master equation of the type discussed in Sect. 15.2; and this equation is *not* of Chapman–Kolmogorov type, i.e. the corresponding classical stochastic process is *not* Markovian.

Let us now return to the case where we consider all three components of the spin σ_κ, considering thus the system at site κ as a quantum open system coupled to the reservoir provided by the other spins; this is the situation considered in the second variation above.

The fourth variation [Thirring, 1983] consists in taking for inital state a pure state with

$$\left.\begin{array}{c} \langle\sigma_\kappa^z\rangle(0) = s \; ; \; \langle\sigma_\kappa^x\rangle(0) = \tfrac{1}{2}\,a\,\sqrt{1-s^2} \; ; \; \langle\sigma_\kappa^y\rangle(0) = \tfrac{1}{2}\,b\,\sqrt{1-s^2} \\[2mm] a^2 + b^2 = 1 \quad \text{and} \quad \prod_{\kappa\in\Lambda}\langle\sigma_\kappa\rangle(0) = \langle\prod_{\kappa\in\Lambda}\sigma_\kappa\rangle(0) \end{array}\right\} . \quad (15.3.14)$$

Then, here again, the relations (15.3.5) prevail, with the change in notation substituting σ_κ for S. The only difference occurs in the fact that now $\sigma_\kappa^z(t) = s$. This implies that when the infinite size limit is taken, the open system at κ approaches the canonical equilibrium state

$$\varrho_\infty = Z_\beta^{-1}\,e^{-\beta\sigma_\kappa^z} \quad \text{with} \quad Z_\beta = \mathrm{Tr}e^{-\beta\sigma_\kappa^z} \quad \text{and} \quad \beta = \tanh^{-1}s \quad (15.3.15)$$

instead of the microcanonical equilibrium state obtained in the previous variations.

For another model of a non-Markovian approach to equilibrium, see [Gates, Gerst, and Kac, 1973].

Model 2: Decay of local thermal deviations from equilibrium. When a local thermal disturbance is imposed on an ordinary macroscopic body in a state of equilibrium, common experience indicates that in due time this disturbance is absorbed, and the system returns to its original equilibrium state. The present model shows that this experience is compatible with the tenets of quantum statistical mechanics.

Scholium 15.3.2. *There exists a macroscopic system $\mathcal{A}$ such that for every local subsystem $\mathcal{A}_\Omega \subset \mathcal{A}$, the complement $\mathcal{A}_R$ of $\mathcal{A}_\Omega$ in $\mathcal{A}$ acts as a thermal reservoir for $\mathcal{A}_\Omega$.*

Specifically, the macrosystem is described by an algebra of observables $\mathcal{A}$, a Hamiltonian dynamics $\{\alpha(t) \mid t \in \mathsf{R}\}$, and a canonical equilibrium state ϕ^{β_o} with respect to the time evolution $\alpha(t)$ and the natural temperature $\beta_o \in (0,\infty)$. The local systems is described by subsalgebras $\mathcal{A}_\Omega \subset \mathcal{A}$, and the corresponding reservoir by $\mathcal{A}_R \subset \mathcal{A}$ so that $\mathcal{A} = \mathcal{A}_\Omega \otimes \mathcal{A}_R$. The initial state of the system is prepared as follows. We take the original Hamiltonian and cut off from it the interaction between the local system and its reservoir. The truncated evolution on the composite system is $\{\tilde{\alpha}(t) = \tilde{\alpha}_\Omega(t) \otimes \tilde{\alpha}_R(t) \mid t \in \mathsf{R}\}$. This evolution is used only to introduce $\tilde{\phi}_\Omega^\beta$ [resp. $\tilde{\phi}_R^{\beta_o}$] the corresponding canonical equilibrium state on $\mathcal{A}_\Omega$ [resp. $\mathcal{A}_R$] for the temperature β [resp. β_o], taken arbitrarily in $(0,\infty)$. The initial state is taken to be the composite state $\tilde{\phi}_\Omega^\beta \otimes \tilde{\phi}_R^{\beta_o}$.

The Scholium claims that it is possible to rig these objects in such a manner that

$$\forall\, A \in \mathcal{A}_\Omega\; :\quad \lim_{t\to\infty} \langle \tilde{\phi}^\beta_\Omega \otimes \tilde{\phi}^{\beta_o}_R ; \alpha(t)[A] \rangle = \langle \phi^{\beta_o} ; A \rangle\;. \tag{15.3.16}$$

Our discussion of Scholium 15.3.2 is divided into two parts: in Part A, we define all the elements that appear in the statement of the Scholium; and in Part B, we give an outline of the proof.

Part A: Construction of the model. Since an infinite time limit is claimed to exist, we expect this limit to be considered only after an infinite size limit has been implemented. This is done now with some details.

We begin with the definition of the global, or macroscopic, objects. Let $\mathcal{F} = \{\Lambda \subset \mathsf{Z} \mid |\Lambda| < \infty\}$ be the collection of all finite subsets of the lattice Z, with $\mathcal{F}$ equipped with the partial ordering given by the set inclusion.

To every site $k \in \mathsf{Z}$ is attached a quantum spin σ_k and thus a copy $\mathcal{A}_k$ of the algebra $\mathcal{M}(2, \mathsf{C})$, the algebra of 2×2 matrices with complex entries. For every $\Lambda \in \mathcal{F}$ we define as usual the algebra $\mathcal{A}_\Lambda = \mathcal{M}(2^{|\Lambda|}, \mathsf{C})$ generated by the elements of the form $A_1 \otimes A_2 \otimes \cdots \otimes A_{|\Lambda|}$ with $A_k \in \mathcal{A}_k$ for each $k \in \Lambda$.

For each pair (Λ_1, Λ_2) of elements of $\mathcal{F}$ that satisfy $\Lambda_1 \subseteq \Lambda_2$ we define a natural embedding of $\mathcal{A}_{\Lambda_1}$ into $\mathcal{A}_{\Lambda_2}$, the injective *–homomorphism J_{21}:

$$\left.\begin{array}{l} A_1 \otimes A_2 \otimes \cdots \otimes A_{|\Lambda_1|} \in \mathcal{A}_{\Lambda_1} \mapsto B_1 \otimes B_2 \otimes \cdots \otimes B_{|\Lambda_2|} \in \mathcal{A}_{\Lambda_2} \\[2mm] \text{where}\quad \forall\, k \in \Lambda_2 : \quad B_k = \begin{cases} A_k & \text{if}\quad k \in \Lambda_1 \\ I_k & \text{if}\quad k \notin \Lambda_1 \end{cases} \end{array}\right\} \tag{15.3.17}$$

These embeddings satisfy the isotony relations $\Lambda_{21}[I_1] = I_2$, and $\Lambda_1 \subset \Lambda_2 \subset \Lambda_3 \models \Lambda_{32} \circ \Lambda_{21} = \Lambda_{31}$. Hence [Kadison and Ringrose, 1983, 1986][Prop. 11.4.1], there exist:

1. a smallest C^*–algebra $\mathcal{A}$ with unit I ;
2. for every $\Lambda \in \mathcal{F}$, an injective *–homomorphism $J_\Lambda : A \in \mathcal{A}_\Lambda \mapsto J_\Lambda[A] \in \mathcal{A}$, with

$$\left.\begin{array}{l} J_\Omega[I_\Omega] = I\, ; \\[3mm] \Lambda_1 \subset \Lambda_2 \models J_{\Lambda_1}[\mathcal{A}_{\Lambda_1}] \subset J_{\Lambda_2}[\mathcal{A}_{\Lambda_2}] \\[3mm] \mathcal{A}_o \equiv \bigcup_{\Lambda \subset \mathsf{Z}} J_\Lambda[\mathcal{A}_\Lambda] \quad \text{is norm–dense in } \mathcal{A}. \end{array}\right\} \tag{15.3.18}$$

In this manner, all finite systems with $\Lambda \subset \mathsf{Z}$ can be viewed with mathematical consistency as subsystems of an infinite system associated to Z. The algebra $\mathcal{A}$ is called the C^*–inductive limit of the collection $\mathcal{A}_\Lambda$ of the algebras associated with the finite subsets $\Lambda \subset \mathsf{Z}$.

To define the dynamics, we start by restricting our attention to the collection $\mathcal{F}_o$ of connected elements in $\mathcal{F}$, i.e. those sets $\Omega \in \mathcal{F}$ of the form

$$\exists\, a, b \in \mathsf{Z} \quad \text{such that} \quad \Omega = \{k \in \mathsf{Z} \mid a \leq k \leq b\}\;. \tag{15.3.19}$$

A Hamiltonian is defined on every local system $\mathcal{A}_\Omega$:

$$H_\Omega = \sum_{k=a}^{b-1} (1 + \zeta)\, \sigma_k^x \sigma_{k+1}^x + (1 - \zeta)\, \sigma_k^y \sigma_{k+1}^y \tag{15.3.20}$$

where $\zeta \in [0, 1)$ is kept fixed throughout. Hence, we are considering a $x - y$ model. This Hamiltonian generates an evolution $\{\alpha_\Omega(t) \mid t \in \mathsf{R}\}$ on $\mathcal{A}_\Omega$; and we denote by $\phi_\Omega^{\beta_o}$ the corresponding canonical equilibrium state at fixed natural temperature $\beta_o \in (0, \infty)$.

The states $\phi_\Omega^{\beta_o}$ and the automorphisms $\alpha_\Omega(t)$ can be extended uniquely by continuity to a state ϕ^{β_o} on $\mathcal{A}$ and a continuous group of automorphisms $\{\alpha(t) \mid t \in \mathsf{R}\}$ of $\mathcal{A}$, such that for all $\Omega \in \mathsf{Z}$ and all $t \in \mathsf{R}$:

$$\phi^{\beta_o} \circ J_\Omega = \phi_\Omega^{\beta_o} \quad ; \quad \alpha(t) \circ J_\Omega = J_\Omega \circ \alpha_\Omega(t). \tag{15.3.21}$$

We omit henceforth the explicit reference to the embeddings J_Ω. The objects $\{\mathcal{A}, \phi^{\beta_o}, \alpha(t)\}$ just defined are the macroscopic objects to which Scholium 15.3.2 refers.

To every contiguous finite subset $\Omega \in \mathcal{F}_o$, we associate the local algebra $\mathcal{A}_\Omega$, generated by the spins in Ω. With $\Omega \in \mathcal{F}_o$ fixed, let Ω^c denote the complement of $\Omega \subseteq \mathsf{Z}$ and $\mathcal{F}_{\Omega^c}$ the collection of all finite subsets of Ω^c. The algebra of observables on the reservoir $\mathcal{A}_R$ is defined as the C^*-inductive limit of the subalgebras associated with these "outer" subsets: the reservoir for the region Ω is the system associated to the spins sitting outside of this region Ω; and we have:

$$\mathcal{A} = \mathcal{A}_\Omega \otimes \mathcal{A}_R \quad . \tag{15.3.22}$$

Suppose now that the interaction between the system Ω and its reservoir R has been cut off. For every finite connected set $\Xi \in \mathcal{F}_o$ containing Ω, we can write without loss of generality Ξ as the disjoint union $\Xi = \Omega_l \cup \Omega \cup \Omega_r$ with $(k_l, k, k_r) \in \Omega_l \times \Omega \times \Omega_r \models k_l < k < k_r$. Let now

$$\tilde{H}_\Xi = H_{\Omega_l} + H_\Omega + H_{\Omega_r} \quad . \tag{15.3.23}$$

Compare with (15.3.19) and note that $\tilde{H}_\Xi$ differs from H_Ξ only by the absence of terms coupling the three contiguous but mutually disjoint regions $\{\Omega_l, \Omega, \Omega_r\}$. An evolution $\{\tilde{\alpha}(t) \mid t \in \mathsf{R}\}$ on the composite system $\mathcal{A}$ is defined from (15.3.23) as the uncut evolution $\{\alpha(t) \mid t \in \mathsf{R}\}$ was defined from (15.3.20). Since (15.3.23) affords no interaction between the local system and the reservoir, $\mathcal{A}_\Omega$ and $\mathcal{A}_R$ are separately stable under $\tilde{\alpha}(t)$. Hence, in particular, the restriction $\tilde{\alpha}_\Omega(t)$ of $\tilde{\alpha}(t)$ to $\mathcal{A}_\Omega$ defines a conservative evolution on $\tilde{\mathcal{A}}_\Omega$. Let then $\tilde{\phi}_\Omega^\beta$ be the canonical equilibrium state on $\mathcal{A}_\Omega$ with respect to the evolution $\tilde{\alpha}_\Omega$ and the arbitrary temperature $\beta \in (0, \infty)$. As for the reservoir, we define similarly the conservative evolution $\{\tilde{\alpha}_R \mid t \in \mathsf{R}\}$ and the canonical equilibrium $\tilde{\phi}_R^{\beta_o}$ with respect to this evolution and the temperature β_o. We have now constructed all the objects appearing in the statement of Scholium 15.3.2.

Part B: Proof of the decay of thermal deviations. We introduce here an auxiliary structure and consider the automorphism γ defined on $\mathcal{A}$ by:

$$\forall \, k \in \mathsf{Z}: \quad \gamma[\sigma_k^z] = \sigma_k^z \quad ; \quad \gamma[\sigma_k^x] = -\sigma_k^x \quad ; \quad \gamma[\sigma_k^y] = -\sigma_k^y \quad . \qquad (15.3.24)$$

The subalgebra of fixed points of γ is denoted

$$\mathcal{A}^\gamma = \{A \in \mathcal{A} \mid \gamma[A] = A\} \quad . \qquad (15.3.25)$$

We have for all $A \in \mathcal{A}$:

$$A = A_+ + A_- \text{ with } A_\pm \equiv \frac{1}{2}(A \pm \gamma[A]) \text{ and } \gamma[A_\pm] = \pm A_\pm \, . \qquad (15.3.26)$$

The algebra $\mathcal{A}^\gamma$ contains all polynomials in $\sigma_k^z, \sigma_m^x, \sigma_n^y$ which are even in the σ^x at different sites, and similarly in the σ^y. For this reason $\mathcal{A}^\gamma$ is referred to as the *even* subalgebra of $\mathcal{A}$. In particular, $\gamma[H_\Omega] = H_\Omega$ and $\gamma[\tilde{H}_\Xi] = \tilde{H}_\Xi$, from which if follows that

$$\left. \begin{array}{c} \forall \, t \in \mathsf{R} \; : \; \alpha(t) \circ \gamma = \gamma \circ \alpha(t) \quad ; \quad \tilde{\alpha}(t) \circ \gamma = \gamma \circ \tilde{\alpha}(t) \\[2mm] \phi^{\beta_\circ} \circ \gamma = \phi^{\beta_\circ} \quad ; \quad \left(\tilde{\phi}^\beta \otimes \tilde{\phi}^{\beta_\circ} \right) \circ \gamma = \tilde{\phi}^\beta \otimes \tilde{\phi}^{\beta_\circ} \end{array} \right\} \; . \qquad (15.3.27)$$

In particular, we consider the restriction $\phi^{\beta_\circ, \gamma}$ and $\alpha^\gamma(t)$ of $\phi^{\beta_\circ}$ and $\alpha(t)$ to $\mathcal{A}^\gamma$ to define the subdynamical system $\{\mathcal{A}^\gamma, \alpha^\gamma, \phi^{\beta_\circ, \gamma}\}$ for which the time-evolution α^γ is *strongly asymptotically abelian*, i.e.

$$\forall \, A, B \in \mathcal{A}^\gamma : \quad \lim_{|t| \to \infty} \| \, [A, \alpha^\gamma(t)[B]] \, \| = 0 \quad . \qquad (15.3.28)$$

This property of the present model was seen in [Emch and Radin, 1971] to follow from a general result proven by [Narnhofer, 1970]. It has several consequences.

First, since we start from a one-dimensional spin-lattice system with a finite-range Hamiltonian, $\phi^{\beta_\circ}$ is an extremal KMS state on $\mathcal{A}$ w.r.t. the evolution α; this entails that $\phi^{\gamma, \beta_\circ}$ is an extreme KMS state on $\mathcal{A}^\gamma$ w.r.t. the evolution α^γ. Together with (15.3.28) this in turn implies that $\phi^{\gamma, \beta_\circ}$ is also extremal invariant w.r.t. α^γ; [Araki and Miyata, 1968] or [Emch, 1972a][Cor.3 to Thm.II.2.12], i.e. $\phi^{\gamma, \beta_\circ}$ cannot be decomposed in a convex combination of time-invariant states, a nice ergodic property.

Furthermore, since $\phi^{\gamma, \beta_\circ}$ is an extremal KMS state, its associated GNS representation $\pi^{\beta_\circ}(\mathcal{A}^\gamma)$ is primary; and it admits a cyclic and separating vector. From the latter property, it follows [Kadison and Ringrose, 1983, 1986][Thm.7.2.3] that every normal state – by no means necessarily a pure state! – on the von Neumann algebra $\mathcal{N}^{\beta_\circ}$ generated by the representation $\pi^{\beta_\circ}(\mathcal{A}^\gamma)$ is a vector state, i.e. for every such ψ^γ there is a vector $\Psi \in \mathcal{H}_{\pi^{\beta_\circ}}$ such that for all $N \in \mathcal{N}^{\beta_\circ} : \langle \psi^\gamma; N \rangle = (N\Psi, \Psi)$. From this and (15.3.28), we obtain that for each such state:

$$\forall A \in \mathcal{A}^\gamma : \quad \lim_{|t| \to \infty} \langle \psi^\gamma; \alpha^\gamma(t)[A] \rangle = \langle \phi^\gamma; A \rangle \quad . \tag{15.3.29}$$

A state ψ on $\mathcal{A}$ is said to be *even* whenever $\psi \circ \gamma = \psi$; for instance – see (15.3.27) – $\phi^{\beta_\circ}$ and $\tilde{\phi}^\beta \otimes \tilde{\phi}^{\beta_\circ}$ are even.

The key to the proof is now the simple fact that an even state ψ on $\mathcal{A}$ is completely determined by its restriction ψ^γ to the algebra $\mathcal{A}^\gamma$. Indeed, with the notation of (15.3.26), we have for every $A \in \mathcal{A}$ $\langle \psi; A_- \rangle = \langle \psi; \gamma[A_-] \rangle = -\langle \psi; A_- \rangle = 0$ and thus $\langle \psi; A \rangle = \langle \psi; A_+ \rangle = \langle \psi^\gamma; A_+ \rangle$. Moreover, since γ commutes with every $\alpha(t)$, we have, for every even state ψ:

$$\forall A \in \mathcal{A} : \quad \langle \psi; \alpha(t)[A] \rangle = \langle \psi^\gamma; \alpha^\gamma(t)[A_+] \rangle . \tag{15.3.30}$$

If, furthermore, ψ is normal on $\mathcal{A}$, so is its restriction ψ^γ to $\mathcal{A}^\gamma$, and (15.3.29), (15.3.30) and (15.3.27) entail that for every even, normal state ψ on $\mathcal{A}$:

$$\forall A \in \mathcal{A} : \quad \lim_{|t| \to \infty} \langle \psi; \alpha(t)[A] \rangle = \langle \phi^{\beta_\circ}; A \rangle \quad . \tag{15.3.31}$$

Since, we know from (15.3.27) that $\tilde{\phi}^\beta \otimes \tilde{\phi}^{\beta_\circ}$ is even, and since one can verify that the difference between the full Hamiltonian (15.3.19) and the cut-off Hamiltonian (15.3.23) is small enough to ensure that $\tilde{\phi}^\beta \otimes \tilde{\phi}^{\beta_\circ}$ is normal on $\mathcal{A}^\gamma$, (15.3.31) contains (15.3.16) as a particular case.   **q.e.d.**

The property (15.3.31) was obtained in [Emch and Radin, 1971]. Mathematically, it is certainly stronger than the claim of Scholium 15.3.2; nevertheless, the formulation of the latter was chosen for the immediacy of its connection with everyday physical experience. An intermediary consequence of (15.3.31) – using again the finite dimensionality of the local algebra $\mathcal{A}_\Omega$ – is that for every even state ψ_Ω on $\mathcal{A}_\Omega$:

$$\lim_{|t| \to \infty} \langle \psi_\Omega \otimes \tilde{\phi}_R^{\beta_\circ}; A \rangle = \langle \phi^{\beta_\circ}; A \rangle$$

thus allowing, for instance, thermal deviations that are not uniform over the region Ω.

Hence, the lessons of the analysis are (a) on the positive side: it exhibits a purely mechanical model that shows a correct thermodynamical behavior; but (b) on the negative side: it depends in an essential way on the asymptotic abelianness property (15.3.28); not many other models that are also amenable to complete mathematical analysis are known to admit a large enough subalgebra on which this property can be verified to hold true. Moreover, (c) no hint is offered on how the approach to equilibrium could be made to be Markovian.

Several interesting variations related to the model presented here have been chiselled; see in particular [Abraham, Baruch, Gallavotti, and Martin-Löf, 1971, Robinson, 1973, Araki and Baruch, 1983, Araki, 1983].

Model 3: Markovian approach to equilibrium. Classical and quantum versions of this model were proposed by [Ford, Kac, and Mazur, 1965]. The model consists of a chain of harmonic osicllators, coupled by a long-range potential; and the argument concludes that: (i) if the chain is in thermal equilibrium, and if an element of the chain is perturbed from its state of equilibrium, then its mechanical interactions with the rest of the chain brings this oscillator back to thermal equilibrium; and (ii) in the long-time, weak-coupling limit this approach is a Markovian diffusion process; It was further proposed that the model be considered as a quantum Brownian motion.

We limit our attention to the quantum model and we follow the spirit of the elaborate reworking of the argument by Davies [1972]. Our presentation of this model is divided into five parts: first, we recall the basics of the Weyl formulation of the canonical commutation relations in the case of one degree of freedom; second, we review an extension of this mathematical formulation to the case of an infinite number of degrees of freedom, and we give the physical interpretation of the construction in terms of an infinite chain of harmonic oscillators; third we define a conservative dynamics of the model; fourth, we discuss the van Hove limit of the restriction of this dynamics to finite subsystems; and fifth, we discuss the Markovian diffusion equation that controls the approach to equilibrium of an oscillator placed in a thermal bath constituted by the other oscillators in the chain.

Part A: the Weyl CCR for one degree of freedom The Weyl form of the canonical commutation relations allows to bypass many of the difficulties associated with unbounded operators; here in particular, this form allows ultimately to control the van Hove limit. We focus on the case of a harmonic oscillator of proper frequency ω, i.e. with energy given by

$$H = \frac{1}{2}(P^2 + \omega^2 Q^2) - \frac{\omega}{2} \qquad \text{with} \qquad [P, Q] = -\mathrm{i}I \qquad (15.3.32)$$

where the constant $\frac{1}{2}\omega$ is inconsequential: it simply ensures that the energy is measured up from its lowest level. As usual, $[A, B] \equiv AB - BA$, except for the fact that we are dealing here with unbounded operators, so that some precautions should be taken to check that such expressions are not used out-side their domains of definition; this is more a matter of accounting hygiene than principle. For purposes of visualization, it is sometimes convenient to introduce, instead of the momemtum and position operators P and Q, the creation and annihilation operators

$$a^* = \tfrac{1}{2}(\sqrt{\omega}Q - \mathrm{i}\tfrac{1}{\sqrt{\omega}}P) \quad ; \quad a_k = \tfrac{1}{2}(\sqrt{\omega}Q + \mathrm{i}\tfrac{1}{\sqrt{\omega}}P) \quad \left.\begin{array}{c} \\ \\ \end{array}\right\} .$$

$$\text{so that} \qquad H = \omega\,a^*a \quad \text{and} \quad [a, a^*] = I \qquad\qquad (15.3.33)$$

As these creation and annihilation operators are also unbounded, the above remark on accounting hygiene holds as well. The Hamiltonian H generates the evolution

$$\begin{pmatrix} P \\ Q \end{pmatrix} \rightarrow \begin{pmatrix} P_t \\ Q_t \end{pmatrix} = \begin{pmatrix} \cos(\omega t) & -\omega \sin(\omega t) \\ \omega^{-1} \sin(\omega t) & \cos(\omega t) \end{pmatrix} \begin{pmatrix} P \\ Q \end{pmatrix} \quad . \tag{15.3.34}$$

The Weyl operators are defined, for any pair $(\xi, \eta) \in \mathsf{R} \times \mathsf{R}$ by:

$$W \begin{pmatrix} \xi \\ \eta \end{pmatrix} = e^{-i(\xi P + \eta Q)} \tag{15.3.35}$$

which satisfy:

$$\left. \begin{aligned} W \begin{pmatrix} \xi_1 \\ \eta_1 \end{pmatrix} W \begin{pmatrix} \xi_2 \\ \eta_2 \end{pmatrix} &= W \begin{pmatrix} \xi_1 + \xi_2 \\ \eta_1 + \eta_2 \end{pmatrix} e^{i\frac{1}{2}(\xi_1 \eta_2 - \xi_2 \eta_1)} \\ \text{and} \qquad\qquad W \begin{pmatrix} \xi \\ \eta \end{pmatrix}^* &= W \begin{pmatrix} -\xi \\ -\eta \end{pmatrix} \end{aligned} \right\} , \tag{15.3.36}$$

so that these operators are unitary and thus, in particular, bounded.

With this notation, the time-evolution is determined by its restriction to the Weyl operators, on which it takes the form:

$$\alpha_t \left[W \begin{pmatrix} \xi \\ \eta \end{pmatrix} \right] \equiv e^{iHt} W \begin{pmatrix} \xi \\ \eta \end{pmatrix} e^{-iHt} = W \begin{pmatrix} \xi_t \\ \eta_t \end{pmatrix} \tag{15.3.37}$$

with

$$\begin{pmatrix} \xi_t \\ \eta_t \end{pmatrix} = \begin{pmatrix} \cos(\omega t) & \omega^{-1} \sin(\omega t) \\ -\omega \sin(\omega t) & \cos(\omega t) \end{pmatrix} \begin{pmatrix} \xi \\ \eta \end{pmatrix} \tag{15.3.38}$$

which suggests the change of variables:

$$\forall\, t \in \mathsf{R} \;:\; \zeta_t \equiv \sqrt{\omega}\,\xi_t + i\,\frac{1}{\sqrt{\omega}}\,\eta_t \;\; \models \;\; \zeta_t = e^{-i\omega t}\,\zeta \quad . \tag{15.3.39}$$

In addition to its formal simplicity, this notation has the advantage to recall the elliptic orbits in the phase-space of the classical harmonic oscillator.

The canonical equilibrium state for each of the harmonic oscillators at natural temperature β can be computed – e.g. by using the KMS condition – to obtain:

$$\left. \begin{aligned} \langle \phi; W \begin{pmatrix} \xi \\ \eta \end{pmatrix} \rangle &= e^{-\frac{1}{4}\Theta |\zeta|^2} \\ \text{with} \qquad \Theta = \coth(\tfrac{1}{2}\beta\omega) \quad \text{and} \quad |\zeta|^2 &= \omega\,\xi^2 + \tfrac{1}{\omega}\,\eta^2 \end{aligned} \right\} . \tag{15.3.40}$$

Part B: The infinite chain of harmonic oscillators We could repeat here the construction of the algebra of the infinite chain of harmonic oscillators as a C*-inductive limit – in a manner analogous to that used in Model 2 for the infinite chain of spins – where now copies of the algebra of the 2×2 matrices are replaced by copies of the algebra generated by the Weyl operators

(15.3.35) and attached to each site. The result is the algebra $\mathcal{A}_Z$ generated by the Weyl operators defined algebraically as follows.

Let $\mathcal{T}_C$ be the complex Hilbert space the elements of which are the functions $f : \theta : [-\pi, \pi] \mapsto f(\theta) \in C$ which are square integrable for the Haar measure $d\theta$. In the sequel, $\langle f, g \rangle$ denotes the scalar product of two elements f and g in $\mathcal{T}_C$. Let now K be the anti-unitary, involutive operaor defined on $\mathcal{T}_C$ by $(Kf)(\theta) = f(-\theta)^*$. With the *real* Hilbert space $\mathcal{T} = \{f\mathcal{T}_C \mid Kf = f\}$, we define

$$
\left.
\begin{aligned}
&W : (f, g) \in \mathcal{T} \times \mathcal{T} \mapsto W \begin{pmatrix} f \\ g \end{pmatrix} \in \mathcal{A}_Z \quad \text{with} \\
&W \begin{pmatrix} f_1 \\ g_1 \end{pmatrix} W \begin{pmatrix} f_2 \\ g_2 \end{pmatrix} = W \begin{pmatrix} f_1 + f_2 \\ g_1 + g_2 \end{pmatrix} e^{i\frac{1}{2}(\langle f_1, g_2 \rangle - \langle f_2, g_1 \rangle)} \\
&\text{and} \quad W \begin{pmatrix} f \\ g \end{pmatrix}^* = W \begin{pmatrix} -f \\ -g \end{pmatrix}
\end{aligned}
\right\}.
\tag{15.3.41}
$$

To have these objects acting in a Hilbert space, we consider the state defined on $\mathcal{A}_Z$ as the extension of

$$
\left.
\begin{aligned}
&\langle \phi ; W \begin{pmatrix} f \\ g \end{pmatrix} \rangle = e^{-\frac{1}{4}\langle \Theta f, f \rangle + \langle \Theta^{-1} g, g \rangle} \\
&\Theta = \tanh(\tfrac{1}{2}\beta C) \\
&\text{for a.a. } \theta \in [-\pi, \pi] \text{ and } \forall f \in \mathcal{T} : (Cf)(\theta) = C(\theta)f(\theta)
\end{aligned}
\right\}.
\tag{15.3.42}
$$

Clearly, (15.3.41) and (15.3.42) are modeled on (15.3.36) and (15.3.40). As we want the multiplication operator C to be a bounded self-adjoint operator with inverse, we require $C(\theta)$ to be real, bounded and not vanishing (except possibly on a set of measure zero). Later on, we impose further conditions on C; see (15.3.50) and (15.3.53).

Upon associating to the state ϕ the GNS representation of the Weyl algebra on $\mathcal{T} \times \mathcal{T}$, we generate the global algebra for the system to be considered.

Having defined the mathematical objects $\mathcal{A}_Z$ and ϕ, we have three immediate tasks: to intepret $\mathcal{A}_Z$ as an algebra of microscopic observables on a macroscopic body; to construct a microscopic evolution that fits our purpose; and to interpret the state ϕ as a canonical equilibrium state with respect to this evolution and the natural temperature β.

With the orthonormal basis $\{e_k \in \mathcal{T} \mid k \in Z\}$ provided by the functions $e_k(\theta) = \frac{1}{\sqrt{2\pi}}e^{ik\theta}$, we define for every $(\xi, \eta) \in R \times R$:

$$
W_k \begin{pmatrix} \xi \\ \eta \end{pmatrix} = W \begin{pmatrix} \xi e_k \\ \eta e_k \end{pmatrix} ;
\tag{15.3.43}
$$

and it follows from this definition that:

1. the operators W_k satisfy the Weyl relations (15.3.36) for one degree of freedom;

2. for $k \neq k'$:

$$\left[W_k \begin{pmatrix} \xi \\ \eta \end{pmatrix} , W_{k'} \begin{pmatrix} \xi' \\ \eta' \end{pmatrix} \right] = 0 \quad ; \tag{15.3.44}$$

3. the operators $\{ W_k \mid k \in \mathsf{Z} \}$ generate the algebra $\mathcal{A}_\mathsf{Z}$.

We can therefore view the algebra $\mathcal{A}_\mathsf{Z}$ as the algebra of observables on an assembly of one-dimensional systems, each of which – say the system described by W_k – sits at a site – say k – of the infinite lattice Z. We cannot yet interpret as harmonic oscillators the individual systems sitting at the sites of the lattice: for this we need to specify the time-evolution, which we do presently.

Part C: The dynamics of the infinite chain. We define the time-evolution on this assembly by its restriction to the Weyl operators W that generate the algebra $\mathcal{A}_\mathsf{Z}$:

$$W \begin{pmatrix} f \\ g \end{pmatrix} \rightarrow W \begin{pmatrix} f_t \\ g_t \end{pmatrix} \tag{15.3.45}$$

with

$$\begin{pmatrix} f_t \\ g_t \end{pmatrix} = \begin{pmatrix} \cos(Ct) & C^{-1}\sin(Ct) \\ -C\sin(Ct) & \cos(Ct) \end{pmatrix} \begin{pmatrix} f \\ g \end{pmatrix} \quad . \tag{15.3.46}$$

Compare with (15.3.38). To interpret the present evolution, we notice first that it is conservative, i.e. (15.3.45–46) extends to a continuous group of automorphisms of $\mathcal{A}_\mathsf{Z}$. Second, with respect to this evolution, the state defined by (15.3.42) is a KMS state for the natural temperature β. Thirdly, to this evolution can be associated the Hamiltonian

$$H = \int_{-\pi}^{\pi} \mathrm{d}\theta \, C(\theta) \, a^*(\theta) \, a(\theta) \tag{15.3.47}$$

where the creation and annihilation operators $a^*(\theta)$ and $a(\theta)$ are formally defined by

$$a^*(\theta) = \frac{1}{\sqrt{2\pi}} \sum_{k \in \mathsf{Z}} e^{ik\theta} a_k^* \quad ; \quad a(\theta) = \frac{1}{\sqrt{2\pi}} \sum_{k \in \mathsf{Z}} e^{-ik\theta} a_k \tag{15.3.48}$$

where a_k^* and a_k are the annihilation and creation operators assigned to the site k by W_k. In the particular case where $\forall \, \theta \in [-\pi, \pi] \; : \; C(\theta) = \omega$, (15.3.47) becomes the Hamiltonian of an assembly of non-interacting harmonic oscillators:

$$H_o = \sum_{k \in \mathsf{Z}} \omega \, a_k^* a_k \quad ; \tag{15.3.49}$$

compare this to (15.3.33). Formally, one rewrite (15.3.47) as

$$H = \sum_k \tilde{C}_{|k-k'|} \, a_k^* \, a_{k'} ,$$

which thus incorporates a translation invariant interaction between harmonic oscillators sitting at the various sites of the lattice Z.

Part D: The van Hove limit of the evolution of local subsystems. We assume henceforth that the multiplication operator C is of the form

$$\left. \begin{array}{c} C^{(\lambda)}(\theta) = \left[\omega^2 + 2\omega B^{(\lambda)}(\theta)\right]^{\frac{1}{2}} \\[2mm] \lim_{\lambda \to 0} \lambda^{-2} B^{(\lambda)} = G(\theta) \quad \text{a.e.} \end{array} \right\} . \qquad (15.3.50)$$

with

In particular, as λ becomes very small:

$$C(\theta) \approx \omega + B^{(\lambda)}(\theta) \quad \text{with} \quad \lim_{\lambda \to 0} B^{(\lambda)} = 0 \quad \text{a.e} \quad . \qquad (15.3.51)$$

To keep tract of the $\lambda-$dependence just introduced in $C^{(\lambda)}$, we write $\alpha^{(\lambda)}{}_t$ for the corresponding evolution given by (3.45-46).

The local subsystem relative to a finite collection $\Omega = \{k_1, \cdots, k_{|\Omega|}\}$ of sites on the lattice Z is described by the algebra $\mathcal{A}_\Omega = \otimes_{k \in \Omega} \mathcal{A}_k$ of observables generated by all Weyl operators of the form

$$W_\Omega \begin{pmatrix} \xi \\ \eta \end{pmatrix} \equiv W \begin{pmatrix} \sum_{k \in \Omega} \xi_k \, e_k \\ \sum_{k \in \Omega} \eta_k \, e_k \end{pmatrix}$$

$$\left. \begin{array}{c} (\xi_{k_1}, \cdots, \xi_{k_{|\Omega|}}) \in \mathsf{R}^{|\Omega|} \\[2mm] (\eta_{k_1}, \cdots, \eta_{k_{|\Omega|}}) \in \mathsf{R}^{|\Omega|} \end{array} \right\} . \qquad (15.3.52)$$

with

Scholium 15.3.3. *With the conventions just introduced, let $\tau = \lambda^2 t$. Then*

1. at fixed $\tau \geq 0$ the following limit exists:

$$\gamma_\tau \left[W_\Omega \begin{pmatrix} \xi \\ \eta \end{pmatrix} \right] = s - \lim_{\lambda \to 0} \alpha^{(0)}{}_{-t} \circ \alpha^{(\lambda)}{}_t \left[W_\Omega \begin{pmatrix} \xi \\ \eta \end{pmatrix} \right]$$

where $s-\lim$ indicates that the limit is taken in the strong-operator topology, i.e. $s - \lim_{\lambda \to 0} X^{(\lambda)} = Y$ means that the limit is controlled in the following sense: for every vector $\Psi \in \mathcal{H}$ and every $\varepsilon > 0$ there exists a $\delta_{\Psi,\varepsilon} > 0$ such that $|\lambda| < \delta_{\Psi,\varepsilon} \models \|(X^{(\lambda)} - Y)\Psi\| < \varepsilon$;

2. For every $\tau \geq 0$ γ_τ is local, i.e. extends to a map $\gamma_\tau : \mathcal{A}_\Omega \to \mathcal{A}_\Omega$ and:

$$\gamma_\tau \left[W_\Omega \begin{pmatrix} \xi \\ \eta \end{pmatrix} \right] = W_\Omega \begin{pmatrix} \xi_\tau \\ \eta_\tau \end{pmatrix} e^{-\frac{1}{4} r_\tau}$$

where, with the identification

$$\begin{pmatrix} \xi \\ \eta \end{pmatrix} \in \mathsf{R}^{|\Omega|} \times \mathsf{R}^{|\Omega|} \longleftrightarrow \zeta \equiv \sqrt{\omega}\, \xi + i \frac{1}{\sqrt{\omega}} \eta \in \mathsf{C}^{|\Omega|} ,$$

we have

$$\zeta_\tau = P\,\mathrm{e}^{-\mathrm{i}G\tau}\zeta \qquad \text{and} \qquad r_\tau = \coth(\tfrac{1}{2}\beta\omega)\left[\,\|\zeta\|^2 - \|\zeta_\tau\|^2\,\right]$$

with P denoting the projector on the subspace $(\mathcal{T}_\mathsf{C})_\Omega \subset \mathcal{T}_\mathsf{C}$ spanned by the vectors $\{e_k \mid k \in \Omega\}$, i.e.

$$Pf = \sum_{k\in\Omega} \langle f, e_k \rangle\, e_k \,.$$

In the sharp form stated above, Scholium 15.3.3 was formulated and proven first in [Davies, 1972]. While the proof turns out to be a somewhat lenghty computation, it is rather straightforward ... likely only once the insight has been gained on what *can* be proven.

Part E: Markovian behavior and diffusion equation Now, we restrict our attention to the case where the subsystem of interest is a single oscillator, i.e. $|\Omega| = 1$; and we assume B_λ as been chosen in such a manner that the multiplication operator G in (15.3.50) is :

$$G(\theta) = D\cot(\theta) + \nu \tag{15.3.53}$$

where D and ν are two real numbers. In this case, the conclusion of Scholium 15.3.3 becomes simply

$$\zeta_\tau = \mathrm{e}^{(\mathrm{i}\nu - |D|)\tau}\,\zeta \qquad \text{and} \qquad r_\tau = \coth(\tfrac{1}{2}\beta\omega)\,|\zeta|^2\,(1 - \mathrm{e}^{-2|D|\tau}) \tag{15.3.54}$$

through which one reads the Markovian semi-group property

$$\forall\, \tau_2 \geq \tau_1 \geq 0 \ : \qquad \gamma_{\tau_2 - \tau_1} \circ \gamma_{\tau_1} = \gamma_{\tau_2} \quad . \tag{15.3.55}$$

In case $D = 0$, γ_τ describes the conservative evolution – compare with (15.3.39) – of an undamped harmonic oscillator of forced proper frequency ν. It is obviously the case $D \neq 0$ that interests us. From now on, we assume that we take a system of coordinates rotating with frequency ν which separates out the purely dissipative evolution $\zeta \to \mathrm{e}^{-|D|\tau}\zeta$. Hence, without loss of generality, we assume henceforth that $\nu = 0$, but *emphatically* $D > 0$.

Let ψ be the initial state of our oscillator, which we assume as usual to be normal. Scholium 15.3.3, together with the special form (15.3.54), gives:

$$\lim_{\tau\to\infty}\, \left\langle \psi;\gamma_\tau\left[W\left(\begin{matrix}\xi\\\eta\end{matrix}\right)\right] \right\rangle = \mathrm{e}^{-\frac{1}{4}\coth(\frac{1}{2}\beta\omega)\,|\zeta|^2} \quad, \tag{15.3.56}$$

i.e. – upon comparing with (15.3.40) – we conclude that in the course of time the state of our oscillator approaches its canonical equilibrium at temperature β, independently of its original state ψ.

To get a better grasp of the way the equilibrium is approached, consider the classical position distribution $\mu(x, \tau)$ defined by its Fourier transform:

$$\tilde{\mu}(k, \tau) \equiv \langle \psi_\tau ; e^{-ikQ} \rangle \quad . \tag{15.3.57}$$

We obtain then the diffusion equation in the harmonic potential

$$\overline{V}(x) = \frac{1}{2}[\overline{\omega}^2]\, x^2 \quad \text{with} \quad \overline{\omega}^2 = 2\,\frac{1}{\beta}\omega \tanh(\frac{1}{2}\beta\omega) \quad , \tag{15.3.58}$$

namely:

$$\partial_\tau \mu(x, \tau) = \overline{D}\, \partial_x^2 \mu(x, \tau) + \beta\overline{D}\left[\overline{V}'(x)\partial_x + \overline{V}(x)\right]\mu(x, \tau) \tag{15.3.59}$$

where $\overline{D} = D\,\beta\overline{\omega}^{2^{-1}}$. In the high-temperature limit $- T \to \infty$ i.e. $\beta \to 0 -$ the effective frequency, potential, and diffusion constant take the particular forms

$$\overline{\omega} \approx \omega \quad ; \quad \overline{V} \approx \frac{1}{2}\omega^2 x^2 \quad ; \quad D \approx D(\beta\omega^2)^{-1} \tag{15.3.60}$$

which are in line with the classical distribution. A similar equation does hold also when one replaces Q by any observable of the form $\xi P + \eta Q$, provided V and D are adjusted accordingly; see e.g. [Emch, 1976b].

The fact that the evolution γ_τ gives raise to diffusion equations played a seminal role in the formulation of the quantum K–flows; this generalizes to the quantum domain the structure known in classical theory as Kolmogorov flows – see Sect. 8.2. Indeed, γ_τ admits a dilation – in the sense of the diagram at the end of Sect. 15.1 – to an evolution $\{\alpha_\tau \mid \tau \in \mathsf{R}\}$ on a von Neumann algebra $\mathcal{N}$, the non-commutative analog of the space $\mathcal{L}^\infty$ of the classical random variables. This dilation can even be chosen to be minimal in the sense that $\{\alpha_\tau[\mathcal{A}_{\{k\}}] \mid \tau \in \mathsf{R}\}$ generates $\mathcal{N}$. The canonical object obtained in this manner satisfies then the axioms of a quantum K–flow, which appears thus in the context of the present model, as a non-commutative version of the classical flow of Brownian motion; for the latter, see Model 8.2.4. The non-commutative version was proposed in its general framework in [Emch, 1976a], and reviewed with particular attention to the present model in [Emch, 1976b]; the abstract non-commutative structure of the generalized K–flows has been further extended, and its general properties thoroughly analyzed; see e.g. [Narnhofer and Thirring, 2000].

The present model is also instructive through one of its limitations. Indeed, while Scholium 15.3.3 asserts the existence of the van Hove limit for *all* local algebras relative to a finite number of sites in the infinite chain, it is also specific enough to establish [Davies, 1972] that the Markov character of the evolution, obtained when we viewed it from a single site, is lost when correlations between as few as two sites are considered. Hence, while the van Hove limit appears to be a recipe – sound in its physical motivation – to focus on the rescaled time scale of the macroscopic regime, its existence alone does *not* ensure Markovicity.

Another aspect of the model surfaces when the van Hove limit is applied to the canonical equilibrium state (15.3.42) as well as to the dynamics; while the latter registers the cumulative effects of the small interaction over long time lapses, the state sees only the vanishing of the coupling constant, and thus approaches in this limit the canonical equilibrium state of the *free* evolution, rather that a KMS state associated to the putative evolution responsible for the approach to equilibrium. Nevertheless (15.3.56) shows that the state so obtained, when restricted to any local algebra, is invariant under the dissipative evolution γ with respect to which it is the universal attractor for all initial normal states.

Model 4: Electric resistivity due to impurities. While electrons are moving without resistance through a perfect rigid crystal, the observed electric resistivity is due to three interactions that the electrons do have, namely (i) with the lattice vibrations; (ii) with one another; and (iii) with random impurities. The first two interactions depend essentially on the temperature, so that at low temperature the third becomes the dominant cause [Kohn and Luttinger, 1957].

The simplest version of the model – the so-called quantum Lorentz gas – consists of a quantum, spinless, non-relativisitic particle moving in a three-dimensional space where it interacts with randomly distributed recoilless particles. The Hamiltonian is of the form

$$H = H_o + \lambda V \tag{15.3.61}$$

and acts in the Hilbert space $\mathcal{H} = \mathcal{L}_{\mathbb{C}}^2(\mathbb{R}^3) = \{\Psi : x \in \mathbb{R}^3 \mapsto \mathbb{C} \mid \int dx |\Psi(x)|^2 \leq \infty\}$. Note for later purposes that the Fourier transform $\Psi \to \tilde{\Psi}$ defined by $\tilde{\Psi}(k) = (2\pi)^{-\frac{3}{2}} \int dx e^{ikx} \Psi(x)$ extends to a unitary operator from $\mathcal{H}$ onto itself. In this momentum representation, the free Hamiltonian takes the form (on a domain of definition):

$$(H\tilde{\Psi})(k) = E_k \tilde{\Psi}(k) \quad \text{with} \quad E_k = \frac{1}{2m}\|k\|^2 ; \tag{15.3.62}$$

or equivalently $H_o = P^2/2m$ where for $j = 1, 2, 3$ $(P_j\Psi)(x) = -i\partial_{x_j}\Psi$.

The interaction with the impurities is modeled here by a stochastic potential,

$$V : x \in \mathbb{R}^3 \mapsto V(x) \in \mathbb{R} \quad . \tag{15.3.63}$$

The impurities are supposed to be independent of one another and their distribution is assumed to be Gaussian, translation invariant, of mean zero and variance γ; namely, for all three-vectors $x, x_1, x_2, \cdots, x_{2n-1}, x_{2n}$, it is entirely characterized by the following conditions:

$$\left. \begin{array}{l} \langle V(x) \rangle = 0 \\ \\ \langle V(x_1)V(x_2) \rangle = \gamma(|x_1 - x_2|) \end{array} \right\} \tag{15.3.64}$$

and

$$\langle V(x_1) \cdots V(x_{2n-1}) \rangle = 0$$

$$\langle V(x_1) \cdots V(x_{2n}) \rangle = \sum \prod \gamma(x_i - x_j)$$

$$\left. \right\} \quad , \qquad (15.3.65)$$

where γ is an integrable function that is of positive type, i.e. such that its Fourier transform $\tilde{\gamma}$ is positive. The sum in the last condition carries over the $(2n - 1)!!$ partitions of $2n$ indices in distinct pairs (where one identifies two pairs of indices that differ only from the order of their indices). For each partition, the product carries over the pairs making up the partition.

Scholium 15.3.4. *Let $U_t^{(\lambda)}$ be the evolution generated by the Hamiltonian (15.3.61), and $U_t^{(o)}$ be the free evolution generated by the Hamiltonian (15.3.62). Then, there exists a contant $\tau_o > 0$ such that, with $\tau \equiv \lambda^2 t$, $0 \le \tau \le \tau_o$ entails that the following limit exists for all Φ and Ψ in $\mathcal{H}$:*

$$\lim_{\lambda \to 0} \langle (U_{-t}^{(o)} U_t^{(\lambda)} \Psi, \Phi) \rangle = (S_\tau \Psi, \Phi)$$

where the semi-group

$$\{ S_\tau = e^{-[\Gamma + i\Delta]\tau} \mid \tau \in \mathsf{R}^+ \}$$

is specified by the multiplication operators $(\Gamma \tilde{\Psi})(k) = \Gamma(k) \tilde{\Psi}(k)$ and $(\Delta \tilde{\Psi})(k) = \Delta(k) \tilde{\Psi}(k)$ with

$$\Gamma(k) = \int d^3 k' \; \delta(E_k - E_{k'}) W(k', k)$$

$$\Delta(k) = (\mathrm{P}) \int d^3 k' \; \frac{1}{E_k - E_{k'}} W(k', k)$$

$$\left. \right\} \quad \text{where} \quad \left\{ \begin{array}{l} E_k = \frac{1}{2m} \|k\|^2 \\[2mm] W(k', k) = (2\pi)^{-\frac{3}{2}} \tilde{\gamma}(k - k') \end{array} \right.$$

(and $(\mathrm{P})\int$ denotes the "principal value" of the integral).

Remarks:

1. Since $\tilde{\gamma}$ is positive, so is Γ, and the semi-group S is dissipative.
2. The presence of the Dirac δ enforces the conservation of the electron energy E, as to be expected from our condition that impurities be recoilless. Accordingly, the electrons approach asymptotically the micro- rather than the macro-canonical equilibrium.
3. The evolution of the extensive, translation-invariant electron observables – those observables that are diagonal in the momentum representation – is governed by the Markovian master equation:

$$\frac{\mathrm{d}}{\mathrm{d}\tau} p_\tau(k) = \int d^3 k' [K(k, k') p_\tau(k') - K(k', k) p_\tau(k)]$$

$$\text{with} \qquad K(k, k') = 2\pi \delta(E_k - E_{k'}) W(k', k)$$

$$\left. \right\} \quad ; \quad (15.3.66)$$

an equivalent form of which is know in the context of the quantum Lorentz gas as the Boltzmann–Peierls equation.

4. The model was rigged in such a way that the proof of Scholium 15.3.4 could follow a mathematically rigourous path, and yet model closely the successive steps in the van Hove's program. Where van Hove proposes to restrict the Dyson perturbation series to a summation over some "most divergent" diagrams, the particular definition of the interaction in the present model induces the average of this series expansion over the Gaussian distribution of the impurities:

$$
\left.
\begin{aligned}
&\langle\, (U^{(o)}_{-t} U^{(\lambda)}_{t} \Psi, \Phi)\, \rangle = \\[4pt]
&\sum_{n=0}^{\infty} (-i\lambda)^n \int_{t \geq t_n \geq \cdots \geq t_1 \geq 0} dt_n \cdots dt_1 \, \langle (V_{t_n} \cdots V_{t_1} \Psi, \Phi) \\[6pt]
&\text{with} \qquad V_t = U^{(0)}_{-t} V U^{(0)}_{t}
\end{aligned}
\right\} \quad ; \quad (15.3.67)
$$

and thus (15.3.64–5) entail that: (i) all odd terms in (15.3.67) vanish exactly; and (ii) for each integer n, the even term of order $2n$ satisfies the following properties:

a) it tends to

$$
\frac{1}{n!} [-\lambda^2 t (\Gamma + i\Delta)]^n \quad \text{as} \quad \lambda \to 0;
$$

b) there is a constant $C > 0$ such that this term is majorized uniformly in λ and t by:

$$
\frac{(2n-1)!!}{n!} (C\tau)^n \approx (2C\tau)^n .
$$

Since $\sum_n (2C\tau)^n$ converges provided $2C\tau < 1$, the limit asserted in Scholium 15.3.4 exists, provided $\tau \in [0, \tau_o]$ with τ_o satisfying $\tau_o < (2C)^{-1}$. Hence, the technical constraint $\tau \leq \tau_o$ is brought about here by the methodology of the proof, namely the control of the convergence of the pertubation series expansion (15.3.67). One would like to think that it is not an actual constraint of the model itself. Moreover, a diagramatic representation of the various terms that contribute to (15.3.67) can be made to model accurately the partitioning proposed in van Hove's diagramatic analysis; see in particular the range on the summation in (15.3.65).

5. The constraint that the configuration space of the model be of dimension $d = 3$ (or larger) can be traced back to the quantum effect, namely the "spreading of the wave packet" which is governed by $|t|^{-\frac{d}{2}}$ and enters in the estimates through its integral for the convergence of which $d \geq 3$ is needed. Again, this appears through the methodology of the proof, although there are some reasons to believe that this constraint may be intrinsic to the model.

6. We spared our Reader the technical derivations involved in proving the existence of the thermodynamical limit for the averaged dynamics in the infinite lattice system. This is hard and somewhat tedious; we chose instead to insist on the core of the argument, as presented above.

A mathematical solution of a lattice model of electrons interacting with re-coilless impurities randomly distributed on Z^3 was obtained in [Martin and Emch, 1975]. Soon afterwards [Spohn, 1977] realized that the methodology of the proof could be carried over to treat some significant generalizations of this model.

In the context of the present discussion, the main improvement was the inclusion of a homogeneous electric field as a term $\lambda^2 E \cdot x$ in the Hamiltonian (15.3.61), so that a transport term $E \cdot \nabla \varrho(k)$ may be brought rigorously in the RHS of the master equation (15.3.65); this term corresponds to the streaming term we saw in the classical Boltzmann equation; see (3.3.3–5). Moreover, the momentum cut-off of the original model was dispensed with; and the random potential was taken to be of the form $\sum v_i V(x-x_i)$ where the v_i are Gaussian random variables, indexed by the lattice, and V is a smooth, but sufficiently localized fixed potential in lieu of the de facto δ–potential used in the original model. Finally, the proofs were conducted in such a manner that they bypass a criticism later leveled [Ho, Landau, and Wilkins, 1993] against a technical legerdemain in the original proof. Working in the opposite direction and looking for a didactic version of the model, [Martin, 1979] proposed a simpler version akin to the one we presented here.

Model 4 is a quantum version of the classical Lorentz gas, properly called after [Lorentz, 1909]. For various aspects of the model, see [Hauge, 1974, Martin, 1979, Spohn, 1991]; and [Chernov, Eyink, Lebowitz, and Sinai, 1993].

Model 5: CP-maps and Bloch equation. In Rem. F.2 in Sect. 8.2, we defined the Bloch equation for spin relaxation. This equation was proposed on an empirical basis by [Bloch, 1946, Bloch and Wangness, 1953]; for the general physical landscape beyond this equation, see [Leggett, Chakravarty, Dorsey, Fisher, Garg, and Zwerger, 1987].

Note that the evolution governed by the Bloch equation does not couple the longitudinal mode – i.e. the algebra generated by $\sigma(z)$ – and the transverse mode – i.e. the evolution as seen from the algebra generated by the observables $\sigma(x)$ and $\sigma(y)$. Hence, when the initial state $\varrho(t=0)$ of the spin satisfies an *initial* random phase assumption according to which it is diagonal in the orthonormal basis $\{\Psi_+, \Psi_-\}$ defined by $\sigma(z)\Psi_\pm = \pm\Psi_\pm$, the transverse mode is washed out, and the Bloch evolution is non-trivial only on its restriction to the algebra generated by $\sigma(z)$. Consequently, in this partic-ular case, the Bloch equation reduces to a Pauli master equation, where the probability $p_\pm(t)$ that the z–component of the spin points in the up- (resp. down-) direction is given by:

$$\frac{d}{dt}p_\pm(t) = -\lambda_\pm p_\pm(t) + \lambda_\mp p_\mp(t) \quad \text{where} \quad \lambda_\pm = \frac{1}{2}(\lambda_\| \mp \varepsilon) \quad . \quad (15.3.68)$$

In this sense, the Markovian evolution described by the full Bloch equation can be viewed as a non-commutative extension of the Pauli master equation.

It describes the general situation where the initial state $\varrho(t = 0)$ of the spin is arbitrary, and thus, in particular, does not necessarily satisfy any initial random phase assumption. It is this generalization we want to study in this subsection.

We mentioned in 8.2.F.2 that the special relation $2T_\| \geq T_\perp$ between the parallel and transverse relaxation times follows from the assumption that the dissipative system described by the Bloch equation admits a conservative extension, i.e. that the semi-group $\{\gamma(t) \mid t \in \mathsf{R}^+\}$ obtained by integrating the Bloch equation satisfies the scheme indicated in the diagram at the end of Sect. 15.1, which is the quantum generalization of the classical situation described by Fig. 8.4(a). Indeed, when this condition is satisfied, we have

$$\forall\, t \in \mathsf{R}^+ \; : \gamma(t) = \mathcal{E} \circ \alpha(t) \circ \mathcal{J} \tag{15.3.69}$$

which entails that every element $\gamma(t)$ in the semi-group $\gamma(\mathsf{R}^+)$ is completely positive, a property we describe presently.

Recall first that an element A of a C^*-algebra $\mathcal{A}$ is said to be positive whenever there exists $X \in \mathcal{A}$ such that $A = X^*X$; in particular when $\mathcal{A} = \mathsf{C}$ – the algebra of complex numbers – this definition gives indeed the charaterization of the positive real numbers $\mathsf{R}^+ \subset \mathsf{C}$. Let now $\mathcal{A}$ and $\mathcal{B}$ be two C^*- algebras; a linear map $\lambda : \mathcal{A} \to \mathcal{B}$ is said to be positive, whenever it preserves positivity. For instance, every *-homomorphism is positive: $\lambda[A] = \lambda[X^*X] = \lambda[X]^*\lambda[X]$. In particular, automorphisms, injections and states are positive maps; and so are conditional expectations (see below for their definition and a list of their basic properties). Hence our maps $\gamma(t)$ are positive: we expect no less from a time-evolution. Our $\gamma(t)$ however satisfy a much more stringent condition as we are about to see.

With $M(n, \mathsf{C})$ denoting the algebra of $n \times n$ matrices with complex entries, consider the natural extension

$$\lambda \otimes I_n : \mathcal{A} \otimes M(n, \mathsf{C}) \to \mathcal{B} \otimes M(n, \mathsf{C}) \tag{15.3.70}$$

where, with $A \in \mathcal{A}$, $M \in M(n, \mathsf{C})$, and $\lambda(A) \in \mathcal{B}$:

$$A \otimes M \to (\lambda \otimes I_n)[A \otimes M] = \lambda[A] \otimes M \quad . \tag{15.3.71}$$

The map λ is said to be $n-$positive whenever $\lambda \otimes I_n$ is a positive map, and λ is said to be *completely positive* if $\lambda \otimes I_n$ is positive for every $n \in \mathsf{Z}^+$.

It is a remarkable feature of non-commutative algebras that a map can be $n-$positive without being $(n + 1)-$ positive. One of the simplest cases is obtained from the case where $\mathcal{A} = \mathcal{B} = M(2, \mathsf{C})$ with $\lambda[A] = (\text{Tr}\, A)I - A$, i.e.

$$\lambda : \begin{pmatrix} a & b \\ c & d \end{pmatrix} = \begin{pmatrix} d & -b \\ -c & a \end{pmatrix} \tag{15.3.72}$$

which is clearly positive, since $\lambda[A]$ and A have the same characterisitc equation. To show that this map however is not $2-$positive, consider the basis:

$$e_1 = \begin{pmatrix} 1 & 0 \\ 0 & 0 \end{pmatrix} \; ; \; e_2 = \begin{pmatrix} 0 & 1 \\ 0 & 0 \end{pmatrix} \; ; \; e_3 = \begin{pmatrix} 0 & 0 \\ 1 & 0 \end{pmatrix} \; ; \; e_4 = \begin{pmatrix} 0 & 0 \\ 0 & 1 \end{pmatrix} \qquad (15.3.73)$$

and the element $A = \sum_{k=1}^{4} e_k \otimes e_k \in M(2, \mathsf{C}) \otimes M(2, \mathsf{C})$. We have then

$$A = \begin{pmatrix} 1 & 0 & 0 & 1 \\ 0 & 0 & 0 & 0 \\ 0 & 0 & 0 & 0 \\ 1 & 0 & 0 & 1 \end{pmatrix} \quad \text{and} \quad (\lambda \otimes I_2)[A] = \begin{pmatrix} 0 & 0 & 0 & -1 \\ 0 & 1 & 0 & 0 \\ 0 & 0 & 1 & 0 \\ -1 & 0 & 0 & 0 \end{pmatrix} \; . \quad (15.3.74)$$

We have indeed $\mathrm{Sp}(A) = \{1, 1, 0, 0\}$, so that A is positive; but $\mathrm{Sp}((\lambda \otimes I_2)[A]) = \{1, 1, 1, -1\}$, so that $(\lambda \otimes I_2)[A]$ is *not* positive. Hence λ is a positive map that is not 2$-$positive. This argument can be generalized to higher dimensions. Nevertheless, every positive map $\lambda : \mathcal{A} \to \mathcal{B}$ is completely positive as soon as at least one of the algebras $\mathcal{A}$ or $\mathcal{B}$ is abelian. Hence, the distinction between positive and completely positive maps is intrinsically a quantum feature.

In the context of interest to us here, we use the following quantum generalization of the concept of conditional expectation. Given a von Neumann algebra $\mathcal{A}$ with a faithful normal state ϕ, a von Neumann sub-algebra $\mathcal{B}$ of $\mathcal{A}$ is said to admit a conditional expectation w.r.t. ϕ if there exists a surjective normal, faithful projection of norm 1:

$$\mathcal{E}_\phi[\,.\,|\,\mathcal{B}] : A \in \mathcal{A} \to \mathcal{E}_\phi[A\,|\,\mathcal{B}] \in \mathcal{B} \qquad (15.3.75)$$

such that for all $A \in \mathcal{A}$:

$$\langle \phi; A \rangle = \langle \phi; \mathcal{E}_\phi[A\,|\,\mathcal{B}] \rangle \quad . \qquad (15.3.76)$$

The import of this somewhat abstract definition is enhanced by the following result, due to [Tomiyama, 1957] and valid for all $A, A_1, A_2 \in \mathcal{A}$ and all $B_1, B_2 \in \mathcal{B}$

1. $0 \le \mathcal{E}_\phi[A\,|\,\mathcal{B}]^* \, \mathcal{E}_\phi[A\,|\,\mathcal{B}] \le \mathcal{E}_\phi[A^*A\,|\,\mathcal{B}]$;
2. $\mathcal{E}_\phi[B_1 A\, B_2\,|\,\mathcal{B}] = B_1 \, \mathcal{E}_\phi[A\,|\,\mathcal{B}]\, B_2$;
3. $A_1 \le A_2 \; \models \; \mathcal{E}_\phi[A_1\,|\,\mathcal{B}] \le \mathcal{E}_\phi[A_2\,|\,\mathcal{B}]$;
4. $\langle \phi; B_1 \mathcal{E}_\phi[A\,|\,\mathcal{B}]\, B_2 \rangle = \langle \phi; B_1 A\, B_2 \rangle$.

Conditional expectations are completely positive maps, as are automorphisms, injections and states. Since the composition of completely positve maps is again a completely positive map, we obtain that our $\gamma(t)$ are completely positive maps.

The concept of completely positive map was introduced by [Stinespring, 1955] as a generalization of the concept of state on a C^*-algebra. A thorough mathematical discussion can be found in [Choi, 1972].

The physical relevance of the concept was recognized and championed by [Lindblad, 1975, 1976]; in particular, he obtained a complete characterization of the generator of semi-groups of completely positive maps. With that characterization in hand, the relation $2T_\parallel \geq T_\perp$ is obtained in straightforward manner [Emch and Varilly, 1979]; see already [Favre and Martin, 1968, Gorini, Kossakowski, and Sudarshan, 1976, Verri and Gorini, 1978].

For an approach to the Bloch equation in the spirit of the van Hove program – specifically, as the weak-coupling/long-time limit of the reduced evolution equation for a spin $\frac{1}{2}$ in interaction with an infinite thermal bath – see [Pulé, 1974, Jaksic and Pillet, 1997].

Conclusions and further readings. The models discussed in the present section prove the compatibility of a Markovian approach to equilibrium and a strict adherence to the tenets of reversible statistical mechanics. In the quantum context of this section, a preliminary mathematical problem was to extend to the quantum realm the theory of stochastic processes, and to find the necessary and sufficient conditions under which a semigroup of completely positive maps can be dilated to such a process. Model 3 showed how this program can be executed for the particular case of quantum Brownian motion. At the end of Rem. 8.2.F.3, we listed references to some of the progress realized in the solution of the general mathematical problem.

Physically, the models indicate how a statistical mechanical theory of irreversible processes compatible with a reversible microscopic dynamics becomes possible when two essential ingredients are brought into play: the thermodynamical limit and the weak-coupling/long-time limit. The first limit is called upon to remove recurrences, whereas the second is intended to define and focus on the time-scale in which the macroscopic transport phenomena are expected to occur. It can be argued with compeling evidences that the question of the proper time scale for the approach to equilibrium has been with statistical mechanics since its inception as a kinetic theory of gases in the second half of the 19th century; see Chap. 3 and references therein. It still remains a lively subject of discussion today; see e.g. [Lebowitz, 1999b]. The remarkable advance during the last half-century is that methods have been developed allowing to control mathematically the above two limits.

Beside the main stream of the problematics discussed in the present chapter, the formalism of the thermodynamical limit developed for the approach to equilibrium in quantum statistical mechanics has also allowed to shed new light on the quantum measurement process and to implement models that bypass some of the most serious of the usual objections; see e.g. [Hepp, 1972, Whitten-Wolfe and Emch, 1976].

We are aware of several possible objections to the scheme proposed in this chapter.

1. The models presented seem to be somewhat artificially rigged for the purpose. This is true, and this is the price we were willing to pay for both the achievement of mathematical rigor and the necessities of a didactic exposition. We also claim that at least the first and fourth models are not too far-fetched as they take stock of, and account for, specific and actual laboratory experiences.

2. We started with a simple computer-assisted discussion of the dog-flea model in Sect. 3.4. A more sophisticated situation is disccused in [Holian, Hoover, and Posh, 1987]. In the present section, we pursued with a diverse variety of models; it can be argued that our selection is either too diverse or too narrow to amount to a comprehensive theory. Here again, consider that the main aim was to settle one specific problem: the logical and mathematical consistency of some basic physical principles; this has been achieved. Consider further that a balance had to be struck between physical reasonableness and technical simplicity. From this platform, the Reader now can – and ought to – reach and pursue the discussion with more advanced texts, such as [Martin, 1979, Spohn, 1980, 1991, Kipnis and Landim, 1999].

3. It may also be argued that these models do not tell us anything about the "real world" inasmuch as one has to deal there with finite systems that evolve over finite times. Except for Model 1 where this criticism has already been answered, the relevance of the use that can be made of limits requires some elaboration, especially in view of the fact that proofs of the existence of limits involve, more often than not, very drastic majorizations. In most cases, trying to get better theoretical estimates is thankless. In connection with the van Hove program, the difficulty is compounded by the fact that – in addition to the infinite volume limit – the van Hove limit involves letting the coupling constant tend to zero and the time tend to infinity with the rescaling from the time t to the time $\tau = \lambda^2 t$.

Nevertheless, on the one hand we should mention that in their modeling of the Bloch equation, already cited at the end of our discussion of model 5, [Jaksic and Pillet, 1997] derived exactly a transport equation, with convergent expansion in powers of the coupling constant λ, and such that the first non-trivial term of this expansion is precisely given by the Bloch equation. On the other hand, graphical estimates of how the actual behavior deviates from the ideal behavior in the van Hove limit have also been obtained recently; see [Facchi and Pascazio, 1999] and references cited there.

From both the analytic side and a capitalization on the advent of powerful computers and computer usage, one is now in possession of getting reasonable control over possible deviations from a limiting behavior. In particular, it appears that recent results converge to vindicate the van Hove program.

The ultimate justification for the consideration of limits is however that the natural scientist must look for trends, and learn to disregard small deviations from the limit which often get buried as the result of uncontrollable interactions with the outside world. Experimental physicists are much aware of the familiar discussion of both systematic and random errors; theoreticians have become conscious that this is also a matter of physical principle: macroscopic systems cannot be considered to be isolated; see [Zeh, 1970].

4. While none of the builders of the models presented in this section succumbed to the twin temptations presented by Boltzmann's Stosszahlansatz and Pauli repeated random phase approximation, all made probabilistic assumptions on the initial states of their models. Our response here is that these assumptions translate the empirical information available to the experimentalists on the preparation of the system at the time they take over. For us, the proper account of partial information is intrinsic to the tenets of statistical mechanics; we gave specific evidences that this account can be carried out rigorously and in a manner that provides a successful explanation of macroscopic behavior from microscopic Hamiltonian premises. Other explanations are surely possible; the challenge is to construct a better one. More than alternate explanations, what the history of the last twenty-five years indicates is that powerful mathematical techniques were and are necessary to overcome the technical difficulties that confine the scope of the analysis of already available physical insights.

5. The formal simplicity of the models discussed in the present section has allowed our access to an understanding of some basic features of the approach to equilibrium; yet the methods themselves have been developed to deal with more sophisticated physical situations; we want to mention here two of these: the theory of the laser [Hepp and Lieb, 1975, Alli and Sewell, 1995, Bagarello and Sewell, 1998]; and the Vlaslov equation of plasma physics [Braun and Hepp, 1977, Sewell, 1997]; see also references cited therein.

6. Moving beyond the models discussed in this section, we should mention that a non-perturbative solution of the old problem of atomic physics – radiative decay – is the purpose of much current research; see [Hübner and Spohn, 1995, Bach, Fröhlich, and Sigal, 2000]; the introduction of each of these papers is written in an informative, and yet largely non-technical style; but the core mathematical techniques that permeate these papers are prohibitive in a text such as ours.

16. The Philosophical Horizon

*Every phrase and every sentence is an end
and a beginning.*

[Eliot, 1943]

16.1 General Issues

The experiments in our conceptual laboratory are now completed. Before we close up, we want to recapitulate some of the findings we have collected along the way and to sum up the lessons we can draw from them.

The scheme we set up is in the tradition of the semantic approach to scientific theories; thus the central theme of the scheme is the separation of the *syntax* from the *semantics* of a theory; the scheme applies to all disciplines, while different local conditions of different sciences may make the scheme easier or more difficult to use. In this regard, thermostatistical physics is among the most difficult topics in physics because of the complicated structures and dynamics it deals with, and the interplay it involves between the micro- and the macro-scopic descriptions. It is no surprise that only a few illustrative examples of model building and usage in the literature of the semantic approach come from this area with the exception of pure thermodynamics which has some of the most simple models.

However, we were motivated to explore further that line of inquiry because we believe that significant improvements of a philosophical approach often arise from serious investigations of the conditions of its application in areas which – at first sight – are less ideally suited to the project; cf. [Wimsatt, 1987, 1994]. Furthermore, thermostatistical physics provides a two-level approach to thermo-phenomena: the purely thermodynamical and the thermostatistical aspects. Some of the laws in the former are even simpler than those of a simple system in mechanics, while some of the systems in the latter comprise perhaps the most complex ones in physics. Since the simple behavior studied in thermodynamics is ontologically the result of the complex behavior studied in statistical mechanics, the question of reductional explanation also becomes an essential ingredient of thermostatistics.

The models we considered include both historical and contemporary ones. From a philosophical perspective, we see the evolution of science, particularly physical sciences, as a pursuit of ever better models to explain (or understand) and predict natural phenomena that are either ready-made in nature for the trained eyes to observe, or are created in the laboratories by imaginative minds. In terms of possible worlds – worlds that the actual world could have been – each consistent theory invented and accepted in the history of science tells us a way our world could have been. The quest is to find out whether that possible world – the world the properties of which are exactly as the model embodies – is indeed the actual world we inhabit; in other words, a quest to know whether the ways in which our world runs are exactly as the model in question tells us it should.

In this sense, the actual world was once thought to be Aristotelian but is not; it was then thought to be Newtonian but is not, etc., and the quest for paradigms goes on. Therefore, for the study of the roles models play in the building of the scientific image of the world, old speculations are just as instructive as the new ones are. Whether one model rather than another describes the actual world is an *empirical question* that scientists of the relevant discipline investigate; and the search for an answer to this question has created conceptual laboratories in the history of science (of physics in particular). The primary task of this book was not the pursuit of the former question, but the study of the latter; compare with articles in [Achinstein and Hannaway, 1985].

The first kind of models we examined consists of thermodynamic models in equilibrium These are models of physical systems the compositions of which are unknown or ignored. One must note that there is a difference between these models and hydrodynamic models in fluid mechanics where the compositions of systems are not ignored but idealized to be of a continuous rather than of a discrete nature.

Here we see the first interesting feature of idealization and approximation in model construction. Although we have already introduced and discussed the notions of idealization and approximation in the main body of the book, we have only done so with respect to specific methods of theory construction, e.g. the theories of ergodicity, especially, in Sect. 9.5. By way of re-examining the subject of idealization and approximation in order to discover their general features, we are now in position to review in some detail this feature which makes the idealization on thermodynamic models an endeavor somewhat different than what one encounters in other models. For the discussion of different aspects of idealization, the Reader is referred to [Jevons, 1874, Barr, 1971, Nowak, 1972, Barr, 1974, Krajewski, 1977, Schwartz, 1978b, Laymon, 1980, Nowak, 1980, Laymon, 1985, McMullin, 1985, Cartwright, 1989].

16.2 Model-Building: Idealization and Approximation

Idealization is an act of conception or reasoning in which properties, conditions, or factors – which are deemed *secondary* or *minor* to the study of certain phenomena – are subtracted, ignored, or made to have vanishing values so that the phenomena appear 'clean' or 'tidy' – at least, clean and tidy enough – so that distinct regularities, or even laws, may be seen emerging from those phenomena or obvious explanations may be given for their observation.

The notion of being secondary or minor carries at least two meanings, one for the *description* of a phenomenon and the other for the *explanation* of it.

With respect to the description, the secondary factors that an idealization may remove are usually *faint* or *hardly noticeable* factors, which means that the removal of such factors should hardly alter the phenomenon so that the resulting description does not risk a misrepresentation of the phenomenon. With respect to the explanation, the secondary or minor factors are usually *irrelevant* or *not quite relevant* factors, which means that the inclusion of the factors can only frustrate the effort of finding an explanation for a phenomenon without the benefit of improving on the explanation. An example of descriptive idealization can be seen in Laplace's model for sound transmission in air (see, Sect. 2.3).

The model assumes that the process of sound transmission is adiabatic because such rapid compressions and expansions as sound 'travels through' air does not leave enough time for heat exchange. This is, from our perspective, an idealization in which a very small amount of heat exchange – which must have taken place despite the great speed of sound – is neglected. Examples of explanatory idealization can be seen in many other models we have discussed in the book; in particular the Ising lattice-model for phase transitions is a good case (see, Sect. 12.2).

In order to explain phase transitions in a ferromagnet (of at least two dimensions), the Ising lattice-model is postulated in which, *inter alia,* the lattice sites are arranged as the spatial repetitions of some simple unit, and the interactions among the sites are of nearest neighbor. This is not likely to be true, given that the model is interpreted literally. But the actual irregularities among the lattice sites, and the presence of interactions that extend beyond nearest-neighbors, are insignificant or irrelevant to the explanation of the *existence* of phase transitions in the magnets, and therefore they are neglected in the idealization. The models resulting from the removal of such factors – the product of idealization – will not be true of the actual world – for otherwise, there would be no factors to be subtracted – but may be close enough to being true of the actual world, so that an understanding and/or a good prediction can be made from the models, e.g. long-range orders emerge from short-range (nearest-neighbor) interactions. This is where approximation enters the picture.

Approximation is a well-worn term in physical sciences, which carries a single unambiguous technical meaning for some authors while for others it carries several distinct meanings. At first one may say that a model is an approximation just in case it results from an idealization as described above. The model approximates a complete and true description of the phenomena as some factors in the phenomena are idealized away and these factors are only secondary in the meaning given above. Note that these are preliminary observations which will be modified as we go through some of the finer features of these two concepts.

The first interesting feature of idealization we mentioned above in connection with the difference between thermodynamics and hydrodynamics falls out of the following observation. One can certainly say that hydrodynamics which idealizes fluids as continuous media deals with approximate models of fluids, provided it is true that at the macroscopic level, whether the fluids are assumed to be continuous or not – though they are in fact not – matters little to the description and explanation of hydrodynamic phenomena. In thermodynamics, the composition of a thermo-system is *truly irrelevant* to the laws of thermodynamics which are true to the system exactly but not approximately. There are certainly idealization involved in the reasoning of thermodynamics, but the result is not an approximation but precise laws at the level of the theory. Pragmatically, this feature is enhanced by the resilience of the fundamental thermodynamic laws and principles against further advances in other parts of physics.

Had Newton said that all mechanical phenomena obey the laws of mechanics that he formulated in his *Principia*, he would have been making a premature statement, because even in the fully improved and polished versions of Lagrange and of Hamilton, Newtonian mechanics involves space *and* time as entirely different concepts, while Einstein showed that they are only the relative concepts of a single absolute: *space-time*. Einstein's relativity theory properly contains the Galilei-Newton mechanics, and the latter is an acceptable approximate description of kinematics only when the velocities to be considered are small against the speed of light. In addition, even at this non-relativistic level, Newtonian physics, involving as it does 2nd order differential equations of motion, describes the evolution as trajectories of points in the phase-space (p, q) of momentum and position. Quantum theory denies the empirical accessibility of points in phase-space: only regions limited by the Heisenberg uncertainty principle, $\Delta p \cdot \Delta q \geq \hbar$ are accessible to experiment; and hence, Newtonian mechanics is a valid approximation only for phenomena where the energy is much larger than the Planck constant $\hbar$. For thermo-physics, this implies that a microscopic model obeying the laws of classical Newtonian mechanics can only describe macroscopic phenomena that happen at high enough temperatures; an example of which can be read in the story of specific heat; see Sect. 10.3.

In contrast, had Planck said that all thermal phenomena in equilibrium obey laws of thermodynamics, he would be telling the truth, whether he truly believed it or not. No amount of later improved understanding of the underlying mechanical processes of thermal phenomena could possibly render the simple theory of thermodynamics, e.g. in the form given by Planck [1903], false or only approximately true; see also [Gibbs, 1902].

Here we see clearly the difference between the type of idealizations that goes into the making of thermodynamic models and the idealizations that go into the making of mechanical models. We shall call the former kind of idealization *ontic* and the latter kind *epistemic*.

These terms are used to indicate that while all idealizations are acts in theory construction – i.e. epistemological in nature – there are facts about the physical world which naturally favor certain acts of idealizations much more than others. There are facts, in other words, about which phenomena are more easily observed than others and variables are more likely to belong together in describing a set of phenomena. With the ontic idealizations, the variables are chosen in such a way that the regularities established among them are so *naturally* insulated from further compositional details that the knowledge of the latter does not affect the truth of the former. The epistemic idealizations do not share such a privilege for they are mostly the result of pragmatic considerations on epistemic accessibility. Linearization or the omission of higher-order terms in a Fourier expansion are but a few typical examples of epistemic idealizations.

Models, in the sense that scientists most readily recognize, play their prominent role in statistical mechanics, the predecessor of which is the kinetic theory of gases.

In the philosophy of science community, statistical mechanics is an often misunderstood science. This is in part because from a methodological point of view, classical statistical mechanics is more similar to quantum theories than any other classical theories, such as mechanics and electrodynamics, and yet it is mostly regarded as a mere, albeit complicated, extension of classical mechanics. The latter view is a profound mistake. Statistical mechanics is among the first attempts of science to construct explanations of observable phenomena from *models* of the unobservable which are supposed to behave exactly like entities the behavior of which is governed by the theory in another branch of physics. It is as if humans, in doing statistical mechanics, are comparing notes with the Maxwell demons – who see the molecules obeying laws of classical mechanics – and try to come up with an explanation of thermodynamic behavior; for Maxwell's demon, see [Earmann and Norton, 1998, 1999].

There are no Maxwell demons to consult, so whatever the demons *might* see can only be gleaned indirectly from such experiments as Perrin's on the Brownian motion (cf. Sect. 3.1).

However much experimental evidence we can gather of that kind, the current model of microscopic composition of bulk matter in its various states is the result of one of the most significant ontic idealizations in physics. The idealization is ontic because there is no compromise or negotiation of the degree of approximation to speak of on this matter. Nature must be structured in such a way that materials in the bulk are Hamiltonian systems of a large number of molecules or else, one might as well forget all together about statistical mechanics – or mechanical explanations of thermo-phenomena.

Epistemic idealizations enter when the details of the Hamiltonians are in question. Here we see examples in our discussion of the kinetic theory in Sect. 3.2. Maxwell's model is better than D. Bernoulli's simply because it is the result of a better idealization, and the idealization is better because it makes less brutal assumptions, such as allowing all molecules to move in all directions in space and do so with different speeds. As a result, the explanation of thermal properties such as pressure from the Bernoulli model is purely mechanical, whilst Maxwell's explanation introduces genuine statistical elements.

While speaking of the epistemic idealization – the results of which are models that approximate the phenomena one tries to understand and/or predict – we want to bring out another dimension of idealization which lies mostly in this category. The *didactic* role of models in the dissemination of scientific knowledge should by no means be neglected. Of this role, there could hardly be a better example than the dog-flea model of the Ehrenfests which we examined in Sect. 3.4.

There may be a subtle danger of overlooking some interesting features of this aspect by thinking of such models as *solely devised for pedagogical reasons*. Many models, including the dog-flea model, which may have been initially constructed for pedagogical reasons become standard models for the intended phenomena. We think this happens for the following reasons. For the uninitiated, the learning of a theory is best facilitated by the highlighting of the essential properties and their lawlike connections of the phenomena that the theory covers. The didactic role of a model which clearly accentuates such features and focusses a student's attention on them is immediate, even if it uses a system which is radically different from the one which exhibits the properties. For the experts, such highlighting of the positive features is perhaps unnecessary; however, if the model is well-made, it should not only get the negative features right but also accentuate them. By negative features we mean the ones that the appropriate idealization should subtract. In other words, the experts would appreciate a model if it turns out naturally to have both the positive and the negative features of the target phenomenon; and especially the latter because it tells how the factors which are taken out by the force of imagination from the description of the target phenomenon can be seen naturally and clearly in a model which uses a radically different system. The dog-flea model is clearly such a model. It shows to the discerning what is **not** pertinent to – and therefore should be idealized away from – the understanding of Boltzmann's model for the kinetic theory of gases. The two-dimensional lattice-models discussed in Sects. 12.2–3 are also such models. We may call the purely didactic models in science as *illustrative* models and the ones we just discussed as *representative* models. To show and impress, one needs mostly illustrative devices which accentuate the positive, but to weigh and judge, one needs a fair representation.

One of the chief aims of epistemic idealization is to arrive at models or claims that are approximate to the intended targets. But what is approximation? How should one measure the distance between two statements? What does it mean to have asymptoticity in certain approximation schemes? We have discussed throughout the book different methods of obtaining approximation in concrete cases (see, for instance, Chaps. 9, and 11–15), it is now time to recapitulate and look at it in general. For the discussion of approximation or approximate truth as general philosophical approaches, see [Popper, 1963, 1976, Hilpinen, 1976, Schwartz, 1978b, Rosenkrantz, 1980, Hartkaemper and Schmidt, 1981, Niiniluoto, 1984, 1986, Oddie, 1986, Niiniluoto, 1987, Weston, 1987, Ramsey, 1992, Weston, 1992].

Upon examining the various cases of approximation in thermostatistical physics, we are led to distinguish two aspects of the notion: one expresses a *static* measure of two assertions as to their closeness, and the other an *asymptotic* measure of a series of assertions approaching a certain given assertion.

The static theory of approximation gives or ought to give a measure of what is known in the philosophy of science literature as *truthlikeness*; see, [Niiniluoto, 1987] and the references therein. It evokes a metric in a space of given physical magnitudes in which the distance between any two statements about certain physical situations is estimated. Once the range of approximation is determined, all statements the distance of which from a given true statement is within the limit should be considered as approximately true.

The asymptotic theory of approximation is more complex; we gave a preliminary discussion in Sect. 1.5, where we differentiated the *controllable* from the *uncontrollable* approximation. Here again is another simple example to drive the point home. With $0 < r < 1$, let

$$f_N : x \in [0, r] \mapsto \sum_{n=0}^{N} x^n \in \mathsf{R}^+ \quad \text{and} \quad f : x \in [0, r] \mapsto \frac{1}{1-x} \in \mathsf{R}^+ \quad .$$

Then

$$\lim_{N \to \infty} f_N = f \quad \text{uniformly} \quad .$$

Specifically

$$\tfrac{1}{1-x} = \sum_{n=0}^{\infty} x^n = \sum_{n=0}^{N} x^n + x^{N+1} \tfrac{1}{1-x}$$

hence

$$\left|\frac{1}{1-x} - \sum_{n=0}^{N} x^n\right| < r^{N+1}\frac{1}{1-r},$$

so that, for *any arbitrarily small* ε

$$N > \frac{\ln[\varepsilon\,(1-r)]}{-\ln r} \quad \models \quad \{\forall x \in [0,r] : |f(x) - f_N(x)| < \varepsilon\} \quad .$$

However, this is only one of the two aspects of the asymptotics. Since approximation is a method of science, it has its *objective* as well as its *subjective* aspect.

The former refers to the objects to which the method applies and the latter refers to the ways in which the method is appropriately used. Suppose that the objects are certain mathematical statements, then whether or not they have controllable approximation is a property of the statements, once the metric approximation (in the static part) is fixed. In this (objective) aspect, the dynamic notion of approximation is very similar to notions such as convergence. But even to a controllable mathematical statement, one may fail to apply a good method of approximation. To obtain the same degree of approximation (as defined in the static part) different approaches may be devised; some may be so costly that to be realistic – achieving the goal within the limit of time and resources – one has to compromise on the degree of approximation. Such methods would be considered as bad approximations, due to the lack of human ingenuity and not a feature of the mathematical statement. As an example, we may look at the [Leibniz, 1682] expansion of π. It is obtained as the evaluation for $x = 1$ of the McLaughlin expansion of arctan x:

$$\frac{\pi}{4} = 1 - \frac{1}{3} + \frac{1}{5} - \frac{1}{7} + \ldots = \sum_{n=0}^{\infty}(-1)^n\frac{1}{2n+1} \quad .$$

To use this expansion to get the value of π in the interval of

$$3\frac{10}{71} < \pi < 3\frac{1}{7},$$

as much as some hundred terms have to be calculated. A good approximation in this sense, however, would be a method that employs the most effective procedure in carrying out the asymptotic approach to the desired degree of approximation. The computer simulation in the case of checking whether the Toda lattice is a chaotic system (Sect. 9.3) is an example of a good approximation. The method of simulation used for it was so good that it was quickly found out that the model is likely to be integrable despite its strong similarity with another known chaotic model – the Henon-Heiles model.

16.3 Model-Building: Simulation

To further illustrate the above point and other related features of model-building in science, let us turn to a special phenomenon in the practice of science which is prevalent in the study of regularities in complex or seemingly random systems consisting of a large number of constituents. The general term for the phenomenon is *simulation*.

With the recent exponential growth in computer power, simulation has asserted its ubiquitous presence in all ways of scientific research. The difficulty and costliness of experiments in certain fields of science also contributed to making the use of stimulation an indispensable step in the process of theory testing.

To illustrate some of the simulation activities that are pursued along this line, we briefly consider an example that is more complex than the usual affairs: GEANT, a program currently employed at CERN for the planning, calibrating, and preliminary testing of a whole range of high-energy experiments which are conducted with its colliders, such as LEP (Large Electron Positron collider) and SPS (Super Proton Synchrotron). The events simulated include Compton collision, photoelectric effect, Cerenkov effect, annihilation of positrons, direct (e+, e-) pair production, muon physics, and some hadronic interactions. Unlike the kind of simulations we have discussed so far, whose purpose is theory testing, such as obtaining truncated and/or ideal approximations to solutions of complex equations, GEANT is used to produce a computational modeling of the experimental instruments, setups, and the processes in and around the colliders, so that the above mentioned (and more) physical proceses may be calibrated and optimized. Let us very briefly describe the components of GEANT so that we see what kind of simulation it really is.

GEANT is capable of creating a virtual environment in which the kinematics of high-energy particles through detectors can be studied in a transparent manner. Its main purpose is to design and/or improve the detectors and to develop and test the tracking and reconstruction of events (including trajectories and error propagations). Its principal application then include transporting virtual particles through an virtual experimental setup for the simulation of detector responses and the graphic representation of the setup and of the particle trajectories. In one track, the MonteCarlo random methods are used to generate different types of particles; they are then taken over by programs for processing and storing event histories. In the other track, programs are used to simulate materials and shapes of the components of the detector. By combining the two tracks, high-energy particle experiments can be simulated by the tracking of the trajectories of the virtual particles through the virtual detectors.

Since calibration and debugging of real experiments in the colliders are the main use of GEANT, this simulation is clearly different from the kind of simulations in which estimations of real solutions of differential equations are its main purpose. What is simulated is not a mathematical or theoretical process – i.e. a derivation or solution, but rather some physical apparati, sctups, and processes. However, there are similarities; for instance, the simulation of a particle's trajectory is not so different from the simulation of a solution of a set of differential equations, since a particle's trajectory is often given by the solution of such equations with the appropriate initial and boundary conditions. The simulation for the setup through which the particles are transported is done by simulating textured geometrical volumes. The material filling up the a volume is represented by the data, such as its atomic number, its atomic weight, and its density. Such is the information with which, after simulational runs of GEANT, evaluations and adjustments of the real experimental setups are given.

However, we emphasize that this kind of simulations neither test any theories nor explain any phenomena; thus, it is not even a model in the usual senses of the term (cf. Sect. 1.3). It may be seen as an example of 'models of experiments' (cf. 5 in Sect. 1.4), for it recreates by computer programs the physical objects and processes in the virtual environment it creates.

Back to our main track, we must recognize that views differ on the role and nature of simulation: how much, or how far, can it substitute for theories; can the results of simulation be used to improve upon or even replace experimental results? As Langer [1999] argues, but see also [Schiesser, 1999], in spite – or perhaps also because – of the prodigious abundance of available data, simulation is no substitute for a theory the role of which remains to be that of separating the relevant information and thrashing out the husk. Nor is simulation the same as the analytical modeling, which is the the primary focus of our book. We discuss below how simulation resembles, and yet differs from, experiments in several crucial aspects. See already [Weissert, 1997] and references therein.

The best way to approach the question is the usual method of conceptual analysis: to examine the typical circumstances in which the method is actually used and to distinguish it from closely related and better understood concepts. The typical scenario of introducing simulation unfolds as follows. One forms a hypothesis about the behavior of a class of systems. It is often in the form of (coupled) differential or difference equations, reflecting the dispositions or tendencies inherent in the systems; under explicitly specified initial and boundary conditions the equations should describe the exact behavior. In our context, the behavior will be the trajectories traced out in a phase space. In other words, different solutions should fall out from the equations as solutions with different sets of initial and boundary conditions. However, the solutions of many of such equations demand computational power which until recently was even beyond the imagination of science. Even today, with the computing ability available from a supercomputer, tasks such as longterm realistic meteorological predictions are still beyond our grasp and are likely to remain so. The case of meteorology was made famous by the Lorentz *butterfly effect*, which emphasized how for certain nonperiodic systems, almost infinitesimal differences in the initial data may result in outcomes as different as if they were picked out randomly from any set of outcomes, cf.

[Lorentz, 1963, Gleick, 1987]. Nevertheless, the computing power now harnessed is such that for some privileged circumstances, enough of a solution can be obtained so that conjectures about the real solution can be made far beyond the stage of wild guesses. Such computational approximations are at the core of the concept of simulation.

Just as the number, π, which as a definite number has a definite decimal expansion – whether or not it is possible to recognize any order in it, (see Sect. 5.3), a large class of (partial) differential equations exists, for which the following is true: given a set of initial (boundary) conditions, they do have solutions, whether or not the solutions can be obtained by analytical means.

To see how the solutions behave, one way – and sometimes the only realistic way – is to use numerical methods and obtain approximate solutions. Whether or not such an approximate solution exists is determined theoretically in view of the numerical method used; this is an *objective* feature of the solution under numerical expansion (see above discussion). With the existence granted, there is still the question of whether a decent approximation can be actually obtained via calculations (the *subjective* aspect). Here enter the powerful computers which in recent years have dramatically increased the chance of asymptotically approaching the real solution via numerical methods. It is now possible to do enough calculations in a reasonably short amount of time – which previously might mean generations – so that one can see some meaningful behavioral patterns in the complex solutions of some of the toughest differential equations. Enough of a solution, that is to say, for us to make *reliable* predictions about what the real solution may be like. Yet, some solutions are stable several examples are discussed in [Hoover, 1999]: see for instance his section 6.7 on the discretization of the Rayleigh–Bénard flow. and some unstable;

To determine which solutions are stable is a problem that, ultimately, can only be solved theoretically; on the answer to that question depends the type of the approximation. The notion of stability again has two different meanings (see Chap. 9): stability with respect to small differences in the initial conditions and stability with respect to the perturbations of the Hamiltonian. This is a subject in close connection with the fertile area of chaos, see [Ott, 1993, Gutzwiller, 1990, Hoover, 1999, Hoppenstaedt, 2000]; and of renormalization, see Chap. 13.

Simulation may be understood in two different ways.

- It may denote the practice of obtaining truncated numerical approximations to solutions of equations, whether or not they are obtainable by analytical means. There is a double approximation in a result of simulation.
  – The approximation coming from the idealization which creates the model from which the differential equations arise;
  – the approximation caused by the simulation itself.

- Simulation may also denote the practice of setting up a *secondary* model for the primary model from which the difficult equations arise. The model of simulation would be a model which involves further idealizations that ensure the solvability or computability of its solutions.

We also have the double approximation implied in the second meaning of simulation; but in this meaning, simulation becomes another kind of model-building activity; only the models of simulation are ones which approximately represent the abstract mathematical structures – themselves models in another sense – rather than physical systems. Here, one should notice an interesting feature of simulation, namely, the discretization of differential equations. The use of numerical methods to approximate the real solution of some differential equation requires that one breaks up the differential at any instant and replace it with a difference at a small time interval.

For examples of this type of simulation, see [Hoover, 1999, Chaps. 5–6]. This move of discretization is an example of the second type of idealizations mentioned above, whose purpose when small enough intervals are taken is to approximate the real solution. However, this idealization does not necessarily remove the results of the simulation further away from the phenomena, because to consider a natural process differentiable (or continuous) is also an idealization – the first type of idealization mentioned above. The compounding of two idealizations does not necessarily make a law or a theory more idealized – i.e. further removed from the phenomena. In this case, since there may be reasons to believe that natural processes are indeed not continuous, the results of simulation might just be closer to what we need to describe the phenomenon than the real solution of the differential equation in question. One must also note that variations (or even complications) arise when the concept of simulation is applied to diverse practices in science. For instance, when hierarchical descriptions of systems (from physics up to biology) and their relationships are concerned, simulation in the second meaning may be involved in the conception of upper-level models in terms of the interactions among their elementary parts; cf. [Pople, 1999, Schweber and Wächter, 2000].

The two meanings of simulation are really two different ways of representing the same practice. We find the second meaning more convenient to depict our conceptual landscape. With simulation as modeling, we may begin to examine its functions in theory building for dynamical systems. As far as modeling is concerned here, an abstract or mathematical system is no different from a physical system. Both involve idealizations which subtract factors the absence of which is thought – hopefully with good reasons – not to affect our understanding of the essential properties of the system in question, be it an abstract system or a concrete one. And some mathematical systems are not any better understood than any physical systems, even though understanding the former involves a different type of epistemic process than the one for understanding the latter. From this observation, we may get a glimpse of the similarities and differences between experiments and simulations.

The similarities are few and superficial and the differences are many and deep. Both simulations and experiments can be used for exploratory as well as testing purposes, although simulations are more often used as tools of exploration, while experiments with physical systems are primarily used for testing theoretical hypotheses about the systems. If one wants to know what the real solution may look like, simulations are run to explore the form of that solution. But if one wants to know whether the solution belongs to a certain category – which is analogous to knowing whether a physical system has certain properties (e.g. whether or not Toda lattice is an integrable model, see Sect. 9.3); or whether the energy gets distributed evenly as among the degrees of freedom in the Fermi–Pasta–Ulam model [Weissert, 1997] – simulations may be run for testing of the hypothesis. However, since the simulated solutions are never the real solution but only (hopefully) closer and closer approximations of it, simulations when used as testing tools are perhaps more appropriate for *falsification* than for *verification*. Just as in real experiments in which scientists usually turn a verifying experiment into a falsifying one by, for instance, trying to refute a null hypothesis rather than to verify the hypothesis from which the null hypothesis is construed, simulators usually try to use a property attributable to the real solution from which it is relatively easy to tell, from some runs of simulation, that the solution could not possibly have that property. In the simulation test of the Toda lattice-model, the property used there was not whether its real solution is integrable – which was initially deemed unlikely – but rather whether it is chaotic, the opposite of integrability. However, one should not be dogmatic about this point, since it is not difficult to imagine or even find instances of simulation in which verification is the intended purpose.

Simulations are experiments in the sense that we do not know the result until the runs in a computer are completed, and if one looks at a computer as a piece of experimental apparatus and the equations together with their initial/boundary conditions as the input, the output of the computer in question can be understood as experimental results. There is some similarity between this operation and the one in which we find out the pH values of certain liquid: one makes the required conditions ready and wait for the results. Whether it is the rapid calculations inside a computer or the rapid chemical reactions inside a pH testing kit, they are in the end physical processes the outcomes of which we have neither foreknowledge nor control.

However, the similarity between experiment and simulation may stop when the simulation is run.

But, a more important difference may be rendered as follows. We may think of an experiment in general as consisting of two procedurally distinct components: devising and executing. If so, simulations as "experiments" on computers differ from real experiments on physical systems. With respect to the first component, there is a fundamental difference between the setting up of a simulation and that of an experiment. In the former, all the relevant computational factors or conditions are absolutely controlled. There is nothing at this stage except planning errors which may bring surprises in the end. This is certainly never true in the experiment: in an actual physical set-up, some not initially identified or planned factors or conditions are always possible, and may bring surprises at the end of an experiment. With respect to the second component – the execution – simulations have a counterpart to the noise factor in experiments on physical systems, which is usually known as the round-off errors. Like noise, such errors are not precisely predictable; nevertheless, bounds for their effects can be estimated so that the results of simulations – just as those of experiments – can be trusted within specified limits.

The differences between simulations and experiments seem to result from a more fundamental disparity between the two. Experiments run on actual physical systems *qua* physical systems, whereas, simulations run on actual physical systems (i.e. computers) *qua* computational algorithms. In other words an "experiment is an active intervention into the course of Nature" [Fuchs and Peres, 2000], while a simulation is not. If one wants to call the latter an intervention, it is one into the procedure of derivation, whose algorithms are known to us. We can neither completely know nor completely control the physical properties of the experimental systems, while we can completely know and control the algorithms of computers (except the round-off errors in the execution phase), which are the only relevant properties in simulations. Computers do have physical properties and they are not always predictable, but such concerns are not relevant to the concept of simulation: one cannot regard as significant any differences in the output of a simulation which result from some unknown conditions in the physical set-up of the computer. Anything significant in the simulation must be completely controlled by the algorithm on which a simulation runs and which ought to be transparent to the person(s) who run the simulation and analyze its results. The Reader will again find ample illustrations of this point in [Hoover, 1999].

16.4 Recapitulation

To conclude what we have discussed so far about idealization and approximation, let us make it clear that although it is frequently accompanied by approximate results, idealization as a tool for model-building is not always connected with the production of approximate results. In fact, it is much better to characterize the purpose of idealization as something other than approximation production. The purpose is rather to conceptually imagine or design a setting (which happens not to obtain in the actual world) in which we can discover true and fundamental laws of nature. Hence, we have the necessary connection between *thought experiments* and idealizations. And as we know, laws of nature, unlike singular facts, may be true in possible worlds other than the actual one, and they may be true in their purest form in those worlds which do not remotely resemble the actual one, cf. [Lewis, 1973, Dretske, 1977, Armstrong, 1983, Lewis, 1986, Van Fraassen, 1989, Carroll, 1994].

The main function of idealization in the whole scheme of theorization in science is, therefore, not merely to help solve technically unsolvable equations but to invent models that *carve nature at its joints*.

Another way of justifying this image of carving nature comes from the distinction between the ontic and the epistemic idealization as examined earlier. To arrive at true and exact laws of nature, the idealization has to be ontic – it has to cut through parts the separation of which is allowed by nature, namely, the de- and the re-composition of the parts are really natural possibilities. Intrinsically, ontic idealization does not have to produce approximate results, especially when the joints are deeply hidden from us. Ontic idealizations produce approximate results only when, as we mentioned above, nature almost does the idealization for us, so to speak, as in the case of astronomy and thermodynamics. But for laws governing, for instance, microscopic processes, ontic idealizations may be far from devices of approximation production. The metaphysical belief behind using such idealizations is that the real mechanisms or causal structures of our overwhelmingly complex phenomenal world are relatively simple. The more complex the phenomena and the deeper the posited mechanisms, the wider the gap between the idealized conditions and the facts, which should not affect the truth of the idealized law or theory in question. If one can conceptually (and sometimes even experimentally) cut the world apart at its joints, one can better understand how it actually operates and predict how not yet existing systems may operate, cf. [Achinstein, 1965, Achinstein and Hannaway, 1985, Brown, 1985, Wimsatt, 1987, 1994, Cartwright, 1997, Morrison, 1997, Liu, 1999] and [Wigner, 1997, 534–549].

We should have realized that statistical mechanics would not be so called if the notion of probability were not essentially involved in its arguments. Without it, the connection between the thermodynamics of bulk matter and the mechanics of molecules which comprise the matter would be crude at best; cf. Bernoulli's model in Chap. 3. In Chaps. 4–6 we have shown the difficult struggle towards a rigorous theory of probability. Such a theory was only possible after one realized that there should be a *clear separation* between the *syntactic* theory of probability, which provides the axiomatics without the interpretation, and the *semantics*.

It was only when Kolmogorov finally freed the syntax of probability from its semantics that the true structure of the concept of probability became available. On the most general level, probability is now understood as a concept of practice that sanctions certain rules of operations – the syntax – and that carries certain meanings which must be at least consistent with the rules – the semantics. Viewing most of the circumstances under which the concept of probability seems to apply appropriately, two readings emerge: one is a *de re* reading and the other is a *de dicto* one; among the former: the limiting frequency and the propensity interpretation/model; and among the latter: De Finnetti's degree-of-belief and Carnap's logical interpretation/model. For details of these models, see Chap. 6.

Less clear is the relationship between these two types of models and the question of which of these two semantics should play the major role in thermostatistical physics. From our discussion in Chaps. 4–6, we have seen the enormous difficulty of making the von Mises semantics rigorous and yet useful in getting a firm grip on the notion of randomness. Such difficulties have recently dampened the enthusiasm of theorists in using it to explicate the meaning of probabilistic concepts or magnitudes in their theories. What is left in the *de re* camp is the propensity interpretation which takes probability as an intrinsic property of chance events and processes. We have chosen in our book not to discuss in depth the large literature of pros and cons of the propensity interpretation because we believe that one of the *de dicto* reading of probability, namely, the de Finnetti semantics, captures more of what physicists have been *doing* in thermostatistics. The arguments for this line are most explicit in the discussion of the Shannon entropy in Sect. 5.3 and the Gibbs ensembles in Sect. 10.1. The lesson in the latter is that the Gibbs measure for the canonical 'ensembles' can be better understood from the de Finnetti perspective in which the only rational or justified degrees of belief one may hold on any observable of a canonical system are that the belief derives from a maximal entropy of the system in the presence of the heat bath. While our Reader should consult these sections for details of the arguments, one general point about them is worth recapitulating here. One should realize that a *de dicto* reading of the concept of probability does not necessarily make it a subjective reading or rendition. The content of probability as degrees of belief is of course subjective, for it expresses the amount of information available to an agent who is assessing the situation. However, the determination of that information, and therefore the probability value, is not subjective at all. It is an objective relation obtaining between the investigator and the system she investigates.

Another way of looking at the various semantics for the Kolmogorov theory of probability is to think of them as different models for the set of axioms; see [Wigner, 1997, 534–549], and the commentary in [Emch, 1993].

There are several aspects of this and similar model-building activities which are worth summarizing here. First, models of this kind can be used to test whether the axiomatic theory is *consistent*. One should note that there are models which are used for this purpose alone, such as the four models presented in Sect. 5.2 for the Kolmogorov axioms. To qualify for playing this role, the models must have all their properties transparently given and none of the properties, either singly or in conjunction with others, should be ambiguous or problematic in any sense. This must be so because otherwise one could not be sure, in case that the model does not satisfy the axioms in question, whether it is a problem with the axioms or one with the model.

This is why, most of the time, models which serve this purpose are purely mathematical models; for a recent discussion of mathematical models, see e.g. [Gershenfeld, 1999]. And this purpose in the model-building activity separates *mathematical physicists* from *theoretical physicists*.

Mathematical physicists relish the challenge of discovering mathematical properties in some physical theories – or theories which have their origins in some observable phenomena – and they precisely articulate the properties by constructing models of the above kind, while theoretical physicists often avoid such properties partly for the reason that these properties are not quite 'real', and thus not relevant to physics. For *property articulation* which is a variation of testing theories for consistency, we have models in Chap. 8. Ergodicity is a fruitful property which not only generates theories for itself but also for other related properties, e.g. the properties in the ergodic hierarchy. It is often a challenge to construct models which have the exact properties defined in highly abstract mathematical terms. The Baker transformation model, the Bermoulli shifts, the Brownian motion, and the Arnold cat, all of them are models that mathematical physicists use to demonstrate the meaning and usage of the properties they define in their theories.

It is quite obvious, however, that the models in Chap. 6 are of an altogether different nature. Since these are results of attempts to relate the abstract theory of probability to the actual usage of the concept in scientific and even daily contexts, the models are complex and involve other notions, e.g. randomness in the limiting frequency model, the precise definitions of which remain controversial till this day. Here, we see a division of labor in the theory of probability. First, the axiomatic theory is tested for its consistency by the construction of purely mathematical models as explained above. Such models tell us little about the 'real' meaning of probability – a widely used and highly entrenched notion in our cultural heritage; see Chap. 4. Second, the axiomatic theory, once proven consistent, is fitted with models that aim at capturing its meanings in practice. This second task is usually a difficult one: the gap between the mathematical (or logical) rigor and the richness of daily usage is usually so big – which is certainly true in the case of probability – that to satisfy fully both sides often appears impossible.

Kolmogorov's syntax is a triumph of mathematical rigor over the morass of loose usage. Today, few people can seriously contemplate any models of probability which do not satisfy the Kolmogorov axioms. Unfortunately, such an outcome is not always achieved, even in closely related areas. Ergodic theory is a case in point. It has been suspected by some physicists and philosophers of science that the mathematically rigorously defined notion(s) of ergodicity might have entirely missed the intuitions and useful insights from which the concept itself was initially conceived. We disagree with such extreme skepticism, and Chaps. 7–9 are devoted to the multifarious aspects of the mathematical theory of ergodicity. The concerns of its philosophical implications, and the predicament we should think we are in, are carefully sifted out in Sect. 9.5.

While models in the ergodic hierarchy address the foundations of statistical physics, the models we met in our study of phase transitions and critical phenomena are concerned with its applications. As such, they are no different from any other models in physics which implement laws or theories under sets of idealized conditions. A rather unique feature of these models is how they are divided into two fundamentally different categories by the taking of the thermodynamic limit. The mean-field models (Sect. 11.3) which do not need thermodynamic limit are more realistic qua the sizes of model-systems, but are neither rigorous nor accurate in predicting testable results, while the Ising-type models (Chap. 12) which do need thermodynamic limit are more realistic qua the structures, the rigor and accuracy of their predictions, but they call for systems beyond those of strictly finite sizes. The thermodynamic limit as an idealization is essential to the rigorous explanation of many micro-explanations of thermo-phenomena, such as phase transitions, critical phenomena, and the approach to equilibrium. Even though it may seem intuitively unappealing, it is not really any more radical an idealization than many others, long-standing and routinely assumed in the history of physics. For instance, it is no more radical a departure from nature than the continuity assumption which lies at the foundations of calculus, the significance of which in physics is generally felt to need no elaboration. Just as taking the thermodynamical limit, making a discontinuous system continuous involves essentially the same kind of limit: the former is a limit to infinite size of the system, while the latter is one of filling up the vacua between the parts, the 'size' of which is also infinite, albeit one case calls for the 'infinitely large' and the other for the 'infinitely small'.

Renormalization (semi-)group method (Chap. 13) is essentially an extension in the use of Ising-type models in dealing with critical phenomena under thermodynamic limit. However, it has enormous philosophical (or methodological) implications which go far beyond the scope of thermostatistical physics. For a recent detailed study of such implications, see [Batterman, 2000]. Batterman believes that the success of the renormalization (semi-) group method in physics provides hope for a general philosophical account of how it is possible to give a reductive account of some higher level phenomena when it is obvious that they are multiply realizable in their lower level mechanisms.

Now that the laboratory is finally closed, we are keenly aware that many worthy subjects in the foundations of thermostatistical physics are left unvisited. But our hope is that this book will have contributed to equip our Readers with the necessary knowledge and tools to explore various areas in the sciences between the micro and the macro phenomena. For instance, for the more philosophically inclined Reader, the issues and controversies concerning the *arrow of time* may be the next stop beyond the present book;

see [Callender, 1998], [Price, 1996], and [Savitt, 1995]; the handling of many of these issues requires a solid grounding along the lines provided in the course of our investigations: the apparent irreversibility prescribed by some macro or phenomenal laws (Chap. 2), the collective effect of a large number of individually reversible processes (Chap. 3), the notion of randomness (Chaps. 5–6), and even the notions of stable and unstable dynamical systems (Chaps. 8–9).

Furthermore, the lessons we learn in Chaps. 7–15, especially the virtue of distinguishing the dealing with interesting and fruitful mathematical concepts from the demand for an explanation of certain physical intuitions, will surely serve well in searching for an explanation of the apparent arrow of time in a world – under certain metaphysical picture – of theoretically reversible processes. For the more scientifically inclined, *chaos* may be the next stop beyond this book, see [Gutzwiller, 1990, Lichtenberg and Lieberman, 1992, Ott, 1993]: we picked up several opportunities to touch on the latter subject, but we chose not to give it the systematic treatment available elsewhere. We trust that our book, especially Chaps. 1–6, will adequately prepare our Reader for further explorations in the territories opened by questioning the logical structure of the concepts and techniques evolved in thermophysics.

A. Appendix: Models in Mathematical Logic

What follows is a collection of elementary definitions and results that reflect only the most basic vocabulary of formal logic. No attempts are made to outline the full mathematical scope of the huge enterprise that logicians call "Model Theory". In a nutshell, the latter concerns itself with the relation between: (a) the properties of sentences specified in a formal language, and (b) the mathematical structures (often, but not necessarily, from algebra) that satisfy these sentences. Since several of the most basic features of this theory undergird the presentation of our main topic in Chap. 1, namely what philosophers of science call the "Theory of Models", we deem it useful to part of our audience that we include a sketch of the mathematical theory in this appendix.

The first feature of the mathematical theory is to distinguish sharply *syntax* from *semantics*, although even some of the best introductory texts – e.g. [Malitz, 1979] from which much of this appendix is borrowed – do not bother to make this distinction explicit, nor even to mention its existence. We found it necessary for our purposes to delineate separately these two topics. We trusted, however, that we could safely ignore here the problems encountered in the discussions of axioms in arithmetic, although our pragmatic attitude in no way reflects on the ultimate importance of the latter.

A.1 Syntax

A first-order language L is viewed here as a fragment of English, so formalized that the sets of well-formed words and sentences are precisely defined; in particular the rules of formation (which we review first) are devised in such a manner that each of the resulting constructs is unambiguous, or *uniquely readable* in a sense that is made precise below.

Definition A.1.1. *Let* Z^+ *and* ω *denote the set of the positive integers and the set of the ordinals, respectively. The* symbols *of the* language L *("first-order predicate calculus") are:*

    the variables: v_n *with* $n \in Z^+$
    the constants: c_α *with* $\alpha \in \omega$

the equality: $\approx$
the connectives: $\wedge$ *(and)*, $\vee$ *(or)*, $\neg$ *(negation)*
the quantifiers: $\exists$ *(there exists)*, $\forall$ *(for all)*
the parentheses: $[\,,\,]$
the n-function symbols: $f_{n,\alpha}$ *with* $n \in \mathbb{Z}^+$ *and* $\alpha \in \omega$
the n-relation symbols: $R_{n,\alpha}$ *with* $n \in \mathbb{Z}^+$ *and* $\alpha \in \omega$.

Remarks A.1.1:

a. Pairs of left-right parentheses is omitted from formulas when no confusion is likely to occur.
b. In some versions of these definitions, suited for instance to a preliminary discussion of arithmetics, one takes a more limited stand: the indexing of the variables and of constants runs over the positive integers; the n-function symbols are the binary function symbols $\oplus$ and $\odot$; and there are no n-relation symbols.

Definition A.1.2.

a. *A type* s *is a set whose elements are either constant symbols, n-function symbols, or n-relation symbols.*
b. *An expression* φ *is a finite sequence of symbols; the type* $\tau(\varphi)$ *of an expression* φ *is the set of constant symbols, function symbols, and relation symbols occurring in* φ; *an expression* φ *is said to be of type* s *whenever* $\tau(\varphi) \subseteq s$.

Definition A.1.3. *Let* s *be a type. The set* Trm_s *of terms of type* s *is the smallest set* T *of expressions satisfying the following three conditions:*

1. $v_n \in T$ *for all* $n \in \mathbb{Z}^+$
2. $c_\alpha \in T$ *for all* $c_\alpha \in s$
3. *for all* $n \in \mathbb{Z}^+$: *if* $t_1, ..., t_n \in T$ *and* $f_{n,\alpha} \in s$, *then*

$$f_{n,\alpha}[t_1, ..., t_n] \in T.$$

Remarks A.1.2:

a. Let v be a variable, c be a constant, f_1 be a 1–function, and f_2 be a 2–function. Then $f_2[f_1[v], c]$ is a term of type $\{f_2, f_1, c\}$.
b. As already noted, one test that these definitions must pass is that they be unambiguous, i.e. in particular, that the rules of formation (the first of which is Def. A.1.3 above) lead to constructs that be uniquely readable; the following result illustrates this demand.

Scholium A.1.1 (readability theorem for terms). *For each term* $t \in \mathrm{Trm}_s$ *exactly one of the following three conditions holds:*

1. *there is a unique* $n \in \mathbb{Z}^+$ *such that* $t = v_n$
2. *there is a unique* $\alpha \in s$ *such that* $t = c_\alpha$

3. *there exist a unique* $n \in \mathbb{Z}^+$ *and a sequence* $\{t_1, ..., t_n, f_{n,\alpha}\}$ *where* $t_1, ..., t_n \in \mathrm{Trm}_s$, *and* $f_{n,\alpha} \in s$ *such that* $t = f_{n,\alpha}[t_1, ..., t_n]$.

An elementary proof, written carefully for the special case of arithmetics – see Rem. A.1.1 – is given in [Malitz, 1979, Thm. 2.10.2] .

Definition A.1.4. *An* atomic formula *is an expression of one of the follow-ing two forms:*

1. $[t_1 \approx t_2]$
2. $R_{n,\alpha}[t_1, ..., t_n]$

where all t_i *appearing in the above two lines are terms.*

Definition A.1.5.

a. *The set* Fm_s *of formulas of type* s *is the smallest set* F *satisfying the following four conditions:*
   1. *every atomic formula of type* s *belongs to* F
   2. *if* $\varphi \in F$, *then* $[\neg \varphi] \in F$
   3. *if* $\varphi \in F$ *and* $\psi \in F$, *then* $[\varphi \vee \psi] \in F$ *and* $[\varphi \wedge \psi] \in F$
   4. *if* $\varphi \in F$ *and* v *is a variable, then* $[\exists v \varphi] \in F$ *and* $[\forall v \varphi] \in F$
b. *If* Σ *is a set of formulas, then the* type $\tau \Sigma$ *of* Σ *is* $\cup \{\tau(\sigma) : \sigma \in \Sigma\}$.

Remark A.1.3: Again, a unique readability theorem holds for formulas; see e.g. [Malitz, 1979, Thm. 2.10.5].

Definition A.1.6.

a. *Let* S *be a sequence* $s_1, s_2, ..., s_n$, *and* $1 \leq i \leq j \leq n$. *The* (i, j)−subse-quence *of* S *is* $s_i, s_{i+1}, ..., s_j$.
b. *Let* φ *be a formula (and hence a sequence). If a* (i, j)−subsequence ψ *of* φ *is a formula, it is said to be the* (i, j)−subformula *of* φ. *A formula* ψ *is said to be a* subformula *of a formula* φ *if it is a* (i, j)−subformula *of* φ *for some* i, j.
c. *The symbol* s *is said to* occur at i *in the formula* φ, *if* s *is the* i−th *term of the sequence* φ.
d. *The variable* v *is said to be* bound at k *in* φ *when both of the following two conditions are satisfied:*
   1. v *occurs at* k *in* φ
   2. *for some* $i < k < j$, *the* (i, j)−subsequence *of* φ *is a formula which takes one of the two forms:* $[\forall v \psi]$ *or* $[\exists v \psi]$.
e. *A variable* v *that occurs at* k *in* φ, *but is* not *bound there is said to be* free *at* k.
f. *We say that* v *occurs* free *in* φ *if for some* k, *v occurs free at* k; *and* v *occurs* bound *in* φ *if for all* k, *v occurs bound at* k.

g. *A formula φ is said to be an* assertion *if no variable occurs free in φ. For two assertions ψ and φ, we write $[\psi \to \varphi]$ as an abreviation for $[[\neg\psi] \vee \varphi]$. We read $[\psi \to \varphi]$ as " if ψ, then φ " or " ψ implies φ "; similarly we write $[\psi \leftrightarrow \varphi]$, for $[[\psi \to \varphi] \wedge [\varphi \to \psi]]$, and we read this " ψ if and only if φ ".*

Examples [from [Malitz, 1979, p. 115]] Consider the formula

$$\varphi = [\exists v_3 [[\forall v_2 [\neg[v_4 \approx v_2]]] \vee [v_3 \approx v_2]]]$$

Then

1. $[\forall v_2 [\neg[v_4 \approx v_2]]]$ is the $(6, 15)$−subsequence of φ, but is *not* a subformula of φ.
2. $[\neg[v_4 \approx v_2]]$ is the $(8, 15)$−subsequence of φ, and it *is* a subformula of φ.
3. v_2 is *bound* at 7 , and at 13; but it is *free* at 21.
4. v_3 is *bound* in both of its occurrences.
5. v_4 is *free* at its single occurrence.
6. φ is *not* an assertion.
7. $[\forall v_2 [\forall v_4 \varphi]]$ *is an assertion.*

At this point, we have completed what we need to say on the "rules of formation" of the language. Yet, we still need to say a few words on the "rules of inference" within the framework of syntax alone.

Definition A.1.7. *Let L be a first-order language in the sense discussed so far.*

a. *An* axiom *is an assertion scheme of L.*
b. *A* rule of inference *$(\Sigma; \sigma)$ is a pair where $\Sigma = \{\sigma_1, ..., \sigma_k\}$ is a $k-$tuple of mutually distinct $L-$assertions, called the* premises *of the rule, and where σ is an $L-$assertion, called the* conclusion *of the rule.*
c. *An* axiom system *$S = (A, B)$ is a pair where A is a set of axioms, and B is a set of rules of inference.*
d. *A* proof *in an axiom system $S = (A, B)$ is a finite sequence $\{\varrho_1, ..., \varrho_n\}$ such that, for each assertion ϱ_i, exactly one of the following holds:*
   1. $\varrho_i \in A$
   2. *there is an inference rule $(\sigma_1, ..., \sigma_k; \varrho_i) \in B$ such that for all $l \leq k$, $\sigma_l \in \{\varrho_j \mid j < i\}$.*
e. *A* theorem *in an axiom system $S = (A, B)$ is the last line of a proof in S. The notation $S \vdash \varrho$ (or $\vdash_S \varrho$) says the assertion ϱ is a theorem in S.*

Remarks A.1.4:

a. It is sometimes useful to split the set A of axioms as the union $A_L \cup A_\Gamma$ of two disjoint sets, where the elements of A_L are called the logical axioms. The definition of the latter however tends to be cumbersome; nevertheless this can be done quite fluently (see below) in the special case

where one restricts one's attention to (first-order) *sentential languages*,
i.e. (first-order) languages involving only assertions (also called then sentences), quantifers, and connectives. One then emphasizes this distinction
by speaking of A_Γ−proof, A_Γ−theorem and by writing $A_\Gamma \vdash \varrho$ in (e)
above.

b. More generally, in devising an axiom system, the rules of inference one
includes in B depend on the choice one has made for the axioms in A,
and on which logical connectives one decides to consider as basic.

Definition A.1.8. *Within a first-order sequential language P, an assertion
scheme is said to be a* logical axiom *whenever it has one of the following three
forms:*

1. $\varphi \to [\psi \to \varphi]$
2. $[\varphi \to [\psi \to \chi]] \to [[\varphi \to \psi] \to [\varphi \to \chi]]$
3. $[[\neg\varphi] \to [\neg\psi]] \to [\psi \to \phi]$

where φ, ψ and χ are assertions.

Scholium A.1.2. *Within a first-order sequential language P, let Σ be a set
of assertions, and φ be an assertion. Then $\Sigma \vdash \varphi$ if and only if there exists
a finite sequence $\{\varphi_1, \ldots, \varphi_n\}$ of assertions φ_i such that $\varphi_n = \varphi$ and, for
each $i \leq n$ one of the following holds:*

1. $\varphi_i \in A_L$
2. $\varphi_i \in A_\Gamma$
3. $\varphi_k \approx [\varphi_j \to \varphi_i]$ *for $j, k < i$.*

Definition A.1.9. *In the above Scholium, the sequence $\{\varphi_1, \ldots, \varphi_n\}$ is
called a* formal proof *of φ from A_Γ.*

Remark A.1.5: From part (3) above it follows that, for any two assertions
φ and ψ : $\{\psi, [\psi \to \varphi]\} \vdash \varphi$. This is commonly known as *Modus Ponens*,
also called the *rule of detachment*. The *discharge of double negation* – namely
$[\neg[\neg\varphi]] \vdash \varphi$ – also follows (see e.g. [Monk, 1976, p. 117]) from the logical
axioms listed in Def. A.1.8.

We now return to the general case of a first-order language.

Definition A.1.10. *We say that a set Σ of assertions is* syntactically inconsistent *if $\Sigma \vdash \sigma$ for all assertions σ; otherwise we say that that Σ is*
consistent. *Finally, we say that Σ is* maximally consistent *if : Σ is syntactically consistent and $\Gamma \supseteq \Sigma$ with Γ syntactically consistent if and only if
$\Gamma = \Sigma$.*

Scholium A.1.3.

a. If Σ is syntactically consistent, and $\Gamma = \{\sigma : \Sigma \vdash \sigma\}$, then Γ is syntactically consistent.

b. *If Σ is maximally consistent, and $\Sigma \vdash \sigma$, then $\sigma \in \Sigma$.*
c. *Σ is inconsistent if and only if $\Sigma \vdash \sigma$ and $\Sigma \vdash \neg\sigma$ for some assertion σ.*

Remark A.1.6: The completeness theorem (to be stated in the semantic part of this summary) provides the meaning for $\vdash$ and the semantics for consistency.

Scholium A.1.4 (Deduction Theorem). *If $\Sigma \cup \{\psi\} \vdash \varphi$, then $\Sigma \vdash [\psi \to \varphi]$.*

Scholium A.1.5 (Lindenbaum Theorem). *For every syntactically consistent set Σ of assertions, there exists a maximally consistent Γ with $\Gamma \supseteq \Sigma$.*

Scholium A.1.6.

a. *If Σ is maximally consistent, and σ is any assertion, then either $\sigma \in \Sigma$ or $[\neg\sigma] \in \Sigma$.*
b. *If Σ is maximally consistent, and ψ, φ are assertions, then $\psi \wedge \varphi \in \Sigma$ if and only if $\psi \in \Sigma$ and $\varphi \in \Sigma$.*

A.2 Semantics

In connection with a study of models in physics it is unfortunate that, in the vernacular of physicists, the word "semantics" is used often with an intended derogatory connotation, suggesting that some kind of specious argument is invoked as an afterthought to save an earlier argument that had somehow misfired.

Yet, the etymological root of the term "semantics' indicates that a *sign* is proposed (or at least is to be proposed) in order to give out the *significance*, or *meaning*, of what has been said. It is surely the latter understanding of the term that is the key to the mathematical notion of semantics: the *interpretation* of the language.

Accordingly, the universe of the interpretation has to be described; therefore the first notions which we are to consider must specify what we call *structures*. The prototype, for the structures to be encompassed by the definition below, is an ordered pair (A, e) where $A \neq \emptyset$ is a set – e.g. Z^+ – and e is a binary relation on A – e.g. $+$.

Definition A.2.1. *Let s be a type in a given language L. A structure of type s is a function $\mathcal{A}$, the domain of which is $s \cup \{\emptyset\}$, and the range of which satisfies the following four conditions:*

1. *$\mathcal{A}(\emptyset)$ is a non-empty set*
2. *for each $c_\alpha \in s$: $\mathcal{A}(c_\alpha) \in \mathcal{A}(\emptyset)$*
3. *for each $R_{n,\alpha} \in s$: $\mathcal{A}(R_{n,\alpha})$ is a n-relation on $\mathcal{A}(\emptyset)$*
4. *for each $f_{n,\alpha} \in s$: $\mathcal{A}(f_{n,\alpha})$ is a n-function on $\mathcal{A}(\emptyset)$.*

The type of a structure $\mathcal{A}$ is written $s\mathcal{A}$. The set $\mathcal{A}(\emptyset)$, also denoted $|\mathcal{A}|$, is called the universe of the structure $\mathcal{A}$. We also write $c_\alpha^{\mathcal{A}}$ for $\mathcal{A}(c_\alpha)$, $R_{n,\alpha}^{\mathcal{A}}$ for $\mathcal{A}(R_{n,\alpha})$, $f_{n,\alpha}^{\mathcal{A}}$ for $\mathcal{A}(f_{n,\alpha})$. If S is a symbol, $S^{\mathcal{A}}$ is called the denotation of S in $\mathcal{A}$.

Remark A.2.1: A group is a structure $\mathcal{G} = (G, \star, Inv, c_0)$, with $|\mathcal{G}| = G$ and $\tau\mathcal{G} = (f_{2,0}, f_{1,0}, c_0)$, where $f_{2,0}^{\mathcal{A}} = \star$ is the binary group operation, $f_{1,0}^{\mathcal{A}} = Inv$ is the unary inverse function, and $c_0^{\mathcal{A}} = c$ is the identity element. From this, we see what the definition of a substructure should be: it should be to a structure what a subgroup is to a group.

Definition A.2.2. *A structure $\mathcal{A}$ is said to be a* substructure *of a structure $\mathcal{B}$, a situation denoted $\mathcal{A} \subseteq \mathcal{B}$, whenever the following five conditions are satisfied:*

1. *$s\mathcal{A} = s\mathcal{B}$ [i.e. the types of the structures $\mathcal{A}$ and $\mathcal{B}$ are the same; cf. Def. A.2.1]*

2. *$|\mathcal{A}| \subseteq |\mathcal{B}|$*

3. *$c_\alpha^{\mathcal{A}} = c_\alpha^{\mathcal{B}}$ for all $c_\alpha \in s\mathcal{A}$*

4. *$f_{n,\alpha}^{\mathcal{A}}(a_1, ..., a_n) = f_{n,\alpha}^{\mathcal{B}}(a_1, ..., a_n)$ for every $f_{n,\alpha} \in s\mathcal{A}$ and all $a_1, ..., a_n \in |\mathcal{A}|$*

5. *$R_{n,\alpha}^{\mathcal{A}}(a_1, ..., a_n)$ if and only if $R_{n,\alpha}^{\mathcal{B}}(a_1, ..., a_n)$ for every $R_{n,\alpha} \in s\mathcal{A}$ and all $a_1, ..., a_n \in |\mathcal{A}|$.*

   b. *When $\mathcal{A}$ is a substructure of $\mathcal{B}$, we say equivalently that $\mathcal{B}$ is an* extension *of $\mathcal{A}$.*

   c. *When $\mathcal{B}$ has a substructure $\mathcal{A}$ the universe of which is a set X, we say that $\mathcal{A}$ is the* restriction *of $\mathcal{B}$ to X*

   d. *The reduct $\mathcal{A} = \mathcal{B}\lceil s_o$ of a structure $\mathcal{B}$ with respect to a type $s_o \subseteq s\mathcal{B}$ is a structure $\mathcal{A}$ such that the following three conditions are satisfied:*
      1. *$s\mathcal{A} = s_o$ i.e. the type of $\mathcal{A}$ is s_o;*
      2. *$|\mathcal{A}| = |\mathcal{B}|$ i.e. $\mathcal{A}$ and $\mathcal{B}$ have the same universe;*
      3. *$S^{\mathcal{A}} = S^{\mathcal{B}}$ i.e. all symbols have the same denotation in $\mathcal{A}$ and in $\mathcal{B}$.*

   e. *A structure $\mathcal{B}$ is said to be an* expansion *of a structure $\mathcal{A}$ if there exists a type s_o such that $\mathcal{A} = \mathcal{B}\lceil s_o$.*

In connection with the last definitions relative to substructures, let us recall that a map $\mu : x \in S_1 \mapsto \mu(x) \in S_2$ between two sets S_1 and S_2 is said to be: an *injection* whenever $\mu(x) = \mu(y)$ implies $x = y$; a *surjection* whenever for every $z \in S_2$ there exists a $x \in S_1$ such that $\mu(x) = z$; a *bijection* whenever it is both an injection and a surjection. In a widely used, yet somewhat cumbersome vocabulary, one refers to an injection as a "$1 - 1$ map" or as a "$1 - 1$ map into"; a surjection as a "map onto"; and a bijection as a "$1 - 1$ map onto".

Definition A.2.3.

a. Let $\mathcal{A}$ and $\mathcal{B}$ be two structures such that $s\mathcal{A} \subseteq s\mathcal{B}$. A map g from $|\mathcal{A}| \to |\mathcal{B}|$ is said to be an embedding whenever the following four conditions are satisfied:

    1. g is an injection

    2. $g(c_\alpha^{\mathcal{A}}) = c_\alpha^{\mathcal{B}}$ for all $c_\alpha \in s\mathcal{A}$

    3. $g(f_{n,\alpha}^{\mathcal{A}}(a_1, ..., a_n)) = f_{n,\alpha}^{\mathcal{B}}(a_1, ..., a_n)$ for all $f_{n,\alpha} \in s\mathcal{A}$ and all $a_1, ..., a_n \in |\mathcal{A}|$

    4. $R_{n,\alpha}^{\mathcal{A}}(a_1, ..., a_n)$ if and only if $R_{n,\alpha}^{\mathcal{B}}(a_1, ..., a_n)$ for all $R_{n,\alpha} \in s\mathcal{A}$ and all $a_1, ..., a_n \in |\mathcal{A}|$.

b. When $s\mathcal{A} = s\mathcal{B}$ and moroever g is bijective, g is said to be an isomorphism, and the two structures $\mathcal{A}$ and $\mathcal{B}$ are said to be isomorphic; we then write $\mathcal{A} \simeq \mathcal{B}$.

Definition A.2.4.

a. An assignment z to $\mathcal{A}$ is a function $z : v \in \mathcal{V} \to z(v) \in |\mathcal{A}|$, where $\mathcal{V}$ denotes the set of variables in L.

b. Let z be an assignment to $\mathcal{A}$, u be a variable, and $a \in |\mathcal{A}|$; the assignment $z_{u,a}$ is defined by

$$z_{u,a}(v) = \begin{cases} a & \text{if } v = u \\ z(v) & \text{otherwise.} \end{cases}$$

c. Let t be a term of type $st \subseteq s\mathcal{A}$ and z be an assignment to $\mathcal{A}$. By induction on the length of t, the assignment z is extended to define $t^{\mathcal{A}} < z >$ by the obvious construction:

$$v_n^{\mathcal{A}} < z > = z(v_n)$$
$$c_\alpha^{\mathcal{A}} < z > = z(c_\alpha)$$
$$(f_{n,\alpha}(t_1, ..., t_n))^{\mathcal{A}} < z > = f_{n,\alpha}^{\mathcal{A}}(t_1^{\mathcal{A}} < z >, ..., t_n^{\mathcal{A}} < z >).$$

Remark A.2.2: The notation $z \begin{pmatrix} u \\ a \end{pmatrix}$ is also used for $z_{u,a}$.

Definition A.2.5. Let φ be a formula and z be an assignment to a structure $\mathcal{A}$. We say that z satisfies φ in $\mathcal{A}$, and we write $\mathcal{A} \models \varphi < z >$, if either:

(1) $\varphi = [t_1 \approx t_2]$ and $t_1^{\mathcal{A}} < z > \approx t_2^{\mathcal{A}} < z >$

(1') $\varphi = [R_{n,\alpha} t_1, ..., t_n]$ and $R_{n,\alpha}^{\mathcal{A}} t_1 < z >, ..., t_n < z >$

(2) $\varphi = [\neg \psi]$ and it is not the case that $\mathcal{A} \models \psi < z >$

(3) $\varphi = [\psi_1 \wedge \psi_2]$ and both $\mathcal{A} \models \psi_1 < z >$ and $\mathcal{A} \models \psi_2 < z >$

(3') $\varphi = [\psi_1 \vee \psi_2]$ and either $\mathcal{A} \models \psi_1 < z >$ or $\mathcal{A} \models \psi_2 < z >$

(4) $\varphi = [\exists v_n \psi]$ and for some $a \in |\mathcal{A}|$, $\mathcal{A} \models \psi < z >_{v_n, a}$

(4') $\varphi = [\forall v_n \psi]$ and for all $a \in |\mathcal{A}|$, $\mathcal{A} \models \psi < z >_{v_n, a}$.

Scholium A.2.1.

a. $\mathcal{A} \models \forall v_n \varphi < z >$ if and only if $\mathcal{A} \models \neg[\exists v_n \neg \varphi] < z >$

b. $\mathcal{A} \models \exists v_n \varphi < z >$ if and only if $\mathcal{A} \models \neg[\forall v_n \neg \varphi] < z >$

c. $\mathcal{A} \models [\varphi \wedge \psi] < z >$ if and only if $\mathcal{A} \models \neg[\neg\varphi \vee \neg\psi] < z >$
d. $\mathcal{A} \models [\varphi \vee \psi] < z >$ if and only if $\mathcal{A} \models \neg[\neg\varphi \wedge \neg\psi] < z >$
e. *If φ is a formula, and z, z' are two assignments to $\mathcal{A}$ such that $z(v) = z'(v)$*
    for all variables v occurring freely in φ, then $\mathcal{A} \models \varphi < z >$ if and only if
    $\mathcal{A} \models \varphi < z' > .$
f. *If σ is an assertion, then either of the following two situations occurs:*
    1. $\mathcal{A} \models \sigma < z >$ *for all assignments z to $\mathcal{A}$*
    2. *no assignment z satisfies σ in $\mathcal{A}$.*

Remark A.2.3: The result (f) above is a trivial consequence of (e); and (e)
justifies the following definitions of "truth", "validity", "theory" and "model"
given below.

Definition A.2.6.

a. *We write $\mathcal{A} \models \sigma$, and we say that an assertion σ is* true *in a structure $\mathcal{A}$*
    (or that $\mathcal{A}$ satisfies *σ), if there is an assignment z in $\mathcal{A}$ that satisfies σ.*
b. *We write $\models \sigma$, and we say that σ is* valid *if for all $\mathcal{A}$ of type $s\mathcal{A} \supseteq s\sigma$:*
    $\mathcal{A} \models \sigma .$

Definition A.2.7.

a. *The* theory *of a structure $\mathcal{A}$ is the set* $\mathrm{Th}\,\mathcal{A} = \{ \sigma \mid \mathcal{A} \models \sigma \}.$
b. *Two structures $\mathcal{A}$ and $\mathcal{B}$ are said to be* elementary equivalent *if $\mathrm{Th}\mathcal{A} =$*
    $\mathrm{Th}\mathcal{B}$, and we write $\mathcal{A} \equiv \mathcal{B}$.
c. *Given a set of assertions Σ, the set* $\mathrm{Mod}\,\Sigma = \{\mathcal{A} \mid \forall \sigma \in \Sigma : \mathcal{A} \models \sigma \}$
    is called the class of models *of Σ. In particular, $\mathcal{A} \models \sigma$ says that $\mathcal{A}$ is*
    a model *of σ.*

Definition A.2.8. *Let Σ be a set of assertions and σ another assertion (in*
the same language L and of the same type s). We say that:

a. *Σ is* satisfiable, *or semantically consistent, if Σ has a model;*
b. *σ is semantically consistent* with *Σ, if $\Sigma \cup \sigma$ is satisfiable;*
c. *σ is independent of Σ, if both σ and $\neg\sigma$ are semantically consistent with*
    Σ;
d. *Σ entails σ, and we write $\Sigma \models \sigma$, if every model of Σ is also a model of*
    σ, i.e. $\mathrm{Mod}\,\Sigma = \mathrm{Mod}\,(\Sigma \cup \{\sigma\}).$

Remarks A.2.4:

a. As an immediate consequence of (c) above, we see that in order to prove
    that σ is independent of Σ, it is sufficient to display two models of Σ,
    one in which σ is true, and one in which σ is false (i.e. $\neg\sigma$ is true).
b. In connection with (d) above, note that Σ entails σ, if and only if $\neg\sigma$ is
    not semantically consistent with Σ.
c. The Reader will have noticed that we now have two notions of consistency:
    a syntactic one and a semantic one; nevertheless, one can show that these
    are equivalent, in the sense of the completeness theorem below.

Scholium A.2.2 (Completeness Theorem). *A set Σ of assertions is syntactically consistent if and only if Σ is satisfiable (i.e. if and only if it is semantically consistent).*

Remarks A.2.5:

a. In particular, we have:
   1. $\vdash \sigma$ if and only if $\models \sigma$
   2. $\Sigma \vdash \sigma$ if and only if $\Sigma \models \sigma$.
b. One can paraphrase (1) as stating that an assertion is a tautology (syntax) if and only if it is valid (semantic). This particular statement is sometimes referred to as "the completeness theorem", while the theorem quoted as such is referred to as the "extended completeness theorem". Some authors however reserve the name "completeness theorem" for the part of (2) above that says only $\{\Sigma \vdash \sigma \text{ if } \Sigma \models \sigma\}$; while referring to the statement $\{\Sigma \models \sigma$ if $\Sigma \vdash \sigma\}$ as the "soundness theorem".

The main stepping stones towards the compactness theorem (Scholium A.2.6) are briefly indicated below.

Definition A.2.9.

a. *We say that a set of assertions Σ is* finitely satisfiable *whenever every finite subset of Σ has a model.*
b. *A constant symbol c is said to be a* witness *for $\exists v\varphi$ in Σ if $\varphi_{v,c} \in \Sigma$.*
c. *We say that a set of assertions Σ is* complete *if for all assertions σ of type $\tau\Sigma$ either $\sigma \in \Sigma$ or $\neg\sigma \in \Sigma$.*

Remark A.2.6: A slightly more general condition for the completeness of Σ is that either $\Sigma \models \sigma$ or $\Sigma \models \neg\sigma$ for all assertions σ of type $\tau\Sigma$.

Scholium A.2.3. *If a set of assertions Σ is finitely satisfiable, then there exists a set of assertions Γ such that the following four conditions are satisfied simultaneously:*

1. $\Gamma \supseteq \Sigma$
2. $\tau\Gamma = \tau\Sigma$
3. Γ *is finitely satisfiable.*
4. Γ *is complete.*

Scholium A.2.4. *If a set Σ of assertions is finitely satisfiable, then there exists a set of assertions Ω such that the following three conditions are satisfied simultaneously:*

1. $\Omega \supseteq \Sigma$
2. *If $\exists v\varphi \in \Omega$ then $\exists v\varphi$ has a witness in Ω.*
3. Ω *is finitely satisfiable.*

Scholium A.2.5. *If a set Σ of assertions is finitely satisfiable and is complete, and if every assertion $\exists v\varphi \in \Sigma$ has a witness in Σ, then Σ has a model.*

Scholium A.2.6 (Compactness Theorem). *Every finitely satisfiable set of assertions Σ has a model.*

The last few results above give a flavor of the subject. As for the time-frame, the proof of the compactness theorem was given in the paper [Gödel, 1930] that stemmed from his dissertation [Gödel, 1929], and he moreover showed there that a proof of the completeness theorem can be derived from this result. A broad and precisely documented account of Gödel's life and work is offered in [Dawson, 1997]. For a very rapid survey of mathematical logic that traces also its main historical landmarks (with specific references) see e.g. *Article* 411 in [Itô, 1987]. More recent and detailed accounts, specifically geared to the subject of Model Theory in mathematical logic, are presented in [Chang and Keisler, 1990, Hodges, 1993] or already [Robinson, 1965, Bell and Slomson, 1971].

For some of the elementary applications to arithmetic, involving the compactness theorem in conjunction with the Loewenheim-Skolem theorems – on which it is sufficient to know here that they compare elementary substructures – in the discussion of non-standard models, see also [Malitz, 1979]. Non-standard analysis was invented in the early 1960s to justify the profusion of infinitesimals needed in a Leibnitz-type calculus; soon its inventor made it to encompass several other mathematical fields; see [Robinson, 1966]. In connection with the compactness theorem, a typical result – see [Nelson, 1976] for a precise formulation in the frame of set theory – is that a standard proof is possible for every standard assertion, the non-standard extension of which has been proven in the non-standard context.

Before we close this Appendix, we still want to make two remarks on the presentation level adopted here. While it was devised to be sufficently general for our purposes in the main body of this text, it is true that we could have taken either a lower road or a higher one.

The first remark concerns a *lower road,* which would have made some passages perhaps a little more intuitive. In particular, the compactness theorem can often be used to reduce a problem on infinite sets to its finitary analog – in spite of the unintuitive tricks for which infinity is notorious (see e.g. [Vilenkin, 1995]). An example where the compactness theorem can be so used is the following matching problem, which we present following Bell and Slomson [1971] (although we must apologize for the somewhat biaised setting by which the problem is known.)

Problem: Let B be a set of boys, each of whom has at most a finite number of girlfriends. If for each integer k any k of the boys have between them at least k girlfriends, is it possible for each boy to marry one of his girlfriends without any of them committing bigamy?

It is easy to see, by elementary means, that the problem admits a positive answer *if the set B is finite, i.e. if* $Card(B) = m < \infty$; indeed the problem is trivial if $m = 1$ and one can then obtain the result, for any finite m, by induction on m. We leave to the Reader either to check that this can be done, or to accept it: we are assuming it here, as our interest is to show how the compactness theorem ensures a positive answer in case B is infinite.

Although Bell and Slomson [1971] give a proof of the compactness theorem in the "full-strength" version in which it is stated above, they also present an elementary proof, adapted to the version of this theorem that obtains in the less sophisticated (zeroth-order) setting of propositional calculus (or SC) ... which gives us an opportunity to indicate briefly what the latter is about. From the beginning, i.e. already from our Def. A.1.1 onwards, SC restricts its attention only to "propositional variables", "connectives" and the punctuation given by "parentheses". In particular, SC does not take into account "quantifiers", thus limiting considerably the scope of logical situations SC can entertain. Yet SC is sufficient for the matching problem at hand. The definition of "formulas" from our Def. A.1.5 is modified, according to the scope of SC, requiring only the weakening of condition (1) to the requirement that a single propositional variable ν be already a formula; condition (2) and (3) being unchanged, and condition (4) becoming without object. Having done this modification, SC can collapse our Defs. A.2.4–2.8 as follows. A "realization" of SC is defined as a map

$$f : \nu \in \mathcal{V} \to \{0, 1\} \tag{A.2.1}$$

or equivalently as its natural extension to the set of all formulas of SC, subject to the two conditions

$$f(\varphi \wedge \psi) = f(\varphi) \wedge f(\psi) \quad \text{and} \quad f(\neg \varphi) = f(\varphi)^* \tag{A.2.2}$$

where $\{0, 1\}$ is equipped with the ordering $0 < 1$ and the complementation $0^* = 1$, $1^* = 0$. A set of formula Σ can then be said to be *satisfiable* if there exists a realization f such that

$$f(\varphi) = 1 \quad \text{for all} \quad \varphi \in \Sigma \quad ; \tag{A.2.3}$$

and Σ is said to be *finitely satisfiable* whenever every finite $\Sigma_o \subset \Sigma$ is satisfiable.

With these preliminaries cleared, the statement of the compactness theorem restricts straightforwardly to the propositional calculus (SC) context, namely: *to verify that Σ is satisfiable, it is sufficient to verify that it is finitely satisfiable.*

We are now ready for the extension of the solution of the above problem in the infinite case.

Upon writing $B = \{b_i \mid i \in I\}$ for the set of all the boys, $G = \{g_j \mid j \in J\}$ for the set of all the girlfriends of the boys in B; and $F(b)$ for the (finite!) set of all girlfriends of the boy b, let indeed L be the propositional language the propositional variables of which are the elements of the set $\{\nu_{i,j} \mid (i, j) \in I \times J\}$ (to be interpreted as "the jth girl is a girlfriend of the ith boy") and Σ be the set of all formulas consisting of:

1. $\forall\, i \in I$, the formula: $\bigvee_{g_j \in F(b_i)} \nu_{ij}$ i.e. $\nu_{io} \vee \cdots \vee \nu_{ij_k}$;
2. $\forall\, (i,j,j') \in I \times J \times J$, with $j \neq j'$, the formula: $\neg(\nu_{ij} \wedge \nu_{ij'})$;
3. $\forall\, (i,i',j) \in I \times J \times J$, with $i \neq i'$, the formula: $\neg(\nu_{ij} \wedge \nu_{i'j})$.

Let now Σ_o be any *finite* subset of Σ, and B_o be the set of all boys b_i such that there exists $j \in J$ for which ν_{ij} occurs in some formula belonging to Σ_o. Since Σ_o is finite, so is B_o; we are therefore in a situation where the problem admits a positive answer. Hence every boy in $b_i \in B_o$ can marry one of his girlfriends, say g_{j_i}, without bigamy being committed. It is then straightforward to verify that

$$f_o(\nu_{ij}) = \begin{cases} 1 \text{ if } \quad j = j_i \\ 0 \text{ otherwise} \end{cases} \tag{A.2.4}$$

defines a realization for which Σ_o is satisfied. *We have now arrived at the crucial point.* Since the above result holds (still separately!) for *every* finite $\Sigma_o \subset \Sigma$, the compactness theorem ensures the existence of a realization f for which Σ is satisfied, allowing each boy $b_i \in B$ to marry, without committing bigamy, the one of his girlfriends, say g_j selected by the condition $f(\nu_{ij}) = 1$.

q.e.d.

Here must we end our short excursion along propositional calculus, the lower road to first-order logic which was the main territory surveyed in this Appendix.

The second of the two remarks announced above is to note that, although first-order logic is the most completely worked out theory in mathematical logics, its domain of applications is limited by its inability to accomodate properly assertions that quantify over properties, such as "for all the colors one can choose from ..."; for such purposes, it becomes necessary to travel on a *higher road:* one must construct a *second-order* logic in which variables for properties are introduced, and quantifiers binding such variables are defined. For modal discourse, with assertions of possibilities and necessities, a *modal logic* has to be developed where operators, such as $\Diamond$ [it is possible that ...] and $\Box$ [it is necessary that ...] have to be defined within the logical system. For these and other adventures in logic, see e.g. for second-order logic: [Ebbinghaus, Flum, and Thomas, 1984]; for modal logic: [Gamut, 1991]; and beyond those: [Gabbay and Guenthner, 1983–1989].

B. Appendix: The Calculus of Differentials

This appendix is divided in three parts. In Sect. B.1, we recall some mathematical definitions and background, and we state the following results: the inverse function theorem (Thm. B.1.1), the Green theorem (Thm. B.1.2) and a corollary of the latter (Thm. B.1.3). These are essential to our presentation of Classical Thermodynamics (see Sects. 2.4). In Sect. B.2 of the present Appendix, we state the extensions of Green's theorem known as the Stokes theorem (Thm. B.2.1) and the Gauss theorem (Thm. B.2.2); we show how these theorems allow an integral presentation of the Maxwell differential equations of classical electromagnetism; the equivalence of these two formulations of the theory thus stands on the same logical footing as does the equivalence between the integral and differential formulations of classical thermodynamics. In Sect. B.3 the general formalism of higher order differentials is presented to emphasize that Stokes' and Gauss' theorems are actually special realizations of a more general result (Thm. B.3.1); to illustrate further the general formalism, we indicate how it leads naturally to a synthetic formulation of Maxwell equations as two relations on $2-$forms in R^4 .

B.1 Green's Theorem

A real-valued function f , defined on an open region $D \subseteq \mathsf{R}^2$, is said to be *differentiable at a point* $(x, y) \in D$ whenever it admits a linear approximation around (x, y) . The geometry behind this definition is sketched in Fig. B.1. Analytically, this means that there exists a linear function

$$\mathrm{d}f_{(x,y)} : (h, k) \in \mathsf{R}^2 \mapsto \mathrm{d}f_{(x,y)}(h, k) \in \mathsf{R} \qquad (\text{B.1.1})$$

such that

$$f(x + h, y + k) - f(x, y) - \mathrm{d}f_{(x,y)}(h, k) = \varepsilon_{(x,y)}(h, k)\sqrt{h^2 + k^2} \qquad (\text{B.1.2})$$

for some function $\varepsilon_{(x,y)}$ satisfying

$$\lim_{(h,k)\to 0} \varepsilon_{(x,y)}(h, k) = 0 . \qquad (\text{B.1.3})$$

$\mathrm{d}f_{(x,y)}$ is said to be the *differential of* f *at* (x, y) . When such a linear

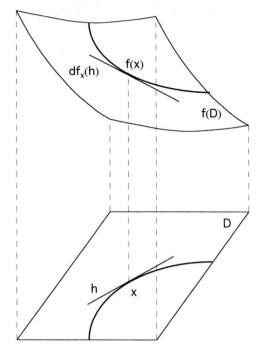

Fig. B.1. Geometrical representation of $\mathrm{d}f$

approximation exists for all $(x, y) \in D$, we say that $\mathrm{d}f$ is the *differential* of f *on* D. In particular, the projection functions

$$\pi^x : (x, y) \in \mathbb{R}^2 \mapsto x \in \mathbb{R} \quad and \quad \pi^y : (x, y) \in \mathbb{R}^2 \mapsto y \in \mathbb{R} \tag{B.1.4}$$

are differentiable on $\mathbb{R}^2$ and we writes simply $\mathrm{d}x$ for $\mathrm{d}\pi^x$ and $\mathrm{d}y$ for $\mathrm{d}\pi^y$. With this notation, we have, for every differentiable f on D, that the following equality holds

$$\mathrm{d}f = (\partial_x f)\, \mathrm{d}x + (\partial_y f)\mathrm{d}y \tag{B.1.5}$$

where $\partial_x f$ and $\partial_y f$ are the *partial derivatives* of f :

$$(\partial_x f)(x, y) = \lim_{h \to 0} \frac{1}{h}[f(x + h, y) - f(x, y)] \tag{B.1.6}$$

and

$$(\partial_y f)(x, y) = \lim_{k \to 0} \frac{1}{h}[f(x, y + k) - f(x, y)] \quad . \tag{B.1.7}$$

Conversely, if f is such that these limits exist, *and* are continuous for all (x, y) in an open region D, then f is differentiable on D; we then say that f is *continuously differentiable on* D.

Finally, we mention that if f, $\partial_x f$, $\partial_y f$, and $\partial_{xy} f$ exist and are continuous in an open region D, then $\partial_{yx} f$ exists also, and $\partial_{yx} f = \partial_{xy} f$ on D.

Let us at this point makes an excursion from real-valued to vector-valued functions, i.e. functions defined on a domain $D \subseteq \mathsf{R}^n$ and taking their values in R^m : $f : (x_1, ..., x_n) \in D \mapsto (f_1(x_1, ..., x_n), ..., f_m(x_1, ..., x_n)) \in \mathsf{R}^m$

[For most of the classical thermodynamics discussed in Chap. 2, we will use either the scalar case $\{n = 2, m = 1\}$, already discussed; or its extension to the case $\{n = 2, m = 2\}$, for which we need the material covered in the present excursion.]

The differential $df_{(x_1,...,x_n)}$ is defined again as the linear approximation of the function f in the neighborhood of the point $(x_1, ..., x_n)$. In terms of partial derivatives, $df_{(x_1,...,x_n)}$ is thus a $n \times m$ matrix:

$$df_{(x_1,...,x_n)}(h_1, ..., h_n) =$$

$$\begin{pmatrix} (\partial_{x_1} f_1)(x_1, ..., x_n) & ... & (\partial_{x_n} f_1)(x_1, ..., x_n) \\ \vdots & \vdots & \vdots \\ (\partial_{x_1} f_m)(x_1, ..., x_n) & ... & (\partial_{x_n} f_m)(x_1, ..., x_n) \end{pmatrix} \begin{pmatrix} h_1 \\ \vdots \\ h_n \end{pmatrix} \tag{B.1.8}$$

or more concisely

$$df = \begin{pmatrix} \partial_{x_1} f_1 & ... & \partial_{x_n} f_1 \\ \vdots & \vdots & \vdots \\ \partial_{x_1} f_m & ... & \partial_{x_n} f_m \end{pmatrix} = (\partial_{x_i} f_j) \quad . \tag{B.1.9}$$

In case $n = m$, the Jacobian of f is defined as the determinant:

$$Jf = \begin{vmatrix} \partial_{x_1} f_1 & ... & \partial_{x_n} f_1 \\ \vdots & \vdots & \vdots \\ \partial_{x_1} f_n & ... & \partial_{x_n} f_n \end{vmatrix} = |\partial_{x_i} f_j| \quad . \tag{B.1.10}$$

Theorem B.1.1. *(Inverse function theorem) Let $f : D \to \mathsf{R}^n$, defined on an open subset $D \subseteq \mathsf{R}^n$, be continuously differentiable. Then, for every $(x_1, ..., x_n) \in D$ for which $(Jf)(x_1, ..., x_n) \neq 0$, there exists an open D_o with $(x_1, ..., x_n) \in D_o \subseteq D$ such that:*

(1) $f : x = (x_1, ..., x_n) \in D_o \mapsto f(x) = (f_1(x_1, ..., x_n), ..., f_n(x_1, ..., x_n)) \in \mathsf{R}^n$ is bijective onto $f(D_o)$, the image of D_o through f;

(2) $f(D_o)$ is open;

(3) there exists a continuously differentiable function $\varphi : f(D_o) \to D_o$ such that
   $(\varphi \circ f)(x) = x \quad \forall \quad x \in D_o$ and $(f \circ \varphi)(u) = u \quad \forall \quad u \in f(D_o)$;

(4) $d\varphi_{f(x)} = [df_x]^{-1} \quad \forall \quad x \in D_o$.

The interpretation of this theorem follows directly from the definition of the differential as a linear approximation. The condition on the Jacobian is that the determinant of the matrix $\mathrm{d}f_x$ [approximating f around x] does not vanish, i.e. that this matrix admits an inverse. Part (4) of the theorem thus asserts in particular that this fact remains true in a neighborhood D_o of x. The other three parts of the theorem then assert that the existence of the inverse of the linear approximation implies the local existence of an inverse for the function f itself, and that the inverse function φ inherits the smoothness of f; in particular its differential is given by (4). For a proof of this theorem, see any of the classical texts in differential calculus, e.g. [Flemming, 1977, Fulks, 1978].

One practical interest of this theorem is that it allows to replace the original variables $(x_1, ..., x_n)$ coordinating D_o by the new variables $(u_1, ..., u_n)$ coordinating $f(D_o)$ obtained by applying the (not necessarily linear) transformation f.

For instance, we constructed for the ideal gas the functions *energy* (2.4.18) and *entropy* (2.4.19) of the original variables *volume* V and *temperature* T. Upon defining

$$f(V, T) = (U(T), S(V, T)),\qquad\qquad (B.1.11)$$

we obtain from (2.4.15–17):

$$\mathrm{d}f_{(V,T)} = \begin{pmatrix} (\partial_V U)(V,T) & (\partial_T U)(V,T) \\ (\partial_V S)(V,T) & (\partial_T S)(V,T) \end{pmatrix} = \begin{pmatrix} 0 & C_V \\ \frac{1}{V}R & \frac{1}{T}C_V \end{pmatrix} \qquad (B.1.12)$$

so that

$$(Jf)(V, T) = -\frac{1}{V}RC_V \neq 0 \quad \forall \quad (V, T) \in D \subseteq \{V > 0, T > 0\} \quad ;$$
$$\qquad\qquad (B.1.13)$$

hence the inverse function theorem allows us to replace the (extrinsic) variables (V, T) by the (intrinsic) variables (U, S) in the description of (reversible) thermodynamical processes.

This ends our brief excursion into the differential calculus of vector-valued functions.

We now return to real-valued functions to extend the primitive notion of the differential of a real-valued function in yet another direction, namely to define a more general object that does not require the existence of a function of which it would be the differential. In brief, we progress from the definition of the "differential of a function" to that of a "differential".

For any pair (P, Q) of continuously differentiable functions defined on an open region $D \subset \mathsf{R}^2$, a *smooth differential*

$$\xi = P\,\mathrm{d}x + Q\,\mathrm{d}y \qquad\qquad (B.1.14)$$

on D is defined by

$$\xi : (x, y; h, k) \in D \times \mathsf{R}^2 \mapsto \xi_{(x,y)}(h, k) = P(x, y)\,h + Q(x, y)\,k \quad . \quad (B.1.15)$$

Note that in the particular case where $\xi = \mathrm{d}f$ for some function f, we have $P = \partial_x f$ and $Q = \partial_y f$. By the smoothness assumption we made on P

and Q, $\partial_{xy}f$ and $\partial_{yx}f$ exist, are continuous, and in fact are equal, so that *in this case*

$$\partial_x Q = \partial_y P \quad . \tag{B.1.16}$$

It is now quite remarkable – and actually essential to a proper understanding of the syntax for the thermodynamics of homogeneous fluids – that a converse theorem holds, provided that D be simply connected. To see this, we need a few more definitions and a major theorem to the effect that:

(i) in the generalization from continuously differentiable functions to smooth differentials, one does not loose the fact that these objects can be integrated along curves; and

(ii) the fundamental theorem of calculus in one variable, namely

$$\int_a^b \mathrm{d}x f(x) = f(b) - f(a) \tag{B.1.17}$$

admits a potent generalization to differentials in two variables: "the" Green Theorem.

A *piece-wise smooth curve* $C \subset \mathsf{R}^2$ is a finite union of abutting smooth curves, i.e. it is a map

$$C : t \in [a,b] \mapsto C(t) = (x_C(t), y_C(t)) \in \mathsf{R}^2 \tag{B.1.18}$$

where the domain $[a,b] \subset \mathsf{R}$ of the parameter t can be written as

$$[a,b] = \cup_{i=1}^n [a_i, a_{i+1}] \tag{B.1.19}$$

with $a_1 = a$ and $a_{n+1} = b$, in such a manner that for each $i = 1, 2, ..., n$ the segment C_i of the curve C corresponding to $[a_i, a_{i+1}]$ is continuously differentiable, with non-vanishing derivative. A *path* Γ is the image of a piece-wise smooth curve:

$$\Gamma = \{ C(t) \mid t \in [a,b] \} \quad . \tag{B.1.20}$$

The path Γ is said to be *simple* if the corresponding curve is non-self-intersecting, i.e. if $a \leq t_1 < t_2 < b$ implies $C(t_1) \neq C(t_2)$; and it is said to be *closed* if $C(a) = C(b)$. A simple closed path is said to be *positively oriented* if it is parametrized in a manner that it is run in a counter-clock manner. A *domain* $D \subset \mathsf{R}^2$ is said to be regular if it is open, is the interior of its closure $\overline{D}$, and its boundary ∂D is a positively oriented, simple, closed path Γ. Let Γ be a path in R^2, and $C(t) = (x(t), y(t))$ be a parametrized curve tracing this path. Let further $\xi = P\mathrm{d}x + Q\mathrm{d}y$ be a smooth differential defined on an open region $D \subset \mathsf{R}^2$ with $\Gamma \subset D$. The *integral* of ξ along Γ is defined by

$$\int_\Gamma \xi = \int_a^b \mathrm{d}t \left[P(x(t), y(t)) \, \dot{x}(t) + Q(x(t), y(t)) \, \dot{y}(t) \right] \tag{B.1.21}$$

where $(\dot{x}(t), \dot{y}(t)) = \dot{C}(t)$ is the derivative of C at t. The notation of the LHS is justified by the fact that defining integral on the RHS depends on Γ only, and not on its parametrization (i.e. on the choice of the curve C used to represent it) provided that the latter preserves the direction in which Γ is traversed.

Fig. B.2. A positively oriented closed simple path

We are now ready to state Green's theorem in the form suitable for our purposes.

Theorem B.1.2. *(Green): Let $D \subset \mathbb{R}^2$ be a regular domain, with boundary Γ and closure $\overline{D}$; and let P and Q be continuously differentiable functions, defined on some open region containing $\overline{D}$. Then*

$$\int_\Gamma P\,\mathrm{d}x + Q\,\mathrm{d}y = \iint_{\overline{D}} \mathrm{d}x\mathrm{d}y\,(\partial_x Q - \partial_y P) \quad . \tag{B.1.22}$$

While we imposed, in the definition of a regular domain that the boundary consist of exactly one simple closed path, one could allow for D to have finitely many holes. The next result however does require this not to be the case.

Theorem B.1.3. *(Corollary to Green's theorem): As D is simply connected, the following four conditions are mutually equivalent:*

(a) throughout D :

$$\partial_x Q = \partial_y P \quad ; \tag{B.1.23}$$

(b) for every closed path $\Gamma \subset D$

$$\int_\Gamma P\,\mathrm{d}x + Q\,\mathrm{d}y = 0 \quad ; \tag{B.1.24}$$

(c) for any two (not necessarily closed!) paths Γ_1 and Γ_2 joining the same two points in D :

$$\int_{\Gamma_1} P\,\mathrm{d}x + Q\,\mathrm{d}y = \int_{\Gamma_2} P\,\mathrm{d}x + Q\,\mathrm{d}y \quad ; \tag{B.1.25}$$

(d) there exists a smooth function f such that

$$\mathrm{d}f = P\,\mathrm{d}x + Q\,\mathrm{d}y \quad . \tag{B.1.26}$$

The Green theorem and its corollary were essentially motivated by the travails that led to the formulation of the principle of thermostatic; its strange history is briefly recounted at the end of Sect. 2.4. Suffice it here to say that Green's contribution was published in 1828, and apparently completely ignored until Thompson discovered it and had it published anew between 1850 and 1854 in Journal für Mathematik [Green, 1828]; not quite coincidentally, 1854 is the year Maxwell won the Smith mathematical competition in response to an essay question submitted by Stokes.

We therefore limit our comments here on the mathematical reach of the theorem. To situate it in mathematical perspective, it is proper to view it as a generalization to higher dimensions of the fundamental theorem of calculus that expresses an integral over a region in terms of its value on the boundary of this region. As was nicely encapsuled in [Fulks, 1978], *"... Green's theorem ... is a generic name for any theorem that gives sufficient conditions for the validity"* of the equality which we wrote as (B.1.22) above. The Reader will find some statement of this theorem in every first-year undergraduate textbook on Calculus or Vector Analysis (where it is usually given in the company of theorems by Gauss and by Stokes which are presented in Sect. B.2). Unless additional restrictive conditions are imposed on D and/or on Γ, the proof of this theorem is however beyond most first-year texts. Fulks gives a precise statement that is general enough for most applications, and proves it; for a more general proof (assuming only that Γ is merely rectifiable, rather than necessarily piece-wise smooth), he refers to [Apostol, 1957] ; see also [Edwards, 1969]. Besides these, an elementary presentation is given by Bressoud [1991], who exhibits the natural place occupied, in the theory of differential forms on manifolds, by the family of related theorems associated with the names of G. Green, C.F. Gauss, M. Ostrogradski and G.G. Stokes.

An awareness to this general setting would benefit the Reader interested in extension of the province of thermodynamics in directions – e.g. the line open by [Carathéodory, 1909] – that lay beyond the confines of Sect. 2.4.

B.2 Stokes' and Gauss' Theorems

We first state two mathematical results; immediately thereafter, we recall some basic notions that enter the formulation of these two results. We then show the sense in which these two theorems are intimately linked to the Green theorem studied earlier in this Appendix. We then illustrate in some detail how electromagnetism (see e.g. [Feynman, Leighton, and Sands, 1964]) – like thermodynamics – can be presented equivalently in differential form or in integral form.

Theorem B.2.1. *(Stokes): Let $D \subset \mathbb{R}^3$ be a smooth, oriented surface, with piece-wise smooth boundary Γ and closure $\overline{D}$; let $\boldsymbol{F}(x, y, z) = (F_1(x, y, z), F_2(x, y, z), F_3(x, y, z))$ be a continuously differentiable vector field, defined on some open region containing $\overline{D}$. Then*

$$\int_\Gamma \mathrm{d}\boldsymbol{r} \cdot \boldsymbol{F} = \iint_{\overline{D}} \mathrm{d}\sigma\, \boldsymbol{n} \cdot \boldsymbol{curl}\, \boldsymbol{F} \quad . \tag{B.2.1}$$

Theorem B.2.2. *(Gauss): Let $M \subset \mathbb{R}^3$ be an open (i.e. solid) region in $\mathbb{R}^3$, with piece-wise smooth boundary ∂M oriented by the outer normal $\boldsymbol{n}$, and let $\overline{M}$ be the closure of M; let further $\boldsymbol{F}(x, y, z) = (F_1(x, y, z), F_2(x, y, z), F_3(x, y, z))$ be a continuously differentiable vector field, defined on some open region containing $\overline{M}$. Then*

$$\iint_{\partial M} \mathrm{d}\sigma\, \boldsymbol{n} \cdot \boldsymbol{F} = \iiint_{\overline{M}} \mathrm{d}V\, \mathrm{div}\, \boldsymbol{F} \quad . \tag{B.2.2}$$

Let us now go over the technical definitions undergirding these statements.

Recall that a smooth surface in $D \subset \mathbb{R}^3$ is said to be orientable if a consistent (i.e. smooth) choice of a unit normal $\boldsymbol{n}$ to the surface can be made (the usual counter-examples are the "one-sided surfaces" known as the *Möbius strip* and the *Klein bottle*); an orientable surface $D \subset \mathbb{R}^3$ is said to be oriented if such a choice of $\boldsymbol{n}$ has been made. The boundary Γ of D is then oriented accordingly, namely in such a manner that the vector product of the normal to the surface and the tangent to the curve is pointing towards the inside of D. Stokes' theorem thus asserts that the line integral over Γ of the tangential component of the vector field $\boldsymbol{F}$ – defined in a manner similar to (B.1.21) – is equal to the surface integral over $\overline{D}$ of the normal component of the vector field $\boldsymbol{curl}\, \boldsymbol{F} \equiv \nabla \times \boldsymbol{F} = (\partial_y F_3 - \partial_z F_2, \partial_z F_1 - \partial_x F_3, \partial_x F_2 - \partial_y F_1)$.

In case $\boldsymbol{F}$ does not depend on z *and* the surface D lies in the (x, y)−plane the Stokes formula (B.2.1) reduces precisely to the Green formula (B.1.22). Hence Stokes' theorem is indeed a natural extension of Green's theorem.

Gauss' theorem involves integrations over manifolds of one dimension higher than the manifolds involved in Stokes' theorem. Note moreover that the manifolds occurring in the LHS of both theorems have no boundaries: Γ is a closed curve, and ∂M is a "closed" surface i.e. has no boundary. So, in particular, the conjunction of both theorems tells us immediately that

$$\iiint_M \mathrm{d}V\, \mathrm{div}\, \boldsymbol{curl}\, \boldsymbol{F} = \iint_{\partial M} \mathrm{d}\sigma\, \boldsymbol{n} \cdot \boldsymbol{curl}\, \boldsymbol{F} = \int_\emptyset \mathrm{d}\boldsymbol{r} \cdot \boldsymbol{F} = 0$$

and since this hold for all M, one obtains the universal identity:

$$\mathrm{div}\, \boldsymbol{curl}\, \boldsymbol{F} = 0 \tag{B.2.3}$$

where the divergence of a vector field $\boldsymbol{G}$ on D is defined as the function $\mathrm{div}\, \boldsymbol{G} \equiv \nabla \cdot \boldsymbol{G} = \partial_x G_1 + \partial_y G_2 + \partial_z G_3$.

In their usual differential formulation the (vacuum) Maxwell equations of electromagnetism are:

$$\text{div } \boldsymbol{E} = \varrho \tag{B.2.4}$$

$$\boldsymbol{curl\ E} = -\partial_t \boldsymbol{B} \tag{B.2.5}$$

$$\text{div } \boldsymbol{B} = 0 \tag{B.2.6}$$

$$\boldsymbol{curl\ B} = \boldsymbol{J} + \partial_t \boldsymbol{E} \tag{B.2.7}$$

where ϱ is a time-dependent scalar field, and $\boldsymbol{E}, \boldsymbol{B}, \boldsymbol{J}$ are time-dependent vector fields; i.e. ϱ is a real-valued function $\varrho : (x, y, z; t) \in D \times I \mapsto \varrho(x, y, z; t) \in \mathsf{R}$ with (x, y, z) running over an open spatial region $D \subset \mathsf{R}^3$, and with the time-parameter t running over an open interval $I \subset \mathsf{R}$; on the same D and I, $\boldsymbol{E}, \boldsymbol{B}, \boldsymbol{J}$ are vector-valued functions of the type $\boldsymbol{X} : (x, y, z; t) \in D \times I \mapsto \boldsymbol{X}(x, y, z; t) \in \mathsf{R}^3$.

Before we enter into the interpretation of these equations, we want to make two remarks.

First, upon using the universal identity (B.2.3), we see that (B.2.4) and (B.2.7) imply:

$$\text{div } \boldsymbol{J} + \partial_t \varrho = 0 \quad . \tag{B.2.8}$$

Second, in the static case, where none of the functions $\varrho, \boldsymbol{E}, \boldsymbol{B}$ and $\boldsymbol{J}$ depends on t, the above Maxwell equations decouple into two mutually independent pairs of equations:

$$\text{div } \boldsymbol{E} = \varrho \quad \text{and} \quad \boldsymbol{curl\ E} = \boldsymbol{0} \tag{B.2.9}$$

$$\text{div } \boldsymbol{B} = 0 \quad \text{and} \quad \boldsymbol{curl\ B} = \boldsymbol{J} . \tag{B.2.10}$$

(B.2.9) and (B.2.10) can be interpreted as the laws of electrostatics and magnetostatics respectively. To see this, we take advantage of the Stokes and Gauss theorems to rewrite these laws in their equivalent integral form. For instance, integrate the first relation in (B.2.9) over "any" (see the assumptions of Gauss' theorem) solid region V to obtain

$$\iiint_V dV \text{ div } \boldsymbol{E} = \iiint_V dV \varrho$$

and replace the LHS of this relation by the surface integral in Gauss' theorem. This gives the first relation in (B.2.11). Conversely, when this relation is supposed to hold for "every" V, the same route can be traveled in the opposite direction to recover the first relation in (B.2.9). Upon using similarly Gauss' theorem for the first relation in (B.2.10), and Stokes' theorem for the second relations in (B.2.9) and (B.2.10), one obtains the remaining three of the integral equations

$$\iint_{\partial V} d\sigma \, \boldsymbol{n} \cdot \boldsymbol{E} = \iiint_V dV \varrho \quad \text{and} \quad \int_\Gamma d\boldsymbol{r} \cdot \boldsymbol{E} = 0 \tag{B.2.11}$$

$$\iint_{\partial V} d\sigma\, \boldsymbol{n} \cdot \boldsymbol{B} = 0 \quad \text{and} \quad \int_{\partial \Sigma} d\boldsymbol{r} \cdot \boldsymbol{B} = \iint_{\Sigma} d\sigma\, \boldsymbol{n} \cdot \boldsymbol{J} \ . \quad \text{(B.2.12)}$$

Although the differential form (B.2.4–7) of the equations of electromagnetism is mathematically equivalent to their integral form (B.2.11–12), the latter equations seem easier to visualize as they describe more immediately the relations one observes in the laboratory.

The first equality in (B.2.11) states that the flux of the vector field $\boldsymbol{E}$ through the boundary ∂V of any solid region $V \subset \mathsf{R}^3$ is equal to the integral over V of the density ϱ. This admits a pragmatic interpretation as the *Gauss' law of electrostatics* where $\boldsymbol{E}$ is the *electric field* and ϱ is the *density of electric charge*.

The second equality in (B.2.11) then states that the circulation of the electric field $\boldsymbol{E}$ around any closed loop Γ vanishes. As the corollary (Thm. B.1.3) of Green's theorem extends to the non-flat situations covered by Stokes' theorem, we can conclude that this is equivalent to the existence of an *electrostatic potential* Φ such that $\boldsymbol{E} = -\boldsymbol{grad}\,\Phi$.

The relations (B.2.11) encompass the basic phenomena of electrostatics, even to the extend that one can derive from them the Coulomb law which asserts that the electric forces between electric charges decrease with the square of the distance between them. The Reader might object that these laws also cover the phenomena of Newtonian gravitation; this is indeed the case, and the inverse square law has indeed the same mathematical origin. There are evidently differences however between electrostatics and gravitation; these go much beyond the quantitative fact that the electrostatic forces are much stronger than the gravitational forces; one qualitative difference is that whereas mass is always non-negative, electrostatics has negative as well as positive charges; another difference is that, for gravitation, nature seems unwilling to offer an equivalent to magnetism.

We briefly indicate how the relations (B.2.12) can be interpreted in magnetostatic terms. The first of these relations, when compared to the corresponding law in electrostatics, is interpreted as the absence of magnetic charges, or as one often says *the absence of magnetic monopoles*. The second of these two laws asserts that, for "any" (see the assumptions of Stokes' theorem) surface Σ, the circulation of the vector field $\boldsymbol{B}$ around the closed loop $\partial \Sigma$ formed by the boundary of Σ is equal to the flux of the vector field $\boldsymbol{J}$ through Σ. This admits a pragmatic interpretation as *Ampère's law of magnetostatics* where $\boldsymbol{B}$ is the *magnetic field* associated to the *density of electric current* $\boldsymbol{J}$.

That electrostatics and magnetostatics are the separate static limits of a common dynamical theory can already be intimated from Ampère's contention that *electric currents are electric charges in motion*. Indeed, there is more in this statement than meets the modern eye; in the course of history, electric charges and electric currents were first studied separately – in electrostatics and magnetism – and there was therefore no reason to worry about the mutual consistency of the arbitrary attributions of signs made for the electric charges and for the direction of the electric currents. It is therefore only a historical freak that we have now to put a minus sign in the conservation law quantifying Ampère's statement, namely:

$$\iint_{\partial V} d\sigma \, \boldsymbol{n} \cdot \boldsymbol{J} = -\frac{d}{dt} \iiint_V dV \varrho \quad . \tag{B.2.13}$$

In fact an ambiguity of magnitude still remains between the two sides of this equation: this is handled by a convenient choice of units. More importantly for our purpose here, we should note, upon using again Gauss' theorem, that the above relation is equivalent to the consequence (B.2.8) of the *time-dependent* Maxwell equations.

The non-static case is then obtained in two steps. The first is *Faraday's law of magnetic induction;* it generalizes the second relation (B.2.11) to assert an *experimental fact*: a time-variation of the flux of a magnetic field $\boldsymbol{B}$ across "any" surface Σ produces an electric field $\boldsymbol{E}$ such that its circulation around the closed loop $\partial\Sigma$ is given by

$$\int_{\partial\Sigma} d\boldsymbol{r} \cdot \boldsymbol{E} = -\frac{d}{dt} \iint_\Sigma d\sigma \, \boldsymbol{n} \cdot \boldsymbol{B} \quad . \tag{B.2.14}$$

This integral relation is equivalent, again as a result of Stokes' theorem, to the Maxwell equation (B.2.5).

It also seems to be an *empirical fact* that both of the first relations in (B.2.9) and (B.2.10) (Bi.e. Gauss' law and the absence of magnetic monopoles) do not require any modification in the non-static case; equivalently, the Maxwell equations (B.2.4) and (B.2.6) are valid in the general non-static situation.

One still has to find some reason to postulate (B.2.7). Upon taking the time-derivative of the Gauss law, we obtain:

$$\iint_{\partial V} d\sigma \, \boldsymbol{n} \cdot \partial_t \boldsymbol{E} = \iiint_V dV \, \partial_t \varrho$$

which, when combined with the conservation law (B.2.13), gives

$$\iint_{\partial V} d\sigma \, \boldsymbol{n} \cdot (\boldsymbol{J} + \partial_t \boldsymbol{E}) = 0$$

or equivalently, through Gauss' theorem

$$\iiint_V dV \, \mathrm{div}\, (\boldsymbol{J} + \partial_t \boldsymbol{E}) = 0$$

which holds for "all" V ; hence:

$$\mathrm{div}\, (\partial_t \boldsymbol{E} + \boldsymbol{J}) = 0$$

so that there exists a (time-dependent) vector field $\boldsymbol{X}$ satisfying:

$$\boldsymbol{curl\ X} = \boldsymbol{J} + \partial_t \boldsymbol{E} \quad . \tag{B.2.15}$$

It was Maxwell's genius to compare (B.2.15) with the second equation in (B.2.10), i.e. with Ampère's law for magnetostatics, and to *postulate* by analogy that the proper dynamical generalization of Ampère's law is to identify the unknown time-dependent vector field $\boldsymbol{X}$ in (B.2.15) with the (now time-dependent) vector field $\boldsymbol{B}$, thus writing (B.2.7).

Through the use of Stokes' theorem, (B.2.7) is equivalent to extending the second of the two relations (B.2.12) so that, in the non-static case, it reads:

$$\int_{\partial \Sigma} \mathrm{d}\boldsymbol{r} \cdot \boldsymbol{B} = \iint_\Sigma \mathrm{d}\sigma\, \boldsymbol{n} \cdot \boldsymbol{J} + \frac{\mathrm{d}}{\mathrm{d}t} \iint_\Sigma \mathrm{d}\sigma\, \boldsymbol{n} \cdot \boldsymbol{E} \quad . \tag{B.2.16}$$

This expression gives the interpretation of (B.2.7): it states that the circulation of the magnetic field $\boldsymbol{B}$ around the closed loop $\partial \Sigma$ [LHS] is equal to the sum of the two terms in the [RHS]: the electric current across the surface Σ and the time-derivative of the flux of the electric field across Σ.

This completes the interpretation of Maxwell differential equations (B.2.4–7), and their consequence (B.2.8), through the following integral equations, the pragmatic significance of which is immediate: Gauss' law [the first equation in (B.2.11)], the Faraday law of magnetic induction [(B.2.14)], the absence of magnetic monopoles [the first equation in (B.2.12)], the Ampère–Maxwell law [(B.2.16)] and the conservation law of electric charges [(B.2.13)].

In closing Sect. B.2, we want to make two historical comments.

The first addresses our mode of presentation. Our main purpose here was to illustrate a mathematical technique, and its use in establishing that the mathematical equivalence of the integral and differential formulations of electrodynamism is structurally very similar to the equivalence between the two aspects of thermodynamics discussed in Sect. 2.4. Our presentation of the Maxwell equations – and in particular of the Ampère–Maxwell law – was made to emphasize the logical distinction between the formal differential equations and the semantics immediately attached to their integral form. However we were not historically faithful in our representing Maxwell's contribution, as it is often represented in textbooks and in anthologies, e.g. [Magie, 1963]: it achieved more than a "mere" formulation of a *mathematical* theory of electrodynamics. A reading of Maxwell's works – for a bibliography, see e.g. [Everitt, 1974]) – shows that he had a comprehensive mastery over a variety of experimental facts pertinent to the behavior of electric and magnetic fields in matter (a distinction must then be made between the electric field $\boldsymbol{E}$ and the electric displacement $\boldsymbol{D}$, and between the magnetic induction $\boldsymbol{B}$ and the magnetic field $\boldsymbol{H}$); he knew about dielectric media, and about paramagnetic, diamagnetic and ferromagnetic media; and he was also aware of experiments to the effect that a magnetic field can induce a rotation in the polarisation plane of light.

The second comment is a brief glimpse beyond the story we used for our illustration, namely that the Maxwell equations unify two separate static theories, electrostatics and magnetostatics, into a new dynamical theory, classical electromagnetism. Indeed, beyond this already remarkable achievement, these equations provided the base for an understanding of *electromagnetic waves,* in vacuum first, but also in matter. From there followed the definitive elaboration of a wave theory of light that supplanted the Descartes–Newton corpuscular theory, and now incorporates the phenomena of light within the much wider theory that electrodynamics has become. It made possible a deeper understanding of the refraction laws of optics and of the diffraction phenomena; the latter were moreover to play later an important role in the discussion of the uncertainty principle in Quantum Mechanics ... even though the wave/particle duality was to steal the show soon thereafter. Finally, as we show in the next section, the Maxwell equations can be written even more naturally in terms of 2–forms in four-dimensional space, a remark that is reflected in the fact that the Maxwell equations, through the invariant role they assign to the speed of light, paved the way to Einstein's theory of Relativity. There is therefore no exageration in viewing Maxwell's equations as opening the door to 20th-century physics.

B.3 Higher Differentials

The purpose of the last section of this appendix is to present a generalization from the ordinary differentials ξ on $D \subseteq \mathsf{R}^2$ (see Sect. B.1) to k–forms on $D \subseteq \mathsf{R}^n$. This generalization allows to see the sense in which the three theorems of Green, Stokes and Gauss are particular expressions of a more general result, namely Thm. B.3.2.

We generalize first the dimension of the domain from the particular case $n = 2$ to general $n < \infty$. The definition of a differentiable function $f : x \in D \mapsto f(x) \in \mathsf{R}$ on an (open) domain $D \subseteq \mathsf{R}^n$ generalizes straightforwardly (see the beginning of this Appendix), and we write:

$$\mathrm{d}f = \sum_{i=1}^{n} \partial_i f \, \mathrm{d}x^i \tag{B.3.1}$$

by which we mean

$$(\mathrm{d}f)(x; h) = \sum_{i=1}^{n} (\partial_i f)(x) \, \mathrm{d}x^i(h) = \sum_{i=1}^{n} (\partial_i f)(x) \, h^i \tag{B.3.2}$$

i.e., for a vector field $X : x \in D \mapsto X(x) = (x, h(x)) \in D \times \mathsf{R}^n$:

$$(\mathrm{d}f)(X) = \sum_{i=1}^{n} (\partial_i f) X^i \quad . \tag{B.3.3}$$

Similarly, we want to view an ordinary differential (or 1–form)

$$\xi = \Xi_1 \, dx^1 + \cdots + \Xi_n \, dx^n \qquad (B.3.4)$$

on an (open) domain $D \subseteq \mathbb{R}^n$ as a linear map

$$\xi : X \in \mathcal{V}(D) \mapsto \xi(X) \in \mathcal{C}(D) \qquad (B.3.5)$$

where $\mathcal{V}(D)$ is the linear space – more precisely: the module – of the vector fields on D, and $\mathcal{C}(D)$ is the linear space of the real-valued functions on D. The function $\xi(X)$ takes, at the point $x \in D$, the value:

$$\xi(X)_x = \sum_{i=1}^{n} \Xi_i(x) \, X^i(x) \quad . \qquad (B.3.6)$$

The second step in the generalization is to consider, for every non-negative integer k, the Cartesian product $\mathcal{V}(D)^k = \mathcal{V}(D) \times \cdots \times \mathcal{V}(D)$ of k copies of $\mathcal{V}(D)$. A k–form on D is defined as a map

$$\xi : \mathcal{V}(D)^k \to \mathcal{C}(D) \qquad (B.3.7)$$

satisfying the following two properties:

- ξ is linear over $\mathcal{C}(D)$ in each of its variables, i.e. for every pair of vector fields (X_i, Y_i), and every $f \in \mathcal{C}(D)$:

$$\left.\begin{array}{l} \xi(X_1, \cdots, X_i + Y_i, \cdots, X_k) = \\ \quad \xi(X_1, \cdots, X_i, \cdots, X_k) + \xi(X_1, \cdots, Y_i, \cdots, X_k) \\[2mm] \xi(X_1, \cdots, f X_i, \cdots, X_k) = f \, \xi(X_1, \cdots, X_i, \cdots, X_k) \end{array}\right\} \quad ; \qquad (B.3.8)$$

- ξ is completely anti-symmetric, i.e. for every permutation p of the indices $1, \cdots, k$:

$$\xi(X_{p_1}, \cdots, X_{p_i}, \cdots, X_{p_k}) = (-1)^p \, \xi(X_1, \cdots, X_i, \cdots, X_k) \qquad (B.3.9)$$

where $(-1)^p$ is the parity of the permutation p, i.e. the parity [$+1$ for even and -1 for odd] of the number of transpositions $\{i, j\} \to \{p_i, p_j\}$ necessary to pass from $\{1, \cdots, i, \cdots, k\}$ to $\{p_1, \cdots, p_i, \cdots, p_k\}$; note that this parity is indeed a property of p alone, although several sequences of transpositions can be followed to achieve the same result.

These properties imply that every k–form should be expressible as a linear combination, with coefficients in $\mathcal{C}(D)$, of some $\binom{n}{k}$ basic k–forms. Recall

$$\binom{n}{k} = \frac{n!}{k! \, (n-k)!}$$

so that, in particular

$$\binom{3}{1} = 3 = \binom{3}{2} \qquad \text{but} \qquad \binom{4}{1} = 4 \ \neq \ 6 = \binom{4}{2} \ .$$

Hence, whereas in three-dimensions the number [3] of linearly independent 2−forms is the same as the number of linearly independent 1−forms, which is also the number of linearly independent vector fields, the situation is quite different in four-dimensions: the number [6] of linearly independent 2−forms is bigger than the number [4] of linearly independent 1−forms. This suggests that 2−forms on 4−dimensional domains may be a proper mathematical universe to describe electromagnetism: $4 = 3 + 1$ is the dimension of space-time, and $6 = 3+3$ is the number of components necessary to include together in the same description the electric field and the magnetic field. We show now how this suggestion can be implemented.

The general theory is based on two operations on k−forms: the exterior product – also called the "wedge" product – and the exterior derivative, which we now define.

The *exterior product* $\xi \wedge \eta$ of a k−form ξ and a l−form η is the $(k+l)$−form

$$\left. \begin{array}{l} \xi \wedge \eta(X_1, \cdots, X_k, X_{k+1}, \cdots, X_{k+l}) = \\[2mm] \sum_{\substack{i_1 < i_2 < \cdots < i_k \\ j_1 < j_2 < \cdots < j_l}} (-1)^p \ \xi(X_{i_1}, \cdots, X_{i_k}) \ \eta(X_{j_1}, \cdots, X_{j_l}) \end{array} \right\} \ . \quad \text{(B.3.10)}$$

This product has the following properties:
- it is antisymmetric in the sense that

$$\xi \wedge \eta = (-1)^{kl} \ \eta \wedge \xi \ ; \tag{B.3.11}$$

- it is linear over $\mathcal{C}(D)$ in each of its factors, i.e.

$$(\xi + \eta) \wedge \zeta = \xi \wedge \zeta + \eta \wedge \zeta \quad \text{and} \quad (f\,\xi) \wedge \zeta = f\,(\xi \wedge \zeta) \ ; \tag{B.3.12}$$

or equivalently, because of (B.3.11):

$$\xi \wedge (\eta + \zeta) = \xi \wedge \eta + \xi \wedge \zeta \quad \text{and} \quad \xi \wedge (f\zeta) = f\,(\xi \wedge \zeta) \ ;$$

- it is associative

$$\xi \wedge (\eta \wedge \zeta) = (\xi \wedge \eta) \wedge \zeta \ . \tag{B.3.13}$$

In particular, the exterior product allows to write every k−form as

$$\xi = \frac{1}{k!} \sum_{i_1 < i_2 < \cdots < i_k} \xi_{i_1, i_2, \cdots, i_k} \ dx^{i_1} \wedge dx^{i_2} \wedge \cdots \wedge dx^{i_k} \tag{B.3.14}$$

where $\xi_{i_1,i_2,\cdots,i_k}$ are smooth functions (thus elements in $\mathcal{C}(D)$) and the dx^i are the ordinary differentials $dx^i : X \in \mathcal{V}(D) \mapsto dx^i(X) = X^i \in \mathcal{C}(D)$.

The second operation extends to higher differentials the operation that passes from a 0−form (i.e. a smooth function) f to a 1−form, namely its differential df. This *exterior derivative*, denoted d, sends any k−form ξ to the $(k+1)$−form $d\xi$ defined by

$$d\xi = \frac{1}{k!} \sum_{i_o < i_1 < i_2 < \cdots < i_k} (\partial_{i_o} \xi_{i_1,i_2,\cdots,i_k}) \, dx^{i_o} \wedge dx^{i_1} \wedge dx^{i_2} \wedge \cdots \wedge dx^{i_k} \quad (\text{B.3.15})$$

where we used the expression (B.3.14) of an arbitrary k−form. Thus d increases by one the rank of the k−forms on which it is applied:

$$d : \xi \in \Lambda^k(D) \mapsto d\xi \in \Lambda^{k+1}(D) \quad\quad\quad (\text{B.3.16})$$

where $\Lambda^k(D)$ denotes the space of all k−forms on $D \subseteq \mathsf{R}^n$, and $1 \leq \cdots \leq k \leq \cdots \leq n$.

This exterior derivative satisfies the following general properties:

- d is additive

$$d(\xi + \eta) = d\xi + d\eta \quad ; \quad\quad\quad (\text{B.3.17})$$

- d satisfies the Leibniz rule

$$d(\xi \wedge \eta) = d\xi \wedge \eta + (-1)^k \, \xi \wedge d\eta \quad\quad\quad (\text{B.3.18})$$

where ξ is a k−form and η is a l−form (with k, l arbitrary);
- $d^2 = 0$, i.e.

$$d(d\xi) = 0 \quad \forall \ \xi \in \Lambda^k(D) \quad \text{and} \quad 1 \leq \cdots \leq k \leq \cdots \leq n \quad . \quad (\text{B.3.19})$$

Note that the above definitions are indeed consistent with $f\xi = f \wedge \xi = \xi \wedge f$: $d(f\xi) = (df) \wedge \xi + f \wedge d\xi = (df)\xi + f \, d\xi$. In particular, (B.3.17) and (B.3.18) together imply that d is linear over R. Moreover (B.3.16–19) characterize d entirely, i.e. imply (B.3.15).

We interrupt here the flow of definitions to propose three illustrations leading naturally to a generalization we need in the main body of this work: the extension of the calculus of differentials to spaces that look only locally like R^n, namely differentiable manifolds.

(i) *curl, grad* and *div* revisited. First, to get a feeling for (B.3.19), consider the familiar case of an arbitrary 1−form ξ on $D \subseteq \mathsf{R}^3$ written as

$$\xi = \Xi_1 \, dx^1 + \Xi_2 \, dx^2 + \Xi_3 \, dx^3 \quad\quad\quad (\text{B.3.20})$$

one then obtains from the above properties of d and $\wedge$:

$$\begin{aligned} d\xi &= H_1 \, dx^2 \wedge dx^3 + H_2 \, dx^3 \wedge dx^1 + H_3 \, dx^1 \wedge dx^2 \quad \text{where} \\ H_i &= \partial_j \Xi_k - \partial_k \Xi_j \quad \text{with} \quad \{i,j,k\} \equiv \{1,2,3\}, \quad\quad \text{i.e.} \quad H = \boldsymbol{curl} \, \Xi \, . \end{aligned}$$
$$(\text{B.3.21})$$

(We used above the notation $\{i, j, k\} \equiv \{1, 2, 3\}$ to mean that $\{i, j, k\}$ stands for all circular permutations of the indices $\{1, 2, 3\}$, namely $\{1, 2, 3\}$ $\{2, 3, 1\}$ and $\{3, 1, 2\}$). Similarly if η is an arbitrary 2−form on $D \subseteq \mathsf{R}^3$ written as

$$\eta = H_1 \, dx^2 \wedge dx^3 + H_2 \, dx^3 \wedge dx^1 + H_3 \, dx^1 \wedge dx^2 \qquad (\text{B.3.22})$$

where, now, H is an arbitrary vector field on $D \subseteq \mathsf{R}^3$, we obtain:

$$d\eta = Z \, dx^1 \wedge dx^2 \wedge dx^3 \quad \text{where} \quad Z = \sum_{i=1}^{3} \partial_i H_i = \operatorname{div} H. \qquad (\text{B.3.23})$$

In particular, when $\xi = df$ we must have $d\xi = d^2 f = 0$ and thus, in line with the above computations:

$$\boldsymbol{curl\, grad} f = \mathbf{0}.$$

The identity (B.2.3)

$$\operatorname{div} \boldsymbol{curl} \,\boldsymbol{\Xi} = 0$$

is obtained in the same way when one starts instead from a 2−form η satisfying $\eta = d\xi$.

(ii) Maxwell equations revisited. In this illustration we make good on an earlier comment we made, and show how 2−forms on R^4 can be used to reformulate the Maxwell equations more concisely. Consider such a general 2−form:

$$\left. \begin{array}{l} \mathcal{F} = (F_1 \, dx^1 \wedge dt \;+\; E_2 \, dx^2 \wedge dt \;+\; E_3 \, dx^3 \wedge dt) \;+ \\ (B_1 \, dx^2 \wedge dx^3 \;+\; B_2 \, dx^3 \wedge dx^1 \;+\; B_3 \, dx^1 \wedge dx^2) \end{array} \right\}. \qquad (\text{B.3.24})$$

One obtains straighforwardly that

$$d\mathcal{F} = (\sum_{\{i,j,k\}} X_i \, dx^j \wedge dx^k \wedge dt) + Z \, dx^1 \wedge dx^2 \wedge dx^3$$

where

$$X_i = (\partial_j E_k - \partial_k E_j) + \partial_t B_i \text{ with } \{i, j, k\} \equiv \{1, 2, 3\}, \text{ and } Z = \sum_{i=1}^{3} \partial_i B_i \,;$$

and where the summation over $\{i, j, k\}$ carries over the three circular permutations of the indices $\{1, 2, 3\}$. This shows that to impose that a 2−form on $D \subseteq \mathsf{R}^4$ is closed, i.e. that

$$d\mathcal{F} = O \quad, \qquad (\text{B.3.25})$$

leads directly (and in fact is mathematically equivalent) to the two Maxwell equations (B.2.5) and (B.2.6) taken together on D. The two Maxwell equations (B.2.4) and (B.2.7) − involving the source terms J and ϱ − are then obtained by a canonical duality that amounts to replacing (B.3.24) by:

$$\tilde{\mathcal{F}} = \left.\begin{array}{l} (B_1 \, \mathrm{d}x^1 \wedge \mathrm{d}t \;+\; B_2 \, \mathrm{d}x^2 \wedge \mathrm{d}t \;+\; B_3 \, \mathrm{d}x^3 \wedge \mathrm{d}t) \;- \\ (E_1 \, \mathrm{d}x^2 \wedge \mathrm{d}x^3 \;+\; E_2 \, \mathrm{d}x^3 \wedge \mathrm{d}x^1 \;+\; E_3 \, \mathrm{d}x^1 \wedge \mathrm{d}x^2) \end{array}\right\} \qquad (\text{B.3.26})$$

and by imposing, instead of (B.3.25):

$$\mathrm{d}\tilde{\mathcal{F}} = \tilde{J} \qquad (\text{B.3.27})$$

with

$$\tilde{J} = \Big(\sum_{\{i,j,k\}} J_i \mathrm{d}x^j \wedge \mathrm{d}x^k \wedge \mathrm{d}t \Big) - \varrho \, \mathrm{d}x^1 \wedge \mathrm{d}x^2 \wedge \mathrm{d}x^3 \qquad (\text{B.3.28})$$

where here again the summation over $\{i, j, k\}$ carries over the three circular permutations of the indices $\{1, 2, 3\}$.

Equivalently, (B.3.27) can be taken as the definition of the source term $\tilde{J}$. From a formal point of view, $\tilde{\mathcal{F}}$ is the most general $2-$form on R^4 with closed dual – the restriction (B.3.25). (B_i, E_i) are then just the names one gives to its components; similarly (J_i, ϱ) are the names one gives to the components of its differential $\mathrm{d}\tilde{\mathcal{F}}$. In this sense, the *syntax* of electromagnetism is intrinsic to R^4; its *semantic* is the interpretation of $(\boldsymbol{E}, \boldsymbol{B})$ and $(\boldsymbol{J}, \varrho)$ as "electric field", "magnetic field"; and "electric current", "electric charge". With this hindsight, one understands better the awe Hertz, as quoted by [Schey, 1997], expressed for the Maxwell equations:

> One cannot escape the feeling that these mathematical formulas have an independent existence and an intelligence of their own, that they are wiser than we are, wiser even than their discoverers, that one gets more out of them than we put into them.

Although without the benefit of the explicit reformulation (B.3.24–28) of the basic equations of electromagnetism, Einstein had the foresight to capitalize on their centrality: he made them the starting point of his special relativity and, furthermore, he saw that he could carry the form of the "electromagnetic tensor" $\mathcal{F}$ readily over into his general relativity; see e.g. [Sachs and Wu, 1977].

(iii) **Hamiltonian mechanics.** Consider the idealization consisting of a material point of mass m, moving on a configuration space R^n. Its instantaneous state is labeled by $(\boldsymbol{p}, \boldsymbol{q}) = (p_1, p_2, \cdots, p_n; q_1, q_2, \cdots, q_n) \in \mathsf{R}^{2n}$, where $\boldsymbol{p}$ denotes its momentum, and $\boldsymbol{q}$ its position. The motion $t \in \mathsf{R} \mapsto (\boldsymbol{p}(t), \boldsymbol{q}(t)) \in \mathsf{R}^{2n}$ is said to obey the laws of Hamiltonian mechanics when there exists a smooth function – called the *Hamiltonian* – $H : (\boldsymbol{p}, \boldsymbol{q}) \in \mathsf{R}^{2n} \mapsto H(\boldsymbol{p}, \boldsymbol{q}) \in \mathsf{R}$ such that:

$$\dot{p}_k = -\partial_{q_k} H \quad ; \quad \dot{q}_k = \partial_{p_k} H \qquad (\text{B.3.29})$$

where $\dot{x}$ denotes the time-derivative of $x(t)$ and ∂_x denotes the partial derivative with respect to x. One may add in H an explicit time-dependence, which we ignore in this elementary exposition, so that H depends of t only implicitly, namely through $\boldsymbol{p}$ and $\boldsymbol{q}$. The evolution of an *observable* – i.e. of a smooth function $f : (\boldsymbol{p}, \boldsymbol{q}) \in \mathsf{R}^{2n} \mapsto f(\boldsymbol{p}, \boldsymbol{q}) \in \mathsf{R}$ – is then given by the chain rule:

$$\dot{f} = \sum_{k=1}^{n} \partial_{p_k} f \; \dot{p}_k + \partial_{q_k} f \; \dot{q}_k \quad . \tag{B.3.30}$$

Upon taking into account (B.3.29), we rewrite (B.3.30) in the two equivalent forms:

$$\dot{f} = X_H f \quad \text{or} \quad \dot{f} = \{f, H\} \tag{B.3.31}$$

where for any smooth functions from R^{2n} to R, the *Hamiltonian vector field* X_H and the *Poisson bracket* $\{f, H\}$ are defined by:

$$\left. \begin{array}{l} X_H = \sum_{k=1}^{n} \; \partial_{p_k} H \; \partial_{q_k} - \partial_{q_k} H \; \partial_{p_k} \\[2mm] \{f, H\} = \sum_{k=1}^{n} \; \partial_{p_k} H \; \partial_{q_k} f - \partial_{q_k} H \; \partial_{p_k} f \end{array} \right\} \quad . \tag{B.3.32}$$

In order to emphasize the core content of these relations, we introduce the 2–form

$$\omega = \sum_{k=1}^{n} \; \mathrm{d}p_k \wedge \mathrm{d}q_k \tag{B.3.33}$$

which is *closed,* i.e.

$$d\omega = 0 \tag{B.3.34}$$

and *non-degenerate,* i.e. with $\mathcal{V}$ denoting the space of all vector fields on R^{2n} :

$$\{X \in \mathcal{V} \text{ and } \{\forall Y \in \mathcal{V} \; : \; \omega(X, Y) = 0 \,\}\} \quad \models \quad X = 0 \quad . \tag{B.3.35}$$

The following relations are then equivalent to the definitions (B.3.32):

$$X_H \rfloor \omega = dH \quad \text{and} \quad \{f, H\} = \omega(X_H, X_f) \tag{B.3.36}$$

where $X \rfloor \omega$ denotes the *contraction* which associates to a vector field X the 1–form $X \rfloor \omega : Y \in \mathcal{V} \mapsto \omega(Y, X) \in \mathcal{C}(\mathsf{R}^{2n})$. The Reader should now think of R^{2n} as the cotangent bundle $T^*\mathsf{R}^n$ of the configuration space R^n. Indeed the next step required by mechanics – at least since Lagrange, who studied spinning tops – is to consider spaces that are more general, in particular configuration spaces that are neither flat nor simply connected, and yet still look locally like R^n. Such extensions of the formalism's scope are also required by geometric optics – an important motivation in Hamilton's original work – and they are at the conceptual center of Einstein's passage from special to general relativity.

Differential and symplectic manifolds. Note: It is not just a happy co-incidence that the nomenclature for differentiable manifolds seems inspired from the familiar objects basic to cartography.

A n-*dimensional manifold* is a set $\mathcal{M}^n$ equiped with an equivalence class of atlases. An *atlas* is a – finite or countably infinite – collection of *charts* (U_j, φ_j) satisfying the following conditions:

1. each U_j is an open subset in R^n and φ_j is a bijective map from U_j to some subset $\varphi_j(U_j) \subseteq \mathcal{M}^n$;
2. for every $m \in \mathcal{M}^n$ there exists at least one chart (U_j, φ_j) such that $m \in \varphi_j(U_j)$;
3. if $x \in U_i$ and $y \in U_j$ are such that $\varphi_i(x) = \varphi_j(y)$, then there exist neighborhoods $V_i \subseteq U_i \subseteq \mathcal{R}^n$ and $V_j \subseteq U_j \subseteq \mathcal{R}^n$ such that $x \in V_i$, $y \in V_j$, $\varphi_i(V_i) = \varphi_j(V_j)$ and the restriction of $\varphi_j^{-1} \circ \varphi_i$, to V_i is a diffeomorphism, i.e. is bijective and differentiable.

Two atlases are said to be *equivalent* whenever their union is also an atlas.

With this, the definition of a differentiable manifold is essentially complete; one says that such a manifold $\mathcal{M}^n$ is modeled on R^n.

Example: The following two charts equip the sphere $\{x \in \mathcal{R}^3 \mid \|x\| = 1\}$ with the structure of a 2–dimensional differentiable manifold. Consider first the plane tangent to the sphere at the South pole; in this plane, draw a circle centered at the South pole and of radius strictly larger than 2, say 2.5; for U_1 take the interior of that circle; for every $x \in U_1$ draw the straightline from x to the North pole; $\varphi_1(x)$ is defined as the point of intersection of this line with the sphere. The second chart (U_2, φ_2) is obtained similarly upon exchanging the roles of the North and South poles.

Notions that are primary to the differential structure of R^n lift to $\mathcal{M}^n$ by a systematic use of charts. A few examples illustrate how to proceed.

A *neighborhood* of a point $m \in \mathcal{M}^n$ is defined as a set of the form $\varphi(V)$ where a chart (U, φ) is chosen such that $V \subseteq U$ be a neighborhood of $x = \varphi^{-1}(m)$. It is customary to assume that for any $m_1, m_2 \in \mathcal{M}^n$ with $m_1 \neq m_2$ there exist neighborhoods $N_1, N_2 \subset \mathcal{M}^n$ such that $m_1 \in N_1$, $m_2 \in N_2$ and $N_1 \cap N_2 = \emptyset$.

Let again $m \in \mathcal{M}^n$ and (U, φ) be a chart such that there exists $x \in U$ with $\varphi(x) = m$; we identify a vector X at x with the equivalence class of all smooth parametrized curves $\gamma : s \in (-\varepsilon, \varepsilon) \mapsto \gamma(s) \in U \subseteq \mathsf{R}^n$ with $\gamma(0) = x$ and $\frac{d\gamma}{ds}(0) = X$. Now lift these curves to curves $\varphi \circ \gamma : s \in (-\varepsilon, \varepsilon) \mapsto \varphi(\gamma(s)) \in \varphi(U) \subseteq \mathcal{M}^n$. The latter collection of curves is defined as the lifting $\varphi^*(X)$ of X and is referred to as a *tangent vector* to $\mathcal{M}^n$ at m. Note that the vectors so defined form indeed a n–dimensional vector space, and that this vector space is independent of the chart used to define it. It is called the *tangent space* to $\mathcal{M}^n$ at $m \in \mathcal{M}^n$, and it is denoted $T_m^* \mathcal{M}^n$. Hence $T^* \mathcal{M}^n \equiv \{T_m^* \mathcal{M}^n \mid m \in \mathcal{M}^n\}$ looks locally like R^{2n} and the resulting $2n$–dimensional manifold is called the *tangent bundle* of $\mathcal{M}$. Note also that, given a chart in which a system of coordinates has been chosen, curves tangent to one another at x allow to interpret X as the differential operator $\sum_k X_k \, \partial_k$; the above construction provides a similar interpretation of $\varphi^*(X)$. A *Riemannian metric* on a manifold $\mathcal{M}$ is a smooth assignment g of an inner product g_m on each $T_m^* \mathcal{M}^n$. A *Riemannian manifold* $(\mathcal{M}^n, g)$ is a manifold on which a Riemannian metric is given.

The notion of k−forms on a manifold is defined by duality from the notion of tangent vector fields. The operations d and $\wedge$ on k−forms lift then naturally to objects defined on a manifold $\mathcal{M}^n$ in a way that parallel there original definition in $\mathcal{R}^n$. We can thus use (B.3.34–35) to define then without ambiguity a *symplectic form* on an even-dimensional manifold $\mathcal{M}^{2n}$ as a 2−form ω that is closed and non-degenerate. A *symplectic manifold* $(\mathcal{M}^{2n}, \omega)$ is a $2n$−dimensional manifold $\mathcal{M}^{2n}$ equipped with symplectic form ω. The following result verifies that we have not straggled too far from the idea that differentiable manifolds should look locally like R^{2n}.

Theorem B.3.1 (Darboux). *Let* $(\mathcal{M}^{2n}, \omega)$ *be symplectic manifold, and* m *be an arbitrary point in* $\mathcal{M}^{2n}$. *Then there exists a local system of coordinates* $\{p_1, \cdots, p_n; q_1, \cdots q_n\}$ *around* m *such that* $\omega = \sum_{k=1}^n \mathrm{d}p_k \wedge \mathrm{d}q_k$.

Note however that this theorem only holds locally: in general, different charts – with different systems of coordinates – become necessary as m runs over $\mathcal{M}^{2n}$.

In the context of diffferentiable manifolds the equivalence of conditions (B.1.23) and (B.1.26) in the corollary to Green's theorem (Thm. B.1.3 above) needs some discussion.

While a k−form η is said to be *closed* if $d\eta = 0$; and it is said to be *exact* if there exists a $(k+1)$−form ξ such that $\eta = d\xi$. Clearly, $d^2 = 0$ implies that every exact form is closed. Poincaré's lemma asserts that the converse, namely that a closed form be exact, holds *provided* some topological conditions are satisfied by the domain D on which the form is defined. For 1−forms on $D \subseteq \mathsf{R}^n$ such a condition is that D be simply connected; in the special case $n = 2$ this was precisely the condition we had seen for the equivalence of (B.1.23) and (B.1.26). For k−forms on $D \subseteq \mathsf{R}^n$ (with $k \geq 2$) the simplest condition is somewhat more restrictive in that it requires D to be star-shaped (i.e. there exists at least one $x_o \in D$ such that for every $x \in D$, the straightline segment joining x_o and x is entirely contained in D). Note that this condition is sufficient; for conditions that are both necessary and sufficient, see for instance [De Rham, 1955, Wilmore, 1959].

To complete our round-up of the theory of k−forms, we return to one of our original motivations, namely to indicate that both Stokes' and Gauss' theorems are in fact particular realisation of the same result.

Theorem B.3.2. *(Generalized Stokes Formula): Let* D^+ *be a compact, oriented* n−*dimensional manifold with boundary* ∂D^+ *; let* ω *be a* $(n-1)$*form on* D. *Then*

$$\int_{\partial D^+} \omega = \int_{D^+} \mathrm{d}\omega \quad . \tag{B.3.37}$$

It is beyond the scope of this appendix to review the theory of integration over general manifolds; see e.g. [Flemming, 1977] ; the Reader who will consult this reference cannot but note that the book carries the above equality, written in bright red characters, across its distinctive yellow front cover, thus testifying to the central importance of the generalized Stokes theorem in the modern presentation of calculus in several variables. Our Reader is simply asked here to read the generalized Stokes formula (B.3.37) in the particular cases where ω stands firstly for a $1-$form and secondly for a $2-$form in $\mathbb{R}^3$: i.e. to note the special expressions taken by $d\omega$ as we gave them in (3.20–21) and (3.22–23); and then to recognize that the particularizations so obtained are indeed the familiar Stokes and Gauss formulas (B.2.1) and (B.2.2).

To go deeper into the theories touched upon in this section, the Reader is advised to visit the following texts; for mechanics [Arnold, 1978]; for differential geometry [Sternberg, 1983]; for symplectic geometry [Guillemin and Sternberg, 1984].

C. Appendix: Recursive Functions

The purpose of this Appendix is to offer a concise orientation into the mathematical theory that formalizes the *intuitive* notions of computable functions (i.e., loosely speaking, functions that can be described in a finite number of steps), and of enumerable sets (i.e., sets or relations that can be listed effectively). The corresponding *formalized* concepts provide the theoretical background for our discussion of the foundations of probability in Part II of this book.

The theory got a running start in the 1930s with the work of Hilbert, Gödel, Kleene, Church, Turing and the Princeton School; see [Soare, 1996, Sieg, 1997] for recent reviews of these early contributions and their motivations, touching also upon their relationships with some aspects of the Hilbert program and Bernays's concerns.

Systematic presentations at the introductory level have been given in several textbooks. We particularly like [Hodel, 1995] and also [Yasuhara, 1971]; it is probably fair to say that a more intuitive and relaxed approach is taken in a variety of other texts, e.g. [Boolos and Jeffrey, 1980, Epstein and Carnielli, 1989]; depending on their backgrounds, taste or inclinations, other Readers may prefer tighter or deeper expositions such as [Hermes, 1969, Tourlakis, 1984, Odifreddi, 1989], although here again this list is but a sample meant to indicate the variety of motivations and/or applications embraced by the theory of recursive functions.

Three main steps have marked the development of the mathematical formalization necessary to delineate the intuitive notion that a function be computable; they can be characterized by the class of the objects on which they focused: (1) the primitive recursive functions; (2) the recursive functions; and (3) the partial recursive functions. For reasons of expository simplicity, we begin with the second class; we next pursue our presentation with a brief discussion of the special case obtained with the primitive recursive functions which were historically proposed first, but soon proved less than adequate; we then present the generalization we ultimately need the functions belonging to the third class.

For each class separately, we first identify the basic (or *"starting"*) functions of that class; we then delineate the allowable operations one can perform to generate step-by-step all the other elements of the class; and we conclude the presentation of this class by listing some of the principal properties of the functions so obtained.

Once each of the three classes of recursive functions has been described, we finally turn to their relevance for the effective listing – or indexing – of sets and relations.

Throughout this Appendix, we use the adjective "computable" (whether applied to functions, sets or relations) only to refer to as-yet undefined, informal, intuitive properties that need to be formalized. Consequently, to emphasize the distinction between the informal and the formalized, we abstain from using the term "computable" in the formal mathematical definitions and properties of the specific objects offered towards the desired formalization; the formal terms are typed in boldface when they are defined.

Definition C.1.1. *Let* $\mathsf{N} = \mathsf{Z}^+\backslash\{0\}$ *denote the set of all natural numbers, i.e. the set of all positive integers* $\{1, 2, \ldots\}$. *The* **starting functions** *for the class of recursive functions are:* $+, \times, \{P_{n,k} \mid (n,k) \in \mathsf{N}^2 \text{ with } k \le n\}$, *and* $K_<$; *where*

$$P_{n,k} : (a_1, \ldots, a_k, \ldots, a_n) \in \mathsf{N}^n \mapsto a_k \in \mathsf{N} \qquad (C.1.1)$$

is referred to as the $n-ary$ **projection function** *to the* $k-coordinate$; *and*

$$K_< : (a_1, a_2) \in \mathsf{N}^2 \mapsto \begin{cases} 0 & \text{if} \quad a_1 < a_2 \\ 1 & \text{if} \quad a_1 \ge a_2 \end{cases} \qquad (C.1.2)$$

is referred to as the **characterisitic function** *for the binary relation* $<$.

Remark C.1.1: In particular, the identity function is a starting function; indeed $P_{1,1}(a) = a \ \forall \ a \in \mathsf{N}$. Note also that a mathematical analyst would call $\chi_R = (1 - K_<)$ the characteristic function of the relation $<$; the convention adopted for instance in [Hodel, 1995], and followed here (see Def. C.1.1), seems indeed better adapted to the purposes of this appendix (see e.g. (C.1.5) below.)

Definition C.1.2. *Let* G *be a* $k-ary$ *function,* $\{H_i \mid 1 \le i \le k\}$ *be* $n-ary$ *functions. The* $n-ary$ *function*

$$F : (a_1, \ldots, a_n) \in \mathsf{N}^n \mapsto G(H_1(a_1, \ldots, a_n), \ldots, H_k(a_1, \ldots, a_n)) \qquad (C.1.3)$$

is called the **composition** *of* G *with the* H_i .

Definition C.1.3. *A* $(n+1)-ary$ *function* G *is said to be* **regular** *if for all* $(a_1, \ldots, a_n) \in \mathsf{N}^n$ *there exists at least one* $x \in \mathsf{N}$ *such that* $G(a_1, \ldots, a_n, x) = 0$. *For every regular* G *we then define the* **minimization** *of* G *as the* $n-ary$ *function:*

$$F : (a_1, \ldots, a_n) \in \mathsf{N}^n \mapsto \mu x[G(a_1, \ldots, a_n, x) = 0] \qquad (C.1.4)$$

$F(a_1, a_2, \ldots, a_n)$ *is the smallest natural number such that* $G(a_1, a_2, \ldots, a_n, x)$ $= 0$.

Remark C.1.2: For instance, the $1-$ary *successor function* $S : a \in \mathsf{N} \mapsto a+1 \in$ N, and the $n-$ary *constant functions* $C_{n,k} : (a_1, \ldots, a_n) \in \mathsf{N}^n \mapsto k \in \mathsf{N}$ can be constructed as follows:

$$S(a) = \mu x[K_<(a, x) = 0] ; \tag{C.1.5}$$

$$\left. \begin{array}{rcl} C_{n,0}(a_1, \ldots, a_n) &=& \mu x[P_{n+1,n+1}(a_1, \ldots, a_n, x) = 0] \\ C_{n,k+1}(a_1, \ldots, a_n) &=& S\left(C_{n,k}(a_1, \ldots, a_n)\right) \end{array} \right\} . \tag{C.1.6}$$

Note that in (C.1.6) we used induction over k. Note also that the operation described in Def. C.1.3 (and refined in Def. C.1.8 below) is known under a plethora of names in the literature; first of all, the word *minimization* that we just used is also spelled *minimalization*; moreover, some authors choose to refer to it as the $\mu-$*operator*, the $\mu-$*operation*, or the Min$-$*operation*; we also should mention, especially in connection with Rems. C.1.3–9 below, that this is also referred to as the *search-operation*.

Remark C.1.3: A warning must be sounded here. The projection function $P_{n+1,n+1} : (a_1, \ldots, a_n, x) \in \mathsf{N}^{(n+1)} \mapsto x \in \mathsf{N}$ is clearly regular, so that the minimization involved in (C.1.6) was indeed properly defined. Still, there are quite innocent-looking functions which are not regular, such as the binary function of addition: $F : (a, x) \in \mathsf{N}^2 \mapsto a + x \in \mathsf{N}$. While it is easy to determine that this particular function is not regular, it is not always possible to recognize in finitely many steps that a given function is not regular. This is a first indication that we will not be able to index effectively the functions that satisfy the following definition.

Definition C.1.4. *A $n-$ary function F is said to be* recursive *if there is a finite sequence of functions* $\{F_1, \ldots, F_n\}$, *ending with $F_n = F$, and such that for every $1 \leq k \leq n$ one of the following conditions hold:*

- *F_k is a starting function (in the sense of Def. C.1.1)*
- *$k > 1$, and F_k is obtained from $\{F_1, \ldots, F_{k-1}\}$:*
   either by composition
   or   by minimization of a regular function.

Remark C.1.4: To show that a $n-$ary function is recursive it is therefore sufficient to show that it satisfies one of the following three conditions:

1. F is a starting function
2. F is the composition of recursive functions
3. F is the minimization of a regular recursive function.

In particular the successor function and the constant functions of Rem. C.1.2 are recursive functions. Hence polynomials $a \mapsto P(a) = \sum_{k=1} c_k a^k$ and exponentials $a \mapsto x^a$ over N are recursive. Many more recursive functions obtain from the following procedure which is partially responsible for the name "recursive" attached to the class of functions we are studying.

Scholium C.1.1. *Let G be a $n-ary$ recursive function, and let H be a $(n+2)-ary$ recursive function. Then the $(n+1)$-ary function F defined by*

$$
\begin{aligned}
F(a_1,\dots,a_n,0) &= G(a_1,\dots,a_n) \\
F(a_1,\dots,a_n,b+1) &= H(a_1,\dots,a_n,b,F(a_1,\dots,a_n,b))
\end{aligned}
\tag{C.1.7}
$$

is recursive. Any function F constructed in this manner is said to have been obtained from H by primitive recursion *starting with G.*

Remark C.1.5: The particular case where H is binary, and G is replaced by a constant, say k in N, the scholium reads in the somewhat more familiar form

$$
\begin{aligned}
F(0) &= k \\
F(b+1) &= H(b,F(b)) \quad.
\end{aligned}
\tag{C.1.8}
$$

In particular, with $k=1$ and $H(n,F(n)) = S(n) \times F(n)$ with S the successor function, this shows that the usual factorial function $F : n \in \mathsf{N} \mapsto n! \in \mathsf{N}$ is recursive.

Primitive recursion seems indeed to be such a powerful operation that one may think of substituting it to the operation of minimization in Def. C.1.4 above, thus avoiding to have to consider regular functions. This may even be accomplished with an axiomatically more economical class of starting functions. One obtains in this manner the first class of functions that were considered as candidates for computability (compare with Def. C.1.4).

Definition C.1.5. *A $n-ary$ function F is said to be* primitive recursive *if there is a finite sequence of functions $\{F_1,\dots,F_n\}$, ending with $F_n = F$, and such that for every $1 \le k \le n$ one of the following conditions hold:*

- *F_k is*
    the successor function S
    the zero function $C_{1,0}$
    a projection function $P_{n,m}$
- *$k > 1$ and F_k is obtained from $\{F_1,\dots,F_{k-1}\}$:*
    either by composition
    or   by primitive recursion.

Remark C.1.6: A reasoning by induction, using Rem. C.1.2 and Scholium C.1.1, shows that every primitive recursive function is recursive. It is also easy to verify that one has not obviously lost too much of the materials included in Def. C.1.4 ; in particular $+$, $\times$ and $K_<$ are primitive recursive. It is in fact quite hard to show that the two classes do *not* coincide, namely that there are recursive functions that are not primitive recursive. A famous example is the binary function known under the name of its discoverer, Ackermann [1928], who presented it as a counterexample against a conjecture by Hilbert to the effect that primitive recursiveness would be the mathematical formalization of the intuitive concept of constructibility. The Ackermann function is defined by

$$A(0, n) = n + 1$$
$$A(m + 1, 0) = A(m, 1)$$
$$A(m + 1, n + 1) = A(m, A(m + 1, n)) \quad . \tag{C.1.9}$$

The proof that this function is not primitive recursive is given in [Hermes, 1969]. The starting point of the proof is the remark that A grows very fast; indeed, an elementary computation gives in its first few steps: $A(1, n) = n + 2$, $A(2, n) = 2n + 3$ $A(3, n) = 2^{n+3} - 3$. In fact, A grows fast enough to provide an upper bound on the growth of primitive recursive functions: one can prove indeed that for every $n-$ary primitive recursive function G there exists a natural number $c \in \mathsf{N}$ such that $G(a_1, \ldots, a_n) < A(c, a_1 + \ldots + a_n)$ $\forall$ $(a_1, \ldots, a_n) \in \mathsf{N}^n$. If A itself were primitive recursive, so would be the function G_1 defined as the diagonal $G_1(a) = A(a, a)$; there would then exist c such that $G_1(x) < A(c, x)$ $\forall$ $x \in \mathsf{N}$ and thus in particular $G_1(c) < A(c, c) = G_1(c)$, a contradiction.

Hence, while the definition of *primitive recursion* presents the advantage of not involving any appeal to regular functions – compare Defs. C.1.4–5, taking into account Rem. C.1.3 – Ackermann's discovery in effect eliminates the class of primitive recursive functions from the candidates for a class encompassing all functions one would reasonably want to regard as computable.

Remark C.1.7: Moreover, we are now going to see a diagonalization argument (akin to the one used in the previous remark on Ackermann function) that allows one to confirm the intuitive thrust of Rem. C.1.3, namely that the class of recursive functions itself still cannot capture in a self-consistent manner the intuitive notion of computability: in particular, we would have liked to be able to exhibit a binary recursive function H that could be used to enumerate, i.e. to index effectively the $1-$ary recursive functions; by this tentative requirement of enumerability, we mean that for every $1-$ary recursive function F there would exist an $e \in \mathsf{N}$, specific to F, with

$$F(a) = H(e, a) \quad \forall \quad a \in \mathsf{N} \quad . \tag{C.1.10}$$

This unfortunately is not possible. The proof is instructive, as it hints at how to get out of the impass epitomized by this negative result.

Scholium C.1.2. *There is no binary recursive function that enumerates all $1-$ary recursive functions.*

Proof. Suppose to the contrary that a binary recursive function H could be adduced with the property that it enumerates, in the sense of (C.1.10), all the $1-$ary recursive functions. Consider then the $1-$ary function

$$F : a \in \mathsf{N} \mapsto H(e, a) + 1 \in \mathsf{N} \quad . \tag{C.1.11}$$

To assume that H is recursive would imply that F is recursive. To assume that H enumerates all $1-$ary recursive functions would then imply in particular that there would exist an e for which (C.1.10) would be satisfied for the function F just constructed. We would therefore have both: $F(e) = H(e, e)$ and $F(e) = H(e, e) + 1$, which is a contradiction.   **q.e.d.**

This suggests the need to enlarge the putative class of computable functions; accordingly one introduces the following definitions.

Definition C.1.6. *For $n \geq 1$ a* n-ary partial function *is a map*

$$F : (a_1, \ldots, a_n) \in \mathcal{D} \mapsto F(a_1, \ldots, a_n) \in \mathsf{N}$$

where $\mathcal{D} \subseteq \mathsf{N}^n$, is called the domain *of F.*

Remark C.1.8: If one wishes to emphasize that a function F has to be viewed as a partial function with $\mathcal{D} = \mathsf{N}^n$, one says that F is total.

The consideration of partial functions that are not total is the essential ingredient in by-passing the difficulties raised by Rem. C.1.3 and Scholium C.1.2. One therefore needs to generalize the notions of composition and of minimization from ordinary functions (see Defs. C.1.2–3) to operations with partial functions; this requires some adjustments which we now outline.

Definition C.1.7. *Let G be a $k-$ary partial function, $\{H_i \mid 1 \leq i \leq k\}$ be $n-$ary partial functions. Let $\mathcal{D}$ be the set of all $(a_1, \ldots, a_n) \in \mathsf{N}^n$ such that $H_i(a_1, \ldots, a_n)$ is defined for each $1 \leq i \leq k$, and such that G is defined for each of the $k-$tuples $(H_1(a_1, \ldots, a_n), \ldots, H_k(a_1, \ldots, a_n)) \in \mathsf{N}^k$. Then the $n-$ary partial function*

$$F : (a_1, \ldots, a_n) \in \mathcal{D} \mapsto G(H_1(a_1, \ldots, a_n), \ldots, H_k(a_1, \ldots, a_n)) \quad \text{(C.1.12)}$$

is called the composition *of G with the H_i.*

Definition C.1.8. *Let G be a $(n + 1)-$ary partial function with domain $\mathcal{D}_G \in \mathsf{N}^{n+1}$; and let $\mathcal{D}$ be the set of all $(a_1, \ldots, a_n) \in \mathsf{N}^n$ such that there exists a $x \in \mathsf{N}$ satisfying all three of the following properties*

1. $(a_1, \ldots, a_n, y) \in \mathcal{D}_G \quad \forall \ y \leq x$
2. $G(a_1, \ldots, a_n, x) = 0$
3. $G(a_1, \ldots, a_n, y) > 0 \quad \forall \ y < x$.

Then the minimization *of G is defined as the $n-$ary function:*

$$F : (a_1, \ldots, a_n) \in \mathcal{D} \mapsto \mu x[G(a_1, \ldots, a_n, x) = 0] \quad . \quad \text{(C.1.13)}$$

Remark C.1.9: Hence F is undefined at $(a_1, \ldots, a_n) \in \mathsf{N}^n$ whenever any of the following situations arises.

- $(a_1, \ldots, a_n, x) \in \mathcal{D}_G \quad \forall \quad x \in \mathsf{N}$ and $G(a_1, \ldots, a_n, x) > 0$;
- there exists $x \in \mathsf{N}$ such that $(a_1, \ldots, a_n, x) \notin \mathcal{D}_G$ and
  $G(a_1, \ldots, a_n, y) > 0 \quad \forall \ y < x$.

The exclusion of these situations is motivated by the fact that any systematic search for the smallest x with $G(a_1, \ldots, a_n, x) = 0$, if attempted in one of these situations, would never terminate.

We are now ready for the needed substitute to Def. C.1.4 .

Definition C.1.9. *A $n-$ary partial function F is said to be* partial recursive *if there is a finite sequence of partial functions $\{F_1, \ldots, F_n\}$ ending with $F_n = F$ and such that for every $1 \leq k \leq n$ one of the following conditions hold:*

- *(a) F_k is a starting function (in the sense of Def. C.1.1)*
- *(b) $k > 1$, and F_k is obtained from $\{F_1, \ldots, F_{k-1}\}$:*
  either by composition of partial functions
  or   by minimization of a partial function.

Remark C.1.10: The analog of Rem. C.1.4 holds for partial recursive functions. The class of partial recursive functions is strictly bigger than the class of recursive functions, namely there are partial recursive functions that are not total, and every total function that is recursive *a fortiori* is partial recursive; in fact a total function is recursive if and only if it is partial recursive. The generalization from recursive to partial recursive functions remedies the objection raised by Scholium C.1.2, as shown by the following result (known as a consequence of the so-called Kleene normal-form theorem, see e.g. [Hodel, 1995]).

Scholium C.1.3. *For every $n \in \mathsf{N}$, there exists a $(n+1)-$ary partial recursive function Φ_n that enumerates all $n-$ary partial recursive functions, i.e. for any $n-$ary partial recursive function F defined on any domain $\mathcal{D} \subseteq \mathsf{N}^n$, there exists an $e \in \mathsf{N}$, specific to F, such that*

$$F(a_1, \ldots, a_n) = \Phi_n(e, a_1, \ldots, a_n) \quad \lor \quad (a_1, \ldots, a_n) \in \mathcal{D} \quad . \qquad (C.1.14)$$

This $e \in \mathsf{N}$ is called the index *of the $n-$ary partial recursive function F.*

Remark C.1.11: Hence, in particular, there exists a binary partial recursive function Φ_1 such that, for every $1-$ary partial recursive function F on $\mathcal{D} \subseteq \mathsf{N}$, there exists an $e \in \mathsf{N}$ with

$$F(a) = \Phi_1(e, a) \quad \forall \quad a \in \mathcal{D} \quad . \qquad (C.1.15)$$

The contradiction encountered in the proof of Scholium C.1.2 is avoided here due to the weaker requirement that Φ_1 be only a partial function so that, for instance, Φ_1 is not required to be defined on its diagonal $\{(e, e) \mid e \in \mathsf{N}\}$.

Remark C.1.12: In fact, the binary partial function Φ_1 enjoys an even more *universal* property: it allows one to enumerate the set of *all* partial recursive functions. One can indeed show [Hodel, 1995] that, for all $n \in \mathsf{N}$ there exists a bijective recursive function $J_n : \mathsf{N}^n \to \mathsf{N}$ such that for every n−ary partial recursive function F on any $\mathcal{D} \subseteq \mathsf{N}^n$, there exists a $e \in \mathsf{N}$ such that:

$$F(a_1, \ldots, a_n) = \Phi_1(e, J_n(a_1, \ldots, a_n)) \quad \forall \quad (a_1, \ldots, a_n) \in \mathcal{D} \quad . \qquad \text{(C.1.16)}$$

For our purposes in Part II, we still need to make precise the allied notions needed to list effectively sets and relations. Here again there are three classes, which we briefly survey from the points of view afforded by the theory of recursive functions.

In the sequel, we found it convenient to reserve the name *set* to designate only the (not necessarily proper) subsets of N, while we use the name n−ary *relation* for the (not necessarily proper) subsets of N^n : sets are 1−ary relations. This notation simply codifies our identifying a n−ary relation with the collection of all n−tuples $(a_1, \ldots, a_n) \in \mathsf{N}^n$ that satisfy this relation. To denote that a particular n−tuple $(a_1, \ldots, a_n) \in \mathsf{N}^n$ satisfies a n−ary relation R we write $R(a_1, \ldots, a_n)$. In agreement with the convention taken in Def. C.1.1 (see also Rem. C.1.1), the characteristic function K_R of a n−ary relation R is the function defined by

$$K_R : (a_1, \ldots, a_n) \in \mathsf{N}^n \quad \mapsto \quad \begin{cases} 0 & \text{if} \quad R(a_1, \ldots, a_n) \\ 1 & \text{otherwise} \end{cases} \qquad \text{(C.1.17)}$$

Definition C.1.10. *A n−ary relation $R \subset \mathsf{N}^n$ is said to be* recursive *(resp.* primitive recursive*) whenever its characteristic function is a n−ary recursive function (resp. a n−ary primitive recursive function.)*

Definition C.1.11. *A n−ary relation $R \subset \mathsf{N}^n$ is said to be* recursively enumerable *whenever there exists a $(n+1)$−ary recursive relation Q such that:*

$$R(a_1, \ldots, a_n) \quad \text{if and only if} \quad \exists \quad x \in \mathsf{N} \quad \text{with} \quad Q(a_1, \ldots, a_n, x) \quad .$$
$$\text{(C.1.18)}$$

Remark C.1.13: Every recursive relation is recursively enumerable, but the converse is not true: there exist recursively enumerable relations that are not recursive (see also Rem. C.1.16 below). Note that R recursive implies $\neg R$ recursive; and thus that both R *and* $\neg R$ are recursively enumerable; the converse of *that* joint implication is also true; hence a relation R is recursive if and only if both R and $\neg R$ are recursively enumerable relations.

Scholium C.1.4. *A non-empty set $S \subseteq \mathsf{N}$ is recursively enumerable if and only if it is the range of a 1−ary recursive function $F : \mathsf{N} \to \mathsf{N}$.*

Scholium C.1.5. *A n−ary relation $R \subseteq \mathsf{N}^n$ is recursively enumerable if and only if it is the domain of a n−ary partial recursive function F.*

Remark C.1.14: The following definition allows us to state a sort of converse to the above scholium.

Definition C.1.12. *The graph $\mathcal{G}_F$ of a $n-$ary partial function F with domain $\mathcal{D} \subseteq \mathsf{N}^n$ is the $(n+1)-$ary relation :*

$$\mathcal{G}_F = \{(a_1, \ldots, a_n, F(a_1, \ldots, a_n)) \in \mathsf{N}^{n+1} \mid (a_1, \ldots, a_n) \in \mathcal{D}\} \quad . \qquad \text{(C.1.19)}$$

Scholium C.1.6. *A $n-$ary partial function F is partial recursive if and only if its graph $\mathcal{G}_F$ is a $(n+1)-$ary recursively enumerable relation.*

Remark C.1.15: We still have to make more explicit the sense in which recursively enumerable relations are indeed *enumerable*, thus formalizing the intuitive requirement that one could effectively list them, or index them (compare with Scholium C.1.3, and Rems. C.1.11–12); indeed, one can prove the existence, for each $n \geq 1$, of a $(n+1)-$ary recursively enumerable relation S_n which *enumerates* all $n-$ary recursively enumerable relations; this means precisely that, at fixed $n \in \mathsf{N}$, there exists for each $n-$ary recursively enumerable relation R, an $e \in \mathsf{N}$ specific to R, such that

$$R(a_1, \ldots, a_n) \quad \text{if and only if} \quad S_n(e, a_1, \ldots, a_n) \quad . \qquad \text{(C.1.20)}$$

e is then called the *index* of the $n-$ary relation R.

Note that (C.1.20) is a much stronger statement than the condition (C.1.18) that enters in Def. C.1.11 itself: the present statement asserts that the function S_n is *the same for all* $n-$ary recursively enumerable relations R. In fact some even stronger statement holds true, namely that all the enumerating relations S_n in (C.1..20) can be expressed in terms of the lowest one of them, the single binary recursively enumerable relation S_1 :

$$S_n(a, a_1, \ldots, a_n) \quad \text{iff} \quad S_1(a, J_n(a_1, \ldots, a_n)) \qquad \text{(C.1.21)}$$

where J_n are the bijective recursive functions $J_n : \mathsf{N}^n \to \mathsf{N}$ we already met in Rem. C.1.12. It is quite remarkable that beyond the mere existence of a binary recursively enumerable relation, S_1, that produces the enumerating condition (C.1.20) through (C.1.21), one can effectively describe such a function in a straightforward manner (this is the relation $SOLN$ introduced in the last section of [Hodel, 1995]) in terms of solutions of equations between recursively specified polynomials with coefficients in N. In this sense, the elements of the collection of recursively enumerable relations are each uniquely specified by a pair $(n, e) \in \mathsf{N} \times \mathsf{N}$ where n tells us that R is a $n-$ary relation, and e gives us its index as defined in (C.1.20): the collection of recursively enumerable relations is indeed enumerable in the precise sense just defined.

Remark C.1.16: We may still add that the binary recursively enumerable relations S_1 of the previous remark also allows to produce (in accordance with Rem. C.1.13) a relation that is recursively enumerable without being recursive. Consider indeed the relation

$$R(a) \quad \text{iff} \quad S_1(a, a) \tag{C.1.22}$$

which is recursively enumerable. If R were recursive, we would have $\neg R$ recursively enumerable, and this relation would be enumerated by S_1, i.e. $\neg R$ would have an index, say $\varepsilon \in \mathbb{N}$:

$$\neg R(a) \quad \text{iff} \quad S_1(\varepsilon, a) \tag{C.1.23}$$

and thus:

$$\neg R(\varepsilon) \quad \text{iff} \quad S_1(\varepsilon, \varepsilon) \quad \text{iff} \quad R(\varepsilon) \quad , \tag{C.1.24}$$

a contradiction, thus proving indeed that there exists a recursively enumerable relation that is not recursive.

In summary, we have garnered in this Appendix some arguments to support a claim to the effect that self-consistent mathematical formalizations have been reached, capturing the essentials from the intuitive notions of computable functions and sets: the *partial recursive functions* (Def. C.1.9), and the *recursively enumerable relations* (Def. C.1.11); see nevertheless [Soare, 1996] who recommends that some distinction be enforced in the usage of these terms.

In addition to the *internal* evidence presented in our sketch of the theory, some *external* evidences could also have been advanced as they were in fact part of the historical development. Indeed, *several other approaches* to the notion of computability of functions have been devised; e.g. Church's λ−calculus, or the reductions based on Turing machines (T), and on register machines (RM) (i.e. machines mimicking quite faithfully ordinary computers, except that they have unlimited memory). *It is a most remarkable fact* that all these approaches ultimately led to requirements equivalent to those of Def. C.1.9 and Def. C.1.11, thus lending credence to the so-called *Church thesis* according to which the class of all effectively computable functions coincides with the class of all partial recursive functions; note that the latter is called a "thesis", not a "theorem": it rather expresses the apparently unescapable feeling that the mathematical formalizations (partial recursive functions and recursively enumerable relations) have fully captured the intuitive notions (effectively computable functions and effectively indexable sets) that were to be formalized; it is certainly not in vain that the name so aptly recognizes the seminal insight provided by one of the prime movers of the theory.

D. Appendix: Topological Essences

We collect in this Appendix the elements of topology needed in the main body of our text. Among the many textbooks that could be recommended for complements and further study at an introductory level, we mention [Kelly, 1955] and [Fairchild and Tulcea, 1971]; both are written in a style that is immediately accessible. For a presentation geared to the needs of functional analysis, see [Schwartz, 1978a] or [Reed and Simon, 1972–1980]. A Reader interested in the origins of the subject may want to consult some of the original texts, such as: [Fréchet, 1906, Kuratowski, 1922, Hausdorff, 1927, Fréchet, 1928, Banach, 1932, Kuratowski, 1933, Sierpinsky, 1934, Alexandroff and Hopf, 1935].

D.1 Basic Definitions

Several equivalent definitions of our subject have been proposed. One of the most concise may be the one proposed in [Kuratowski, 1922]:

Definition D.1.1. *A topological space is a set X equipped with a closure operation $A \to \overline{A}$ which assigns to each subset $A \subseteq X$ a subset $\overline{A} \subseteq X$ subject to the following four axioms:*

$$(1)\ \overline{\emptyset} = \emptyset\ ; \quad (2)\ A \subseteq \overline{A}\ ; \quad (3)\ \overline{\overline{A}} = \overline{A}\ ; \quad (4)\ \overline{(A \cup B)} = \overline{A} \cup \overline{B}\ .$$

$\overline{A}$ *is then called the* closure *of* A.

Example D.1.1: The "coarse topology" $\mathcal{T}_1$ on an arbitrary set X is defined by:

$$\forall\, A \subseteq X\ :\ \overline{A} = \begin{cases} \emptyset & \text{if}\ \ A = \emptyset \\ X & \text{otherwise} \end{cases}.$$

Example D.1.2: The "discrete topology" $\mathcal{T}_2$ on an arbitrary set X is defined by:

$$\forall\, A \subseteq X\ :\ \overline{A} = A\ \ .$$

Example D.1.3: The topology $\mathcal{T}_3$ on an arbitrary set X is defined by:

$$\forall\, A \subseteq X \; : \; \overline{A} = \begin{cases} A & \text{if} \;\; 0 \le \operatorname{card}(A) < \infty \\ X & \text{otherwise} \end{cases}.$$

In the above three Examples, when X is chosen to be the same set, we have: if $\operatorname{card} X > 1$, then $(X, \mathcal{T}_1)$ is different from $(X, \mathcal{T}_2)$ and $(X, \mathcal{T}_3)$; and if $\operatorname{card} X = \infty$, then $(X, \mathcal{T}_2)$ is different from $(X, \mathcal{T}_3)$.

The next definition lists a few of the objects derived from Def. D.1.1:

Definition D.1.2. *A subset $F \subseteq X$ is said to be* closed *whenever $F = \overline{F}$. A subset $U \subseteq X$ is said to be* open *whenever its complement $U^c = (X - U)$ is closed. A subset $V \subseteq X$ is said to be a* neighborhood *of a point $x \in X$ if there exists an open set U such that $x \in U \subseteq V$.*

Following [Hausdorff, 1927], we review some of the alternate definitions of a topological space. These definitions are equivalent to the original one in the sense that each of these definitions chooses one of the derived notions introduced in Def. D.1.2 and gives it primacy by axiomatizing it; this axiomatization is devised so that: (i) it captures the properties of the chosen notion when it appears as a derived notion in the other alternate definitions; and (ii) the axioms are sufficient to derive all the properties of the notions appearing as primary in the other alternate definitions.

The first of these alternate definitions is to give primacy to the closed sets.

Definition D.1.3. *A* topological space *is a set X in which one family $\mathcal{C}$ of subsets $F \subseteq X$ has been chosen, satisfying the following axioms:*

1. *$\emptyset \in \mathcal{C}$ and $X \in \mathcal{C}$;*
2. *$F_1, F_2 \in \mathcal{C}$ entails $F_1 \cup F_2 \in \mathcal{C}$;*
3. *for any collection $\{F_\lambda \in \mathcal{C} \mid \lambda \in \Lambda\} \; : \; \bigcap_{\lambda \in \Lambda} F_\lambda \in \mathcal{C}$.*

Definition D.1.4. *With $\mathcal{C}$ as in the above definition, the subsets $F \subseteq X$ which belong to $\mathcal{C}$ are said to be* closed. *The closure $\overline{A}$ of an arbitrary subset $A \subseteq X$ now enters as the smallest closed subset containing A; finally the* open *sets and the* neighborhoods *are defined as in Def. D.1.2.*

The complementarity beween the closed and the open sets suggests the following.

Definition D.1.5. *A* topological space *is a set X in which one family $\mathcal{O}$ of subsets $U \subseteq X$ has been chosen, satisfying the following axioms:*

1. *$\emptyset \in \mathcal{O}$ and $X \in \mathcal{O}$;*
2. *$U_1, U_2 \in \mathcal{O}$ entails $U_1 \cap U_2 \in \mathcal{O}$;*
3. *for any collection $\{U_\lambda \in \mathcal{O} \mid \lambda \in \Lambda\} \; : \; \bigcup_{\lambda \in \Lambda} U_\lambda \in \mathcal{O}$.*

Definition D.1.6. *With $\mathcal{O}$ as in the above definition, the subsets $U \subseteq X$ which belong to $\mathcal{O}$ are called* open. *The* closed *subsets $F \subseteq X$ are then introduced as the subsets, the complements of which are open. The* closure $\overline{A}$ *of an arbitrary subset $A \subseteq X$, and the* neighborhoods *of a point are again defined as in Def. D.1.4.*

Yet another alternate definition axiomatizes the neighborhoods [Hausdorff, 1927] (already in 1914 ed.):

Definition D.1.7. *A* topological space *is a set X equipped with a function $\mathcal{N}$ that assigns to each $x \in X$ a family $\mathcal{N}(x)$ of subsets of X subject to the following axioms:*

1. *for each $V \in \mathcal{N}(x)$: $x \in V$;*
2. *for every pair (V_1, V_2) of elements of $\mathcal{N}(x)$: $V_1 \cap V_2 \in \mathcal{N}(x)$;*
3. *for every subset $V \in \mathcal{N}(x)$ and any V' with $V \subseteq V'$: $V' \in N(x)$;*
4. *for each $V \in \mathcal{N}(x)$ there exists $W \in \mathcal{N}(x)$ such that : $y \in W \models V \in \mathcal{N}(y)$.*

As for the earlier alternate definitions, this one is completed by describing how the other notions derive from it.

Definition D.1.8. *Given an assignment $\mathcal{N}$ as above, the elements V of $\mathcal{N}(x)$ are called the* neighborhoods *of x. Now, a subset $U \subseteq X$ is said to be* open *whenever $x \in U$ entails that there exists $V \in \mathcal{N}(x)$ with $V \subseteq U$. The* closed *sets and the* closure *operation are then defined as in Def. D.1.6.*

To the three Examples, given immediately after Def. D.1.1, let us add one more, which illustrates the convenience of starting with Def. D.1.7.

Example D.1.4: Let $X = \mathbb{R}$; and $|x|$ denote the absolute value of x. For each $x \in X$, and each $\varepsilon > 0$, let

$$V_\varepsilon(x) = \{y \in X \mid |y - x| < \varepsilon\}; \qquad (D.1.1)$$

then the assignment

$$\mathcal{N}(x) = \{V_\varepsilon(x) \mid \varepsilon > 0\} \qquad (D.1.2)$$

satisfies the four axioms of Def. D.1.7; for reasons to be specified in Sect. D.2, this is called the *natural metric topology* – or simply the *natural topology* – on $\mathbb{R}$.

The Examples provided so far can be used to illustrate the following facts:

1. while any arbitrary union of open sets is open, it is *not true* that the intersection of an infinite family of open sets is necessarily open [even if the family has countably many members]; finite intersections of open sets, nevertheless, are always open;
2. a similar situation holds for closed sets, with the roles of intersection and union interchanged;

3. while an open set is clearly a neighborhood of any of its points, the converse is *not true;* namely neighborhoods are not necessarily open;
4. there may exist sets that are neither open nor closed;
5. some sets can be both open and closed; in particular, whatever the topology, both of the two subsets $\emptyset$ and X are open as well as closed.

The last of the above remarks suggests to single out the following definition.

Definition D.1.9. *The topological space $(X, \mathcal{T})$ is said to be* connected *whenever $\emptyset$ and X are the only subsets of X that are both open and closed.*

This nomenclature comes from the following easily checked facts.

Theorem D.1.1. *Let $(X, \mathcal{T})$ be a toplogical space; then the following conditions are equivalent:*

1. *$(X, \mathcal{T})$ is connected;*
2. *$X = U \cup V$, with U and V non-empty open subsets of X, entails $U \cap V \neq \emptyset$;*
3. *$X = E \cup F$, with E and F non-empty closed subsets of X, entails $E \cap F \neq \emptyset$.*

Note that R, when equipped with its metric topology, is connected.

There are still several other ways to define a topology on a set. One approach is to axiomatize the notion of convergence [Fréchet, 1906]. The following definitions should give an idea on the properties that need to be distinguished.

Definition D.1.10. *Let $(X_1, \mathcal{T}_1)$ and $(X_2, \mathcal{T}_2)$ be two topological spaces; $D \subseteq X_1$; $y_o \in X_2$ and let $f : x \in D \mapsto f(x) \in X_2$. We say that y_o is the* limit *of $f(x)$ as x approaches x_o if, for every neigborhood V_{y_o} of y_o in X_2, there is a neighborhood V_{x_o} of x_o in X_1 such that $x \in V_{x_o} \cap D \models f(x) \in V_{y_o}$; we then write $\lim_{x \to x_o} f(x) = y_o$. If, in addition, $x_o \in D$ and $y_o = f(x_o)$, we say that f is* continuous *at $x_o \in X_1$. A function is said to be* continuous *on $D_o \subseteq D$ whenever f is continuous for all $x_o \in D_o$.*

Notes:

1. In case $X = Y = \mathsf{R}$ with its *metric* topology, the above definition embodies the $(\varepsilon, \delta)-$definition we all met in our freshman calculus class: f is continuous at x_o, if for every $\varepsilon > 0$ there exists a $\delta_\varepsilon > 0$ such that $|x - x_o| < \delta_\varepsilon \models |f(x) - f(x_o)| < \varepsilon$ with δ_ε depending in general on x_o; when the same δ_ε can be found so that the above holds for all $x_o \in D$ simultaneously, we say that f is *uniformly* continuous.
2. A sequence $\{u_n \in X_2 \mid n \in \mathsf{Z}^+\}$ can be viewed as a particular case of the above, i.e. as a function $f : \frac{1}{n} \in D \mapsto u_n \in X_2$ with $D \subset X_1 \equiv \{0\} \cup D \equiv \{0\} \cup \{\frac{1}{n} \mid n \in \mathsf{Z}^+\}$ and X_1 equipped with the metric topology it inherits from R; and $x_o = 0$. Then the existence and value of the limit $u_o = \lim_{n \to \infty} u_n$ are defined through the definition of $u_o = \lim_{\frac{1}{n} \to 0} f(\frac{1}{n})$. See also Def. D.2.3 below.

Theorem D.1.2. *Let* $(X_1, \mathcal{T}_1)$, $(X_2, \mathcal{T}_2)$ *and* $f : X_1 \to X_2$ *be as in Def. D.1.10 (with* $D = X_1$ *) ; and for any subset* $B \subseteq X_2$ *, let* $f^{-1}(B) = \{x \in X_1 \mid f(x) \in B\}$ *. Then the following four conditions are equivalent:*

1. f *is continuous on* X_1 ;
2. $\forall\, A \subseteq X_1 :\quad f(\overline{A}\,) \subseteq \overline{f(A)}$;
3. $\forall\, F \subseteq X_2$ *with* F *closed* : $\quad f^{-1}(F) \subseteq X_1$ *is closed* ;
4. $\forall\, O \subseteq X_2$ *with* O *open* : $\quad f^{-1}(O) \subseteq X_1$ *is open* .

Warning: Conditions (3) and (4) above must be distinguished from the following two conditions:

5. $\forall\, F \subseteq X_1$ *with* F *closed* : $\quad f(F) \subseteq X_2$ *is closed* ;
6. $\forall\, O \subseteq X_1$ *with* O *open* : $\quad f(O) \subseteq X_2$ *is open* .

When condition 5 (resp. 6) is satisfied, f is said to be *closed* (resp. *open*).

Example D.1.5: Let $X_1 = \mathsf{R}$ equipped with $\mathcal{T}_1$, the discrete topology defined in Example D.1.2, and $X_2 = \mathsf{R}$ equipped with $\mathcal{T}_2$, the metric topology defined in Example D.1.4; and let $f : x \in X_1 \mapsto x \in X_2$. Then f is continuous. Note that f is also a bijection.

Example D.1.6: Let $(X_1, \mathcal{T}_1)$, $(X_2, \mathcal{T}_2)$ and f be as in Example D.1.5. Despite the fact that f is a continuous bijection, f^{-1} is *not* continuous; and f is neither open nor closed.

These two Examples, when considered in conjunction, motivate the following two definitions and the accompanying theorems.

Definition D.1.11. *Let* $(X_1, \mathcal{T}_1)$, $(X_2, \mathcal{T}_2)$, $f : X_1 \to X_2$ *and* f^{-1} *be as in Def. D.1.10.* f *is said to be a* homeomorphism *whenever all of the following three conditions are satisfied:*

1. f *is bijective between* X_1 *and* X_2 ;
2. f *is continuous on* X_1 ;
3. f^{-1} *is continuous on* X_2 .

Two topological spaces are said to be homeomorphic *if there exists a homeomorphism from one onto the other.*

The relation of being homeomorphic is transitive and symmetric; the importance of the second of these attributes is emphasized by the counterexample provided by the two topological spaces of Examples D.1.5–6: they are not homeomorphic, despite the fact that there exists a bijective continuous map from the first to the second.

Theorem D.1.3. *Let* $(X_1, \mathcal{T}_1)$, $(X_2, \mathcal{T}_2)$ *, be two topological spaces, and* $f : X_1 \to X_2$ *be a bijection; then the following conditions are equivalent:*

1. f *is a homeomorphism;*
2. f *is continuous and open;*
3. f *is continuous and closed.*

Theorem D.1.4. *Let $\mathcal{T}_1$ and $\mathcal{T}_2$ be two topologies defined on the same set X, and suppose that the identity map from $(X, \mathcal{T}_1)$ to $(X, \mathcal{T}_2)$ is continuous. Then the following condtions are equivalent:*

1. *$\forall\, A \subseteq X \;:\quad \overline{A}^1 \subseteq \overline{A}^2$;*
2. *with $\mathcal{C}_k$ denoting the collection of closed sets w.r.t. $\mathcal{T}_k \;:\quad \mathcal{C}_1 \supseteq \mathcal{C}_2$;*
3. *with $\mathcal{O}_k$ denoting the collection of open sets w.r.t. $\mathcal{T}_k \;:\quad \mathcal{O}_1 \supseteq \mathcal{O}_2$;*
4. *with $\mathcal{N}_k(x)$ denoting the collection of neighborhoods of x w.r.t. $\mathcal{T}_k$, $\forall\, x \in X \;:\quad \mathcal{N}_1(x) \supseteq \mathcal{N}_2(x)$.*

Definition D.1.12. *Whenever any – and thus all – of the above four conditions are satisfied, we say that the topology $\mathcal{T}_1$ is finer than the topology $\mathcal{T}_2$; we also say that $\mathcal{T}_1$ is stronger than $\mathcal{T}_2$, or that $\mathcal{T}_2$ is coarser (or weaker) than $\mathcal{T}_1$; we write then $\mathcal{T}_1 \geq \mathcal{T}_2$.*

Note that the discrete topology – see Example D.1.2 – is the finest of all topologies one can place on a set X and that the coarse topology – see Example D.1.1 – is the coarsest. In particular, with $X = \mathsf{R}$ and $\mathcal{T}_k$ as in Examples (1.k) with $k = 1, 2, 3, 4 : \mathcal{T}_1 \leq \mathcal{T}_3 \leq \mathcal{T}_4 \leq \mathcal{T}_2$.

Back to the general case, the partial ordering $\mathcal{T} \leq \mathcal{T}'$ equips the set of all topologies on X with the structure of a complete lattice; if $\{\mathcal{T}_\lambda \mid \lambda \in \Lambda\}$ is any collection of topologies on X, then among the topologies $\mathcal{T}$ on X finer than all $\mathcal{T}_\lambda$, there is a coarsest one, denoted $\mathcal{T}_\Lambda = \sup_{\lambda \in \Lambda} \mathcal{T}_\lambda$.

D.2 Examples from Functional Analysis

Example D.1.4 suggests several generalizations. The most immediate is:

Example D.2.1: For any finite, fixed, positive integer n, replace $X = \mathsf{R}$ in Example D.1.4 by $X = \mathsf{R}^n$; and (D.1.1) by

$$V_\varepsilon(x) = \{y \in X \mid \|y - x\| < \varepsilon\} \quad \text{where} \quad \|y - x\| = \sqrt{\sum_{k=1}^{n} (y_k - x_k)^2} \quad \text{(D.2.1)}$$

which reduces indeed to (D.1.1) in case $n = 1$. It is easy to check that this assignment satisfies the axioms of Def. D.1.7 and thus endows R^n with the structure of a topological space. R^n is in fact the $n-$dimensional archetype of a special class of topological spaces, singled out in the following definition.

Definition D.2.1. *A real normed space is a linear (or vector) space X on R, equipped with a norm, i.e. with a map*

$$\| \cdot \| : x \in X \mapsto \|x\| \in \mathsf{R}^+$$

satisfying the following axioms:

1. $\forall \; x \in X \quad with \quad x \neq 0 : \quad \|x\| > 0 \quad ;$
2. $\forall \; (c, x) \in \mathsf{R} \times X \; : \quad \|cx\| = |c| \, \|x\| \quad ;$
3. $\forall \; (x, y) \in X \times X \; : \quad \|x + y\| \leq \|x\| + \|y\| \quad .$

Note that (1) and (2) together imply that $\|x\| = 0$ if and only if $x = 0$. Complex normed spaces are defined similarly by replacing R by C. In neither case do we require that a normed space be finite-dimensional. We give examples of infinite-dimensional normed spaces below, but first, we want to note that, in fact, even the linear space structure is not essential to the definition of a topology that expresses how close points are from one another. Indeed, Def. D.2.1 suggests the following generalization:

Definition D.2.2. *A metric space* (X, d) *is a set* X *equipped with a distance, i.e. with a map*

$$d : (x, y) \in X \times X \mapsto d(x, y) \in \mathsf{R}^+$$

satisfying the following axioms:

1. $\forall \; (x, y) \in X \times X \; : \quad d(y, x) = d(x, y) \quad ;$
2. $\forall \; (x, y) \in X \times X : \quad d(x, y) = 0 \quad$ *if and only if* $\quad x = y;$
3. $\forall \; (x, y, z) \in X \times X \times X \; : \quad d(x, z) \leq d(x, y) + d(y, z) \quad .$

Clearly every normed space becomes a metric space under the trivial identification

$$d(x, y) = \|y - x\| . \tag{D.2.2}$$

This is referred to as the *distance canonically associated to the norm* of the normed space (X, d).

In particular, note that, upon replacing (D.1.1) by

$$V_\varepsilon(x) = \{y \in X \mid d(x, y) < \varepsilon\} \tag{D.2.3}$$

we equip any metric space (X, d) with the structure of a topological space; unless explicit mention is made to the contrary, this topological structure is assumed whenever we refer to a metric space.

Definition D.2.3. *Let* (X, d) *be a metric space; a sequence* $\{x_n \in X \mid n \in \mathsf{Z}^+\}$ *is said to be a* Cauchy *sequence in* (X, d) *whenever it satisfies the following condition : for every* $\varepsilon > 0$ *there exists some* $N_\varepsilon \in \mathsf{Z}^+$ *such that*

$$\forall \; m, n > N_\varepsilon \; : \quad d(x_n, x_m) < \varepsilon .$$

Given a metric space (X, d), *a Cauchy sequence in* X *is said to converge in* X *if there exists* $x_o \in X$ *satisfying the following condition: for every* $\varepsilon > 0$ *there exists some* $N_\varepsilon \in \mathsf{Z}^+$ *such that*

$$\forall \; n > N_\varepsilon \; : \quad d(x_n, x_o) < \varepsilon .$$

A metric space (X, d) *is said to be* complete *provided every Cauchy sequence in* X *converges.*

Note that, for each finite n, R^n equipped with the distance (D.2.2), is a complete metric space. Upon restricting this natural metric on R^2 to the unit disk $\{x \in \mathsf{R}^2 \mid \|x\| \leq 1\}$, we obtain a complete metric space that is not a normed space. To obtain an example of a metric space that is not complete, it suffices to puncture this unit disk at the origin.

Completeness becomes an issue that requires to be checked with care when considering infinite-dimensional linear spaces, the elements of which are functions.

Example D.2.2: Let $(X, \mathcal{T})$ be a topological space, and let $\mathcal{C}_\mathsf{R}(X)$ be the set of all bounded continuous functions from X to R. With $(f+g)(x) \equiv f(x)+g(x)$ and $(cf)(x) \equiv cf(x)$, $\mathcal{C}_\mathsf{R}(X)$ becomes an infinite-dimensional linear space. To say that f is bounded means that

$$\|f\| \equiv \sup_{x \in X} |f(x)|$$

is finite. It is easy to verify that $\|.\| : f \in \mathcal{C}_\mathsf{R}(X) \mapsto \|f\| \in \mathsf{R}$ is a norm. It is also true that the distance associated to this norm equips $\mathcal{C}_\mathsf{R}(X)$ with the structure of a *complete* metric space. To verify the latter, one first proves: R being complete and $\{f_n\}$ Cauchy $\models \exists f : x \in X \to \mathsf{R}$ such that $\forall \varepsilon > 0$ $\exists N_\varepsilon$ with: $x \in X$ and $n > N_\varepsilon$ $\models |f(x) - f_n(x)| < \varepsilon$. A 3ε-argument, involving the continuity of f_n, establishes finally that f is continuous and bounded; and that the Cauchy sequence $\{f_n\}$ converges to f.

This Example motivates the following definition.

Definition D.2.4. *A* Banach space *is a normed (real or complex) linear space that is complete with respect to the distance canonically associated to its norm.*

Example D.2.3: Let $(\Omega, \mathcal{F}, \mu)$ be a measure space; we do *not* require that the measure be finite, i.e. that $\mu(\Omega) < \infty$, so that R with its Lebesgue measure is covered as a particular case. Then the $\mathcal{L}^p$ spaces defined below are Banach spaces:

for $1 \leq p < \infty$:

$$\mathcal{L}^p(\Omega, \mathcal{F}, \mu) \doteq \{f : \Omega \to \mathsf{R} \mid \|f\|_p \equiv \left[\int_\Omega d\mu(x) |f(x)|^p\right]^{\frac{1}{p}} < \infty\} \left.\right\} \quad \text{(D.2.4)}$$

for $p = \infty$:

$$\mathcal{L}^\infty(\Omega, \mathcal{F}, \mu) \doteq \{f : \Omega \to \mathsf{R} \mid \|f\|_\infty \equiv \text{ess sup}_{x \in \Omega} |f(x)| < \infty\} \left.\right\} \quad \text{(D.2.5)}$$

In each of these spaces, for $\|\cdots\|$ to be a norm, we must have $\|f\| = 0$ if and only if $f = 0$, and thus we identify in each of these spaces, any two functions that differ on at most a set of measure zero. Hence, the elements of these spaces are in fact equivalence classes of functions; the notation $\doteq$ in (D.2.4–5) is meant to draw attention to this fact. (D.2.4) subsumes the theory of Lebesgue integration; (D.2.5) needs a complementary explanation. Given any measure space $(\Omega, \mathcal{F}, \mu)$, let $\mathcal{N} \equiv \{N \in \mathcal{F} \mid \mu(N) = 0\}$. A function f, defined almost everywhere on $(\Omega, \mathcal{F}, \mu)$, is said to be *essentially bounded* if there exists $N \in \mathcal{N}$ such that the restriction of f to $\Omega - N$ is bounded. The *essential least upper bound* of $|f|$ is written:

$$\text{ess sup}_{x \in \Omega} |f(x)| \equiv \inf_{N \in \mathcal{N}} \sup_{x \in \Omega - N} |f(x)| \quad .$$

Definition D.2.5. *The dual X^* of a complex Banach space X is the Banach space of all bounded linear functionals $\varphi : f \in X \mapsto \langle \phi; f \rangle \in \mathsf{C}$ equipped with the following operations of addition, multiplication by scalars, and norm:*

1. *$\phi_1 + \phi_2 : f \in X \mapsto \langle \phi_1 + \phi_2; f \rangle \in \mathsf{C}$ with $\langle \phi_1 + \phi_2; f \rangle = \langle \phi_1; f \rangle + \langle \phi_2; f \rangle$;*
2. *$c\phi : f \in X \mapsto \langle c\phi; f \rangle \in \mathsf{C}$ with $\langle c\phi; f \rangle = c\langle \phi; f \rangle$;*
3. *$\|\phi\| = \sup_{f \in X; \|f\| \leq 1} |\langle \phi; f \rangle|$.*

The dual of a real Banach space is defined similarly by replacing C by R in Def. D.2.8.

Example D.2.4: With $\mathcal{L}^1$, $\mathcal{L}^2$, and $\mathcal{L}^\infty$, as in Example D.2.3: $\mathcal{L}^\infty$ is the dual of $\mathcal{L}^1$, while $\mathcal{L}^2$ is self-dual; in particular:

$$\langle g, f \rangle \equiv \int_\Omega d\mu(x) \, g(x) \, f(x) \tag{D.2.6}$$

is well-defined and finite, whenever

$$f \in \mathcal{L}^1(\Omega, \mathcal{F}, \mu) \quad \text{and} \quad g \in \mathcal{L}^\infty \Omega, \mathcal{F}, \mu) \tag{D.2.7}$$

or

$$f \text{ and } g \in \mathcal{L}^2(\Omega, \mathcal{F}, \mu) \quad . \tag{D.2.8}$$

Our main text also requires us to single out the following class of Banach spaces.

Definition D.2.6. *A complex Hilbert space is a complex linear space $\mathcal{H}$ equipped with a scalar product, i.e. a map*

$$(\cdot, \cdot) : (\Phi, \Psi) \in \mathcal{H} \times \mathcal{H} \mapsto \langle \Phi, \Psi \rangle \in \mathsf{C}$$

satisfying the following axioms:

1. *$\forall \Psi \in \mathcal{H} : \langle \Psi, \Psi \rangle \geq 0$ and with $\Psi \in \mathcal{H} \to \|\Psi\| \equiv \langle \Psi, \Psi \rangle^{\frac{1}{2}}$, $\mathcal{H}$ is a Banach space;*
2. *$\forall \, (\Phi, \Psi) \in \mathcal{H} \times \mathcal{H} : \quad \langle \Psi, \Phi \rangle = \langle \Phi, \Psi \rangle^* $;*
3. *$\forall \, \Phi_1, \Phi_2, \Psi \in \mathcal{H} : \quad \langle \Phi_1 + \Phi_2, \Psi \rangle = \langle \Phi_1, \Psi \rangle + \langle \Phi_2, \Psi \rangle$;*
4. *$\forall \, \Phi, \Psi \in \mathcal{H} \text{ and } \forall \, c \in \mathsf{C} : \quad \langle c\Phi, \Psi \rangle = c \langle \Phi, \Psi \rangle$.*

Real Hilbert spaces are defined similarly by replacing C in Def. D.2.6 by R and the complex conjugation * by the identity map.

Example D.2.5: The typical Examples of infinite-dimensional Hilbert spaces are: the space l^2 of square-summable sequences; and the space $\mathcal{L}^2$ of square-integrable functions on R^n. These spaces are in fact isomorphic, as follows from the following construction. Let $\{\Psi_k \mid k \in \mathsf{Z}^+\}$ be an orthonormal basis in $\mathcal{L}^2$; the map $\Psi \in \mathcal{L}^2 \mapsto \psi_k \equiv \langle \Psi, \Psi_k \rangle \in l^2$ provides indeed an isometric, bijective map between these two spaces. We cannot leave even such a brief introduction to Hilbert spaces without mentioning one of their most important properties, namely that their self-duality is reflected in the Cauchy–Schwarz inequality

$$|\langle \Phi, \Psi \rangle|^2 \leq \|\Phi\| \, \|\Psi\| \,. \tag{D.2.9}$$

While the metric of a metric space specifies its topology, the converse is not true: several metrics may lead to the same topology; moreover, a specific metric may turn out to be cumbersome when one is merely interested in topological aspects. The following definition marks this shift of emphasis.

Definition D.2.7. *A topological space $(X, \mathcal{T})$ is said to be* metrizable *if there exists a metric d on X such that the topology $\mathcal{T}_d$ associated to d is equivalent to the original topology $\mathcal{T}$, i.e. $\mathcal{T}_d \leq \mathcal{T} \leq \mathcal{T}_d$, which we write $\mathcal{T}_d = \mathcal{T}$.*

Naturally, with such an existential definition, one needs some criterion to recognize that a space is metrizable. We now proceed towards such a criterion for the case where X is a vector space; Example D.2.6 illustrates this situation.

Definition D.2.8. *Let X be a real vector space. A map*

$$p : x \in X \mapsto p(x) \in \mathsf{R}^+$$

is said to be a semi-norm *if it satisfies the following axioms:*

1. $\forall \, (c, x) \in \mathsf{R} \times X : \; p(cx) = |c| \, p(x)\,;$
2. $\forall \, (x, y) \in X \times X : \; p(x, y) \leq p(x) + p(y)\,.$

A family $\{p_\lambda \mid \lambda \in \Lambda\}$ of semi-norms on X is said to separate *points in X whenever*

$$\{\, p_\lambda(x) = 0 \quad \forall \quad \lambda \in \Lambda \} \quad \models \quad x = 0\,.$$

A locally convex space *is a vector space equipped with a family of semi-norms that separate points of X. The natural topology on the locally convex space $(X, \{p_\lambda \mid \lambda \in \Lambda\})$ is defined by the neighborhood assignment $\mathcal{N}$ that associates, to every $x_o \in X$, every $\varepsilon > 0$, and every $n-$tuple of indices $(\lambda_1, \cdots, \lambda_n)$ with $\lambda_k \in \Lambda$, the set*

$$V_{(\lambda_1, \cdots, \lambda_n)}(x_o) = \{x \in X \mid \forall \, k = 1, \cdots, n : p_{\lambda_k}(x - x_o) < \varepsilon \} \quad .$$

Complex locally convex spaces are defined similarly by replacing R by C in Def. D.2.8. Note that the topology of a locally convex space can be defined more concisely – but less constructively – than done in Def. D.2.8.

Theorem D.2.1. *The natural topology on a locally convex space is the coarsest topology for which*

1. $\forall \lambda \in \Lambda$: *the semi-norm* $p_\lambda : x \in X \to p_\lambda(x) \in \mathsf{R}^+$ *is continuous;*
2. $+ : (x, y) \in X \times X \to x + y \in X$ *is continuous.*

Note that if the family of semi-norms is countable, we can read $\lambda \in \Lambda$ as $n \in \mathsf{Z}^+$; and verify that

$$d(x, y) \equiv \sum_{n=1}^{\infty} 2^{-n} \frac{p_n(y - x)}{1 + p_n(y - x)} \tag{D.2.10}$$

defines a metric on X. This is the idea behind the criterion for metrizability of a topological vector space announced immediately after Def. D.2.7::

Theorem D.2.2. *A locally convex space is metrizable if and only if its topology can be prescribed by a countable family of semi-norms.*

Example D.2.6: Let $\mathcal{S}(\mathsf{R})$ be the real vector space of all smooth functions $f : x \in \mathsf{R} \mapsto f(x) \in \mathsf{R}$, satisfying the following conditions:

$$\forall \, (m, n) \in \mathsf{Z}^+ \times \mathsf{Z}^+ \ : \ p_{(m,n)}(f) \equiv \sup_{x \in \mathsf{R}} \left| x^m \frac{\mathrm{d}^n f}{\mathrm{d} x^n}(x) \right| < \infty \, .$$

Since $\{p_{(m,n)} \mid (m, n) \in \mathsf{Z}^+ \times \mathsf{Z}^+\}$ define a countable family of semi-norms that separate points in R, $\mathcal{S}(\mathsf{R})$ is a metrizable vector space. In addition, it is complete as a metric space when equipped with the metric defined by (D.2.10) [upon identifying the countable set index sets $\mathsf{Z}^+ \times \mathsf{Z}^+$ and Z^+]; this is to say that $\mathcal{S}(\mathsf{R})$ is a *Fréchet* space, namely it is a complete metrizable locally convex space. The dual $\mathcal{S}(\mathsf{R})^*$ of $\mathcal{S}(\mathsf{R})$ – i.e. the space of all continuous linear functionals $T : f \in \mathcal{S}(\mathsf{R}) \mapsto \langle T; f \rangle$ – is known as the Schwartz space of *tempered distributions*.

We consider now topological spaces equipped with algebraic structures.

Definition D.2.9. *A Banach* $^*-$*algebra* $\mathcal{A}$ *is a complex Banach space on which an associative product* $(A, B) \in \mathcal{A} \times \mathcal{A} \mapsto AB \in \mathcal{A}$ *and an involution* $A \in \mathcal{A} \mapsto A^* \in \mathcal{A}$ *are defined in such a manner that, for all* $A, B, C \in \mathcal{A}$ *and* $a, b, c \in \mathsf{C}$:

1. $(A + B)C = AC + BC$; $(cA)B = c(AB) = A(cB)$;
2. $(A^*)^* = A$; $(aA + bB)^* = a^* A^* + b^* B^*$; $(AB)^* = B^* A^*$;
3. $\|AB\| \le \|A\| \, \|B\|$.

When, in addition $\|A^* A\| = \|A\|^2$, *we say that* $\mathcal{A}$ *is a* C^*-*algebra.*

Note: For every C*−algebra $\mathcal{A}$: $A \in \mathcal{A} \models \|A^*\| = \|A\|$.

Examples D.2.7:

1. The Banach space $\mathcal{L}^\infty(\Omega, \mathcal{F}, \mu)$ (see Example D.2.3) when equipped with the operations $fg : x \in \Omega \mapsto f(x)g(x) \in \mathsf{C}$ and $f^* : x \in \Omega \mapsto f(x)^* \in \mathsf{C}$ becomes a C*−algebra, which is commutative, i.e. $f, g \in \mathcal{A} \models fg = gf$.

2. Let $\mathcal{B}(\mathcal{H})$, be the set of all bounded linear operators from a complex Hilbert space $\mathcal{H}$ into itself, equipped with the operations $A + B : \Psi \in \mathcal{H} \mapsto A\Psi + B\Psi \in \mathcal{H}$, $\|A\| = \sup_{\Psi \in \mathcal{H}; \|\Psi\| \leq 1} \|A\Psi\|$, $AB : \Psi \in \mathcal{H} \mapsto A(B\Psi) \in \mathcal{H}$ and A^* defined by $\forall \Phi, \Psi \in \mathcal{H} : (A^*\Phi, \Psi) = (\Phi, A\Psi)$. $\mathcal{B}(\mathcal{H})$ is a C*−algebra.

3. When $2 \leq \dim\mathcal{H} < \infty$, $\mathcal{H} = \mathsf{C}^n$ and $\mathcal{B}(\mathsf{C}^n)$ is the C*−algebra of all $n \times n$ matrices with complex entries. In case $n = 2$ this is the algebra generated by the Pauli matrices (12.2.8), and it is not commutative. Hence, commutativity is violated in every C*−algebra $\mathcal{B}(\mathcal{H})$ with $2 \leq \dim\mathcal{H} \leq \infty$.

4. Let $\mathcal{H}$ be an infinite dimensional complex Hilbert space; $\mathcal{F}(\mathcal{H}) = \{A \in \mathcal{B}(\mathcal{H}) \mid \dim(A\mathcal{H}) < \infty\}$; and $\mathcal{C}(\mathcal{H})$ be the completion of $\mathcal{F}(\mathcal{H})$ in the Banach space $\mathcal{B}(\mathcal{H})$. Then, the identity operator I is not an element of $\mathcal{C}(\mathcal{H})$ and thus $\mathcal{C}(\mathcal{H}) \subset \mathcal{B}(\mathcal{H})$ with $\mathcal{C}(\mathcal{H}) \neq \mathcal{B}(\mathcal{H})$. $\mathcal{C}(\mathcal{H})$ is nevertheless a sub-C*−algebra of $\mathcal{B}(\mathcal{H})$.

Definition D.2.10. *A* W*−algebra $\mathcal{W}$ *is a* C*−algebra *dual to a Banach space.*

Examples D.2.8:

1. $\mathcal{L}^\infty(\Omega, \mathcal{F}, \mu)$ is a W*−algebra: it is the dual of the Banach space $\mathcal{L}^1(\Omega, \mathcal{F}, \mu)$.

2. $\mathcal{B}(\mathcal{H})$ is a W*−algebra: it is the dual of the Banach space $\mathcal{T}(\mathcal{H})$ obtained as the completion of $\mathcal{F}(\mathcal{H})$ with respect to the norm $\|A\|_T \equiv \mathrm{Tr}(A^*A)^{\frac{1}{2}}$.

3. While $\mathcal{T}(\mathcal{H})^* = \mathcal{B}(\mathcal{H})$ we have, when $\dim\mathcal{H} = \infty$: $\mathcal{T}(\mathcal{H}) \subsetneq \mathcal{T}(\mathcal{H})^{**} = \mathcal{B}(\mathcal{H})^*$; and $\mathcal{B}(\mathcal{H})^* = \mathcal{T}(\mathcal{H}) \oplus \mathcal{C}(\mathcal{H})^\perp$ where $\mathcal{C}(\mathcal{H})^\perp$ is the set of bounded linear functionals on $\mathcal{B}(\mathcal{H})$ that vanish on $\mathcal{C}(\mathcal{H})$

4. When $\dim\mathcal{H} = \infty$, the C*−algebra $\mathcal{C}(\mathcal{H})$ (see Example D.2.7.4) is *not* a W*−algebra.

Definition D.2.11. *A von Neumann algebra is a sub-*-algebra* $\mathcal{N} \subseteq \mathcal{B}(\mathcal{H})$ *such that* $\mathcal{N}'' = \mathcal{N}$ *where, for any subset* $\mathcal{X} \subset \mathcal{B}(\mathcal{H})$, *the commutant of* $\mathcal{X}$ *is the set* $\mathcal{X}' = \{A \in \mathcal{B}(\mathcal{H}) \mid \forall X \in \mathcal{X} : AX - XA = 0\}$. *A von Neumann algebra* $\mathcal{N}$ *is said to be a (von Neumann) factor whenever* $\mathcal{N} \cap \mathcal{N}' = \mathsf{C}I \equiv \{cI \mid c \in \mathsf{C}\}$.

Examples D.2.9:

1. $\mathcal{B}(\mathcal{H})' = \mathsf{C}I$ and $(\mathsf{C}I)' = \mathcal{B}(\mathcal{H})$; hence $\mathcal{B}(\mathcal{H})$ and $\mathsf{C}I$ are von Neumann algebras.

2. The sub-*-algebras of $\mathcal{M}, \mathcal{N} \subset \mathcal{B}(\mathbb{C}^4)$ defined respectively as the sets of all matrices of the form

$$\left\{ \begin{pmatrix} a & b & & \\ c & d & & \\ & & a & b \\ & & c & d \end{pmatrix} \mid a, b, c, d \in \mathsf{C} \right\} \quad \text{resp.} \quad \left\{ \begin{pmatrix} \alpha & & \beta & \\ & \alpha & & \beta \\ \gamma & & \delta & \\ & \gamma & & \delta \end{pmatrix} \mid \alpha, \beta, \gamma, \delta \in \mathsf{C} \right\}$$

satisfy the relations $\mathcal{M}' = \mathcal{N}$ and $\mathcal{N}' = \mathcal{M}$. Therefore $\mathcal{M}$ and $\mathcal{N}$ are von Neumann algebras; and, since $\mathcal{M} \cap \mathcal{N} = \mathbb{C}I$, they are factors. Moreover, with V denoting the (involutive) transposition matrix

$$V = \begin{pmatrix} 1 & 0 & 0 & 0 \\ 0 & 0 & 1 & 0 \\ 0 & 1 & 0 & 0 \\ 0 & 0 & 0 & 1 \end{pmatrix} = V^* \quad ,$$

$M \in \mathcal{M} \mapsto V^* M V \in \mathcal{N}$ is bijective; and $\mathcal{M}$ and $\mathcal{N}$ are unitarily equivalent representations of the von Neumann algebra $\mathcal{B}(\mathbb{C}^2)$. Factors that are conjugate to their commutant are a characteristic feature of the algebra of observables describing a pure thermodynamical phase of a quantum dynamical system (i.e. extremal KMS states); see e.g. Sects. 10.2 or 15.3.

3. Every von Neumann algebra contains the identity operator. Hence, when $\dim \mathcal{H} = \infty$, the C^*-algebra $\mathcal{C}(\mathcal{H})$ (see Example D.2.7.4) is *not* a von Neumann algebra (compare to Example D.2.8.4).

Remarks:

1. The operation $\mathcal{X} \subseteq \mathcal{B}(\mathcal{H}) \to \mathcal{X}'' \subseteq \mathcal{B}(\mathcal{H})$ satisfies $\mathcal{X} \subseteq \mathcal{X}''$ and $\mathcal{X}'' = (\mathcal{X}'')''$, which is very much reminiscent of the basic closure operation defining a topology (see Def. D.1.1). Indeed, let us consider on $\mathcal{B}(\mathcal{H})$ the five topologies defined by the convergence w.r.t. the following semi-norms for the uniform topology ($\mathcal{T}_u$), the (semi-)norm $A \to \|A\|$;
for the strong topology ($\mathcal{T}_s$), the semi-norms $A \to \|A\Phi\|$ where Φ runs over $\mathcal{H}$;
for the weak topology ($\mathcal{T}_w$), the semi-norms $A \to |(A\Phi, \Psi)|$ where Φ and Ψ run over $\mathcal{H}$;
for the ultrastrong topology ($\mathcal{T}_{us}$) the semi-norms $A \to [\sum_{k \in \mathsf{Z}^+} \|A\Phi_k\|^2]^{\frac{1}{2}}$ where $\{\Phi : k \in \mathsf{Z}^+ \mapsto \mathcal{H}\}$ runs over all square-summable sequences, i.e. all sequences satisfying $\sum_{k \in \mathsf{Z}^+} \|\Phi_k\|^2 < \infty$;
for the ultraweak topology ($\mathcal{T}_{uw}$) the semi-norms $A \to |\sum_{k \in \mathsf{Z}^+} (A\Phi_k, \Psi_k)|$ where $\{\Phi : k \in \mathsf{Z}^+ \mapsto \mathcal{H}\}$ and $\{\Psi : k \in \mathsf{Z}^+ \mapsto \mathcal{H}\}$ run over all square-summable sequences.
When $\dim \mathcal{H} = \infty$, these five topologies are different, with the uniform topology being the strongest, and the weak topology the weakest; but $\mathcal{T}_{uw}$ and $\mathcal{T}_s$ are not comparable. In the notation of Def. D.1.12: $\mathcal{T}_u > \mathcal{T}_{us} > \mathcal{T}_s > \mathcal{T}_w$ and $\mathcal{T}_u > \mathcal{T}_{us} > \mathcal{T}_{uw} > \mathcal{T}_w$. While the above partial

order relations are strict, we have nevertheless that if a sub-*-algebra $\mathcal{A}$ of $\mathcal{B}(\mathcal{H})$ is such that $\{A\Psi \mid A \in \mathcal{A}, \Psi \in \mathcal{H}\}$ is norm-dense in $\mathcal{H}$, then $\mathcal{A}''$ is the closure of $\mathcal{A}$ in any of the four topologies: $\mathcal{T}_{us}$, $\mathcal{T}_{uw}$, $\mathcal{T}_s$ $\mathcal{T}_w$; and $\mathcal{A}''$ is closed in $\mathcal{T}_u$.

2. In the same way as every abstract C^*−algebra can be realized as a concrete sub-C^*−algebra of some $\mathcal{B}(\mathcal{H})$, every abstract W^*−algebra can be realized as a concrete von Neumann sub-algebra of some $\mathcal{B}(\mathcal{H})$. The essence of this representation theorem is captured in the following result.

Scholium D.2.1 (GNS construction). *Let $\mathcal{A}$ be a C^*−algebra with unit I, and φ be a positive linear functional on $\mathcal{A}$ with $\langle \varphi; I \rangle = 1$. Then, there exists a Hilbert state $\mathcal{H}$ a vector $\Phi \in \mathcal{H}$ with $\|\Phi\| = 1$, and a representation $\pi : A \in \mathcal{A} \mapsto \pi(A) \in \mathcal{B}(\mathcal{H})$ such that $\{\pi(A)\Phi \mid A \in \mathcal{A}\}$ is norm-dense in $\mathcal{H}$, and for all $A \in \mathcal{A}$: $(\pi(A)\Phi, \Phi) = \langle \varphi; A \rangle$. Moreover, this representation is unique up to unitary equivalence.*

The *proof* is based on the following two observations. First, φ is positive linear bounded; hence for every $A, B \in \mathcal{A}$: $0 \le \langle \varphi; B^*A^*AB \rangle \le \|A^*A\| \langle \varphi; B^*B \rangle$, so that the closed subspace $\mathcal{K} = \{K \in \mathcal{A} \mid \langle \varphi; K^*K \rangle = 0\}$ is a left-ideal of $\mathcal{A}$. Second, the map $(A, B) \to \langle \varphi; B^*A \rangle$ is a positive sesquilinear map from $\mathcal{A} \times \mathcal{A}$ to $\mathbb{C}$. With $\Phi_A \in \mathcal{H}_o \equiv \mathcal{A}/\mathcal{K}$ denoting the equivalence class of $A \in \mathcal{A}$, the above two observations entail: (a) $(\Phi_A, \Phi_B) \in \mathcal{H}_o \times \mathcal{H}_o \mapsto \langle \varphi; B^*A \rangle \in \mathbb{C}$ is well defined, sesquilinear, and *positive definite;* (b) $\pi_o(A) : \Phi_B \to \Phi_{AB}$ is well-defined, and satisfies $\|\pi_o(A)\Phi_B\|^2 \le \|A^*A\| \|\Phi_B\|^2$. (a) entails that we have defined a scalar product that equips $\mathcal{H}_o$ with the structure of pre-Hilbert space. $\mathcal{H}$ is its completion, with $\Phi = \Phi_I$. (b) entails that π_o can be uniquely extended by continuity to a representation $\pi : \mathcal{A} \to \mathcal{B}(\mathcal{H})$, i.e. π satisfies $\pi(A + B) = \pi(A) + \pi(B)$; $\pi(cA) = c\pi(A)$; $\pi(AB) = \pi(A)\pi(B)$; $\pi(A^*) = \pi(A)^*$; and $\|\pi(A)\| \le \|A\|$. The uniqueness of the triple $\{\mathcal{H}, \Phi, \pi\}$ up to unitary equivalence follows immediately.   **q.e.d.**

In quantum theory, φ appears as a state on an algebra $\mathcal{A}$ of observables; thus the GNS construction associates to φ a representation of $\mathcal{A}$ as an algebra of operators acting on a Hilbert space tailored to the physical circumstances described by φ.

As factors play an important role in relativisitic quantum field theory and statistical mechanics, it is interesting to know that they can be classified according to a *dimension function* defined canonically on the lattice $\mathcal{P}$ of the projectors in $\mathcal{N}$. To see this, we first consider a generalization of the usual notion of trace defined for $\mathcal{B}(\mathcal{H})$.

Definition D.2.12. *Let $\mathcal{N}$ be a von Neumann algebra, and $\mathcal{N}^+ = \{A^*A \mid A \in \mathcal{N}\}$ be the cone of its positive elements. A trace for $\mathcal{N}$ is a map $\tau : A \in \mathcal{N}^+ \mapsto \langle \tau; A \rangle \in [0, +\infty]$ such that for all $A, B \in \mathcal{N}^+$, all $c \in \mathbb{R}^+$, and all $C \in \mathcal{N}$: $\langle \tau; A + B \rangle = \langle \tau; A \rangle + \langle \tau; B \rangle$, $\langle \tau; cA \rangle = c\langle \tau; A \rangle$, and $\langle \tau; C^*C \rangle = \langle \tau; CC^* \rangle$. A trace τ is said to be* faithful *if $\langle \tau; A \langle = 0$ with $A \in \mathcal{N}^+ \models A = 0$.*

A trace is said to be normal *if, for every increasing net $\{A_\nu\} \subset \mathcal{N}^+$ with lowest upper bound $A \in \mathcal{N}^+$, $\langle \tau; A \rangle$ is the lowest upper bound of $\{\langle \tau; A_\nu \rangle\}$. A trace τ is said to be* semi-finite *if, for every $A \in \mathcal{N}^+$ with $A \neq 0$, there exists $B \in \mathcal{N}^+$ with $A - B \in \mathcal{N}^+$ and $B \neq 0$ such that $\langle \tau; B \rangle > 0$. The von Neumann algebra $\mathcal{N}$ is said to be* semi-finite *if, for every $A \in \mathcal{N}^+$ with $A \neq 0$, there exists a semi-finite, normal trace τ for $\mathcal{N}$ such that $\langle \tau; A \rangle \neq 0$.*

Definition D.2.13. *Given a von Neumann algebra $\mathcal{N}$, let $\mathcal{P} = \{P \in \mathcal{N} \mid P^* = P = P^2\}$, equipped with the natural partial order $P \subseteq Q$. P and Q in $\mathcal{P}$ are said to be* equivalent, *which we denote $P \sim Q$ whenever there exists $U \in \mathcal{N}$ such that $U^*U = P$ and $UU^* = Q$. If, given P and Q in $\mathcal{P}$ there exists some $R \in \mathcal{P}$ such that $R \sim P$ and $R \subseteq Q$, we write $P \leq Q$; we further write $P < Q$ when $P \leq Q$ and $P \nsim Q$. An element $P \in \mathcal{P}$ is said to be* finite *if $\{Q \in \mathcal{P}, Q \subseteq P, Q \sim P\} \models P = Q$; otherwise P is said to be* infinite.

Scholium D.2.2 (Classification of factors). *Let $\mathcal{N}$ be a factor; and $\mathcal{P} = \{P \in \mathcal{N} \mid P^* = P = P^2\}$. Then:*

1. *for any $P, Q \in \mathcal{P}$, one has necessarily either $P \leq Q$ or $Q \leq P$, i.e. $\mathcal{P}$ is totally ordered by the relation $\leq$ of Def. D.2.13;*
2. *there exists a faithful normal trace τ for $\mathcal{N}$, with τ semi-finite if $\mathcal{N}$ is supposed to be semi-finite. These properties charactirize τ uniquely, up to a multiplicative constant;*
3. *the restriction $\Delta : p \in \mathcal{P} \mapsto \Delta(P) = \langle \tau; P \rangle$ is a dimension function, i.e. it satisfies:*
   a. *$\forall P \in \mathcal{P} : \Delta(P) \geq 0$ with $\Delta(P) = 0$ iff $P = 0$;*
   b. *$P \perp Q \quad \models \quad \Delta(P + Q) = \Delta(P) + \Delta(Q)$;*
   c. *$P \sim Q \quad \models \quad \Delta(P) = \Delta(Q)$;*
   d. *$\Delta(P) = \infty$ iff P is infinite;*
   e. *$\{P \text{ finite}, P < Q\} \quad \models \quad \Delta(P) < \Delta(Q)$.*
4. *Up to a multiplicative constant, $\{\Delta(P) \mid P \in \mathcal{P}\}$ must be one of the following five sets: $\{0, 1, 2, 3, \cdots, n < \infty\}$, resp. $\{0, 1, 2, 3, \cdots, \infty\}$, $[0, 1]$, $[0, \infty]$, $\{0, \infty\}$.*

Definition D.2.14. *According to the range of its dimension function (see Scholium D.2.2), the factor $\mathcal{N}$ is then said to be of type I_n, resp. I_∞, II_1, II_∞, III. One thus distinguishes between factors that are discrete $\{I_n, I_\infty\}$ vs. continuous $\{II_1, II_\infty, III\}$; finite $\{I_n, II_1\}$ vs. infinite $\{I_\infty, II_\infty, III\}$; semi-finite $\{I_n, I_\infty, II_1, II_\infty\}$ vs. purely infinite $\{III\}$.*

Examples D.2.10 The familiar von Neumann algebras $\mathcal{B}(\mathcal{H})$ are factors of type I_n or I_∞ depending on whether $\dim\mathcal{H} = n$ or ∞. These are the algebras traditionally used to describe quantum systems with finitely many degrees of freedom. Factors of type II_1 were discussed by [Von Neumann, 1981] in connection with "continuous geometries", a subject he had pursued on-and-off since the mid-1930s; a special factor of this type appears in the mathematical treatment of the quantum version of the Arnold CAT due to [Benatti, Narnhofer, and Sewell, 1991, Emch, Narnhofer, Thirring, and Sewell, 1994]. Factors of type III are ubiquitous in quantum field theory [Haag and Kastler, 1964, Buchholz, D'Antoni, and Fredenhagen, 1987, Haag, 1996].

The above classification of factors (discrete vs continuous, etc) can be extended to general von Neumann algebras, at the cost of moving away from the immediately visualizable notion of relative dimension of projectors, as in part (3) of Scholium D.2.2.

Bibliographical notes. The study of "rings of operators" – as von Neumann algebras were first known – was initiated by John von Neumann in collaboration with Francis Murray (then a freshly minted Ph.D. graduate from Marshall Stone). Their results appeared in a series of papers published between 1936 and 1943 starting with [Murray and Neumann, 1936]. For general references on C^*−algebras, W^*−algebras, von Neumann algebras, and their representations, see: [Dixmier, 1957, 1964, Sakai, 1971, Pedersen, 1979, Kadison and Ringrose, 1983, 1986]; and for a presentation geared towards applications to physics, [Emch, 1972a, Bratteli and Robinson, 1979, 1987, Emch, 1984, Haag, 1996].

D.3 Separability and Compactness

First of all, we demonstrate with the help of an example that the notion of convergence based on sequences (see Def. D.2.3) must be extended (from its original context in metric spaces) to cover the case of general topological spaces.

Example D.3.1: Let $X = [0, 1]$ and equip it with the topology in which the non-empty open sets are those subsets $U \subseteq X$ whose complements contain at most a countable infinity of points. Note that $\{1\}^c = [0, 1)$ so that $\{1\}$ is not open, and thus $[0, 1)$ is not closed. Hence $\overline{[0, 1)} = [0, 1]$, and thus

$$1 \in \overline{[0, 1)} \quad . \tag{D.3.1}$$

It would therefore seem natural to conjecture the existence of a sequence $\{x_n \mid n \in \mathsf{Z}^+\} \subset [0, 1)$ that converges to 1. This however is not possible. Indeed, on the one hand: $\{x_n \mid n \in \mathsf{Z}^+\}$ is countable, so $\{x_n \mid n \in \mathsf{Z}^+\}^c$ is open, and thus

$$\{x_n \mid n \in \mathsf{Z}^+\} \quad \text{is closed} \quad . \tag{D.3.2}$$

But on the other hand: $\{x_n \mid n \in \mathsf{Z}^+\}^c \supset [0, 1)^c = \{1\}$, i.e.

$$1 \notin \{x_n \mid n \in \mathsf{Z}^+\}. \tag{D.3.3}$$

From (D.3.2–3) we read that $\{x_n \mid n \in \mathsf{Z}^+\}$ is a closed set not containing 1; hence $\{x_n \mid n \in \mathsf{Z}^+\}$ cannot converge to 1, contrary to the conjecture suggested by (D.3.1).

Definition D.3.1. *A set Λ is said to be* directed *if it is equipped with a partial ordering $\leq$ such that for any two elements λ_1, λ_2 of Λ, there exists a element $\lambda \in \Lambda$ such that, for $k = 1, 2$: $\lambda_k \leq \lambda$. A* net *in a space X is a map $x : \lambda \in \Lambda \mapsto x_\lambda \in X$ where Λ is a directed set; we say then that the net is indexed by Λ. A net $y : \mu \in M \mapsto y_\mu \in X$ is said to be a* subnet *of a net $x : \lambda \in \Lambda \mapsto x_\lambda \in X$ if there is a map $i : M \to \Lambda$ such that:*

- $\forall \mu \in M$: $y_\mu = x_{i(\mu)}$
- $\forall \lambda \in \Lambda$ *there exists* $\mu_\lambda \in M$ *such that :* $\mu \in M$ *with* $\mu \geq \mu_\lambda \models i(\mu) \geq \lambda$.

Example D.3.2: Z^+ with its usual (total) ordering is a directed set; and a net indexed by Z^+ is what we already know as a *sequence*. Similarly, the notion of subnet generalizes that of subsequence. Hence to obtain a genuine extension of the concept of sequence, it is essential that the definition of a net does *not* require its indexing set Λ to be countable. This generality allows to beat the unpleasant situation uncovered by Example D.3.1: there, the contradiction is eliminated by the fact that the complement of a net in $[0, 1)$ is not necessarily closed.

Nets that are indexed by an interval in R, or a cone in R^n, are frequently encountered nets which are not sequences. Another useful example of a directed set is provided by an increasing family of closed subspaces (or projectors) in a Hilbert space, where the partial ordering is given by the inclusion of subspaces.

We are now ready for the necessary adjustment in the definition of convergence.

Definition D.3.2. *A net $\{x_\lambda \mid \lambda \in \Lambda\}$ in a topological space X is said to* converge *to a point $x \in X$ if for every neighborhood V of x, there exists some $\lambda_V \in \Lambda$ such that: $\lambda \in \Lambda$ with $\lambda \geq \lambda_V \models x_\lambda \in V$. We then say that x is a* limit point *of the net. A point $x \in X$ is said to be a* cluster point *of a net $\{x_\lambda \in X \mid \lambda \subset \Lambda\}$ if the latter admits a subnet $\{y_\mu \in X \mid \mu \in M\}$ converging to x.*

Note that:

1. the above definition carries over from sequences to nets the familiar definitions of convergence, limit points, and cluster points;
2. it can be shown that for each $A \subset X$ the following two conditions on $x \in X$ are equivalent:
   a) $x \in \overline{A}$;
   b) there exists a net $\{x_\lambda \in A \mid \lambda \in \Lambda\}$ converging to x.
   Hence, by substituting nets for sequences, one defeats contradictions of the type met in Example D.3.1;
3. nothing in the above definition guarantees that a net has either a limit point, or that it has only one. We shall review conditions under which these situations can be remedied.

Theorem D.3.1. *Let* $(X, \mathcal{T})$ *be a topological space. Then the following two conditions are equivalent:*

1. *each net* $\{x_\lambda \mid \lambda \in \Lambda\}$ *converges to at most* one *point;*
2. *for each* pair *of distinct* points x_1 *and* x_2 *in* X, *there exists a pair of disjoint* neighborhoods $V_1 \ni x_1$ *and* $V_2 \ni x_2$.

Definition D.3.3. *A topological space satisfying any one – and thus both – of the equivalent conditions of Thm. D.3.1 is said to be a* Hausdorff space; *in particular, the second of the above conditions is known as the* second separability axiom *or* axiom T_2.

Incidentally, the notation T_2 should alert the Reader that separability can be defined in other – non–equivalent – ways; these, however, would be of little interest in analysis, as a consequence of the following result.

Theorem D.3.2 (Tychonoff–Urysohn). *Every metrizable topological space is a Hausdorff space.*

Example D.3.3: Example D.1.2 is a Hausdorff space; it is indeed immediate to check that the discrete topology satisfies the separability condition T_2. In fact, this space is metrizable; consider indeed the metric:

$$d(x, y) = \begin{cases} 0 & \text{if} \quad x = y \\ 1 & \text{if} \quad x \neq y \end{cases} .$$

Example D.3.4: If card $X = \infty$, the topological space of Example D.1.3 is not a Hausdorff space. Indeed, for $k = 1, 2$, let $x_k \in X$ with $x_1 \neq x_2$; let further $V_k \in \mathcal{N}(x_k)$; by Def. D.1.2 there exist U_k open such that $x_k \in U_k \subseteq V_k$ and thus $V_k^c \subseteq U_k^c$; but, by the definition of the topology in this example: U_k open means $\text{card}(U_k^c) < \infty$. Hence $\text{card}(V_k^c) < \infty$. Suppose now that it were possible to choose these V_k such that V_1 and V_2 are disjoint i.e. $V_1 \cap V_2 = \emptyset$, i.e. $V_1^c \cup V_2^c = X$, and hence

$$\text{card } X = \text{card}(V_1^c \cup V_2^c) \leq \text{card}(V_1^c) + \text{card})(V_2^c) < \infty \quad ,$$

a contradiction with our initial assumption that card $X = \infty$; hence this topological space is not a Hausdorff space; consequently, by virtue of Thm. D.3.2, this space is not metrizable.

To define compactness, we need the following auxilliary concepts.

Definition D.3.4. *A collection* $\{U_\lambda \subseteq X \mid \lambda \in \Lambda\}$ *is said to be a* cover *of* $A \subseteq X$ *whenever*

$$A \subseteq \bigcup_{\lambda \in \Lambda} U_\lambda ;$$

the cover is said to be an open cover *whenever, in addition,*

$$\forall \lambda \in \Lambda : U_\lambda \text{ is open} .$$

A cover $\{U_\lambda \subseteq X \mid \lambda \in \Lambda\}$ *is said to be a* finite cover *whenever* $\operatorname{card}(\Lambda) < \infty$.
A cover $\{V_\mu \subseteq X \mid \mu \in M\}$ *is said to be a* subcover *of* $\{U_\lambda \subseteq X \mid \lambda \in \Lambda\}$
whenever $M \subseteq \Lambda$ *and* $\forall \, \mu \in M \, : V_\mu = U_\mu$.

Example D.3.5: With R equipped with its usual metric topology, then

$$\{V_n = (-n, n) \mid n \in \mathsf{Z}^n\}$$

is an open cover of R, and it admits no finite subcover.

Example D.3.6: Another cover of R with these properties is

$$\{V_\varepsilon(x) = (x - \varepsilon, x + \varepsilon) \mid \varepsilon > 0, \, x \in \mathsf{R}\} \quad .$$

Example D.3.7: In contrast with the above, let now $A = [0, 1] \subset \mathsf{R}$, with R
equipped with its metric topology; then

$$\{\, V^A{}_\varepsilon(x) = (x - \varepsilon, x + \varepsilon) \mid \varepsilon > 0, \, x \in \mathsf{A}\,\}$$

is clearly an open cover of A, and it admits a finite subcover.

Example D.3.8: As a warning, however, consider $B = (0, 1] \subset \mathsf{R}$, with R
equipped with its metric topology; then

$$\left\{ W_n = \left(\frac{1}{n+1}, 1 \right] \mid n \in \mathsf{Z}^+ \right\}$$

is clearly an open cover of B, but it admits no finite subcover.

Example D.3.9: Lest the Reader be inclined to think that the lack of being
closed is the only culprit, consider again R, still equipped with its metric
topology; it is then easy to produce subsets $A \subseteq R$ that are closed, and
yet open covers of A which do not admit finite subcovers; see e.g. R itself in
Examples D.3.5–6; or $[0, \infty) \equiv \mathsf{R}^+ \subset \mathsf{R}$.

The following definition brings order in this collection of examples which,
at first sight, may appear to be somewhat lawless.

Definition D.3.5. *A Hausdorff space* $(X, \mathcal{T})$ *is said to be* compact *whenever every open cover of* X *admits a finite subcover. Let* A *be a subset of
a topological space* $(X, \mathcal{T})$; *we denote by* $(A, \mathcal{T}_A)$ *the topological space obtained by defining the sets that are open relative to* A *as the elements of*
$\mathcal{O}_A \equiv \{U \cap A \mid U \in \mathcal{O}\}$ *where* $\mathcal{O}$ *is the collection of open sets of* X . *A subset* $A \subseteq X$ *is said to be* compact *whenever the topological space* $(A, \mathcal{T}_A)$ *is
compact.*

In reference to Examples D.3.5–9, note that R R^+ and $(0, 1]$ are not compact,
while $[0, 1]$ is compact. More generally, we have the following result:

Theorem D.3.3 (Heine–Borel). *For* $A \subset \mathsf{R}^n$ *the following conditions are
equivalent:*

1. *A is compact;*
2. *A is both closed and bounded.*

Note that compactness is a robust notion, as evidenced by any of the results collected below:

Theorem D.3.4. *Let $(X, \mathcal{T})$ be an arbitrary topological space, then*

1. $\{A, B \subset X \text{ and } A, B \text{ compact}\} \models (A \cup B) \text{ compact}$;
2. $\{A \subseteq X \text{ and } A \text{ compact}\} \models A \text{ closed}$;
3. $\{B \subseteq A \subseteq X, \ B \text{ closed and } A \text{ compact}\} \models B \text{ compact}$;
4. *Let $(X_k, \mathcal{T}_k)$ $(k = 1, 2)$ be two topological spaces, $f : X_1 \to X_2$ be continuous, and $A \subseteq X_1$; then: A compact in $X_1 \models f(A)$ compact in X_2.*

We shall also need occasionally – see Thm. D.4.4 – the following allied concept:

Definition D.3.6. *A topological space $(X, \mathcal{T})$ is said to be* locally compact *if every $x \in X$ has at least one compact neighborhood.*

The following results are reported here to convey that life – and analysis in particular – is much easier when carried on compact spaces, especially when a metric is available.

Theorem D.3.5.

1. *Let $(X_k, \mathcal{T}_k)$ $(k = 1, 2)$ be two topological spaces, with X_1 compact ; and $f : X_1 \to X_2$ be a continuous injection; then $g : y \in f(X) \mapsto f^{-1}(x) \in X$ is continuous;*
2. *in particular, with X_k as above, every continuous bijection $f : X_1 \to X_2$ is a homeomorphism – compare with Def. D.1.11;*
3. *let $(X, \mathcal{T})$ be a compact topological space, and $f : x \in X \mapsto f(x) \in \mathbb{R}$, the latter space being equipped with its usual metric topology; then f is bounded – below and above – and its bounds are reached, i.e. there exist (finite!) x_m and x_M in X such that $\forall x \in X : f(x_m) \leq f(x) \leq f(x_M)$.*

Theorem D.3.6 (Bolzano–Weierstrass). *Let $(X, \mathcal{T})$ be a topological space that is metrizable. Then the following conditions on $A \subseteq X$ are equivalent:*

1. *A is compact;*
2. *every sequence $\{x_n \in A \mid n \in \mathbb{Z}^+\}$ admits a converging subsequence that converges to some $a \in A$.*

Theorem D.3.7 (uniform continuity). *Let (X_k, d_k) $(k = 1, 2)$ be two metric spaces, with X_1 compact. Then the following conditions on $f : X_1 \to X_2$ are equivalent:*

1. *f is continuous, i.e. for every $x_1 \in X_1$ and every $\varepsilon > 0$, there exists some $\delta_\varepsilon(x_1)$ such that $\forall x \in X_1$ with $d_1(x, x_1) < \delta_\varepsilon(x_1) : d_2(f(x), f(x_1)) < \varepsilon(x_1)$;*
2. *f is uniformly continuous, i.e. for every $\varepsilon > 0$, there exists some δ_ε such that $\forall x, x_1 \in X_1$ with $d_1(x, x_1) < \delta_\varepsilon : d_2(f(x), f(x_1)) < \varepsilon$.*

D.4 The Baire Essentials

We now return to the basic topology of open and closed subsets, as we need a few more elements of the nomenclature. X stands throughout for an arbitrary topological space.

Definition D.4.1. *The* interior *of a subset* $A \subseteq X$, *denoted* $\mathrm{Int}(A)$, *is the set of all its* interior points, *i.e.* $\mathrm{Int}(A) = \{x \in X \mid \exists\, O \text{ open with } x \in O \subseteq A\}$.
    A subset $A \subset X$ *is said to be* dense *in* X *whenever* $\overline{A} = X$.
    A subset $A \subset X$ *is said to be* nowhere dense *in* X *whenever* $\mathrm{Int}(\overline{A})$ *is empty. i.e. equivalently when the complement* $\overline{A}^c \equiv (X - \overline{A})$ *of its closure is dense in* X.
    A subset $A \subseteq X$ *is said to be* meager *(or to be of the first Baire category) if it is a countable union of nowhere dense subsets of* X; *otherwise the set is said to be of the second Baire category. The complement of a meager set is said to be a* residual *set.*

Note that a residual set is thus a countable intersection of open dense subsets of X.

Example D.4.1: Given R with its natural metric topology, consider the set Q of all rational numbers, and its complement, the set $\mathsf{Q}^c = \mathsf{R} - \mathsf{Q}$ of all irrational numbers. Both Q and Q^c are dense in R, but Q is meager. Note that Q has no interior point. This property motivated the following definition, due to Baire [1899].

Definition D.4.2. *A* Baire space *is a topological space* $(X, \mathcal{T})$ *such that every meager set* $A \subset X$ *has empty interior.*

The above concept may be approached in different ways:

Theorem D.4.1. *Let* $(X, \mathcal{T})$ *be a topological space. Then the following four conditions are equivalent:*

1. $(X, \mathcal{T})$ *is a Baire space;*
2. *for every family* $\{F_n \mid n \in Z^+\}$ *of closed nowhere dense subsets* $F_n \subset X$, *the set* $\cup_{n \in Z^+} F_n$ – *which is not necessarily closed* – *retains the property of having an empty interior;*
3. *for every family* $\{O_n \mid n \in Z^+\}$ *of open, dense subsets* $O_n \subset X$, *the set* $\cap_{n \in Z^+} O_n$ – *which is not necessarily open* – *retains the property of being dense in* X.
4. *each non-empty open subset of* X *is of the second Baire category.*

    Moreover:

Theorem D.4.2. *The following two conditions on a subset* A *of a Baire space* $(X, \mathcal{T})$ *are equivalent:*

1. *A is meager;*
2. *A is contained in a countable union $\cup_{n \in \mathbb{Z}^+} F_n$ of closed nowhere dense subsets $F_n \subset X$.*

In connection with the second of the above conditions recall that, for all n, we have that F_n is closed, with $\text{Int}(F_n) = \emptyset$.

The following two results indicate that Baire spaces are common features in the topological landscape:

Theorem D.4.3 (Baire–Hausdorff). *Every topological space homeomorphic to a complete metric space is a Baire space.*

Theorem D.4.4. *Every locally compact Hausdorff space is a Baire space.*

In view of the very richness of the definition of a Baire space, it seems reasonable to consider the following concept.

Definition D.4.3. *A property P defined on the points of a Baire space X is said to be* generic *whenever the set $X_P \equiv \{x \in X \mid x \vdash P\}$ – i.e. the set of points of X which have property P – contains a residual set of X. A property E is said to be* not generic *whenever the set $X_E \equiv \{x \in X \mid x \vdash E\}$ is meager.*

Note that being "generic" is not an attribute of any specific point, but is an attribute of a property. For instance, the property of being an irrational number is generic in $\mathbb{R}$, while the property of being rational is not generic.

E. Appendix: Models vs. Models

Very different concepts are covered by the common noun 'model' as it is routinely used by practitioners in various disciplines, even among the physical scientists. In particular, a fault-line between two such acceptions seems to have developed through theoretical and mathematical physics; some of its manifestations are explored in this Wigner Symposium lecture [Emch, 1993] for the purpose of helping clarify the significance of these differences.

E.1 Models in Wigner's Writings and in the Third Wigner Symposium

The problem that I address in this lecture became more sharply focused in my mind when I was engaged in preparing the annotations for the Wigner papers that are reprinted as Volume VI of his Collected Works [Wightman and Mehra, 1995]. To sketch the issue in broad strokes, I would say that, while in most of the articles we included in that volume the noun *'model'* describes attempts to produce fully controlled mathematical constructs, the very same noun has a significantly different meaning when used in conference interventions we reprinted in the section on Nuclear Physics: there it describes tentative semi-intuitive schemes devised to explore the complexity of what physicists often call the 'real world'. One could be tempted to dismiss these differences on account of the rhetorical context: well-prepared lectures or carefully written papers as apposed to improvised contributions and spontaneous discussions; however, this would amount to an escape from the problem of dealing with the intentional underpinnings of Wigner's discourse.

Another manifestation of this problem has been exemplified, already in the lectures by Professor Doebner and by Professor Schaeffer during the first sessions of this Third Wigner Symposium: both speakers used routinely, and yet properly, the word 'models' to present their contributions, while pursuing nevertheless quite different methodologies. Afterwards, during conversations held in the marvelously urbane atmosphere at *Christ Church*, we agreed also that there was here a genuine discrepency, the cause of which was worth elucidating: how did it come about that we all could legitimately hope to

understand each other while using the same vocable to serve so radically
different purposes? In this lecture I intend to give this case a fair trial, hence
the title 'Models vs. Models'.

E.2 A Search for Precedents

To help focus our attention, we start with Bridgman [1927]. Accordingly, we
make here three presuppositions: (i) a primary task of science is to relate the
world of appearances (laboratory observations or data collections) with the
world of conceptions (theories or ideas); (ii) these relations – and the world's
elements they relate – are to be stated, specified or identified in terms of
operations (laboratory manuals and logical rules); and (iii) not only should
these relations proceed from and to *both* experiments and mental demands,
but this mode of description in terms of operations must also inform the
design of the admissible evidences, from the laboratory as well as from the
scholar's den. How close to a positivist manifesto this may be is not our
concern here (although it surely would be if we were to concentrate solely
on understanding Wigner's use of the term 'model'); rather, our contention
is simply that these presuppositions contain enough consensual elements to
provide a controlled entry point to our discussion of the role that models
play in our own current professional activities as theoretical or mathematical
physicists.

   This framework offers the advantage of limiting our search. It allows us to
ignore – in the beginning of this investigation – the most extreme meanings
of the noun 'model': *e.g.* at one end of the spectrum, the 'model' of the
fashion designer's world would not be likely to inform our query, nor would
the role-'model' of the socio-psycho-educational industry; at the other end
of this spectrum of meanings, this framework also excludes such constructs
as the 'model' electric train of yesteryear and, perhaps more arguably, the
sleek 'models' that splash from the wind-tunnel into the showrooms. We do
not intend to intimate that none of these modeling activities has a place in
a comprehensive investigation, but simply that our trying to encompass them
in the seminal core could hinder our defining of this core in a manner that
elucidates the dichotomy in the usages presented at the outset of this lecture.

   Even so, the field of enquiry still open to us remains quite wide [Suppes,
1960, Freudenthal, 1961, Hesse, 1970, Suppes, 1988, 1993]. These writings
offer an introduction into the range of the basic problems one has to face
when dealing with the notion of model in the sciences.

   Hesse marks the territory with a variety of no-threspassing posters, warning of the necessity to be specific; she starts by having two imaginary pro-
taganists, *Campbellian* and *Duhemist* argue that a definition of the notion
of model should not aim at so much faithfulness as to make a model undistinguishable from a theory; while that point would not be accepted in some

camps, it certainly marks a facile pitfall that has proven fatal to some inattentive explorers. With her two protagonists, she then proceeds on a sequence of one-to-one trials of the notion of model *as* or *against* each of the following temptingly close alternate concepts: heuristic device, representation, analogy, conceptual construct, material construct, metaphor; and the varied uses of each as didactive devices, communication tools and providers of consistency arguments. While the landscape is thus painted quite broadly and with exquisitely transparent colors, the light in which it basks is somewhat too soft for the delineation of the specific problem outlined in Sect. 1. Nevertheless, the most pragmatic distinctions she presents, especially towards the end of her book, prove useful to us here; also, in this connection, a more recent reflection on the implementation and role of the metaphor in mathematical machines is presented in [Haken, Karlqvist, and Svedin, 1993].

Corroborating some of the pragmatic arguments that surfaced in Hesse's book, precious insight can be gained from several of the contributions in the collection edited by Freudenthal [1961], most particularly from the classifications proposed therein by Leo Apostel (*Towards the Formal Study of Models in the Non-Formal Sciences*) and by H.J. Groenewold (*The Model in Physics*). Among other interesting ideas discussed there, these authors propose to distinguish models by their functions in the progress of research projects, from its initiation to the communication of its results: experimentation, substitution, simplification, representation, explanation. It would be interesting to examine again this classification in terms of the praxis of the different types of modeling among a wider spectrum of researchers than may be encompassed here: from medicine, epidemiology, engineering, on to biology, chemistry, physics, and all the way to mathematics, logics, and the philosophy of science.

For another systematic approach to the definition and role of models in the philosophy of science, explicit mention should be made of the remarkable selection from Suppes' works already listed above [Suppes, 1993]. Chapter 6 in [Suppes, 1988] is most immediately relevant to some of our present concerns as his interest is to study the structure of a theory in terms of its models. For this, he delineates a notion of isomorphism of models; a theory of representations that goes beyond the use of the term in ordinary discourse; and a notion of invariance to assert the meaningfulness of these constructs. Rather than minutely rehearsing what Suppes says, it ought to be sufficient for our purpose to illustrate the definiteness and the limitations of his scheme by his own choosing as "example of a simple and beautiful representation theorem ... Cayley's theorem that every group is isomorphic to a group of transformations" and his own commentary "One source of the concept of group, as it arouse in the nineteenth century, came from the consideration of ... transformations. It is interesting that the elementary axioms for groups are sufficient to characterize transformations in this abstract sense, namely in the sense that any model of the axioms, *i.e.* any group, can be shown to be isomorphic to a group of transformations." [Suppes, 1988]. He also generalizes his notion of isomorphism of models to a notion of homomorphism, opening thus a tantalizing alley towards a more dynamical meaning for his modeling. Yet, it is patent from the tenor of his arguments, his examples, and his results, that the *theories* his analysis best applies to are relatively stable mathematical theories, rather than more tentative physical theories. In particular, the attempt to make statements about *all* models of a theory is bound to seem a bit outlandish to many physical scientists who are all too aware of the flux of theories they have to contend with. Suppes' line is pursued by De Costa and French [1990] who succeed in pushing into further recesses the limits of this reflection.

In this connection, it may not be too absurdly redundant to recall for an audience of theoretical and mathematical physicists, that a whole branch of mathematical logic, at the confluence of Symbolic Logic and Universal Algebra, is known as Model Theory [Chang and Keisler, 1990]. There, in the vast perspective provided by Carnap and Tarski, one regards the construction of models as the explicit presentation of a bridge between the *postulates* and the *realm of objects*; or said somewhat differently, a *model m* of a *theory T* is a *possible realization* in which all valid sentences of T are satisfied. A model is viewed there as a semantic, *i.e.* a valuation in $\{0, 1\}$, defined on the pair $\mathcal{S} \times \mathcal{A}$ constituted by a syntax $\mathcal{S}$, *i.e.* a collection of finite sentences that are valid with respect to stated syntactic rules; and a collection of structures $\mathcal{A}$. Given Σ, a subcollection of sentences in $\mathcal{S}$, a structure $A \in \mathcal{A}$ is said to be a model for Σ exactly when $m(\phi, A) = 1$ for all $\phi \in \Sigma$. It is remarkable that even such an elementary scheme offers immediately some interesting results, two of which – most particularly the first – are well-worth mentioning in connection with the case we plan to make in Sect. 3 below: they are the so-called compactness and soundness/completeness theorems. Let us define $\Sigma \subset \mathcal{S}$ as *satisfiable* if and only if it has at least one model; and as *finitely satisfiable* if and only if every finite subset of Σ is satisfiable. The *compactness theorem* for first-order logics then asserts that every finitely satisfiable Σ is satisfiable! The second theorem states that a proposition $\phi \in \mathcal{S}$ is provable, within the rules of this syntax $\mathcal{S}$, from a set of axioms $\Sigma \subset \mathcal{S}$, if ('completeness') and only if ('soundness') for every structure $A \in \mathcal{A}$ satisfying Σ, one has $m(\phi, A) = 1$.

Lest we remain suspended in these lofty heights, let us now remember another of Bridgman's admonitions:

> We have ... a pragmatic matter, namely that we have observed after much experience that if we want to do certain kinds of things with our concepts, our concepts had better be constructed in certain ways ... when we push our analysis in the limit ... operations are not ultimately sharp or irreducible any more ... We always run into a haze eventually, and all our concepts are describable only in a spiraling approximation
>
> [Bridgman, 1955]

I wish to retain here three directives from this practical wisdom: (i) our models should be adapted to the purpose of simplification of the issues at hand; (ii) such simplifications almost always involve approximations; and (iii) these approximations, if not necessarily or immediately compatible among themselves, at least should be controllable enough to be accountable in future passes of the spiraling theory and/or experiment.

Towards such a dynamical approach to modeling for theories in gestation, I found support in my reading of [Redhead, 1980]. To avoid a painfully bulky ontological search that may run out of control, I too feel expedient to make at the onset some kind of commitment to a realistic position, albeit as mildly as I can conceive, assuming thereby that there exist somewhere – accessible for at least partial viewings or observations – some things, objects, or systems, in connection with which one can hope to devise theories and which one could explore further with the help of models, thus rendering the 'real world' more 'actual'. Thus, I try never to use hierarchical scales nor to expound the argument beyond the case at hand, namely sorting out the types of models according to their uses in theoretical or mathematical physics; Therefore I do not address such questions as upgrading a model to the status of theory, or downgrading a theory to the status of model. Nonetheless, I argue along with Redhead for the capital importance to be placed on the quality of the trust one is ready to vest in approximations.

E.3 The Case

The notion of approximation is, indeed, the touchstone which I propose to use to discriminate between models as they are used in contemporary theoretical and mathematical physics.

In order to maintain a descriptive, rather than normative, stance I do not attempt to draw strict boundaries but rather proceed from concrete pieces of evidence: my specific examples of models are drawn from the common lore of standard textbooks in physics; your own favorite will do; should you want to check the original motivations, an effort was made to list the primary sources in my own survey of the foundations [Emch, 1984].

Thus I place squarely in the *first* type of models, the shell model of nuclear physics, the Einstein–Debye model for the specific heat of solids, the BCS model of superconductivity, the Feynmann path integrals, the quark model of elementary particles, and the various cosmological models for the birth of the Universe.

In a representative sample of what I have in mind for the *second* type of models we need to consider, I include the Lenz–Ising model for ferromagnetism, a few models in non-equilibrium statistical mechanics, some of the models for the quantum measurement process, most of the models of constructive quantum field theory, the Schwartzschild model for a space-time that accomodates a black hole, and the Dyson–Lenard and Lieb–Thirring models for the stability of matter.

Most of these models have wide currency, and each does involve more or less overt approximations.

To qualify for the first of the above two classes, the models definitely had to have *explanatory* value: being 'realistic' – here the physicist's vernacular is used rather than the philosopher's – is a primary consideration for all of these models, although a dynamics is provided by the fact that physicists will want to argue, here as elsewhere, about whether the models are 'realistic enough', whether they cover a big enough and natural enough range of interesting situations; Wigner is reported to have said something to the effect: "Give me three parameters and I build you a dog, four and I make it wag its tail."

To be admissible in the second class of models, the approximation defining the model had to be explicitly declared at the beginning, the accent being severely placed on these models being exactly solvable – possibly only in terms of precisely controlled limits, but without any mute restrictions that are deemed 'too evident' to be declared – so that these models can be used to *prove the consistency* of the theory from which they proceed. This aspect of the function of these models is thus quite close to the role a model is supposed to play in mathematical logics.

No such undue worry about mathematical rigor slows down the builders of the models of the first type. These colleagues feel secure, as George Uhlenbeck used to say, that as long as nothing explodes, everything should be roughly right, while if some terrible mathematical malversations were committed the physicist would be alerted *a posteriori* by the gross features of the solution itself. What replaces mathematical rigor, to keep such modeling within the bound of the exact sciences, is a strong 'physical intuition' that is surely very elusive when one tries to define it for the skeptic who has not experienced it! The confidence borne out of successful habits seems to deserve more than a small part of the credit. Feynman had been so annointed, but as Mark Kac had warned him – and his Mesmerized audiences – what should the rest of us do once he is gone ...

Once a model of the second kind has been proposed, physical common sense is given, at best, perfunctory credit in what passes for its successful manipulation. Once such a model is stated, much of the skills required for benefiting from it are different in essential manners from the skills involved in exploiting a model of the first kind. Ising, after having solved within the best standards of mathematical rigor of his time the one-dimensional version of the model that now bears his name, indulged in an extraneous bout of 'physical' exhuberance and opined that higher dimensional versions of the model were not to be considered as hopeful candidates for models of magnetism. It was apparently Lenz who had proposed the model to his student.

This dichotomy is thus real and patent. Committed to the attention of the philosophers of science, the recognition of this complementarity might also benefit the science students struggling through their initiation rites. More-over, at the very pragmatic level where scientific constructs become recognizable scientific contributions, it seems as if two thermodynamical phases are presently allowed to separate, even among the scientific journals in our community, each of these phases identifying itself by which one of these two types of models it deems acceptable.

E.4 Closing Statements

One principal remaining question is whether the coexistence of these two modeling activities, in view of the pragmatic distances often inforced by their practitioners, simply have to be accepted on account of the increasing specialization of science.

An optimistic observer may wish rather for a deeper explanation. One may speculate on whether a more constructive explanation could germinate from the ideas of structural stability, whereby one can state qualitatively whether certain behaviors arc stable against perturbations of their structure and thus, in particular, against the removing of some of the approximations used to predict that behavior. Notwithstanding some disappointing statements to the effect that Hamiltonian systems are not generic in the theory of dynamical systems, one might hope that a more careful specification of the categories – *i.e.* the objects and their morphisms – of scientific theories will have to be developed so that, some day soon, this branch of mathematics could help delineate predictively the frontiers to the domain of applicability of specific models envisaged by physicists, be they of the theoretical or of the mathematical persuasions.

A further hope, that some restoring mechanism will prevail, may also be entertained from the fact that models too have histories.

Some models may happen to become obsolete, not only for the obvious reason that their cause has been settled, but also more radically because one cannot – or simply does not – argue in their style anymore. The quaint hardware of the *Gedanken* experiments of the Bohr–Einstein debates have little if any equivalent nowadays in the primary literature: when trying to locate such arguments in the current discourse, one is almost invariably cornered into resorting to secondary sources. While there surely are notable exceptions, the very problems to be addressed in this lecture required us to focus deliberately on the main and developing trends of the current professional literature in theoretical and mathematical physics; in this respect at least, I think that the delineation proposed here between the two classes of models we use is a fair description of the contemporary praxis.

Perhaps more disconcertingly it can happen that, in the course of their history, some models see even their intended features *qua models* change so much as to move them from one class to the other. Two of the examples mentioned in the beginning of the previous section can also serve to illustrate this point. The immediate success of the BCS model was clearly due to its explanatory value: more than a half-dozen of the disparate phenomena collectively referred to as superconductivity had found a unifying mechanism in the interaction of Cooper pairs. However, one soon suspected – and somewhat later actually proved – that the laboratory predictions based on the BCS Hamiltonian may be controlled exactly in the thermodynamical limit: the BCS model had become exactly solvable; furthermore, it turned out to be an exact model establishing the self-consistency of our mathematical and physical descriptions of a more general phenomenon: the spontaneous symmetry breaking occurring in the many-body systems of statistical mechanics and quantum field theory. All the while, the BCS model's importance was fading as an explanatory model for superconductivity at significantly higher temperatures than those one had to work with in the heroic times. The history of the Lenz–Ising model followed the opposite path. After the initial disappointement brought by the solution of its one-dimensional avatar, some of the most important features of the two-dimensional model yielded to a rigorous analysis that showed the compatibility of phase transitions with the framework of a statistical mechanics that incorporates the thermodynamical limit. At this point the model belonged resolutely to the second type in our classification. Yet, minor semantic modifications of the model were made, that extended its modeling range much beyond magnetism. This expansion went hand-in-hand with new predictions of 'universal' relations between various critical exponents, predictions that turned out to have excellent explanatory content for otherwise quite diverse laboratory circumstances.

Hence, the analysis of the specific roles models play in theoretical and mathematical physics also throws light on yet another twist of Nature: when it yields, it does sometimes choose paths that straddle our classifications efforts.

Acknowledgments

I am indebted to Michael Redhead (Cambridge) and Abner Shimony (Boston) for their stimulating and enlightening remarks; to them I wish to express my gratitude, as well as to Chuang Liu and Rick Smith (Gainesville) for their help regarding some of the materials I wanted to explore in this Wigner Symposium lecture.

Note added in proof. As our book was going to print, two papers appeared that illustrate the poles of the modeling world in mathematical physics: [Pillay, 2000] and [Schweber and Wächter, 2000]. The logician views a model as the "real thing" while that of which it is a model (e.g. a set of axioms) is regarded as an "idealization". The theoretical physicist, in contrast, sees in the model a "manageable approximation" of a "complex theory". For the logician, the model should have some universal character. For the theoretical physicist, its nature is relative to the techniques available, so much so that the massive advent of computer simulation is viewed as a revolution; a threshold was passed through which "more" became "different"; Schweber and Wächter cite [Anderson, 1992] who, in turn, refers to Marx: quantitative differences became qualitative ones. It is interesting to compare this with the light in which Pillay [2000] presents the compactness theorem (for a statement of the latter see our Scholium A.2.6).

References

D.B. Abraham, E. Baruch, G. Gallavotti, and A. Martin-Löf. Dynamics of local perturbations in the XY model. I. approach to equilibrium. *Stud. in Appl. Math.*, 50:121–131, 1971. see already *ibid.* Thermalization of a magnetic impurity in the isotropic XY model, *Phys. Rev. Lett.*, 25 (1970) 1449-1450.

D.B. Abraham and A. Martin-Löf. The transfer matrix for a pure phase in the two–dimensional Ising model. *Commun. Math. Phys.*, 32:245–268, 1973.

D.B. Abraham and P. Reed. Interface profile in the two–dimensional Ising model. *Commun. Math. Phys.*, 49:35–46, 1976.

R. Abraham and J.E. Marsden. *Foundations of mechanics (2nd ed.).* Benjamin/Cummings, Reading MA, 1978.

R. Abraham and J. Robbin. *Transversal mappings and flows.* Benjamin/Cummings, Reading MA, 1967.

L. Accardi, A. Frigerio, and J.T. Lewis. Quantum stochastic processes. *Publ. RIMS Kyoto University*, 18:97–133, 1982.

L. Accardi and G. Pistone. de Finetti's theorem, sufficiency, and Dobrushin's theory. In *Exchangeability in probability and statistics*, pages 125–156. North–Holland, Amsterdam, 1982.

L. Accardi and W. Von Waldenfelds, editors. *Quantum probability and applications III.* Springer, Berlin, 1988.

P. Achinstein. Models, analogies and theories. *Philosophy of Science*, 31:328–350, 1964.

P. Achinstein. Theoretical models. *Brit. J. Phil. Sci.*, 16:102–119, 1965.

P. Achinstein. *Concepts of science: a philosophical analysis.* The Johns Hopkins University Press, Baltimore MD, 1968.

P. Achinstein and O. Hannaway, editors. *Observation, experiment and hypothesis in modern physical science.* MIT Press, Cambridge MA, 1985.

W. Ackermann. Zum Hilbertschen Aufbau der reellen Zahlen. *Math. Ann.*, 99: 118–133, 1928.

H.P. Agnew and A.P. Morse. Extensions of linear functionals with applications to limits, integrals, measures, and densities. *Ann. Math. (2)*, 39:20–30, 1938.

C.C. Agosta, I.F. Silvera, H.T.C. Stoof, and B.J. Verhaar. Trapping of neutral atoms with resonant microwave radiation. *Phys. Rev. Lett.*, 62:2361–2364, 1989. C.C. Agosta and I.F. Silvera, Proposed trapping of atomic hydrogen with microwave radiation, *in* Spin–polirized quantum systems, S. Stringari, ed., World Scientific Publishers, Singapore, 1989, pp. 254–257.

M. Aizenmann. Translational invariance and instability of phase coexistence in the two–dimensional Ising system. *Commun. Math. Phys.*, 73:83–94, 1980.

S. Albeverio, Yu.G. Kondratiev, and Yu.V. Kozitsky. Critical properties of a quantum hicrarchical model. *Lett. Math. Phys.*, 40:287–291, 1997.

D.J. Aldous. Exchangeability and related topics. In P.L. Hennequin, editor, *Ecole d'été de probabilités de Saint–Flour XIII* , pages 1–198. LNM 1117, Springer, Berlin, 1985.

P.S. Aleksandrov, N.I. Akhiezer, B.V. Gnedenko, and A.N. Kolmogorov. Sergei Natanovich Bernstein. *Russ. Math. Surveys*, 24:169–176, 1969.

V.M. Alekseev and M.V. Yakobson. Symbolic dynamics and hyperbolic dynamical systems. *Physics Reports*, 75:287–325, 1981.

P. Alexandroff and H. Hopf. *Topologie I*. Springer, Berlin, 1935. reprinted, Chelsea, Bronx NY, 1972.

R. Alicki, M. Bozejko, and W.A. Majewski, editors. *Quantum probability*. Banach Center Publications, Vol. 43, Polish Academy of Sciences, Warsaw, 1998.

R. Alicki and M. Fannes. Defining quantum dynamical entropy. *Lett. Math. Phys.*, 32:75–82, 1994.

R. Alicki and H. Narnhofer. Comparison of dynamical entropies for non–commutative shifts. *Lett. Math. Phys.*, 33:241–247, 1995.

J.F. Allen and A.D. Misener. Flow of liquid Helium II. *Nature*, 141:75, 1938.

G. Alli and G.L. Sewell. New methods and structures in the theory of the multimode Dicke laser model. *J. Math. Phys.*, 36:5598–5626, 1995.

I. Amemiya and H. Araki. A remark on Piron's paper. *Publ. RIMS*, A 2:423–427, 1967.

D.J. Amit. *Field theory, the renormalization group and critical phenomena (rev. 2nd ed.)*. World Scientific, Singapore, 1984.

G. Amontons. Discours sur quelques propriétés de l'air et le moyen d'en connaitre la température dans tous les climats de la terre. *Mém. Acad. Roy. Sci.*, June: 18, 1702.

A.M. Ampère. Idées de M. Ampère sur la chaleur et la lumière. *Bibliothèque Universelle de Genève, Sciences et Arts*, 49:225–235, 1832.

A.M. Ampère. Notes sur la chaleur et la lumière considérées comme résultant de mouvements vibratoires. *Ann. de Chimie*, 58:432–444, 1835.

D.L. Anderson. *The discovery of the electron*. van Nostrand, Princeton NJ, 1964.

M.H. Anderson, J.R. Ensher, M.R. Matthews, C.E. Wieman, and E.A. Cornell. Observation of Bose–Einstein condensation in a dilute atomic vapor. *Science*, 269:198–201, 1995.

P.W. Anderson. Considerations on the flow of superfluid Helium. *Rev. Mod. Phys.*, 38:298–310, 1966. Macroscopic coherence and superfluidity, Contemporary Physics, 1(1969) 47–54; reprinted in [Anderson, 1994][pp. 264–271].

P.W. Anderson. *Basic notions of condensed matter physics*. Benjamin-Cummings, Menlo Park CA, 1984.

P.W. Anderson. More is different. *Science*, 177:393–395, 1992. reprinted in [Anderson, 1994][pp.1–4].

P.W. Anderson. *A Career in theoretical physics*. World Scientific, Singapore, 1994.

M.R. Andrews, C.G. Townsend, H.J. Miesner, D.S. Durfee, D.M. Kurn, and W. Ketterle. Observation of interference between two Bose condensates. *Science*, 275: 637–639, 1997.

T. Andrews. On the continuity of the gaseous and liquid states of matter. *Phil. Trans. Roy. Soc. (London)*, 159:575–590, 1869.

J. Andries, M. Fannes, P. Tuyls, and R. Alicki. The dynamical entropy of the quantum Arnold cat map. *Rev. Math. Phys.*, 8:167–184, 1996.

D.V. Anosov. Geodesic flows on compact Riemannian manifolds of negative curvature. *Proc. Steklov Math. Inst.*, 90:1–235, 1967. Engl. transl. by S. Feder, AMS 1969.

L. Apostel. Towards the formal study of models in the non-formal sciences. In Freudenthal [1961], pages 1–37.

T.M. Apostol. *Mathematical analysis*. Addison–Wesley, Reading MA, 1957.

H. Araki. Gibbs states of a one dimensional quantum lattice. *Commun. Math. Phys.*, 14:120–157, 1969. see also [Araki, 1970].

H. Araki. One dimensional quantum lattice system. In Michel and Ruelle [1970], pages 75–86.

H. Araki. On the XY model on two–sided infinite chain. *Publ. RIMS Kyoto*, No. 435: 1–33, 1983.

H. Araki and E. Baruch. On the dynamics and ergodic properties of the XY model. *J. Stat. Phys.*, 31:327–345, 1983.

H. Araki and H. Miyata. On KMS boundary condition. *Publ. RIMS Kyoto*, A4: 373–385, 1968.

H. Araki and E.J. Woods. Representations of the canonical commutation relations describing a nonrelativistic free Bose gas. *J. Math. Phys.*, 4:637–662, 1963.

D.M. Armstrong. *What is a law of nature?* Cambridge University Press, Cambridge, 1983.

V.I. Arnold. Small denominators II. Proof of a theorem of A.N. Kolmogorov on the preservation of conditionally–periodic motions under small perturbations of the Hamiltonian. *Usp. Math. Nauk*, 18:13–40, 1963. Russian Math. Surveys **18**(1963) 9-36.

V.I. Arnold. *Mathematical methods of classical mechanics*. Springer, New York, 1978.

V.I. Arnold and A. Avez. *Ergodic problems of classical mechanics*. Benjamin, New York, 1968.

K.J. Arrow. *Social choice and individual values; Monograph 12; Cowles Foundation for Research in Economics at Yale University*. Yale University Press, New Haven CT; Originally published by Wiley, New York, 1951,1963.

K.J. Arrow. Exposition of the theory of choice under uncertainty. In C.B. McGuire and R. Radner, editors, *Decision and Organization, Studies in mathematical and managerial economics, vol.12*. North–Holland, Amsterdam, 1972.

R.A. Ash. *Real analysis and probability*. Academic Press, New York, 1972.

J. Ashkin and W.E. Lamb. The propagation of order in crystal lattices. *Phys. Rev.*, 64:159–178, 1943.

A. Aspect, J. Dalibard, and G. Roger. Experimental test of Bell's inequalities using time–varying analyzers. *Phys. Rev. Lett.*, 49:1804 1807, 1982. see also: A. Aspect, Trois tests experimentaux des inegalités de Bell par mesure de corrélation de polarisation de photons, Thèse, Orsay, 1983; and J.F. Clauser, M.A. Horn, A. Shimony and R.A. Holt, Proposed experiment to test local hidden–variable theories, Phys. Rev. Lett. 23, 880-884, 1969.

Augustine of Hippo. *De Musica*. n.a., Colonia Julia Carthago, 389.

A. Bach. Boltzmann's distribution of 1877. *Arch. Exact Sci.*, 41:1–40, 1990.

V. Bach, J. Fröhlich, and I.M. Sigal. Return to equilibrium. *J. Math. Phys.*, 41: 3985–4060, 2000.

L. Bachelier. *Théorie de la spéculation*. Gauthier-Villars, Paris; also published in *Ann. ENS* 17, 21–86, 1900.

F. Bacon. *Novum organum*. (transl. by R. Ellis and James Spedding; with preface and notes; Routledge, London, 1905), 1560.

F. Bagarello and G.L. Sewell. New structures in the theory of the laser model II. Microscopic dynamics and a nonequilibrium entropy principle. *J. Math. Phys.*, 39:2730–2747, 1998.

L. Bahar and J. Spencer. Paul Erdös (1913–1996). *Notices AMS*, 45:64–66, 1998.

D.H. Bailey, J.M. Borwein, P.B. Borwein, and S. Plouffe. The quest for pi. *Math. Intelligencer*, 19:50–57, 1997.

T.A. Bak, editor. *Statistical mechanics, foundations and applications.* Benjamin, New York, 1967.

K.M. Baker. *Condorcet, from natural philosophy to social mathematics.* University Chicago Press, Chicago IL, 1975.

G.A. Baker, Jr. One–dimensional order–disorder model which approaches phase transition. *Phys. Rev.*, 122:1477–1484, 1961.

G.A. Baker, Jr. Certain general order–disorder models in the limit of long–range interaction. *Phys. Rev.*, 126:2075–2078, 1962.

G.A. Baker, Jr. Ising model with a scaling interaction. *Phys. Rev. D*, 5:2622–2633, 1972. See also: G. A. Baker, Jr. and G.R. Golner, Spin–spin correlations in an Ising model for which scaling is exact, Phys. Rev. Lett. 31 (1973) 22–25.

G.A. Baker, Jr. *Quantitative theory of critical phenomena.* Academic Press, San Diego CA, 1990.

R.M. Balian. *From microphysics to macrophysics, methods and applications of statistical physics; 2 vol.* D. ter Haar and J.F. Gregg, transl.,Springer, New York, 1991,1992.

W. Balzer, C.U. Moulines, and J.D. Sneed. *An architectonic for science: the structuralist program.* D. Reidel, Dordrecht, 1987.

S. Banach. Sur les fonctionelles linéaires I, II. *Studia Math.*, 1:211–216, 223–239, 1929.

S. Banach. *Théorie des opérations linéaires.* Monografje Matematyczne, Warsaw, 1932. 2nd ed.: Chelsea, New York, 1978.

J. Bardeen. Wave functions for superconducting electrons. *Phys. Rev.*, 80:567–574, 1950. see also: Zero-point vibrations and superconductivity, *ibid.* 79 (1950) 167–168; Relation between lattice vibrations and London theories of superconductivity *ibid.* 81 (1951) 829–834.

J. Bardeen. Theory of the Meissner effect in superconductors. *Phys. Rev.*, 97: 1724–1725, 1955.

J. Bardeen. Theory of superconductivity. In Flügge [1956], pages 274–369. Compare with *ibid.* Physica Suppl. 24 (1958) 27–34.

J. Bardeen. Electron-phonon interactions and superconductivity. *Nobel acceptance speech*, 1972. Reprinted in [Nobel Foundation, 1998] V: 54–69.

J. Bardeen, L.N. Cooper, and J.R. Schrieffer. Theory of superconductivity. *Phys. Rev.*, 108:1175–1204, 1957. Announcement in *ibid.* 106 (1956) 162–164.

J. Bardeen and J.R. Schrieffer. Recent developments in superconductivity. *Progr. Low Temp. Physics*, III:170–287, 1961.

J. Barone and A. Novikov. A history of the axiomatic formulation of probability from Borel to Kolmogorov. *Arch. Hist. Exact Sci.*, 18:123–190, 1977/78.

W.F. Barr. A syntactic and semantic analysis of idealizations in science. *Philosophy of Science*, 38:258–272, 1971.

W.F. Barr. A pragmatic analysis of idealizations in physics. *Philosophy of Science*, 41:48–64, 1974.

J. Barrow-Green. *The three–body problem.* Amer. Math. Soc., Providence RI, 1997.

R.W. Batterman. Irreversibility and statistical mechanics: a new approach. *Phil. Sci.*, 57:395–419, 1990.

R.W. Batterman. Why equilibrium statistical mechanics works; universality and the renormalization group. *Phil. Sci.*, 65:183–208, 1998.

R.W. Batterman. Multiple realizability and universality. *Brit. J. Phil. Sci.*, 51: 115–145, 2000.

M. Baxter and A. Rennie. *Financial calculus.* Cambridge University Press, Cambridge, 1996.

R.J. Baxter. *Exactly solved models in statistical mechanics.* Academic Press, London, 1982.

T. Bayes. An essay towards solving a problem in the doctrine of chances. *Phil. Trans. Roy. Soc. London*, 53:370–418, 1763. See also *ibid.* **54** (1764) 296–335 [1765]; and for a modern edition: E.S. Pearson and M.G. Kendall, *'Studies in the History of Statistics and Probability'*, Griffin, London, 1970, pp. 134–153.

A.F. Beardon. *The geometry of discrete groups*. Springer, New York, 1983.

C. Beccaria. *Dei delitti e delle pene*. s.n., Livorno, 1764. Engl. transl. D. Young, Hackett, Indianapolis, 1986.

J.G. Bednorz and K.A. Müller. Perovskite–type oxides – the new approach to high-T superconductivity. *Nobel acceptance speech*, 1987. Reprinted in [Nobel Foundation, 1998] VI: 424 – 457.

V.P. Belavkin. Quantum Ito B*-algebras, their classification and decomposition. In Alicki et al. [1998], pages 63–70.

J.L. Bell and A.B. Slomson. *Models and ultraproducts*. North–Holland, Amsterdam, 1971.

J.S. Bell. On the Einstein–Podolsky–Rosen paradox. *Physics*, 1:195–200, 1964. 'On the problem of hidden variables in quantum mechanics', Rev. Mod. Phys. 38, 447–452, 1966; *Speakable and unspeakable in quantum mechancs*, Cambridge University Press, Cambridge, 1987.

J. Bellissard. K–theory of C*-algebras in solid state physics. In T.C. Dorlas, N.M. Hugenholtz, and M. Winnink, editors, *Statistical mechanics and field theory*, pages 99–157. Springer, Berlin, 1986.

F. Benatti, H. Narnhofer, and G.L. Sewell. A noncommutative version of the Arnold cat map. *Lett. Math. Phys.*, 21:157–192, 1991.

P.L. Bender and C.J. Gorter. A few remarks on the two–field model for superconductors. *Physica*, 18:597–604, 1952.

G. Benettin, G. Gallavotti, G. Jona-Lasinio, and A. Stella. On the Onsager–Yang value of the spontaneous magnetization. *Commun. Math. Phys.*, 30:45–54, 1973.

G. Benfatto and G. Gallavotti. *Renormalization group*. Princeton University Press, Princeton NJ, 1995.

C.H. Bennet. The thermodynamics of computation – a review. *Intern. J. Theor. Phys.*, 21:905–940, 1982. see also: *Demons, engines, and the 2nd law*, Scientific American **257** (1987) 108–116.

J.O. Berger, B. Boukai, and Y. Wang. Unified frequentist and Bayesian testing of a precise hypothesis. *Statistical Science*, 12:133–148, 1997. see also: D.V. Lindley, 'Comment', *ibid.*, 149–152; T.A. Louis, 'Comment', *ibid.*, 152–154; D. Hinkley, 'Comment', *ibid.*, 155–156; J.O. Berger et al., 'Rejoinder', *ibid.*, 156–160; S.E. Feinberg, 'Introduction to R.A. Fisher's "On inverse probability and likelihood"', *ibid.*, 161; J. Aldrich, 'R.A. Fisher and the making of maximum likelihood 1912–22', *ibid.*, 162–176; A.W.F. Edwards, 'What did Fisher mean by "inverse probability" in 1912-1922', *ibid.*, 177–182.

I. Berggren, J. Bornwein, and P. Bornwein. *Pi: A source book*. Springer, New York, 1999.

T.D. Berlin and M. Kac. The spherical model of a ferromagnet. *Phys. Rev.*, 86: 821–835, 1952.

Claude Bernard. *Introduction à la médecine expérimentale*. J.–B. Baillière, Paris, 1865.

D. Bernoulli. *Hydrodynamica*. Dulsecker, Argentoratum (=Strasbourg), 1738.

D. Bernoulli. Essai d'une nouvelle analyse de la mortalité causée par la petite vérole et des avantages de l'inoculation pour la prévenir. *Mem. Acad. Sci. Paris pour 1760,*, pages 1–43, 1760–1765. (Read 16 April 1760, it was published only in 1765, however with a new and loaded *'Introduction apologétique'*).

Jakob Bernoulli. *Ars Conjectandi (Opus posthumum)*. Imp. Thurnis. Fraetr., Basileae (=Basel), 1713.

F. Bernstein. Ueber eine Anwendung der Mengenlehre auf ein aus der Theorie des säkularen Störungen herrührendes Problem. *Math. Ann.*, 71:417–439, 1912.

S. Bernstein. On the axiomatic foundation of the theory of probability (in Russian). *Mitt. Math. Ges. Charkov = Comm. Soc. Math. Kharkov*, 15:209–274, 1917.

G. Bertin and L.A. Radicati. The bifurcation from the Maclaurin to the Jacobi sequence as a second–order phase transition. *Astrophys. J.*, 206:815–821, 1976.

J. Bertrand. *Calcul des probabilités*. Gauthier-Villars, Paris, 1888.

E.W. Beth. Semantics of physical theories. In Freudenthal [1961], pages 48–51.

E.W. Beth. Towards an up-to-date philosophy of the natural sciences. *Methodos*, 1:178–185, 1949.

P. Billingsley. *Ergodic theory and information*. Wiley, New York, 1965.

M.A. Biondi, A.T. Forrester, M.O. Garfunkel, and C.B. Satterthwaite. Experimental evidence for an energy gap in superconductors. *Rev. Mod. Phys.*, 30: 1109–1136, 1958.

M.A. Biondi, M.P. Garfunkel, and A.O. Coubrey. Millimeter–wave absorbtion in superconducting Aluminium. *Phys. Rev.*, 101, 1956. See also: Biondi, Garfunkel and Coubrey, 'Microwave measurements of the energy gap in Aluminium,' Phys. Rev. 108 (1957) 495-497; Biondi and Garfunkel, 'Millimeter wave absorption in superconducting Aluminium, I. Temperature dependence of the energy gap, II. Calculation of skin depth,' Phy. Rev. 116 (1959) 853–861, 862–867.

G. Birkhoff. *Lattice theory*. Amer.Math.Soc., New York, Providence, 1940. rev. 1949; prelim. 3rd ed. 1963; 3rd ed. 1967.

G. Birkhoff and J. Von Neumann. The logic of quantum mechanics. *Ann. of Math.*, 37:823–843, 1936.

G.D. Birkhoff. Proof of the ergodic theorem. *Proc. Nat. Acad. Sci. USA*, 17: 656–660, 1931.

G.D. Birkhoff. What is the ergodic theorem. *Amer. Math. Monthly*, 49:222–226, 1942.

G.D. Birkhoff and P.A. Smith. Structure analysis of surface transformations. *Journ. de Math. Pures et Appl.*, 7:345–379, 1928.

D.J. Bishop and J.D. Reppy. Study of the superfluid transition in two–dimensional ^{4}He films. *Phys. Rev. Lett.*, 40:1727–1730, 1978.

J. Black. *Lectures on the elements of chemistry*. J. Robison, ed.; Longman and Rees, Edinburgh, 1803.

W. Black. *An arithmetical and medical analysis of the diseases and the mortality of the human species*. Dilly, London, 1789.

P.M. Bleher and Ya. G. Sinai. Critical indices for the Dyson's asymptotically hierarchical models. *Commun. Math. Phys.*, 45:247–278, 1975. See already *ibid.* 'Investigation of the critical point in models of the type of Dyson's hierarchical model,' *Commun. Math. Phys.*, 33 (1973) 23–42.

P.M. Bleher and P. Major. Critical phenomena and universal exponents in statistical physics: on Dyson hierarchical model. *Ann. Prob.*, 15:431–477, 1987.

G.S. Blevins, W. Gordy, and W.M. Fairbank. Superconductivity and millimeter wave frequencies. *Phys. Rev.*, 100:1215–1216, 1955.

F. Bloch. Ueber die Quantenmechanik der Elektronen in Kristallgittern. *Zeit. d. Phys.*, 52:555–600, 1928.

F. Bloch. Nuclear induction. *Phys. Rev.*, 70:460–474, 1946.

F. Bloch and R.K. Wangness. The dynamical theory of nuclear induction. *Phys. Rev.*, 89:728–739, 1953.

J.R. Blum and D.L. Hanson. On the isomorphism problem for Bernoulli schemes. *Bull. Am. Math. Soc.*, 69:221–223, 1963.

T. Bodineau. The Wulff construction in three and more dimensions. *Commun. Math. Phys.*, 207:197–229, 1999.

T. Bodineau. *A microscopic derivation of 3D equilibrium crystal shapes.* IAMP Congress, London, 2000.

N.N. Bogoliubov. On the theory of superfluidity. *J. Phys. USSR*, 11:23–32, 1947.

N.N. Bogoliubov. On a new method in the theory of superconductivity. *Nuovo Cim.*, 7:794–805, 1958.

N.N. Bogoliubov and D.V. Shirkov. *Introduction to the theory of quantized fields.* Interscience, New York, 1959.

P. Bohl. Ueber ein in der Theorie der säkulare Störungen vorkommendes Problem. *Crelle Journ. f. Mathematik*, 135:189–283, 1909.

G. Bohlmann. Lebensversicherung–Mathematik. In *Encyclopädie der mathematischen Wissenschaften, Bd. I-2, Hft. 6*, pages 852–917. Teubner, Leipzig, 1901.

D. Bohm. A suggested interpretation of the quantum theory in terms of "hidden variables" I and II. *Phys. Rev.*, 85:166–179, 180–193, 1952.

N. Bohr. *Studier over Metallernes Elektrontheori.* Thaning and Appel, København, 1911. Reprinted w. transl. and notes in *Collected works*, L. Rosenfeld and I.R. Nielsen, eds. North–Holland, Amsterdam, 1972, 1:167–290, 291–395.

L. Boltzmann. Studien über das Gleichgewicht der lebendigen Kraft zwischen bewegten materiellen Punkten. *Sitzungsber. Akad. Wiss., Wien*, 58:517–560, 1868.

L. Boltzmann. Lösung eines mechanischen Problems. *Wien. Ber.*, 58:1035–1044, 1869.

L. Boltzmann. Einige allgemeine Sätze über Wärmegleichgewicht unter Gasmolekülen. *Sitzungsber. Akad. Wiss., Wien*, 63:679–711, 1871a.

L. Boltzmann. Ueber das Wärmegleichgewicht zwischen mehratomigen Gasmolekülen. *Wien. Ber.*, 63:397–418, 1871b.

L. Boltzmann. Weitere Studien über das Wärmegleichgewicht unter Gasmolekülen. *Sitzungsber. Akad. Wiss., Wien*, 66:275–370, 1872.

L. Boltzmann. Bemerkungen über einige Probleme der mechanischen Wärmetheorie; Ueber die Beziehung zwischen den zweiten Hauptsätze der mechanischen Wärmetheorie und der Wahrscheinlichkeitsrechnung resp. den Sätzen über das Wärmegleichgewicht; Weitere Bemerkungen über einige probleme der mechanischen Wärmetheorie. *Sitzungsber. Akad. Wiss., Wien*, 75, 76, 78: 62–100, 373–435, 7–46, 1877, 1878.

L. Boltzmann. Ueber die Magnetizierung eines Ringes. *Wien. Anz.*, 17:12–13, 1880. 'Ueber die absolute Geschwindigkeit der Elektizität im Elektrischen Strome', Phil. Mag. **9** (1880) 308–309.

L. Boltzmann. Referat über die Abhandlung von J.C. Maxwell "Ueber Boltzmanns Theorie betreffend die mittlere Verteilung der lebendige Kraft in einem System materieller Punkte". *Wied. Ann. Beiblätter*, 5:403–417, 1881.

L. Boltzmann. Ableitung der Stefan'schen Gesetzes betreffend die Abhängigkeit der Wärmestrahlung von der Temperatur aus der elektromagnetischen Lichttheorie. *Ann. d. Physik*, 22:291–294, 1884. see already: L. Boltzmann, Ueber eine von Hrn. Bartoli entdeckte Beziehung der Wärmestrahlung zum zweiten Haupsatze, Ann. d. Physik 22(1884) 31–39.

L. Boltzmann. Ueber die Eigenschaften monozyklischer Systeme. *J. reine u. angew. Math.*, 98:68–94, 1884/5.

L. Boltzmann. Ueber die mechanischen Analogien des zweiten Hauptsätzen der Thermodynamik. *J. reine u. angew. Math.*, 100:201–212, 1887.

L. Boltzmann. III. Teil der Studien über Gleichgewicht der lebendigen Kraft. *Sitz.math–phys. Klasse, Köning. Bayer. Akad. Wiss. zu München*, 22:329–358, 1892.

L. Boltzmann. Ueber die mechanische Analogie des Wärmegleichgewichtes zweier sich berührender Körper. *Sitzungsber. Akad. Wiss., Wien*, 103:1125–1134, 1894. (with G.H. Bryan), also: Proc. Phys. Soc. London, **13** (1895) 485–493.

L. Boltzmann. On certain questions of the theory of gases. *Nature*, 51:413–415, 1895a.

L. Boltzmann. On the minimum theorem in the theory of gases. *Nature*, 52:221, 1895b.

L. Boltzmann. Entgegnung auf die wärmetheoretischen Betrachtungen des Hrn. E. Zermelo. *Ann. der Physik*, 57:773–784, 1896. continued in: 'Zu Hrn. Zermelos Abhandlung "Ueber die mechanische Erklärungen irreversibler Vorgäge" ', *ibid* **60** (1897) 392–398.

L. Boltzmann. *Vorlesungen über Gastheorie*. J.A. Barth, Leipzig, 1896–98, 1896–1898. = *Lectures on gas theory,* S.G. Brush, transl., UC Press, Berkeley CA, 1964.

L. Boltzmann. *Vorlesungen über die Principe der Mechanik*. Barth, Leipzig, 1897. For some aftertaste, see: 'Ueber die Prinzipien der Mechanik: zwei akademische Antrittreden', S. Hirzel, Leipzig, 1903; collected in *Populäre Schriften*, Barth, Leipzig, 1905, pp. 308 and 330.

L. Boltzmann. Ueber die sogenannte H-Kurve. *Math. Ann.*, 50:325–333, 1898.

R. Bombelli. *L'algebra : parte maggiore dell'arimetica; divisa in tre libri*. Rossi, Bologna, 1572.

V.L. Bonch-Bruevich and S.V. Tyablikov. *The Green function method in statistical mechanics*. North–Holland, Amsterdam, 1962.

F. Bonetto, J.L. Lebowitz, and L. Rey-Bellet. Fourier law: a challenge to Theorists. In Fokas et al. [2000], pages 128–150.

G.S. Boolos and R.C. Jeffrey. *Computability and logic (2nd ed.)*. Cambridge University Press, Cambridge, 1980.

E. Borel. *Leçons sur la théorie des fonctions*. Gauthier-Villars, Paris, 1898. 2nd ed. 1914.

E. Borel. *Eléments de la théorie des probabilités*. Hermann, Paris, 1909a.

E. Borel. Les probabilités dénombrables et leurs applications arithmétiques. *Rend. Circ. Mat. Palermo*, 27:247–271, 1909b.

E. Borel. Mécanique statistique et irreversibilité. *J. Phys.*, 3 (5e ser.):189–196, 1913.

E. Borel. Exposé français de l'article sur la mécanique statistique [de P. et T. Ehrenfest (1911)]. *Encyclopédie des Sc. math. 4, vol.I, suppl. II*, pages 272–292, 1914a.

E. Borel. *Introduction géometrique à quelques théories physiques*. Gauthier-Villars, Paris, 1914b.

E. Borel. *Le hasard*. Alcan, Paris, 1914c.

E. Borel. *Oeuvres (4 vols.)*. Editions du CNRS, Paris, 1972.

D. Bornwein, J.M. Bornwein, and P. Maréchal. Surprise maximization. *Amer. Math. Monthly*, 107:517–527, 2000.

S.N. Bose. Plancks Gesetz und Lichtquantenhypothese. *Zeits. d. Phys.*, 26:178–181, 1924.

N. Bourbaki. *Intégration*. Hermann, Paris, 1952.

R. Boyle. *New experiments physico-mechanical, touching the spring of the air, and its effects;* and *A defense of the doctrine touching the spring and weight of air.* T. Robinson, Oxford, London, 1660, 1662.

C.C. Bradley, C.A. Suckett, J.J. Tollet, and R.G. Hulet. Evidence of Bose-Einstein condensation in an atomic gas with attractive interactions. *Phys. Rev. Lett.*, 75:1687–1690, 1995. *ibid.* Bose-Einstein condensation of Lithium: Observation of limited condensate number, 78, 985-989, 1995.

L. Bradley. *Smallpox inoculation, an eighteenth–century mathematical controversy; translations and commentary.* University Publishers, Nottingham, 1971.

R.B. Braithwaite. Models in the empirical sciences. In Nagel et al. [1962], pages 224–231.

O. Bratteli and D.W. Robinson. *Operator algebras and quantum statistical mechanics.* Springer, New York, 1979, 1987.

W. Braun and K. Hepp. The Vlasov dynamics and its fluctuations in the 1/N limit of inteacting classical particles. *Commun. Math. Phys.*, 56:101–113, 1977.

D.M. Bressoud. *Second–year calculus.* Springer, New York, 1991.

P.W. Bridgman. *The logic of modern physics.* Macmillan, New York, 1927. Arno Press, New York, 1980.

P.W. Bridgman. *Reflections of a physicist.* Philosophical Library, New York, 1955.

L. Brillouin. *Science and information theory. 2d ed.* Academic Press, New York, 1962. 1st ed. 1952; see also 'Maxwell's demon cannot operate: information and entropy, I', J. Appl. Phys. **22** (1951) 334–337; *ibid. II*, 338-343; also: 'The negentropy principle of information', *ibid.* **24** (1953) 1152–1163.

Britannica. *Principles of thermodynamics, statistical thermodynamics: the Boltzmann factor and the partition function.* Britanica.com, http://www.-britanica.com/bcom/eb/article/2/ 0,5716,115212+2,00.html, online.

T.A. Brody. *The philosophy behind physics.* Springer, Berlin, 1993.

L. E.J. Brouwer. Beweis der Invarianz der Dimensionzahl. *Math. Annalen*, 70:161–165, 1910. See also; *ibid.* **71**, *305–313*, 1912; **71**, *314–319*, 1912; **72**, *55–56*, 1912.

F.E. Browder, editor. *The mathematical heritage of Henri Poincaré.* Amer. Math. Soc., Providence RI, 1983.

A. Brown, M.W. Zemansky, and H.A. Boorse. Behavior of the heat capacity of superconducting Niobium below 4.5°K. *Phys. Rev.*, 86:134–135, 1952.

J.R. Brown. *Ergodic theory and topological dynamics.* Academic Press, New York, 1976.

J.R. Brown. Explaining the success of science. *Ratio*, 27:49–66, 1985.

R. Brown. A brief account of microscopic observations made ... on the particles contained in the pollen of plants; and on the general existence of active molecules in organic and inorganic bodies; Additional remarks. *Philosophical Magazine*, N.S. 4; 6:161–173; 161–166, 1828, 1829.

A.A. Brudno. The complexity of the trajectories of a dynamical system. *Russian Math. Surveys*, 33:197–198, 1978.

A.A. Brudno. Entropy and the complexity of the trajectories of a dynamical system. *Trans. Moscow Math. Soc.*, 2:127–151, 1983.

S.G. Brush. Foundations of statistical mechanics 1845–1915. *Arch. Hist. Exact Sci.*, 4:145–183, 1967.

S.G. Brush. *The kind of motion we call heat, 2 vols.* North–Holland, Amsterdam, 1976.

S.G. Brush. *Statistical physics and the atomic theory of matter, from Boyle and Newton to Landau and Onsager.* Princeton University Press, Princeton NJ, 1983.

H.A. Buchdahl. *The concepts of classical thermodynamics.* Cambridge University Press, Cambridge, 1966.

D. Buchholz, C. D'Antoni, and K. Fredenhagen. The universal structure of local algebras. *Commun. Math. Phys.*, 111:123–135, 1987.

J.R. Buchler, J.M. Perdang, and E.A. Spiegel, editors. *Chaos in astrophysics.* D. Reidel, Dordrecht, 1985.

M. Bunge, editor. *Delaware seminar in the foundations of physics.* Springer, New York, 1967.

S.H. Burbury. Boltzmann's minimum function. *Nature*, 51:78, 1894.

K. Burnett, M. Edwards, and C.W. Clark. The theory of Bose–Einstein condensation of dilute gas. *Physics Today (Dec.99)*, 52:37–42, 1999.

P. Busch, J.P. Lahti, and P. Mittelstaedt. *The quantum theory of measurement.* Springer, Berlin, 1991.

G. Callender. The view from no-when. *Brit. J. Phil. Sci.*, 49:135–157, 1998.

C. Calude. *Information and randomness: an algorithmic perspective.* Springer, Berlin, Heidelberg, 1994.

C. Calude. What is a random string? In *The foundation debate – complexity and constructivity in mathematics and physics*, pages 101–113. Vienna Circle Yearbook, Kluwer, Dordrecht, 1995.

C. Calude and I Chitescu. A combinatorial characterization of P. Martin–Löf tests. *Intern'l J. Comput. Math.*, 17:53–64, 1988.

A. Camus. *Caligula.* NRF, Gallimard, Paris, 1947.

D.M. Cannell. George Green: An enigmatic mathematician. *Amer. Math. Monthly*, 106:137–151, 1999. *ibid.*, George Green, mathematician and physicist, 1793–1841, Athlone Press, London, 1993.

F.P. Cantelli. Sulla legge dei grandi numeri. *Mem. Acad. Lincei*, 11:330–349, 1916. See also: Sulla probabilità come limite della frequenza, Rend. Acad. Lincei **26** (1917), 39–45.

F.P. Cantelli. Una teoria astratta del calcolo delle probabilità. *Giornale dell' Istituto Italiano degli Attuari*, 3:257–263, 1932.

G. Cantor. Ein Betrag zur Mannigfaltigkeitslehre. *J. reine u. angew. Math.*, 84: 242–258, 1878.

C. Carathéodory. Untersuchen über die Grundlagen der Thermodynamik. *Math. Ann.*, 67:355–386, 1909. See also *ibid.* 'Ueber die Bestimmung der Energie und der absoluten Temperatur mit Hilfe von reversiblen Prozessen', Sitz. Ber. Preuss. Akad. Wiss. Phys. Math. Kl. (1925) 39–47. For a middle-of-the-road discussion, see [Falk and Jung, 1959].

C. Carathéodory. Ueber den Wiederkehrsatz von Poincaré. *Sitz. Preuss. Akad. Wiss.*, pages 579–584, 1919.

H. Cardano. *Practica arithmetice, mensurandi singularis.* Bernardini Calusci, Mediolani, 1539. (also *Liber de ludo aleae* (1564) = *Book on games of chance*, S.H. Gould, transl. *in* O. Ore: *Cardano, The gambling scholar*, Princeton University Press, Princeton NJ, 1953).

J. Cardy. *Scaling and renormalization in statistical physics.* Cambridge University Press, Cambridge, 1996.

H.J. Carmichael. *Statistical methods in quantum optics I: Master equations and Fokker-Planck equations.* Springer–Verlag, New York, 1999.

R. Carnap. *Logical foundations of probability.* University of Chicago Press, Chicago IL, 1950.

R. Carnap. *The continuum of inductive methods.* University of Chicago Press, Chicago IL, 1952.

R. Carnap. Replies and expositions. In Schlipp [1963], pages 966–998.

N.L. Sadi Carnot. *Réflexions sur la puissance motrice du feu, et sur les machines propres à développer cette puissance.* Bachelier, Paris, 1824. English transl. in [Mendoza, 1960].

V. Carraud. *Pascal et la philosophie.* Presses Universitaires de France, Paris, 1992.

J.W. Carroll. *Laws of nature.* Cambridge University Press, Cambridge, 1994.

E. Cartan and A. Einstein. *Letters on absolute parallelism, 1929–1932*. Princeton University Press, Princeton NJ, 1979. R. Debever, ed.; incl. original text and English transl. by J. Leroy and J. Ritter; the excerpt in our Scetion 9.1 is from a letter from AE to EC dated 13.II.30 and is directed there to the introduction of the cosmological constant.

N. Cartwright. *How the laws of physics lie*. Clarendon Press, Oxford, 1983.

N. Cartwright. *Nature's capacities and their measurement*. Clarendon Press, Oxford, 1989.

N. Cartwright. Models: the blueprints for laws. *Philosophy of Science*, 64:S292–S303, 1997.

J. Case. The modeling and analysis of financial time series. *AMS Monthly*, 105:401–411, 1998.

F. Castella, L. Erdos, F. Frommlet, and P.A. Markowich. Fokker–Planck equations as scaling limits of reversible quantum systems. *J. Stat. Phys.*, 100:543–561, 2000.

C. Cercignani, R. Illner, and M. Pulvirenti. *The mathematical theory of dilute gases*. Springer, New York, 1994.

C. Cercignani and D.H. Sattinger. *Scaling limits and models in physical processes*. Birkhäuser, Basel, 1998. for an update on the BBGKY hierarchy, see also C. Cercignani, V.I. Gerasimenko and D. Ya Petrina, Many–particle dynamics and kinetic equations, Kluwer, Dordrecht, 1997.

R. Cerf. Large deviations for the three dimensional supercritical percolation. *Astérisque*, 267:1–177, 2000.

F. Cernushi and H. Eyring. An elementary theory of condensation. *J. Chem. Phys,*, 7:547–551, 1939.

F. Cesi, G. Guadagni, F. Martinelli, and R.H. Schonmann. On the 2D stochastic Ising model in the phase coexistence region near the critical point. *J. Stat. Mech.*, 85:55–102, 1996.

G.J. Chaitin. On the length of programs for computing finite binary sequences. *J. ACM*, 13:547–569, 1966. *ibid.* **16** (1969) 145-159; also for a collection of more recent papers by this author: *Information, randomness and incompleteness (2nd ed.)*, World Scientific, Singapore, 1990; and for a systematic exposition: [Chaitin, 1987].

G.J. Chaitin. Information–theoretic computational complexity. *IEEE Trans. In form. Theory*, IT 20:10–15, 1974. reproduced in G.J. Chaitin, *Information, randomness and incompleteness* (2nd ed.), World Scientific, Singapore, 1990.

G.J. Chaitin. *Algorithmic information theory*. Cambridge University Press, Cambridge, 1987.

D.G. Champernowne. The construction of decimals normal in the scale of ten. *J. London Math. Soc.*, 8:254–260, 1933.

S. Chandrasekhar. *Ellipsoidal figures of equilibrium*. Yale University Press, New Haven CN, 1969.

S. Chandrasekhar. *Newton's Principia for the Common Reader*. Clarendon Press, Oxford, 1995.

C.C. Chang and H.J. Keisler. *Model theory*. North-Holland, Amsterdam, 1990.

S. Chapman and T.G. Cowling. *The mathematical theory of non–uniform gases*. Cambridge University Press, Cambridge, 1932.

P.R. Chernoff. Mathematical obstructions to quantization. *Hadronic Physics*, 4:879–898, 1981.

N.I. Chernov, G.L. Eyink, J.L. Lebowitz, and Y. Sinai. Steady state electric conductivity in the periodic Lorentz gas. *Commun. Math. Phys.*, 154:569–601, 1993.

628   References

C. Chevalley. *Pascal: contingence et probabilités*. Presses Universitaires de France, Paris, 1995.

F. Chevy, K.W. Madison, and J. Dalibard. Measurement of the angular momentum of a rotating Bose–Einstein condensate. *Phys. Rev. Lett.*, 85:2223–2227, 2000.

M.D. Choi. Positive linear maps on C*-algebras. *Can. J. Math.*, 24:520–529, 1972.

K.L. Chung. Probability and Doob. *Math. Monthly*, 105:28–35, 1998.

A. Church. On the concept of random sequence. *Bull. Amer. Math. Soc.*, 47: 130–135, 1940.

B.P. Clapeyron. Mémoire sur la puissance motrice de la chaleur. *Journ. Ecole Polytechnique*, XIV, 23:153–190, 1834.

R. Clausius. Ueber die Art der Bewegung, welche wir Wärme nennen. *Ann. d. Physik, ser.2*, 100:353–390, 1857.

R. Clausius. Ueber die mittlere länge der Wege, welche bei der Molecülen zurück gelegt werden, nebst einige anderen Bermerkungen über die mechanische Wärmetheorie. *Ann. der Physik, ser.2*, 105:239–258, 1858.

R. Clausius. Ueber verschiedene für die Anwendungen bequeme Formen der Haupgleichungen der mechanishen Wärmetheorie. *Ann. d. Physik*, 125:353–400, 1865.

R. Clausius. *The mechanical theory of heat*. W.R. Browne, transl., MacMillan, London, 1879.

R.J. Clausius. Ueber die bewegende Kraft der Wärme und die Gesetze welche sich daraus für die Wärmelehre selbst ableiten lassen. *Ann. d. Physik*, 79:368–397 and 508–524, 1850.

D.L. Cohen. *Measure theory*. Birkhäuser, Boston MA, 1980.

E.G.D. Cohen, editor. *Fundamental problems in statistical mechanics*. North–Holland, Amsterdam, 1962.

E.G.D. Cohen and W. Thirring, editors. *The Boltzmann equation, theory and applications*. Springer, New York, 1973.

L.A. Colding. Undersögelse on de almindelige naturkraefter og deres gjensidige afhaengighed og isaerdeleshed om den ved visse faste legemers gnidnig udviklede varne. *Dansk. Vid. Selsk.*, 2:121–146, 1851. (see: 'On the History of the Principle of the Conservation of Energy', Philosophical Magazine **27** 56-64, 1864).

P. Collet and J.P. Eckmann. *A renormalization group analysis of the hierarchical model in statistical mechanics*. LNP 74, Springer, Berlin, 1978.

A. Connes. Entropie de Kolmogoroff–Sinai et mécanique statistique quantique. *C. R. Acad. Sci. Paris*, 301:1–6, 1985.

A. Connes. *Géometrie non–commutative*. InterEditions, Paris, 1990.

A. Connes, H. Narnhofer, and W. Thirring. Dynamical entropy of C*-algebras and von Neumann algebras. *Commun. Math. Phys.*, 112:691–719, 1987.

A. Connes and M. Rieffel. Yang–Mills for non–commutative two-tori. *Contemp. Math.*, 62:237–266, 1987.

A. Connes and E. Størmer. Entropy for automorphisms of II_1 von Neumann algebras. *Acta Math.*, 134:289–306, 1975.

L.N. Cooper. Bound electron pairs in a degenerate Fermi gas. *Phys. Rev.*, 104: 1189–90, 1956.

L.N. Cooper. Microscopic quantum interference effects in the theory of superconductivity. *Nobel acceptance speech*, 1972. Reprinted in [Nobel Foundation, 1998] V: 73–93.

A.H. Copeland. Admissible numbers in the theory of probability. *Amer. J. Math.*, 50:153–162, 1928. See also: 'Admissible numbers in the theory of geometrical probability', *ibid.* **53** (1931), 153–162; 'The theory of Probability from the point of view of Admissible Numbers', Ann. Math. Statist. **3** (1932), 1435–1456; 'Consistency of the conditions determining Kollectivs', Trans. Amer. Math. Soc. **42**.

A.H. Copeland and P. Erdös. Note on normal numbers. *Bull. Amer. Math. Soc.*, 52:857–860, 1946.

W.S. Corak, B.B. Goodman, C.B. Satterthwaite, and A. Wexler. Atomic heats of normal and superconducting Vanadium. *Phys. Rev.*, 102:656–661, 1956.

W.S. Corak and C.B. Satterthwaite. Atomic heats of normal and superconducting Tin. *Phys. Rev.*, 102:662–666, 1956.

I. Cornfeld, S.V. Fomin, and Ya. G. Sinai. *Ergodic theory.* Springer, Berlin, 1982.

S.L. Cornish, N.R. Claussen, J.L. Roberts, E.A. Cornell, and C.E. Wieman. Stable 85Rb Bose-Einstein condensates with widely tunable interactions. *Phys. Rev. Lett.*, 85:1795–, 2000.

D. Costantini and M.C. Galavotti, editors. *Probability, dynamics and causality.* Kluwer, Dordrecht, 1997. reprinted from Erkenntnis, 45, 2/3, 1996.

A.A. Cournot. *Exposition de la théorie des chances et des probabilités.* Hachette, Paris, 1843.

J.J. Cross. Integral theorems in Cambridge mathematical physics. In P.M. Harman, editor, *Wranglers and physicists: studies on Cambridge physics in the nineteenth century*, pages 112–148, esp. 121–129. Manchester University Press, Manchester, 1985.

M.J. Cunningham. Very high frequency absorption in superconductors. *Phys. Rev.*, 101:1431–1432, 1956.

J.T. Cushing and E. McMullin, editors. *Philosophical consequences of quantum theory.* University of Notre Dame Press, Notre Dame IN, 1989.

A.I. Dale. Bayes or Laplace? An examination of the origin and early applications of Bayes' theorem. *Arch. Hist. Exact Sci.*, 27:23–47, 1982.

A.I. Dale. On Bayes' theorem and the inverse Bernoulli theorem. *Historia Mathematica*, 15:348–360, 1988. See already [Dale, 1982].

J. D'Alembert. Sur l'application du calcul des probabilités à l'inoculation de la petite vérole; (read on 12 nov. 1760 to the Académie Royale). *Opuscules Mathématiques, II,*, 11:26–95, 1760.

E. Dalfovo, S. Giorgini, L.P. Pitaevski, and S. Stringari. Theory of Bose–Einstein condensation in trapped gases. *Rev. Mod. Phys.*, 71:463–512, 1999.

T. Danzig. *Henri Poincaré, critic of crisis, reflections on his universe of discourse.* Scribner, New York, 1954.

L. Daston. Review of "Simeon-Denis Poisson et la science de son temps; M. Métivier, P. Costabel, P. Dugas, eds., Ecole Polytechnique, 1981". *Historia Mathematica*, 14:198–200, 1987.

L. Daston. *Classical probability in the Enlightenment.* Princeton University Press, Princeton NJ, 1988.

L. Daston. How probabilities came to be objective and subjective. *Historia Mathematica*, 21:330–344, 1994.

E.B. Davies. Diffusion for weakly coupled quantum oscillators. *Commun. math. Phys.*, 27:309–325, 1972.

E.B. Davies. Markovian master equations I, II, III. *Commun. Math. Phys.*, 39: 91–110, 1974. Math. Ann. **219**, 147-158 (1976); Ann. Inst. Henri Poincaré **11**, 265–273 (1975).

E.B. Davies. *Quantum theory of open systems.* Academic Press, London, 1976.

K.B. Davis, M.O. Mewes, M.R. Andrews, N.J. Van Druten, D.S. Durfee, D.M. Kurn, and W. Ketterle. Bose-Einstein condensation in a gas of Sodium atoms. *Phys. Rev. Lett.*, 75:3969–3973, 1995.

C.J. Davisson and L.H. Germer. Diffraction of electrons by a crystal of nickel. *Phys. Rev.*, 30:705–740, 1927.

R.H. Daw and E.S. Pearson. Abraham de Moivre's 1733 derivation of the normal curve: A bibliographical note. *Biometrika*, 59:677–680, 1973.

J.W. Dawson. *Logical dilemmas – The life and work of Kurt Gödel*. A K Peters, Wellesley MA, 1997.

M. J.A.C. De Condorcet. Application de l'analyse à cette question de déterminer la probabilité d'un arrangement régulier et l'effet d'une intention de le produire. *Mém. Acad. Sci. pour 1781,*, pages 720–728, 1784. see also: 'Réflexions sur la méthode de déterminer la probabilité des événements futurs d'après l'observation des événements passés', *ibid.* 1783, 539–553 (publ. 1786); *'Essai sur l'application de l'analyse à la probabilité des décisions rendues à la pluralité des voix'*, Paris, 1795; *'Eléments du calcul des probabilités et ses applications aux jeux de hasard, à la loterie et aux jugements des hommes'*, Paris, An XIII (1805).

N.C.A. De Costa and S. French. The model-theoretic approach in the philosophy of S science. *Philosophy of Science*, 57:248–265, 1990.

B. De Finetti. La prévision, ses lois logiques, et ses sources subjectives. *Ann. Inst. H. Poincaré*, 7:1–68, 1937.

B. De Finetti. *Probability, induction and statistics*. Wiley, New York, 1972. *La probabilità e la statistica nei rapporti cum l'induzioni, secondo i diversi punti di vista; Atti corso CIME su induzione e statistica, Varenna, 1959.*

S.R. De Groot and P. Mazur. *Non–equilibrium thermodynamics*. North–Holland, Amsterdam, 1962.

A. De Moivre. *Annuities upon lives; or, The valuation of annuities upon any number of lives, as also, of reversions to which is added, an appendix concerning the expectations of life, and probabilities of survivorship*. W.P. for F. Fayram and B. Mott, London, 1725.

A. De Moivre. *Approximatio ad summam terminorum Binomii $\overline{a+b}^{n}$ in seriem expansi ...* p.s., Londoni, 1733. Engl. transl. by De Moivre: *A Method of approximating the sum of the terms of the binomial $\overline{a+b}^{n}$ expanded into a series from whence are deducted some practical rules to evaluate the degree of ascent which is to be given experiments;* included, with some additions, in the 2nd and 3rd editions of his 'Doctrine of chances'.

A. De Moivre. *The doctrine of chances; or, A method of calculating the probabilities of events in play (2nd ed.)*. H. Woodfall for the author, London, 1738. 1st ed.:W. Pearson for the author, London, 1718; 3rd ed.: A. Millar, London, 1756.

J. De Nobel and P. Lindenfeld. The discovery of superconductivity. *Physics Today (Sep.96)*, 49:40–42, 1996.

G. De Rham. *Variétés différentielles*. Hermann, Paris, 1955.

B.D. Deaver and W.M. Fairbank. Experimental evidence for quantized flux in superconducting cylinders. *Phys. Rev. Lett.*, 7:43–46, 1961.

P. Debye. Zur Theorie des spezifischen Wärme. *Ann. d. Physik*, 39:789–839, 1912.

R. Dedekind. Ueber die von drei Moduln erzeugte Dualgruppe. *Math. Ann.*, 53: 371–408, 1900.

J.D. Deuschel and Stroock. *Large deviations*. Academic Press, Boston, 1989.

J.D. Deuschel, D.W. Stroock, and H. Zessin. Microcanonical distribution for lattice gases. *Commun. Math. Phys.*, 139:83–101, 1991.

R.L. Devaney. Dynamics of simple maps. In Devaney and Keene [1989], pages 1–24.

R.L. Devaney. *A first course in chaotic dynamical systems, theory and experiment*. Addison-Wesley, Reading MA, 1992.

R.L. Devaney and L. Keene, editors. *Chaos and fractals*. AMS, Proc. Symp. Appl. Math. 39, Providence RI, 1989.

C. Di Castro and G. Jona Lasinio. On the microscopic foundations of the scaling laws. *Phys. Lett.*, 29A:322–323, 1969.

P. Diaconis and D. Freedman. de Finetti's generalizations of exchangeability. In *Studies in inductive logic and probability II*, pages 233–249. University of California Press, Berkeley CA, 1980a.

P. Diaconis and D. Freedman. Finite exchangeable sequences. *Ann. Prob.*, 8:745–764, 1980b. see also [Diaconis and Freedman, 1980a].

F. Diacu. Poincaré and the three–body problem. *Historia Mathematica*, 26:175–178, 1999. see also [Barrow-Green, 1997].

F. Diacu and P. Holmes. *Celestial encounters: The origin of chaos and stability.* Princeton University Press, Princeton NJ, 1996.

L.E. Dickson. *History of the theory of numbers, vol.I.* Chelsea, New York, 1971. (Textually unaltered reprint of the 1923 edition).

D. Diderot and J. d'Alembert. *Encyclopédie,* vol. I–VII, Briasson et al., Paris, 1751-1757; vol. VIII–XVII, S. Faulche, Neufchastel, 1765; suppl. & Table analytique et raisonnée des matières, 1–6, Rey. Amsterdam & Panckoucke, Paris, 1776–1780, 1751–1780.

A. Dinghas. Ueber einen geometrischen Satz von Wulff für die Gleichtgewichtsform von Kristallen. *Z. Krist.*, 105:301–314, 1944.

P. A.M. Dirac. The fundamental equations of quantum mechanics. *Proc. Roy. Soc., London,* A109:642–656, 1925. 'Quantum mechanics and a preliminary investigation of the hydrogen atom', *ibid.* A110(1926) 561–579; 'The elimination of the nodes in quantum mechanics', A111 (1926) 281–305.

J. Dixmier. *Les algèbres d'opérateurs dans l'espace Hilbertien (algèbres de von Neumann).* Gauthier-Villars, Paris, 1957. Engl. transl. by F. Jellett: North-Holland, Amsterdam, 1981.

J. Dixmier. *Les C*-algèbres et leurs représentations.* Gauthier-Villars, Paris, 1964. 2nd ed. 1969; Engl. transl.: North-Holland, Amsterdam, 1981.

R.L. Dobrushin. The existence of a phase transition in the two– and three–dimensional Ising models. *Theor. Prob. Appl.*, 10:209–230, 1965.

R.L. Dobrushin. Gibbs state describing coexistence of phases for a three–dimensional Ising model. *Theory Prob. Appl.*, 17:582–600, 1972.

R.L. Dobrushin and O. Hryniv. Fluctuations of the phase boundary in the 2D Ising model. *Commun. Math. Phys.*, 189:395–445, 1997.

R.L. Dobrushin, R. Kotecki, and S. Shlosman. *The Wulff construction: a global shape from local interaction; Translations of Mathematical Monographs, 104.* American Mathematical Society, Providence, 1992.

R.L. Dobrushin, R. Kotecki, and S. Shlosman. A microscopic justification of the Wulff construction. *J. Stat. Phys.*, 72:1–14, 1993.

R.L. Dobrushin and B. Tirozzi. The central limit theorem and the problem of equivalence of ensembles. *Commun. Math. Phys.*, 54:173–192, 1977.

R. Doll and M. Näbauer. Experimental proof of magnetic flux quantization in a superconducting ring. *Phys. Rev. Lett.*, 7:51–52, 1961.

C. Domb and M.S. Green, editors. *Phase transitions and critical phenemena.* Academic Press, London and New York, 1972.

T.C. Dorlas. *Statistical mechanics: fundamentals and model solutions.* Institute of Physics Pub., Bristol; Philadelphia PA, 1999.

A.C. Doyle. *The hound of the Baskervilles.* McClure & Phillips, New York, 1902.

J.M. Doyle, J.C. Sandberg, I.A. Yu, C.I. Cesar, D. Kleppner, and T. J. Greytak. Hydrogen in the submillikelvin regime: sticking probability on superfluid ^{4}He. *Phys. Rev. Lett.*, 67:603–606, 1991.

A.J. Dragt. Eternity, chaos, Lie algebras, integrability and accelerator design. In J.R. Buchler, J.R. Ipser, and C.A. Williams, editors, *Integrability in dynamical systems*, pages 83–85. Ann. NY Acad. Sc. 536, New York, 1985.

M. Dresden. *H.A. Kramers, Between tradition and revolution*. Springer, New York, 1987.

M. Dresden. Kramers's contributions to statistical mechanics. *Physics Today (Sep.88)*, 41:26–33, 1988.

F. Dretske. Laws of nature. *Philosophy of Science*, 44:248–268, 1977.

D.A. Dubin. *Solvable models in algebraic statistical mechanics*. Clarendon, Oxford, 1974.

P. Dubreil. L'histoire des nombres mystérieux. In F. Le Lionnais, editor, *Les grands courants de la pensée mathématique*, pages 99–113. Editions des Cahiers du Sud, Marseille, 1948.

P. Dubreil and M.L. Dubreil-Jacotin. *Leçons d'algèbre moderne*. Dunod, Paris, 1961. Engl. transl. by A. Geddes: Lectures in modern algebra, Hafner, New York, 1967.

R. Dugas. *La théorie physique au sens de Boltzmann, et ses prolongements modernes*. Griffon, Neuchatel–Suisse, 1959.

P. Duhem. *The aim and structure of physical theory*. Princeton University Press, Princeton NJ, 1954. La théorie physique : son objet, sa structure; M. Rivière, Paris, 1914.

P.L. Dulong and A.T. Petit. Sur quelques points importants de la théorie de la chaleur. *Ann. Chim. Phys.*, 10:395–413, 1819.

H.S. Dumas, K.R. Meyer, and D.S. Schmidt, editors. *Hamiltonian dynamical systems; history, theory and applications*. Springer, New York, 1995.

N Dunford and J.T. Schwartz. *Linear Operators* (2nd printing). Interscience, New York, 1964.

F.J. Dyson. Existence of a phase transition in a one–dimensional Ising ferromagnet. *Commun. Math. Phys.*, 12:91–107, 1969. 'Non–existence of spontaneous magnetization in a one–dimensional Ising ferromasgnet', *ibid.* 212–215; 'An Ising ferromagnet with discontinuous long–range order', *ibid.* 21 (1971) 269–283.

J. Earman. *A primer on determinism*. Reidel, Dordrecht, 1986.

J. Earman. *World enough and space-time*. A Bradford Book, The MIT Press, Cambridge MA, 1989.

J. Earman. *Bayes or burst? A critical examination of Bayesian confirmation theory*. MIT Press, Cambridge MA, 1996.

J. Earman and M. Rédei. Why ergodic theory does not explain the success of equilibrium statistical mechanics. *Brit. J. Phil. Sci.*, 47:63–78, 1996.

J. Earmann and J.D. Norton. Exorcist xiv: the wrath of Maxwell's demon. Part I: from Maxwell to Szilard; Part II: from Szilard to Landauer and beyond. *Stud. Hist. Mod. Phys.*, 29; 30:435–471; 1–40, 1998, 1999.

H.D. Ebbinghaus, J. Flum, and W. Thomas. *Mathematical logic*. Springer, New York, 1984.

J.P. Eckmann and M. Hairer. Non–equilibrium statistical mechanics of strongly anharmonic chains of oscillators. *Commun. Math. Phys.*, 212:105–164, 2000.

J.P. Eckmann, C.A. Pillet, and L. Rey-Bellet. Entropy production in non–linear, thermally driven Hamiltonian systems. *J. Stat. Phys.*, 95:305–331, 1999a.

J.P. Eckmann, C.A. Pillet, and L. Rey-Bellet. Non–equilibrium statistical mechanics of anharmonic chains coupled to two heat baths at different temperatures. *Commun. Math. Phys.*, 201:657–697, 1999b. See also [Eckmann, Pillet, and Rey-Bellet, 1999a].

J.P. Eckmann and D. Ruelle. Ergodic theory of chaos and strange attractors. *Revs. Mod. Phys.*, 57:617–656, 1985.

H.M. Edwards. *Advanced calculus.* Houghton Mifflin, Boston MA, 1969.

P. Ehrenfest. Welche Züge der Lichtquantenhypothese spielen in der Theorie der Wärmestrahlung eine wesentliche Rolle. *Ann. d. Physik*, 36:91–118, 1911.

P. Ehrenfest and T. Ehrenfest. Ueber zwei bekannte Einwände gegen das Boltzmannsche H–Theorem. *Phys. Zeitschrift*, 8:311–314, 1907.

P. Ehrenfest and T. Ehrenfest. *Begriffliche Grundlagen der statistischen Auffassungen in der Mechanik, Encyklopädie der mathematischen Wissenschaften, Band 4, Teil 32.* Teubner, Leipzig, 1911. Engl. transl.: *The conceptual foundations of the statistical approach in mechanics,* Cornell University Press, Ithaca NY, 1959.

A. Einstein. Ueber die von molekularkinetischen Theorie der Wärme geforderte Bewegung von in Flüssigkeiten suspendierten Teilchen. *Ann. d. Physik*, 17: 549–560, 1905a.

A. Einstein. Ueber einen die Erzeugung und Verwandlungen des Lichtes betreffenden heuristischen Gesichtpunkt. *Ann. d. Physik*, 17:132–148, 1905b.

A. Einstein. Zur Theorie der Lichterzeugung und Lichtabsorption. *Ann. d. Physik*, 20:199–206, 1906.

A. Einstein. Die Plancksche Theorie der Strahlung und die Theorie der spezifischen Wärme. *Ann. d. Physik*, 22:180–190, 1907.

A. Einstein. Ueber die Entwicklung unserer Anschauungen über das Wesen und die Konstitution der Strahlung. *Phys. Zeits.*, 10:817–825, 1909a.

A. Einstein. Zum gegenwärtigen Stand des Strahlungsproblems. *Phys. Zeits.*, 10: 185–193, 1909b.

A. Einstein. Elementare Betrachtungen über die thermische Molekularbewegung in festen Körper. *Ann. d. Physik*, 35:679–694, 1911.

A. Einstein. Quantentheorie des einatomigen idealen Gases. *Sitz. Ber. Akad. Wissens., Berlin*, pages 261–267, 1924. *ibid.* (1925) 3–14.

A. Einstein, B. Podolsky, and N. Rosen. Can quantum–mechanical description of physical reality be considered complete? *Phys. Rev.*, 47:777–780, 1935.

T.S. Eliot. *Four quartets.* Harcourt & Brace, New York, 1943.

G. Ellis and R. Tavakol. Mixing properties of compact $K = -1$ FLRW models. In D. Hobill, A. Burd, and A. Coley, editors, *Deterministic chaos in general relativity*, pages 237–250. Plenum, New York, 1994.

R.S. Ellis. *Entropy, large deviations, and statistical mechanics.* Springer, New York, 1985. see also: P. Dupuis and R.S. Ellis, A weak convergence approach to the theory of large deviations, Wiley, New York, 1997.

G.G. Emch. Coarse–graining in Liouville space and master equation. *Helv. Phys. Acta*, 37:532–544, 1964. *ibid.* Sur la généralité des équations maitresses quantiques, Helv. Phys. Acta 37 (1964) 270–283.

G.G. Emch. Non–markovian model for the approach to equilibrium. *J. Math. Phys.*, 7:1198–1206, 1966a.

G.G. Emch. Rigourous results in non–equilibrium statistical mechanics. In W.E. Brittin, editor, *Lectures in theoretical physics, Vol.VIII A. Statistical physics and solid state physics*, pages 55–99. University Colorado Press, Boulder CO, 1966b.

G.G. Emch. *Algebraic methods in statistical mechanics and quantum field theory.* Wiley, New York, 1972a.

G.G. Emch. The C*-algebraic approach to phase transitions. In Domb and Green [1972], pages 1: 137–175.

G.G. Emch. Positivity of the K–entropy on non–abelian K–flows. *Z. Wahrscheinlichkeitstheorie verw. Gebiete*, 29:241–252, 1974.

G.G. Emch. Generalized K–flows. *Commun. math. Phys.*, 49:191–215, 1976a.

G.G. Emch. Non–equilibrium quantum statistical mechanics. *Acta Phys. Austr.*, Suppl. XV:79–131, 1976b. P. Urban, ed., Springer, Wien.

G.G. Emch. An algebraic approach for spontaneous symmetry breaking in quantum statistical mechanics. In P. Kramer and M. Dal Cin, editors, *Groups, systems and many-body physics*, pages 246–284. Friedr. Vieweg & Son, Braunschweig, 1977.

G.G. Emch. Prequantization and KMS structures. *J. Theor. Phys.*, 20:981–904, 1981. G.G. Emch and S.T. Ali, 'Geometric quantization: modular reduction theory and coherent states', J. Math. Phys. 27 (1986) 2936–2943; G.G. Emch, 'KMS structures in geometric quantization', Contemporary Mathematics, 62 (1987) 175–186.

G.G. Emch. Stochasticity in non–equilibrium statistical mechanics. In S. Albeverio, Ph. Combe, and M. Sirugue-Collin, editors, *Stochastic processes in quantum theory and statistical physics*, pages 149–153. LNP 173, Springer, Berlin, 1982.

G.G. Emch. *Mathematical and conceptual foundations of 20th–century physics*. North–Holland, Amsterdam, 1984.

G.G. Emch. Models vs. models. *Lecture presented at the Third Wigner Symposium, Oxford*, 1993. The Proceedings, supposed to appear in *Intern'l J. Mod. Phys. B*, are still to be published; the original paper is published for the first time as Appendix E of the present book.

G.G. Emch, S. Albeverio, and J.P. Eckmann. Quasi–free generalized K–flows. *Rep. Math. Phys.*, 13:73–85, 1978.

G.G. Emch and M. Guenin. Gauge invariant formulation of the BCS model. *J. Math. Phys.*, 7:915–921, 1966.

G.G. Emch and S. Hong. Time–like geodesic flows on Lorentz manifolds. *Ergod. Th. & Dynam. Sys.*, 7:175–192, 1987.

G.G. Emch and J.M. Jauch. Structures logiques et mathématiques en physique quantique. *Dialectica*, 19:259–279, 1965.

G.G. Emch and H. J.F. Knops. Pure thermodynamical phases as extremal KMS states. *J. Math. Phys.*, 11:3008–3018, 1970.

G.G. Emch, H.J.F. Knops, and E.J. Verboven. On partial weakly clustering states with an application to the Ising model. *Commun. Math. Phys.*, 8:300–314, 1968. Also: EKV, 'On the extension of invariant partial states in statistical mechanics', *ibid.* 7 (1968) 164–172.

G.G. Emch, H.J.F. Knops, and E.J. Verboven. Breaking of Euclidean symmetry with an application to the theory of crystallization. *J. Math. Phys.*, 11:1655–1668, 1970.

G.G. Emch, H. Narnhofer, W. Thirring, and G.L. Sewell. Anosov actions on non-commutative algebras. *J. Math. Phys.*, 35:5582–5599, 1994.

G.G. Emch and C. Piron. Symmetry in quantum theory. *J. Math. Phys.*, 4:469–473, 1963.

G.G. Emch and C. Radin. Relaxation of local thermal deviations from equilibrium. *J. Math. Phys.*, 12:2041–2046, 1971.

G.G. Emch and J.C. Varilly. On the standard form of the Bloch equation. *Lett. Math. Phys.*, 3:113–116, 1979.

A. Emch-Dériaz. L'inoculation justifiée – or was it? *Eighteenth–Century Life*, 7: 65–72, 1982. rev. in The 18th–Century, a current bibliography, 12 III-165, AHS Press, New York, 1992; see also: AEDz: *Tissot, Physician of the Enlightenment*, P. Lang, New York and Bern, 1992.

R.L. Epstein and W.A. Carnielli. *Computability, computable functions, logic, and the foundations of mathematics*. Wadsworth & Brooks/Cole Advanced Books, Pacific Grove CA, 1989.

P. Erdös and M. Kac. On the Gaussian law of errors in the theory of additive functions. *Proc. Nat. Acad. Sci. USA*, 25:206–207, 1939. The Gaussian law of errors in the theory of additive number theoretic functions, Amer. J. Math. **62** (1940), 738-742.

M.H. Ernst, L.K. Haines, and J.R. Dorfman. Theory of transport coefficients for moderately dense gases. *Rev. Mod. Phys.*, 41:296–316, 1969.

J.W. Essam and M.E. Fisher. Padé approximant studies of the lattice gas and Ising ferromagnet below critical point. *J. Chem. Phys.*, 38:802–812, 1963.

L. Euler. De fractionibus continuis. *Nova Acta Acad. Sci. Petrop. 1737*, 9:98–137, 1744a.

L. Euler. *Methodus inveniendi lineas curvas maximi minimive proprietate gaudentes*. Bousquet, Lausanne, Genève, 1744b.

L. Euler. Recherches générales sur la mortalité et la multiplication du genre humain'; *and* 'Sur les rentes viagères. *Hist. Acad. Berlin pour 1760,*, pages 144–164 *and* 165–175, 1767.

C. W.F. Everitt. Maxwell. In C.C. Gillispie, editor, *Dictionary of Scientific Bibliography*, pages **9**:198–230. Scribner, New York, 1974.

H. Ezawa. The representation of the canonical variables in the limit of infinite space volume: the case of the BCS model. *J. Math. Phys.*, 5:1078–1090, 1964.

P. Facchi and S. Pascazio. Deviations from the exponential law and van Hove's $\lambda^2 t$ limit. *Physica A*, 271:133–146, 1999.

A. Fagot. *Médecine et probabilité: Actes de la journée du 15 décembre 1979*. Didier-érudition, Paris, 1982.

G.B. Fahrenheit. Experimentas circa gradum caloris liquorum nonnullorum ebullientium instituta. *Phil. Trans. Roy. Soc. London*, 33:1, 1724.

W.W. Fairchild and C.I. Tulcea. *Topology*. Saunders, Philadelphia PA, 1971.

M. Falcioni, U.M.B. Marconi, and A. Vulpiani. Ergodic properties of high–dimensional symplectic maps. *Phys. Rev. A*, 44:2263–2270, 1991.

K.J. Falconer. *Fractal geometry – Mathematical foundations and applications*. Wiley, New York, 1990.

G. Falk and H. Jung. *Axiomatik der Thermodynamik*. Handbuch der Physik III/2, Springer, 1959.

M. Fannes and A. Verbeure. Correlation inequalities and equilibrium states. *Commun. Math. Phys.*, 55:125–131, 1977. *ibid.* 57, 165-171.

M. Fannes and A. Verbeure, editors. *Micro, meso and macroscopic approaches in physics*. Plenum, New York, 1994.

I. Farquhar. *Ergodic theory in statistical mechanics*. Wiley, New York, 1964.

C. Favre and Ph. Martin. Dynamique quantique des systèmes amortis non-markoviens. *Helv. Phys. Acta*, 41:333–361, 1968.

W. Feller. *An introduction to probability theory and its applications, vols. 1 (3rd ed.) and 2 (rev. 2nd.ed.)*. Wiley, New York, 1968, 1971.

E. Fermi, J. Pasta, and S. Ulam. Studies of nonlinear problems. Technical report, Los Alamos, 1940. see E. Fermi, Collected works, vol. II, pp. 978–988, University of Chicago Press, Chicago IL, and Accademia Nazionale dei Lincei, Rome, 1965.

R.P. Feynman. Atomic theory of the $\lambda-$ transition in Helium. *Phys. Rev.*, 91: 1291–1301, 1953. see also [Feynman, 1958].

R.P. Feynman. Application of quantum mechanics to liquid Helium. *Prog. Low Temp. Phys.*, 1:17–53, 1955. for a didactic account, see [Feynman, 1972].

R.P. Feynman. Excitations in liquid Helium. *Physica Suppl.*, 24:18–26, 1958.

R.P. Feynman. *Statistical Mechanics*. Addison–Wesley, Reading MA, 1972.

R.P. Feynman, R.B. Leighton, and M. Sands. *The Feynman lectures on physics, Vol.III*. Addison-Wesley, Reading MA, 1964.

A. Fine. Do correlations need to be explained? In Cushing and McMullin [1989], pages 175–194.

T.L. Fine. *Theories of probability*. Academic Press, New York, 1973.

B.S. Finn. Laplace and the speed of sound. *Isis*, 55:7–19, 1964.

M.E. Fisher. The nature of critical points. In W.E. Britten, editor, *Lectures in theoretical physics VII-C (Boulder, 1964)*, pages 1–159. University of Colorado Press, Denver, 1965.

M.E. Fisher. The theory of condensation and the critical point. *Physics*, 3:255–283, 1967a. See also [Fisher, 1967b].

M.E. Fisher. The theory of equilibrium critical phenomena. *Rep. Progr. Phys.*, 30: 615–730, 1967b.

M.E. Fisher. Rigourous inequalities for critical–point correlation exponents. *Phys. Rev.*, 180:594–600, 1969.

M.E. Fisher. The renormalization group in the theory of critical behavior. *Rev. Mod. Phys.*, 46:597–616, 1974.

M.E. Fisher. Scaling, universality and renormalization group theory. In F.J.W. Hahne, editor, *Critical phenomena: proceedings of the summer school held at the University of Stellenbosch, South Africa, January 18-29, 1982*, pages 1–139. Springer, Berlin, 1983.

M.E. Fisher. Phases and phase diagrams: Gibbs's legacy today. In D.G. Caldi and G.D. Mostow, editors, *Proceedings of the Gibbs symposium, Yale University, 1989*, pages 39–72. Amer. Math. Soc. and Amer. Inst. Phys., Providence RI, 1990.

M.E. Fisher. Renormalization group theory: its basis and formulation in statistical physics. *Rev. Mod. Phys.*, 70:653–681, 1998. Based on a lecture delivered on 2 March 1996 at the Boston Colloquium for the Philosophy of Science: "A historical examination and philosophical reflections on the foundations of quantum field theory".

R.A. Fisher. *Statistical methods and scientific inference*. Oliver & Boyal, Edimburgh, 1956.

R.A. Fisher. *Statistical methods for research workers, (13th ed.)*. Hafner Publ., New York, 1968.

R. Fitzgerald. An optimal spoon stirs up vortices in a Bose–Einstein condensate. *Physics Today (Aug. 00)*, 53:19–21, 2000.

H. Flaschka. The Toda lattice – Existence of integrals. *Phys. Rev. B*, 9:1924–1925, 1974. see also: 'The Toda lattice – Inverse scattering solution', *Progr. Theoret. Phys.* **51**(1974) 703-715.

W. Flemming. *Functions of several variables (2nd ed.)*. Springer, New York, 1977.

S. Flügge, editor. *Encyclopedia of physics/Handbuch der Physik, XV*. Springer, Berlin, 1956.

A. Fokas, A. Grigorian, T. Kibble, and B. Zegarlinski, editors. *Mathematical physics 2000*. Imperial College Press, London, 2000.

A.D. Fokker. Die mittlere Energie rotierender Dipole im Strahlungsfeld. *Ann. d. Physik*, 43:810–820, 1914.

G.W. Ford, M. Kac, and P. Mazur. Statistical mechanics of assemblies of coupled oscillators. *J. Math. Phys.*, 6:504–515, 1965.

J. Ford. Irreversibility in non–linear oscillator systems. *Pure and Appl. Chem.*, 22: 401–408, 1970.

J. Ford and G.H. Lunsford. Stochastic behavior of resonant nearly linear oscillators systems in the limit of zero non–linear coupling. *Physical Review A*, 1:59–70, 1970.

J. Ford, S.D. Stoddard, and J.S. Turner. On the integrabilibty of the Toda lattice. *Progr. Theoret. Phys.*, 50:1547–1560, 1973.

J. B.J. Fourier. Mémoire sur la statique, contenant la démonstration du principe des vitesses virtuelles, et la théorie des moments. *J. Ecole Polytechnique*, Ve cahier, 1798. reprinted in Oeuvres de Fourier, G. Darboux, ed., Gauthier-Villars, Paris, 1890; Vol. 2, pp. 477-521.

J. B.J. Fourier. *Théorie analytique de la chaleur*. Firmin Didot, Paris, 1822. Re-edited by G. Darboux, Gauthier-Villars, 1888.

R. Fox, editor. *The caloric theory of gases from Lavoisier to Regnault*. Clarendon Press, Cambridge, 1971.

R. Fox, editor. *Edition critique avec introduction et commentaire, augmentée de documents d'archives et de divers manuscrits de Carnot*. Librairie Philosophique J. Vrin, Paris, 1978.

M. Fréchet. Sur quelques points du calcul fonctionnel. *Rend. Circ. Mat. Palermo*, 22:1–74, 1906.

M. Fréchet. Sur l'intégrale d'une fonctionelle étendue à un ensemble abstrait. *Bull. Soc. Math. France*, 43:248–265, 1915.

M. Fréchet. *Les espaces abstraits*. Gauthier-Villars, Paris, 1928.

L. Freudenthal, editor. *The concept and the role of the model in mathematics and natural and social sciences*. D. Reidel, Dordrecht, 1961.

Dale G. Fried, Thomas C. Killian, Lorenz Willmann, David Landhuis, Stephen C. Moss, Daniel Kleppner, and Thomas J. Greytak. Bose–Einstein condensation of atomic Hydrogen. *Phys. Rev. Lett.*, 81:3811–3814, 1998.

K.S. Friedman. A partial vindication of ergodic theory. *Phil. Sci.*, 43:151–162, 1976.

K.S. Friedman and A. Shimony. Jaynes's maximum entropy prescription and probability theory. *J. Stat. Phys.*, 3:381–384, 1971.

F.G. Frobenius. Ueber Matrizen aus positiven Elementen. *Sitzungsber. Kön. Preuss. Akad. Wiss. Berlin*, pages 471–476, 1908. for a generalization, see *ibid.* II, 514–518, 1909; reprinted in Gesammelte Abhandlungen (J.P. Serre, ed.), Springer, Berlin, 1968, vol.III, pp. 404-409; 410–414.

H. Fröhlich. Theory of the superconductive state.I. The ground state at the absolute zero of temperature. *Phys. Rev.*, 79:845–856, 1950. 'Interaction of electrons with lattice vibration', Proc. Roy. Soc. (London) A215 (1952) 291–298.

J. Fröhlich and T. Spencer. The Kosterlitz–Thouless transition in two–dimensional abelian spin systems and the Coulomb gas. *Commun. Math. Phys.*, 81:527–602, 1981.

C.A. Fuchs and A. Peres. Quantum theory needs no "interpretation". *Physics Today (Mar. 2000)*, 53:70–71, 2000.

W. Fulks. *Advanced Calculus, 3rd ed.* Wiley, New York, 1978.

R. Fürth, editor. *A. Einstein, Investigations on the theory of Brownian movement*. A.D. Cooper, transl., Dover, New York, 1956.

D. Gabbay and F. Guenthner, editors. *Handbook of philosophical logic: 1. Elements of classical logic; 2. Extensions of classical logic; 3. Alternatives to classical logic; 4. Topics in the philosophy of language*. Reidel, Dordrecht, 1983–1989.

P. Gács. Randomness and probability – complexity of description. In *Encyclopedia of statistical sciences. vol.7*, pages 551–555. Wiley, New York, 1986. See already: 'Exact expressions for some random tests', *Zeitschr. f. Math. Logik u. Grundlagen Math.* **26** (1960) 385–394; 'On the relation between descriptional complexity and algorithmic probability', *Theor. Comp. Sci.* **22** (1983) 71–93.

L. Galgani. Ordered and chaotic motions in Hamiltonian systems and the problem of equipartition of the energy. In Buchler et al. [1985], pages 245–257.

L. Galgani. Relaxation times and the foundationss of classical statistical mechanics in the light of modern perturbation theory. In G. Gallavotti and P.F. Zweifel, editors, *Non–linear evolution and chaotic phenomena*, pages 147–159. Nato ASI Series B 176, 1987.

L. Galgani, A. Giorgilli, A. Martinoli, and S. Vanzini. On the problem of energy equipartition for large systems of the Fermi–Pasta–Ulam type; analytical and numerical estimates. *Physica D*, 59:334–348, 1992.

Galileo. *Sidereus Nuncius*. Baglioni, Venice, 1610. Engl. transl. w. comments by A. Van Helden, University of Chicago Press, Chicago IL, 1989.

G. Gallavotti. The phase separation line in the two–dimensional Ising model. *Commun. Math. Phys.*, 27:103–136, 1972.

G. Gallavotti. Ergodicity, ensembles, irreversibility in Boltzmann and beyond. *J. Stat. Phys.*, 78:1571–1589, 1995.

G. Gallavotti. *Statistical mechanics, a short treatise*. Springer, Berlin, 1999.

G. Gallavotti. Fluctuations and entropy driven space–time intermittency in Navier–Stokes fluids. In Fokas et al. [2000], pages 48–58.

G. Gallavotti. Non–equilibrium in statistical and fluid mechanics: ensembles and their equivalence. Entropy driven intermittency. *J. Math. Phys.*, 41:4061–4081, 2000b.

G. Gallavotti and E. G.D. Cohen. Dynamical ensembles in non–equilibrium statistical mechanics. *Phys. Rev. Lett.*, 74:2694–2697, 1995a.

G. Gallavotti and E. G.D. Cohen. Dynamical ensembles in stationary states. *J. Stat. Phys.*, 80:931–970, 1995b.

G. Gallavotti and H. Knops. The hierarchical model and the renormalization group. *Nuovo Cim.*, 5:341–368, 1975. See also: H. Van Beyeren, G. Gallavotti and H. Knops, 'Conservation laws in the hierarchical model', Physica 78 (1974) 541–548.

G. Gallavotti and A. Martin-Löf. Surface tension in the Ising model. *Commun. Math. Phys.*, 25:87–126, 1972.

J.A. Gallian. *Contemporary abstract algebra, (4th ed.)*. Heath, Lexington MA, 1998.

L. T.F. Gamut. *Logic, language, and meaning, vol.2*. University of Chicago Press, Chicago IL, 1991.

F.R. Gantmacher. *The theory of matrices*. Chelsea, New York, 1959. Transl. from the Russian by K.A. Hirsch, and reprinted several times.

D.J. Gates, I. Gerst, and M. Kac. Non–Markovian diffusion in idealized Lorentz gases. *Arch. Rat. Mech. and Anal.*, 31:106–135, 1973.

C.F. Gauss. *Disquisitiones arithmeticae*. G. Fleisher, Lipsiae (Leipzig), 1801. Engl. transl.: Yale University Press, New Haven, 1966; Springer, New York, 1986.

C.F. Gauss. *Theoria motus corporum coelestium in sectionibus conisis solem ambientum*. F. Perthes et I.H. Besser, Hamburg, 1809. = 'Theory of the motion of the heavenly bodies moving about the Sun in conic sections', C.H. Davy, transl., Little & Brown, Boston MA, 1857; reprinted, Dover, New York, 1963.

C.F. Gauss. Disquisitio de elementis ellipticis palladis. *reprinted in [Gauss, 1880]*, 5:1–24, 1810.

C.F. Gauss. Bestimmung der Genauigkeit der Beobachtungen. *Zschr. f. Astronomie*, 1, 1816. Reprinted in [Gauss, 1880], vol. 4, pp. 109–117.

C.F. Gauss. *Theoria combinationis observationum erroribus minimis obnoxiae*. Dieterich, Göttingen, 1823. Reprinted in [Gauss, 1880], vol. 4 pp.1–93; see already [Gauss, 1810, 1816].

C.F. Gauss. *Werke*. Kön. Ges. Wiss., Göttingen, 1880.

J. Gavarret. *Principes généraux de statistique médicale, ou Développement des règles qui doivent présider à son emploi*. Librairie Fac. Med., Paris, 1840. See also: *Lois générales de l'électricité dynamique, analyse et discussion des principaux phénomènes physiologiques et pathologiques qui s'y rapportent*, Bachelier, Paris, 1843; *Physique médicale, de la chaleur produite par les corps vivants*,

Masson, Paris, 1855; *Physique biologique, les phénomènes physiques de la vie*, Masson, Paris, 1869.

L.-J. Gay-Lussac. Recherches sur la dilatation des gaz et des vapeurs. *Annales de Chimie*, 43:137–175, 1802.

L.-J. Gay-Lussac. Premier essai pour déterminer les variations de température qu'éprouvent les gaz en changeant de densité, et conservation de leur capacité pour le calorique. *Mémoires de Physique et de Chimie de la Société d'Arceuil*, 1:180–204, 1807.

I.M. Gel'fand and S.V. Fomin. Geodesic flows on manifolds of constant negative curvature. *Usp. Math. Nauk*, 47:118–137, 1952. Transl. Amer. Math. Soc. bf 2 (1955) 49–67.

I.M. Gel'fand and N. Ya. Vilenkin. *Generalized functions, vol.4: applications of harmonic analysis*. Academic Press, New York, 1964.

M. Gell-Mann and J.B. Hartle. Alternative decohering histories in quantum mechanics. In K.K. Phua and Y. Yamaguchi, editors, *Proceedings of the 25th International Conference on High Energy Physics, Singapore, 1990*, page xxx. World Scientific, Singapore, 1991,.

M. Gell-Mann and F.E. Low. Quantum electrodynamics at small distances. *Phys. Rev.*, 95:1300–1312, 1954.

H.O. Georgii. *Gibbs measures and phase transitions*. de Gruyter, Berlin, 1988.

H.O. Georgii. Large deviations and maximum entropy principle for interacting random fields on Z^d. *Ann. Prob.*, 21:1845–1875, 1993.

N. Gershenfeld. *The nature of mathematical modeling*. Cambridge University Press, Cambridge, 1999.

I. Giaever. Energy–gap in superconductors measured by electron tunneling. *Phys. Rev. Lett.*, 5:147–148, 464–466, 1960. I. Giaever and K. Megerle, 'Study of superconductors by electron tunneling', Phys. Rev. 122 (1961) 1101–1111; J. Nicol, S. Shapiro and P.H. Smith, 'Direct measurment of the superconducting energy gap', Phys. Rev. Lett. 5 (1961) 461.

I. Giaever. Electron tunneling and superconductivity. *Nobel acceptance speech*, 1973. Reprinted in [Nobel Foundation, 1998] V: 137–153.

E. Gibbon. *The history of the decline and fall of the Roman Empire*. Strahan and Cadell, London, 1776/1788. (We used the Bury edition, Methuen, London, 1909).

J.W. Gibbs. Graphical methods in the thermodynamics of fluids; A method of geometrical representation of the thermodynamical properties of substances by means of surfaces. *Trans. Connecticut Academy*, II:309–342; 382–404, 1873.

J.W. Gibbs. *Elementary principles of statistical mechanics, developed with special reference to the rational foundations of thermodynamics*. Scribner, New York, 1902.

J.W. Gibbs. *The scientific papers of J. Willard Gibbs, Volume I, Thermodynamics*. Dover, New York, 1961. Original edition: Longmanns, Green and Co., 1906.

J. Giedymin. *Science and convention: essays on Henri Poincaré's philosophy of science and the conventionalist tradition*. Pergamon Press, Oxford, 1982.

R. Giere. *Explaining science: a cognitive approach*. University of Chicago Press, Chicago IL, 1988.

R.N. Giere and A.W. Richardson. *Origins of logical empiricism*. Minnesota Studies in the Philosophy of Science vol. 16, University of Minnesota Press, Minneapolis MN, 1996.

R. Giles. *Mathematical foundations of thermodynamics*. Macmillan, New York, 1964.

J. Ginibre. Simple proof and generalization of Griffiths second inequality. *Phys. Rev. Lett.*, 23:828–830, 1969. see also 'General formulation of Griffiths inequalities', Commun. Math. Phys. 16 (1970) 310–328.

J. Ginibre. On some recent work of Dobrushin. In Michel and Ruelle [1970], pages 163–175. see also 'Existence of phase transitions in quantum lattice systems', Commun. Math. Phys. 14(1969) 205-234.

A. Giorgilli. Energy equipartition and Nekhoroshev–type estimates for large systems. In Dumas et al. [1995], pages 147–161.

A.M. Gleason. Measures on the closed subspaces of a Hilbert space. *J. Math. Mech.*, 6:885–894, 1957.

J. Gleick. *Chaos: making a new science*. Viking, New York, 1987.

J. Glimm and A. Jaffe. *Quantum physics; a functional integral point of view*. Springer, New York, 1981.

R.E. Glover III and M. Tinkham. Transmission of superconducting films at millimeter microwave and far–infrared frequencies. *Phys. Rev.*, 104:844–845, 1956.

R.E. Glover III and M. Tinkham. Conductivity of superconducting films for photon energies between 0.3 and 40 kT_c. *Phys. Rev.*, 108:243–256, 1957.

K. Gödel. Ueber die Vollständigkeit des Logikkalküls. *Doctoral Diss. University of Vienna*, 1929.

K. Gödel. Ueber die Vollständigkeit der Axiome des logischen Funktionenkalküls. *Monathefte f. Math. u. Phys.*, 37:349–360, 1930.

R. Goldblatt. Orthomodularity is not elementary. *J. Symbolic Logic*, 49:401–404, 1984.

N. Goldenfeld. *Lectures on phase transitions and the renormalization group*. Addison-Wesley, Reading MA, 1992.

G.S. Goodman. Statistical independence and normal numbers: an aftermath to Mark Kac's Carus monograph. *Amer. Math. Monthly*, 106:112–126, 1999.

V. Gorini, A. Kossakowski, and E. C.G. Sudarshan. Completely positive dynamical semigroups on N–level systems. *J. Math. Phys.*, 17:821–825, 1976.

D.L. Goroff. *Henri Poincaré and the birth of chaos theory: An introduction to the English translation; in Henri Poincaré, New Methods of Celestial Mechanics, Vol. 1*. AIP, New York, 1993.

C.J. Gorter. The two–fluid model for superconductors and Helium II. *Prog. Low Temp. Phys.*, 1:1–16, 1955.

C.J. Gorter and H. Casimir. On supraconductivity. I. *Physica*, 1:306–320, 1934. *ibid.* 'Zur Thermodynamik des Supraleitenden Zustände', *Physik. Zschr.* 35 (1934) 963–966.

D. Goswami and K. Sinha. Minimal quantum dynamical semigroup on a von Neumann algebra. *Quantum Probability and Related Topics*, 2:221–239, 1999. see also *ibid.*, 'Hilbert modules and stochastic dilation of a quantum dynamical semigroup on a von Neumann algebra', Commun. Math. Phys., **205** (1999), 377–403.

H. Grad. *Principles of the kinematic theory of gases, in Handbuch der Physik*. Springer, Berlin, 1958.

H. Grad. Asymptotic theory of the Boltzmann equation. *Phys. Fluids*, 6:147–181, 1963.

H. Grad. Levels of description in statistical mechanics and thermodynamics. In Bunge [1967], pages 49–76.

I. Grattan-Guiness. Why did George Green write his Essay of 1828 on electricity and magnetism? *Amer. Math. Monthly*, 102:387–396, 1995.

G. Green. *An essay on the application of mathematical analysis to the theories of electricity and magnetism*. p.s., Nottingham, 1828. Re–published (under

Thomson's sponsorship) in Journal für Mathematik, 39 (1850) 73-89; 44 (1852) 356-374; 47 (1854) 161-221.

D.M. Greenberger and A. Zeilinger. *Fundamental problems in quantum theory*. Annals NYS 755, New York, 1995.

F.P. Greenleaf. *Invariant means on topological groups*. Van Nostrand–Reinhold, New York, 1969.

J. Gregory. *Letter to Collins, in: 'James Gregory tercentary memorial volume'*. H.W. Turnbull, ed., Royal Society of Edimburg, 1939, 1670.

A. Griffin, D.W. Snoke, and S. Stringari, editors. *Bose–Einstein condensation*. Cambridge University Press, Cambridge, 1995.

R.B. Griffiths. Peierls proof of spontaneous magnetization in a two–dimensional Ising ferromagnet. *Phys. Rev.*, 136A:437–439, 1964.

R.B. Griffiths. Ferromagnets and simple fluids near critical points: some thermodynamical inequalities. *J. Chem. Phys.*, 43:1958–1968, 1965. See already: 'Thermodynamical inequality near the critical point for ferromagnets and fluids', Phys. Rev. Lett. 14 (1965) 623-624.

R.B. Griffiths. Correlations in Ising Ferromagnets. I. *J. Math. Phys.*, 8:478–483, 1967. II. *ibid.* pp. 484–489.

R.B. Griffiths. Rigourous results and theorems. In Domb and Green [1972], pages 8–109.

R.B. Griffiths. Consistent histories and the interpretation of quantum mechanics. *J. Stat. Phys.*, 36:219–272, 1984. for an update, see 'Consistent quantum realism: a reply to Bassi and Ghirardi', *ibid.*, 99, 1409-1425, 2000.

R.B. Griffiths and J.L. Lebowitz. Random spin systems: some rigorous results. *J. Math. Phys.*, pages 1284–1292, 1968.

H.J. Gronewold. On the principles of elementary quantum mechanics. *Physica*, 12: 405–460, 1946.

E.P. Gross. Structure of quantized vortex in boson systems. *Nuov. Cim.*, 20: 454–477, 1961. 'Hydrodynamics of the superfluid condensate', *J. Math. Phys.*, 4(1963) 195–207.

F. Guerra, L. Rosen, and B. Simon. The $P(\phi)_2$ Euclidean quantum field theory as classical statistical mechanics. *Ann. Math.*, 101:111–259, 1975.

V. Guillemin and S. Sternberg. *Symplectic techniques in physics*. Cambridge University Press, Cambridge; New York, 1984.

B.M. Gurevitch. The entropy of horocycle flows. *Dokl. Akad. Nauk*, 136:768–770, 1961. Sov. Math. Dokl. **2** (1961) 124–130.

M.C. Gutzwiller. *Chaos in classical and quantum mechanics*. Springer, New York, 1990.

G. Györgyi, H. Ling, and G. Schmidt. Torus structure in higher–dimensional Hamiltonian systems. *Phys. Rev. A*, 40:5311–5318, 1989.

R. Haag. The mathematical structure of the Bardeen–Cooper–Schrieffer model. *Nuovo Cim.*, 25:287–298, 1962.

R. Haag. *Local quantum physics, 2nd ed.* Springer, Berlin, 1996.

R. Haag, N. Hugenholtz, and M. Winnink. On the equilibrium states in quantum statistical mechanics. *Commun. Math. Phys.*, 5:215–236, 1967.

R. Haag and D. Kastler. An algebraic approach to quantum field theory. *J. Math. Phys.*, 5:848–861, 1964.

A. Haar. Der Massbegriff in der Theorie der kontinuierchen Gruppen. *Ann. Math.(2)*, 34:147–169, 1933.

I. Hacking. *The emergence of probability; a philosophical study of early ideas about probability, induction and statistical inference*. Cambridge University Press, Cambridge, 1975. (paperback editions, 1984, 1985).

I. Hacking. Was there a probabilistic revolution 1800–1930? In M. Heidelberger, L. Krüger, and M. Rheiwald, editors, *Probability since 1800*, pages 487–506. ZiF report Nr. 25, Bielfeld, 1983.

J. Hadamard. Les surfaces à courbures opposées et leurs lignes géodésiques. *J. Math. pures appl.*, 4:27–73, 1898.

A. Hahn. Ueber lineare Gleichungsysteme in lineare Raumen. *J. reine u. angew. Math.*, 157:214–229, 1927.

E. Hairer and G. Wanner. *Analysis by its history*. Springer, New York, 1996.

H. Haken, A. Karlqvist, and U. Svedin. *The machine as metaphor and tool.* Springer, Berlin, 1993.

T. Hall. *Carl Friedrich Gauss: a biography*. MIT Press, Cambridge, 1970. A. Froderberg, transl.

E. Halley. An estimate of the degrees of mortality of mankind, drawn from curious tables of the births and funerals at the city of Breslaw; with an attempt to ascertain the price of annuities upon lives'; *and* 'Some further considerations on the Breslaw bills of mortality. *Phil. Trans. Roy. Soc. London*, 17:596–610 and 654–656, 1693. Edited with an introduction by Lowell J. Reed, The John Hopkins Press, Baltmore MD, 1942.

P. Halmos. *Measure theory*. van Nostrand, New York, 1950.

P. Halmos. *Lectures on ergodic theory*. Chelsea, New York, 1958.

P. Halmos. *Lectures on Boolean algebras*. van Nostrand, New York, 1963.

M. Hamermesh. *Group theory and its application to physical problems*. Addison-Wesley, Reading MA, 1962.

R.W. Hamming. *Coding and information theory (2nd ed.)*. Prentice–Hall, Englewood Cliffs NJ, 1986.

T.W. Hänsch and A.L. Schawlow. Cooling of gas by laser radiation. *Opt. Commun.*, 13:68, 1975.

G.H. Hardy and S.S. Ramanujan. The normal number of prime factors of a number n. *Quart. J. Pure Appl. Math.*, 48:76–92, 1917.

G.H. Hardy and E.M. Wright. *An introduction to the theory of numbers (5th ed.)*. Clarendon Press, Oxford, 1979.

A. Hartkaemper and H.-J. Schmidt, editors. *Structure and approximation in physical theories*. Plenum, New York, 1981.

D. Hartley. *Observations on man, his frame, his duty, and his expectations.* S. Richardson, London, 1749. facsimile: G. Olms, Hildesheim, 1967 or Garland, New York, 1971.

E.H. Hauge. What can one learn from Lorentz models. In G. Kirczenow and J. Marro, editors, *Transport phenomena*, pages 337–367. LNP 31, Springer, Heidelberg, 1974.

F. Hausdorff. *Mengenlehre*. De Gruyter, Berlin, 1927. 1st ed.: Grundzüge der Mengenlehre, Veit, Leipzig, 1914; Engl. transl. from the 3rd ed. (1937) by J.R. Aumann et al,: Set theory, Chelsea, New York, 1957.

D.M. Hausman. *The inexact and separate science of economics*. Cambridge University Press, Cambridge, 1992.

L.C. Hebel and C.P. Slichter. Nuclear relaxation in superconducting Aluminium. *Phys. Rev.*, 107:901–903, 1957. Hebel and Slichter, Nuclear relaxation in normal and superconducting Aluminium, Phys. Rev. 113 (1959) 1504–1519.

C.E. Hecht. The possible superfluid behaviour of hydrogen atoms gases and liquids. *Physica*, 25:1159–1161, 1959.

G. Hedlund. On the metrical transitivity of the geodesics on closed surfaces of constant negative curvature. *Ann. Math.*, 35:787–808, 1935. See also: The dynamics of geodesic flows, Bull. Amer. Math. Soc. **45** (1939) 241–246; A new proof for a metrically transitive system, Amer. J. Math. **62** (1940) 233–242.

D.A. Heijhal. *The Selberg trace formula for PSL(2, R)*. LNM 548, Springer, New York, 1976.

S.P. Heims. Approach to equilibrium. *Am. J. Phys.*, 33:722–727, 1965.

W. Heisenberg. Zur Theorie der Ferromagnetismus. *Zeits. f. Physik*, 49:619–636, 1928.

W. Heisenberg. *The physical principles of the quantum theory*. University of Chicago Press, Chicago IL, 1930.

E. Helfand. Approach to a phase transition in a one–dimensional model. *J. Math. Phys.*, 5:127–132, 1964a.

E. Helfand. Statistical mechanics of systems with long–range interaction. In H.J. Frisch and J.L. Lebowitz, editors, *Equilibrium theory of classical fluids*, pages III. 41–70. Benjamin, New York, 1964b.

A. Hellmich, G. Röpke, A. Schnell, and H. Stein. Onset of superfluidity in nuclear matter. In Griffin et al. [1995], pages 584–594.

P.C. Hemmer. On the van der Waals theory of the vapor–liquid equilibrium. IV. The pair–correlation function and equation of state for long–range forces. *J. Math. Phys.*, 5:75–84, 1965.

C. Hempel. *Aspects of scientific explanation*. The Free Press, New York, 1965.

M. Hénon. Integrals of the Toda lattice. *Phys. Rev. B*, 9:1921–1923, 1974.

M. Hénon. Numerical exploration of Hamiltonian systems. In G. Iooss, R.H.G. Helleman, and R. Stora, editors, *Chaotic behavior of deterministic systems, Les Houches, 1981*, pages 53–170. North–Holland, Amsterdam, 1983.

M. Hénon and C. Heiles. The applicability of the third integral of motion: some numerical experiments. *Astron. J.*, 69:73–79, 1964.

K. Hepp. *Théorie de la renormalisation*. LNP 2, Springer, Berlin, 1969.

K. Hepp. Quantum theory of measurement and macroscopic observables. *Helv. Phys. Acta*, 45:237–248, 1972.

K. Hepp and E. Lieb. On the superradiant phase transition for molecules in a quantized radiation field: The Dicke maser model. *Annals of Phys. (N.Y.)*, 76: 360–404, 1973. *ibid*. Phase transition in reservoir driven open systems with applications to lasers and superconductors, *Helv. Phys. Acta*, 46 (1973) 573-602; The equilibrium statistical mechanics of matter interacting with the quantized radiation field, *Phys. Rev.*, A8 (1973) 2517-2525.

K. Hepp and E.H. Lieb. The laser: a reversible quantum dynamical system with irreversible classical macroscopic motion. In *Dynamical systems*, pages 178–208. LNP 38, Springer, Heidelberg, 1975. see already [Hepp and Lieb, 1973].

H. Hermes. *Enumerability, decidability, computability*. Grundlehren der mathematischen Wissenschaften, 127, Springer, New York, 1969.

Ch. Hermite. Sur la fonction exponentielle. *C. R. Acad. Sci. Paris*, 77:18–24, 74–79, 226–233, 285–293, 1873.

H. Hertz. *Die Prinzipien der Mechanik in neuem Zusammenhange dargestellt*. Leipzig, 1894. The principles of mechanics presented in a new form, Macmillian, London and New York, 1899.

P. Hertz. Ueber die mechanische Grundlagen der Thermodynamik. *Ann. d. Physik (4)*, 33:225–274, 537–552, 1910.

H.F. Hess. Evaporation cooling of magnetically trapped and compressed spin–polarized hydrogen. *Phys. Rev.*, B 34:1476–1479, 1986.

M. Hesse. *Models and analogies in science*. University Notre Dame Press, Notre Dame IN, 1970.

E. Hewitt and L.J. Savage. Symmetric measures on cartesian products. *Trans. Amer. Math. Soc.*, 80:470–501, 1955.

T. Hida. *Stationary stochastic processes*. Princeton University Press, Princeton NJ, 1970.

D. Hilbert. Ueber die stetige Abildung einer Linie auf einer Flächenstück. *Math. Annalen*, 38:459–460, 1891.

D. Hilbert. *Grundlagen der Geometrie*. Teubner, Leipzig, 1899. This is the 1st ed.; by 1922 the 5th ed. had already appeared (same publisher); Engl. transl. by L. Unger (from the 10th German ed.) Foundations of geometry, Open Court, LaSalle IL, 1971.

D. Hilbert. Sur les problèmes futurs des mathématiques. In *Comptes Rendus du Deuxième Congrès International des Mathématiciens, Paris, 1900*, pages 58–114. Gauthier-Villars, Paris, 1902. See also: Mathematical developments arising from Hilbert problems, F.E. Browder (ed.), Proc. Symp. Pure Math. XXVIII, 1976.

D. Hilbert. Axiomatisches Denken. *Math. Ann.*, 78:405–415, 1918.

R. Hilpinen. Approximate truth and truthlikeness. In M. Przlecki, K. Szanianwski, and R. Wojicki, editors, *Formal Methods in the Methodology of Empirical Sciences*, pages 19–42. Reidel, Dordrecht, 1976.

T.G. Ho, L.J. Landau, and A.J. Wilkins. On the weak coupling limit for a Fermi gas in a random potential. *Rev. Math. Phys.*, 5:209–298, 1993.

R.E. Hodel. *An introduction to mathematical logic*. PWS Publ., Boston MA, 1995.

W. Hodges. *Model theory*. Cambridge University Press, Cambridge, 1993. A shorter model theory, Cambridge University Press, Cambridge, 1997.

P.C. Hohenberg. Existence of long–range order in one and two dimensions. *Phys. Rev.*, 158:383–386, 1967.

P.C. Hohenberg and P.C. Martin. Microscopic theory of superfluid Helium. *Ann. Phys.(NY)*, 34:291–359, 1965.

B.L. Holian, W.G. Hoover, and H.A. Posh. Resolution of Loschmidt's paradox: the origin of irreversible behavior in reversible atomistic dynamics. *Phys. Rev. Lett.*, 59:10–13, 1987.

S.S. Holland. The current interest in orthomodular lattices. In J.C. Abbott, editor, *Trends in lattice theory*, pages 41–126. Van Nostrand Reinhold, New York, 1970.

P.J. Holmes. Nonlinear oscillations and the Smale horseshoe map. In Devaney and Keene [1989], pages 25–39.

G. Holton. *Einstein, history, and other passions*. AIP Press, Woodbury NY, 1995.

W.G. Hoover. *Time reversibility, computer simulation, and chaos*. World Scientific, Singapore, 1999.

E. Hopf. On the time average theorem in dynamics. *Proc. Nat. Acad. Sci. USA*, 18:93–100, 1932.

E. Hopf. *Ergodentheorie*. Springer, Berlin, 1937.

E. Hopf. Statistik der geodätischen Linien in Mannigfaltigkeiten negativer Krümung. *Ber. Verh. Sächs. Akad. Wiss. Leipzig*, 91:261–304, 1939.

F.C. Hoppenstaedt. *Analysis and simulation of chaotic systems, 2nd ed.* Springer, Berlin, 2000.

J. Horgan. Icon and Bild: a note on the analogical structure of models - the role of models in experiment and theory. *Brit. J. Phil. Sci.*, 45:599–604, 1994.

C. Howson and P. Urbach. *Scientific reasoning: the Bayesian approach (2nd ed.)*. Open Court, Chicago IL, 1993.

K. Huang. *Statistical mechanics*. Wiley, New York, 1963, 1987.

K. Huang. Bose–Einstein condensation and superfluidity. In Griffin et al. [1995], pages 31–50.

M. Hübner and H. Spohn. Radiative decay: nonperturbative approaches. *Revs. Math. Phys.*, 7:363–387, 1995.

T. Hudetz. Quantum dynamical entropy revisited. In Alicki et al. [1998], pages 241–251.

R.L. Hudson. Analogs of de Finetti's theorem and the interpretative problems of quantum mechanics. *Found. Phys.*, 11:805–808, 1981.

R.L. Hudson and G.R. Moody. Locally normal symmetric states and an analogue of de Finetti's theorem. *Z. Wahrscheinlichkeitstheorie verw. Gebiete*, 33:343–351, 1976.

R.L. Hudson and K.R. Parthasarathy. Quantum Ito formula and stochastic evolution. *Commun. Math. Phys.*, 93:301–323, 1984.

A.L. Hughes. On the emission velocities of photoelectrons. *Phil. Trans. Roy. Soc. (London)*, 212:205–226, 1912.

R. I.G. Hughes. Bell's theorem, ideology, and structural explanation. In Cushing and McMullin [1989], pages 195–207.

R. I.G. Hughes. *The structure and interpretation of quantum mechanics*. Harvard University Press, Cambridge MA, 1989b.

R. I.G. Hughes. The Bohr atom, models, and realism. *Philosophical Topics*, 18: 71–84, 1990.

D. Hume. *A treatise of human nature*. John Noon, London, 1740.

D. Hume. *An enquiry concerning human understanding*. M. Cooper, London, 1748.

L. Hurd, C. Grebogi, and E. Ott. On the tendency toward ergodicity with increasing number of degrees of freedom in Hamiltonian systems. In J. Seimenis, editor, *Hamiltonian mechanics*, pages 123–129. Plenum, New York, 1994.

C. Huygens. *Tractatus de ratiociniis in aleæ ludo*, in F. Van Schooten: *Exercitationum mathematicarum, Libri V*. Elzevir, Amstelodami, 1657. French and Dutch transl. *in: Oeuvres complètes,* Société hollandaise des sciences, M. Nijhoff, La Haye, 1888-1979; **14:** 49–91.

F. Iachello. The bosonization in nuclear physics. In Griffin et al. [1995], pages 418–437.

R.S. Ingarden, editor. *Marian Smoluchovski, his life and scientific work*. Polish Scientific Publishers, Warsaw, 1999. 1st ed. PWN, Warsaw, 1986.

E. Ising. Beitrag zur Theorie des Ferromagnetismus. *Zeits. f. Physik*, 31:253–258, 1925.

R.B. Israel. *Convexity in the theory of lattice gases*. Princeton University Press, Princeton NJ, 1979.

Kiyosi Itô. *Encyclopedic dictionary of mathematics*. MIT Press, Cambridge MA, 1987.

B. Iversen. *Hyperbolic geometry*. Cambridge University Press, London, 1992.

V. Jaksic and C.A. Pillet. From resonances to master equations. *Ann. Inst. Poincaré*, 67:425–445, 1997.

V. Jaksic and C.A. Pillet. Ergodic properties of classical dissipative systems I. *Acta Math.*, 181:245–282, 1998.

M. Jammer. *The conceptual development of quantum mechanics*. McGraw Hill, New York, 1966.

R. Jancel. *Foundations of classical and quantum mechanics*. Pergamon Press, Oxford, 1969. Transl. from the French ed., Gauthier-Villars, 1963.

J.M. Jauch. *Foundations of quantum mechanics*. Addison–Wesley, Reading MA, 1968.

E.T. Jaynes. Foundations of probability theory and statistical mechanics. In Bunge [1967], pages 77–101.

E.T. Jaynes. *Papers on probability, statistics and statistical physics*. Kluwer, Dordrecht, 1989. First ed. 1983.

J.B. Jeans. On the partition of energy between matter and ether. *Phil. Mag.*, 10: 91–98, 1905. ibid. p. 98; also: ' Comparison between two theories of radiation' Nature 72 (1905) 293.

T. Jech. *Set theory*. Academic Press, New York, 1978.

H. Jeffreys. *The theory of probability*. Oxford University Press, London, 1939.

F. Jelinek. BCS-spin model, its thermodynamic representations and automorphisms. *Commun. Math. Phys.*, 9:169–175, 1968.

W.S. Jevons. *The principles of science: a treatise on logic and scientific method.* Macmillan, London, 1874.

K. Johansson. Separation of phases in one–dimensional gases. *Commun. Math. Phys.*, 169:521–561, 1995. see already 'Condensation of a one–dimensional lattice gas' *ibid.* 141 (1991) 41–61; 'Separation of phases at low temperatures in a one–dimensional continuous gas' *ibid.* 141 (1991) 259–278.

P. Jordan and E.P. Wigner. Ueber das Paulische Aequivalenzverbot. *Zeits. f. Physik*, 47:631–651, 1928.

B.D. Josephson. Possible new effects in superconductive tunnelling. *Phys. Lett.*, 1: 251–253, 1962. For an expository review, see: The discovery of tunneling supercurrents, Nobel acceptance speech, 1973, Reprinted in [Nobel Foundation, 1998] V: 157–164.

B.D. Josephson. Inequality for the specific heat. I. Derivation; II. Application to critical phenomena. *Proc. Phys. Soc.*, 92:269–275; 276–284, 1967.

J.P. Joule. On the caloric effects of magneto-electricity and the mechanical value of heat. *Phil. Mag. (Ser. 3)*, 23:263–276, 347–355, 435–443, 1843.

J.P. Joule. On the changes of temperature produced by the rarefaction and condensation of air. *Phil. Mag. (Ser. 3)*, 26:369–383, 1845.

J.P. Joule. Expériences sur l'identité entre le calorique et la force mécanique. Détermination de l'équivalent par la chaleur dégagée pendant la friction du mercure. *C. R. Acad. Sci. (Paris)*, 25:309–311, 1847.

J.P. Joule. Sur l'équivalent mécanique du calorique. *C. R. Acad. Sci. (Paris)*, 28: 132–135, 1849.

J.P. Joule. On the mechanical equivalent of heat. *Phil. Trans. Roy. Soc. London*, 140:61–82, 1850.

J.P. Joule and W. Thomson. On the thermal effects of fluids in motion, II *and* IV. *Phil. Trans. Roy. Soc. London*, 144 *and* 152:321–364 *and* 579–589, 1854 and 1862.

G.S. Joyce. Critical properties of the spherical model. In Domb and Green [1972], pages 2:375–442.

M. Kac. A discussion of the Ehrenfest model (Preliminary report). *Bull. Amer. Math. Soc.*, 52:621, 1946.

M. Kac. On the notion of recurrence in discrete stochastic processes. *Bull. Amer. Math. Soc.*, 53:1002–1010, 1947a.

M. Kac. Random walk and the theory of Brownian motion. *Amer. Math. Monthly*, 54:369–391, 1947b.

M. Kac. On the partition function of a one–dimensional gas. *Phys. Fluids*, 2:8–12, 1959a.

M. Kac. *Probability and related topics in physical sciences*. Interscience, New York, 1959b.

M. Kac. *Statistical dependence in probability, analysis, and number theory, Carus Math. Monographs, no.12*. Wiley, New York, 1959c.

M. Kac. Statistical mechanics of some one–dimensional systems. In *Studies in mathematical analysis and related topics: essays in honor of George Pólya*, pages 165–169. Stanford University Press, Stanford, 1962.

M. Kac. Mathematical mechanisms of phase transitions. In *Statistical physics (Brandeis Summer Institute 1966)*, pages 241–305, Vol.1. Gordon and Breach, New York, 1968.

M. Kac. Autobiographical note. In K. Baclawski and M.D. Donsker, editors, *Mark Kac: probability, number theory, and statistical physics; Selected papers*, pages ix–xxiii. MIT Press, Cambridge MA, 1979.

M. Kac. *Enigmas of chance*. Harper & Row, New York, 1985. Paperback ed.: University of California, Berkeley CA, 1987.

M. Kac and E. Helfand. Study of several lattice systems with long–range forces. *J. Math. Phys.*, 4:1078–1088, 1963.

M. Kac and C.J. Thompson. Spherical model and the infinite spin dimensionality limit. *Physica Norvegica*, 5:163–168, 1971.

M. Kac, G.E. Uhlenbeck, and P.C. Hemmer. On the van der Waals theory of the vapor–liquid equilibrium. I. Discussion of a one–dimensional model. *J. Math. Phys*, 4:216–228, 1963. II. 'Discussion of the distribution function' *ibid.* 4 (1963) 229-247; III. 'Discussion of the critical region', *ibid.* 5 (1964) 60–74.

L.P. Kadanoff. Scaling laws for Ising models near T_c. *Physics*, 2:263–272, 1966. See also [Kadanoff, Götze, Hamblen, Hecht, Lewis, Palciauskas, Rayl, and Swift, 1967].

L.P. Kadanoff, W. Götze, D. Hamblen, R. Hecht, E.A.S. Lewis, V.V. Palciauskas, M. Rayl, and J. Swift. Static phenomena near critical points: theory and experiment. *Rev. Mod. Phys.*, 39:395–431, 1967.

R.V. Kadison and J.R. Ringrose. *Fundamentals of the theory of operator algebras: I. Elementary theory; II. Advanced theory*. Wiley, New York, 1983, 1986.

Yu. Kagan. Kinetics of Bose–Einstein condensate formation in an interacting Bose gas. In Griffin et al. [1995], pages 202–225.

T. Kahan. Un document historique de l'académie des sciences de Berlin sur l'activité d'Albert Einstein (1913). *Archives Internationales d'Histoire des Sciences*, 15: 337–342, 1962. this paper offers a French translation as well as the complete German original of the letter; see also [Seelig, 1956] pp.144-145; the first and second parts of our quote are given in both English and German by [Jammer, 1966][p.44]; (we proposed our own translations).

V.P. Kalashnikov and D.N. Zubarev. The derivation of time–irreversible generalized master equation. *Physica*, 56:345–364, 1971.

T. Kamac. Subsequences of normal sequences. *Israel J. Math.*, 16:121–149, 1973.

H. Kamerlingh Onnes. Further experiments with liquid Helium. D. On the change of electric resistence at very low temperature, etc. The disappearance of resistence of Mercury. *Commun. Phys. Lab. University Leiden*, 122b, 1911. see already: *ibid.* 119, 120, 122 (1911).

H. Kamerlingh Onnes. Investigations into the properties of substances at low temperatures, which have led, amongst other things, to the preparation of liquid Helium. *Nobel acceptance speech*, 1913. Reprinted in [Nobel Foundation, 1998] I: 306-336.

P. Kapitza. Viscosity of liquid Helium below the λ–point. *Nature*, 141:74, 1938.

H.A. Kastrup. The contributions of Emmy Noether, Felix Klein and Sophus Lie to the modern concept of symmetries in physical systems. In M.G. Doncel, A. Hermann, L. Michel, and A. Pais, editors, *Symmetries in physics (1600–1980)*, pages 115–163. Bellaterra (distrib. by WSP), Barcelona, 1987.

T. Kato. *Perturbation theory for linear operators*. Springer, New York, 1966.

B. Kaufman. Crystal statistics II. Partition function evaluation by spinor analysis. *Phys. Rev.*, 76:1232–1243, 1949. see also: B. Kaufman and L. Onsager, 'Crystal statistics III. Short range order in a binary Ising lattice', Phys. Rev. 76(1949) 1244-1252.

S. Kazumasa and S. Tomohei. Hamiltonian systems with many degrees of freedom: asymptotic motion and intensity of motion in phase spece. *Phys. Rev. E*, 54: 4685–4700, 1996.

W.H. Keesom and A.P. Keesom. On the anomaly in the specific heat of liquid Helium. *Proc. Kon. Akad. Wetens., Amsterdam*, 35:736–742, 1932.

W.H. Keesom and P.H. Van Laer. Measurements of the atomic heats of Tin in the superconductive and the non–superconducting state. *Physica*, 5:193–201, 1938.

W.H. Keesom and M. Wolfke. Two different liquid states of Helium. *Proc. Kon. Akad. Wetens., Amsterdam*, 31:90–94, 1928.

D.G. Kelly and S. Shermann. General Griffiths inequalities on correlations in Ising ferromagnets. *J. Math. Phys.*, 9:466–484, 1968. see also S. Sherman, Cosets and ferromagnetic correlation inequalities, *Commun. Math. Phys.* 14 (1969) 1–4.

J.L. Kelly. *General topology*. van Nostrand, Princeton NJ, 1955.

J.G. Kemeny. Carnap's theory of probability and induction. In Schlipp [1963], pages 711–738.

T. Kennedy, E.H. Lieb, and B.S. Shastry. The XY model has long–range order for all spins and all dimensions greater than one. *Phys. Rev. Lett.*, 61:2582–2584, 1984.

W. Ketterle. Experimental studies of Bose–Einstein condensation. *Physics Today (Dec.99)*, 52:30–35, 1999.

J.M. Keynes. *A treatise on probability*. Macmillan, London, 1921. Harper & Row, New York, 1962.

A. Khinchin. Ueber einem Satz der Wahrscheinlichkeitsrechnung. *Fundam. Math.*, 6:9–20, 1924.

A. Khinchin. Zur Birkhoffs Lösung des Ergodenproblems. *Math. Annalen*, 107: 485–488, 1932.

A. Khinchin. Zur mathematischen Begründung der statistischen Mechanik. *Z. f. Angewandte Mathematik und Mechanik*, 13:101–103, 1933.

A. Khinchin. *Mathematical foundations of statistical mechanics*. Transl. from the Russian by G. Gamov; Dover, New York, 1949.

A.I. Khinchin. *Mathematical foundations of information theory*. R.A. Silverman and M.D. Friedman, transl.; Dover, New York, 1957.

A.Ya. Kipnis, B.E. Yavelov, and J.S. Rowlinson. *Van der Waals and molecular science*. Clarendon, London, 1996.

C. Kipnis and C. Landim. *Scaling limits of interparticle systems*. Springer, Berlin, 1999.

W. Kirchherr, M. Li, and P. Vitanyi. The miraculous universal distribution. *Math. Intelligencer*, 19:7–15, 1997.

J.G. Kirkwood, E.K. Mann, and B.J. Adler. Radial distribution functions and the equation of state of a fluid composed of hard spherical molecules. *J. Chem. Phys.*, 18:1040–1047, 1959.

J.G. Kirkwood and E. Monroe. Statistical mechanics of fluids. *J. Chem. Phys.*, 9: 514–526, 1941.

T.H. Kjeldsen. A contextual historical analysis of the Kuhn–Tucker theorem in nonlinear programming: the impact of World War II. *Historia Mathematica*, 27:331–361, 2000.

M.J. Klein. Entropy and the Ehrenfest urn model. *Physica*, 22:569–575, 1956.

M.J. Klein. *Paul Ehrenfest: The making of a theoretical physicist*. North–Holland, Amsterdam, 1970.

M.J. Klein. The development of Boltzmann's statistical ideas. In Cohen and Thirring [1973], pages 53–106.

M. Kline. *Mathematical thought from ancient to modern times*. Oxford University Press, New York, 1972.

R.S. Knox. *Theory of excitons*. Academic Press, New York, 1980.

D.E. Knuth. *The art of computer programming, vol. 2 (2nd ed.)*. Addison-Wesley, Reading MA, 1981.

K.W.F. Kohlrausch and E. Schroedinger. Das Ehrenfestsche Model der $H-$Kurve. *Phys. Zeitschrift*, 27:306–613, 1926.

W. Kohn and J.M. Luttinger. Quantum theory of electrical transport phenomena. *Phys. Rev.*, 108:590–611, 1957. *ibid.* 109 (1958) 1892-1909.

A.N. Kolmogorov. Das Gesetz des iterierten Logarithmus. *Math. Annalen*, 101: 126–135, 1929.

A.N. Kolmogorov. Ueber die analytischen Methoden in der Wahrscheinlichkeitsrechnung. *Math. Ann.*, 104:415–458, 1931.

A.N. Kolmogorov. *Grundbegriffe der Wahrscheinlichkeitsrechnung*. Springer, Berlin, 1933. Foundations of the theory of probability, (N. Morrison, transl.), Chelsea, New York, 1956.

A.N. Kolmogorov. On conservation of conditionally–periodic motions for a small change in Hamilton's function. *Dokl. Akad. Nauk SSSR*, 98:525–530, 1954. in Russian; see H. Dahlby, transl. (LA-TR-71-67), in G. Casati and J. Ford, eds., LNP 93, Springer, Berlin, 1979, pp. 51-56.

A.N. Kolmogorov. *The general theory of dynamical systems and classical mechanics*. in Proc. 1954 International Congress of Mathematicians, pp. 315–333, North–Holland, Amsterdam, 1957. Engl. transl. in Appendix (pp.741–757) in [Abraham and Marsden, 1978].

A.N. Kolmogorov. A new metric invariant of transitive systems and automorphisms of Lebesgue spaces. *Dokl. Akad. Nauk SSSR*, 119:861–864, 1958. see also Math. Rev. **21** No. 2035a.

A.N. Kolmogorov. On the entropy per time unit as a metric invariant of automorphisms. *Dokl. Akad. Nauk SSSR*, 124:754–755, 1959.

A.N. Kolmogorov. On tables of random numbers. *Sankya, The Indian Journal of Statistics, Ser. A*, 25:369–376, 1963. See also: 'Three approaches for defining the concept of information quantity', *Problems Inform. Transmission* **1** (1965) 3–11; 'Logical basis for information theory and probability theory', *IEEE Trans. Inform. Theory* **14** (1968), 662–664; 'On the logical foundations of probability theory', *in:* Probabiblity theory and mathematical statistitics, Proc. 4th USSR-Japan Symposium, 1982, K. Ito and J.V. Prokhorov, eds., LNM 1021, Springer, Berlin, 1983, pp. 1–5; 'Combinatorial foundations of information theory and the calculus of probability', *Uspekhi Mat. Nauk* **38** (1983) 27–36 = *Russian Math. Surveys* **38** (1983) 29–40; V.A. Uspensky, 'Kolmogorov and mathematical logic', *J. Symbol. Logic* **57** (1992) 385–412.

B.O. Koopman. Hamiltonian systems and transformations in Hilbert space. *Proc. Nat. Acad. Sci. USA*, 17:315–318, 1931.

B.O. Koopman. The axioms and algebra of intuitive probability. *Ann. of Math.*, 41:269–292, 1940.

B.O. Koopman and J. Von Neumann. Dynamical systems of continuous spectra. *Proc. Nat. Acad. Sci. USA*, 18:255–263, 1932.

J.M. Kosterliz and D.J. Thouless. Ordering, metastability and phase transitions in two–dimensional systems. *J. Phys. C*, 6:1181–1203, 1973.

L.G. Kraft. *A device for quantizing, grouping and coding amplitude modulated pulses*. MS Thesis, EE Dept., MIT, Cambridge MA, 1949.

W. Krajewski. *Correspondence principle and the growth of knowledge*. Reidel, Dordrecht, 1977.

H.A. Kramers and G.H. Wannier. Statistics of the two-dimensional ferromagnet I & II. *Phys. Rev.*, 60:252–262 & 263–276, 1941.

P. Kroes. Structural analogies between physical systems. *Brit. J. Phil. Sci.*, 40: 145–154, 1989.

A.K. Krönig. Grundzüge einer Theorie der Gase. *Ann. der Physik, ser.2*, 99:315–322, 1856.

N.S. Krylov. *Works on the foundations of statistical physics*. Princeton University Press, Princeton NJ, 1979. Transl. by A.B. Migdal, Ya.G. Sinai and Yu.L. Zeeman, with a preface by A.S. Wightman, a foreword by A.B. Migdal and Ya.G. Sinai, a personal biography by V.A. Fock, a scientific biography by A.B. Migdal and V.A. Fock, and an update by Ya. G. Sinai.

R. Kubo. Statistical mechanics of irreversible processes. *J. Phys. Soc. Japan*, 12: 570–586, 1957.

H.W. Kuhn and A.W. Tucker. Non–linear programming. *2nd Berkeley Symp. Math. Stat. Prob., J. Neyman, ed.*, pages 481–492, 1951.

T.S. Kuhn. Energy conservation as an example of simultaneous discovery. In M. Clagett, editor, *Critical Problems in the History of Science*, pages 321–356. University of Wisconsin Press, Madison, 1959. reprinted by T.S.K. in his "The Essential Tension", The University of Chicago Press, Chicago IL, 1977, pp. 86-104.

L. Kuipers and H. Niederretter. *Uniform distribution of sequences*. Wiley, New York, 1974.

B. Kümmerer. Markov dilations on W*-algebras. *J. Funct. Anal.*, 63:139–177, 1985a.

B. Kümmerer. On the structure of Markov dilations on W*-algebras. In L. Accardi and W. Von Waldenfelds, editors, *Quantum probability and applications II*, pages 318–331. Springer, Berlin, 1985b.

B. Kümmerer. Survey on a theory of non–commutative Markov processes. In Accardi and Von Waldenfelds [1988], pages 228–244.

K. Kuratowski. Sur l'opération $\overline{A}$ de l'analysis situs. *Fund. Math.*, 3:182–199, 1922.

K. Kuratowski. *Topologie*. Fundaszu Kultury Narodowej, Warsawa–Lwow, 1933. Rev. 1948-1952; Engl. transl. by A. Kirkov and by J. Jaworowski, Academic Press, 1966-1968.

A.G. Kurosh. *The theory of groups*. Chelsea, New York, 1955-56.

A.G. Kurosh. *Lectures on general algebra*. Chelsea, New York, 1963.

H. Kyburg. Propensities and probabilities. *Brit. J. Phil. Sci.*, 25:358–375, 1974.

H. Kyburg and H. Smokler. *Studies in subjective probability*. Wiley, New York, 1964.

S.E. Lacroix. *Traité élémentaire du calcul des probabilités*. Bachelier, Paris, 1816 and 1833.

E. Ladenburg. Untersuchungen über die entladenden Wirkung des ultravioletten Lichtes auf negativ geladene Metallplatten im Vakuum. *Ann. d. Physik*, 12: 558–578, 1903.

M. Laezkovich. On Lambert's proof of the irrationality of pi. *Amer. Math. Monthly*, 104:439–445, 1997.

J.L. Lagrange. *Recherches sur la nature et la propagation du son*. Turin mémoire, 1759. Reproduced in [Lagrange, 1867], pp.1: 131–132.

J.L. Lagrange. Sur la résolution des équations numériques. *Nouv. Mém. Acad. Berlin pour 1767*, 23:311–352, 1769. Additions, *ibid.* **24** (1768), 111–180 (publ. 1770); Traité de la résolution des équations numériques de tous les degrés (Nvelle ed. rev. et aug. par l'auteur), Courcier, Paris, 1808.

J.L. Lagrange. *Mécanique analytique*. La Veuve Desaint, Paris, 1788.

J.L. Lagrange. Mémoire sur une question concernant les annuités. *Mem. Acad. Berlin pour 1792 & 1793,*, pages 235–246, 1798.

J.L. Lagrange. *Leçons sur la théorie des fonctions*. Courcier, Paris, 1806.

J.L. Lagrange. *Oeuvres*. J.-A. Serret, ed., Gauthier-Villars, Paris, 1867.

J.H. Lambert. Mémoire sur quelques propriétés remarquables des quantités transcendentes circulaires et logarithmiques. *Hist. Acad. Berlin pour 1761,*, pages 265–322, 1768. (Presented in 1767).

L.D. Landau. The theory of superfluidity of helium II. *J. Phys. USSR*, 5:71–90, 1941. On the theory of superfluidity, Phys. Rev. 75 (1949) 884–885.

N.P. Landsman. *Mathematical topics between classical and quantum mechanics.* Springer, New York, 1998.

O.E. Lanford. Entropy and equilibrium states in classical mechanics. In Lenard [1973], pages 1–113.

O.E. Lanford. On a derivation of the Boltzmann equation. *Astérisque*, 40:117–137, 1976. See already: 'Time–evolution of large classical systems' *in* 1974 Battelle Rencontres, J. Moser, ed., LNP **38**, Springer, Berlin, 1975.

L.J. Lange. An elegant continued fraction for π. *Amer. Math. Monthly*, 106:456–458, 1999.

J. Langer. Computing physics: Are we taking it too seriously? or not seriously enough? *Physics Today (Jul.99)*, 52:11–13, 1999.

L. Lanz, G.R. Lanz, and A. Scotti. Considerations on the problem of deriving the master equation and characterizing macroscopic observables. In Bak [1967], pages 32–42.

P.S. Laplace. Mémoire sur la probabilité des causes par les événements. *Mem. Acad. Sci. Paris*, 1774. *in* [Laplace, 1878-1912], **8**:27 65.

P.S. Laplace. Mémoire sur les probabilités. *Mem. Acad. Sci. Paris*, 1781. *in* [Laplace, 1878-1912], **9**:381–485.

P.S. Laplace. *Essai philosophique sur les probabilités.* Inaugural lecture at ENS, Paris, 1795. Became the *Introduction* (pp.i-cxlii) *in* [Laplace, 1781].

P.S. Laplace. *Exposition du système du monde.* Imprimerie du Cercle-Social, Paris, An IV, 1796.

P.S. Laplace. Sur la vitesse du son dans l'air et dans l'eau. *Ann. de Chimie et Phys.*, 3:238–241, 1816.

P.S. Laplace. *Théorie analytique des probabilités (3ème ed.).* Courcier, Paris, 1820. (1st ed. 1812).

P.S. Laplace. *Traité de mécanique céleste.* Bachelier, Paris, 1823.

P.S. Laplace. *Oeuvres complètes (14 vol.).* J. Bertrand and J.B. Dumas, eds., Gauthier-Villars, Paris, 1878-1912.

E.N. Lassettre and J.P. Howe. Thermodynamic properties of binary solid solutions on the basis of the nearest neighbor approximation. *J. Chem. Phys.*, 9:747–754, 1941.

D. Lavis. The role of statistical mechanics in classical physics. *Brit. J. Phil. Sci.*, 28:255–279, 1977.

A.L. Lavoisier. *Traité élementaire de chimie.* Cuchet, Paris, 1789.

P.D. Lax. Integrals of nonlinear equations of evolution and solitary waves. *Commun. Pure Applied Math.*, 21:467–490, 1968.

R. Laymon. Idealization, explanation, and confirmation. In P.D. Asquith and R.N. Giere, editors, *PSA 1980: 1*, pages 336–350. Philosophy of Science Association, East Lansing MI, 1980.

R. Laymon. Idealizations and the testing of theories by experimentation. In P. Achinstein and O. Hannaway, editors, *Observation, Experiment and Hypothesis in Modern Physical Science*, pages 147–173. MIT Press, Cambridge MA, 1985.

W.H. Leatherdale. *The role of analogy, model and metaphor in science.* North–Holland, Amsterdam, 1974.

H. Lebesgue. *Leçons sur l'intégration.* Gauthier-Villars, Paris, 1928. (There was a 1st ed. (1904), but Kolmogorov uses the 1928 ed.).

J.L. Lebowitz. Bounds on the correlations and analycity properties of ferromagnetic Ising spin systems. *Commun. Math. Phys.*, 28:313–321, 1972.

J.L. Lebowitz. Macroscopic dynamics and macroscopic laws. In G.W. Horton, L.E. Reichl, and V.G. Szebehely, editors, *Long–time predictions in dynamics*, pages 3–19. Wiley, New York, 1983.

J.L. Lebowitz. Microscopic origin of irreversible macroscopic behavior. *Physica*, A263:516–527, 1999a.

J.L. Lebowitz. Statistical mechanics: a selective review of two central issues. *Rev. Mod. Phys.*, 71:S346–S357, 1999b.

J.L. Lebowitz and A. Martin-Löf. On the uniqueness of the equilibrium state for Ising spin systems. *Commun. Math. Phys.*, 25:276–282, 1972.

J.L. Lebowitz and A.E. Mazel. Improved Peierls argument for high–dimensional Ising models. *J. Stat. Phys.*, 90:1051–1059, 1998.

J.L. Lebowitz, A.E. Mazel, and E. Presutti. Rigorous proof of a liquid–vapor transition in a continuum particle system. *Phys. Rev. Lett.*, 80:4701–4704, 1998. See also [Lebowitz, Mazel, and Presutti, 1999].

J.L. Lebowitz, A.E. Mazel, and E. Presutti. Liquid–vapor phase transitions for systems with finite–range interactions. *J. Stat. Phys.*, 94:955–1025, 1999. See also [Lebowitz, Mazel, and Presutti, 1998].

J.L. Lebowitz, A.E. Mazel, and Y.M. Suhov. An Ising interface between two walls: competition between two tendencies. *Rev. Math. Phys.*, 8:669–687, 1996.

J.L. Lebowitz and E.W. Montroll. *Nonequilibrium phenomena I: the Boltzmann equation.* North–Holland, Amsterdam, 1983.

J.L. Lebowitz and O. Penrose. Modern ergodic theory. *Physics Today (Feb.73)*, 26: 23–29, 1973.

J.L. Lebowitz and H. Spohn. A Gallavotti–Cohen–type symmetry in the large deviation functional for stochastic dynamics. *J. Stat. Phys.*, 95:333–365, 1999.

T.D. Lee and C.N. Yang. Statistical theory of equations of state and phase transitions II. Lattice gas and Ising model. *Phys. Rev.*, 87:410–419, 1952.

S. Leeds. Malament and Zabell on Gibbs phase averaging. *Phil. Sci.*, 56:325–340, 1989.

H.S. Leff and A.F. Rex, editors. *Maxwell's demon: entropy, information, computing.* Princeton University Press, Princeton NJ, 1990.

A.M. Legendre. *Nouvelles méthodes pour la determination des orbits des comètes.* F. Didot, Paris, 1805.

A.M. Legendre. *Eléments de géometrie.* F. Didot, Paris, An II, 1794. Essai sur la théorie des nombres, Duprat, Paris, An VI (1798), Nvelle ed. 1830.

A.J. Leggett. Diatomic molecules and Cooper pairs. In A. Pekalski and R. Przystawa, editors, *Modern trends in the theory of condensed matter*, pages 13–35. Springer, Berlin, 1980.

A.J. Leggett. Broken gauge symmetry in a Bose condensate. In Griffin et al. [1995], pages 452–462.

A.J. Leggett. Superfluidity. *Rev. Mod. Phys.*, 71:S318–S323, 1999.

A.J. Leggett, S. Chakravarty, A.T. Dorsey, M.P.A. Fisher, A. Garg, and W. Zwerger. Dynamics of the dissipative two–state system. *Rev. Mod. Phys.*, 59:1–85, 1987.

G.W. Leibniz. *Dissertatio de Arte Combinatoria.* Frick & Seubold, Lipsiae (Leipzig), 1666. in *Mathematische Schriften*, C.I. Gerhardt, ed.; reprinted in 1962 by G. Olms, Hildesheim.

G.W. Leibniz. De vera proportione circuli ad quadratum circumscriptum in numens rationalibus expressa. *Acta Erudit. Lips. (Leipzig)*, 1:41–45, 1682.

G.W. Leibniz. *Essais sur l'entendement humain; in Philosophische Schriften 5, p. 448.* C.I. Gerhardt, ed.; reprinted in 1978 by G. Olms, Hildesheim, 1703–1705. (publ. 1765). See also: *New essays concerning human understanding;*

transl. w. notes by Alfred Gideon Langley (3d ed.); Open Court, LaSalle IL, 1949.

A. Lenard, editor. *Statistical mechanics and mathematical problems.* LNP 20, Springer, New York, 1973.

J.M.H. Levelt-Sengers. Compressibility, Gas. In R.M. Besançon, editor, *The Encyclopedia of Physics*, pages 118–119. Reinhold, New York, 1966.

B.G. Levi. Work on atom trapping and cooling gets warm reception in Stockholm. *Physics Today (Dec.97)*, 50:17–19, 1997.

B.G. Levi. At long last, a Bose–Einstein condensate is formed in Hydrogen. *Physics Today (Dec.98)*, 51:17–19, 1998a.

B.G. Levi. Two–dimensional electron gases continue to exhibit intriguing behavior. *Physics Today (Dec.98)*, 51:22–23, 1998b.

L.A. Levin. On the notion of random sequence. *Soviet Math. Dokl.*, 14:1413–1416, 1973.

P. Lévy. *Calcul des probabilités.* Gauthier-Villars, Paris, 1925.

D. Lewis. *Counterfactuals.* Harvard University Press, Cambridge MA, 1973.

D. Lewis. *On the plurality of worlds.* Blackwell, Oxford, 1986.

G.N. Lewis. The conservation of photons. *Nature*, 118:874–875, 1926.

J.T. Lewis, C.E. Pfister, and W.G. Sullivan. Large deviations and the thermodynamic formalism: a new proof of the equivalence of ensembles. In Fannes and Verbeure [1994], pages 181–192.

M. Li and P. Vitanyi. *An introduction to Kolmogorov complexity and its applications (2nd ed.).* Springer, New York, 1997.

A.J. Lichtenberg and M.A. Lieberman. *Regular and chaotic dynamics (2nd ed.).* Springer, New York, 1992.

E.H. Lieb. Exact solution of the problem of the entropy of the two–dimensional ice. *Phys. Rev. Lett.*, 18:692–694, 1967.

E.H. Lieb. Some of the early history of exactly soluble models. *Intern. J. Mod. Phys. B*, 11:3–10, 1997.

E.H. Lieb and D.C. Mattis. *Mathematical Physics in one dimension.* Academic Press, New York, 1966.

E.H. Lieb, R. Seiringer, and J. Yngvason. Bosons in a trap: a rigourous derivation of the Gross–Pitaevskii energy functional (13 pages). *Phys. Rev.*, A.61–043602, 2000a. ArXiv:math-ph/9908027 26 Aug 1999; for a summary: The ground state energy and density of interacting Bosons in a trap, arXiv:math-phys/9911026 19 Nov 1999.

E.H. Lieb, R. Seiringer, and J. Yngvason. A rigorous derivation of the Gross–Pitaevskii energy functional for a two–dimensional Bose gas. *arXiv:cond–mat/0005026*, v2, 3 May, 2000b.

E.H. Lieb and F.Y. Wu. Two–dimensional ferroelectric models. In Domb and Green [1972], pages 1: 331–490.

E.H. Lieb and J. Yngvason. A guide to entropy and the second law of thermodynamics. *Notices of the Amer. Math. Soc.*, 45:571–581, 1998. The full version is 'The physics and mathematics of the second law of thermodynamics', Phys. Rep. 310 (1999) 1–96 [for Fig.8, see *ibid.* 314 (1999) 699]; see also their: 'A fresh look at entropy and the second law of thermodynamics', Physics Today (Apr.00) 53 (2000) 32–37.

G. Lindblad. Completely positive maps and entropy inequalities. *Commun. Math. Phys.*, 40:147–151, 1975.

G. Lindblad. On the generators of quantum dynamical semi–groups. *Commun. Math. Phys.*, 48:119–130, 1976.

G. Lindblad. Non–Markovian stochastic processes and their entropy. *Commun. Math. Phys.*, 65:281–294, 1979.

G. Lindblad. Dynamical entropy for quantum probability. In Accardi and Von Waldenfelds [1988], pages 183–191.

G. Lindblad. Determinism and randomness in quantum dynamics. *J. Phys. A*, 26: 7193–7211, 1993.

F. Lindemann. Ueber die zahl π. *Math. Ann.*, 20:213–225, 1882.

J. Liouville. Note sur l'irrationalité du nombre e. *Journ. de Math.*, 5:192 and 193–199, 1840.

J. Liouville. Mémoire sur des classes très étendues' ... and 'Nouvelle démonstration d'un théorème sur les irrationnelles algébriques inséré dans les comptes rendus de la dernière séance. *C. R. Acad. Sci. Paris*, 18:883–885 and 910–911, 1844. See also: 'Sur des classes très étendues de quantités dont la valeur n'est ni algébrique, ni même réductible à des irrationnelles algébriques', Journ. de Math. **16** (1851), 133–142.

J. Lissajous. Mémoire sur l'étude optique des mouvements vibratoires. *Ann. Chem.*, 51:147–231, 1857.

C. Liu. Approximation, idealization, and laws of nature. *Synthese*, 118:229–256, 1999.

E.A. Lloyd. *The structure and confirmation of evolutionary theory*. Greenwood Press, Westport CT, 1988.

P. Lochak. Stability of Hamiltonian systems over exponentially long times: the near–linear case. In Dumas et al. [1995], pages 221–229.

J. Locke. *An essay concerning human understanding*. T. Basset, London, 1690. see P.H. Nidditch, Clarendon Press, Oxford, 1975.

A. Lomnicki. Nouveaux fondements de la théorie des probabilités. *Fundam. Math.*, 4:34–71, 1923.

F. London. Macroscopical interpretation of superconductivity. *Proc. Roy. Soc. (London)*, 152A:24–33, 1935.

F. London. The λ−phenomenon of liquid Helium and the Bose–Einstein degeneracy. *Nature*, 141:643–644, 1938a.

F. London. On the Bose–Einstein condensation. *Phys. Rev.*, 54:947–954, 1938b.

F. London. The state of liquid Helium neart absolute zero. *J. Phys. Chem.*, 43: 49–69, 1939.

F. London. *Superfluids, vol. 1*. Wiley, New York, 1950.

F. London and H. London. The electromagnetic equations of the supraconductor. *Proceedings of the Royal Society (London)*, A149:71–88, 1935. Supraleitung und Diamagnetismus, Physica 2 (1935) 341–354.

E.N. Lorentz. Deterministic nonperiodic flow. *Journal of the Atmospheric Sciences*, 20:130–141, 1963.

H.A. Lorentz. The motion of electrons in metallic bodies. *Proc. Akad. Wetenschappen, Amsterdam*, 7:438–453, 585–593, 684–691, 1905.

H.A. Lorentz. *The theory of electrons and its applications to the phenomena of light and radiant heat*. Teubner, Leipzig, 1909. 2nd ed. Stechert, New York, 1915; Dover ed., 1952.

J. Loschmidt. Ueber den Zustand des Wärmegleichgewichtes eines Systems von Körpern mit Rücksicht auf die Schwerkraft I,II,III. *Sitzungsber. Akad. Wiss., Wien*, 73, 75:128–142, 366–372, 287–298, 1876,1877.

P.C.A. Louis. *Recherches anatomiques, pathologiques et thérapeutiques, 2 vols.* J.–B. Baillière, Paris, 1829.

D. Loveland. A new interpretation of von Mises' concept of a random sequence. *Z. Math. Logik u. Grundlagen d. Math.*, 12:279–294, 1966. See also: 'A variant of the Kolmogorov concept of complexity', *Inform. Control* **15** (1969) 510–526.

I.J. Lowe and R.E. Nordberg. Free–induction decay in solids. *Phys. Rev.*, 107: 46–61, 1957.

G.B. Lubkin. Nobel prize in physics to Lee, Osheroff and Richardson for discovery of superfluidity in ^{3}He. *Physics Today (Sep.96)*, 49:17–19, 1996.

G. Ludwig. *Die Grundstukturen einer physikalischen Theorie, 2nd ed.* Springer, Berlin, 1990. see also [Schröter, 1998].

S. Lundqvist. Presentation speech. *Nobel Prize*, 1972. Reprinted in [Nobel Foundation, 1998] V: 47–49.

G.H. Lunsford and J. Ford. On the stability of periodic orbit for nonlinear oscillator systems in regions exhibiting stochastic behavior. *J. Math. Phys.*, 13:700–705, 1972.

Shang-Keng Ma. *Modern theory of critical phenomena.* W.A. Benjamin, Reading MA, 1976.

E. Mach. *Die Geschichte und die Wurzel des Satzes der Erhaltung der Arbeit.* Prague, Engl. transl. P.E.B. Jourdain, "Conservation of Energy", Open Court Publ., Chicago IL, 1911, 1872.

E. Mach. *Die Mechanik in ihrer Entwicklung historisch–kritish dargestellt.* Brockhaus, Leipzig, 1883. Hilbert refers to the 4th ed., 1901. See also The science of Mechanics: a critical and historical account of its development (6th ed.) [transl. by T.J. McCormack, from the 9th German ed. (1933)] The Open Court, LaSalle IL and London, 1974.

E. Mach. *Beiträge zur Analyse der Empfingdungen.* G. Fisher, Jena, 1886. Engl. translation from the 1st ed., by C.M. Williams; rev. from the 5th German ed., by S. Waterlow, Open Court, Chicago IL, 1914; reprinted as *The Analysis of sensations*, New York, Dover, 1959.

E. Mach. On the principle of the conservation of energy. *The Monist*, 5:no. 1, October, 1894.

G.W. Mackey. On a theorem by Stone and von Neumann. *Duke Math. J.*, 16: 313–325, 1949.

K.W. Madison, F. Chevy, W. Wohlleben, and J. Dalibard. Vortex lattices in a stirred Bose–Einstein condensate. *Phys. Rev. Lett.*, 84:806–809, 2000.

W.F. Magie. *A source book in physics.* Harvard University Press, Cambridge MA, 1963.

K. Mahler. Arithmetische Eigenschaften einer Klasse von Dezimalbrüchen. *Kon. Nederl. Akad. Wetens.*, 40:421–428, 1937. 'Ueber die Dezimalbruchtentwicklung gewisser Irrationalzahlen', Mathematica B, Zuften **6** (1937), 22–26.

K. Mahler. On the approximation of π. *Kon. Nederl. Akad. Wetens.*, 56:30–42, 1953.

L.E. Maistrov. *Probability theory: a historical sketch.* Academic Press, New York, 1974. This annotated Engl. transl. is by S. Kotz; original publication in Russian: 1967.

D.B. Malament and S.L. Zabell. Why Gibbs phase averages work - the role of ergodic theory. *Phil. Sci.*, 47:339–349, 1980.

J. Malitz. *Introduction to mathematical logic.* Springer, New York, 1979.

R. Mane. *Ergodic theory and differentiable dynamics.* Springer, Berlin, 1983.

E. Maor. *e – the story of a number.* Princeton University Press, Princeton NJ, 1994.

P.M. Marcus and E. Maxwell. The two–fluid model of superconductivity with application to isotope effects. *Phys. Rev.*, 91:1035–1042, 1953.

E. Mariotte. *Essay de la Nature de l' Air.* in 'Oeuvres', Leiden, 1717, 1676 and 1679.

L. Markus and K.R. Meyer. *Generic Hamiltonian systems are neither integrable nor ergodic; Memoirs No. 144.* Amer. Math. Soc., Providence RI, 1974.

P.C. Martin. A microscopic approach to superfluidity and superconductivity. *J. Math. Phys.*, 4:208–215, 1964.

P.C. Martin and J. Schwinger. Theory of many–particle systems. *Phys. Rev.*, 115: 1342–1373, 1959.

Ph.A. Martin. *Modèles en mécanique statistique des processus irreversibles.* LNP 103, Springer, Berlin, 1979.

Ph.A. Martin and G.G. Emch. A rigorous model sustaining van Hove's phenomenon. *Helv. Phys. Acta*, 48:59–78, 1975.

P. Martin-Löf. The definition of random sequences. *Information and Control*, 9: 602–619, 1966.

P. Martin-Löf. Complexity oscillations in infinite binary sequences. *Z. Wahrsch. verw. Geb.*, 19:225–230, 1971.

F. Maseres. *The doctrine of permutations and combinations being an essential and fundamental part of the doctrine of chances: as it is delivered by Mr. James Bernoulli, in his excellent treatise on the doctrine of chances, entitled Ars conjectandi, and by the celebrated Dr. John Wallis, of Oxford, in a tract intitled from the subject, and published at the end of his Treatise on algebra: in the former of which tracts is contained, a demonstration of Sir Isaac Newton's famous binomial theorem, in the cases of integral powers, and of the reciprocals of integral powers: together with some other useful mathematical tracts.* B. and J. White, London, 1795.

A. Matsubara, T. Atai, S. Hotta, J.S. Kothonen, and T. Mizusaki. Quest for Kosterlitz–Thouless transition in two–dimensional atomic hydrogen. In Griffin et al. [1995], pages 478–486.

J. Rosser Matthews. *Quantification and the quest for medical certainty.* Princeton University Press, Princeton, NJ, 1995.

B.T. Matthias, T.H. Geballe, and V.B. Compton. Sperconductivity. *Rev. Mod. Phys.*, 35:1–22, 1963. Errata, p. 414.

P. Mattila. *The geometry of sets and mesures in Euclidean spaces.* Cambridge University Press, Cambridge, 1995.

E. Maxwell. Isotope effect in the superconductivity of Mercury. *Phys. Rev.*, 78:477, 1950.

J.C. Maxwell. Illustrations of the dynamical theory of gases: Part 1. On the motions and collisions of perfectly elastic spheres; Part 2. On the process of diffusion of two or more kinds of moving particles among one another. *Phil. Mag.*, 19, 20:19–32, 21–37, 1860. See also: [Maxwell, 1867].

J.C. Maxwell. On the dynamical theory of gases. *Phil. Trans. Roy. Soc. London*, 157:49–88, 1867.

J.C. Maxwell. van der Waals on the continuity of the gaseous and liquid states. *Nature*, 10:477–480, 1874.

J.C. Maxwell. On the dynamical evidence of the molecular constitution of bodies. *Nature*, 11:357–359; 374–377, 1875. *J. Chem. Soc.* 13, 418–438.

J.C. Maxwell. Report on Dr. Andrews's paper "On the gaseous state of matter". *Roy. Soc. Archives*, 1875/6.

J.C. Maxwell. Diffusion. *Encyclopedia Britanica, 9th ed.*, 7:214–, 1878. reprinted in *The scientific papers of James Clark Maxwell*, W.D. Niven, ed.,Cambridge University Press, 1890; Dover, New York, 1952.

J.C. Maxwell. *The electric researches of the Honorable Henry Cavendish, F.R.S. written between 1771 and 1781.* Cambridge University Press, London, 1879a.

J.C. Maxwell. On Boltzmann's theorem on the average distribution of energy in a system of material points. *Trans. Cambridge Phil. Soc.*, 12:547–570, 1879b.

J.R. Mayer. Bemerkungen über die Kräfte der unbelebten Natur. *Ann. d. Chemie und Pharm.*, 42:233–240, 1842. See also 'Remarks on the forces of inorganic nature', *Phil. Mag. (Ser. 4)*, **24**, 371, (1862).

R.M. Mazo. *Statistical mechanical theories of transport processes.* Pergamon, Oxford, 1967.

P. Mazur and J. Van der Linden. Asymptotic form of the structure function for real systems. *J. Math. Phys.*, 4:271–277, 1963.

P.J. McKenna. Large torsional oscillations in suspension bridges: Fixing an old approximation. *Amer. Math. Monthly*, 106:1–18, 1999.

E. McMullin. What do physical models tell us? In B. Van Rootselaar and J.F. Staal, editors, *Logic, methodology and philosophy of science III*, page 392. North Holland, Amsterdam, 1968.

E. McMullin. Structural explanation. *American Philosophical Quarterly*, 15:139–147, 1978.

E. McMullin. Galilean idealization. *Studies in the History and Philosophy of Science*, 16:247–273, 1985.

M.L. Mehta. *Random matrices and the statistical theory of energy levels, 2nd ed.* Academic Press, Boston MA, 1991. 1st ed., 1967.

W. Meissner and R. Ochsenfeld. Ein neuer Effect bei Eintritt der Supraleitung. *Naturwiss.*, 21:787–788, 1933. W. Meissner, 'Bericht über neuere Arbeiten zur Supraleitfähigkeit', Physik. Zschr. 35 (1934) 931–938.

M. Mendès-France. Nombres normaux. Applications aux fonctions pseudo-aléatoires. *J. d'Anal. Math.*, 20:1–56, 1967.

E. Mendoza, editor. *Reflections on the motive power of fire, by Sadi Carnot; and other papers on the second law of thermodynamics, by E. Clapeyron and R. Clausius;, with an introduction.* Dover, New York, 1960.

N.D. Mermin. Absence of ordering in certain classical systems. *J. Math. Phys.*, 8:1061–1064, 1967. See also: Phys. Rev. 176 (1968) 250–254.

N.D. Mermin. Hidden variables and the two theorems of John Bell. *Rev. Mod. Phys.*, 65:803–815, 1993.

N.D. Mermin and H. Wagner. Absence of ferromagnetism and antiferromagnetism in one- and two–dimensional isotropic Heisenberg models. *Phys. Rev. Lett.*, 17:1133–1136, 1966.

L.D. Meshalkin. A case of isomorphisms of Bernoulli schemes. *Dokl. Akad. Nauk*, 128:41–44, 1959.

A. Messager and S. Miracle-Sole. Correlation functions and boundary conditions in the Ising ferromagnet. *J. Stat. Phys.*, 17:245–262, 1977.

A. Messager, S. Miracle-Solé, and J. Ruiz. Convexity properties of surface tension and equilibrium crystals. *J. Stat. Phys.*, 67:449–470, 1992.

C. Mette. *Invariantentheorie als Grundlage des Konventionalismus: Ueberlegungen zur Wissenschaftstheorie Jules Henri Poincarés.* Die Blaue Eule, Essen, 1986.

L. Michel and D. Ruelle, editors. *Systèmes à un nombre infini de degrés de liberté.* Editions du CNRS, Paris, 1970.

R.A. Millikan. New tests of Einstein's photoelectric equation. *Phys. Rev.*, 6:55, 1915. see also his: 'Some new values of the positive potentials assumed by metals in a high vacuum under the influence of ultra–violet light', Phys. Rev. 3 (1910) 287–288; 'A direct determination of h', Phys. Rev. 4 (1914) 73–75; 'A direct photoelectric determination of the Planck constant h', Phys. Rev. 7 (1916) 355-388.

R.A. Minlos and Ya.G. Sinai. The phenomenon of "phase separation" at low temperature in some lattice models of a gas I. *Math. USSR–Sb.*, 2:335–395, 1967. II. *Trans. Moscow Math. Soc.* 19, 121–196, 1968.

R.A. Minlos, S. Shlosman, and Yu.M. Suhov, editors. *On Dobrushin's way. From probability theory to statistical physics.* American Mathematics Society, Providence RI, 2000.

S. Miracle-Solé. Surface tension, step free energy and facets in the equilibrium crystal shape. *J. Stat. Phys.*, 79:183–214, 1995.

S. Miracle-Solé and J. Ruiz. On the Wulff construction as a problem of equivalence of ensembles. In Fannes and Verbeure [1994], pages 295–302.

J.D. Monk. *Mathematical logic*. Springer, New York, 1976.

E.W. Montroll. Statistical mechanics of nearest neighbor systems. *J. Chem. Phys.*, 9:706–721, 1941.

E.W. Montroll. Some remarks on the integral equations of statistical mechanics. In Cohen [1962], pages 230–249.

E.W. Montroll, R.B. Potts, and J.C. Ward. Correlations and spontaneous magnetization of the two–dimensional Ising model. *J. Math. Phys*, 4:308–322, 1963.

D.J. Moore. On state spaces and property lattices. *Stud. Hist. Phil. Mod. Phys.*, 30:61–83, 1999.

M. Morrison. Physical models and biological contexts. *Philosophy of Science*, 64: S315–S324, 1997.

R.W. Morse. Ultrasonic attenuation in metals at low temperature. In K. Mendelssohn, editor, *Progress in Cryogenics*, pages 1: 220–259. Heywood, London, 1959.

J. Moser. On invariant curves of area–preserving mapping of an annulus. *Nachr. Akad. Wiss. Göttingen, Math. Phys. Kl.*, IIa, 1962. see also: *On the theory of quasi–periodic motion,* SIAM Review **8**(1966) 145–172; 'Stable and random motion in dynamical systems, with special emphasis on classical mechanics', Ann. Math. Studies no. 77, Princeton University Press, 1973; 'Dynamical systems, theory and applications', LNP **38**, Springer, New York, 1975.

F.J. Murray and J. Von Neumann. On rings of operators. *Ann. of Math.*, 37:116–229, 1936. see also: FJM & JvN, Trans. Amer. Math, Soc. 41 (1937) 208–248; JvN, Ann. of Math 41 (1940) 94–161; FJM & JvN, Ann. of Math. 44 (1943) 7716–808.

B. Nachtergaele. *Interfaces and their excitations in quantum lattice models*. IAMP Congress, London, 2000. see already: N. Datta, A. Messager, B. Nachtergaele, 'Rigidity of interfaces in Falicov–Kimball model', *J. Stat. Phys.*, 99, 461–555, 2000.

E. Nagel. *The structure of science*. Harcourt, Brace and World, New York, 1961.

E. Nagel, P. Suppes, and A. Tarski, editors. *Logic, methodology and philosophy of science: Proceedings of the 1960 international congress*. Stanford University Press, Stanford CA, 1962.

J.F. Nagle. Lattice statistics of hydrogen bounded crystals. I. The residual entropy of ice. *J. Math. Phys.*, 7:1484–1491, 1966.

S. Nakajima. On quantum theory of transport phenomena. *Progr. theor. Phys.*, 20: 948–959, 1958.

W. Narkiewicz. *Number theory* (S. Kanemitsu, transl.). World Scientific, Singapore, 1977.

H. Narnhofer. On Fermi lattice systems with quadratic Hamiltonians. *Acta Phys. Austr.*, 31:349–353, 1970.

H. Narnhofer. Quantized Arnold cat maps can be entropic K–systems. *J. Math. Phys.*, 33:1502–1510, 1992.

H. Narnhofer, I. Peter, and W. Thirring. How hot is the de Sitter space? *Intern. J. Mod. Phys.*, B 10:1507–1520, 1996.

H. Narnhofer and W. Thirring. From relative entropy to entropy. *Fizika*, 17:257–265, 1985.

H. Narnhofer and W. Thirring. Equivalence of modular K and Anosov dynamical systems. *Rev. Math. Phys.*, 12:445–459, 2000.

C.L. M.H. Navier. Statistique appliquée à la médecine. *C.R. Acad. Sci. Paris*, 1: 247–250, 1835.

N.N. Nekhoroshev. Exponential estimates of the stability time of near–integrable Hamiltonian systems. *Russian Math. Surveys*, 32:1–65, 1977. II, Trudy Sem. Petrova **5** (1979) 5–50.

D.R. Nelson and J.M. Kosterlitz. Universal jump in the superfluid density of two–dimensional superfluids. *Phys. Rev. Lett.*, 39:1201–1205, 1977.

E. Nelson. *Dynamical theories of Brownian motion*. Princeton University Press, Princeton NJ, 1967.

E. Nelson. Internal set theory. *Bull. Amer. Math. Soc.*, 83:1165–1198, 1976.

W. Nernst. Ueber neuere Probleme der Wärmetheorie. *Berlin. Ber.*, page 86, 1911.

Y. Nestrenko. Modular functions and transcendence problems. *C. R. Acad. Sci. Paris, Ser. I.*, 322:909–914, 1996.

M. Netto. Beitrag zur Mannigfaltigkeitslehre. *J. reine u. angew. Math.*, 86:263–268, 1879.

I. Newton. *Philosophiæ naturalis principia mathematica*. Roy. Soc., London, 1687. I.B. Cohen and A. Whitman, transl., University California Press, Berkeley CA, 1999.

I. Newton. Scala graduum caloris. *Phil. Trans. Roy. Soc. London*, 22:824–829, 1701. Engl. transl. in 'Isaac Newton's Papers and Letters on Natural Philosophy', I.B. Cohen, ed., p.259-268, Harvard University Press, Cambridge MA., 1958.

I. Newton. Methodus differentialis. In *Analysis per quantitatum series, fluxiones, ac differencias: cum enumeratione linearum tertii ordinalis*, pages 93–101. ex o.p. (W.W. Jones), Londoni, 1711. (Manuscript mentionned in a letter to Oldenburg, October 1676).

I. Niiniluoto. *Is science progressive?* Reidel, Dordrecht, 1984.

I. Niiniluoto. Theories, approximations and idealizations. In R.B. Marcus, G.J.W. Dorn, and P. Weingartner, editors, *Logic, methodology and philosophy of science*, pages 255–289. North Holland, Amsterdam, 1986.

I. Niiniluoto. *Truthlikeness*. Reidel, Dordrecht, 1987.

R. Nillsen. Normal numbers without measure theory. *Amer. Math. Monthly*, 107: 639–644, 2000.

Nobel Foundation. *Nobel Lectures in Physics*. World Scientific, Singapore, 1998.

Th. Nordström. Presentation speech. *Nobel*, 1913. Reprinted in [Nobel Foundation, 1998] I: 303–305.

L. Nowak. Laws of science, theories, measurement. *Philosophy of Science*, 39: 533–548, 1972.

L. Nowak. *The structure of idealization: towards a systematic interpretation of the Marxian idea of science*. Reidel, Dordrecht, 1980.

H. Nowotny. *After the breakthrough: the emergence of high–temperature superconductivity as a research field*. Cambridge University Press, Cambridge, 1997.

P. Nozières. Some comments on Bose–Einstein condensation. In Griffin et al. [1995], pages 15–30.

G. Oddie. *Likeness to truth*. Reidel, Dordrecht, 1986.

P. Odifreddi. *Classical recursion theory: the theory of functions and sets of natural numbers*. North-Holland, Elsevier Science Publ., Amsterdam, 1989.

M. Ohya and D. Petz. *Quantum entropy and its use*. Springer, Berlin, 1993.

J.M. Ollagnier. *Ergodic theory and statistical mechanics*. Springer, Berlin, 1985.

R. Omnes. Logical reformulation of quantum mechanics. *J. Stat. Phys.*, 53:893–932, 1988.

L. Onsager. Crystal statistics. I. A two–dimensional model with an order–disorder transition. *Phys. Rev.*, 65:117–149, 1944.

L. Onsager. Discussion remark, IUPAP 1948 Florence meeting. *Nuovo Cim. Suppl.*, 6 (Suppl.2):261, 1949a.

L. Onsager. Discussion remark, IUPAP 1948 Florence meeting. *Nuovo Cim.*, 6 (Suppl.2):249, 1949b.

L. Onsager. Magnetic flux through a superconducting ring. *Phys. Rev. Lett.*, 7:50, 1961.

D.S. Ornstein. Bernoulli shifts with the same entropy are isomorphic. *Adv. in Math.*, 4:337–352, 1970.

D.S. Ornstein. *Ergodic theory, randomness, and dynamical systems.* Yale University Press, New Haven, 1974.

L.S. Ornstein. *Application of the statistical mechanics of Gibbs to molecular theoretic questions.* (in Dutch) Dissertation, Leiden, 1908.

D.D. Osheroff, R.C. Richardson, and D.M. Lee. Evidence for a new phase of solid He^3. *Phys. Rev. Lett.*, 28:885–888, 1972. D.D. Osheroff, W.J. Gully, R.C. Richardson and D.M. Lee, New magnetic phenomena in liquid He^3, below 3 mK, *ibid.* 29 (1972) 920–923.

E. Ott. *Chaos in dynamical systems.* Cambridge University Press, Cambridge, 1993.

J.C. Oxtoby and S.M. Ulam. Measure–preserving homeomorphisms and metrical transitivity. *Ann. Math.*, 42:874–920, 1941.

A. Pais. *Subtle is the Lord: the science and the life of Albert Einstein.* Clarendon Press, Oxford, 1982.

V. Pareto. *Manuel d'économie politique.* Giard and Brière, Paris, 1909. 2nd ed.: Marcel Giard, Paris, 1927.

V. Pareto. *Manuale di economia politica con una introduzione alla scienza sociale.* Societa Editrice Libraria, Milano, 1919.

G. Parisi. *Statistical field theory.* Addison–Wesley, Redwood City CA, 1988.

W. Parry. *Entropy and generators in ergodic theory.* Benjamin, New York, 1969.

K.R. Parthasarathy. *Probability measures on metric spaces.* Academic Press, New York, 1967.

K.R. Parthasarathy. *Introduction to Probability and Measure.* Springer, New York, 1977.

B. Pascal. *Traité du triangle arithmétique, avec quelques autres petits traités sur les mêmes matières.* G. Desprez, Paris, 1665. *Oeuvres complètes,* J. Mesnard, ed., Desclée de Brouwer, Paris, 1964, **II:** 1166-1195. Mesnard makes a compelling case that the *Triangle* was first printed in 1654, and only distributed in 1665.

B. Pascal. *Pensées.* édition de Port–Royal, 1670. (We used the text and numbering of the Brunschwicg edition, Collections des Grands Ecrivains de la France, Hachette, Paris, 1908–1924).

A.Z. Patashinskii and V.I. Pokrovskii. *Fluctuation theory of phase transitions.* Pergamon, Oxford, 1979.

W. Pauli. Ueber das H-theorem vom Anwachsen der Entropie vom Standpunkt der neuen Quantenmechanik. In P. Debye, editor, *Probleme der modernen Physik, Arnold Sommerfeld zum 60. Geburtstage, gewidnet von seinen Schulern,* pages 30–45. Hirzel, Leipzig, 1928.

L. Pauling. The structure and entropy of ice and of other crystals with some randomness of atomic arrangement. *J. Am. Chem. Soc.*, 57:2680–2684, 1935.

M. Pavicic. Bibliography on quantum logic and related structures. *Int. J. Theor. Phys.*, 31:373–461, 1992.

G. Peano. Sur une courbe, qui remplit toute une aire plane. *Math. Annalen*, 36: 157–160, 1890.

P.A. Pearce and C.J. Thompson. The spherical limit for n–vector correlations. *J. Stat. Phys.*, 17:189–196, 1977.

K. Pearson. Historical note on the origin of the normal curve of errors. *Biometrika*, 16:402–404, 1924.

K. Pearson. James Bernoulli's theorem. *Biometrika*, 17:201–210, 1925.

G. Pedersen. *C*-algebras and their automorphism groups*. Academic Press, London, 1979.

R. Peierls. On Ising's model of ferrromagnetism. *Proc. Camb. Phil. Soc.*, 32:477–481, 1936.

O. Penrose and L. Onsager. Bose–Einstein condensation and liquid Helium. *Phys. Rev.*, 104:576–584, 1956.

J. Perrin. *Brownian movement and molecular reality.* tranl. from: Annales de chimie et de physique, 8ème série, Sept. 1909, Taylor and Francis, London, 1910.

O. Perron. Ueber Matrizen. *Math. Ann.*, 64:248–263, 1907.

I.J. Peter and G.G. Emch. Quantum Anosov flows: a new family of examples. *J. Math. Phys.*, 39:4513–4539, 1998.

V.V. Petrov. *Limit theorems of probability theory: Sequences of independent random variables*. Clarendon Press, Oxford, 1995.

G.F. Peverone. *Due brevi e facili trattati: il primo d'arithmetica, l'altro di geometria*. G. di Tournes, Lione(=Lyons), 1558.

P. Pfeuty and G. Toulouse. *Introduction to the renormalization group and critical phenomena*. Wiley, New York, 1977.

C.E. Pfister. Large deviations and phase separation in the two–dimensional Ising model. *Helv. Phys. Acta*, 64:953–1054, 1991.

C.E. Pfister and Y. Velenik. Interface, surface tension and reentrant pinning in the 2D Ising model. *Commun. Math. Phys.*, 204:269–312, 1999.

E. Picard. *Sadi Carnot, bibliographie et manuscript*. Gauthier-Villars, Paris, 1927.

A. Pillay. Model theory. *Notices Amer. Math. Soc.*, 47:1373–1381, 2000.

S.A. Pirogov. Peierls argument for the anisotropic Ising model. In Minlos et al. [2000], page 195.

S.A. Pirogov and Ya. G. Sinai. Phase diagrams in classical lattice systems. *Theor. Math. Phys.*, 25,26:1185–1192, 39–49, 1975, 1976. reproduced in Ya.G. Sinai, Mathematical problems of statistical mechanics, World Scientific, Singapore, 1991.

C. Piron. Axiomatique quantique. *Helv. Phys. Acta*, 37:439–468, 1964.

C. Piron. *Foundations of quantum mechanics*. Benjamin, Reading MA, 1976. For an adversorial view, see chap. 20 in [Brody, 1993].

L.P. Pitaevsky. Vortex lines in an imperfect Bose gas. *Sov. Phys. JETP*, 13:451–454, 1961.

M. Plancherel. Sur l'incompatibilité de l'hypothèse ergodique et des équations de Hamilton. *Archives des Sciences Physiques et Naturelles, Genève (4)*, 33:254–255, 1912.

M. Plancherel. Beweis der Unmöglichkeit ergodischer mechanischer Systeme. *Ann. d. Physik*, 42:1061–1063, 1913.

M. Planck. Ueber eine Verbesserung der Wienschen Spektralgleichung. *Verh. d. Deutsch. Phys. Ges.*, 2:202–204, 1900a.

M. Planck. Zur Theorie des Gesetzes der Energieverteilung in Normalspektrum. *Verh. d. Deutsch. Phys. Ges.*, 2:237–245, 1900b. *same title*, Ann. d. Physik 4 (1900) 553–563.

M. Planck. *Treatise on thermodynamics*. Longmans, Green, London, 1903.

M. Planck. Ueber einen Satz der statistischen Dynamik und seine Erweiterung in der Quantentheorie. *Sitz. d. Preuss. Akad. Wiss.*, pages 324–341, 1917.

M. Planck. Ueber die Begrundung des zweites Haupsatzes der Thermodynamik. *Sitzungber. Preuss. Akad. Wiss. Phys. Math. Kl.*, pages 453–463, 1926.

M. Planck. *Die Entstehung und bisherige Entwicklung der Quantentheorie (Nobel Prize Lecture, June 2, 1920);* reprinted in *Physikalische Abhandlungen und Vorträge,*vol. 3, pp. 121–134. Vieweg, Braunschweig, 1958. Engl. transl. *A Survey of Physics*, Methuen, London, 1922; Dover, New York, 1960.

H. Poincaré. Théorie des groupes Fuchsiens. *Acta Math.*, 1:1–62, 1882.

H. Poincaré. Sur l'équilibre d'une masse fluide animée d'un mouvement de rotation. *Acta Mathematica*, 7:259–380, 1885. see also his Leçons sur les hypothèses cosmogoniques, 2nd ed., Hermann, Paris, 1913 [esp. pp. LX-LXI, 53–58, 132–189].

H. Poincaré. Sur le problème des trois corps et les équations de la dynamique. *Acta Math.*, 13:1–270, 1890.

H. Poincaré. *Les méthodes nouvelles de la mécanique céleste.* Gauthier-Villars, Paris, 1892–1907. Engl. transl. with introduction by D.L. Goroff: AIP, New York, 1993.

H. Poincaré. La théorie cinétique des gas. *Revue générale des sciences pures et appliquées*, 5:513–521, 1894. Reprinted in Oeuvres, vol 10, pp. 246–263.

H. Poincaré. *La science et l'hypothèse.* Flammarion, Paris, 1903.

H. Poincaré. *La valeur de la science.* Flammarion, Paris, 1904.

H. Poincaré. *Leçons de mécanique céleste.* Gauthier-Villars, Paris, 1905–1910.

H. Poincaré. *Science et méthode.* Flammarion, Paris, 1908a.

H. Poincaré. Sur les petits diviseurs dans la théorie de la lune. *Bull. Astr.*, 25: 321–360, 1908b.

H. Poincaré. *Calcul des probabilités, 2nd éd., revue et augmentée par l'auteur.* Gauthier-Villars, Paris, 1912. The 1st ed. indicates: Leçons professées pendant le deuxième semestre 1893–1894, redigées par A. Quiquet, G. Carre, Paris, 1896.

S.D. Poisson. Essai sur le calcul des variations. *Journ. Ec. Polytechn.*, 8:266–344, 1809.

S.D. Poisson. Sur la vitesse du son'; 'Sur la chaleur des gaz et des vapeurs. *Ann. de Chimie et Phys.*, 23:5–16; 337–352, 1823.

S.D. Poisson. *Recherches sur la probabilité des jugements en matière criminelle et en matière civile.* Bachelier, Paris, 1837. (see already C.R. Acad. Sci. Paris **1** (1835) 473-494).

L. Pontrjagin. *Topological groups.* Princeton University Press, Princeton NJ, 1939.

J.A. Pople. Nobel lecture: Quantum chemical models. *Rev. Mod. Phys.*, 71:1267–1274, 1999.

K.R. Popper. *Logik der Forschung.* Springer, Wien, 1935. The logic of scientific discovery, Hutchinson, London, 1959; Basic Books, New York, 1959; Harper and Row, New York, 1968.

K.R. Popper. *Conjectures and refutations.* Routledge, London, 1963.

K.R. Popper. A note on verisimilitude. *Brit. J. Phil. Sci.*, 27:145–159, 1976.

K.R. Popper. *A world of propensities.* Theomemes, Bristol, 1990.

T. Poston. *Fuzzy geometry.* PhD thesis, Warwick, 1971a.

T. Poston. Fuzzy geometry. *Manifold*, 10:25–33, 1971b.

H. Price. *Time's arrow and archimedes' point: new directions for the physics of time.* Oxford University Press, Oxford, 1996.

I. Prigogine. *Non–equilibrium statistical mechanics.* Wiley, New York, 1962.

I. Prigogine. *The end of certainty: time, chaos, and the new laws of nature.* The Free Press (Simon and Schuster), New York, 1997.

I. Prigogine and P. Résibois. On the kinetics of the approach to equilibrium. *Physica*, 27:629–646, 1961.

J.V. Pulé. The Bloch equations. *Commun. Math. Phys.*, 38:241–256, 1974.

H. Pulte. Jacobi's criticism of Lagrange: The changing role of mathematics in the foundations of classical mechanics. *Historia Math.*, 25:154–184, 1998.

P. Quay. A philosophical explanation of the explanatory function of ergodic theory. *Phil. Sci.*, 45:47–59, 1978.

A. Quetelet. *Sur l'homme et le développement de ses facultés, ou Essai de physique sociale.* C. Muquardt, J.-B. Baillière, J. Isaakoff, Bruxelles, Paris, St–Petersburg, 1835 and 1869.

C. Radin. Low temperature and the origin of crystalline symmetry. *Internat. J. Mod. Phys.*, B1:1157–1191, 1987.

C. Radin. Global order from local sources. *Bull. Amer. Math. Soc.*, 25:335–364, 1991.

F.P. Ramsey. Truth and probability, 1926. In R.B. Braithwaite, editor, *The foundations of mathematics and other logical essays*, pages 287–292. Routledge and Keagan Paul, Trench, Trubner, London, 1931. Also in F.P. Ramsay, Philosophical Papers, edited by D.H. Mellor, Cambridge University Press, Cambridge, 1990, pp. 52-94.

J.L. Ramsey. Towards an expanded epistemology for approximations. In M. Forbes D. Hull and K Okruhlik, editors, *PSA 1992 vol. 1*, pages 154–164. Philosophy of Science Association, East Lansing MI, 1992.

M. Randeira. Crossover from BCS theory to Bose–Einstein condensation. In Griffin et al. [1995], pages 355–392.

J. W.S. Rayleigh. *The theory of sound.* Macmillan, London, 1877–78.

J. W.S. Rayleigh. Remarks upon the law of complete radiation. *Phil. Mag.*, 49: 539–540, 1900. see also: The constant of radiation as calculated from molecular data, Nature 72 (1905) 243.

M. Rédei. Krylov's proof that statistical mechanics cannot be founded on classical mechanics and interpretation of classical statistical mechanics. *Philosophia naturalis*, 29:268–284, 1992.

M. Rédei. *Quantum logic in algebraic approach.* Kluwer, Dordrecht, 1998.

M. Rédei. Unsolved problems in mathematics: J.von Neumann's address to the ICM, Amsterdam, 1954. *Math. Intelligencer*, 21:7–12, 1999. see also: Von Neumann's concept of quantum logic and quantum probability, in "John von Neumann and the foundations of quantum mechanics", M. Rédei and M. Stölzner, eds., Kluwer, Dordrecht, 2001.

A.G. Redfield. Nuclear spin relaxation in superconducting Aluminium. *Phys. Rev. Lett.*, 3:85–86, 1959.

M. Redhead. Models in physics. *Brit. J. Phil. Sci.*, 31:145–163, 1980.

M. Reed and B. Simon. *Methods of modern mathematical physics, 4 vols.* Academic Press, New York, 1972–1980.

E. Regazzini. De Finetti's reconstruction of the Bayes–Laplace paradigm. In Costantini and Galavotti [1997], pages 19–36. reprinted from Erkenntnis, 45, 2/3, 1996.

H.V. Regnault. Recherches sur la dilatation des vapeurs. *Ann. chimie et physique (3ème ser.; Mém. Acad. Sci. (Paris)*, 4; 21:5–67; 15–120, 1842; 1847.

H.V. Regnault. Recherches sur la chaleur spécifique des fluides élastiques. *C. R. Acad. Sci. (Paris)*, 36:676–689, 1853.

K. Reich. *Carl Friedrich Gauss: 1777–1855 (2nd ed.).* Moos Verlag, Munchen, 1985.

H. Reichenbach. *Experience and prediction: an analysis of the foundations and structure of knowledge.* University of Chicago Press, Chicago IL, 1938. See also Wahrscheinlischkeitstheorie, Sijthoff's, Leiden, 1935; Les fondements logiques du calcul des probabilités, Ann. Inst. H. Poincaré, 1937.

I.C. Reynolds, B. Serin, W.H. Wright, and I.B. Nesbitt. Superconductivity of isotopes of Mercury. *Phys. Rev.*, 78:487, 1950.

O.W. Richardson and K.T. Compton. The photoelectric effect. *Phil. Mag.*, 24: 575–594, 1912.

F. Riesz and B. Sz.-Nagy. *Leçons d'analyse fonctionnelle*. Gauthier-Villars and Akadémiai Kiadó, Paris and Budapest, 1955.

R.M. Rilke. *Die Weise von Liebe und Tod des Cornets Christoph Rilke*. Juncker, Berlin, 1906.

A. Robinson. *Introduction to model theory and to the metamathematics of algebra*. North-Holland, Amsterdam, 1965.

A. Robinson. *Non–standard analysis*. North–Holland, Amsterdam, 1966. 2nd ed. 1974.

D.W. Robinson. Return to equilibrium. *Commun. Math. Phys.*, 31:171–189, 1973.

V.A. Rohlin. Lectures on ergodic theory. *Russian Math. Surveys*, 22:1–52, 1967.

V.A. Rohlin and Ya. G. Sinai. Construction and properties of invariant measurable partitions. *Dokl. Akad. Nauk SSSR*, 141:1038–1041, 1961. = Sov. Math. **2**, 1611–1614 (1961).

R. Rosenkrantz. Measuring truthlikeness. *Synthese*, 45:463–488, 1980.

A. Rosenthal. Beweis der Unmöglichkeit ergodischer Gassysteme. *Ann. d. Physik*, 42:796–806, 1913.

W. Rudin. *Principles of mathematical analysis*. McGraw–Hill, New York, 1964.

I. Rudnick. Critical surface density of the superfluid component in ^{4}He films. *Phys. Rev. Lett.*, 40:1454–1455, 1978.

D. Ruelle. Statistical mechanics of a one–dimensional lattice gas. *Commun. Math. Phys.*, 9:267–278, 1968.

D. Ruelle. *Statistical mechanics: rigorous results*. Benjamin, New York, 1969.

D. Ruelle. *Chance and chaos*. Princeton University Press, Princeton NJ, 1991.

G.S. Rushbrooke. On the thermodynamics of the critical region for the Ising problem. *J. Chem. Phys.*, 39:842–843, 1963.

L. Russo. The infinite cluster method in the two–dimensional Ising model. *Commun. Math. Phys.*, 67:251–266, 1979.

R.K. Sachs and H. Wu. *General relativity for mathematicians*. Springer, New York, 1977.

S. Sakai. *C*-algebras and W*-algebras*. Springer, New York, 1971. reprinted, 1998.

W. Salmon. Four decades of scientific explanation. In P. Kitcher and W.C. Salmon, editors, *Scientific explanation. Minnesota Studies in the Philosophy of Science, vol. 13*, pages 3–219. University of Minnesota Press, Minneapolis MN, 1989.

A. Sard. The measure of the critical values of a differentiable map. *Bull. Amer. Math. Soc.*, 48:233–234, 1942.

J.L. Sauvageot and J.P. Thouvenot. Une nouvelle définition de l'entropie dynamique des systèmes non–commutatifs. *Commun. Math. Phys.*, 145:411–423, 1992.

L.J. Savage. *The foundations of statistics*. Wiley, New York, 1954.

S.F. Savitt, editor. *Time's arrows today, recent physical and philosophical work on the direction of time*. Cambridge University Press, Cambridge, 1995. See already [Zeh, 1981].

M.R. Schafroth. Bemerkungen zur Frölichschen Theorie der Supraleitung. *Helv. Phys. Acta*, 24:645–662, 1951. Nuovo Cim. 9 (1952) 291.

M.R. Schafroth. Superconductivity of a charged boson gas. *Phys. Rev.*, 96:1149, 1954. Theory of superconductivity, *ibid.* 1442.

M.R. Schafroth. Theoretical aspects of superconductivity. In F. Seitz and D. Turnbull, editors, *Solid state physics.10*, pages 293–498. Academic Press, New York, 1960.

E. Scheibe. The Physicists' conception of progress. In H. Schnädelbach, editor, *Rationalität, Philosophishe Beiträge*, pages 91–116. Suhrkamp, Frankfurt, 1988. See also [Ludwig, 1990, Schröter, 1998, Scheibe, 2001].

E. Scheibe. A new theory of reduction in physics. In J. Earman, A.I. Janis, G.J. Massey, and N. Rescher, editors, *Philosophical problems of the internal and external worlds: essays on the philosophy of Adolf Grünbaum*, pages 249–271. University of Pittsburgh Press, Pittsburgh PA, 1993.

E. Scheibe. *Die Reduktion Physikalischer Theorien: ein Beitrag zur Einheit der Physik. Teil I: Grundlagen und elementare Theorie*. Springer, Berlin, 1997.

E. Scheibe. *Die Reduktion Physikalischer Theorien: ein Beitrag zur Einheit der Physik. Teil II: Inkommensurabilität und Grenzfallreduktion*. Springer, Berlin, 1999.

E. Scheibe. *Between Rationalism and Empiricism: Selected Papers in the Philosophy of Physics, B. Falkenburg, ed.* Springer, Heidelberg, 2001.

H.M. Schey. *Div, grad, curl and all that – an informal text on vector calculus (3rd ed.)*. W.W. Norton, New York, 1997.

W.E. Schiesser. "Computer in physics" prompts model debate. *Physics Today (Dec. 99)*, 52:15, 79, 1999.

P.A. Schlipp, editor. *The philosophy of Rudolf Carnap*. Open Court, La Salle IL, 1963.

T. Schmidt. On normal numbers. *Pacific J. Math.*, 10:661–678, 1960.

I. Schneider. *Die Entwicklung der Wahrscheinlichkeitstheorie von den Anfangen bis 1933: Einfuhrungen und Texte*. Akademie-Verlag/Wissenschaftliche Buchgesellschaft, Berlin/Darmstadt, 1988.

C.P. Schnorr. A survey of the theory of random sequences. In *Basic problems in methodology and linguistics*, pages 193–210. Reidel, Dordrecht, 1977. See already: 'A unified approach to the definition of random sequences', *Math. Syst. Th.* **5** (1971) 246–258; 'Process complexity and effective random tests', *J. Comp. Sci.* **7** (1973) 376–388.

J.R. Schrieffer. *Superconductivity*. Benjamin, New York, 1964.

J.R. Schrieffer. Macroscopic quantum phenomena from pairing in superconductors. *Nobel acceptance speech*, 1972. Reprinted in [Nobel Foundation, 1998] V: 97–108.

J. Schröter. Das Ludwigsche Konzept physikalischer Theorien. *Preprint, Paderborn*, 1998.

T.D. Schultz, D.C. Mattis, and E.H. Lieb. Two–dimensional Ising model as a soluble problem of many fermions. *Rev. Mod. Phys.*, 36:856–871, 1964.

L. Schwartz. *Théorie des distributions*. Hermann, Paris, 1957.

L. Schwartz. *Analyse (2e partie): Topologie générale et analyse fonctionelle*. Hermann, Paris, 1978a.

R.J. Schwartz. Idealization and approximations in physics. *Philosophy of Science*, 45:595–603, 1978b.

B. Schwarzschild. Physics Nobel Prize goes to Tsui, Stormer and Laughlin for the fractional quantum Hall effect. *Physics Today (Dec.98)*, 51:17–19, 1998.

S. Schweber and M. Wächter. Complex systems, modelling and simulation. *Stud. Hist. Phil. Mod. Phys.*, 31B:583–609, 2000.

S.S. Schweber. *QED and the men who made it*. Princeton University Press, Princeton NJ, 1994. See in particular the *Epilogue: Some reflections on renormalization theory*.

C. Seelig. *Albert Einstein, a documentary biography*. Staples Press, London, 1956.

T. Seidenfeld. Why I am not an objective Bayesian; some reflections prompted by Rosenkrantz. *Theory and Decision*, 11:413–440, 1979.

B. Serin. Superconductivity, experimental part. In Flügge [1956], pages 210–272.

B. Serin, I.C. Reynolds, and I.B. Nesbitt. Superconductivity of isotopes of Mercury. *Phys. Rev.*, 78:813–814, 1950.

J. Serrin, editor. *New perspectives in thermodynamics.* Springer, Berlin, 1986.

G.L. Sewell. Stability, equilibrium and metastability in statistical mechanics. *Phys. Rep.*, 57:307–342, 1980.

G.L. Sewell. *Quantum theory of collective phenomena.* Clarendon Press, Oxford, 1986.

G.L. Sewell. Macroscopic quantum electrodynamics of a plasma model: derivation of the Vlaslov kinetics. *Lett. Math. Phys.*, 40:203–222, 1997.

G.L. Sewell. *Quantum mechanics and its emergent macrophysics.* Princeton University Press, Princeton NJ, forthcoming.

C.E. Shannon. The mathematical theory of communication. *Bell Syst. Techn. Journ.*, 27:379–423, 623–656, 1948.

C.E. Shannon and W. Weaver. *The mathematical theory of communication.* The University of Illinois Press, Urbana IL, 1949.

O.B. Sheynin. S.D. Poisson's work in probability. *Arch. Hist. Exact Sci.*, 18:245–300, 1977/78.

A.B. Shidlovskii. *Transcendental numbers.* N. Koblitz, transl., de Gruyter, Berlin, 1989.

P. Shields. *The theory of Bernoulli shifts.* University of Chicago Press, Chicago IL, 1973.

A. Shimony. Coherence and the axioms of confirmation. *J. Symbolic Logic*, 20:1–28, 1955.

A. Shimony. Events and processes in the quantum world. In R. Penrose and C.J. Isham, editors, *Quantum concepts in space and time*, pages 182–203. Clarendon Press, Oxford, 1986.

A. Shimony. *Search for a naturalistic world view (2 vol.).* Cambridge University Press, Cambridge, 1993.

W. Sieg. Step by recursive step: Church's analysis of effective calculability. *Bull. Symbolic Logic*, 3:154–180, 1997.

W. Sierpinsky. Sur la valeur asymptotique d'une certaine somme. *Bull. Intern'l Acad. Sci. et Lettres, Cracovie*, A:9–11, 1910.

W. Sierpinsky. *Introduction to general topology.* University Toronto Press, Toronto, 1934.

I.F. Silvera and M. Reynolds. A hybrid microwave–static magnetic trap for spin-polarized hydrogen. *Physica*, B 169:449–450, 1991.

I.F. Silvera and J. T.M. Walravan. Stabilization of atomic Hydrogen at low temperature. *Phys. Rev. Lett.*, 44:164–168, 1980. see also: I.F. Silvera and V.V. Goldman, 'Atomic Hydrogen in contact with Helium surface: Bose condensation, adsorption isotherms and stability', Phys. Rev. Lett. **45**, 915–918, 1980; B.W. Statt and A.J. Berlinsky, 'The theory of spin relaxation and recombination in spin–polarized atomic Hydrogen', Phys. Rev. Lett. **45**, 2105–2109, 1980.

I.F. Silvera and J. T.M. Walravan. Spin polarized atomic Hydrogen. *Prog. Low Temp. Phys.*, X:139–370, 1986.

T. Simpson. *Miscellaneous tracts ... on mechanics, physical astronomy and speculative mathematics.* J. Nourse, London, 1757. esp. pp.64–75, a slightly expanded version of his Phil. Trans. Roy. Soc. London 49, 82–93, 1755.

Ya. G. Sinai. On the concept of entropy of a dynamical system. *Dokl. Akad. Nauk SSSR*, 124:768–771, 1959. = Math. Rev. **21** No. 2036a.

Ya. G. Sinai. Dynamical systems with countably Lebesgue spectra. *Izvestia Math. Nauk*, 25:899–924, 1961. Transl. Amer. Math. Soc., ser. 2, **39** 83–110 (1961).

Ya. G. Sinai. On the foundation of the ergodic hypothesis for a dynamical system of statistical mechanics. *Soviet Math. Dokl.*, 4:1818–1822, 1963.

Ya. G. Sinai. Dynamical systems with elastic reflections. *Russian Math. Surveys*, 25:137–189, 1970.

Ya. G. Sinai. *Introduction to ergodic theory*. Princeton University Press, Princeton NJ, 1976.

Ya. G. Sinai. *Theory of phase transitions.Rigorous results*. Pergamon, Oxford, 1982.

Ya. G. Sinai and N. I Chernov. Ergodic properties of certain systems of two–dimensional disks and three–dimensional balls. *Russian Math. Surveys*, 42: 181–207, 1987.

L. Sklar. Statistical explanation and ergodic theory. *Phil. Sci.*, 40:194–212, 1973.

L. Sklar. *Physics and chance; philosophical issues in the foundations of statistical mechanics*. Cambridge University Press, Cambridge, 1993.

B. Skyrms and W.L. Harper, editors. *Causation, chance, and credence, vol. 1*. Kluwer, Dordrecht, 1988.

C. Smith and M.N. Wise. *Energy and Empire: a biographical study of Lord Kelvin*. Cambridge University Press, Cambridge, 1989.

M. Smoluchowski. Sur le chemin moyen parcouru par les molécules d'un gaz et sur son rapport avec la théorie de la diffusion. *Bull. Acad. Sci. Cracovie, Cl. Sci. Math. Natur.*, pages 202–213, 1906a. Reprinted with Engl. Transl. in (Ingarden 1999), pp. 29–42.

M. Smoluchowski. Zur kinetischen Theorie der Brownschen Molekularbewegung und der Suspensionen. *Ann. d. Physik*, 21:756–780, 1906b.

M. Smoluchowski. Drei Vorträge über Diffusion, Brownsche Molekularbewegung und Koagulation von Kolloidteilschen. *Phys. Zschr.*, 17:557–571; 585–599, 1916. Reprinted with Engl. Transl. in (Ingarden 1999), pp. 43–77, 78–107, 108–127.

J.D. Sneed, editor. *The logical structure of mathematical physics*. D. Reidel, Dordrecht, 1971.

R.L. Soare. Computability and recursion. *Bull. Symb. Logic*, 2:284–321, 1996.

R.J. Solomonoff. A formal theory of inductive inference. *Inform. Control*, 7:1–22, 224–254, 1964. See already: M.L. Minsky, 'Problems of formulation for artificial intelligence' *in:* Mathematical problems in the biological sciences, Proc. Symp. Applied Math. **14**, R.E. Bellman, ed., AMS providence RI, 1962, pp.35, 43.

R.M. Solovay. A model of set theory in which every set of reals is Lebesgue measurable. *Ann. of Math.*, 92:1–56, 1970.

W. Somerset Maugham. *The moon and sixpence*. G.H. Doran, New York, 1919.

Sophocles. *Oedipus the King*. n.a., Athens, ca. 425 BC. F. Storr, transl., Heritage, New York, 1956.

H. Spohn. Derivation of the transport equations for electrons moving through random impurities. *J. Stat. Phys.*, 17:385–412, 1977.

H. Spohn. Kinetic equations from Hamiltonian dynamics: Markovian limits. *Rev. Mod. Phys.*, 53:569–615, 1980.

H. Spohn. *Large scale dynamics of interacting particles*. Springer, Berlin, 1991.

H. Spohn and R. Dümcke. The proper form of the generator in the weak coupling limit. *Z. f. Physik B*, 34:419–422, 1979.

H.E. Stanley. *Introduction to phase transitions and critical phenomena*. Clarendon Press, Oxford, 1971.

J. Stefan. Ueber die Beziehung zwischen der Wärmestrahlung und der Temperatur. *Wiener Ber.*, 79:391–428, 1879.

W. Stegmüller. *Theorienstruktruen under Theoriendynamik*. Springer, Berlin, 1973. Engl. transl: The structure and dynamics of theories; Springer, New York, 1976.

H. Steinhaus. Les probabilités dénombrables et leur rapport à la théorie de la mesure. *Fundam. Math.*, 4:286–310, 1923.

H. Steinhaus. La théorie et les applications des fonctions indépendentes au sens stochastique. *Activités Scientifiques et Industrielles*, 738:57–73, 1938.

S. Sternberg. *Lectures on differential geometry (2nd ed.)*. Chelsea, New York, 1983.

M. Stifel. *Arithmetica Integra*. Apud Iohan. Petreium, Norimbergae (=Nurenberg), 1544.

W.F. Stinespring. Positive functions on C*-algebras. *Proc. Amer. Math. Soc.*, 6: 211–216, 1955.

D.R. Stinson. *Cryptography: theory and practice*. CRC Press, Boca Raton, 1995.

D.S. Stoddard and J. Ford. Numerical experiment on the stochastic behavior of a Lennard–Jones gas system. *Phys. Rev. A*, 8:1504–1512, 1973.

M.H. Stone. Linear transformations in Hilbert space III. *Proc. Nat. Acad. Sci. USA*, 16:172–175, 1930.

M.H. Stone. *Linear transformation in Hilbert space and their applications to analysis*. Am. Math. Soc., Providence, 1932.

M.H. Stone. The theory of representation for Boolean algebras. *Trans. Amer. Math. Soc.*, 40:37–111, 1940. Applications of the theory of Boolean algebras to general topology *ibid.* 41 (1937) 375–481.

H. T.C. Stoof. Condensate formation in a Bose gas. In Griffin et al. [1995], pages 226–280.

E. Størmer. Symmetric states on infinite tensor products of C^*–algebras. *J. Funct. Anal.*, 3:48–68, 1969.

E. Størmer. Entropy of endomorphisms and relative entropy in finite von Neumann algebras. *J. Funct. Anal.*, 171:34–52, 2000.

D.J. Struik. *A concise history of mathematics (4th ed.)*. Dover, 1948. 1987.

D.J. Struik. *A sourcebook in mathematics, 1200-1800*. Harvard University Press, Cambridge MA, 1969.

E.C.G. Stuekelberg and A. Petermann. La normalisation des constantes dans la théorie des quanta. *Helv. Phys. Acta*, 26:499–520, 1953.

C Stutz and B. Williams. Ernst Ising. *Physics Today (Mar.99)*, 52:106–107, 1999.

S.J. Summers. Bell's inequalities and algebraic structure. In S. Doplicher, R. Longo, J.R. Roberts, and L. Zsido, editors, *Operator algebras and quantum field theory*, pages 633–646. International Publishers, distributed by AMS, Providence RI, 1997.

F. Suppe, editor. *The structure of scientific theories*. University of Illinois Press, Urbana IL, 1977.

F. Suppe. *The semantic conception of theories and scientific realism*. University of Illinois Press, Urbana IL, 1989.

P. Suppes. A comparison of the meaning and uses of models in mathematics and the empirical sciences. *Synthese*, 12:287–301, 1960.

P. Suppes. Models of data. In Nagel et al. [1962], pages 252–261.

P. Suppes. What is a scientific theory? In R. Colodny, editor, *Philosophy of science today*, pages 55–67. University of Pittsburgh Press, Pittsburgh, PA, 1972.

P. Suppes. Representation theory and the analysis of structure. *Philosophia Naturalis*, 25:254–268, 1988.

P. Suppes. *Models and methods in philosophy of science: selected essays*. Kluwer, Dordrecht, 1993.

K. Svozil. *Quantum logics*. Springer-Verlag, New York, Singapore, 1998.

R.J. Swenson. Derivation of generalized master equations. *J. Math. Phys.*, 3:1017–1022, 1962. Generalized master equation and t-matrix expansion, *ibid.* 4 (1963) 544–551.

D. Szasz. On the K–property of some planar hyperbolic billards. *Commun. Math. Phys.*, 145:595–604, 1992.

D. Szasz. Ergodicity of classical billard balls. *Physica A*, 194:86–92, 1993.

D. Szasz. Boltzmann's ergodic hypothesis, a conjecture for centuries? *Studia Scient. Math. Hungar.*, 31:299–322, 1996.

L. Szilard. Ueber die Entropieverminderung in einer thermodynamischen System bei Eingriffen intelligenter Wesen. *Zschr. f. Physik*, 53:840–856, 1929. = On the decrease of entropy in a thermodynamical system by the intervention of an intelligent being, in *Collected works of Leo Szilard*, B.T. Feld and G. Weiss Szilard, eds., MIT Press, Cambridge MA, 1972, pp. 120-129; see also [Leff and Rex, 1990] pp. 124–133.

H. Takahashi. A simple method for treating the statistical mechanics of one–dimensional substances. *Proc. phys. math. Soc. Japan*, 24, 1942. reprinted in [Lieb and Mattis, 1966], pp. 25–27.

M. Takesaki. *Tomita's theory of modular algebras and its applications*. LNM 128, Springer, New York, 1970.

M. Takesaki. States and automorphisms of operator algebras – standard representations and KMS boundary condition. In Lenard [1973], pages 205–246.

N. Tartaglia. *General trattato di numeri et misure*. T. di Nauo, Vinegia, 1556.

D. Teets and K. Whitehead. The discovery of Ceres: how Gauss became famous. *Mathematics Magazine (Apr. 99)*, 72:83–93, 1999.

H.N.V. Temperley and E.H. Lieb. Relations between the 'percolation' and the 'colouring' problem and other graph-theoretical problems associated with regular planar lattices. Some exact results for the 'percolation' problem. *Proc. Roy. Soc.*, A322:251–280, 1971.

D. Ter Haar. *Elements of statistical mechanics*. Rinehart, New York, 1954. Note: Ter Haar is also one of the translators of [Balian, 1991,1992].

D. Ter Haar and C.D. Green. The statistical aspect of Boltzmann's H-theorem. *Proc. Phys. Soc. A*, 66:153–159, 1953.

W. Thirring. The mathematical structure of the BCS model and related models. In A. Cruz and T.W. Preist, editors, *The many–body problem*, pages 125–160. Plenum, London, 1969.

W. Thirring. *Quantum mechanics of large systems*. E.M. Harrell, transl.; Springer, New York, Wien, 1983.

W. Thirring and A. Wehrl. On the mathematical structure of the BCS model. *Commun. Math. Phys.*, 4:303–314, 1967. W. Thirring, *ibib.* II., 7 (1968) 181–199.

C.J. Thompson. *Mathematical statistical mechanics*. Macmillan, New York, 1972.

W. Thomson. On an absolute thermometric scale, founded on Carnot's theory of the motive power of heat, and calculated from Regnault's observations. *Phil. Mag. (3)*, 33:313–317, 1848.

W. Thomson. A mathematical theory of magnetism. *Phil. Trans. Roy. Soc. London*, 141:243–285, 1851a.

W. Thomson. On the dynamical theory of heat, with numerical results deduced from Mr. Joule's equivalent of the thermal unit, and M. Regnault's observations on steam. *Trans. Roy. Soc. Edimburgh*, 3:48–52, 1851b. see also: *ibid.***20** 261-288 (1853); *Phil. Mag.* **4**, 8–21, 105–117 and 168–176 (1852); and 'Deux mémoires sur la théorie dynamique de la chaleur', *Liouville Journ. de Math.***12**, 209-241, 1852.

W. Thomson. On a universal tendency in nature to the dissipation of mechanical energy. *Proc. Roy. Soc. Edinburgh*, 3:139–142, 1852; see also: *Phil. Mag.* **4**, 304–306 (1852).

D.J. Thouless. The BCS model Hamiltonian as an exactly soluble problem in statistical mechanics. In J.E. Bowcock, editor, *Methods and problems in statistical mechanics, in honour of R.E. Peierls*, pages 29–35. Elsevier, New York, 1970.

M. Tinkham. Energy gap interpretation of experiments on infrared transmission through superconducting films. *Phys. Rev.*, 104:845–846, 1956.

L. Tisza. Sur la supraconductibilité thermique de l'Helium II liquide et la statistique de Bose–Einstein. *C.R. Acad. Sci. Paris*, 207:1035–1037, 1938a.

L. Tisza. Transport phenomena in Helium II. *Nature*, 141:913, 1938b.

M. Toda. Vibration of a chain with non–linear interaction – Exact solution in terms of Jacobian elliptic functions. *J. Phys. Soc. Japan*, 22:431–436, 1967. see also: Wave in non–linear lattice, *Progr. Theoret. Phys.(Suppl.)* **45** (1970) 174–200.

M. Toda. Integrability of the trajectories of the lattice with cubic non–linearity. *Phys. Lett.*, 48A:335–336, 1974.

I. Todhunter. *A History of the mathematical theory of probability from the time of Pascal to that of Laplace*. Macmillan, Cambridge, 1865. Thoemmes, Bristol, 1993.

M. Tomita. *Standard forms of von Neumann algebras*. Fifth Functional Analysis Symposium of the Mathematical Society of Japan, Sendai, 1967.

J. Tomiyama. On the projection of norm 1 in W*-algebras. *Proc. Japan Acad.*, 33: 608–612, 1957. see also: II. Tohoku Math. J. 10 (1958) 204–209; III. *ibid.* 11 (1959) 125–129.

L. Tonks. The complete equation of state of one, two and three–dimensional gases of hard elastic spheres. *Phys. Rev.*, 50:955–963, 1936.

G.J. Tourlakis. *Computability*. Reston Publ., Prentice-Hall, Reston, VA, 1984.

C. Truesdell. *The rational mechanics of flexible or elastic bodies 1638–1788, Introduction to Leonhardi Euleri Opera Omnia, ser.2, vol. 11*. Orell Füssli, Zurich, 1960. [esp. Section 28].

C. Truesdell. Ergodic theory in classical statistical mechanics. In P. Caldirola, editor, *Ergodic theories*, pages 21–56. Academic Press, New York, 1961.

C. Truesdell. *Six lectures on modern natural philosophy*. Springer, New York, 1966.

C. Truesdell. *The tragicomical history of thermodynamics, 1822–1854*. Springer, New York, 1980. See also [Truesdell, 1966, Truesdell and Wang, 1984, Serrin, 1986, Truesdell and Bharatha, 1989].

C. Truesdell. The influence of elasticity on analysis: the classical heritage. *Bull. Amer. Math. Soc.*, 9:293–310, 1983.

C. Truesdell. *An idiot's fugitive essays on science*. Springer, New York, 1984.

C. Truesdell and S. Bharatha. *The concept and logic of classical thermodynamics as a theory of heat engines. Rigorously constructed upon foundations laid by S. Carnot and F. Reech*. Springer, New York, 1989.

C. Truesdell and R.G. Muncaster. *Fundamentals of Maxwell's kinetic theory of a simple monatomic gas treated as a branch of rational mechanics*. Academic Press, New York, 1980.

C. Truesdell and C.C. Wang. *Rational thermodynamics, 2nd ed.* Springer, New York, 1984. 1st ed. 1969.

H.G. Tucker. *A graduate course in probability*. Academic Press, New York, 1967.

P. Turán. On a theorem by Hardy and Ramanujan. *J. London Math. Soc.*, 7:274–276, 1934. See also: Ueber einige Verallgemeinungen einem Satz von Hardy und Ramanujan, *ibid* 11 (1936), 125–133.

P. Tuyls. Comparing quantum dynamical entropies. In Alicki et al. [1998], pages 411–420.

J. Uffink. Can the maximum entropy principle be explained as a consistency requirement. *Stud. Hist. Phil. Mod. Phys.*, 26:223–261, 1995.

J. Uffink. The constraint rule of the maximum entropy principle. *Stud. Hist. Phil. Mod. Phys.*, 27:47–79, 1996a.

J. Uffink. Nought but molecules in motion: essay–review. *Stud. Hist. Phil. Mod. Phys.*, 27:373–387, 1996b.

G.E. Uhlenbeck. The classical theories of the critical phenomena. In *Critical phenomena, Proccedings of a conference, Washington, April 1965*, pages 3–6. National Bureau of Standards, Washington, D.C., 1966.

G.E. Uhlenbeck. Summarizing remarks. In Bak [1967], pages 574–582.

G.E. Uhlenbeck. An outline of statitical mechanics. In *Fundamental problems in statistical mechanics II*, pages 1–27. (E.G.D. Cohen, ed.), North Holland, Amsterdam, 1968.

S.M. Ulam. On the ergodic behavior of dynamical systems. *LA-2055*, May 10, 1955. The text of this lecture is included in *Analogies between analogies*, A.R. Bednarek and F. Ulam, eds., Los Alamos Series in Basic and Applied Sciences, University California Press, Berkeley CA, 1990, pp. 155-162.

J.G. Valatin. Comments on the theory of superconductivity. *Nuovo Cim.*, 7:843–857, 1958.

H. Van Beijeren. Interface sharpness in the Ising model. *Commun. Math. Phys.*, 40:1–6, 1975.

J.D. Van der Waals. *Over de continuiteit van der gas- en vloeistoftoestand (Dissertation)*. A. W, Stijhoff, Leiden, 1873. see also [Van der Waals,̄ 1881, 1890].

J.D. Van der Waals. *Die Kontinuität des gasförmigen und flussigen Zustandes*. Barth, Leipzig, 1881. this transl. of [Van der Waals, 1873] by F. Roth contains new materials drawn from later papers by Van der Waals; see also the 1899 edition.

J.D. Van der Waals. *The continuity of the liquid and gaseous states of matter*. Taylor and Francis, London, 1890. transl. from [Van der Waals, 1881] by R. Threlfall and J. F. Adair.

B. Van Fraassen. A formal approach to the philosophy of science. In S. Morgenbesser, editor, *Paradigms and paradoxes*, pages 303–366. Basic Books, New York, 1967.

B. Van Fraassen. On the extension of Beth's semantics of physical theories. *Philosophy of Science*, 37:325–334, 1970.

B. Van Fraassen. *The scientific image*. Clarendon Press, Oxford, 1983.

B. Van Fraassen. *Laws and symmetry*. Clarendon Press, Oxford, 1989.

B.C. Van Fraassen. *Quantum mechanics: an empiricist view*. Clarendon Press, Oxford, 1991.

L. Van Hove. Quelques propriétés générales de l'intégrale de configuration d'un système de particules avec interaction. *Physica*, 15:951–961, 1949.

L. Van Hove. Sur l'intégrale de configuration pour les systèmes de particules à une dimension. *Physica*, 16:137–143, 1950.

L. Van Hove. Sur le problème des relations entre les transformations unitaires de la mécanique quantique et les transformations canoniques de la mécanique classique. *Mem. Acad. Roy. Belgique, cl. sc.*, 37:610–620, 1951. Sur certaines représentations unitaires d'un groupe infini de transformations, ibid. 36 (1951) 1–102.

L. Van Hove. Quantum mechanical perturbations giving rise to a statistical transport equation'; 'The approach to equilibrium in quantum statistics'; 'The ergodic behaviour of quantum many–body systems. *Physica*, 21,23,25:517–540; 441–480; 268–276, 1955, 1957, 1959.

L. Van Hove. The problem of master equations. In P. Caldirola, editor, *Ergodic theories*, pages 155–169. Academic Press, New York, 1961.

L. Van Hove. Master equation and approach to equilibrium for quantum systems. In Cohen [1962], pages 157–172.

N.G. Van Kampen. Fundamental problems in statistical mechanics of irreversible processes. In Cohen [1962], pages 173–202.

M. Van Lambalgen. The axiomatization of randomness. *J. Symbol. Logic*, 55: 1143–1167, 1990. See already: 'Von Mises' definition of random sequences reconsidered', *ibid.* **52** (1987) 725–755; 'Algorithmic information theory', *ibid.* **54** (1989) 1389–1400.

V.S. Varadarajan. *Geometry of quantum theory (2nd ed.)*. Springer, New York, 1985.

M. Verri and V. Gorini. Quantum dynamical semigroups and multipole relaxation of a spin in isotropic surroundings. *J. Math. Phys.*, 19:1803–1807, 1978.

N.I. Vilenkin. *In search of infinity*. Birkhauser, Boston MA, 1995.

J. Ville. *Etude critique de la notion de collectif.* Gauthier-Villars, Paris, 1939. See already: 'Sur la notion de collectif', *C. R. Acad. Sci. Paris* **203** (1936) 26–27.

W.F. Vinen. Detection of single quanta of circulation in rotating Helium II. *Nature*, 181:1524–1525, 1958. *ibid. Proc. Roy. Soc. (London)*, A 260, 218-236, 1961; for a discussion, see [Vinen, 1961].

W.F. Vinen. Votex lines in liquid Helium II. *Progr. Low Temp. Phys.*, III:1–57, 1961.

G. Vitali. *Sul problemo della measura dei gruppi di una retta*. Bologna, 1905.

D. Voiculescu. The analogues of entropy and of Fisher's information measure in free probability theory I. *Commun. math. phys.*, 155:71–92, 1993.

D. Voiculescu. Dynamical approximation entropies and topological entropy in operator algebras. *Commun. Math. Phys.*, 170:249–281, 1995.

D. Voiculescu. *Free probability theory*. Amer. Math. Soc., Providence RI, 1997.

D. Voiculescu. The analogues of entropy and of Fisher's information measure in free probability theory V: noncommutative Hilbert transform. *Invent. Math.*, 132:189–227, 1998. See also: The noncommutative Hilbert transform approach to free entropy, in *Quantum probability*, Banach Center Publications, Vol. 43, Polish Acad. Sci., Warsaw, 1998, pp. 421–427.

P. Volkmann. *Einführung in der Studium der theoretischen Physik, insbesondere in das der analytischen Mechanik mit einer Einleitung in die Theorie der physikalischen Erkenntnis*. Teubner, Leipzig, 1900.

H. Von Helmoltz. *Ueber die Erhaltung der Kraft. Eine physikalische Abhandlung*. G. Reimer, Berlin, 1847.

R. Von Mises. Grundlagen der Wahrscheinlichkeitsrechnung. *Math. Z.*, 5:52–99, 1919.

R. Von Mises. *Wahrscheinlichkeit, Statistik und ihre Wahrheit*. J. Springer, Wien, 1928. 2nd Engl. transl.: Probability, Statistics and Truth, MacMillan, New York, 1957; reprinted 1961; prepared by H. Geiringer from the 3rd and 'definitive' German ed., J. Springer, Wien, 1951; the 2nd German ed. had appeared in 1936. See also 4th German ed. (supervised by H. Geiringer) Springer, Wien, New York, 1972.

R. Von Mises. *Wahrscheinlichkeitsrechnung und ihre Anwendung in der Statistik und theoretische Physik*. Deutike, Leipzig u. Wien, 1931.

R. Von Mises. *Kleines Lehrbuch der Positivismus, Einfuhrung in die empiristische Wissenschaftsauffassung*. W.P. van Stockum/University of Chicago, Den Haag/Chicago, IL, 1939. Transl. by J. Bernstein and R.G. Newton, as: Positivism, Harvard University Press, Cambridge MA, 1951.

R. Von Mises. On the foundations of probability and statistics. *Ann. Math. Stat.*, 12:191–205, 1941.

J. Von Neumann. Thermodynamik quantenmechanischer Gesamtheiten. *Gött. Nach.*, 1 No. 11:273–291, 1927. see already: Mathematische Begründung der Quantenmechanik, *ibid.* 1–57; Wahrscheinlichkeitstheoretischer Aufbau der Quantenmechanik, *ibid.* 245-272.

J. Von Neumann. Die Eindeutigkeit der Schrödingerschen Operatoren. *Math. Ann.*, 104:570–578, 1931.

J. Von Neumann. *Mathematische Grundlagen der Quantenmechanik.* Springer, Berlin, 1932a. Mathematical Foundations of Quantum Mechanics (R.T. Beyer, transl.), Princeton University Press, Princeton NJ, 1955.

J. Von Neumann. Proof of the quasi–ergodic hypothesis. *Proc. Nat. Acad. Sci. USA*, 18:70–82, 1932b. See also 'Physical applications of the ergodic hypothesis', *ibid.* 263–266.

J. Von Neumann. The uniqueness of Haar's measure. *Mat. Sb.*, 1:721–734, 1936.

J. Von Neumann. Various techniques used in connection with random digits. *J. Res. Nat. Bur. Stand. Appl. Math. Series*, 3:36–38, 1951. Collected Works, vol. V, MacMillan, New York, 1963, 768-770.

J. Von Neumann. Continuous geometries with transition probabilities. *Mem. AMS*, 34, No. 252:1–210, 1981. Prepared and edited by I. Halperin from a 1937 manuscript; see also: Von Neumann's Collected Works, Vol. IV: Continuous geometries and other topics, A.H. Taub ed., Pergamon, London, 1961.

J. Von Neumann. *Quantum logic (strict- and probability logic), unpublished manuscript, J. von Neumann Archives.* Library of Congress, Washington DC, ca 1937. reviewed by A.H. Taub, in J. von Neumann, Collected works, vol. IV, Pergamon Press, 1961.

J. Von Plato. *Creating modern probability: its mathematics, physics and philosophy in historical perspective.* Cambridge University Press, Cambridge, 1994.

P. B.M. Vranas. Epsilon-ergodicity and the success of equilibrium statistical mechanics. *Phil. Sci.*, 65:688–708, 1998.

S. Wagon. Is π normal? *Math. Intelligencer*, 7:65–67, 1985.

A. Wald. Sur la notion de collectif dans le calcul des probabilités. *C. R. Acad. Sci. Paris*, 202:180–183, 1936. See: Die Wiederspruchfreiheit des Kollektivbegriffes der Wahrscheinlichkeitsrechnung, Ergebnisse eines Kolloquium (Vienna) 8 (1937), 38–72; Die Wiederspruchfreiheit des Kollektivbegriffes, Actualités Sci. Indust. **735** (1938), 79–99.

J. Wallis. *Arithmetica Infinitorium* in *Operum mathematicorum.* Lichfield, Oxford, 1655.

J. Wallis. *A discourse of combinations, alterations and aliquot parts;* attached to the English edition of his *Algebra.* John Playford, London, 1685. Also reprinted in [Maseres, 1795].

P. Walters. *An introduction to ergodic theory.* Springer, New York, 1982.

W. Walton and E. Sitwell. *Façade.* Oxford University Press, London, 1951. Music by William Walton. Poems by Edith Sitwell.

G.H. Wannier. *Statistical physics.* Wiley, Dover, New York, 1966, 1987.

E. Waring. *On the principles of translating algebraic quantities into probable relations and annuities.* J. Nicholson, Cambridge, 1792.

M W. Wartofsky. *Models: representation and the scientific understanding.* D. Reidel, Dordrecht, 1979.

F.H. Weber. Die specifische Wärme des Kohlenstoffs. *Ann. d. Physik*, 147:311–319, 1872.

P. Weiss. L'hypothèse du champ moléculaire et la propriété ferromagnétique. *Journ. phys. théor. appl.*, 6:661–690, 1907.

T.P. Weissert. *The genesis of simulation in dynamics: pursuing the Fermi-Pasta-Ulam problem.* Springer, New York, 1997.

T. Weston. Approximate truth. *Journal of phil. logic*, 16:203–227, 1987.

T. Weston. Approximate truth and scientific realism. *Philosophy of Science*, 59: 53–74, 1992.

A. Wexler and W.S. Corak. Superconductivity of Vanadium. *Phys. Rev.*, 85:85–90, 1952.

H. Weyl. Ueber die Gibbssche Erscheinung und verwandte Konvergenzphänomene. *Rend. Circ. Mat. Palermo*, 30:377–407, 1910.

H. Weyl. Ueber die Gleichverteilung von Zahlen mod Eins. *Math. Ann.*, 77:313–352, 1916.

B. Whitten-Wolfe and G.G. Emch. A mechanical quantum measuring process. *Helv. Phys. Acta*, 49:45–55, 1976.

B. Widom. Surface tension and molecular correlations near the critical point; equation of state in the neighborhood of the critical point. *J. Chem. Phys.*, 43: 3892–3897; 3898–3905, 1965.

C.E. Wieman, D.E. Pritchard, and D.J. Wineland. Atom cooling, trapping, and quantum manipulation. *Rev. Mod. Phys.*, 71:S253–S262, 1999.

W. Wien. Temperatur und Entropie der Strahlung. *Ann. d. Physik*, 52:132–165, 1894.

W. Wien. Ueber die Energievertheilung im Emissionspektrum eines schwartzen Körpers. *Ann. d. Physik*, 58:662–669, 1896.

N. Wiener and A. Wintner. On the ergodic dynamics of almost periodic systems. *Amer. J. Math.*, 63:794–819, 1941.

A.S. Wightman. Introduction to the problems. In G. Velo and Wightman A. S, editors, *Regular and chaotic motions in dynamical systems*, pages 1–26. Plenum Press, New York, 1985.

A.S. Wightman and J. Mehra, editors. *The collected works of Eugene P. Wigner*. Springer, Berlin, 1995.

E.P. Wigner. Events, laws of nature, and invariance principles. *Mimeographical notes*, ca. 1980. Reproduced in (Wightman and Mehra 1995), pp. VI: 334–342.

E.P. Wigner. *Philosophical reflections and syntheses*. Springer, Berlin, 1997. Annotated by G.G. Emch; this is a paperback reprint of vol. 6 in [Wightman and Mehra, 1995].

T.J. Wilmore. *An introduction to differential geometry*. Oxford University Press, London, 1959.

K.G. Wilson. Renormalization group and critical phenomena. *Phys. Rev. B*, 4: 3174–3183; 3184–3205, 1971.

K.G. Wilson. The renormalization group: critical phenomena and the Kondo problem. *Rev. Mod. Phys.*, 47:773–840, 1975.

K.G. Wilson and M.E. Fisher. Critical exponents in 3.99 dimensions. *Phys. Rev. Lett.*, 28:240–243, 1972.

K.G. Wilson and G. Kogut. The renormalization group and the ε expansion. *Physics Report C*, 12:75–199, 1974.

W.C. Wimsatt. False models as means to truer theories. In M. Nitecki and A. Hoffman, editors, *Neutral models in biology*, pages 23–55. Oxford University Press, Oxford, 1987.

W.C. Wimsatt. The ontology of complex systems: levels, perspectives, and causal thickets. In M. Matthen and R.X. Ware, editors, *Biology and society: reflections on methodology*, pages 207–274. Canadian Journal of Philosophy: University Calgary Press, Calgary, Alta, 1994.

D. Wineland and H. Dehmelt. Proposed 10^{14} $\delta\nu < \nu$ laser fluorescence spectroscopy on Tl^+ mono–ion oscillator III. *Bull. Amer. Phys. Soc.*, 20:637, 1975.

M. Winnink. On the equilibrium states of infinite quantum systems at $T \neq 0$. In Bak [1967], pages 59–71.

J.A. Wolf. *Spaces of constant negative curvature.* McGraw–Hill, New York, 1967.

J.P. Wolfe, J.L. Lin, and D.W. Snoke. Bose–Einstein condensation of a nearly ideal gas: excitons in Cu_2O. In Griffin et al. [1995], pages 281–329.

G. Wulff. Zur Frage der Geschwindigkeit des Wachstums und der Auflösung der Krystallflächen. *Z. Kryst.*, 34:449–530, 1901.

C.N. Yang. Spontaneous magnetization of a two-dimensional Ising model. *Phys. Rev.*, 85:808–816, 1952.

C.N. Yang. Concept of off–diagonal long range order and the quantum phases of liquid He and of superconductors. *Rev. Mod. Phys.*, 34:694–704, 1962.

C.N. Yang and T.D. Lee. Statistical theory of equations of state and phase transitions I. Theory of condensation. *Phys. Rev.*, 87:404–409, 1952.

A. Yasuhara. *Recursive function theory and logic.* Academic Press, New York, 1971.

K. Yoshida. Paramagnetic susceptibility in superconductors. *Phys. Rev.*, 110:769–770, 1958.

S. Yosida. *Functional analysis.* Springer, New York, 1974.

L.S. Young. Developments in chaotic dynamics. *Notices of the AMS*, 45:1318–1328, 1998.

S.L. Zabell. Symmetry and its discontents. In Skyrms and Harper [1988], pages 155–190.

M. Zahradnik. Contour methods and Pirogov–Sinai theory for continuous spin lattice models. In Minlos et al. [2000], pages 197–220.

E.C. Zeeman. Topology of the brain and visual perception. In M.K. Fort, Jr, editor, *Topology of 3–manifolds*, pages 240–256. Prentice–Hall, Englewood Cliffs NJ, 1961.

H. Zeh. On the interpretation of measurement in quantum mechanics. *Found. Phys.*, 1:69–76, 1970. Discussed in E.P. Wigner, Review of the quantum mechanical measurement problem, 1984, reprinted in (Wightman and Mehra, 1995) pp.VI: 225–244.

H.D. Zeh. *The physical basis of the direction of time.* Springer, Berlin, 1981. 4th ed., 2001.

A. Zeilinger. Experiment and foundations of quantum physics. *Rev. Mod. Phys.*, 71:S283–S287, 1999.

M.W. Zemansky. *Temperatures, very low and very high.* van Nostrand, Princeton NJ, 1964.

E. Zermelo. Ueber einem Satz der Dynamik und die mechanische Wärmetheorie'; 'Ueber mechanische Erklärungen irreversibler Vorgänge: Eine Antwort auf Hrn. Boltzmann's "Entgegnung ...". *Ann. d. Physik*, 57, 59:485–494; 793–80, 1896.

F. Zernike. Propagation of order in cooperative phenomena. *Physica*, 7:565–585, 1940.

R. Zwanzig. Ensemble method in the theory of irreversibility. *J. Chem. Phys.*, 33:1338–1341, 1960. see also 'Statistical mechanics of irreversible processes', in Lectures in theoretical physics 3, Boulder CO, 1960, W.E. Brittin, ed. pp. 106-141.

R. Zwanzig. On the identity of three generalized master equations. *Physica*, 30: 1109–1123, 1964.

Citation Index

Subject Index

Printing (Computer to Film): Saladruck Berlin
Binding: Stürtz AG, Würzburg